Lecture Notes in Computer Science 9948

Commenced Publication in 1973
Founding and Former Series Editors:
Gerhard Goos, Juris Hartmanis, and Jan van Leeuwen

Akira Hirose · Seiichi Ozawa
Kenji Doya · Kazushi Ikeda
Minho Lee · Derong Liu (Eds.)

Neural
Information Processing

23rd International Conference, ICONIP 2016
Kyoto, Japan, October 16–21, 2016
Proceedings, Part II

 Springer

Editors

Akira Hirose
The University of Tokyo
Tokyo
Japan

Seiichi Ozawa
Kobe University
Kobe
Japan

Kenji Doya
Okinawa Institute of Science and
  Technology Graduate University
Onna
Japan

Kazushi Ikeda
Nara Institute of Science and Technology
Ikoma
Japan

Minho Lee
Kyungpook National University
Daegu
Korea (Republic of)

Derong Liu
Chinese Academy of Sciences
Beijing
China

ISSN 0302-9743   ISSN 1611-3349 (electronic)
Lecture Notes in Computer Science
ISBN 978-3-319-46671-2   ISBN 978-3-319-46672-9 (eBook)
DOI 10.1007/978-3-319-46672-9

Library of Congress Control Number: 2016953319

LNCS Sublibrary: SL1 – Theoretical Computer Science and General Issues

Printed on acid-free paper

This Springer imprint is published by Springer Nature
The registered company is Springer International Publishing AG
The registered company address is: Gewerbestrasse 11, 6330 Cham, Switzerland

Preface

This volume is part of the four-volume proceedings of the 23rd International Conference on Neural Information Processing (ICONIP 2016) held in Kyoto, Japan, during October 16–21, 2016, which was organized by the Asia-Pacific Neural Network Society (APNNS, http://www.apnns.org/) and the Japanese Neural Network Society (JNNS, http://www.jnns.org/). ICONIP 2016 Kyoto was the first annual conference of APNNS, which started in January 2016 as a new society succeeding the Asia-Pacific Neural Network Assembly (APNNA). APNNS aims at the local and global promotion of neural network research and education with an emphasis on diversity in members and cultures, transparency in its operation, and continuity in event organization. The ICONIP 2016 Organizing Committee consists of JNNS board members and international researchers, who plan and run the conference.

Currently, neural networks are attracting the attention of many people, not only from scientific and technological communities but also the general public in relation to the so-called Big Data, TrueNorth (IBM), Deep Learning, AlphaGo (Google DeepMind), as well as major projects such as the SyNAPSE Project (USA, 2008), the Human Brain Project (EU, 2012), and the AIP Project (Japan, 2016). The APNNS's predecessor, APNNA, promoted fields that were active but also others that were leveling off. APNNS has taken over this function, and further enhances the aim of holding technical and scientific events for interaction where even those who have extended the continuing fields and moved into new/neighboring areas rejoin and participate in lively discussions to generate and cultivate novel ideas in neural networks and related fields.

The ICONIP 2016 Kyoto Organizing Committee received 431 submissions from 38 countries and regions worldwide. Among them, 296 (68.7 %) were accepted for presentation. The first authors of papers that were presented came from Japan (100), China (78), Australia (22), India (13), Korea (12), France (7), Hong Kong (7), Taiwan (7), Malaysia (6), United Kingdom (6), Germany (5), New Zealand (5) and other countries/regions worldwide.

Besides the papers published in these four volumes of the Proceedings, the conference technical program includes

- Four plenary talks by Kunihiko Fukushima, Mitsuo Kawato, Irwin King, and Sebastian Seung
- Four tutorials by Aapo Hyvarinen, Nikola Kazabov, Stephen Scott, and Okito Yamashita,
- One Student Best Paper Award evaluation session
- Five special sessions, namely, bio-inspired/energy-efficient information processing, whole-brain architecture, data-driven approach for extracting latent features from multidimensional data, topological and graph-based clustering methods, and deep and reinforcement learning
- Two workshops: Data Mining and Cybersecurity Workshop 2016 and Workshop on Novel Approaches of Systems Neuroscience to Sports and Rehabilitation

The event also included exhibitions and a technical tour.

Kyoto is located in the central part of Honshu, the main island of Japan. Kyoto formerly flourished as the imperial capital of Japan for 1,000 years after 794 A.D., and is presently known as "The City of Ten Thousand Shrines." There are 17 sites (13 temples, three shrines, and one castle) in Kyoto that form part of the UNESCO World Heritage Listing, named the "Historic Monuments of Ancient Kyoto (Kyoto, Uji and Otsu Cities)." In addition, there are three popular, major festivals (Matsuri) in Kyoto, one of which, "Jidai Matsuri" (The Festival of Ages), was held on October 22, just after ICONIP 2016.

We, the general chair, co-chair, and Program Committee co-chairs, would like to express our sincere gratitude to everyone involved in making the conference a success. We wish to acknowledge the support of all the sponsors and supporters of ICONIP 2016, namely, APNNS, JNNS, KDDI, NICT, Ogasawara Foundation, SCAT, as well as Kyoto Prefecture, Kyoto Convention and Visitors Bureau, and Springer. We also thank the keynote, plenary, and invited speakers, the exhibitors, the student paper award evaluation committee members, the special session and workshop organizers, as well as all the Organizing Committee members, the reviewers, the conference participants, and the contributing authors.

October 2016

Akira Hirose
Seiichi Ozawa
Kenji Doya
Kazushi Ikeda
Minho Lee
Derong Liu

Organization

 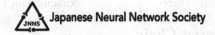

General Organizing Board

JNNS Board Members

Honorary Chairs

Shun-ichi Amari   RIKEN
Kunihiko Fukushima   Fuzzy Logic Systems Institute

Organizing Committee

General Chair

Akira Hirose   The University of Tokyo, Japan

General Co-chair

Seiichi Ozawa   Kobe University, Japan

Program Committee Chairs

Kenji Doya   OIST, Japan
Kazushi Ikeda   NAIST, Japan
Minho Lee   Kyungpook National University, Korea
Derong Liu   Chinese Academy of Science, China

Local Arrangements Chairs

Hiroaki Nakanishi   Kyoto University, Japan
Ikuko Nishikawa   Ritsumeikan University, Japan

Members

Toshio Aoyagi   Kyoto University, Japan
Naoki Honda   Kyoto University, Japan
Kazushi Ikeda   NAIST, Japan

Shin Ishii	Kyoto University, Japan
Katsunori Kitano	Ritsumeikan University, Japan
Hiroaki Mizuhara	Kyoto University, Japan
Yoshio Sakurai	Doshisha University, Japan
Yasuhiro Tsubo	Ritsumeikan University, Japan

Financial Chair

Seiichi Ozawa	Kobe University, Japan

Member

Toshiaki Omori	Kobe University, Japan

Special Session Chair

Kazushi Ikeda	NAIST, Japan

Workshop/Tutorial Chair

Hiroaki Gomi	NTT Communication Science Laboratories, Japan

Publication Chair

Koichiro Yamauchi	Chubu University, Japan

Members

Yutaka Hirata	Chubu University, Japan
Kay Inagaki	Chubu University, Japan
Akito Ishihara	Chukyo University, Japan

Exhibition Chair

Tomohiro Shibata	Kyushu Institute of Technology, Japan

Members

Hiroshi Kage	Mitsubishi Electric Corporation, Japan
Daiju Nakano	IBM Research - Tokyo, Japan
Takashi Shinozaki	NICT, Japan

Publicity Chair

Yutaka Sakai	Tamagawa University, Japan

Industry Relations

Ken-ichi Tanaka   Mitsubishi Electric Corporation, Japan
Toshiyuki Yamane   IBM Research - Tokyo, Japan

Sponsorship Chair

Ko Sakai   University of Tsukuba, Japan

Member

Susumu Kuroyanagi   Nagoya Institute of Technology, Japan

General Secretaries

Hiroaki Mizuhara   Kyoto University, Japan
Gouhei Tanaka   The University of Tokyo, Japan

International Advisory Committee

Igor Aizenberg   Texas A&M University-Texarkana, USA
Sabri Arik   Istanbul University, Turkey
P. Balasubramaniam   Gandhigram Rural Institute, India
Eduardo Bayro-Corrochano CINVESTAV, Mexico
Jinde Cao   Southeast University, China
Jonathan Chan   King Mongkut's University of Technology, Thailand
Sung-Bae Cho   Yonsei University, Korea
Wlodzislaw Duch   Nicolaus Copernicus University, Poland
Tom Gedeon   Australian National University, Australia
Tingwen Huang   Texas A&M University at Qatar, Qatar
Nik Kasabov   Auckland University of Technology, New Zealand
Rhee Man Kil   Sungkyunkwan University (SKKU), Korea
Irwin King   Chinese University of Hong Kong, SAR China
James Kwok   Hong Kong University of Science and Technology,
                          SAR China
Weng Kin Lai   Tunku Abdul Rahman University College, Malaysia
James Lam   The University of Hong Kong, SAR Hong Kong
Kittichai Lavangnananda   King Mongkut's University of Technology, Thailand
Min-Ho Lee   Kyungpoor National University, Korea
Soo-Young Lee   Korea Advanced Institute of Science and Technology,
                          Korea
Andrew Chi-Sing Leung   City University of Hong Kong, SAR China
Chee Peng Lim   University Sains Malaysia, Malaysia
Chin-Teng Lin   National Chiao Tung University, Taiwan
Derong Liu   The Institute of Automation of the Chinese Academy of
                          Sciences (CASIA), China

Chu Kiong Loo	University of Malaya, Malaysia
Bao-Liang Lu	Shanghai Jiao Tong University, China
Aamir Saeed Malik	Petronas University of Technology, Malaysia
Danilo P. Mandic	Imperial College London, UK
Nikhil R. Pal	Indian Statistical Institute, India
Hyeyoung Park	Kyungpook National University, Korea
Ju. H. Park	Yeungnam University, Republic of Korea
John Sum	National Chung Hsing University, Taiwan
DeLiang Wang	Ohio State University, USA
Jun Wang	Chinese University of Hong Kong, SAR Hong Kong
Lipo Wang	Nanyang Technological University, Singapore
Zidong Wang	Brunel University, UK
Kevin Wong	Murdoch University, Australia
Xin Yao	University of Birmingham, UK
Li-Qing Zhang	Shanghai Jiao Tong University, China

Advisory Committee Members

Masumi Ishikawa	Kyushu Institute of Technology
Noboru Ohnishi	Nagoya University
Shiro Usui	Toyohashi University of Technology
Takeshi Yamakawa	Fuzzy Logic Systems Institute

Technical Program Committee

Abdulrahman Altahhan	Tetsuo Furukawa
Sabri Arik	Kuntal Ghosh
Sang-Woo Ban	Anupriya Gogna
Tao Ban	Hiroaki Gomi
Matei Basarab	Shanqing Guo
Younes Bennani	Masafumi Hagiwara
Ivo Bukovsky	Isao Hayashi
Bin Cao	Shan He
Jonathan Chan	Akira Hirose
Rohitash Chandra	Jin Hu
Chung-Cheng Chen	Jinglu Hu
Gang Chen	Kaizhu Huang
Jun Cheng	Jun Igarashi
Long Cheng	Kazushi Ikeda
Zunshui Cheng	Ryoichi Isawa
Sung-Bae Cho	Shin Ishii
Justin Dauwels	Teijiro Isokawa
Mingcong Deng	Wisnu Jatmiko
Kenji Doya	Sungmoon Jeong
Issam Falih	Youki Kadobayashi

Juyang Weng
Bin Xu
Tetsuya Yagi
Nobuhiko Yamaguchi
Hiroshi Yamakawa
Toshiyuki Yamane
Koichiro Yamauchi
Tadashi Yamazaki
Pengfei Yan
Qinmin Yang
Xiong Yang

Zhanyu Yang
Junichiro Yoshimoto
Zhigang Zeng
Dehua Zhang
Li Zhang
Nian Zhang
Ruibin Zhang
Bo Zhao
Jinghui Zhong
Ding-Xuan Zhou
Lei Zhu

Contents – Part II

Neuromorphic Hardware

Sensory Perception

Pattern Recognition

Social Networks

Brain-Machine Interface

Computer Vision

Machine Learning

Non-parametric e-mixture of Density Functions

Hideitsu Hino[1(✉)], Ken Takano[2], Shotaro Akaho[3], and Noboru Murata[2]

[1] University of Tsukuba, 1-1-1 Tennodai, Tsukuba, Ibaraki 305-8573, Japan
hinohide@cs.tsukuba.ac.jp
[2] Waseda University, Okubo 3-4-1, Shinjuku-ku, Tokyo 169-8555, Japan
[3] AIST, 1-1-1 Umezono, Tsukuba, Ibaraki 305-8568, Japan

Abstract. Mixture modeling is one of the simplest ways to represent complicated probability density functions, and to integrate information from different sources. There are two typical mixtures in the context of information geometry, the m- and e-mixtures. This paper proposes a novel framework of non-parametric e-mixture modeling by using a simple estimation algorithm based on geometrical insights into the characteristics of the e-mixture. An experimental result supports the proposed framework.

Keywords: Mixture model · Information geometry · Non-parametric method

1 Introduction

Suppose we have a *target* dataset $\mathcal{D}^{(0)} = \{x_k^{(0)}\}_{k=1}^{n_0}$, which is composed of n_0 samples $x_k^{(0)} \in \mathbb{R}^d$ generated from a probability distribution with a probability density function (pdf) p_0. We also have N *auxiliary* datasets $\{\mathcal{D}^{(i)}\}_{i=1}^N$. The i-th dataset $\mathcal{D}^{(i)} = \{x_k^{(i)}\}_{k=1}^{n_i}$ is composed of n_i samples $x_k^{(i)} \in \mathbb{R}^d$ generated from a probability distribution with a pdf p_i. We consider the situation that the *target* dataset has a much smaller amount of data than auxiliary datasets and we wish to obtain more feasible estimate of p_0 taking advantage of informative auxiliary datasets. This situation is often seen, for example, in classification problems of EEG [1] and audio signals [2]. We consider representing the *target* distribution p_0 as a mixture of other auxiliary pdfs p_i by weighting data in auxiliary datasets.

Constructing a mixture model of probability distributions is a standard approach for integrating information of different sources. There are two typical mixtures in the context of information geometry [3], the m- and the e-mixture. The m-mixture is a convex combination of auxiliary pdfs $p_i(x)$ defined as $p^m(x; \boldsymbol{\theta}) = \sum_{i=1}^N \theta_i p_i(x)$, $\sum_{i=1}^N \theta_i = 1$, $\theta_i \geq 0$, where $\boldsymbol{\theta} = \{\theta_i\}_{i=1}^N$ is a mixture ratio vector of the pdfs $p_i(x), i = 1, \ldots, N$. The Gaussian Mixture Model (GMM) is an example of the m-mixtures [4]. The e-mixture $p^e(x)$ has the following particular form:

$$p^e(x; \boldsymbol{\theta}) = \exp\left\{ \sum_{i=1}^N \theta_i \log p_i(x) - b(\boldsymbol{\theta}) \right\}, \quad \sum_{i=1}^N \theta_i = 1, \tag{1}$$

© Springer International Publishing AG 2016
A. Hirose et al. (Eds.): ICONIP 2016, Part II, LNCS 9948, pp. 3–10, 2016.
DOI: 10.1007/978-3-319-46672-9_1

where $b(\boldsymbol{\theta})$ is the normalization term. Figure 1 shows the e- and the m-mixture of two Gaussian distributions. Solid lines indicate two Gaussian distributions. Dashed and dotted lines indicate the m- and the e-mixture of two Gaussians, respectively, with a uniform mixture ratio. From the definitions of mixture models and from Fig. 1, we can see that the m-mixture of Gaussian does not remain to be a Gaussian, while the e-mixture results in Gaussian again. This property holds for any distributions in the exponential family. It is a favorable characteristic when we want to model the distributions within the exponential family.

In a variety of research fields, the m-mixtures are applied along with the EM algorithm [5]. The e-mixtures, however, are applied by only a few authors in spite of their good properties [6]. This is because estimating e-mixtures is sometimes computationally intractable due to logarithmic and exponential functions. It

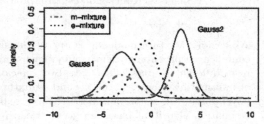

Fig. 1. An example of the m- and e-mixture of two Gaussian distributions.

needs to select an appropriate distribution family for auxiliary pdfs which can be calculated in the e-mixture form.

In this paper, from a viewpoint of information geometry, we propose a novel framework for estimating e-mixtures of non-parametric models by using a weighted empirical distribution function of all the given data

$$p(x) = \sum_{k=1}^{K} w_k \delta(x - x_k), \tag{2}$$

where $\mathbf{w} = \{w_k\}_{k=1}^{K}$ is a weight vector, i.e. each element w_k of $\mathbf{w}$ represents sampling probability of a datum $x_k \in \mathbb{R}^d$, and $\delta(\cdot)$ is the Dirac delta function. When a sufficient number of data are given, non-parametric models such as an empirical distribution function can express the underlying distribution more precisely than parametric models. We estimate the weighted empirical distribution function in terms of the e-mixture, i.e. our objective is determining the weight $\mathbf{w}$ of all the given data.

2   Preliminary on Information Geometry

We consider a statistical space $\mathcal{S}$ which is composed of arbitrary probability density functions $\{p(x)\}$, where x is a random variable. A point in the space $\mathcal{S}$ corresponds to a pdf, and the closeness between two points in $\mathcal{S}$ is measured by certain divergence measure. The KL-divergence [7] is an example of the divergence between two probability distributions p and q, which is defined as

$$D_{KL}(p, q) = \int p(x) \log p(x)/q(x) dx. \tag{3}$$

Let us consider a relationship among three pdfs p, q, and r,

$$D_{KL}(p,q) - D_{KL}(p,r) - D_{KL}(r,q) = \int \{p(x) - r(x)\}\{\log r(x) - \log q(x)\}dx. \quad (4)$$

When the right hand side of the Eq. (4) equals 0, $\{p(x) - r(x)\}$ and $\{\log r(x) - \log q(x)\}$ are *orthogonal* in the statistical space $\mathcal{S}$. This property, namely, $D_{KL}(p,q) - D_{KL}(p,r) - D_{KL}(r,q) = 0$ is known as the *Pythagorean relation* [3].

Mixture models are regarded as subspaces of $\mathcal{S}$ spanned by finite number of pdfs. The following theorem characterizes the e-mixture of pdfs $\{p_i\}_{i=1}^N$.

Theorem 1 (characterization of the e-mixture [8]**).** *For any mixture parameter* $\boldsymbol{\theta} = \{\theta_i\}_{i=1}^N$, *a sum of KL-divergence* $\sum_{i=1}^N \theta_i D_{KL}(q, p_i)$ *weighted by* $\boldsymbol{\theta}$ *is minimized at the e-mixture as*

$$\arg\min_{q \in \mathcal{P}} \sum_{i=1}^N \theta_i D_{KL}(q, p_i) = p^e(x; \boldsymbol{\theta}), \quad (5)$$

where $\mathcal{P}$ *is the set of probability density functions.*

3 Problem Formulation

To realize flexible modeling, our algorithm estimates a mixture in a non-parametric manner. We construct an empirical distribution p_0 from the target dataset $\mathcal{D}^{(0)} = \{x_k^{(0)}\}_{k=1}^{n_0}$ as $p_0(x) = \frac{1}{n_0} \sum_{k=1}^{n_0} \delta(x - x_k^{(0)})$. Similarly, we construct *auxiliary empirical distributions* $p_i, i = 1, \ldots, N$ from auxiliary datasets $\mathcal{D}^{(i)} = \{x_k^{(i)}\}_{k=1}^{n_i}$ as $p_i(x) = \frac{1}{n_i} \sum_{k=1}^{n_i} \delta(x - x_k^{(i)})$. The e-mixture p^e of these auxiliary empirical distributions $p_i, i = 1, \ldots, N$ can be written, with an abuse of delta functions, as

$$p^e(x; \boldsymbol{\theta}) = \exp\left\{ \sum_{i=1}^N \theta_i \log \frac{1}{n_i} \sum_{k=1}^{n_i} \delta(x - x_k^{(i)}) - b(\boldsymbol{\theta}) \right\}, \quad (6)$$

where $\boldsymbol{\theta} = \{\theta_i\}_{i=1}^N$ is a mixture ratio vector of the auxiliary pdfs p_i. The expression (6) is not mathematically formal because of the log of delta functions. Thus, we define a mixture model which satisfies Eq. (5) as an e-mixture of non-parametric models avoiding the mathematically incorrect expression (6). With an appropriate parameter $\boldsymbol{\theta}$, we define a non-parametric e-mixture $\hat{p}^e$ as

$$\hat{p}^e(x; \boldsymbol{\theta}) = \arg\min_{q \in \mathcal{P}} \sum_{i=1}^N \theta_i D_{KL}(q, p_i), \quad (7)$$

where set of pdfs $\mathcal{P}$ is defined by using weighted empirical distributions as

$$\mathcal{P} = \left\{ p(x) = \sum_{k=1}^K w_k \delta(x - y_k); \sum_{k=1}^K w_k = 1, w_k \geq 0 \right\}. \quad (8)$$

The variable y_k is the k-th element of the set $\mathcal{D}$ of the whole datasets, which is defined by $\mathcal{D} = \mathcal{D}^{(0)} \cup \mathcal{D}^{(1)} \cup \mathcal{D}^{(2)} \cup \cdots \cup \mathcal{D}^{(N)}$. Here, $K = |\mathcal{D}|$ is the number of data points in $\mathcal{D}$ and $\mathbf{w} = \{w_k\}_{k=1}^{K}$ is a weight vector of weighted empirical distributions. Although the weight vector depends on the mixture ratio $\boldsymbol{\theta}$, for the sake of simplicity, we write the weight vector $\mathbf{w}$ instead of $\mathbf{w}(\boldsymbol{\theta})$.

Our purpose is to estimate $\hat{p}^e$, namely, to determine weight vector $\mathbf{w}$ on the set $\mathcal{D}$ of all the given data. A subspace $\mathcal{E}$ of e-mixtures of empirical distributions, which is composed of weighted empirical distributions satisfying Eq. (7) with arbitrary mixture ratios $\boldsymbol{\theta}$ is defined as

$$\mathcal{E} = \left\{ p \,\middle|\, p(\cdot; \boldsymbol{\theta}) = \arg\min_{q \in \mathcal{P}} \sum_i \theta_i D_{KL}(q, p_i), \sum_i \theta_i = 1, \theta_i \geq 0 \right\}. \qquad (9)$$

We perform our proposed e-mixture estimation algorithm on this subspace with satisfying requisite conditions for the e-mixture described in Sect. 3.1.

Finally, we note that Eq. (8) does not indicate the m-mixture of auxiliary pdfs because each weight w_k in Eq. (8) contains $\boldsymbol{\theta}$ implicitly. Since the e- and m-mixture have different restrictions with respect to $\boldsymbol{\theta}$, restrictions on weight are also different depending on mixture models. Equation (8) is the m-*representation of the e-mixture*, and we develop an algorithm to optimize the weight vector $\mathbf{w}$ in Eq. (8) for the e-mixture estimation.

3.1 Requisite Conditions for e-mixture

For estimating the e-mixture in a non-parametric manner, following two conditions are imposed:

 (i) $\hat{p}^e$ is the e-mixture of auxiliary empirical distributions $p_i, i = 1, \ldots, N$,
(ii) $\hat{p}^e$ is the projection of p_0 onto the subspace $\mathcal{E}$.

Let us consider the first condition. According to Theorem 1, the probability density function which minimizes the weighted KL-divergence is written in the form of the e-mixture. Therefore, if we obtain the weight $\mathbf{w}$ of $\hat{p}^e$ in Eq. (7), which minimizes the weighted KL-divergence in Eq. (5), $\hat{p}^e$ is regarded as the e-mixture of auxiliary pdfs with given mixture ratios $\boldsymbol{\theta}$.

For the second condition, we consider the subspace $\mathcal{E}$, which includes $p_i, i = 1, \ldots, N$. Our objective is finding the closest pdf $\hat{p}^e \in \mathcal{E}$ from p_0 in the sense of the KL-divergence. Then, any pdf $q \in \mathcal{E}$ satisfies the Pythagorean relation, when we consider $p = p_0, q = q$, and $r = \hat{p}^e$ in Eq. (4). Also, because each auxiliary pdf p_i is in $\mathcal{E}$, the following equation holds

$$D_{KL}(p_0, \hat{p}^e) + D_{KL}(\hat{p}^e, p_i) - D_{KL}(p_0, p_i) = 0. \qquad (10)$$

This equation claims that the geodesics from p_0 to $\hat{p}^e$ and that from $\log \hat{p}^e$ to $\log p_i$ are orthogonal. For applying these theorems to our non-parametric algorithm, a non-parametric KL-divergence estimator is required.

Divergence estimators based on the k-nearest neighbor method are well investigated [9,10], and those methods are extended to deal with weighted observations [11], which we denote $\hat{D}_{KL}(\mathcal{D}_{\mathbf{w}}, \mathcal{D}'_{\mathbf{v}})$. See [11] and the journal version of this paper for details.

4 Proposed Algorithm

Since the e-mixture of non-parametric models is determined by the weight $\mathbf{w}$ and the mixture ratio $\boldsymbol{\theta}$, we denote the e-mixture by $\hat{p}^e_{\mathbf{w}(\boldsymbol{\theta})}$ henceforth. Note that the weight vector $\mathbf{w}$ depends on the mixture ratio $\boldsymbol{\theta}$ implicitly. Now, we introduce an alternating optimization algorithm for $\mathbf{w}$ and $\boldsymbol{\theta}$. Step 1 computes the weight $\mathbf{w}$ given a fixed mixture ratio $\boldsymbol{\theta}$. Step 2 computes the mixture ratio $\boldsymbol{\theta}$ given a fixed weight $\mathbf{w}$. These two steps are computed alternately until $\mathbf{w}$ is converged. For initialization, we start with uniform weights $\{w_k = 1/K\}_{k=1}^K$ and uniform ratios $\{\theta_i = 1/N\}_{i=1}^N$.

Step1: The e-mixture is estimated with a given mixture ratio $\boldsymbol{\theta}$ based on Theorem 1. Namely, we compute the weight $\mathbf{w}$ which minimizes the weighted KL-divergence with a fixed mixture ratio $\boldsymbol{\theta}$:

$$\min_{\mathbf{w}} \mathcal{L}(\mathbf{w}), \quad \mathcal{L}(\mathbf{w}) \equiv \sum_{i=1}^N \theta_i \hat{D}_{KL}(\mathcal{D}_{\mathbf{w}}, \mathcal{D}_{\mathbf{u}}^{(i)}) = \mathbf{w}^{\mathrm{T}}(\mathbf{g} + \mathbf{f}(\mathbf{w})) \qquad (11)$$

where $\mathbf{u}$ denotes the uniform weight, i.e., $\mathcal{D}_{\mathbf{u}}^{(i)} = \{x_k^{(i)}, 1/n_i\}_{k=1}^{n_i}$, and

$$\mathbf{g} = \{g_k\}_{k=1}^K, \ g_k = \sum_{i=1}^N \theta_i \log \varepsilon_\alpha(y_k, \mathcal{D}_{\mathbf{u}}^{(i)}), \ \mathbf{f}(\mathbf{w}) = \{f_k\}_{k=1}^K, \ f_k = -\log \varepsilon_\alpha(y_k, \mathcal{D}_{\mathbf{w}}^{-k}).$$

Here $\varepsilon_\alpha(y_k, \mathcal{D}_{\mathbf{u}}^{(i)})$ is the α-quantile distance from y_k to the sample in $\mathcal{D}_{\mathbf{u}}^{(i)}$. We obtain the weight vector $\mathbf{w}$ by minimizing equation $\mathcal{L}(\mathbf{w})$ with a gradient projection method

$$\mathbf{w}^{(s+1)} \leftarrow \Pi\left(\mathbf{w}^{(s)} - \eta_s\left(\mathbf{g} + \mathbf{f}(\mathbf{w}^{(s)})\right)\right), \qquad (12)$$

where η_s is a learning rate and Π is the projection operator for the weight $\mathbf{w}^{(s+1)}$, i.e. $\sum_{k=1}^K w_k^{(s+1)} = 1, w_k^{(s+1)} \geq 0$. We set $\eta_s = 0.9^s \frac{1}{20 \times K}$ in our experiment.

Step2: We next determine the mixture ratio $\boldsymbol{\theta}$ so that $\hat{p}^e_{\mathbf{w}(\boldsymbol{\theta})}$ becomes the projection of p_0 onto the subspace $\mathcal{E}$. Figure 2 shows the geometrical interpretation of Step 2. If $\hat{p}^e_{\mathbf{w}(\boldsymbol{\theta})}$ is the projection of p_0 onto $\mathcal{E}$, the dotted line between p_0 and $\hat{p}^e_{\mathbf{w}(\boldsymbol{\theta})}$, and the solid line between $\log \hat{p}^e_{\mathbf{w}(\boldsymbol{\theta})}$ and $\log p_i$ are orthogonal, namely, p_i, $\hat{p}^e$ and p_0 satisfies the Pythagorean relation. Based on this geometrical intuition, we update the mixture ratio θ_i according to the violation of the Pythagorean relation $r_i \equiv D_{KL}(p_0, \hat{p}^e_{\mathbf{w}(\boldsymbol{\theta})}) + D_{KL}(\hat{p}^e_{\mathbf{w}(\boldsymbol{\theta})}, p_i) - D_{KL}(p_0, p_i)$, which takes 0 when the mixture ratio θ_i is optimal. Two top triangles in Fig. 2 show relations among p_0, $\hat{p}_{\mathbf{w}(\boldsymbol{\theta})}^e$ and $p_i, i = \{1, 4\}$. The upper left panel shows the case of the acute-angled triangle where r_1 is positive. In this case, θ_1 is smaller than optimal and $\hat{p}_{\mathbf{w}(\boldsymbol{\theta})}^e$ should be closer to p_1. Conversely, the upper right panel shows the case of the obtuse-angled triangle, and θ_4 is larger than the optimal.

In order to reflect this geometrical comprehension, we now introduce a weakly increasing piecewise linear function ϕ of r_i. The function ϕ controls θ, reflecting the degree of violation of the Pythagorean relation r_i. The function ϕ increases (decreases) θ_i when the r_i satisfies $r_i > 0$ ($r_i < 0$). With this function ϕ, we update the mixture ratio $\boldsymbol{\theta} = \{\theta_i\}_{i=1}^N$ according to the violation of the Pythagorean relation

$$\theta_i \leftarrow \theta_i \times \phi(\hat{r}_i), \qquad (13)$$

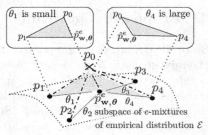

Fig. 2. Step 2 of the algorithm. Each empirical distribution p_i is on the curved surface with dotted lines which is the subspace of e-mixtures of empirical distributions $\mathcal{E}$.

where $\hat{r}_i$ is estimated by $\hat{r}_i = \hat{D}_{KL}(\mathcal{D}_{\mathbf{u}}^{(0)}, \mathcal{D}_{\mathbf{w}}) + \hat{D}_{KL}(\mathcal{D}_{\mathbf{w}}, \mathcal{D}_{\mathbf{u}}^{(i)}) - \hat{D}_{KL}(\mathcal{D}_{\mathbf{u}}^{(0)}, \mathcal{D}_{\mathbf{u}}^{(i)})$, and ϕ is defined as $\phi(\hat{r}_i) = c \times \hat{r}_i + 1(\underline{r} \leq \hat{r}_i \leq \overline{r}), c \times \underline{r} + 1(\hat{r}_i < \underline{r}), c \times \overline{r} + 1(\hat{r}_i > \overline{r})$, where c is a positive constant. Note that the function ϕ satisfies $\phi(0) = 1$. In our experiments we use $\underline{r} = \frac{-0.95}{c}, \overline{r} = \frac{0.95}{c}$, and $c \in (0, 1)$. These two steps search the closest e-mixture of auxiliary pdfs from $\mathcal{D}^{(0)}$ by assigning weights for samples on the set $\mathcal{D}$.

5   Experimental Result

To show how the proposed method works, we used synthetic data to demonstrate how our proposed algorithm works. Experimental results with more synthetic and real-world datasets will be provided in the journal version of the paper. We show our non-parametric e-mixture estimation algorithm works when the underlying distributions are Gaussian. Suppose we have a set of auxiliary datasets $\{\mathcal{D}^{(i)}\}_{i=1}^N$. Each dataset has $2,000$ data points sampled from a Gaussian distribution $\mathcal{N}(\boldsymbol{\mu}_i, \boldsymbol{\Sigma}_i)$. We are also given a target dataset $\mathcal{D}^{(0)}$, which contains data points of size $n_0 = 200$, from the e-mixture $p^e = \mathcal{N}(\boldsymbol{\mu}^e, \boldsymbol{\Sigma}^e)$ of auxiliary pdfs $\{\mathcal{N}(\boldsymbol{\mu}_i, \boldsymbol{\Sigma}_i)\}_{i=1}^N$. The mean vector and covariance matrix of the e-mixture p^e are calculated by $\boldsymbol{\mu}^e = \boldsymbol{\Sigma}^e \left(\sum_{i=1}^N \theta_i \boldsymbol{\Sigma}_i^{-1} \boldsymbol{\mu}_i \right)$, and $(\boldsymbol{\Sigma}^e)^{-1} = \sum_{i=1}^N \theta_i \boldsymbol{\Sigma}_i^{-1}$, respectively. For illustration purpose, we consider a 2-component 2-dimensional ($N = 2, d = 2$) Gaussian mixture. Our aim is in generating 2,000 points from the non-parametric e-mixture distribution constructed from $\mathcal{D}^{(0)}$, $\mathcal{D}^{(1)}$, and $\mathcal{D}^{(2)}$. We set the coefficient $c = 0.5$ in ϕ.

Top panels of Fig. 3-A show datasets $\mathcal{D}^{(1)}$, $\mathcal{D}^{(2)}$, and 2,000 points sampled from p^e. The bottom panels show $\mathcal{D}^{(0)}$, the e-mixture $\hat{p}_{\mathbf{u}}^e$ with the uniform weight

u, and a result of the non-parametric e-mixture estimation, respectively. The size of each mark in the bottom panels represents weight for each sample.

Estimated θ by iterations is shown in Fig. 3-B. Horizontal dotted lines are true values of mixture ratio $\theta = \{0.8, 0.2\}$. Contours of the empirical covariance matrices centered as the estimated means with the progress of the algorithm are shown in Fig. 3-C. Solid ellipse express the contours of the original distributions $p_1 = \mathcal{N}(\boldsymbol{\mu}_1, \boldsymbol{\Sigma}_1)$ and $p_2 = \mathcal{N}(\boldsymbol{\mu}_2, \boldsymbol{\Sigma}_2)$. Grey solid ellipse is the ground truth e-mixture p^e, and all dotted ellipses express the contours of the estimated e-mixtures $\hat{p}^e_{\mathbf{w}(\boldsymbol{\theta})}$ by iterations. Each ellipse contain 90 % probability mass of distribution.

From the experimental result, we can see that the proposed algorithm can approximate the ground truth distribution for the target dataset as the non-parametric e-mixture constructed by the given datasets with appropriate weights.

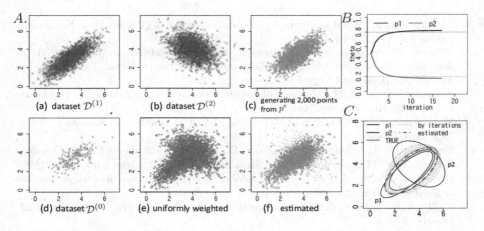

Fig. 3. A: Top panels show scatter plots of datasets $\mathcal{D}^{(1)}$, $\mathcal{D}^{(2)}$ and 2,000 points sampled from p^e. Bottom panels plot target dataset $\mathcal{D}^{(0)}$, uniformly weighted e-mixture $\hat{p}^e_{\mathbf{w}(\boldsymbol{\theta})}$, and the estimated $\hat{p}^e_{\mathbf{w}(\boldsymbol{\theta})}$. B: Mixture ratios θ_1 and θ_2 by iterations. C: Contours of the estimated distributions by iterations.

6   Conclusion and Future Work

We proposed a non-parametric e-mixture estimation algorithm based on the geometric characterization of e-mixtures. Firstly, we discussed the relationship between a certain pdf and the closest e-mixture of auxiliary pdfs with given mixing ratios in terms of the KL divergence as stated in Theorem 1. Secondly, we gave a representation of a non-parametric e-mixture model by using a weighted empirical distribution. Thirdly, we introduced an estimator of the KL divergence between weighted empirical distributions. Then our problem is reduced to finding the optimal weights of the weighted empirical distribution of all the given data which satisfies the condition of Theorem 1. Consequently, we provided a way to use samples included in auxiliary datasets to better expressing the target distribution.

One of the virtues of the e-mixture model is that negative mixture ratios can be considered because mixture ratios are in the exponential function. In this paper, we derived multiplicative update algorithm for estimating mixture ratios, which does not support negative mixture ratios. Therefore another interesting direction of our future work might be design of an e-mixture estimation algorithm which can dealt with negative mixture ratios.

Acknowledgement. Part of this work was supported by JSPS KAKENHI No. 2512-0009, 25120011, and 16K16108.

References

1. Tu, W., Sun, S.: A subject transfer framework for EEG classification. Neurocomputing **82**, 109–116 (2011)
2. Silva, J., Narayanan, S.S.: Information divergence estimation based on data-dependent partitions. In: Proceedings on IEEE International Conference on Acoustics, Speech and Signal Processing, pp. 429–432 (2001)
3. Amari, S., Nagaoka, H.: Methods of Information Geometry. American Mathematical Society, Providence (2000)
4. McLachlan, G., Peel, D.: Finite Mixture Models. Probability and Statistics. Wiley, New York (2000)
5. Dempster, A.P., Laird, N.M., Rubin, D.B.: Maximum likelihood from incomplete data via the EM algorithm. J. Roy. Stat. Soc. Ser. B (Methodol.) **39**, 1–38 (1997)
6. Genest, C., Zidek, J.V.: Combining probability distributions: a critique and an annotated bibliography. Stat. Sci. **1**, 114–135 (1986)
7. Kullback, S., Leibler, R.A.: On information and sufficiency. Ann. Math. Stat. **22**, 79–86 (1951)
8. Murata, N., Fujimoto, Y.: Bregman divergence and density integration. J. Math Ind. **1**, 97–104 (2009)
9. Wang, Q., Kulkarni, S.R., Verdú, S.: Divergence estimation of continuous distributions based on data-dependent partitions. IEEE Trans. Inf. Theor. **51**, 3064–3074 (2005)
10. Wang, Q., Kulkarni, S.R., Verdú, S.: Divergence estimation for multidimensional densities via k-nearest-neighbor distances. IEEE Trans. Inf. Theor. **55**, 2392–2405 (2009)
11. Hino, H., Murata, N.: Information estimators for weighted observations. Neural Netw. **1**, 260–275 (2013)

An Entropy Estimator Based on Polynomial Regression with Poisson Error Structure

Hideitsu Hino[1]([⊠]), Shotaro Akaho[2], and Noboru Murata[3]

[1] University of Tsukuba, 1-1-1 Tennoudai, Tsukuba, Ibaraki 305–8573, Japan
hinohide@cs.tsukuba.ac.jp
[2] National Institute of Advanced Industrial Science and Technology,
1-1-1 Umezono, Tsukuba, Ibaraki 305–8568, Japan
[3] Waseda University, 3-4-1 Ohkubo, Shinjuku-ku, Tokyo 169–8555, Japan

Abstract. A method for estimating Shannon differential entropy is proposed based on the second order expansion of the probability mass around the inspection point with respect to the distance from the point. Polynomial regression with Poisson error structure is utilized to estimate the values of density function. The density estimates at every given data points are averaged to obtain entropy estimators. The proposed estimator is shown to perform well through numerical experiments for various probability distributions.

Keywords: Entropy · Regression · Density estimation · Poisson error structure

1 Introduction

Let X be a p-dimensional random variable with a probability density function (pdf) $f(X)$. The differential entropy [1,2] of this distribution with pdf $f(x)$ is defined by

$$H(f) = - \int f(x) \ln f(x) \mathrm{d}x. \tag{1}$$

We consider estimating the entropy $H(f)$ in non-parametric manner using a set of observations $\mathcal{D} = \{x_i\}_{i=1}^n$, where $x_i, i = 1, \ldots, n$ are the independent realizations of X with pdf $f(x)$. There are a large number of non-parametric entropy estimation methods. The simplest approach is firstly estimating the pdf using the observed dataset $\mathcal{D}$ by using, for example the kernel density estimator [3], then substitute the estimate $\hat{f}(x)$ into the definition of the entropy. The entropy can be estimated by numerical integration, though, it is known that numerical integration for multivariate function is unstable and time consuming, it is recommended in [4] to use empirical expectation with respect to $\mathcal{D}$ as

$$\hat{H}(\mathcal{D}) = -\frac{1}{n} \sum_{i=1}^{n} \ln \hat{f}(x_i). \tag{2}$$

© Springer International Publishing AG 2016
A. Hirose et al. (Eds.): ICONIP 2016, Part II, LNCS 9948, pp. 11–19, 2016.
DOI: 10.1007/978-3-319-46672-9_2

One of the most popular methods for differential entropy estimation is the method based on the k-nearest neighbor method [5–10]. In this work, we derive a nonparametric entropy estimator based on the second order expansion of probability mass function and polynomial regression with Poisson error structure. The proposed method is experimentally shown to work well for estimating the differential entropy of various probability distributions.

2   Preliminary and Notation

We consider the problem of estimating the value $f(z)$ of the probability density function at the *inspection point* $z \in \mathbb{R}^p$ using the set of observation $\mathcal{D} = \{x_i\}_{i=1}^n$. Let the p-dimensional ball with radius ε centered at z be $b(z; \varepsilon) = \{x \in \mathbb{R}^p | \|z - x\| < \varepsilon\}$, which has volume $|b(z; \varepsilon)| = c_p \varepsilon^p$, where $c_p = \pi^{p/2}/\Gamma(p/2 + 1)$ is a volume element of the p-dimensional unit ball and $\Gamma(\,\cdot\,)$ is the gamma function. The probability mass of the ball is defined by

$$q_z(\varepsilon) = \int_{x \in b(z;\varepsilon)} f(x)\mathrm{d}x. \tag{3}$$

Expanding the integrand, we obtain

$$\begin{aligned} q_z(\varepsilon) &= \int_{x \in b(z;\varepsilon)} \{f(x) + (x - z)^\top \nabla f(z) + O(\varepsilon^2)\}\, \mathrm{d}x \\ &= |b(z; \varepsilon)| \left(f(z) + O(\varepsilon^2)\right) = c_p \varepsilon^p f(z) + O(\varepsilon^{p+2}). \end{aligned}$$

Assume that the radius ε of the ball is enough small and ignore the second order term. Then, approximating the left hand side of the above equation by the proportion of the number of samples fallen in the ball to the whole sample size n, we obtain a first order approximation of the value of pdf as

$$\hat{f}(z; \varepsilon) = \frac{k_\varepsilon}{n c_p \varepsilon^p}, \tag{4}$$

where k_ε is the number of samples in $\mathcal{D}$ inside the ε-ball [11–13]. Conversely, when we fix the number of sample points from the inspection point to k, we obtain the k-NN density estimator $\hat{f}^{nn}(z; k) = k/(n c_p \varepsilon_k^p)$, where ε_k is the distance between the inspection point to the k-th nearest point. Denoting the values of k-NN estimator at $x_i \in \mathcal{D}$ using $\mathcal{D} \backslash \{x_i\}$, namely, without using $\{x_i\}$, by $\hat{f}_i^{nn}(x_i; k)$, we obtain the k-NN based entropy estimator [6] by

$$\hat{H}^{nn}(\mathcal{D}; k) = -\sum_{i=1}^n \ln \hat{f}_i^{nn}(x_i; k). \tag{5}$$

3   Second Order Method

In our previous work [14], we derived nonparametric entropy estimators based on the second order expansion of the integrand of Eq. (3).

Proposition 1. *The probability mass of the ε-ball around z is expanded as*

$$q_z(\varepsilon) = c_p f(z)\varepsilon^p + \frac{n}{4(p/2+1)} c_p \mathrm{Tr}\nabla^2 f(z)\varepsilon^{p+2} + O(\varepsilon^{p+4}). \tag{6}$$

Approximating the left hand side of Eq. (6) by k_ε/n, and dividing the equation by $c_p\varepsilon^p$, we obtain

$$\frac{k_\varepsilon}{nc_p\varepsilon^p} = f(z) + C\varepsilon^2 + O(\varepsilon^4), \tag{7}$$

where $C = \frac{n\mathrm{Tr}\nabla^2 f(z)}{4(p/2+1)}$. Introducing the *response variable* $Y_\varepsilon = \frac{k_\varepsilon}{nc_p\varepsilon^p}$ and the *explanatory variable* $X_\varepsilon = \varepsilon^2$, and ignoring higher order term with respect to ε, we obtain a linear equation

$$Y_\varepsilon \simeq f(z) + CX_\varepsilon. \tag{8}$$

This Eq. (8) can be regarded as a linear regression model with respect to $(X_\varepsilon, Y_\varepsilon)$. These variables vary with the different values of ε. Taking a set of radii $\mathcal{E} = \{\varepsilon_i\}_{i=1}^m$ and regarding the pairs $\{(X_\varepsilon, Y_\varepsilon)\}_{\varepsilon \in \mathcal{E}}$ observed samples, we can estimate $f(z)$ and C by minimizing the squared error

$$R = \frac{1}{m} \sum_{\varepsilon \in \mathcal{E}} (Y_\varepsilon - f(z) - CX_\varepsilon)^2, \tag{9}$$

which is nothing but the fitting of simple linear model. Namely, the intercept of the linear model is the estimate of the value of the pdf at z. Let $\hat{f}_i^s(x_i)$ be the estimate obtained by solving Eq. (9) without using a sample x_i. Then, by leave-one-out estimate, we obtain a nonparametric entropy estimator

$$\hat{H}^s(\mathcal{D}) = -\frac{1}{n} \sum_{i=1}^n \ln \hat{f}_i^s(x_i), \tag{10}$$

which we call the Simple Regression Entropy Estimator (SRE) [14].

In [14], another entropy estimator is also proposed, by substituting the relation Eq. (8) to the empirical estimate of the differential entropy (2) and fitting linear model. Suppose ε is fixed, and consider Eq. (8) at the inspection point $x_i \in \mathcal{D}$. Here $Y_\varepsilon = \frac{k_\varepsilon}{nc_p\varepsilon^p}$ and $C = \frac{n\nabla^2 f(x_i)}{4(p/2+1)}$ depend on the inspection point x_i, we denote them as Y_ε^i and C^i, respectively. To derive an entropy estimator based on Eq. (2), we consider the minus of the logarithm of $Y_\varepsilon^i = f(x_i) + C^i X_\varepsilon$. By averaging this quantity with respect to all sample points $x_i \in \mathcal{D}$, we obtain

$$-\frac{1}{n}\sum_{i=1}^{n}\ln Y_{\varepsilon}^{i} = -\frac{1}{n}\sum_{i=1}^{n}\ln\left\{f(x_i)+C^iX_{\varepsilon}\right\}$$

$$= -\frac{1}{n}\sum_{i=1}^{n}\ln f(x_i) - \frac{1}{n}\sum_{i=1}^{n}\ln\left(1+\frac{C^iX_{\varepsilon}}{f(x_i)}\right)$$

$$\simeq -\frac{1}{n}\sum_{i=1}^{n}\ln f(x_i) - \frac{1}{n}\left(\sum_{i=1}^{n}\frac{C^i}{f(x_i)}\right)X_{\varepsilon}.$$

The last equation is from the first order Taylor expansion of $\ln(1+x)$. The first term of the above equation is the empirical estimate (2) of the entropy. So, by defining $\bar{Y}_{\varepsilon} = -\frac{1}{n}\sum_{i=1}^{n}\ln Y_{\varepsilon}^i$, $H(\mathcal{D}) = -\frac{1}{n}\sum_{i=1}^{n}f(x_i)$, and $\bar{C} = -\frac{1}{n}\sum_{i=1}^{n}\frac{C^i}{f(x_i)}$, we obtain a relationship

$$\bar{Y}_{\varepsilon} = H(\mathcal{D}) + \bar{C}X_{\varepsilon}. \tag{11}$$

In the same manner as in SRE, this equation is valid for each of sample points $\{(X_{\varepsilon},\bar{Y}_{\varepsilon})\}_{\varepsilon\in\mathcal{E}}$. By fitting a linear model, we obtain an entropy estimator $\hat{H}^d(\mathcal{D})$ as the estimated intercept, which is called the Direct Regression Entropy Estimator (DRE). The SRE and DRE take the higher order information of pdf into account, though they do not consider the characteristic of the error structure for the observation, namely, the number of samples within a ball should be treated as a counting variable.

4   Proposed Method

In this section, we derive a novel entropy estimator based on linear regression with a Poisson error structure. The left hand side of the Eq. (6) is approximated by k_{ε}/n again. Then, multiply both sides of equation by n to obtain

$$k_{\varepsilon} \simeq c_p n f(z)\varepsilon^p + c_p n \frac{n}{4\Gamma(p/2+1)}\mathrm{Tr}\nabla^2 f(z)\varepsilon^{p+2}. \tag{12}$$

This equation is regarded as a regression of k_{ε} on $(\varepsilon^p, \varepsilon^{p+2})$. Namely, for a certain $\varepsilon > 0$, the explanatory variable is defined by $X = (\varepsilon^p, \varepsilon^{p+2})$ and response variable is defined by $Y = k_{\varepsilon}$, which are linked by a simple generalized linear model $Y = \beta^{\top}X$. Since k_{ε} is a counting data, the Poisson error structure is a natural choice. We note that it is common to adopt the logarithmic link function for Poisson regression as a link function in the generalized linear model, though, in our formulation, the identity link function is natural. The logarithmic link function can avoid negative values for the response variable while the identity link cannot. However, our aim is not in prediction but in fitting to obtain the coefficient as the estimate of pdf value, and the estimated pdfs in all of our experiments were non-negative.

More concretely, for a set of radii $\mathcal{E} = \{\varepsilon_i\}_{i=1}^{m}$, we calculate the corresponding pairs of explanatory and response variable $\{(Y_i, X_i)\}_{i=1}^{m}$ where $X_i = (\varepsilon_i^p, \varepsilon_i^p)$ and

Y_i is the number of samples within the ε_i-ball centered at the inspection point. Then, we maximize the likelihood function

$$L(\boldsymbol{\beta}) = \prod_{i=1}^{m} \frac{e^{-X_i^\top \boldsymbol{\beta}}(X_i^\top \boldsymbol{\beta})^{Y_i}}{Y_i!} \tag{13}$$

of a Poisson distribution with respect to $\boldsymbol{\beta} = (\beta_1, \beta_2)$. The ML estimate $\hat{\beta}_1$ for the first coefficient β_1 is divided by $c_p n$ to obtain the estimate of the pdf value at z as $\hat{f}(z) = \hat{\beta}_1/(c_p n)$. This procedure is done for each $X \in \mathcal{D}$ and by leave-one-out method, we obtain the proposed entropy estimator in the same manner as in SRE. The notable characteristic of the proposed method, compared to conventional SRE, is in utilizing the Poisson error structure. We call the proposed entropy estimator EPI (Entropy Estimator with Poisson-noise structure Identity-link regression) henceforth.

5   Numerical Experiments

We apply some conventional and proposed entropy estimators for samples from various distributions to see the accuracies of the estimators.

5.1   Univariate Case

We evaluate the performance of entropy estimation by the absolute error

$$AE = |H(f) - \hat{H}(\mathcal{D})| \tag{14}$$

between the ground truth entropy $H(f)$ and the estimates $\hat{H}(\mathcal{D})$. We used the following 15 distributions: (1) Normal, (2) Skewed, (3) Strongly Skewed, (4) Kurtotic, (5) Bimodal, (6) Skewed Bimodal, (7) Trimodal, (8) Claw, (9) 4th Power Exponential, (10) Logistic, (11) Laplace, (12) t with df=5, (13) Mixed t, (14) Exponential, and (15) Cauchy, which are shown in Fig. 1. Details of these distributions are shown in [14]. We compare the proposed method to four existing methods, (a) KDE: pdf is estimated by kernel density estimation, and the estimate $\hat{f}(x)$ is substituted in Eq. (2). The kernel band width is optimized by using the unbiased cross-validation method [15]. (b) kNN: entropy is estimated by the k-NN method [6]. The number k is fixed to $k = 3$ following the empirical results reported in [16]. (c, d) SRE, DRE: the methods explained in Sect. 3.

From the ground truth distributions, we sampled datasets 100 times and perform entropy estimation, and the performance is estimated by average and standard deviations of AE as shown in Table 1. The number of sample in each dataset is 500. We note that we performed the same experiments with different sample sizes, and observed similar tendencies. From this result, it is seen that the proposed method significantly outperforms other four methods in 2 out of 15 distributions, and marks the second best method in 9 methods.

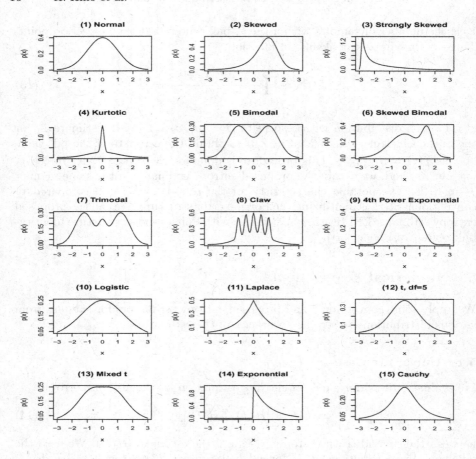

Fig. 1. Plots of 15 probability density functions for generating samples.

5.2 Multivariate Case

For seeing the effect of dimensions on the estimation accuracy, we performed a set of experiments with multidimensional distributions. It is difficult to calculate the ground truth entropy values for general multidimensional distributions, hence we use Gaussian distributions with three different covariance structures:

Isometric: covariance matrix is a p dimensional unit matrix.
Band: diagonal elements of covariance matrix is one, and its upper and lower elements are 0.3.
Full Correlation: Each element of the covariance matrix is set to

$$[\Sigma_p]_{ij} = 0.9^{|i-j|+1}, \quad 0 \le i, j \le p. \tag{15}$$

Table 1. Averages of absolute errors of entropy estimations for different seven methods. Sample size n is set to 500. The minimum AE results are shown in boldface, the second best method is shown with †, and when the minimum is statistically significant in t-test with $\alpha = 0.05$ compared to the second best result, the result is shown with ∗.

	KDE	kNN	SRE	DRE	EPI
Type 1	**0.028**(0.0194)	0.040(0.0292)	0.066(0.0334)	0.030(0.0233)	†0.029(0.0192)
Type 2	†0.029(0.0246)	0.039(0.0310)	0.061(0.0329)	0.032(0.0226)	**0.027**(0.0239)
Type 3	0.149(0.1891)	†0.035(0.0247)	0.087(0.0295)	0.139(0.0259)	∗**0.020**(0.0163)
Type 4	0.219(0.2491)	**0.040**(0.0313)	0.149(0.0409)	0.088(0.0381)	†0.041(0.0301)
Type 5	0.022(0.0161)	0.033(0.0226)	0.022(0.0161)	**0.017**(0.0122)	†0.021(0.0154)
Type 6	0.026(0.0206)	0.037(0.0259)	0.027(0.0200)	**0.023**(0.0170)	†0.024(0.0172)
Type 7	0.022(0.0189)	0.032(0.0228)	**0.018**(0.0127)	†0.019(0.0144)	0.020(0.0131)
Type 8	0.154(0.1289)	0.038(0.0354)	∗**0.025**(0.0185)	0.055(0.0289)	†0.036(0.0257)
Type 9	0.022(0.0150)	0.034(0.0254)	0.025(0.0191)	†0.021(0.0159)	**0.020**(0.0136)
Type 10	0.053(0.0617)	0.041(0.0318)	0.091(0.0412)	†0.038(0.0229)	∗**0.033**(0.0256)
Type 11	0.156(0.2368)	0.049(0.0353)	0.131(0.0426)	**0.036**(0.0260)	†0.039(0.0322)
Type 12	0.094(0.1175)	†0.041(0.0345)	0.108(0.0411)	**0.040**(0.0299)	0.042(0.0304)
Type 13	0.251(0.3148)	0.044(0.0334)	0.086(0.0381)	**0.035**(0.0257)	†0.041(0.0327)
Type 14	0.505(0.4753)	**0.045**(0.0334)	0.073(0.0465)	0.127(0.0479)	†0.048(0.0361)
Type 15	1.903(1.0509)	0.477(0.0933)	∗**0.169**(0.0740)	0.509(0.1055)	†0.310(0.1006)

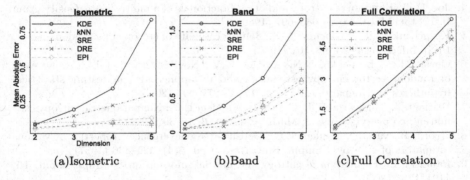

Fig. 2. Averages of absolute errors of entropy estimation when p is varied from 2 to 5. The number of samples is fixed to $n = 300$.

We varied the sample size to $n = 100, 300, 500, 700$, though, we didn't see systematic difference, and we show the case with $n = 300$ in Fig. 2. From Fig. 2, we can see that for all of three distributions, the proposed method shows moderate increase in estimation error as the increase of dimension, and it is a strong candidate of the non-parametric entropy estimator among classical kNN, KDE, and other recently proposed methods.

6   Conclusion

We proposed a non-parametric entropy estimator based on the second order expansion of probability mass function, and polynomial fitting with respect to the distance from the inspection point. By modeling the error structure by Poisson distribution, we obtained comparable or superior estimation accuracies to conventional method. It is also shown that the proposed method works well for multi-dimensional cases. Our future work includes investigation of statistical properties of the proposed estimator, including the optimal choice of the radii $\mathcal{E}$. We are also planning to apply the proposed estimator to various real-world problems which require accurate entropy estimation.

Acknowledgement. Part of this work was supported by JSPS KAKENHI No. 25120009, 25120011, and 16K16108.

References

1. Cover, T.M., Thomas, J.A.: Elements of information theory. Wiley, Hoboken (1991)
2. Shannon, C.E.: A mathematical theory of communication. Bell Syst. Tech. J. **27**(3), 379–423 (1948)
3. Wand, M.P., Jones, M.C.: Kernel Smoothing. Chapman & Hall/CRC, London (1994)
4. Joe, H.: Estimation of entropy and other functionals of a multivariate density. Ann. Inst. Stat. Math. **41**(4), 683–697 (1989)
5. Kozachenko, L.F., Leonenko, N.N.: Sample estimate of entropy of a random vector. Prob. Inf. Transm **23**, 95–101 (1987)
6. Goria, M.N., Leonenko, N.N., Mergel, V.V., Novi Inverardi, P.L.: A new class of random vector entropy estimators and its applications in testing statistical hypotheses. J. Nonparametric Stat. **17**(3), 277–297 (2005)
7. Beirlant, J., Dudewicz, E.J., Györfi, L., Meulen, E.C.: Nonparametric entropy estimation: an overview. Int. J. Math. Stat. Sci. **6**, 17–39 (1997)
8. Györfi, L., van der Meulen, E.C.: Density-free convergence properties of various estimators of entropy. Comput. Stat. Data Anal. **5**(4), 425–436 (1987)
9. Paninski, L.: Estimation of entropy and mutual information. Neural Comput. **15**, 1191–1253 (2003)
10. Pérez-Cruz, F.: Estimation of information theoretic measures for continuous random variables. In: NIPS, pp. 1257–1264 (2008)
11. Loftsgaarden, D.O., Quesenberry, C.P.: A nonparametric estimate of a multivariate density function. Ann. Math. Stat. **36**(3), 1049–1051 (1965)
12. Mack, Y.P., Rosenblatt, M.: Multivariate k-nearest neighbor density estimates. J. Multivar. Anal. **9**(1), 1–15 (1979)
13. Moore, D.S., Yackel, J.W.: Consistency properties of nearest neighbor density function estimators. Ann. Stat. **5**(1), 143–154 (1977)
14. Hino, H., Koshijima, K., Murata, N.: Non-parametric entropy estimators based on simple linear regression. Comput. Stat. Data Anal. **89**, 72–84 (2015)
15. Rudemo, M.: Empirical choice of histograms and kernel density estimators. Scand. J. Stat. **9**(2), 65–78 (1982)

16. Khan, S., Bandyopadhyay, S., Ganguly, A.R., Saigal, S., Erickson, D.J., Pro-
topopescu, V., Ostrouchov, G.: Relative performance of mutual information estimation methods for quantifying the dependence among short and noisy data. Phys.
Rev. E: Stat., Nonlin., Soft Matter Phys. **76**(2 Pt 2), 026209 (2007)

A Problem in Model Selection of LASSO and Introduction of Scaling

Katsuyuki Hagiwara[(⊠)]

Faculty of Education, Mie University, 1577 Kurima-Machiya-cho,
Tsu 514-8507, Japan
hagi@edu.mie-u.ac.jp

Abstract. In this article, we considered to assign a single scaling parameter to LASSO estimators for investigating and improving a problem
of excessive shrinkage at a sparse representation. This problem is important because it directly affects a quality of model selection in LASSO.
We derived a prediction risk for LASSO with scaling and obtained an
optimal scaling parameter value that minimizes the risk. We then showed
the risk is improved by assigning the optimal scaling value. In a numerical example, we found that an estimate of the optimal scaling value is
larger than one especially at a sparse representation; i.e. excessive shrinkage is relaxed by expansion via scaling. Additionally, we observed that
a risk for LASSO is high at a sparse representation and it is minimized
at a relatively large model while this is improved by the introduction
of an estimate of the optimal scaling value. We here constructed a fully
empirical risk estimate that approximates the actual risk well. We then
observed that, by applying the risk estimate as a model selection criterion, LASSO with scaling tends to obtain a model with low risk and high
sparsity compared to LASSO without scaling.

Keywords: LASSO · Sparse modeling · Shrinkage · Thresholding ·
Scaling

1 Introduction

In recent years, sparse modeling is an important topic in machine learning and
statistics. Especially, LASSO (Least Absolute Shrinkage and Selection Operator)
is a popular method that has been extensively studied [2,3,7–9]. In LASSO,
estimators are obtained by minimizing a cost function that is defined by squared
error sum plus ℓ_1 regularizer; i.e. it is an ℓ_1 penalized least squares method.

LASSO has a nature of soft-thresholding that implements thresholding and
shrinkage of coefficients. These two properties are simultaneously controlled by
a single regularization parameter. LASSO is known to yield a sparse representation due to the thresholding property, by which the regularization parameter
apparently determines tradeoff between bias and variance. If the parameter value
is too large then effective coefficients are possible to be removed. This causes a

© Springer International Publishing AG 2016
A. Hirose et al. (Eds.): ICONIP 2016, Part II, LNCS 9948, pp. 20–27, 2016.
DOI: 10.1007/978-3-319-46672-9_3

large bias. Conversely, if the parameter value is too small then fruitless coefficients remain. This causes a large variance. However, it is not straightforward since the parameter also determines an amount of shrinkage. If the parameter value is large then both of a threshold level and an amount of shrinkage are large. Therefore, the amount of shrinkage is automatically large at a small model. This leads to a possibility of excessive shrinkage that causes a high risk at a sparse representation. The important point is that, due to this problem, a model selection method based on prediction error tends to choose a larger model even when there is a sparse representation. This drawback in model selection is actually pointed out in [5] as a dilemma between sparsity and generalization. Note that this problem of excessive shrinkage is not entirely solved by a post-estimation of coefficients since it relates to a quality of model selection. There are several methods to solve the problem of excessive shrinkage. [3] has proposed SCAD (Smoothly Clipped Absolute Deviation) penalty instead of ℓ_1 penalty. [8] has proposed adaptive LASSO that employs a kind of weighted ℓ_1 penalty. An ℓ_1 penalty term is modified by different ways (functions) in SCAD and adaptive LASSO while shrinkage amounts are taken to be small for the least squares estimators with large absolute values. In these works, however, relationship between excessive shrinkage and model selection is not focused.

On the other hand, to relax excessive shrinkage, it is natural to consider the expansion of LASSO estimators by scaling. Actually, it has been shown that a simple scaling works well in case of soft-thresholding [4]; i.e. orthogonal design case. In this article, we consider to apply a scaling method to LASSO for investigating an excessive shrinkage problem of LASSO and model selection property which is not considered in SCAD and adaptive LASSO. For simplicity, we consider to assign a single common scaling parameter to LASSO estimators and derive a risk that is defined by an expected error between target output and output estimate. To clarify a problem of excessive shrinkage in LASSO, we investigate the difference of risks between LASSO with scaling and without scaling. In a numerical experiment, we then focus on risk curves of LASSO with scaling and without scaling, by which we compare their qualities of model selection via the estimates of risks.

In Sect. 2, we formulate a regression problem and LASSO. In Sect. 3, we explain a scaling method in LASSO, in which we derived a risk for LASSO with a single common scaling parameter and give some analyses on the risk. In this section, furthermore, we give a fully empirical model selection criterion for LASSO with scaling. In Sect. 4, we show numerical investigations on a problem of excessive shrinkage and effectiveness of scaling in a toy example. Also, we confirm the effectiveness of the given model selection criterion. Section 5 is devoted for conclusions and future works.

2   Regression Problem and LASSO

Let $x = (x_1, \ldots, x_m)$ and y be explanatory variables and a response variable, for which we have n samples: $\{(x_{i,1}, \ldots, x_{i,m}, y_i) : i = 1, \ldots, n\}$. We define

$\boldsymbol{x}_j = (x_{1,j}, \ldots, x_{n,j})' \in \mathbb{R}^n$ for $j = 1, \ldots, m$, where $'$ stands for the transpose operator. We assume that $m \le n$ holds and $\boldsymbol{x}_1, \ldots, \boldsymbol{x}_m$ are linearly independent. We also define $\mathbf{X} = (\boldsymbol{x}_1, \ldots, \boldsymbol{x}_m)$ and $\boldsymbol{y} = (y_1, \ldots, y_n)'$. Let $\varepsilon_1, \ldots, \varepsilon_n$ be i.i.d. samples from $N(0, \sigma^2)$; i.e. normal distribution with mean 0 and variance σ^2. Thus, $\boldsymbol{\varepsilon} \sim N(\mathbf{0}_n, \sigma^2 \mathbf{I}_n)$, where $\mathbf{0}_n$ is an n-dimensional zero vector and $\mathbf{I}_n$ is an $n \times n$ identity matrix. We define $\boldsymbol{\varepsilon} = (\varepsilon_1, \ldots, \varepsilon_n)'$ and assume $\boldsymbol{y} = \boldsymbol{\mu} + \boldsymbol{\varepsilon}$. We therefore have $\boldsymbol{\mu} = \mathbb{E}\boldsymbol{y}$, where $\mathbb{E}$ is the expectation with respect to the joint probability distribution of $\boldsymbol{y}$. We consider a regression problem by $\mathbf{X}\boldsymbol{b}$, where $\boldsymbol{b} = (b_1, \ldots, b_m)$ is a coefficient vector. Let $\widehat{\boldsymbol{b}} = (\widehat{b}_1, \ldots, \widehat{b}_m)$ be an estimator of $\boldsymbol{b}$. LASSO is a method for obtaining coefficient estimators that minimize ℓ_1 regularized cost function defined by

$$C_\lambda(\boldsymbol{b}) = \|\boldsymbol{y} - \mathbf{X}\boldsymbol{b}\|^2 + \lambda\|\boldsymbol{b}\|_1, \tag{1}$$

where $\|\cdot\|$ is the Euclidean norm and $\|\boldsymbol{b}\|_1 = \sum_{k=1}^{n} |b_j|$. $\lambda \ge 0$ is a regularization parameter. The second term of the right hand side of (1) is called ℓ_1 regularizer. Let $\widehat{\boldsymbol{b}}_\lambda = (\widehat{b}_{1,\lambda}, \ldots, \widehat{b}_{m,\lambda})$ be a LASSO solution. Since the LASSO is known to be yield a sparse representation under an appropriate choice of λ, some of elements in $\widehat{\boldsymbol{b}}_\lambda$ are exactly zeros. We define $B_\lambda = \{i : \widehat{b}_{i,\lambda} \ne 0\}$ and $\widehat{k}_\lambda = |B_\lambda|$.

3   LASSO with Scaling

3.1   LASSO with Scaling

We now consider to assign a positive scaling parameter to LASSO estimator component-wisely. We define $\boldsymbol{\alpha} = (\alpha_1, \ldots, \alpha_m)$, $\alpha_k \ge 0$. Let $\mathbf{A}$ be a diagonal matrix whose diagonal vector is $\boldsymbol{\alpha}$. LASSO with component-wise scaling estimator is given by $\mathbf{A}\widehat{\boldsymbol{b}}_\lambda$; i.e. $\alpha_k \widehat{b}_{k,\lambda}$ for $k \in \{1, \ldots, m\}$. Since our main purpose in this paper is to clarify whether the introduction of scaling improves a problem excessive shrinkage in LASSO or not, for simplicity, we here consider $\mathbf{A} = \alpha \mathbf{I}_m$, $\alpha \ge 0$; i.e. assignment of a single scaling parameter common for all LASSO estimators. We define $\widehat{\boldsymbol{\mu}}_\lambda = \mathbf{X}\widehat{\boldsymbol{b}}_\lambda$ that is a LASSO output vector. The output vector with a single scaling parameter is given by $\alpha\widehat{\boldsymbol{\mu}}_\lambda$

3.2   Derivation of Risk Under a Single Scaling Parameter

A prediction capability of LASSO estimator with scaling is measured by a risk defined by

$$R_n(\lambda, \alpha) = \frac{1}{n}\mathbb{E}\left[\|\alpha\widehat{\boldsymbol{\mu}}_\lambda - \boldsymbol{\mu}\|^2\right]. \tag{2}$$

$R_n(\lambda, 1)$ is a risk of LASSO estimator.

Since $\boldsymbol{y} = \boldsymbol{\mu} + \boldsymbol{\varepsilon}$ and $\boldsymbol{\varepsilon} \sim N(\mathbf{0}, \sigma^2 \mathbf{I}_n)$, we have

$$\begin{aligned}
\mathbb{E}(\alpha\widehat{\boldsymbol{\mu}}_\lambda - \boldsymbol{y})'(\boldsymbol{y} - \boldsymbol{\mu}) &= \alpha\mathbb{E}(\widehat{\boldsymbol{\mu}}_\lambda - \mathbb{E}\widehat{\boldsymbol{\mu}}_\lambda)'(\boldsymbol{y} - \boldsymbol{\mu}) + \mathbb{E}(\alpha\mathbb{E}\widehat{\boldsymbol{\mu}}_\lambda - \boldsymbol{y})'(\boldsymbol{y} - \boldsymbol{\mu}) \\
&= \alpha\mathbb{E}(\widehat{\boldsymbol{\mu}}_\lambda - \mathbb{E}\widehat{\boldsymbol{\mu}}_\lambda)'(\boldsymbol{y} - \boldsymbol{\mu}) + \mathbb{E}(\boldsymbol{\mu} - \boldsymbol{y})'(\boldsymbol{y} - \boldsymbol{\mu}) \\
&= \alpha\mathbb{E}(\widehat{\boldsymbol{\mu}}_\lambda - \mathbb{E}\widehat{\boldsymbol{\mu}}_\lambda)'(\boldsymbol{y} - \boldsymbol{\mu}) - n\sigma^2.
\end{aligned} \tag{3}$$

We thus have

$$R_n(\lambda, \alpha) = \frac{1}{n}\mathbb{E}\left[\|\alpha\widehat{\boldsymbol{\mu}}_\lambda - \boldsymbol{y}\|^2\right] + \frac{1}{n}\mathbb{E}\left[\|\boldsymbol{y} - \boldsymbol{\mu}\|^2\right] + 2\mathbb{E}(\alpha\widehat{\boldsymbol{\mu}}_\lambda - \boldsymbol{y})'(\boldsymbol{y} - \boldsymbol{\mu})$$

$$= \frac{1}{n}\mathbb{E}\|\alpha\widehat{\boldsymbol{\mu}}_\lambda - \boldsymbol{y}\|^2 - \sigma^2 + \frac{2\alpha}{n}\mathbb{E}(\widehat{\boldsymbol{\mu}}_\lambda - \mathbb{E}\widehat{\boldsymbol{\mu}}_\lambda)'(\boldsymbol{y} - \boldsymbol{\mu}). \tag{4}$$

In Theorem 1 of [9] for LASSO estimate,

$$\mathbb{E}(\widehat{\boldsymbol{\mu}}_\lambda - \mathbb{E}\widehat{\boldsymbol{\mu}}_\lambda)'(\boldsymbol{y} - \boldsymbol{\mu}) = \sigma^2\mathbb{E}\widehat{k}_\lambda \tag{5}$$

has been derived by applying Stein's lemma [6]. We thus have

$$R_n(\lambda, \alpha) = \frac{1}{n}\mathbb{E}\|\alpha\widehat{\boldsymbol{\mu}}_\lambda - \boldsymbol{y}\|^2 - \sigma^2 + \frac{2\alpha\sigma^2}{n}\mathbb{E}\widehat{k}_\lambda. \tag{6}$$

Therefore, C_p type model selection criterion or SURE (Stein's unbiased risk estimate) for LASSO with scaling is given by

$$r_n(\lambda, \alpha, \sigma^2) = \frac{1}{n}\|\alpha\widehat{\boldsymbol{\mu}}_\lambda - \boldsymbol{y}\|^2 - \sigma^2 + \frac{2\alpha\sigma^2}{n}\widehat{k}_\lambda. \tag{7}$$

As a special case of $\alpha = 1$, SURE for LASSO is obtained by

$$r_n(\lambda, 1, \sigma^2) = \frac{1}{n}\|\widehat{\boldsymbol{\mu}}_\lambda - \boldsymbol{y}\|^2 - \sigma^2 + \frac{2\sigma^2}{n}\widehat{k}_\lambda. \tag{8}$$

By setting the derivative of (6) with respect to α to zero, the minimizing scaling value of $R_n(\lambda, \alpha)$ is given by

$$\alpha_{\text{opt}} = \frac{\mathbb{E}\widehat{\boldsymbol{\mu}}_\lambda' \boldsymbol{y} - \sigma^2\mathbb{E}\widehat{k}_\lambda}{\mathbb{E}\|\widehat{\boldsymbol{\mu}}_\lambda\|^2} = \frac{\mathbb{E}\widehat{\boldsymbol{\mu}}_\lambda' \boldsymbol{\mu}}{\mathbb{E}\|\widehat{\boldsymbol{\mu}}_\lambda\|^2}. \tag{9}$$

Through a simple calculation using (6) and (9), we have

$$R_n(\lambda, 1) - R_n(\lambda, \alpha_{\text{opt}}) = \frac{1}{n}(\alpha_{\text{opt}} - 1)^2\mathbb{E}\|\widehat{\boldsymbol{\mu}}_\lambda\|^2. \tag{10}$$

Therefore, LASSO with the optimal scaling value improves the risk compared to naive LASSO at any λ. Actually, the right hand side of (10) has shown to be $O(n^{-1}\log n)$ in case of an orthogonal design [4].

3.3   Empirical Model Selection Criterion

We need an estimate of σ^2 and an appropriate value of α if we apply (7) as a model selection criterion for choosing λ.

Firstly, for estimating the noise variance in a regression problem, [1] has recommended to apply

$$\widehat{\sigma}^2 = \frac{\boldsymbol{y}'(\mathbf{I}_n - \mathbf{H}_\gamma)^2 \boldsymbol{y}}{\text{trace}[(\mathbf{I}_n - \mathbf{H}_\gamma)^2]}, \tag{11}$$

where trace is the trace of a matrix and $\mathbf{H}_\gamma = \mathbf{X}(\mathbf{X}'\mathbf{X} + \gamma\mathbf{I}_n)^{-1}\mathbf{X}'$ with $\gamma > 0$. $\mathbf{H}_\gamma$ can be viewed as the hat matrix under the ℓ_2 regularization with a regularization parameter γ. In general, the ℓ_2 regularization is introduced for better generalization and stabilization. We need to carefully select the parameter value for the former reason. However, since the purpose to introduce γ here is to stabilize an estimate of the noise variance, especially when m is large. Therefore, we just set γ to a small value, say, 10^{-6} in applications. By replacing σ^2 with the above $\widehat{\sigma}^2$ in (7), we have a fully empirical model selection criterion for LASSO by

$$r_n(\lambda, 1, \widehat{\sigma}^2) = \frac{1}{n}\|\widehat{\boldsymbol{\mu}}_\lambda - \boldsymbol{y}\|^2 - \widehat{\sigma}^2 + \frac{2\widehat{\sigma}^2}{n}\widehat{k}_\lambda. \tag{12}$$

On the other hand, we need an appropriate value for α in constructing a model selection criterion for LASSO with scaling. We here consider to employ

$$\widehat{\alpha}_{\text{opt}} = \frac{\widehat{\boldsymbol{\mu}}'_\lambda \boldsymbol{y} - \widehat{\sigma}^2 \widehat{k}_\lambda}{\|\widehat{\boldsymbol{\mu}}_\lambda\|^2} \tag{13}$$

as an estimate of α_{opt}. Since $r_n(\lambda, \alpha, \sigma^2)$ is a quadratic function of α, (13) is the minimizer of $r_n(\lambda, \alpha, \widehat{\sigma}^2)$. And (13) apparently seems to be a natural estimate of (9). By replacing σ^2 and α with the above $\widehat{\sigma}^2$ and $\widehat{\alpha}_{\text{opt}}$ in (7), we have a fully empirical model selection criterion for LASSO with scaling by

$$r_n(\lambda, \widehat{\alpha}_{\text{opt}}, \widehat{\sigma}^2) = \frac{1}{n}\|\widehat{\alpha}_{\text{opt}}\widehat{\boldsymbol{\mu}}_\lambda - \boldsymbol{y}\|^2 - \widehat{\sigma}^2 + \frac{2\widehat{\sigma}^2}{n}\widehat{\alpha}_{\text{opt}}\widehat{k}_\lambda. \tag{14}$$

4   Numerical Experiment

We refer to LASSO with scaling by LASSO-S. Note that LASSO-S modifies non-zero coefficients that are obtained by LASSO. Therefore, LASSO-S is a two stage estimation procedure in this meaning.

For $u \in \mathbb{R}$, we define $g_{b,\tau}(u) = \exp\left\{(u - b)^2/(2\tau)\right\}$, where $b \in \mathbb{R}$ and $\tau > 0$. Let u_i, $i = 1, \ldots, n$ be equidistant points in $[-5, 5]$. We define the (i, j) entry of $\mathbf{X}$ by $x_{i,k_j} = g_{k_j,\tau}(u_i)$, where $k_j \in \{u_1, \ldots, u_n\}$ and $j = 1, \ldots, m$. Therefore, we consider a curve fitting problem using a linear combination of m Gaussian basis functions whose centers are some of input data points. We generate y_i by $y_i = \sum_{k=1}^3 \beta_k g_{d_k,\tau}(u_i) + \varepsilon_i$, where $\varepsilon_i \sim N(0, \sigma^2)$. Thus, we consider the case where there exists a small true representation; i.e. exact and very sparse case. In the numerical experiment here, we set $n = 500$, $m = 40$, $\tau = 0.1$, $(\beta_1, \beta_2, \beta_3) = (-4, 3, -2)$ and $(d_1, d_2, d_3) = (\lceil m/4 \rceil, \lceil 2m/4 \rceil, \lceil 3m/4 \rceil)$. We set $\sigma^2 = 1$. We here employ a modification of the least angle regression (LARS) for calculating LASSO path [2]; i.e. LASSO solution at each regularization parameter value.

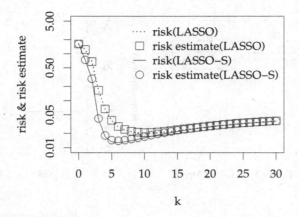

Fig. 1. Averages of risk and risk estimate at each step of LARS-LASSO.

Since the regularization parameter corresponds to the number of un-removed coefficients, we here observe the relationship between the number of un-removed coefficients and risk. Since we know the true representation, we can calculate both the actual risk and the fully empirical risk estimate here. We repeat this for 1000 times and calculate averages of risk and risk estimate. The averages of risk and risk estimate of LASSO and LASSO-S are depicted in Fig. 1. The horizontal axis is the number of steps in LARS-LASSO. λ decreases as the number of steps increases. Note that λ is common for both of LASSO and LASSO-S at each step. In this figure, the solid and dotted line indicate averages of risks for LASSO and LASSO-S respectively. The open square and circle indicate averages of risk estimates for LASSO and LASSO-S respectively. We can find several important insights in this figure.

– In both of LASSO and LASSO-S, risk estimates are well consistent with actual risks. This implies that the risk estimates can be model selection criteria in both methods.
– Although the number of steps in LARS-LASSO is not exactly equal to the number of un-removed coefficients, the latter tends to increase as the former increases since the regularization parameter decreases as the step increases. Therefore, the optimal number of un-removed coefficients that minimize the risk of LASSO-S is smaller than that of LASSO. In other words, LASSO-S prefers a sparse representation in terms of the risk. Moreover, the minimum of risk for LASSO-S is smaller than that for LASSO. Since the true number of non-zero coefficients is very small here, these results show the effectiveness of scaling. From another viewpoint, we can say that this result shows a problem of excessive shrinkage at a sparse representation in LASSO.

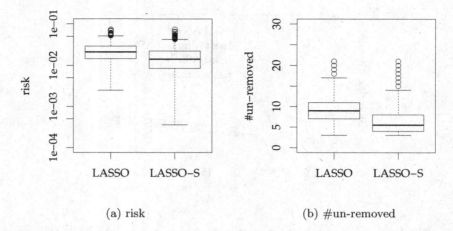

(a) risk   (b) #un-removed

Fig. 2. Averages of risk and the number of un-removed coefficients at the optimal step in terms of risk estimate.

– We can see that the risk estimate around its minimum value is relatively flat for LASSO. On the other hand, the minimum value of risk estimate is clearly identified for LASSO-S. Additionally, the above arguments on the properties of risk apply to the risk estimate since it approximates the actual risk well. Therefore, we can expect that model selection via risk estimate yields a model with low risk and high sparsity in LASSO-S compared to naive LASSO.

In Fig. 2, we show the boxplots of (a) risk and (b) the number of un-removed coefficients at the minimum of the risk estimate; i.e. at the selected model via risk estimate. This result supports that LASSO-S can obtain a model with low risk and high sparsity compared to naive LASSO. Therefore, we can say that

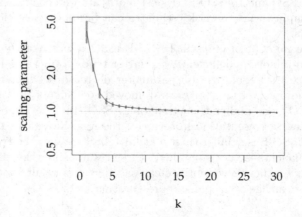

Fig. 3. Average of $\widehat{\alpha}_{\text{opt}}$ at each step of LARS-LASSO.

scaling is effective in model selection using prediction error based criterion; i.e. risk estimate. In Fig. 3, we show the average of $\widehat{\alpha}_{opt}$ at each LAR-LASSO step, in which vertical gray line at each k indicates the standard deviation centered at the average. In this figure, we can see that $\widehat{\alpha}_{opt}$ is larger than 1 for small k, thus, for large regularization parameter value that corresponds to the case where the number of un-removed coefficients is small; i.e. a sparse case. Especially, it is notable at a very small k. On the other hand, as in Fig. 1, risk at a sparse representation for LASSO-S is smaller than that for LASSO. These facts imply that expansion of LASSO estimator is effective or is needed for improving prediction capability at a sparse representation. Or, from another viewpoint, we can say that a problem of excessive shrinkage at a sparse representation in LASSO is serious in model selection.

5   Conclusions and Future Works

In this article, we showed that a problem of excessive shrinkage at a sparse representation in LASSO is improved by introducing scaling of LASSO estimators. The important point argued here is that, due to this problem in LASSO, a larger model tends to be selected by prediction error based criterion even when a sparse representation exists. Contrastly, a sparse representation is properly obtained by an empirical model selection criterion under the introduction of scaling. However, effect of scaling here may not be satisfactory in real applications. This is because a scaling parameter is common for all coefficients; i.e. the optimal amount of scaling may depend on the corresponding variable. For example, adaptive LASSO employs component-wise regularization parameters and succeeds in applications. Along this line, we will consider a component-wise scaling as a future work.

References

1. Carter, C.K., Eagleson, G.K.: A comparison of variance estimators in nonparametric regression. J. R. Statist. Soc. B **54**, 773–780 (1992)
2. Efron, B., Hastie, T., Johnstone, I., Tibshirani, R.: Least angle regression. Ann. Statist. **32**, 407–499 (2004)
3. Fan, J., Li, R.: Variable selection via nonconcave penalized likelihood and its oracle properties. J. Amer. Statist. Assoc. **96**, 1348–1360 (2001)
4. Hagiwara, K.: On scaling of soft-thresholding estimator. Neurocomputing **194**, 360–371 (2016)
5. Leng, C.L., Lin, Y., Wahba, G.: A note on the lasso and related procedures in model selection. Stat. Sin. **16**, 1273–1284 (2006)
6. Stein, C.: Estimation of the mean of a multivariate normal distribution. Ann. Stat. **9**, 1135–1151 (1981)
7. Tibshirani, R.: Regression shrinkage and selection via the lasso. J. R. Statist. Soc. Ser. B. **58**, 267–288 (1996)
8. Zou, H.: The adaptive lasso and its oracle properties. J. Amer. Statist. Assoc. **101**, 1418–1492 (2006)
9. Zou, H., Hastie, T., Tibshirani, R.: On the degree of freedom of the LASSO. Ann. Statist. **35**, 2173–2192 (2007)

A Theoretical Analysis of Semi-supervised Learning

Takashi Fujii[1], Hidetaka Ito[2], and Seiji Miyoshi[2(✉)]

[1] Graduate School of Science and Engineering, Kansai University,
3-3-35 Yamate-cho, Suita-shi, Osaka 564-8680, Japan
k735312@kansai-u.ac.jp
[2] Faculty of Engineering Science, Kansai University,
3-3-35 Yamate-cho, Suita-shi, Osaka 564-8680, Japan
{h.ito,miyoshi}@kansai-u.ac.jp

Abstract. We analyze the dynamical behaviors of semi-supervised learning in the framework of on-line learning by using the statistical-mechanical method. A student uses several correlated input vectors in each update. The student is given a desired output for only one input vector out of these correlated input vectors. In this model, we derive simultaneous differential equations with deterministic forms that describe the dynamical behaviors of order parameters using the self-averaging property in the thermodynamic limit. We treat the Hebbian and Perceptron learning rules. As a result, it is shown that using unlabeled data is effective in the early stages for both of the two learning rules. In addition, we show that the two learning rules have qualitatively different dynamical behaviors. Furthermore, we propose a new algorithm that improves the generalization performance by switching the number of input vectors used in an update as the time step proceeds.

Keywords: On-line learning · Semi-supervised learning · Statistical-mechanical method · Thermodynamic limit · Optimal scheduling

1 Introduction

Learning can be roughly classified into batch learning and on-line learning [1]. In batch learning, some given examples are used more than once. In this paradigm, a student gives correct answers after training if the student has an adequate degree of freedom. However, it is necessary to have a long amount of time and a large memory in which many examples are stored. In contrast, examples used once are discarded in on-line learning. In this case, a student cannot give correct answers for all examples used in training. However, there are some merits: for example, a large memory for storing many examples is not necessary and it is possible to follow a time-variant teacher [2,3].

Recently, it has become possible to easily obtain a large number of data through the Internet, reducing the cost of collecting data. In supervised learning, however, the cost of giving correct labels to each of the collected data is high.

© Springer International Publishing AG 2016
A. Hirose et al. (Eds.): ICONIP 2016, Part II, LNCS 9948, pp. 28–36, 2016.
DOI: 10.1007/978-3-319-46672-9_4

Therefore, for cost reduction it is useful to give labels to only some data and to give the same labels to similar data. Such learning is receiving a lot of attention and is called semi-supervised learning [4,5].

In this paper, we analyze the dynamical behaviors of semi-supervised learning in the framework of on-line learning by using the statistical-mechanical method [6]. We consider that both a teacher machine and a student machine are simple perceptrons and that K correlated input vectors are used in an update. Each input vector is N-dimensional. A model where correct labels are given to all input vectors has already been analyzed [7]. In this paper, we analyze a case where a correct label is given to only one of the K input vectors as a model of semi-supervised learning. We derive simultaneous differential equations with deterministic forms that describe the dynamical behaviors of order parameters using the self-averaging property in the thermodynamic limit. We treat two well-known learning rules, that is, Hebbian and Perceptron learning [8]. Furthermore, we propose a new algorithm that improves the generalization performance by switching the number K of input vectors used in an update as the time step proceeds.

2   Model

In this paper, we consider that both a teacher machine and a student machine are simple perceptrons with the connection weights $\boldsymbol{B}$ and $\boldsymbol{J}^m$, respectively, where m denotes the time step. For simplicity, the connection weights of the teacher and student are simply called the teacher and student, respectively. The teacher $\boldsymbol{B} = (B_1, \cdots, B_N)^\top$ and student $\boldsymbol{J}^m = (J_1^m, \cdots, J_N^m)^\top$ are N-dimensional vectors. Here, $\top$ denotes transposition. Each component B_i of $\boldsymbol{B}$ is independently drawn from a distribution with a mean of zero and a variance of unity and fixed. Each component J_i^0 of the initial value $\boldsymbol{J}^0$ of $\boldsymbol{J}^m$ is independently drawn from a distribution with a mean of zero and a variance of unity. The cosine of the angle θ^m between $\boldsymbol{B}$ and $\boldsymbol{J}^m$ is R^m. In the case of simple perceptrons, the outputs of the teacher and student are $\mathrm{sgn}(\boldsymbol{B} \cdot \boldsymbol{x}_k^m)$ and $\mathrm{sgn}(\boldsymbol{J}^m \cdot \boldsymbol{x}_k^m)$, respectively. Here, $\mathrm{sgn}(\cdot)$ is the sign function and $\boldsymbol{x}_k^m$ is one of the K input vectors $\boldsymbol{x}_k^m = (x_{k1}^m, \cdots, x_{kN}^m)^\top$, $k = 1, \cdots, K$, that are used in an update. Each component x_{ki}^m of $\boldsymbol{x}_k^m$ is drawn from a distribution with a mean of zero and a variance of $1/N$. In this paper, the thermodynamic limit $N \to \infty$ is also treated. Therefore,

$$\|\boldsymbol{B}\| = \sqrt{N}, \ \|\boldsymbol{J}^0\| = \sqrt{N}, \ \|\boldsymbol{x}_k^m\| = 1, \tag{1}$$

where $\|\cdot\|$ denotes the L2-norm. The cosine of the angle between $\boldsymbol{x}_k^m$ and $\boldsymbol{x}_{k' \neq k}^m$ is assumed to be a^2. Generally, since the norm $\|\boldsymbol{J}^m\|$ of the student changes as the time step proceeds, the ratio l^m of the norm to $\|\boldsymbol{J}^0\|$ is introduced and is called the length of the student. That is,

$$l^m = \frac{\|\boldsymbol{J}^m\|}{\|\boldsymbol{J}^0\|} = \frac{\|\boldsymbol{J}^m\|}{\sqrt{N}}. \tag{2}$$

In the update at each time step, the K input vectors $\{\boldsymbol{x}_k^m\}$ are used. Here, the output of the teacher for $\boldsymbol{x}_1^m$ is used as the desired output vectors for all $\boldsymbol{x}_k^m$, $k = 1, 2, \cdots, K$. Therefore,

$$\boldsymbol{J}^{m+1} = \boldsymbol{J}^m + f_1^m \boldsymbol{x}_1^m + \sum_{k=2}^{K} g_k^m \boldsymbol{x}_k^m. \tag{3}$$

Here, f_1^m and g_k^m are the update functions when using the input vector $\boldsymbol{x}_1^m$ and the input vectors $\boldsymbol{x}_k^m$, $k = 2, \cdots, K$, respectively. f_1^m and g_k^m are determined by the learning rule. Hebbian and Perceptron learning are well-known learning rules for simple perceptrons [8]. The update functions are as follows.

Hebbian learning

$$f_1^m = \eta \, \mathrm{sgn}(v_1^m), \tag{4}$$
$$g_k^m = \eta \, \mathrm{sgn}(v_1^m), \tag{5}$$

Perceptron learning

$$f_1^m = \eta \, \Theta(-u_1^m v_1^m)\mathrm{sgn}(v_1^m), \tag{6}$$
$$g_k^m = \eta \, \Theta(-u_k^m v_1^m)\mathrm{sgn}(v_1^m). \tag{7}$$

Here, η denotes the learning rate. $\Theta(\cdot)$ denotes the step function. v_j^m and u_j^m, $j = 1, 2, \cdots, K$, are defined by

$$v_j^m = \boldsymbol{B} \cdot \boldsymbol{x}_j^m, \quad u_j^m = \frac{\boldsymbol{J}^m \cdot \boldsymbol{x}_j^m}{l^m}. \tag{8}$$

Thus, v_j^m, $v_{j'}^m$, u_j^m, and $u_{j'}^m$, $j \neq j'$, have a Gaussian distribution with a mean of zero and a covariance matrix of

$$\Sigma_4 = \begin{pmatrix} 1 & a^2 & R^m & a^2 R^m \\ a^2 & 1 & a^2 R^m & R^m \\ R^m & a^2 R^m & 1 & a^2 \\ a^2 R^m & R^m & a^2 & 1 \end{pmatrix}. \tag{9}$$

3   Theory

3.1   Generalization Error

One purpose of statistical learning theory is to theoretically obtain the generalization error ϵ_g, which is the mean of errors ϵ over the distribution of a new input vector $\boldsymbol{x}$. The generalization error ϵ_g can be calculated as follows [8]:

$$\epsilon_g = \langle \epsilon \rangle = \int d\boldsymbol{x} P(\boldsymbol{x}) \epsilon(\boldsymbol{x}) = \int du dv P(u,v) \epsilon(u,v) = \frac{1}{\pi} \cos^{-1} R. \tag{10}$$

Here,

$$P(u,v) = \frac{1}{2\pi\sqrt{|\Sigma_2|}} \exp\left(-\frac{(u,v)\Sigma_2^{-1}(u,v)^{\mathrm{T}}}{2}\right), \quad \Sigma_2 = \begin{pmatrix} 1 & R \\ R & 1 \end{pmatrix}. \tag{11}$$

3.2 Simultaneous Differential Equations for Dynamical Behaviors of Order Parameters

Equation (10) shows that ϵ_g is a function of the order parameter R. Therefore, we discuss the dynamical behavior of R. We derive simultaneous differential equations with deterministic forms that describe the dynamical behaviors of order parameters by self-averaging in the thermodynamic limit. To simplify the analysis, the following auxiliary order parameter is introduced:

$$r^m = R^m l^m. \tag{12}$$

Here, R^m is the cosine of the angle between the teacher $\boldsymbol{B}$ and student $\boldsymbol{J}^m$. Thus,

$$R^m = \frac{\boldsymbol{J}^m \cdot \boldsymbol{B}}{\|\boldsymbol{J}^m\|\|\boldsymbol{B}\|} = \frac{1}{l^m N} \sum_{i=1}^{N} J_i^m B_i. \tag{13}$$

To obtain a differential equation for r, multiplying both sides of Eq. (3) by the teacher $\boldsymbol{B}$ and using Eqs. (1), (2), (8), (12), and (13), we obtain

$$Nr^{m+1} = Nr^m + f_1^m v_1^m + \sum_{k=2}^{K} g_k^m v_k^m. \tag{14}$$

Thereafter, we introduce the continuous time t, which is the time step m normalized by the dimension N, and use it to represent the learning process. If the student is updated Ndt times in an infinitely small time dt, we can obtain Ndt equations as follows:

$$Nr^{m+1} = Nr^m + f_1^m v_1^m + \sum_{k=2}^{K} g_k^m v_k^m, \tag{15}$$

$$\vdots$$

$$Nr^{m+Ndt} = Nr^{m+Ndt-1} + f_1^{m+Ndt-1} v_1^{m+Ndt-1} + \sum_{k=2}^{K} g_k^{m+Ndt-1} v_k^{m+Ndt-1}. \tag{16}$$

Summing all these equations, when r is changed by dr through Ndt updates, we obtain

$$N(r+dr) = Nr + Ndt\langle f_1 v_1 \rangle + (K-1)\langle g_k v_k \rangle. \tag{17}$$

Note that the effect of the stochastically generated input vector on the right-hand side of Eq. (17) has been replaced by the mean for the input vector. From Eq. (17), we obtain a differential equation that describes the dynamical behavior of r in a deterministic form as

$$\frac{dr}{dt} = \langle f_1 v_1 \rangle + (K-1)\langle g_k v_k \rangle. \tag{18}$$

Next, we obtain a differential equation for l. Squaring both sides of Eq. (3) and proceeding in the same manner as for r, we derive the following differential equation for l:

$$\frac{dl}{dt} = \frac{1}{2l}\langle f_1^2 \rangle + \langle f_1 u_1 \rangle + \frac{1}{2l}(K-1)\langle g_k^2 \rangle + (K-1)\langle g_k u_k \rangle$$
$$+ \frac{1}{2l}(K-1)(K-2)\langle g_k g_{k'} \rangle a^2 + \frac{1}{l}(K-1)\langle f_1 g_k \rangle a^2. \qquad (19)$$

If r and l are obtained, we can obtain R from Eq. (12). In each learning rule, if we can obtain the eight sample averages in Eqs. (18) and (19) as functions of r and l, Eqs. (18) and (19) are simultaneous differential equations closed in r and l. By solving these equations, we can deterministically discuss the macroscopic dynamical behaviors.

3.3 Hebbian Learning

In the case of Hebbian learning, the three sample averages $\langle f_1 v_1 \rangle$, $\langle f_1 u_1 \rangle$, and $\langle f_1^2 \rangle$ were analytically calculated by executing the Gaussian integrations as follows [8]:

$$\langle f_1 v_1 \rangle = \eta \sqrt{\frac{2}{\pi}}, \quad \langle f_1 u_1 \rangle = \eta \frac{2R}{\sqrt{2\pi}}, \quad \langle f_1^2 \rangle = \eta^2. \qquad (20)$$

Performing similar integration to that carried out to obtain Eq. (20), we analytically calculate the remaining five sample averages as follows:

$$\langle g_k v_k \rangle = \eta \frac{2a^2}{\sqrt{2\pi}}, \quad \langle g_k u_k \rangle = \eta \frac{2a^2}{\sqrt{2\pi}} R, \quad \langle g_k^2 \rangle = \langle f_1 g_k \rangle = \langle g_k g_{k'} \rangle_{k \neq k'} = \eta^2. \qquad (21)$$

Substituting Eqs. (20) and (21) into Eqs. (18) and (19), we analytically solve the simultaneous differential equations as follows:

$$r = \eta \sqrt{\frac{2}{\pi}} \left(1 + (K-1)a^2\right) t, \qquad (22)$$

$$l^2 = \eta^2 \frac{2}{\pi} \left(1 + (K-1)a^2\right)^2 t^2 + \eta^2 K \left(1 + (K-1)a^2\right) t + 1, \qquad (23)$$

where $l(0) = 1$ and $r(0) = 0$ are used as initial conditions. From Eqs. (10), (12), (22), and (23), ϵ_g is analytically obtained as follows:

$$\epsilon_g = \frac{1}{\pi} \cos^{-1} \frac{\eta \sqrt{\frac{2}{\pi}} \left(1 + (K-1)a^2\right) t}{\sqrt{\eta^2 \frac{2}{\pi} \left(1 + (K-1)a^2\right)^2 t^2 + \eta^2 K \left(1 + (K-1)a^2\right) t + 1}}. \qquad (24)$$

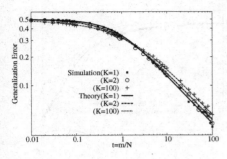

Fig. 1. Learning curves (Hebbian).

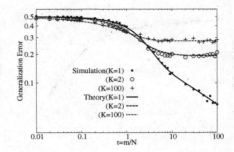

Fig. 2. Learning curves (Perceptron).

3.4 Perceptron Learning

In the case of Perceptron learning, the three sample averages $\langle f_1 v_1 \rangle$, $\langle f_1 u_1 \rangle$, and $\langle f_1^2 \rangle$ were analytically calculated by executing the Gaussian integrations as follows [8]:

$$\langle f_1 v_1 \rangle = \eta \frac{1-R}{\sqrt{2\pi}}, \quad \langle f_1 u_1 \rangle = \eta \frac{R-1}{\sqrt{2\pi}}, \quad \langle f_1^2 \rangle = \frac{\eta^2}{\pi}\cos^{-1}R. \tag{25}$$

Performing similar integration to that carried out to obtain Eq. (25), we analytically calculate $\langle g_k^2 \rangle$, $\langle g_k u_k \rangle$, and $\langle g_k v_k \rangle$ as follows:

$$\langle g_k^2 \rangle = \frac{\eta^2}{\pi}\cos^{-1}a^2 R, \quad \langle g_k u_k \rangle = \eta \frac{a^2 R - 1}{\sqrt{2\pi}}, \quad \langle g_k v_k \rangle = \eta \frac{a^2 - R}{\sqrt{2\pi}}. \tag{26}$$

Because $\langle f_1 g_k \rangle$ and $\langle g_k g_{k'} \rangle$ cannot be analytically calculated, it is necessary to numerically execute the following integrals:

$$\langle f_1 g_k \rangle = \eta^2 \int dv_1 du_1 du_k P(v_1, u_1, u_k) \Theta(-u_1 v_1)\mathrm{sgn}(v_1)\Theta(-u_k v_1)\mathrm{sgn}(v_1), \tag{27}$$

$$\langle g_k g_{k'} \rangle = \eta^2 \int dv_1 du_k du_{k'} P(v_1, u_k, u_{k'}) \Theta(-u_k v_1)\mathrm{sgn}(v_1)\Theta(-u_{k'} v_1)\mathrm{sgn}(v_1). \tag{28}$$

4 Results and Discussion

4.1 Learning Curves

Figures 1 and 2 show the theoretical results and the corresponding simulation results for each learning rule. In the theoretical calculations, for Perceptron learning, we numerically executed Eqs. (27), (28) by Simpson's rule and Eqs. (18) and (19) by the Runge-Kutta method. The conditions are $\eta = 1$ and $a = 0.8$. In the computer simulations, the dimension N is 10^3 and the generalization error ϵ_g

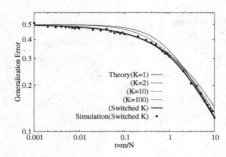
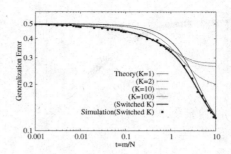

Fig. 3. Effect of switching (Hebbian).   **Fig. 4.** Effect of switching (Perceptron).

was measured through tests using 10^4 random input vectors at each time step. In Figs. 1 and 2, the curves represent the theoretical results and the symbols represent the simulation results. The obtained theory quantitatively agrees well with the results of the computer simulations. However, there is a slight disagreement between the theoretical results and computer simulations since the computer simulations were performed in a finite-dimensional space where the self-averaging property does not hold.

Figures 1 and 2 show that the generalization errors with large K are small in the early stages for both learning rules. However the generalization errors with small K are small in the final stages of learning. That is, the use of many unlabeled data is effective in the early stages of learning. In Hebbian learning, the generalization error continues to decrease regardless of the value of K. On the other hand, in Perceptron learning, the generalization error with $K \geq 2$ has a lower bound while the generalization error with $K = 1$ continues to decrease. These results illustrate that there are qualitative differences between Hebbian learning and Perceptron learning.

4.2   Switching the Number K of Inputs

So far, we have considered on-line learning using K inputs in an update. In this case, the larger the number K of input vectors, the smaller the generalization error ϵ_g in the early stages for both of the two learning rules as shown in Figs. 1–2. That is, it is effective to use unlabeled data in the early stages of learning. However, the smaller the value of K, the smaller ϵ_g becomes as the time step proceeds. Therefore, there is a possibility that the learning curve is improved by switching the number K of input vectors to $K - 1$ at the time step when the rates of decrease in ϵ_g for K and $K - 1$ equal each other.

From Eq. (10), ϵ_g monotonically decreases with increasing R. Therefore, a rapid decrease in ϵ_g is equivalent to a rapid increase in R. Thus, we switch the number K of input vectors to $K - 1$ when the time T satisfies

$$R(T, K) = R(T', K - 1), \tag{29}$$

$$\left.\frac{dR(t, K)}{dt}\right|_{t=T} = \left.\frac{dR(t, K - 1)}{dt}\right|_{t=T'} \tag{30}$$

for some time T'.

At the moment of switching, we multiply the student J by $l(T', K - 1)/l(T, K)$. In the present implementation, we numerically construct, in advance, look-up tables for R, $\frac{dR}{dt}$, and l versus t for each of $K = K_{\text{init}}, \ldots, 1$. These tables are then used to optimize the switching schedule in accordance with Eqs. (29) and (30).

Figures 3 and 4 show the effect of the proposed switching algorithm for each learning. Here, the conditions are $\eta = 1$, $a = 0.8$, and $K_{\text{init}} = 100$. In the computer simulations, the dimension N is 10^4 and the generalization error ϵ_g was measured through tests using 10^4 random input vectors at each time step. These figures show that ϵ_g that obtained for the proposed algorithm is smaller than ϵ_g using a constant K. This indicates the effectiveness of the proposed method. The generalization errors ϵ_g for the proposed algorithm asymptotically approach those with $K = 1$ as the time step proceeds.

5 Conclusion

In this paper, we have analyzed the generalization performance of semi-supervised learning in the framework of on-line learning with a statistical-mechanical method. It has been shown that using unlabeled data is effective in early stages for two types of learning. There are qualitative differences between Hebbian learning and Perceptron learning. Furthermore, we have proposed a new algorithm that improves the generalization performance by switching K as the time step proceeds. From the viewpoint of applications, the proposed switching algorithm can obtain the best generalization performance even if the update is stopped at any moment. In other words, the best generalization performance is obtained using the smallest number of labeled data with help of many unlabeled data.

References

1. Saad, D. (ed.): On-Line Learning in Neural Networks. Cambridge University Press, Cambridge (1998)
2. Urakami, M., Miyoshi, S., Okada, O.: Statistical mechanics of on-line learning when a moving teacher goes around an unlearnable true teacher. J. Phys. Soc. Jpn. **76**, 044003 (2007)
3. Hirama, T., Hukushima, K.: On-line learning of an unlearnable true teacher through mobile ensemble teachers. J. Phys. Soc. Jpn. **77**, 094801 (2008)
4. Chapelle, O., Scholkopf, B., Zien, A.: Semi-Supervised Learning. MIT Press, Cambridge (2006)

5. Zhu, X., Goldberg, A.B.: Introduction to Semi-Supervised Learning: Synthesis Lectures on Artificial Intelligence and Machine Learning. Morgan & Claypool Publishers, city of publication (2009)
6. Engel, A., Broeck, C.: Statistical Mechanics of Learning. Cambridge University Press, Cambridge (2001)
7. Nakao, K., Narukawa, Y., Miyoshi, S.: Statistical mechanics of on-line learning using correlated examples. IEICE Trans. Inf. Syst. **E94–D**(10), 1941–1944 (2011)
8. Nishimori, H.: Statistical Physics of Spin Glasses and Information Processing: An Introduction. Oxford University Press, Oxford (2001)

Evolutionary Multi-task Learning for Modular Training of Feedforward Neural Networks

Rohitash Chandra[✉], Abhishek Gupta, Yew-Soon Ong, and Chi-Keong Goh

Rolls Royce @ NTU Corporate Lab, Nanyang Technological University,
Nanyang View, Singapore, Singapore
c.rohitash@gmail.com

Abstract. Multi-task learning enables learning algorithms to harness shared knowledge from several tasks in order to provide better performance. In the past, neuro-evolution has shownpromising performance for a number of real-world applications. Recently, evolutionary multitasking has been proposed for optimisation problems. In this paper, we present a multi-task learning for neural networks that evolves modular network topologies. In the proposed method, each task is defined by a specific network topology defined with a different number of hidden neurons. The method produces a modular network that could be effective even if some of the neurons and connections are removed from selected trained modules in the network. We demonstrate the effectiveness of the method using feedforward networks to learn selected n-bit parity problems of varying levels of difficulty. The results show better training and generalisation performance when the modules for representing additional knowledge are added by increasing hidden neurons during training.

Keywords: Evolutionary multitasking · Neuro-evolution · Modular design · Multi-task learning

1 Introduction

Neuro-evolution employs evolutionary algorithms for training neural networks [1] which can be classified into direct [1,2], and indirect encoding strategies [3]. In direct encoding, every connection and neuron is specified directly and explicitly in the genotype [1,2]. Direct encoding has also been used in the evolution of feedforward networks for pattern recognition problems using conventional evolutionary algorithms [4], memetic based approaches [5] and cooperative coevolution [6–8].

In indirect encoding, the genotype specifies rules or some other structure for generating the network. Neuro-evolution of augmenting topologies (NEAT) has been a popular indirect encoding that begins evolution with the simplest network topology and adapts nodes and weights together during evolution [3]. Performance of direct and indirect encodings vary for specific problems, while indirect encodings seem very intuitive and have biological motivations, they in several cases have shown not to outperform direct encoding strategies [7,9]. Neuro-evolution has also been applied to evolve modular neural networks [10].

© Springer International Publishing AG 2016
A. Hirose et al. (Eds.): ICONIP 2016, Part II, LNCS 9948, pp. 37–46, 2016.
DOI: 10.1007/978-3-319-46672-9_5

Modular neural networks are motivated from repeating structures in nature [10]. They were introduced for visual recognition tasks that were trained by genetic algorithms and produced generalisation capability [10]. More recently, a modular neural network was presented where the performance and connection costs were optimised through neuro-evolution which achieved better performance when compared to fully connected neural networks [11]. They have also been designed with the motivation to learn new tasks without forgetting old ones [12]. It was shown that modular networks learn new tasks faster from knowledge of previous tasks. Modular neural network architectures have been beneficial for hardware implementations [13]. In particular, they enable smaller networks to be used as building blocks for a larger network. Such transfer of knowledge from simpler to progressively more complex tasks can be achieved by the process of multi-task learning [14].

The notion of multi-task learning has been prevalent in the machine learning community for at least the past two decades [14]. The main stimulus for the approach has been the potential for exploiting relevant information available in related tasks by simultaneous learning using a shared representation. Keeping this motivation in mind, we hereafter extend this idea to neuro-evolution where the implicit parallelism of population-based search is fully unleashed via *evolutionary multitasking* - a novel paradigm that has only recently been conceived [15–17]. Multitasking, from the standpoint of *optimization*, has shown to facilitate autonomous knowledge exchange (in the form of implicit genetically encoded information transfer) between essentially self-contained (but possibly similar) tasks that are processed (evolved) concurrently in a unified solution representation space, thereby often leading to improved convergence characteristics.

In this paper, we present multi-task learning for neural networks that evolves modular network topologies. In the proposed method, each of the tasks is defined by specific network topologies that can also be viewed as modules which feature building blocks for transfer. The method produces a modular network that is effective or feasible even if some of the neurons and connections are removed from selected modules in the trained network. We demonstrate the effectiveness of the method using feedforward neural networks (FNNs) that learn different instances of the n-bit parity problem.

The rest of the paper is organised as follows. Section 2 presents the proposed method and Sect. 3 presents experiments and results. Section 4 concludes the paper with a discussion of future work.

2   Evolutionary Multi-task Learning

Although neural networks have been successfully applied to several real-world problems, their design of topology is typically based on trial-and-error which is often time consuming. While neuro-evolution has been popular, the potential for multi-task learning though neuro-evolution has not been explored.

Essentially, neuro-evolution employs an evolutionary algorithm (EA) that features operators such as selection, crossover and mutation for evolving the weights of the feedforward network. The EA is composed of a population of p

individuals (chromosomes) that are real-coded to represent the weights and bias in the network. The individuals are mapped into the network using direct encoding techniques where all the weights are encoded in a consecutive order, hence each individual consists of total number of weights and biases in the network.

The proposed method, where multitasking is used for neuro-evolution considers different tasks as neural network topologies defined by different number of hidden neurons. Since, we are employing direct encoding for weight representation in the evolutionary algorithm, different tasks in the multi-tasking method results in varied length real-parameter chromosomes in the evolutionary algorithm. Hence, we employ the strategies for creating new offspring using genetic operators proposed in evolutionary multi-tasking [15]. We focus on tasks as network topologies that are applied to the same learning problem.

Algorithm 1. Multi-Tasking Neuro-Evolution for FNNs

Step 1: Define different tasks given by number of hidden neurons
Task 1 (p hidden neurons); Task 2 (q hidden neurons); Task 3 (r hidden neurons)
Step 2: Initialise *popsize* individuals in the unified search space
(i) Randomly associate every individual with *any one* of Task (1, 2, or 3)
(ii) Evaluate individuals for their associated task only
for each generation until termination **do**
   (i) Select and create new offspring via crossover and mutation
   (ii) Associate each offspring with *any one* of Task (1, 2, or 3) via *imitation*
   (ii) Evaluate offspring for their associated task only
   (iii) Select elitist *popsize*/3 indviduals from each of Task (1, 2, and 3) for next
   generation
end for

Algorithm 1 gives details for the proposed evolutionary multi-task learning method. Without loss of generality, we assume there exists three tasks at a time. The algorithm follows a similar workflow as the multi-factorial evolutionary algorithm proposed in [15].

The critical step for successful evolutionary multi-task learning is the formulation of the unified solution representation scheme encompassing the search spaces of all tasks. Such unification can be achieved in a straightforward manner in a multi-dimensional real space for real-parameter optimization (as is the case with training neural network weights). The main challenge is with regard to accommodating for the heterogeneity in search space dimensionality of constitutive tasks. Accordingly, following the methodology proposed in [15], if the three tasks in Algorithm 1 possess dimensionality D_1, D_2, and D_3, respectively, then the dimensionality of the unified search space is given by $D_{multitask} = \max\{D_1, D_2, D_3\}$. Thus, a candidate solution in the unified space is characterized by a vector of $D_{multitask}$ elements. While evaluating an individual associated with the jth task, we simply refer to (or extract) D_j relevant elements from the list of $D_{multitask}$ elements. The motivation behind employing such a search space merging scheme (instead of simplistic concatenation) is to facilitate the underlying EA in providing implicit information transfer across related tasks. This transfer, at least within the present framework, is achieved during the crossover operation as discussed in [16].

At this juncture, an appropriate choice of D_j elements to be extracted for the jth task is crucial for effective multi-task learning. This must be customized by accounting for the properties of the multi-task learning instance at hand. For the present example, we assume p (hidden neurons in Task 1) $< q$ (hidden neurons in Task 2) $< r$ (hidden neurons in Task 3); which implies that $D_1 < D_2 < D_3$. Thus, we enforce that all weights connected to the first p hidden neurons (regardless of Task [1, 2, or 3]) refer to the same $D_p = D_1$ elements among $D_{multitask}$. Similarly, all the weights connected to the next $q - p$ hidden neurons (regardless of Task [2 or 3]) refer to the same $D_{q-p} = D_2 - D_1$ elements among $D_{multitask}$. Finally, all the weights connected to the last $r - q$ hidden neurons in Task 3 refer to the last $D_{r-q} = D_3 - D_2$ elements among $D_{multitask}$. Notice that in the proposed procedure $D_p + D_{q-p} + D_{r-q} = D_{multitask}$. The encoding process is further illustrated in Fig. 1.

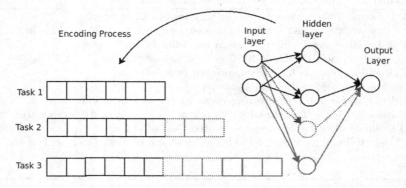

Fig. 1. Encoding FNN as tasks in evolutionary multi-task learning. Note that the colours associated with the synapses in the network are linked to their encoding that are given as different tasks. Task 1 employs a network topology with 2 hidden neurons while the rest of the tasks add extra hidden neurons to it.

To conclude the description of the algorithm, we highlight the association of every individual with *any one* task in the multi-task learning environment. Doing so primarily leads to a saving in the computational cost as evaluating every individual exhaustively for every task is likely to be expensive. While in the initial population the associations are randomly assigned (whilst ensuring uniform representation for all tasks), an assignment strategy based on the memetic concept of *vertical cultural transmission* [18] is adopted in subsequent generations for the genetically modified offspring. In particular, it is prescribed that an offspring randomly *imitates* the association of any one of the parents from which it is created. For further details, refer to [15].

2.1   N-bit Parity Problem

The *n*-bit parity problem essentially is defined as the number of even parity bits in a binary string. Typically, n is the size of the string and all the possible permutations are used to make a dataset.

The n-bit parity problem has been used in the literature to demonstrate the effectiveness of a particular neural network architecture [19] or training algorithm [20]. For instance, the n-bit parity problem was solved using a neural network that allowed direct connections between the input layer and the output layer and trained with linear programming [19]. The effectiveness of certain neuro-evolution methods has also been demonstrated using n-bit problem [20–22].

3 Experiments and Results

This section presents the experimental results for evolutionary multi-task learning (EMTL) of feedforward network topologies for the n-bit parity problem. Although 3 or 4 bit parity problems have also been used in the literature [21, 22], we specifically chose 4, 6 and 8-bit parity problems in order to demonstrate the effectiveness of the proposed method in order to cater for increasing levels of difficulty.

3.1 Design of Experiments

We design the experiments where we compare the performance for the evolutionary single-task learning (ESTL) approach given different number of hidden neurons with EMTL.

The ESTL is implemented through an EA that employs a population size (*popsize*) of 30 individuals. Accordingly, a population of 90 individuals is employed for EMTL which executes three self-contained tasks at once. The algorithm terminates once a total of at least 30 000 function evaluations are completed for each task. Furthermore, note that identical crossover and mutation operators are employed for both methods. In particular, a simulated binary crossover (SBX) operator (with distribution index of 2) [23] and polynomial mutation (with distribution index of 5) are used [24]. The choice of low distribution indices encouraging enhanced exploration is preferred as we employ an elitist selection strategy ensuring preservation of high quality genetic material.

The mean squared error (MSE) given in Eq. 1 and the classification performance are used to evaluate the performance of the respective methods.

$$MSE = \frac{1}{N} \sum_{i=1}^{N} (y_i - \hat{y}_i)^2 \qquad (1)$$

where y_i and $\hat{y}_i$ are the observed and predicted output, respectively. N is the length of the data.

3.2 Results

The results for 4, 6 and 8 bit problems that compare EMTL with ESTL are given in Tables 1, 2, 3, 4, 5 and 6, for MSE and classification performance, respectively.

In Tables 1, 3 and 5, EMTL achieves better performance than ESTL for all the cases for the respective problems. The same trend is also given for the

Table 1. MSE (std-dev) of EMTL vs. ESTL for the 4-bit problem

Method	Task 1	Task 2	Task 3
EMTL(4, 5, 6)	0.0653 (0.0591)	0.0472 (0.0433)	0.0361 (0.0326)
ESTL (4, 5, 6)	0.1557 (0.0364)	0.1020 (0.0232)	0.0699 (0.0285)
EMTL(6, 7, 8)	0.0071 (0.0219)	0.0058 (0.0178)	0.0043 (0.0132)
ESTL(6, 7, 8)	0.0699 (0.0285)	0.0427 (0.0336)	0.0325 (0.0309)

Table 2. Classification performance (std-dev) of EMTL vs. ESTL for the 4-bit problem

Method	Task 1	Task 2	Task 3
EMTL(4, 5, 6)	86.6667 % (3.1714)	90.4167 % (4.2590)	95.4167 % (5.4272)
ESTL (4, 5, 6)	73.3333 % (15.3912)	87.0833 % (6.1267)	90.4167 % (6.5104)
EMTL(6, 7, 8)	91.8750 % (4.6857)	94.3750 % (3.0041)	98.1250 % (2.9131)
ESTL(6, 7, 8)	90.4167 % (6.5104)	94.1667 % (4.9057)	94.7917 % (5.2119)

Table 3. MSE (std-dev) of EMTL vs. ESTL for the 6-bit Problem

Method	Task 1	Task 2	Task 3
EMTL(4, 5, 6)	0.0977 (0.0240)	0.0789 (0.0264)	0.0639 (0.0224)
ESTL (4, 5, 6)	0.1343 (0.0392)	0.1153 (0.0235)	0.0949 (0.0254)
EMTL(6, 7, 8)	0.0642 (0.0221)	0.0459 (0.0148)	0.0492 (0.0269)
ESTL(6, 7, 8)	0.0949 (0.0254)	0.0695 (0.0304)	0.0643 (0.0272)

classification performance in Tables 2, 4 and 6. The results clearly demonstrate that the EMTL is better than ESTL given that the overall optimisation time is the same for different strategies (given by number of hidden neurons) for all the respective problems.

Figures 2 and 3 show the convergence trend for the respective strategies for the 8-bit parity problem. In the case, the mean MSE for 30 experimental runs is shown at different stages of evolution. It is clear that EMTL converges with higher quality solutions.

3.3 Discussion

The experiments were designed to observe the behaviour of evolutionary multi-task learning when compared to single-task learning. We observed interesting results where the multi-task learning performed better. This is due to the way the problem was decomposed using a dynamic programming approach where the main task is divided into smaller tasks as building blocks that transfer to larger tasks given by larger network topologies. The results show that the building blocks of knowledge in smaller tasks were transferred and effectively refined when presented with more weights and neurons in the larger tasks.

Table 4. Classification performance (std-dev) of EMTL vs. ESTL for the 6-bit problem

Method	Task 1	Task 2	Task 3
EMTL(4, 5, 6)	85.3125 % (8.8064)	90.5208 % (6.1129)	93.1250 % (4.0538)
ESTL (4, 5, 6)	81.1458 % (10.0337)	83.4375 % (9.2174)	86.1458 % (8.1214)
EMTL(6, 7, 8)	92.7604 % (4.5597)	95.6250 % (2.6081)	95.0000 % (4.5211)
ESTL(6, 7, 8)	86.1458 % (8.1214)	92.0833 % (7.4533)	92.5521 % (4.7087)

Table 5. MSE (std-dev) of EMTL vs. ESTL for the 8-bit problem

Method	Task 1	Task 2	Task 3
EMTL(5, 6, 7)	0.1108 (0.0249)	0.0949 (0.0257)	0.0861 (0.0232)
ESTL (5, 6, 7)	0.1349 (0.0279)	0.1223 (0.0279)	0.1269 (0.0322)
EMTL(7, 8, 9)	0.0843 (0.0274)	0.0661 (0.0178)	0.0618 (0.0161)
ESTL(7, 8, 9)	0.1269 (0.0322)	0.1093 (0.0275)	0.0936 (0.0217)

One can argue that the design of experiments are biased towards the multi-task learning approach as it was given more computational budget (90 000) when compared to single task learning approach (30 000). Note that this is because the multi-task learning approach is solving 3 tasks at the same time exhibiting

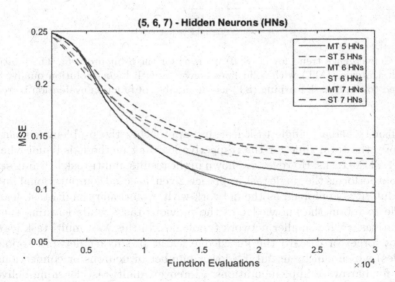

Fig. 2. Convergence trend for (5, 6, 7) neurons for the 8-bit problem. Note that the multi-task learning strategies (MT) with solid lines converge with higher solution quality when compared to single-task learning (ST) as the number of function evaluation increases.

Table 6. Classification performance (std-dev) of EMTL vs. ESTL for the 8-bit problem

Method	Task 1	Task 2	Task 3
EMTL(5, 6, 7)	86.5234 % (4.3712)	88.6198 % (4.6375)	89.0885 % (5.3569)
ESTL (5, 6, 7)	82.0313 % (6.4415)	81.9010 % (9.5784)	84.1406 % (6.1272)
EMTL(7, 8, 9)	90.5859 % (5.4334)	92.9427 % (3.2954)	93.5417 % (2.3810)
ESTL(7, 8, 9)	84.1406 % (6.1272)	85.7813 % (6.8659)	88.4115 % (5.0280)

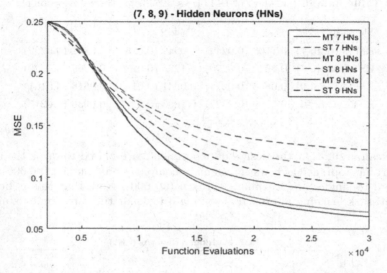

Fig. 3. Convergence trend for (7, 8, 9) neurons for the 8-bit problem. The multi-task learning strategies (MT) with solid lines converge with higher solution quality when compared to single-task learning (ST) as the number of function evaluation increases.

parallelism, whereas, single task learning approaches the problem sequentially. Therefore, the total time for the respective learning methods is equal when we consider all the tasks. However, as shown in the results, multi-task learning significantly outperforms the single task approach given an equal computational budget.

Modularity is enforced as the network with evolutionary multi-task learning was able to retain the knowledge of the previous tasks while learning the new tasks, i.e. tasks with smaller network topologies. In this way, multi-task learning does not forget or discard the knowledge learnt in smaller network topologies (modules) which can be useful if a lower number of neurons or connections are needed for hardware implementations. Moreover, multi-task learning delivers a neuro-evolution strategy which employs direct encoding along with the advantages of indirect encoding where the number of hidden neurons and associated connection are evolved during evolution.

In the case of hardware implementation [13], it is essential to have a modular trained neural network that is operational with a degree of error for *safe mode operation* when some of the neurons or links in selected modules are damaged during an event.

4   Conclusions and Future Work

We presented an evolutionary multi-task learning method that produced modular neural networks which have the property to retain knowledge in some of the modules when others are affected. This can be particularly helpful for hardware implementations where certain defects in the modules do not affect the entire knowledge representation and hence the network can operate in *safe mode.*

The results showed that the multi-task learning approach was helpful in achieving better solution quality when compared to a related method. This was due to the way the training was done where the larger problem was solved in parts using a dynamic programming approach.

In future work, the approach can be extended to tasks that consist of different problems rather than neural network topologies. There is scope for using other evolutionary techniques such as coevolution with application to problems that have building blocks or require multi-task learning. The design proposed method can also be helpful for hardware implementation.

References

1. Angeline, P., Saunders, G., Pollack, J.: An evolutionary algorithm that constructs recurrent neural networks. IEEE Trans. Neural Netw. **5**(1), 54–65 (1994)
2. Moriarty, D.E., Miikkulainen, R.: Forming neural networks through efficient and adaptive coevolution. Evol. Comput. **5**(4), 373–399 (1997)
3. Stanley, K.O., Miikkulainen, R.: Evolving neural networks through augmenting topologies. Evol. Comput. **10**(2), 99–127 (2002)
4. Sexton, R.S., Dorsey, R.E.: Reliable classification using neural networks: a genetic algorithm and backpropagation comparison. Decis. Support Syst. **30**(1), 11–22 (2000)
5. Cant-Paz, E., Kamath, C.: An empirical comparison of combinations of evolutionary algorithms and neural networks for classification problems. IEEE Trans. Syst. Man Cybern. B Cybern. **35**(5), 915–933 (2005)
6. Garcia-Pedrajas, N., Hervas-Martinez, C., Munoz-Perez, J.: COVNET: a cooperative coevolutionary model for evolving artificial neural networks. IEEE Trans. Neural Netw. **14**(3), 575–596 (2003)
7. Gomez, F., Schmidhuber, J., Miikkulainen, R.: Accelerated neural evolution through cooperatively coevolved synapses. J. Mach. Learn. Res. **9**, 937–965 (2008)
8. Chandra, R.: Competition and collaboration in cooperative coevolution of Elman recurrent neural networks for time-series prediction. IEEE Trans. Neural Netw. Learn. Syst. **26**, 3123–3136 (2015)
9. Heidrich-Meisner, V., Igel, C.: Neuroevolution strategies for episodic reinforcement learning. J. Algorithms **64**(4), 152–168 (2009). Reinforcement Learning

10. Happel, B.L., Murre, J.M.: Design and evolution of modular neural network archi-tectures. Neural Networks **7**(6–7), 985–1004 (1994). Models of Neurodynamics and Behavior
11. Clune, J., Mouret, J.-B., Lipson, H.: The evolutionary origins of modularity. Proc. R. Soc. of London B: Biol. Sci. **280**(1755) (2013)
12. Ellefsen, K.O., Mouret, J.-B., Clune, J.: Neural modularity helps organismsevolve to learn new skills without forgetting old skills. PLoS Comput. Biol. **11**(4), 1–24 (2015)
13. Misra, J., Saha, I.: Artificial neural networks in hardware: a survey of two decades of progress. Neurocomputing **74**(13), 239–255 (2010). Artificial Brains
14. Caruana, R.: Multitask learning. Mach. Learn. **28**(1), 41–75 (1997)
15. Gupta, A., Ong, Y.S., Feng, L.: Multifactorial evolution: toward evolutionary mul-titasking. IEEE Trans. Evol. Comput. **20**(3), 343–357 (2016)
16. Gupta, A., Ong, Y.-S., Feng, L., Tan, K.C.: Multiobjective multifactorial optimiza-tion in evolutionary multitasking. IEEE Trans, Cybernetics (2016, Accepted)
17. Ong, Y.-S., Gupta, A.: Evolutionary multitasking: a computer science view of cognitive multitasking. Cognitive Comput., 1–18 (2016)
18. Chen, X., Ong, Y.-S., Lim, M.-H., Tan, K.C.: A multi-facet survey on memetic computation. IEEE Trans. Evol. Comput. **15**(5), 591–607 (2011)
19. Liu, D., Hohil, M.E., Smith, S.H.: N-bit parity neural networks: new solutions based on linear programming. Neurocomputing **48**(14), 477–488 (2002)
20. Mangal, M., Singh, M.P.: Analysis of pattern classification for the multidimen-sional parity-bit-checking problem with hybrid evolutionary feed-forward neural network. Neurocomputing **70**(79), 1511–1524 (2007). Advances in Computational Intelligence and Learning, 14th European Symposium on Artificial Neural Net-works 2006
21. Mirjalili, S., Hashim, S.Z.M., Sardroudi, H.M.: Training feedforward neural net-works using hybrid particle swarm optimization and gravitational search algorithm. Appl. Math. Comput. **218**(22), 11125–11137 (2012)
22. Chandra, R., Frean, M.R., Zhang, M.: Crossover-based local search in cooperative co-evolutionary feedforward neural networks. Appl. Soft Comput. **12**(9), 2924–2932 (2012)
23. Deb, K., Agrawal, R.B.: Simulated binary crossover for continuous search space. Complex Syst. **9**(2), 115–148 (1995)
24. Deb, K., Deb, D.: Analysing mutation schemes for real-parameter genetic algo-rithms. Int. J. Artif. Intell. Soft Comput. **4**(1), 1–28 (2014)

On the Noise Resilience of Ranking Measures

Daniel Berrar[(⊠)]

School of Arts and Sciences, College of Engineering,
Shibaura Institute of Technology, 307 Fukasaku,
Minuma-ku, Saitama 337-8570, Japan
dberrar@shibaura-it.ac.jp

Abstract. Performance measures play a pivotal role in the evaluation
and selection of machine learning models for a wide range of applications. Using both synthetic and real-world data sets, we investigated the
resilience to noise of various ranking measures. Our experiments revealed
that the area under the ROC curve (AUC) and a related measure, the
truncated average Kolmogorov-Smirnov statistic (taKS), can reliably discriminate between models with truly different performance under various types and levels of noise. With increasing class skew, however, the
H-measure and estimators of the area under the precision-recall curve
become preferable measures. Because of its simple graphical interpretation and robustness, the lower trapezoid estimator of the area under the
precision-recall curve is recommended for highly imbalanced data sets.

Keywords: Ranking · Classification · Noise · Robustness · ROC curve ·
AUC · H-measure · taKS · Precision-recall curve

1 Introduction

Reliable performance assessment is a central task in machine learning. There
exist a plethora of performance measures and graphical tools for the comparison
of predictive models [10,14], and practitioners are faced with a non-trivial question: how should the "right" performance measure be chosen? We posit that two
main factors should guide this choice. First, a measure should be easily intelli-
gible, ideally with a neat graphical interpretation. Second, a measure should be
able to reliably discriminate between models with a truly different performance,
even under various types and levels of noise.

Here, we are considering ranking measures that summarize the performance
of a binary scoring classifier, i.e., a mapping that assigns a real-valued score to
each instance, which expresses the expectation that an instance is a member
of the positive class. In particular, we are interested in the question how well
these measures can discriminate between truly different models under various
conditions. For example, when the classes are extremely skewed, ROC curves
and the AUC generally do not discriminate well between models with different
performance [2,4]. Extremely imbalanced classes are very common in real-world

© Springer International Publishing AG 2016
A. Hirose et al. (Eds.): ICONIP 2016, Part II, LNCS 9948, pp. 47–55, 2016.
DOI: 10.1007/978-3-319-46672-9_6

applications, for instance, fraud prediction tasks [12]. In this case, precision-recall curves might paint a clearer picture of the performance difference.

In this study, we investigate the reliability of ranking measures under different types and levels of noise. Related comparative studies can be found in [6,13]. The present study extends our previous work [1] by including estimators of the area under the precision-recall curve and new experiments on the resilience to attribute noise. Our results confirm that the AUC and a related metric, taKS, are robust measures under various types and levels of noise. However, for increasingly imbalanced data sets, the H-measure and the estimators of the area under the precision-recall curve are preferable alternatives.

2   Basic Notation and Ranking Measures

We now briefly describe the investigated ranking measures, beginning with some basic notation. We consider binary prediction tasks where a data set D contains n instances $\mathbf{x}_i$, $i = 1..n$, and each instance belongs to exactly one class y, i.e., $(y, \mathbf{x}_i)$, $y \in \{0, 1\}$; here, 1 denotes the positive class and 0 denotes the negative class. Let $X = \mathbf{a}_1 \times ... \times \mathbf{a}_m$ denote the instance space over the set of attributes $\mathbf{a}_1...\mathbf{a}_m$. A scoring classifier is a mapping $C : X \to \mathbb{R}$ that produces a class membership score for each instance, for example, a posterior probability $P(y = 1 | X = \mathbf{x}_i)$. These scores allow a total ordering of the instances, possibly with ties. Based on the scores, the instances are ranked in decreasing order from the most to the least likely positive. Let n_+ be the number of positive instances and n_- be the number of negative instances in D, with $n_+ > 0$ and $n_- > 0$. Let $h_+(t_i)$ denote the hits, i.e., the number of positive instances at or above the threshold t_i, $i = 1..k$, where $k = n+1$ is the total number of possible thresholds. Accordingly, let $h_-(t_i)$ denote the number of negative instances. The recall—or equivalently, the true positive rate (TPR)—at threshold t_i is $r(t_i) = \text{TPR}(t_i) = \frac{h_+(t_i)}{n_+}$. The precision is $p(t_i) = \frac{h_+(t_i)}{i-1}$ for $i > 1$ and 0 for $i = 1$. The false positive rate at threshold t_i is $\text{FPR}(t_i) = \frac{h_-(t_i)}{n_-}$.

One of the most widely used summary statistics for the ranking performance is the area under the ROC curve (AUC) [5], which is equivalent to the Wilcoxon rank-sum statistic [7]: given any randomly selected positive and negative instance, the AUC is the probability that the model assigns a higher score to the positive instance (i.e., ranks it before the negative one). The rank-sum interpretation leads to an elegant method of calculating the AUC [9],

$$\text{AUC} = \frac{S_- - 0.5 n_- (n_- + 1)}{n_+ n_-}, \tag{1}$$

where S_- is the sum of the ranks of the negative instances. Alternatively, a ROC curve can be summarized by the area under the ROC convex hull (AUCH) [7,15]. From an empirical ROC curve, the AUCH can be calculated with the trapezoidal rule and the points $(x_i, y_i) = (0,0)$, the minimum set of points spanning the concavities, and the point $(1,1)$,

$$\text{AUCH} = \sum_{i=1}^{k-1} y_i(x_{i+1} - x_i) + 0.5(y_{i+1} - y_i)(x_{i+1} - x_i) \ . \tag{2}$$

The H-measure was proposed as an alternative to the AUC [8],

$$\text{H-measure} = 1 - \frac{\int Q(T(c), c)u(c)\mathrm{d}c}{\pi_+ \int_0^{\pi^-} cu(c)\mathrm{d}c + \pi_- \int_{\pi_-}^1 (1 - c)u(c)\mathrm{d}c} \ , \tag{3}$$

where c_+ and c_- are the costs associated with the misclassification of a positive and negative instance, respectively; π_+ and π_- are the prior probabilities of positive and negative instances, respectively; and $c = c_+/(c_+ + c_-)$; $T(c) = \arg\min_t\{c\pi_+(1 - \text{TPR}(t)) + (1 - c)\pi_-\text{FPR}(t)\}$; $Q(t, c) = \{c\pi_+(1 - \text{TPR}(t)) + (1 - c)\pi_-\text{FPR}(t)\}(c_+ + c_-)$; $u(c) = c(1 - c)/\int_0^1 c(1 - c)\mathrm{d}c$.

The Kolmogorov-Smirnov (KS) statistic (Eq. 4) is the maximum absolute difference between two cumulative distributions; here,

$$\text{KS} = \max_t\{|\text{TPR}(t_i) - \text{FPR}(t_i)|\} \ . \tag{4}$$

A recently proposed variant, the truncated average Kolmogorov-Smirnov (taKS) statistic [1], calculates the average distance between the TPR- and FPR-curves, excluding the start- and endpoints $(0, 0)$ and $(1, 1)$,

$$\text{taKS} = \frac{1}{k - 2} \sum_{i=2}^{k-1} (\text{TPR}(t_i) - \text{FPR}(t_i)) \ . \tag{5}$$

This metric was derived from the area between the TPR- and FPR-curves, which corresponds to AUC minus 0.5 in the case that no ties exist.

Precision-recall graphs plot the precision as a function of the recall. However, the best method of calculating the area under the precision-recall curve is not obvious [2], since the precision normally does not change monotonically with increasing recall. As a result, different estimators of the area under the precision-recall curve (AUCPR) are conceivable. A frequently used summary measure is the average precision (Eq. 6),

$$\text{AUCPR}_{\text{avg}} = \sum_{i=1}^{k} p(t_i)[r(t_i) - r(t_{i-1})] \ , \tag{6}$$

with $r(t_0) = 0$. The area under an empirical precision-recall curve can be approximated by a trapezoidal estimator [2],

$$\text{AUCPR}_{\text{minmax}} = \sum_{i=1}^{k-1} \frac{p_{\min}(t_i) + p_{\max}(t_{i+1})}{2}[r(t_{i+1}) - r(t_i)] \ , \tag{7}$$

where $p_{\min}(t)$ and $p_{\max}(t)$ denote the minimum and maximum precision, respectively, for a recall at threshold t. Here, we consider two further estimators of the area under the precision-recall curve. The *upper trapezoid* (Eq. 8) is defined as

$$\text{AUCPR}_{\max} = \sum_{i=1}^{k-1} \frac{p_{\max}(t_i) + p_{\max}(t_{i+1})}{2} [r(t_{i+1}) - r(t_i)] \,, \tag{8}$$

and the *lower trapezoid* (Eq. 9) is defined as

$$\text{AUCPR}_{\min} = \sum_{i=1}^{k-1} \frac{p_{\min}(t_i) + p_{\min}(t_{i+1})}{2} [r(t_{i+1}) - r(t_i)] \,, \tag{9}$$

with $\text{AUCPR}_{\min} \leq \text{AUCPR}_{\text{minmax}} \leq \text{AUCPR}_{\max}$. The upper trapezoid was shown to overestimate the area under the precision-recall curve [2,3]. By contrast, the lower trapezoid is likely to underestimate the area. However, this does not necessarily preclude them from being useful metrics for model selection; what matters is that these measures can reliably identify the better ranker.

3   Experiments

To investigate how resilient the ranking measures are to different levels and types of noise, we adopted an approach similar to the methodology described in [1]. The main differences are: (i) we now use a slightly more precise error criterion (Eq. 10); (ii) we consider now estimators of the area under the precision-recall curve (Eqs. 7, 8, 9); and (iii) we carry out new experiments on attribute noise (Sect. 3.2). We consider again two scoring classifiers, C_1 and C_2, where C_1 is the truly better one. Then, we progressively add noise that increasingly blurs the differences in performance. A measure M_i can be considered more robust than another measure M_j if M_i is less affected by the added noise. We used both synthetic and real-world data sets from the UCI repository [11]. All experiments were carried out in R 3.2.3 [16]. The R code for the estimators of the area under the precision-recall curve is provided at https://github.com/dberrar/AUCPR.

3.1   Synthetic Data Sets

We generated a vector of 100 real values that were randomly sampled from $U(0,1)$ and used these values as the ranking scores for a hypothetical test set comprising 100 instances. We assigned the positive class label to all instances with a ranking score of at least 0.5 and the negative class labels to the remaining cases. The resulting matrix contains 100 rows (each row representing one test instance) and 2 columns. The first column contains the class labels, and the second column contains the ranking scores, which can be thought of as predicted class posterior probabilities. Then, we selected 10 instances at random and replaced their scores by a random number from $U(0,1)$. We can expect that half of the new scores now lead to an incorrect classification for a threshold of 0.5. Thus, the expected accuracy is 95 %. We used this corrupted result as the output of a hypothetical classifier C_1. To generate the output of C_2, we proceeded analogously, except that we replaced the scores of 10 additional, randomly selected cases by a random number from $U(0, 1)$. The expected accuracy of C_2 is therefore only 90 %. Thus, the output of C_1 is clearly better than that of C_2. Without noise, all ranking measures can clearly identify C_1 as the preferable model, but with increasing levels of noise, this task becomes more and more challenging. We considered the following three types of noise:

1. **Misclassification noise** affects the class labels. In this experiment, we assessed how resilient the ranking measures are to class label errors. The noise level ranged from 0 to 100, where 0 means that no class label was altered and 100 means that 100 % of all class labels were randomly determined by a coin flip. For a noise level of 0, the error rates of all measures are expected to be 0, whereas for a noise level of 100, the expected error rates are 100 % because there are no differences anymore between the outputs of C_1 and C_2.
2. **Probability noise** affects the ranking scores. In this experiment, we assessed how sensitive the ranking measures are to class posterior probabilities that are not well calibrated. For a noise level of i, we sampled 100 real numbers from $U[-i, +i]$ and added these numbers to the 100 ranking scores. The noise level ranged from $i = 0$ (no noise) to $i = 0.5$ in a stepsize of 0.005.
3. **Class frequency noise** affects the class ratio. In this experiment, we assessed how sensitive the ranking measures are to class skew. We changed the class skew by progressively deleting $x\%$ of the positive instances, where x ranged from 0 (i.e., no changes) to 95 (i.e., 95 % of randomly selected positive instances were deleted).

For each noise type and level, we generated each model $v = 10,000$ times, and each time, we evaluated their outputs with the ranking measures. For each measure, we counted how many times it indicated that C_2 was at least as good as C_1. Let $M(C)$ denote the value of a performance measure M for a model C. We defined the error rate of M as

$$\epsilon(M) = \frac{1}{v} \sum_{i=1}^{v} \delta(M), \text{with} \quad \delta(M) = \begin{cases} 1 \text{ if } M(C_2) \geq M(C_1) \\ \\ 0 \text{ if } M(C_2) < M(C_1) \end{cases} \tag{10}$$

We then plotted $\epsilon(M)$ for each noise level. For example, if the resulting error curve of AUC is consistently below that of KS, then we can conclude that AUC is more robust to noise than KS.

3.2 Real-World Data Sets

We investigated how sensitive the performance measures are to *attribute noise*, i.e., noise that affects the attributes of a data set. We considered again two classifiers, C_1 and C_2, where C_1 is the truly better model. The choice of the learning algorithm was assumed to have no influence on the experimental outcome of interest. To build C_1, we opted for naive Bayes learning because it produces class posterior probabilities. The performance of C_1 was assessed in 10-fold stratified cross-validation. Let V_i denote the ith validation set, $i = 1..10$, consisting of m cases $\mathbf{x}_{ij}$, $j = 1..m$, and let s_{ij} denote the ranking score of the jth validation case, with $s_{ij} = P(y = 1|\mathbf{x}_{ij})$. The ranking scores of C_2 are the same as those of C_1, except that for each validation set V_i, randomly selected 10 % of the ranking scores were replaced by random numbers uniformly sampled from $[\min(s_{ij}), \max(s_{ij})]$. For small validation sets ($\frac{m}{10} < 2$), two randomly selected scores were replaced. While it is possible that the resulting ranking by C_2 is better than that of C_1 for a particular validation set, it is unlikely that the corrupted classifier performs better over all 10 validation sets. Consequently, we expect that model C_1 can be clearly identified as being superior to C_2 based on the average cross-validated ranking performance.

Next, we added attribute noise to the data set D as follows. We selected q attributes at random and then randomly permuted the values within each selected attribute, where q ranged from 0 to the total number of attributes in D. Thus, with increasing q, the relationship between the target classes and the attributes becomes increasingly blurred. When q equals the total number of attributes, D is effectively a random data set, and C_1 and C_2 are expected to have the same performance. For each value of q, we repeated the cross-validation 1,000 times, each time selecting and permuting q attributes at random. We counted how often C_2 achieved a ranking performance at least as high as that of C_1 and calculated the error rate $\epsilon(M)$ for each ranking measure M (cf. Eq. 10). For $q = 0$, $\epsilon(M)$ is expected to be 0, as C_1 is clearly the better classifier. When q equals the total number of attributes, $\epsilon(M)$ is expected to be 0.5.

4   Results

Figure 1 shows the results of the experiments with the synthetic data. Under class label noise (Fig. 1a), the measures based on precision-recall curves are the

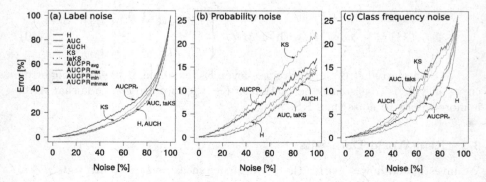

Fig. 1. Error rates of the ranking measures for (a) noise affecting the class labels, (b) noise affecting the ranking scores, and (c) noise affecting the class frequency.

least robust, while the most robust measures are taKS and AUC. As expected, their differences are negligible because these measures are linearly related to each other in the absence of ties. The AUC and taKS are also the most robust measures under probability noise (Fig. 1b). By contrast, under class frequency noise (Fig. 1c), taKS and AUC are the least robust. From a noise level of already 20 %, which corresponds to a class ratio of approximately 4 : 5, the error curves of the measures based on the precision-recall curve and the H-measure are below the error curves of AUC and taKS. Overall, under class frequency noise, the H-measure is the most robust. The error curves of the average precision and the three trapezoidal estimators largely overlap in all three experiments. For a class

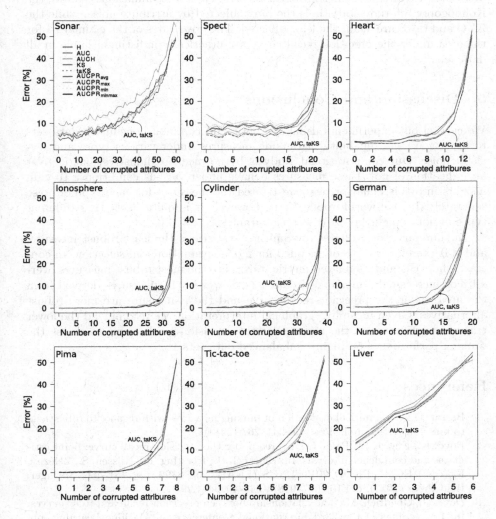

Fig. 2. Robustness to attribute noise.

frequency noise level of 80 % (corresponding to an approximate class ratio of 1 : 5) and more (Fig. 1c), the average error rates are 17.12 % ($AUCPR_{avg}$), 17.34 % ($AUCPR_{max}$), 15.82 % ($AUCPR_{min}$), and 16.18 % ($AUCPR_{minmax}$). Hence, the lower trapezoid, $AUCPR_{min}$, is slightly more robust than the other estimators under class frequency noise.

In the experiments involving real-world data sets (Fig. 2), the most striking observation is that all measures are remarkably resilient to attribute noise for all data sets, with the exception of the data set Liver. For the data sets Spect, Heart, Ionosphere, Cylinder, German, and Pima, the error rates begin to increase when about 70 % of the attributes are corrupted. Apparently, around 30 % of the attributes are already sufficient for an effective class discrimination. Overall, the Kolmogorov-Smirnov statistic is the most affected by attribute noise, while the AUC and taKS are again the least affected. The error rates of the estimators for the area under the precision-recall curve are practically indistinguishable in all data sets.

5   Discussion and Conclusions

We carried out experiments using synthetic and common benchmark data sets to investigate the reliability of ranking measures under various types and levels of noise. Our computational study is by no means exhaustive, but it gives an indication of how these measures behave under various conditions. To gain insights into this behavior, we investigated the measures for each type of noise separately. In real-world applications, however, it is quite likely that different types of noise might be at play concomitantly.

In summary, the Kolmogorov-Smirnov statistic is the least robust measure, and it is therefore not recommended for model evaluation and selection. In contrast, the AUC and its related metric, taKS, are the most robust measures overall. For increasingly imbalanced data sets, however, the measures derived from the area under the precision-recall curve and the H-measure are more robust alternatives. Given its simple graphical interpretation, we recommend the lower trapezoid estimator of the area under the precision-recall curve to assess the ranking performance on highly imbalanced data sets.

References

1. Berrar, D.: An empirical evaluation of ranking measures with respect to robustness to noise. J. Artif. Intell. Res. **49**, 241–267 (2014)
2. Boyd, K., Eng, K.H., Page, C.D.: Area under the precision-recall curve: point estimates and confidence intervals. In: Blockeel, H., Kersting, K., Nijssen, S., Železný, F. (eds.) ECML PKDD 2013. LNCS (LNAI), vol. 8190, pp. 451–466. Springer, Heidelberg (2013). doi:10.1007/978-3-642-40994-3_29
3. Davis, J., Goadrich, M.: The relationship between precision-recall and ROC curves. In: Proceedings of the 23rd International Conference on Machine Learning, pp. 233–240. ACM (2006)

4. Drummond, C.: Machine learning as an experimental science, revisited. In: Proceedings of the 21st National Conference on Artificial Intelligence: Workshop on Evaluation Methods for Machine Learning, pp. 1–5. AAAI Press (2006)
5. Fawcett, T.: ROC graphs: notes and practical considerations for researchers. Technical Report HPL-2003-4, HP Laboratories, pp. 1–38 (2004)
6. Ferri, C., Hernández-Orallo, J., Modroiu, R.: An experimental comparison of performance measures for classification. Pattern Recogn. Lett. **30**, 27–38 (2009)
7. Flach, P.: ROC analysis. In: Sammut, C., Webb, G. (eds.) Encyclopedia of Machine Learning, pp. 869–874. Springer, US (2010)
8. Hand, D.: Measuring classifier performance: a coherent alternative to the area under the ROC curve. Mach. Learn. **77**, 103–123 (2009)
9. Hand, D., Till, R.: A simple generalisation of the area under the ROC curve for multiple class classification problems. Mach. Learn. **45**, 171–186 (2001)
10. Hernández-Orallo, J., Flach, P., Ferri, C.: A unified view of performance metrics: translating threshold choice into expected classification loss. J. Mach. Learn. Res. **13**, 2813–2869 (2012)
11. Lichman, M.: UCI Machine Learning Repository (2013). http://archive.ics.uci.edu/ml
12. Oentaryo, R., Lim, E.P., Finegold, M., Lo, D., Zhu, F., Phua, C., Cheu, E.Y., Yap, G.E., Sim, K., Nguyen, M.N., Perera, K., Neupane, B., Faisal, M., Aung, Z., Woon, W.L., Chen, W., Patel, D., Berrar, D.: Detecting click fraud in online advertising: a data mining approach. J. Mach. Learn. Res. **15**(1), 99–140 (2014)
13. Parker, C.: On measuring the performance of binary classifiers. Knowl. Inf. Syst. **35**, 131–152 (2013)
14. Prati, R.C., Batista, G., Monard, M.C.: A survey on graphical methods for classification predictive performance evaluation. IEEE Trans. Knowl. Data Eng. **23**(11), 1601–1618 (2011)
15. Provost, F., Fawcett, T.: Robust classification for imprecise environments. Mach. Learn. **42**(3), 203–231 (2001)
16. R Core Team: R: A Language and Environment for Statistical Computing. R Foundation for Statistical Computing, Vienna (2015). https://www.R-project.org/

BPSpike II: A New Backpropagation Learning Algorithm for Spiking Neural Networks

Satoshi Matsuda$^{(\boxtimes)}$

Nihon University,
1-2-1, Izumi-cho, Narashino, Chiba 275-8575, Japan
matsuda.satoshi@nihon-u.ac.jp,
matsuda.satoshi0127@gmail.com

Abstract. Using gradient descent, we propose a new backpropagation learning algorithm for spiking neural networks with multi-layers, multi-synapses between neurons, and multi-spiking neurons. It adjusts synaptic weights, delays, and time constants, and neurons' thresholds in output and hidden layers. It guarantees convergence to minimum error point, and unlike SpikeProp and its extensions, does not need a one-to-one correspondence between actual and desired spikes in advance. So, it is stably and widely applicable to practical problems.

Keywords: Spiking neural networks · Learning algorithm · Backpropagation

1 Introduction

Many learning algorithms have been presented for spiking neural networks [1,2,4–10]. SpikeProp and its extensions [1,4,6,9,10], emply backpropagation and so guarantee the convergence to minimum error point, but need to specify a one-to-one correspondence between actual and desired spikes in advance. Specifying such a correspondence in advance is sometimes very difficult or impossible in practice. On the other hand, Resume [8] and SPAN [7] do not need to specify such a correspondence in advance, but cannot guarantee the convergence to minimum error point.

In this paper, we propose a new backpropagation learning algorithm, called BPSpike II, which adjusts all the parameters; synaptic weights, synaptic delays, synaptic time constants, and neuron thresholds in multiple layered networks and multiple spiking neurons, although only the synaptic weights learning is shown here because of page limitation. BPSpike II not only guarantees the convergence to minimum error point but also does not need to specify one-to-one correspondence between actual spikes and desired spikes in advance.

© Springer International Publishing AG 2016
A. Hirose et al. (Eds.): ICONIP 2016, Part II, LNCS 9948, pp. 56–65, 2016.
DOI: 10.1007/978-3-319-46672-9_7

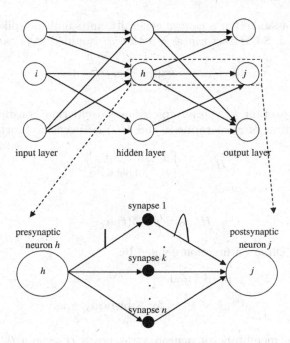

Fig. 1. Spiking neural networks with multiple layers and multiple synapses between neurons. The upper one shows the network structure that consists of multiple layers. Although it is shown only one hidden layer, any number of hidden layers are possible. The lower one zooms in on the broken lined portion in the upper. Two neurons are connected by a series of time delayed synaptic links.

2 Definitions

2.1 Spiking Neural Networks

We employ spiking neural networks with multiple layers and multiple synapses between neurons shown in Fig. 1, and the spike response model [3].

A membrane potential, $u_j(t)$, of neuron j is given by

$$u_j(t) = \sum_{h \in P_j} \sum_{s \in S_h} \sum_{k \in L_{hj}} w_{hj}^k y_{hjs}^k(t) + \eta_j(t - t_{j,last}), \qquad (1)$$

where P_j is a set of presynaptic neurons of neuron j, S_h is a spike train emitted by neuron h, L_{hj} is a set of synapses from neuron h to j (multi-synapses connection), w_{hj}^k is a weight of synapse k from neuron h to j, $y_{hjs}^k(t)$ is an unweighted postsynaptic potential evoked by the s-th spike of neuron h at synapse k to neuron j, $\eta_j(z)$ is a refractoriness function of neuron j, $t_{j,last}$ is the most recent firing time of spike emitted by neuron j prior to current time t, that is,

$$t_{j,last} = \hat{t}_j(t) = \max_{s \in S_j} \{t_{js} < t\}, \qquad (2)$$

and t_{js} is the firing time of the s-th spike emitted by neuron j.

Note that we assume that a neuron generally emits multipe spikes during our observation period. So, refractoriness function η_j is added in (1) and given by

$$\eta_j(z) = -\rho \cdot \exp(-\frac{z}{\tau_j^{ref}})H(z), \tag{3}$$

where ρ is some constant, τ_j^{ref} is a refractoriness decay time constant and determines the decay time of refractoriness, and $H(\cdot)$ is Heaviside function given by

$$H(z) = \begin{cases} 1 & \text{if } z > 0 \\ 0 & \text{otherwise} \end{cases} \tag{4}$$

and approximated by

$$H(z) = \int_{-\infty}^{z} \delta(t)dt, \tag{5}$$

where $\delta(t)$ is the Dirac's δ function defined by

$$\delta(t) = 0 \quad \text{if } t \neq 0$$
$$\int_{-\epsilon}^{\epsilon} \delta(t)dt = 1 \qquad \text{for arbitrarily small } \epsilon. \tag{6}$$

Whenever the membrane of neuron j reaches a threshold θ_j from below, neuron j emits a spike and that moment defines the firing time $t^{(f)}$ of the spike, that is,

$$t^{(f)}: \quad u_j(t^{(f)}) = \theta_j \quad \text{and} \quad \frac{du_j(t)}{dt}\Big|_{t=t^{(f)}} > 0, \tag{7}$$

We employ $x_j(t) = u_j(t) - \theta_j$ in stead of $u_j(t)$ in order for the proposed learning algorithm to work well also for neuron thresholds learning.

A postsynaptic potential, $y_{hjs}^k(t)$, evoked by the s-th spike of neuron h at synapse k to neuron j is given by

$$y_{hjs}^k(t) = \epsilon_{hj}^k(t - t_{hs} - d_{hj}^k) \tag{8}$$
$$\epsilon_{hj}^k(z) = \frac{ez}{\tau_{hj}^k} \exp(-\frac{z}{\tau_{hj}^k})H(z) \tag{9}$$

where $\epsilon_{hj}^k(z)$ is the spike response function, sometimes called α function, of synapse k from neuron h to j, d_{hj}^k is a delay time of synapse k from neuron h to j, τ_{hj}^k is a postsynaptic potential decay time constant of synapse k from neuron h to j and determines the rise and decay time of the postsynaptic potential.

2.2   Convolved Spike Train and Error Function

A spike and a spike train emitted by neuron at $t^{(f)}$ is represented by

$$\delta(t - t^{(f)}) \tag{10}$$

and

$$F = \{t_1, t_2, \cdots, t_p\} \quad \text{or} \quad s(t) = \sum_f \delta(t - t_f), \tag{11}$$

respectively, where t_f is a firing time of the f-th spike emitted by neuron.

In order to calculate the difference between an actual and a desired spike trains without explicitly specifying one-to-one correspondence between spikes in the actual and disired spike trains in advance, we convolve each spike train with a kernel function [7]. As a kernel function, we employ α function,

$$\alpha(z) = \frac{ez}{\tau} \exp(-\frac{z}{\tau}). \tag{12}$$

Then, a spike, $\delta(t - t_f)$, and a spike train, $s(t) = \sum_f \delta(t - t_f)$, are convolved as

$$\tilde{\delta}(t - t_f) = \alpha(t - t_f)H(t - t_f) \tag{13}$$

and

$$\tilde{s}(t) = \sum_f \tilde{\delta}(t - t_f) = \sum_f \alpha(t - t_f)H(t - t_f) \tag{14}$$

respectively. Note that $\epsilon(\cdot)$ given by (9) is also α function.

By gradient descent, the proposed learning algorithm changes a parameter to reduce the square error E given by

$$E(t) = \frac{1}{2} \sum_{j \in J} \left[\tilde{o}_j^d(t) - \tilde{o}_j^a(t) \right]^2, \tag{15}$$

where J is a set of output neurons and $\tilde{o}_j^a(t)$ and $\tilde{o}_j^d(t)$ are a convolved actual and desired spike trains emitted by output neuron j, respectively,

$$\tilde{o}_j^a(t) = \sum_{f \in S_j^a} \tilde{\delta}(t - t_{jf}^a) = \sum_{f \in S_j^a} \alpha(t - t_{jf}^a)H(t - t_{jf}^a)$$

$$\tilde{o}_j^d(t) = \sum_{g \in S_j^d} \tilde{\delta}(t - t_{jg}^d) = \sum_{g \in S_j^d} \alpha(t - t_{jg}^d)H(t - t_{jg}^d)$$

S_j^a and S_j^d are an actual and desired spike trains emitted by neuron j, rspectively, and t_{js}^a and t_{js}^d are an actual and a desired firing times of s-th spike emitted by output neuron j, respectively.

Note that SpikeProp [1] and its extensions [4,6,9,10] also employs the gradient descent but their error function is

$$E_{SpikeProp} = \frac{1}{2} \sum_{j \in J} \sum_{s \in S_j} (t_{js}^a - t_{js}^d)^2. \tag{16}$$

Thus, they need to specify a one-to-one correspondence between an actual and desired spikes in advance. However, it is sometimes very difficult or impossible in practice.

Using the error function $E(t)$ given by (15), we do not need to specify a one-to-one correspondence between actual and desired spikes.

In the next section, for multiple layered networks and multiple spiking neurons, a learning algorithm for synaptic weights are given. The learning algorithms for synaptic delays, time constants, and neurons thresholds are not given in this paper because of page limitaion, but will be shown soon somewhere.

3   Learning Algorithm for Synaptic Weights

3.1   Synaptic Connections to Output Layers

(1) Online Learning.
Using gradient descent, the change of weight, $\Delta w_{hj}^k(t)$, of the k-th synapse from hidden neuron h to output neuron j, is given by

$$\Delta w_{hj}^k(t) = -\gamma_w \frac{\partial E(t)}{\partial w_{hj}^k} = -\gamma_w \sum_{s \in S_j^a} \left[\frac{\partial E(t)}{\partial \tilde{o}_j^a(t)} \frac{\partial \tilde{o}_j^a(t)}{\partial t_{js}^a} \frac{\partial t_{js}^a}{\partial x_j(t)} \frac{\partial x_j(t)}{\partial w_{hj}^k} \Big|_{t=t_{js}^a} \right]$$

$$= -\gamma_w \sum_{s \in S_j^a} \left[\delta_{js}(t) \frac{\partial x_j(t)}{\partial w_{hj}^k} \Big|_{t=t_{js}^a} \right], \tag{17}$$

where γ_w ia a learning rate and

$$\delta_{js}(t) = \frac{\partial E(t)}{\partial \tilde{o}_j^a(t)} \frac{\partial \tilde{o}_j^a(t)}{\partial t_{js}^a} \frac{\partial t_{js}^a}{\partial x_j(t)}. \tag{18}$$

And, we have

$$\frac{\partial E(t)}{\partial \tilde{o}_j^a(t)} = \frac{1}{2} \frac{\partial}{\partial \tilde{o}_j^a(t)} \sum_{j' \in J} [\tilde{o}_{j'}^d(t) - \tilde{o}_{j'}^a(t)]^2 = -[\tilde{o}_j^d(t) - \tilde{o}_j^a(t)] \tag{19}$$

$$\frac{\partial \tilde{o}_j^a(t)}{\partial t_{js}^a} = \frac{\partial}{\partial t_{js}^a} \sum_{s' \in S_j^a} \tilde{\delta}_j(t - t_{js'}^a) = \frac{\partial}{\partial t_{js}^a} \tilde{\delta}_j(t - t_{js}^a) = \frac{\partial}{\partial t_{js}^a} \left[\alpha_j(t - t_{js}^a) H(t - t_{js}^a) \right]$$

$$= H(t - t_{js}^a) \frac{\partial \alpha_j(t - t_{js}^a)}{\partial t_{js}^a} = \left(\frac{1}{\tau} - \frac{1}{t - t_{js}^a} \right) \alpha(t - t_{js}^a) H(t - t_{js}^a) \tag{20}$$

$$\frac{\partial t_{js}^a}{\partial x_j(t)} = \frac{\partial x_j^{-1}(\theta)}{\partial x_j(t)} \Big|_{x_j^{-1}(\theta)=t_{js}^a} = \frac{1}{\frac{\partial x_j(t)}{\partial x_j^{-1}(\theta)} \Big|_{x_j^{-1}(\theta)=t_{js}^a}} = \frac{1}{-\frac{\partial x_j(t)}{\partial t} \Big|_{t=t_{js}^a}} \tag{21}$$

$$= \left\{ \sum_{h' \in H_j} \sum_{s' \in S_{h'}^a} \sum_{k' \in L_{h'j}} \left[w_{h'j}^{k'} y_{h'js'}^{k'}(t_{js}^a) \left(\frac{1}{\tau_{h'j}^{k'}} - \frac{1}{t_{js}^a - t_{h's'}^a - d_{h'j}^{k'}} \right) \right] \right.$$

$$\left. + \frac{1}{\tau_j^{ref}} \eta_j(t_{js}^a - t_{j,s-1}^a) \right\}^{-1}, \tag{22}$$

where by letting $t_{j0} = -\infty$, we have $\eta_j(t_{j1}^a - t_{j0}^a) = 0$ and so the second terms in (22) of $\frac{\partial t_{j1}^a}{\partial x_j(t)}$ disappear.

Note that since the membrane potential given by (1) does not have the inverse function, (21) does not generally hold. However, following [1], since $du_j/dt > 0$ holds around t_{js}^a, by restricting $du_j/dt > 0$, we have (21).

Thus, by (18), (19), (20), and (22), we can compute $\delta_{js}(t)$.

Last, we have

$$\frac{\partial x_j(t)}{\partial w_{hj}^k}\bigg|_{t=t_{js}^a} = \frac{\partial}{\partial w_{hj}^k}\bigg\{\sum_{h' \in H_j}\sum_{s' \in S_{h'}^a}\sum_{k' \in L_{h'j}} w_{h'j}^{k'} y_{h'js'}^{k'}(t) + \eta_j(t - t_{j,last}^a)\bigg\}\bigg|_{t=t_{js}^a}$$

$$= \sum_{s' \in S_h^a} y_{hjs'}^k(t_{js}^a) + \frac{1}{\tau_j^{ref}}\eta_j(t_{js}^a - t_{j,s-1}^a)\frac{\partial t_{j,s-1}^a}{\partial x_j(t)}\frac{\partial x_j(t)}{\partial w_{hj}^k}\bigg|_{t=t_{j,s-1}^a} \qquad (23)$$

Thus, $\frac{\partial x_j(t)}{\partial w_{hj}^k}\big|_{t=t_{js}^a}$ is defined recursively. However, by letting $t_{j0} = -\infty$, $\eta_j(t_{j1}^a - t_{j0}^a) = 0$ and the second term of $\frac{\partial x_j(t)}{\partial w_{hj}^k}\big|_{t=t_{j1}^a}$ can be negligble. So, $\frac{\partial x_j(t)}{\partial w_{hj}^k}\big|_{t=t_{j1}^a}$ is fixed without recursion.

Then, by (17), (23), and δ_{js}, we can obtain $\Delta w_{hj}^k(t)$.

⟨**Example 1**⟩

Assume that postsynaptic neuron, j, is connected from presynaptic neuron, h, by a single synapse link, and every neuron emits only a single spike. Thus, the spike trains of these neurons are

$$F_h^a = \{t_{h1}^a\}, \ F_j^a = \{t_{j1}^a\}, \ \text{and} \ F_j^d = \{t_{j1}^d\}. \qquad (24)$$

In addition, we assume that

$$t_{j1}^d < t_{j1}^a \quad \text{and} \quad w_{mj}^1 > 0. \qquad (25)$$

Then, our learning algorithm becomes

$$\Delta w_{hj}^1(t) = \gamma_w e^2 \frac{\tau_{hj}^1}{\tau^3} \frac{(t_{j1}^a - t_{h1}^a)(t - t_{j1}^a - \tau)}{w_{hj}^1 \cdot (t_{j1}^a - t_{h1}^a - d_{hj}^1 - \tau_{hj}^1)} \exp(-\frac{t - t_{j1}^a}{\tau})H(t - t_{j1}^a)$$

$$\times \bigg[(t - t_{j1}^d)\exp(-\frac{t - t_{j1}^d}{\tau}) - (t - t_{j1}^a)\exp(-\frac{t - t_{j1}^a}{\tau})\bigg] \qquad (26)$$

$$\begin{cases} = 0 \ \text{if} \ t < t_{j1}^a, \ t_{j1}^d = t_{j1}^a, \ t = t_0, \ t = t_{j1}^a + \tau, \ t \to \infty \\ > 0 \ \text{if} \ t < t_{j1}^a + \tau \ \& \ t < t_0, \ t > t_{j1}^a + \tau \ \& \ t > t_0 \\ < 0 \ \text{if} \ t > t_{j1}^a + \tau \ \& \ t < t_0, \ t < t_{j1}^a + \tau \ \& \ t > t_0, \end{cases} \qquad (27)$$

$$\because \ t_{j1}^d + \tau < t_0 < t_{j1}^a + \tau \ \text{and} \ t_{h1}^a < t_{j1}^a < t_{h1}^a + \tau_{hj}^1 + d_{hj}^1$$

where

$$t_0 = \frac{t_{j1}^d \exp(\frac{t_{j1}^d}{\tau}) - t_j^a \exp(\frac{t_{j1}^a}{\tau})}{\exp(\frac{t_{j1}^d}{\tau}) - \exp(\frac{t_{j1}^a}{\tau})} \ \bigg(\geq \max\{t_{j1}^a, t_{j1}^d\} \bigg) \qquad (28)$$

The time course of $\Delta w^1_{hj}(t)$ shown in (26) for $t^d_{j1} = 1.0$, $t^a_{j1} = 1.5$, $\tau^1_{hj} = 1.0$, and $d^1_{hj} = 0$ is shown in Fig. 2.

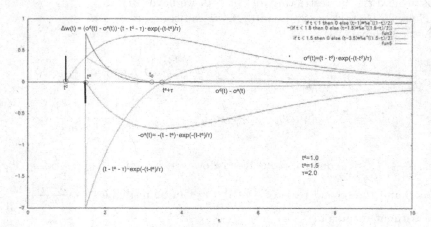

Fig. 2. The time course of $\Delta w^1_{hj}(t)$ shown in (26): disired and actual spikes are shown in bold black short vertical lines at $t^d_{j1} = 1.0$ and $t^a_{j1} = 1.5$, respectively, $\tilde{o}^d(t)$ in blue, $-\tilde{o}^a(t)$ in red, $\tilde{o}^d(t) - \tilde{o}^a(t)$ in green, $(t - t^a_{j1} - \tau) \exp(-\frac{t-t^a_{j1}}{\tau})$ in purple, and $\Delta w^1_{hj}(t)$ in black lines, respectively. (Color figure online)

(2) Offline Learning.

The online learning shown in the previous subsection is not easy to implement. We present an offline learning algorithm, which is easily implemented. It adjusts the synaptic weights all at once after observation period.

$$\Delta w^k_{hj} = \int_0^\infty \Delta w^k_{hj}(t)dt = -\gamma_w \sum_{s \in S^a_j} \left\{ \frac{\partial x_j(t)}{\partial w^k_{hj}}\Big|_{t=t^a_{js}} \int_0^\infty \delta_{js}(t)dt \right\}. \quad (29)$$

Then, we have

$$\int_0^\infty \delta_{js}(t)dt = \int_0^\infty \frac{\partial E(t)}{\partial \tilde{o}^a_j(t)} \frac{\partial \tilde{o}^a_j(t)}{\partial t^a_{js}} \frac{\partial t^a_{js}}{\partial x_j(t)} dt$$

$$= -\int_0^\infty \left\{ \sum_{g \in S^d_j} \left[\alpha_j(t - t^d_{jg})H(t - t^d_{jg}) \right] \right.$$

$$\left. - \sum_{f \in S^a_j} \left[\alpha_j(t - t^a_{jf})H(t - t^a_{jf}) \right] \right\}$$

$$\times H(t - t^a_{js}) \frac{\partial \alpha_j(t - t^a_{js})}{\partial t^a_{js}} \frac{\partial t^a_{js}}{\partial x_j(t)} dt$$

$$= -\sum_{g \in S^d_j} \frac{\partial t^a_{js}}{\partial x_j(t)} \int_{\max\{t^d_{jg}, t^a_{js}\}}^\infty \alpha_j(t - t^d_{jg}) \frac{\partial \alpha_j(t - t^a_{js})}{\partial t^a_{js}} dt$$

$$+ \sum_{f \in S_j^a} \frac{\partial t_{js}^a}{\partial x_j(t)} \int_{\max\{t_{jf}^a, t_{js}^a\}}^{\infty} \alpha_j(t - t_{jf}^a) \frac{\partial \alpha_j(t - t_{js}^a)}{\partial t_{js}^a} \, dt \tag{30}$$

$$= -\frac{1}{\tau} \left(\frac{e}{2}\right)^2 \frac{\partial t_{js}^a}{\partial x_j(t)} \left\{ \sum_{g \in S_j^d} \left[(t_{jg}^d - t_{js}^a) \exp\left(-\frac{|t_{jg}^d - t_{js}^a|}{\tau}\right) \right] \right.$$

$$\left. - \sum_{f \in S_j^a} \left[(t_{jf}^a - t_{js}^a) \exp\left(-\frac{|t_{jf}^a - t_{js}^a|}{\tau}\right) \right] \right\}. \tag{31}$$

Thus, by (22), we obtain $\int_0^{\infty} \delta_{js}(t) dt$. So, by (29) and (23), we have Δw_{hj}^k.

⟨**Example 2**⟩
Under the same assumption as Example 1, our offline learning algorithm becomes

$$\Delta w_{hj}^k = \gamma_w \frac{\tau_{hj}^1}{\tau} \left(\frac{e}{2}\right)^2 \frac{(t_j^d - t_j^a)(t_j^a - t_h^a - d_{hj}^1)}{w_{hj} \cdot (t_j^a - t_h^a - d_{hj}^1 - \tau_{hj}^1)} \exp\left(-\frac{|t_j^d - t_j^a|}{\tau}\right) \tag{32}$$

$$\begin{cases} = 0 \ \text{if} \ t_{j1}^d = t_{j1}^a \\ > 0 \ \text{if} \ t_{j1}^d < t_{j1}^a \\ < 0 \ \text{if} \ t_{j1}^d > t_{j1}^a. \end{cases} \quad \because \ t_h^a + d_{hj}^1 < t_j^a < t_h^a + d_{hj}^1 + \tau_{hj}^1 \tag{33}$$

The results shown in (33) is reasonable.

3.2 Synaptic Connections to Hidden Layers

(1) Online Learning.
The change of weight, $\Delta w_{ih}^l(t)$, of the l-th synapse from input or hidden neuron i to hidden neuron h, is given by

$$\Delta w_{ih}^l(t) = -\gamma_w \frac{\partial E(t)}{\partial w_{ih}^l} = -\gamma_w \sum_{p \in S_h} \left[\frac{\partial E(t)}{\partial t_{hp}^a} \frac{\partial t_{hp}^a}{x_h(t)} \frac{\partial x_h(t)}{\partial w_{ih}^l} \bigg|_{t=t_{hp}^a} \right]$$

$$= -\gamma_w \sum_{p \in S_h} \left[\delta_{hp}(t) \frac{\partial x_h(t)}{\partial w_{ih}^l} \bigg|_{t=t_{hp}^a} \right], \tag{34}$$

where

$$\delta_{hp}(t) = \frac{\partial E(t)}{\partial t_{hp}^a} \frac{\partial t_{hp}^a}{\partial x_h(t)} = \sum_{j \in J} \sum_{s \in S_j} \left\{ \frac{\partial E(t)}{\partial \tilde{o}_j^a(t)} \frac{\partial \tilde{o}_j^a(t)}{\partial t_{js}^a} \frac{\partial t_{js}^a}{\partial x_j(t)} \frac{\partial x_j(t)}{\partial t_{hp}^a} \bigg|_{t=t_{js}^a} \right\} \frac{\partial t_{hp}^a}{\partial x_h(t)}$$

$$= \sum_{j \in J} \sum_{s \in S_j} \left\{ \delta_{js}(t) \frac{\partial x_j(t)}{\partial t_{hp}^a} \bigg|_{t=t_{js}^a} \right\} \frac{\partial t_{hp}^a}{\partial x_h(t)}. \tag{35}$$

Then,

$$
\begin{aligned}
\left.\frac{\partial x_j(t)}{\partial t_{hp}^a}\right|_{t=t_{js}^a} &= \sum_{k'\in L_{hj}} w_{hj}^{k'} \left.\frac{\partial y_{hjp}^{k'}(t)}{\partial t_{hp}^a}\right|_{t=t_{js}^a} \\
&\quad + \left.\frac{\partial \eta_j(t-t_{j,s-1}^a)}{\partial t_{j,s-1}^a}\right|_{t=t_{js}^a} \frac{\partial t_{j,s-1}^a}{\partial x_j(t)} \left.\frac{\partial x_j(t)}{\partial t_{hp}^a}\right|_{t=t_{j,s-1}^a} \\
&= \sum_{k'\in L_{hj}} w_{hj}^{k'}\, y_{hjp}^{k'}(t_{js}^a)\left(\frac{1}{\tau_{hj}^{k'}} - \frac{1}{t_{js}^a - t_{hp}^a - d_{hj}^{k'}}\right) \\
&\quad + \frac{1}{\tau_j^{ref}}\, \eta_j(t_{js}^a - t_{j,s-1}) \frac{\partial t_{j,s-1}^a}{\partial x_j(t)} \left.\frac{\partial x_j(t)}{\partial t_{hp}^a}\right|_{t=t_{j,s-1}^a}.
\end{aligned}
\tag{36}
$$

The same as before, the above recursion terminates finally.

And, the same as $\frac{\partial t_{js}^a}{\partial x_j(t)}$ shown in (22), we have

$$
\begin{aligned}
\frac{\partial t_{hp}^a}{\partial x_h(t)} &= \left\{ \sum_{i'\in I_h} \sum_{q\in S_{i'}^a} \sum_{l'\in L_{i'h}} w_{i'h}^{l'} y_{i'hq}^{l'}(t_{hp}^a)\left(\frac{1}{\tau_{i'h}^{l'}} - \frac{1}{t_{hp}^a - t_{i'q}^a - d_{i'h}^{l'}}\right)\right. \\
&\quad \left. + \frac{1}{\tau_h^{ref}}\, \eta_h(t_{hp}^a - t_{h,p-1}^a) \right\}^{-1}.
\end{aligned}
\tag{37}
$$

Then, by (35), (36), (37), and δ_{js}, we can obtain $\delta_{hp}(t)$.

Last, the same as $\left.\frac{\partial x_j(t)}{\partial w_{hj}^l}\right|_{t=t_{js}^a}$ shown in (23), we have

$$
\left.\frac{\partial x_h(t)}{\partial w_{ih}^l}\right|_{t=t_{hp}^a} = \sum_{q\in S_i} y_{ihq}^l(t_{hp}^a) + \frac{1}{\tau_h^{ref}}\eta_h(t_{hp}^a - t_{h,p-1}^a)\frac{\partial t_{h,p-1}^a}{\partial x_h(t)}\left.\frac{\partial x_h(t)}{\partial w_{ih}^l}\right|_{t=t_{h,p-1}^a}
\tag{38}
$$

The same as before, the above recursion terminates finally. Thus, by (34), (38), and $\delta_{hp}(t)$, we obtain $\Delta w_{ih}^l(t)$.

(2) Offline Learning.

The same for synaptic connections to output neurons, an offline learning is practical also for synaptic connections to hidden neurons. We have

$$
\begin{aligned}
\Delta w_{ih}^l &= \int_0^\infty \Delta w_{ih}^l(t)dt = -\gamma_w \int_0^\infty \sum_{p\in S_h}\left\{\delta_{hp}(t)\left.\frac{\partial x_h(t)}{\partial w_{ih}^l}\right|_{t=t_{hp}^a}\right\}dt \\
&= -\gamma_w \sum_{p\in S_h}\left\{\left.\frac{\partial x_h(t)}{\partial w_{ih}^l}\right|_{t=t_{hp}^a}\int_0^\infty \delta_{hp}(t)dt\right\},
\end{aligned}
\tag{39}
$$

and

$$\int_0^\infty \delta_{hp}(t)dt = \int_0^\infty \sum_{j\in J} \sum_{s\in S_j} \left\{ \delta_{js}(t) \frac{\partial x_j(t)}{\partial t_{hp}^a}\Big|_{t=t_{js}^a} \right\} \frac{\partial t_{hp}^a}{\partial x_h(t)} dt$$

$$= \sum_{j\in J} \sum_{s\in S_j} \left\{ \frac{\partial x_j(t)}{\partial t_{hp}^a}\Big|_{t=t_{js}^a} \frac{\partial t_{hp}^a}{\partial x_h(t)} \int_0^\infty \delta_{js}(t)dt \right\}. \tag{40}$$

Thus, by (36), (37), and (31), we have obtained $\int_0^\infty \delta_{hp}(t)dt$. So, by (39) and (38), we have Δw_{ih}^l.

4 Conclusion

We have presented a new backpropagation learning algorithms, called BPSpike II, which adjusts synaptic weights in multiple layered networks and multiple spiking neurons. BPSpike II can also adjust synaptic delays, synaptic time constants, and neuron thresholds. Its performance evaluations with simulations are now in progress.

References

1. Bohte, S.M., Kok, J.N., La Poutre, H.: Error-backpropagation in temporally encoded networks of spiking neurons. Neurocomputing **48**, 17–37 (2002)
2. Florian, R.V.: The chronotron: a neuron that learns to fire temporally precise spike patterns. PLoS ONE **7**(8), e40233 (2012)
3. Gerstner, W., Kistler, W.: Spiking Neuron Models. Cambridge University Press, Cambridge (2002)
4. Ghosh-Dastidar, S., Adeli, H.: A new supervised learning algorithm for multiple spiking neural networks with application in epilepsy and seizure detection. Neural Netw. **22**, 1419–1431 (2009)
5. Guetig, R., Sompolinsky, H.: The tempotron: a neuron that learns spike timing-based decisions. Nat. Neurosci. **9**(3), 420–428 (2006)
6. Matsuda, S.: BPSpike:a backpropagation learning for all parameters in spiking neural networks with multiple layers and multiple spikes. In: IJCNN 2016 (2016)
7. Mohemmed, A., Schliebs, S., Matsuda, S., Kasabov, N.: Training spiking neural networks to associate spatio-temporal input-output spike patterns. Int. J. Neural Syst. **22**(4), 1250012 (2012)
8. Ponulak, Filip: Supervised learning in spiking neural networks with ReSuMe method, Doctoral Dissertation. Poznan University of Technology, Poznan, Poland (2006)
9. Schrauwen, B., van Campenhout, J.: Improving spikeProp: enhancements to an error-backpropagation rule for spiking neural networks. In: Proceeduings of 15th ProRISC Workshop (2004)
10. Yan, X., Zeng, X., Han, L., Yang, J.: A supervised multi-spike learning algorithm based on gradient descent for spiking neural networks. Neural Netw. **43**, 99–113 (2013)

Group Dropout Inspired by Ensemble Learning

Kazuyuki Hara[1]([envelope]), Daisuke Saitoh[2], Takumi Kondou[2], Satoshi Suzuki[3],
and Hayaru Shouno[3]

[1] College of Industrial Technology, Nihon University, 1-2-1 Izumi-cho,
Narashino-shi, Chiba 275-8575, Japan
hara.kazuyuki@nihon-u.ac.jp
[2] Graduate School of Industrial Technology, Nihon University, Chiba, Japan
[3] Graduate School of Informatics and Engineering,
The University of Electro-Communications, 1-5-1 Chofugaoka,
Chofu-shi, Tokyo 182-8585, Japan

Abstract. Deep learning is a state-of-the-art learning method that is
used in fields such as visual object recognition and speech recognition.
This learning uses a large number of layers and a huge number of units
and connections, so overfitting occurs. Dropout learning is a kind of
regularizer that neglects some inputs and hidden units in the learning
process with a probability p; then, the neglected inputs and hidden units
are combined with the learned network to express the final output. We
compared dropout learning and ensemble learning from three viewpoints
and found that dropout learning can be regarded as ensemble learning
that divides the student network into two groups of hidden units. From
this insight, we explored novel dropout learning that divides the student
network into more than two groups of hidden units to enhance the benefit
of ensemble learning.

Keywords: Dropout learning · Ensemble learning · Soft-committee
machine · Teacher-student formulation

1 Introduction

Deep learning [1,2] is attracting much attention in visual object recognition,
speech recognition, object detection, and many other fields. It provides automatic
feature extraction and can achieve outstanding performance [3].

Deep learning uses a very deep layered network and a huge amount of data, so
overfitting is a serious problem. To avoid overfitting, dropout learning [3] is used
for regularization. Dropout learning consists of two processes. At learning time,
some hidden units are neglected with a probability p, and this process reduces
the network size; therefore, overfitting is avoidable. At test time, learned hidden
units and hidden units that have not been learned are summed up and multiplied
by p to calculate the network output. We found that this method can be regarded
as ensemble learning.

© Springer International Publishing AG 2016
A. Hirose et al. (Eds.): ICONIP 2016, Part II, LNCS 9948, pp. 66–73, 2016.
DOI: 10.1007/978-3-319-46672-9_8

Ensemble learning improves the performance of a single network by using many networks. Bagging and the Ada-boost algorithm are well known [4]. We had theoretically analyzed ensemble learning using linear or non-linear perceptrons [5,6].

In this paper, we first analyze dropout learning regarded as ensemble learning. On-line learning [7,8] is used to learn a network. From the results, we find that dropout learning can be regarded as ensemble learning, except for when using different sets of hidden units in dropout learning. We then propose a novel dropout learning called group dropout. The proposed method divides the hidden units in the student network into several groups, and then each group learns from the teacher independently. After the learning, group outputs are averaged to calculate the student output.

2   Model

In this paper, we use a teacher-student formulation and assume the existence of a teacher that produces the desired output for the student network. By introducing the teacher, we can directly measure the similarity of the student weight vector compared to that of the teacher. First, we formulate a teacher network (referred to as "teacher") and a student network (referred to as "student") and then introduce the gradient descent algorithm.

The teacher and student are a soft-committee machine with N input units, hidden units, and an output, as shown in Fig. 1. The teacher consists of K hidden units, and the student consists of K' hidden units. Each hidden unit is a perceptron. The kth hidden weight vector of the teacher is $\boldsymbol{B}_k = (B_{k1}, \ldots, B_{kN})$, and the k'th hidden weight vector of the student is $\boldsymbol{J}_{k'}^{(m)} = (J_{k'1}^{(m)}, \ldots, J_{k'N}^{(m)})$, where m denotes the number of learning iterations. In the soft-committee machine, all hidden-to-output weights are fixed to be $+1$ [8]. This network calculates the majority vote of the hidden outputs.

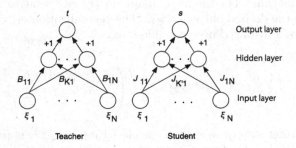

Fig. 1. Network structures of teacher and student

We assume that both the teacher and the student receive N-dimensional input $\boldsymbol{\xi}^{(m)} = (\xi_1^m, \ldots, \xi_N^{(m)})$, the teacher outputs $t^{(m)} = \sum_{k=1}^{K} t_k^{(m)} = \sum_{k=1}^{K} g(d_k^{(m)})$, and the student outputs $s^{(m)} = \sum_{k'=1}^{K'} s_{k'}^{(m)} = \sum_{k'=1}^{K'} g(y_{k'}^{(m)})$. Here, $g(\cdot)$ is the output function of a hidden unit, $d_k^{(m)}$ is the inner potential of the kth hidden unit of the teacher calculated using $d_k^{(m)} = \sum_{i=1}^{N} B_{ki}\xi_i^{(m)}$, and $y_{k'}^{(m)}$ is the inner potential of the k'th hidden unit of the student calculated using $y_{k'}^{(m)} = \sum_{i=1}^{N} J_{k'i}^{(m)}\xi_i^{(m)}$.

We assume that the ith elements $\xi_i^{(m)}$ of the independently drawn input $\boldsymbol{\xi}^{(m)}$ are uncorrelated random variables with zero mean and unit variance; that is, the ith element of the input is drawn from a probability distribution $P(\xi_i)$. The thermodynamic limit of $N \to \infty$ is also assumed. The statistics of the inputs in the thermodynamic limit are $\left\langle \xi_i^{(m)} \right\rangle = 0$, $\left\langle (\xi_i^{(m)})^2 \right\rangle \equiv \sigma_\xi^2 = 1$, and $\left\langle \|\boldsymbol{\xi}^{(m)}\| \right\rangle = \sqrt{N}$, where $\langle \cdots \rangle$ denotes the average and $\|\cdot\|$ denotes the norm of a vector. For each element B_{ki}, $k = 1 \sim K$ is drawn from a probability distribution with zero mean and $1/N$ variance. With the assumption of the thermodynamic limit, the statistics of the teacher weight vector are $\langle B_{ki} \rangle = 0$, $\left\langle (B_{ki})^2 \right\rangle \equiv \sigma_B^2 = 1/N$, and $\langle \|\boldsymbol{B_k}\| \rangle = 1$. This means that any combination of $\boldsymbol{B}_l \cdot \boldsymbol{B}_{l'} = 0$. The distribution of inner potential $d^{(m)}$ follows a Gaussian distribution with zero mean and unit variance in the thermodynamic limit.

For the sake of analysis, we assume that each element of $J_{k'i}^{(0)}$, which is the initial value of the student vector $\boldsymbol{J}_{k'}^{(0)}$, is drawn from a probability distribution with zero mean and $1/N$ variance. The statistics of the k'th hidden weight vector of the student are $\left\langle J_{k'i}^{(0)} \right\rangle = 0$, $\left\langle (J_{k'i}^{(0)})^2 \right\rangle \equiv \sigma_J^2 = 1/N$, and $\left\langle \|\boldsymbol{J}_{k'}^{(0)}\| \right\rangle = 1$ in the thermodynamic limit. This means that any combination of $\boldsymbol{J}_l^{(0)} \cdot \boldsymbol{J}_{l'}^{(0)} = 0$. The output function of the hidden units of the student $g(\cdot)$ is the same as that of the teacher. The statistics of the student weight vector at the mth iteration are $\left\langle J_{k'i}^{(m)} \right\rangle = 0$, $\left\langle (J_{k'i}^{(m)})^2 \right\rangle = (Q_{k'k'}^{(m)})^2/N$, and $\left\langle \|\boldsymbol{J}_{k'}^{(m)}\| \right\rangle = Q_{k'k'}^{(m)}$. Here, $(Q_{k'k'}^{(m)})^2 = \boldsymbol{J}_{k'}^{(m)} \cdot \boldsymbol{J}_{k'}^{(m)}$. The distribution of the inner potential $y_{k'}^{(m)}$ follows a Gaussian distribution with zero mean and $(Q_{k'k'}^{(m)})^2$ variance in the thermodynamic limit.

Next, we introduce the stochastic gradient descent (SGD) algorithm for the soft-committee machine. For the possible inputs $\{\boldsymbol{\xi}\}$, we want to train the student to produce the desired outputs $t = s$. The generalization error is defined as the squared error ε averaged over possible inputs:

$$\varepsilon_g^{(m)} = \left\langle \varepsilon^{(m)} \right\rangle = \frac{1}{2}\left\langle (t^{(m)} - s^{(m)})^2 \right\rangle = \frac{1}{2}\left\langle \left(\sum_{k=1}^{K} g(d_k^{(m)}) - \sum_{k'=1}^{K'} g(y_{k'}^{(m)}) \right)^2 \right\rangle, \tag{1}$$

At each learning step m, a new uncorrelated input, $\boldsymbol{\xi}^{(m)}$, is presented, and the current hidden weight vector of the student $\boldsymbol{J}_{k'}^{(m)}$ is updated using

$$\boldsymbol{J}_{k'}^{(m+1)} = \boldsymbol{J}_{k'}^{(m)} + \frac{\eta}{N}\left(\sum_{l=1}^{K} g(d_l^{(m)}) - \sum_{l'=1}^{K'} g(y_{l'}^{(m)}) \right) g'(y_{k'}^{(m)})\boldsymbol{\xi}^{(m)}, \tag{2}$$

where η is the learning step size and $g'(x)$ is the derivative of the output function of the hidden unit $g(x)$.

3   Ensemble Learning and Dropout Learning

In this section, we compare dropout learning and ensemble learning to clarify the effect of random selection of hidden units.

3.1   Ensemble Learning

Ensemble learning is performed by using many learners (referred to as "students") to achieve better performance [5]. In ensemble learning, each student learns from the teacher independently, and each student output $s_{k'_{en}}$ is averaged to calculate the ensemble output s_{en}. We assume that the teacher and the students are the soft-committee machines.

$$s_{en} = \sum_{k'_{en}=1}^{K_{en}} C_{k'_{en}} s_{k'_{en}} = \sum_{k'_{en}=1}^{K_{en}} C_{k'_{en}} \sum_{k'=1}^{K'} g(y_{k'}) \tag{3}$$

Here, $C_{k'_{en}}$ is a weight for averaging. K_{en} is the number of students.

There are three cases of setting the number of hidden units in the student: (1) $K' < K$, (2) $K' = K$, and (3) $K' > K$. In the case of $K' < K$, this is an unlearnable and insufficient case because the degree of complexity of the student is smaller than that of the teacher. In the case of $K' = K$, this is a learnable case because the degree of complexity of the student is the same as that of the teacher. In the case of $K' > K$, this is a learnable and redundant case because the degree of complexity of the student is larger than that of the teacher [10]. Therefore, if we set $K' = K$, the network performance will be the best of them.

3.2   Dropout Learning

Dropout learning is used in deep learning to prevent overfitting [3]. A small amount of data compared with the size of a network may cause overfitting [10]. In the state of overfitting, the learning error and the test error become different.

The learning equation of dropout learning for the soft-committee machine can be written as follows.

$$J_{k'}^{(m+1)} = J_{k'}^{(m)} + \frac{\eta}{N} \left(\sum_{l=1}^{K} g(d_l^{(m)}) - \sum_{l' \notin D^{(m)}}^{(1-p)K'} g(y_{l'}^{(m)}) \right) g'(y_{k'}^{(m)}) \xi^{(m)}, \tag{4}$$

Here, $D^{(m)}$ shows a set of the hidden units that randomly selected with respect to the probability p from all the hidden units at the mth iteration. Note that the second term in the bracket of RHS of Eq. (4) is a soft-committee machine composed of not selected hidden units. The hidden units in $D^{(m)}$ are not subject to learning then, the size of the student becomes small, and a shrunken student may avoid overfitting. This effect is the dropout learning opportunity. After the learning, the student's output $s^{(m)}$ is calculated by using the sum of learned hidden outputs and hidden outputs that have not been learned multiplied by p.

$$s^{(m)} = p * \left\{ \sum_{l' \notin D^{(m)}}^{(1-p)K'} g(y_{l'}^{(m)}) + \sum_{l' \in D^{(m)}}^{pK'} g(y_{l'}^{(m-1)}) \right\} \tag{5}$$

This equation is regarded as the ensemble of a learned soft-committee machine (the first term of RHS) and that of a not learned (the second term of RHS) when the probability is $p = 0.5$. However, in the deep learning, a set of hidden units in $D^{(m)}$ is changed in every iteration where the same set of hidden units are used in the ensemble learning. Therefore, dropout learning is regarded as ensemble learning using a different set of hidden units in every iteration. Therefore, we refer to dropout learning as "random dropout" in this paper.

3.3   Comparison Between Random Dropout and Ensemble Learning

We compared dropout learning and ensemble learning from three viewpoints: (1) selecting the hidden units in a group randomly or using the same hidden units, (2) dividing the student into two or more groups that contains a part of hidden units in the student, and (3) averaging the outputs of learned networks and those of an un-learned networks or averaging only the output of learned networks. Dropout learning involves selecting the hidden units in a group randomly, dividing the student into two groups, and averaging the output of learned hidden units and that of un-learned networks. Ensemble learning involves using the same hidden units in a group throughout the learning, dividing the students into more than two groups, and averaging the output of learned networks.

In comparison, we used two soft-committee machines with 50 hidden units for ensemble learning. The output function $g(x)$ is a sigmoid-like function $\mathrm{erf}(x/\sqrt{2})$. For dropout learning, we used 100 hidden units. We set $p = 0.5$; then, dropout learning selected 50 hidden units in $D^{(m)}$ with 50 unselected hidden units remaining. Therefore, dropout learning and ensemble learning had the same architectures. The number of input units was $N = 1000$, and the learning step size was set to $\eta = 0.01$.

Figure 2 shows the results. The results are obtained by taking the average of the results of ten trials. 10^4 independent data are used in the learning, and 10^3 independent data are used to calculate the MSE. The horizontal axes are time $t = m/N$, and the vertical axes are the MSE calculated for N data. In Fig. 2(a), "Single" shows the soft-committee machines with 50 hidden units, and "Ensemble" shows the results given by ensemble learning. Test errors are used in these figures. In Fig. 2(b), "Test" shows the MSE given by the test data, and "Learning" shows the MSE given by the learning data. Dropout learning is used. From Fig. 2(a), the ensemble learning achieved an MSE smaller than that of the single network. However, from Fig. 2(b), dropout learning achieved an MSE smaller than that of ensemble learning. Therefore, dropout learning outperforms the ensemble learning because it uses randomly selected hidden units and averages the outputs of learned networks and those of un-learned networks.

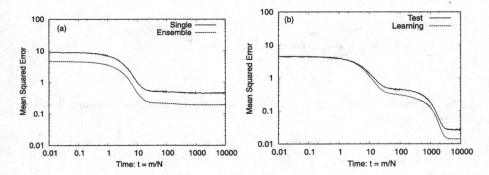

Fig. 2. Results of comparison between dropout learning and ensemble learning. (a) is ensemble learning of two networks, and (b) is dropout learning with respect to $p = 0.5$.

4   Group Dropout

In the dropout learning, two groups of the learned hidden units and those of the not-learned hidden units are used. This means that the dropout learning uses only two students for ensemble. In ensemble learning, using many students will achieve better performance. Therefore, the effect of the ensemble in the dropout learning is not enough. Therefore, we propose "group dropout", which involves dividing the student into more than two groups that composed of hidden units of the student and applying ensemble learning. In group dropout, the number of hidden units in a group are the same as that in the teacher network.

We divide the student (with K' hidden units) into $K_{gr} = K'/K$ groups (See Fig. 3. Here, $K' = 4$ and $K = 2$). These groups learn from the teacher independently, and then we calculate the ensemble output s_{en} by averaging the group outputs s_{gr} as:

$$s_{gr} = \frac{1}{K_{gr}} \sum_{k'_{gr}=1}^{K_{gr}} s_{k'_{gr}} = \frac{1}{K_{gr}} \sum_{k'_{gr}=1}^{K_{gr}} \sum_{k'=1}^{K} g(y_{k'_{gr}k'}). \tag{6}$$

Here, $s_{k'_{gr}}$ is the output of a group with K hidden units, and $y_{k'_{gr}k'}$ is the k'th hidden output in the k'_{gr}th group. We set the number of hidden units in a group to K, and this makes the learning of the group network learnable.

Figure 4 shows the results. The results are given by taking the average of the results of 10 trials. In these figures, the horizontal axes show the learning time $t = m/N$. The vertical axes show the MSE. The simulation conditions are the same as those of Fig. 2. The line labeled "Half Dropout" shows the results of the ensemble of two groups of the students, and it uses the same hidden units in a group during the learning. "Half Dropout" is the same as "Random Dropout" except for using the same hidden units in a group and the ensemble of learned soft-committee machines.

As can be seen from Fig. 4, the residual error of "without Dropout" stays large and can be considered as overfitting the data. The residual error of

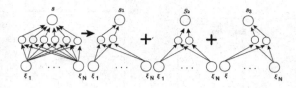

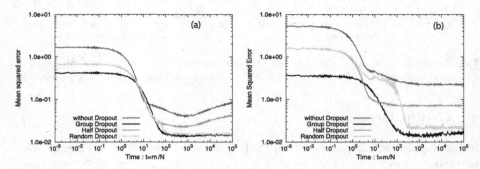

Fig. 3. Student is divided into three groups to learn by ensemble learning. Teacher composes of two hidden units.

Fig. 4. Effect of proposed method. (a) shows when $K' = 8$, and (b) shows when $K' = 30$. $K = 2$ for both. Results are given by the average of ten trials.

"Half dropout" is in the middle of "without Dropout" and "Random Dropout" or "Group Dropout", so dividing the student into two groups and applying ensemble learning does not perform well. The residual error of "Random Dropout" and "Group Dropout" converge to a small value, so these are considered as not overfitting the data. Therefore, both "Random Dropout" and "Group Dropout" can work as a regularizer. As can be seen from these figures, when the number of hidden units is small (Fig. 4(a)), the MSE of the group dropout and that of the random dropout are identical. However, when the number of hidden units is large (Fig. 4(b)), the MSE of the group dropout outperforms that of the random dropout.

The group dropout differs from the random dropout for three reasons. First, the dropout learning divided the student into two groups; however, the proposed method divided the student into more than two groups. Second, the dropout learning randomly selected the hidden units to be neglected at each learning step; however, the proposed method uses the same hidden units in the group. Third, the dropout learning is the ensemble learning of learned hidden units and un-learned hidden units; however, the group dropout is the ensemble of only learned hidden units.

5 Conclusion

In this paper, we analyzed the dropout learning regarded as ensemble learning. We have shown that the dropout learning can be regarded as ensemble learning

except for when using a different set of hidden units in every learning iteration. From this analysis, it is clarified that using a different set of hidden units
outperforms ensemble learning. We next proposed using group dropout, which
divides the student into several groups with some hidden units that are identical
to those of the teacher, and perform ensemble learning. The proposed method
outperforms the dropout learning when the number of hidden units in a group is
larger than that of the teacher. To clarify the effect of averaging the outputs of
learned networks and that of un-learned networks, and to explore group dropout
using different hidden units in a group are our future works.

Acknowledgments. The authors thank Professor Masato Okada and Assistant
Professor Hideitsu Hino for insightful discussions.

References

1. LeCun, Y., Bengio, Y., Hinton, G.: Deep learning. Nature **521**, 436–444 (2015)
2. Hinton, G.E., Osindero, S., Teh, Y.W.: A fast learning algorithm for deep belief
   nets. Neural Comput. **18**, 1527–1554 (2006)
3. Krizhevsky, A., Sutskever, I., Hinton, G.E.: ImageNet classification with deep convolutional neural networks. In: Advanced in Neural Information Processing System
   25 (2012)
4. Freund, Y., Schapire, R.E.: A decision-theoretic generalization of on-line learning
   and an application to boosting. J. Comput. Syst. Sci. **55**, 119–139 (1997)
5. Hara, K., Okada, M.: Ensemble learning of linear perceptrons: on-line learning
   theory. J. Phys. Soc. Jpn. **74**(11), 2966–2972 (2005)
6. Miyoshi, S., Hara, K., Okada, M.: Analysis of ensemble learning using simple perceptron based on online learning theory. Phys. Rev. E **71**, 036116 (2005)
7. Biehl, M., Schwarze, H.: Learning by on-line gradient descent. J. Phys. A Math.
   Gen. Phys. **28**, 643–656 (1995)
8. Saad, D., Solla, S.A.: On-line learning in soft-committee machines. Phys. Rev. E
   52, 4225–4243 (1995)
9. Wager, S., Wang, S., Liang, P.: Dropout training as adaptive regularization. In:
   Advance in Neural Information Processing System 26 (2013)
10. Bishop, C.M.: Pattern Recognition and Machine learning. Springer, New York
    (2006)

Audio Generation from Scene Considering Its Emotion Aspect

Gwenaelle Cunha Sergio and Minho Lee[✉]

School of Electronics Engineering, Kyungpook National University,
1370 Sankyuk-Dong, Taegu 702-701, South Korea
gwena.cs@gmail.com, mholee@gmail.com

Abstract. Scenes can convey emotion like music. If that's so, it might be possible that, given an image, one can generate music with similar emotional reaction from users. The challenge lies in how to do that. In this paper, we use the Hue, Saturation and Lightness features from a number of image samples extracted from videos excerpts and the tempo, loudness and rhythm from a number of audio samples also extracted from the same video excerpts to train a group of neural networks, including Recurrent Neural Network and Neuro-Fuzzy Network, and obtain the desired audio signal to evoke a similar emotional response to a listener. This work could prove to be an important contribution to the field of Human-Computer Interaction because it can improve the interaction between computers and humans. Experimental results show that this model effectively produces an audio that matches the video evoking a similar emotion from the viewer.

Keywords: Neuro-Fuzzy Network · Recurrent Neural Network · Emotion · Mean Opinion Score · HSL color space · Music elements

1 Introduction

The need to communicate is intrinsic to any living organism, and as such, humans have always felt the need to communicate with each other. Studies into origin of language suggests how, for the need to register history, spoken language evolved into a system of writing in the form of cave paintings [1].

An important area of research in this regard is to find a method to convert art of one modality into the art of another modality. Currently there are some applications that generate music given an image, such as Photosounder [2], SonicPhoto [3] and Paint2Sound [4]. Photosounder is the "first audio editor/synthesizer to have an entirely image-based approach to sound creation and editing" and it can transform sounds into images as well as transform images into sounds. The other two applications only transform pictures to sound, and not the other way around.

However, a common limitation of available applications is that they do not consider emotion when generating the music. Thus, the goal of this paper is to implement an algorithm that takes into consideration that important aspect of music and scenes.

© Springer International Publishing AG 2016
A. Hirose et al. (Eds.): ICONIP 2016, Part II, LNCS 9948, pp. 74–81, 2016.
DOI: 10.1007/978-3-319-46672-9_9

To do that, we will try to find a relationship between the Hue, Saturation and Lightness (HSL) components of a scene, and their variation through time, and music components such as tone, pitch, rhythm and loudness and analyze the performance of the proposed method by comparing the spectrograms of the target and estimates audio output, which is a visual representation of the music, for intuitive understanding.

This paper is organized in 5 sections starting with the introduction. Section 2 discusses the relationship between emotion, music and picture, followed by Sect. 3, which explains the proposed method and evaluation method. Section 4 introduces the experimental results and Sect. 5 is the conclusion and future work.

2   State of the Art: Emotion in Music and Picture

2.1   Emotion from Pictures

In the engineering field, Zhang [5,6] aims to embed emotion into a machine that can analyze images and learn more complex emotions by interacting with humans and analyzing EEG signals and image features. In [7], the authors extract texture information from images in the IAPS dataset, using Wiccest and Gabor features and use those to train a support vector machine (SVM) network to classify them into different emotional valences. After the training, a collection of masterpieces is applied to the system, obtaining a performance a little higher than 50 %, and showing that machines have a potential to derive emotion from paintings.

The research in [8] is also relevant to this topic with Mikels collecting "descriptive emotional category data on subsets of the IAPS in an effort to identify images that elicit one discrete emotion more than others". His research reveals that the analyzed images have multiple levels of emotion and that the IAPS set used is a good asset when investigating discrete emotions.

2.2   Emotion from Music

Daniel J. Levitin [9] wrote in his first best-selling book [10] that "the essence of music performance is being able to convey emotion" and that's the reason why most of us listen to music, for that emotional experience. According to him, in western music, major scales are associated with happy emotions and minor scales with sad ones and a fast tempo is generally regarded as happy, whilst a slow one is regarded as sad.

In a study, Kim [11] extracts emotion indicators given a video, that is, given musical and visual components. It extracts Tempo, Melody and Loudness from the musical component and Hue, Saturation, Lightness (or Intensity) and Orientation from the visual component. It also uses EEG signals as the Valence and Arousal indicators, and, in the end, it uses Adaptive Neuro-Fuzzy Inference System (ANFIS) to classify the emotion as positive or negative one.

In a similar study, Lee [12] developed a system based on 3D fuzzy visual and EEG features that recognizes a person's emotional state through concurrently analyzing a movie clip as visual stimuli and EEG signals from human subjects' brain.

3   Proposed Method: Scene to Music

3.1   Overall Architecture

The proposed method, see Fig. 1, is a hybrid system composed of an Adaptive Neuro-Fuzzy Inference System (ANFIS) and two Recurrent Neural Networks (RNN). The model's input is a 3×1 descriptor representing the scene's Hue, Saturation and Lightness (HSL) features and the output is also a 3×1 descriptor representing the audio's features: Tempo, Loudness and Rhythm (TLR).

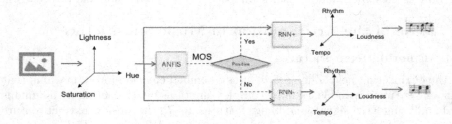

Fig. 1. Hybrid System

The ANFIS is used to classify the emotion extracted from a scene's features into one of 2 groups such as positive and negative ones. Such task was assigned to a Fuzzy System because it is more reasonable to model emotion according to fuzzy mathematics [13]. The ANFIS includes adaptive learning algorithm to identify the membership function parameters and rules based on the Sugeno type fuzzy inference systems (FIS). The FIS parameters are generated with the *genfis1* function in Matlab with 2 generalized bell membership functions, since there are only 2 possible emotions. The input to this sysmte is the HSL features extracted from the scene and the target output is MOS of that particular scene obtained by surveying 5 subjects and asking them to rate the videos as 1 when having a very negative emotion or 9 when having a very positive one.

The RNNs are used to estimate the sound features, and the one that will be used depends on whether the ANFIS classified the scene as having a positive or negative emotion. This network was chosen because the scene's variation through time is also an important feature that needs to be considered and music has dynamic properties that cannot be ignored. The RNN used here considers 3 time delays and 1 hidden layer, if one does not consider the output layer as such, with 12 neurons. The input to this network is the HSL features from the scene and the target output is the TLR features from which the music can be later generated and which should convey a similar emotion as the scene.

3.2   Feature Extraction from Scene and Music

This work uses a modified version of Kim's work [11] to extract the scene and audio features, the modification being that it does not use EEG signals to do so nor the scene's orientation. Additionally, when obtaining the audio features, instead of obtaining those from the entire audio file, it segments the audio and extracts the features from those segments while also keeping the original audio segment in another matrix for future use.

In [11], the sound from the scene is extracted and converted into a spectrogram by using the Short Time Fourier Transform (STFT). After that, *tempo* can be estimated by calculating where the strongest autocorrelation occurs in the signal away from the origin. The *loudness* feature is estimated through a mapping of the sound pressures in decibels, which is done by a time-frequency decomposition that reflects the human ear response compensation. Finally, *rhythm* is "reflected in terms of tone information", which includes the audio's frequency characteristics and the gradient of frequency variation over the period of time, both parameters of which have been calibrated and tested by many previous experiments.

As for the scene features, those are extracted using 3D fuzzy GIST [14] based on a 3D tensor data that includes the L*C*H color space (or Lightness, Chroma and Hue respectively). These tensor data are M×M×T, where M×M is the dimension of a single frame and T is the number of frames in the sequence, which can also be regarded as the duration of the scene. The color informations are then clustered into three clusters using the Fuzzy C-Means (FCM) algorithm. As a result, each frame will have a 3×1 descriptor for each color feature (H, S and L), or, a final 9×1 descriptor which is obtained by concatenating the descriptors from the color space.

A modification was done to this feature extraction method as to make the model easier to visualize and also to shorten the training time of the neural networks. Instead of concatenating three 3×1 descriptors as they were, the average of the values in this descriptor was taken, obtaining a single value for each color component. Those values were then concatenated, resulting in a 3×1 descriptor representing the Hue, Saturation and Lightness.

3.3 Music Generation

Figure 2 illustrates how music was generated in this model. The original audios in the training video dataset were segmented and had their features extracted in order to form the dataset that was used during training. While those features were stored, so was their original equivalent audios to be used later on.

The first step in the music generation is to obtain the estimated audio features from each frame. Then, the Mean Absolute Error (MAE) of the difference between this descriptor d and each and all of the features in the dataset used during training is calculated. The following equation shows how to calculate the MAE of the difference between d and one set of features in the dataset f:

$$MAE = \frac{1}{n} \sum_{j=1}^{n} |d_j - f_j| \tag{1}$$

This equation has to be applied to all the features and the one that results in the smallest MAE is then chosen to represent this specific frame. This process

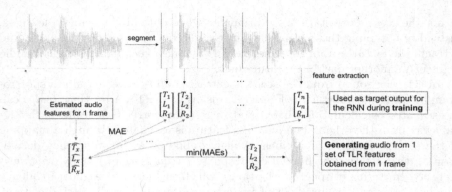

Fig. 2. Diagram for Music Generation

is repeated to all the frames in the scene, concatenating the obtained audios as they are chosen, so that at the end there will be one music representing the full scene.

3.4   Evaluation Method

To evaluate this model we will give to it as input a scene excerpt used during the training stage, to which we know its target audio, and obtain the estimated audio output. We will then compare the spectrogram of those two audios and check on similarities between them, both visually and quantitatively by using the MAE function. The smaller the value obtained with that function, the more similarity there are between the two signals. The Mean Opinion Score (MOS) will also be used as an evaluation method as another quantitative measure.

4   Experimental Results

4.1   Lindsey Stirling Dataset

To train our model we used a dataset composed of excerpts from 8 instrumental music videos by American violinist, dancer, performance artist, singer and composer Lindsey Stirling[1]. As shown in Fig. 3, the music videos chosen were Roundtable Rival, Crystallize, Beyond the Veil, Elements, Take Flight, Phantom of the Opera, Lord of the Rings Medley and Moon Trance, all of which can be visualized in this YouTube playlist: https://www.youtube.com/playlist?list=PLg5IYs6I5_xPkTWQ6P_YOiTTh7IBlc7ZH.

---

[1] https://www.youtube.com/user/lindseystomp.

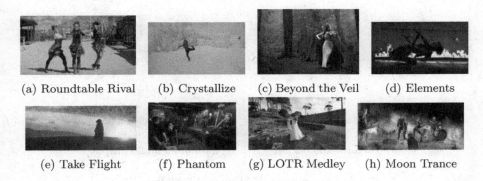

(a) Roundtable Rival   (b) Crystallize   (c) Beyond the Veil   (d) Elements

(e) Take Flight   (f) Phantom   (g) LOTR Medley   (h) Moon Trance

Fig. 3. Dataset

4.2 Results

The ANFIS network tested with the dataset used during training performed very well, with approximately 90 % accuracy. Figures 4(a), (b) and (c) show the learning curve of the ANFIS network and target and estimated MOS for the HSL input respectively. In Figs. 4(b) and (c), the blue dots represent samples with negative emotion and the red ones, positive. It can be seen that the network had some problems when trying to classify the negative samples that were too closely related to the positive ones. It is highly likely that those samples fall in a neutral emotion region.

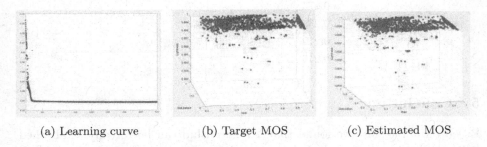

(a) Learning curve   (b) Target MOS   (c) Estimated MOS

Fig. 4. Results for ANFIS with HSL input: performance of ∼ 90 % (Color figure online)

We used a 11 s video segment from "Lord of the Rings Medley", see Fig. 3(g), to test our work. Figure 5 shows the spectrogram of the target and estimated output from the hybrid system in that order. The MAE for those signals was calculated as 0.212 and the MOS collected was 7.0.

Figure. 5 shows that while most of the spectrum were estimated with a reasonable accuracy, there were some gaps that failed to approximate. This might be solved with further fine tuning of the networks' parameters or by introducing deep learning. Table 1 shows some extended results, including MAE and MOSs, of other sample scenes that were tested with the model. Most of the samples gave

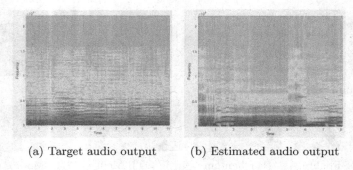

(a) Target audio output   (b) Estimated audio output

Fig. 5. Spectrogram, MAE = 0.212, MOS = 7.0

back a reasonable result except for samples 4 and 6, which tended to be in a more neutral region of the spectrum. It was also noticed some discrepancy in opinion between the subjects regarding a few samples, meaning that the resulting audio needs to be clearer as to convey an emotion similar to the scene.

Table 1. Extended results

Sample	MAE	Expected MOS	Obtained MOS
1	0.200	8.0	6.3
2	0.154	3.8	2.3
3	0.258	6.4	5.0
4	0.176	6.4	2.7
5	0.273	6.6	6.7
6	0.254	6.0	4.3
7	0.212	7.2	7.0
8	0.249	5.2	6.3

5   Conclusion

Based on our results we found that a relationship can be roughly established between scene and music by relating the Hue, Saturation and Lightness of an image to the tone, pitch, rhythm and loudness of a sound. However, this work still has a room to improve in the future.

In future works we would like to generate new original songs from the audio features instead of just comparing and associating it to known songs. We also plan on balancing the dataset more evenly between the positive and negative emotions mentioned. Another extension would be to add a second dimension to the emotion axis called arousal, where the positive and negative emotions will both be further classified as having a high or low arousal, that is, if the sample made the subject feel a strong or a weak emotion. What that means is that this model will be considering a total of 4 groups to classify an emotion. A neutral state should also be taken into consideration.

Acknowledgments. This work was supported by the Industrial Strategic Technology Development Program (10044009) funded by the Ministry of Trade, Industry and Energy (MOTIE, Korea).

References

1. Mark, J.J.: Writing: Definition (2011). http://www.ancient.eu/writing/
2. Rouzic, M.: Photosounder (2008). http://photosounder.com/
3. White, D.: Sonicphoto (2011). http://www.skytopia.com/software/sonicphoto/
4. Singh, J.F.: Paint2sound (2012). http://flexibeatz.weebly.com/paint2sound.html
5. Zhang, Q., Jeong, S., Lee, M.: Autonomous emotion development using incremental modified adaptive neuro-fuzzy inference system. Neurocomputing **96**, 33–44 (2012). Elsevier
6. Zhang, Q., Lee, M.: Emotion development system by interacting with human eeg and natural scene understanding. Cogn. Syst. Res. **14**, 37–49 (2012). Elsevier
7. Yanulevskaya, V., Gemert, J.V., Roth, K., Herbold, A., Sebe, N., Geusebroek, J.: Emotional valence categorization using holistic image features. In: 15th IEEE International Conference on Image Processing (ICIP), pp. 101–104 (2008)
8. Mikels, J.A., Fredrickson, B.L., Larkin, G.R., Lindberg, C.M., Maglio, S.J., Reuter-Lorenz, P.A.: Emotional category data on images from the international affective picture system. Behav. Res. Meth. **37**(4), 626–630 (2005)
9. Levitin, D.J.: Dr. daniel j. levitin: Neuroscientist, musician, author (2015). http://daniellevitin.com/publicpage/
10. Levitin, D.J.: This Is Your Brain on Music: The Science of a Human Obsession. Dutton Penguin Books Ltd., New York (2006)
11. Kim, T.: The acoustic and visual emotive signals classification in movies using brain cognitive signal and fuzzy clustering and adaptive neuro-fuzzy inference system (2014)
12. Lee, G., Kwon, M., Kavuri, S., Lee, M.: Emotion recognition based on 3D fuzzy visual and eeg features in movie clips. Neurocomputing **144**, 560–568 (2014). Elsevier
13. Kwon, I.K., Lee, S.Y.: Design of emotional space modeling using neuro-fuzzy. Adv. Sci. Technol. Lett. **46**, 6–9 (2014)
14. Lee, Giyoung, Kwon, Mingu, Sri, S.K., Lee, M.: Emotion recognition based on 3D fuzzy visual and eeg features in movie clips. Neurocomputing **144**, 560–568 (2014). Elsevier

Semi Supervised Autoencoder

Anupriya Gogna$^{(\boxtimes)}$ and Angshul Majumdar

Indraprasatha Institute of Information Technology-Delhi, New Delhi, India
{anupriyag, angshul}@iiitd.ac.in

Abstract. Autoencoders are self-supervised learning tools, but are unsupervised in the sense that class information is not required for training; but almost invariably they are used for supervised classification tasks. We propose to learn the autoencoder for a semi-supervised paradigm, i.e. with both labeled and unlabeled samples available. Given labeled and unlabeled data, our proposed autoencoder automatically adjusts – for unlabeled data it acts as a standard autoencoder (unsupervised) and for labeled data it additionally learns a linear classifier. We use our proposed semi-supervised autoencoder to (greedily) construct a stacked architecture. We demonstrate the efficacy our design in terms of both accuracy and run time requirements for the case of image classification. Our model is able to provide high classification accuracy with even simple classification schemes as compared to existing models for deep architectures.

Keywords: Autoencoder · Feature extraction · Classification · Semi-supervised learning

1 Introduction

Autoencoders (AE) [1] are known to be self-supervised models, i.e. the input and the output are the same. They are unsupervised in the sense that they do not require any label/class information for training. However, they are almost always used for supervised classification in various domains [2–3]. They are essentially 3-layer structures wherein a hidden (middle) layer representation is learned such that it can be used to reconstruct the input. The hidden layer representation, derived under suitable constraints, provide greater insight into data assisting in classification tasks.

Conventionally, feature extraction tools have relied on shallow architectures like kernel machines [4] and principal component analysis (PCA) [5]. In recent works, deep architectures, constructed by stacking together AE (or Restricted Boltzmann machines), have been shown to yield more abstract representation which in turn improved classification accuracy [6–8]. This is credited to the fact that such a model mimics our brain's natural processing and helps better represent complex data.

Deep architectures are hierarchical structures composed of multiple layers, each representing data at a different level of abstraction; the representation learned from one layer acting as input to the next. In this work, we design an autoencoder based deep architecture targeting performance improvement in classification task – specifically image classification.

Training a deep architecture involves two steps. In the first stage – pre training – each layer (AE) is individually trained using either the raw data (for first stage) or

A. Hirose et al. (Eds.): ICONIP 2016, Part II, LNCS 9948, pp. 82–89, 2016.
DOI: 10.1007/978-3-319-46672-9_10

hidden layer representation of previous layer (for subsequent layers) as input. This is a fully unsupervised step. The next stage – fine tuning – the decoder portion of the stacked autoencoder is removed and the targets are attached to the deepest/bottleneck layer. A suitably chosen objective function (like classification accuracy) is then optimized using gradient decent algorithm.

A major challenge in way of achieving high accuracy of (image) classification is the presence of unwarranted factor, such as illumination variation, rotation, scaling or noise, inflicting the raw images which alter the inter-class variability. Traditionally each of the effects was treated separately. In most practical scenarios, several of these effects are present simultaneously, in such cases such tailored schemes (to account for each effect simultaneously) are not effective.

The current trend is to learn automatic feature representation techniques that are implicitly robust to such effects; AE being an important tool for it; AE [1] focuses on learning the (hidden layer) representation such that input can be constructed from the same in least square sense; it ensures that the feature set provides a good representation of the input data. However, for classification task, discrimination amongst feature vectors of different classes is as desirable as the fidelity of feature vector to information content of the input data.

In this work, we propose a variant of standard AE - semi-supervised AE (SS-AE) - aimed at improving inter-class discrimination amongst feature representation to assist in better classification. Our SS-AE makes use of available label information by adding an additional label based penalty to the base objective function. Our modified formulation ascertains that all feature vectors belonging to same label mapping are similar; this use of additional information helps in mitigating influence of unwarranted factors as well. The improved robustness of features assists in achieving high accuracy, with even simple classification schemes.

There have been few previous works on supervised AE [9]; however fully labeled training data is almost never available. In most practical problems, the amount of unlabeled data is much more than the available supervised information. There are previous works on semi-supervised neural networks [10–12]; to the best of our knowledge there is only a single study on semi-supervised AE. Authors in [13] propose a semi-supervised model wherein the least square objective function is replaced by a weighted least square formulation; weights capturing the relevance of each input variable (node) towards final classification task. It leads to only slight improvement in prediction accuracy and the weight matrix needs to be pre-computed manually.

On the other hand, our model benefits from its capability to automatically adjust itself to given (limited) amount of label information; ranging from unsupervised standard AE to fully supervised design. We use our SS-AE to build a stacked architecture, following a greedy approach.

In addition to the advantage highlighted above, incorporating label information within the AE framework, enables our model to be trained in a single feed forward pass; unlike the standard AE based designs [1, 6]. Even in the works proposed in [9, 13] only the pre-training stage – weight initialization-benefits from available label information, and two stage learning is still required. The single pass learning in our model ensures reduced computation cost.

We report the performance of our model on the image classification problem. The results clearly indicate that our model's capability to generate discriminatory features achieves good classification accuracy with even simple (off the shelf) classification schemes.

2 Semi-supervised Autoencoder Design

Autoencoders consist of an encoding and a decoding unit [1]. The encoding part learns the feature representation (f) – hidden layer variables – of the input data sample (d) under a (usually non-linear) parameterized mapping (Φ) as in (1); W_e representing the encoding weight matrix (including bias term) to be learned.

$$f = \Phi(W_e \times d) \tag{1}$$

The decoding counterpart reconstructs the input in least square sense, using (2); W_d being the decoding weight matrix.

$$\overline{d} = W_d \times f \tag{2}$$

The non-linear mapping, such as sigmoid or hyperbolic tangent, between input and hidden layer ensures that hidden layer variables are valued between 0 and 1. Such as constraint need not be imposed on reconstructed input which can take any real values and thus decoder can be linear.

The parameter set for an AE (W_e, W_d) is learned by optimizing a suitable cost function. Traditionally, the reconstruction error (3) between input and the reconstructed signal, in terms of Euclidean distance, is employed as the cost function.

$$\min_{W_e, W_d} \left\| d - \overline{d} \right\|_2^2 = \min_{W_e, W_d} \left\| d - W_d \times \Phi(W_e \times d) \right\|_2^2 \tag{3}$$

2.1 Proposed Model

It has been shown that variants of the basic AE, designed for a specific application (such as Denoising AE [14]), outperform the standard model. Motivated by the finding, we propose a semi-supervised autoencoder, utilizing available label information, for improved classification accuracy.

Similar to a traditional AE, our design consists of an encoding unit - which maps the input data to hidden layer variables and a decoding unit - which uses the hidden layer information to construct the output. The main difference from traditional AE is that for labeled training samples, we additionally learn a linear map from the hidden layer to the targets/labels.

Learning weights by optimizing (3), ensures maximizing the lower bound on the mutual information between the input and learnt feature set [15]. We propose to modify the formulation in (3) by introducing additional constraint of label consistency.

The weights are recovered such that the feature vectors map to the class labels, wherever available, under a linear mapping.

Considering all the available (N) training samples we can rewrite (3) as follows,

$$\min_{W_d, W_e} \sum_{i=1}^{N} \|d - W_d \Phi(W_e d)\|_2^2 \tag{4}$$

To incorporate our proposed idea, we modify the base formulation in (4) as in (5), where d_l denotes the labeled and d_u the unlabeled training samples.

$$\min_{W_d, W_e, M} \sum_{l=1}^{S} \left(\|d_l - W_d \Phi(W_e d_l)\|_2^2 + \lambda \|c_l - M\Phi(W_e d_l)\|_2^2 \right) + \sum_{u=S+1}^{N} \|d_u - W_d \Phi(W_e d_u)\|_2^2 \tag{5}$$

In above equation, along with the data consistency (reconstruction) term we use the label information, wherever available, by mapping the feature vector to the corresponding class label vector via a linear map M. Class label vector, is constructed such that its ith element set to 1 if the corresponding training sample belongs to class i, rest being zero. The regularization parameter λ controls the relative contribution of the two for labeled data.

Use of additional class information helps improve the discriminative capability of the autoencoder over standard design (4). The reconstruction term in (5) ensures that the learned feature vector captures the relevant information from the training data in the l_2-norm sense. The label consistent term forces the features (of labeled training data) of every class to map to a distinct class label; we assume a linear mapping defined by M. Such a constraint improves discriminative capability of the encoded feature vectors and also mitigates impact of unwarranted factors which improves subsequent classification accuracy. The mapping matrix M is also learned during the optimization process.

Such label consistency terms have been used in the past for discriminative dictionary learning [16, 17] and restricted Boltzman machine [12]. This is the first work that introduces such label consistent formulation into the AE framework. In addition, we restrict our hidden layer dimension to be less than that of input/output layer to assist in dimensionality reduction. However, to provide greater freedom in our model, unlike conventionally employed tied weights, we do not impose any such constraints.

2.2 Algorithm Design

Recasting our model (5) in terms of matrices we get

$$\min_{W_d, W_e, M} \|D_L - W_d \Phi(W_e D_L)\|_F^2 + \lambda \|C_L - M\Phi(W_e D_L)\|_F^2 + \|D_U - W_d \Phi(W_e D_U)\|_F^2 \tag{6}$$

where, matrix D_L/D_U contain labeled/unlabeled training samples, and C_L contains corresponding label vectors. Introducing proxy variables Z_L, Z_U we can rewrite (6) as

$$\min_{W_d, W_e, M} \|D_L - W_d Z_L\|_F^2 + \lambda \|C_L - M Z_L\|_F^2 + \|D_U - W_d Z_U\|_F^2$$

$$s.t \ \Phi(W_e D_L) = Z_L, \ \Phi(W_e D_U) = Z_U \tag{7}$$

The constraints can be incorporated using Lagrangian as in (8)

$$\min_{W_d, W_e M, Z_L, Z_U} \|D_L - W_d Z_L\|_F^2 + \lambda \|C_L - M Z_L\|_F^2 + \|D_U - W_d Z_U\|_F^2$$

$$+ \mu \left(\|\Phi(W_e D_L) - Z_L\|_F^2 + \|\Phi(W_e D_U) - Z_U\|_F^2 \right) \tag{8}$$

We split (8) into multiple sub problems (9–13), each minimizing over one variable.

$$\min_{W_d} \|D_L - W_d Z_L\|_F^2 + \|D_U - W_d Z_U\|_F^2 \tag{9}$$

$$\min_{Z_L} \|D_L - W_d Z_L\|_F^2 + \lambda \|C_L - M Z_L\|_F^2 + \mu \|\Phi(W_e D_L) - Z_L\|_F^2 \tag{10}$$

$$\min_{Z_U} \|D_U - W_d Z_U\|_F^2 + \mu \|\Phi(W_e D_U) - Z_U\|_F^2 \tag{11}$$

$$\min_{M} \lambda \|C_L - M Z_L\|_F^2 \tag{12}$$

$$\min_{W_e} \|\Phi(W_e D_L) - Z_L\|_F^2 + \|\Phi(W_e D_U) - Z_U\|_F^2 \tag{13}$$

Equation (13) can be rewritten to cast it as a simple least square expression (14).

$$\min_{W_e} \left\| W_e D_L - \bar{\bar{Z}}_L \right\|_F^2 + \left\| W_e D_U - \bar{\bar{Z}}_U \right\|_F^2, where \Phi^{-1} Z_* = \bar{\bar{Z}}_* \tag{14}$$

Each sub problem (9–12, 14) is a least square expression which can be solved alternately until convergence, using any conjugate gradient solver. Because of space constraint, we do not show the detailed derivation; complete algorithm for our semi-supervised design (SS-AE) is given in Fig. 1.

3 Experimental Setup and Results

We conduct experiments on MNIST digit dataset and its variants [18], CIFAR-10 image [19] and USPS digit [20] datasets. The images are scaled to [0,1] and we do not perform any other pre-processing.

Our SS-AE is used to construct a 3-layer structure, with the representation learnt from the final layer acting as extracted feature vector. Each layer is greedily optimized employing the regularization parameter ($\lambda = 1e + 1$) computed using l-curve technique [21]. We compare our approach against with existing stacked Autoencoder design [22] and deep belief net [23] (using RBM), also stacked as a 3-layer structures. The derived

$$\boxed{\begin{aligned}
&\textit{Initialize variables; Set regularization parameter;}\\
&\textit{while } k < \textit{max\_iter or obj\_fun(k) - obj\_fun(k-1) < 1e-7}\\[4pt]
&W_d \leftarrow \min_{W_d} \left\| \begin{bmatrix} D_L & D_U \end{bmatrix} - W_d \begin{bmatrix} Z_L & Z_U \end{bmatrix} \right\|_F^2\\[4pt]
&Z_L \leftarrow \min_{Z_L} \left\| \begin{bmatrix} D_L & \sqrt{\lambda}C_L & \sqrt{\mu}\Phi(W_e D_L) \end{bmatrix}^T - \begin{bmatrix} W_d & \sqrt{\lambda}M & \sqrt{\mu}I \end{bmatrix}^T Z_L \right\|_F^2\\[4pt]
&Z_U \leftarrow \min_{Z_U} \left\| \begin{bmatrix} D_U & \sqrt{\mu}\Phi(W_e D_U) \end{bmatrix}^T - \begin{bmatrix} W_d & \sqrt{\mu}I \end{bmatrix}^T Z_U \right\|_F^2\\[4pt]
&M \leftarrow \min_M \lambda \left\| C_L - M Z_L \right\|_F^2\\[4pt]
&W_e \leftarrow \min_{W_e} \left\| \begin{bmatrix} \overline{\overline{Z}}_L & \overline{\overline{Z}}_U \end{bmatrix} - W_e \begin{bmatrix} D_L & D_U \end{bmatrix}^T Z_U \right\|_F^2\\[4pt]
&\textit{end while}
\end{aligned}}$$

Fig. 1. Algorithm for proposed design of a discriminative autoencoder

Table 1. Classification accuracy for fully supervised model (in %)

Dataset	KNN			SRC			SVM		
	SS-AE	DBN	SAE	SS-AE	DBN	SAE	SS-AE	DBN	SAE
USPS	**95.35**	94.67	89.84	**95.71**	95.47	91.93	95.37	77.63	91.88
CIFAR-10	32.08	**34.77**	21.34	36.75	**40.02**	27.41	33.12	**36.36**	23.01
MNIST	**97.26**	97.05	96.11	**98.20**	88.43	97.29	**98.22**	88.44	97.40
MNIST-Rot	**85.16**	84.71	80.71	**90.21**	79.47	84.89	**86.53**	76.59	79.83
MNIST-Back	**78.21**	77.16	70.97	**85.40**	75.09	76.94	**85.52**	75.22	74.99
MNIST-Rand	**87.67**	86.36	81.11	**92.07**	79.67	85.49	**90.32**	78.59	85.34
MNIST-RotBack	**53.17**	50.47	44.6	**63.77**	49.68	50.96	**58.97**	48.53	49.14

Table 2. Classification accuracy using for semi-supervised autoencoder (in %)

	Using KNN					Using SRC				
	20	40	60	80	100	20	40	60	80	100
USPS	95.02	95.22	95.27	95.32	95.35	95.49	95.55	95.62	95.67	95.71
CIFAR-10	31.05	31.22	31.41	31.76	32.08	35.37	35.50	35.75	36.28	36.75
MNIST	97.07	97.12	97.19	97.25	97.26	98.11	98.15	98.18	98.20	98.20
MNIST-Rot	84.83	84.90	84.98	85.11	85.16	89.81	89.88	90.07	90.13	90.21
MNIST-Back	77.97	77.99	78.14	78.18	78.21	85.13	85.16	85.26	85.35	85.40
MNIST-Rand	87.38	87.42	87.44	87.58	87.67	91.92	91.93	91.95	91.03	92.07
MNIST-RotBack	52.87	52.97	53.01	53.07	53.17	63.59	63.62	63.67	63.73	63.77

feature set from each design is fed into standard classifiers – multiclass SVM with RBF kernel, KNN (K-nearest neighbor) and SRC (Sparse Classifier) [24].

Table 1, gives the classification accuracy for all the approaches, considering fully supervised model. It can be observed that the performance of our model is consistently better than other methods for MNIST and USPS datasets. For CIFAR-10 dataset, although DBN gives better results, our design performs considerably better than the existing standard autoencoder based structure, SAE.

Table 2 gives the variation in classification accuracy for our proposed design as a function of the percentage of label information using KNN and SRC. As expected, with the increase in label information the classification accuracy improves. Classification accuracy obtained with simple (KNN) classifier indicates that for difficult classification problems (like MNIST-back) the impact of label information is more pronounced than simpler cases (MNIST). Results for SVM also follow similar trend; howvever, they are not shgown because of spoace constraint.

4   Conclusion

In this work, we proposed a semi-supervised autoencoder for feature extraction. We use our model to build a deep architecture such that the representation (hidden) layer variable of the final layer acts as the final feature vector.

We augment the reconstruction loss cost function design with additional label consistency term derived from available label information. The autoencoder weights are derived such that the representation layer variables, in addition to maintaining (input) data fidelity, display inter-class discrimination. This is ensured by use of add-on regularization factor which promotes mapping of feature vectors to their respective class labels wherever available. Use of semi-supervised learning helps improve robustness of our model by forcing the extracted features to be invariant to image variations such as illumination effects or scaling/rotation. We use our SS-AE model to build a deep network. Improved robustness ensures that high classification accuracy is obtained even with standard off-the-shelf classifiers.

Experiments conducted on image classification datasets validate our claim that our model yields higher classification accuracy, than standard approaches, while simultaneously achieving lower computational complexity.

References

1. Hinton, G.E., Zemel, R.S.: Autoencoders, minimum description length, and Helmholtz free energy. Adv. Neural Inf. Process. Syst. **6**, 3–10 (1994)
2. Vincent, P., Larochelle, H., Bengio, Y., Manzagol, P-A.: Extracting and composing robust features with denoising autoencoders. In: Proceedings of the 25th International Conference on Machine Learning, pp. 1096–1103. ACM (2008)
3. Chen, Y., Lin, Z., Zhao, X., Wang, G., Gu, Y.: Deep learning-based classification of hyperspectral data. IEEE J. Sel. Top. Appl. Earth Obs. Remote Sens. **7**(6), 2094–2107 (2014)
4. Maldonado, S., Weber, R., Basak, J.: Simultaneous feature selection and classification using kernel-penalized support vector machines. Inf. Sci. **181**(1), 115–128 (2011)

5. Kong, H., Li, X., Wang, L., Teoh, E.K., Wang, J.-G., Venkateswarlu, R.: Generalized 2D principal component analysis. In: Proceedings of 2005 IEEE International Joint Conference on Neural Networks, IJCNN 2005, vol. 1, pp. 108–113. IEEE (2005)
6. Bengio, Y.: Learning deep architectures for AI. Found. Trends® Mach. Learn. **2**(1), 1–127 (2009)
7. Hinton, G.E., Osindero, S., Teh, Y.-W.: A fast learning algorithm for deep belief nets. Neural Comput. **18**(7), 1527–1554 (2006)
8. Ciresan, D., Meier, U., Schmidhuber, J.: Multi-column deep neural networks for image classification. In: 2012 IEEE Conference on Computer Vision and Pattern Recognition (CVPR), pp. 3642–3649. IEEE (2012)
9. Gao, S., Zhang, Y., Jia, K., Lu, J., Zhang, Y.: Single sample face recognition via learning deep supervised autoencoders. IEEE Trans. Inf. Forensics Secur. **10**(10), 2108–2118 (2015)
10. Huang, G., Song, S., Gupta, J.N.D., Wu, C.: Semi-supervised and unsupervised extreme learning machines. IEEE Trans. Cybern. **44**(12), 2405–2417 (2014)
11. Ranzato, M., Szummer, M.: Semi-supervised learning of compact document representations with deep networks. In: Proceedings of the 25th International Conference on Machine Learning, pp. 792–799. ACM (2008)
12. Larochelle, H., Mandel, M., Pascanu, R., Bengio, Y.: Learning algorithm for the classification restricted boltzmann machine. J. Mach. Learn. Res. **13**(1), 643–669 (2012)
13. Almousli, H., Vincent, P.: Semi supervised autoencoders: better focusing model capacity during feature extraction. In: Lee, M., Hirose, A., Hou, Z.-G., Kil, R.M. (eds.) ICONIP 2013. LNCS, vol. 8226, pp. 328–335. Springer, Heidelberg (2013)
14. Vincent, P., Larochelle, H., Bengio, Y., Manzagol, P.-A.: Extracting and composing robust features with denoising autoencoders. In: Proceedings of the 25th International Conference on Machine Learning, pp. 1096–1103. ACM (2008)
15. Lemme, A., Reinhart, R.F., Steil, J.J.: Online learning and generalization of parts-based image representations by non-negative sparse autoencoders. Neural Netw. **33**, 194–203 (2012)
16. Jiang, Z., Lin, Z., Davis, L.S.: Learning a discriminative dictionary for sparse coding via label consistent K-SVD. In: 2011 IEEE Conference on Computer Vision and Pattern Recognition (CVPR), pp. 1697–1704. IEEE (2011)
17. Shrivastava, A., Pillai, J.K., Patel, V.M., Chellappa, R.: Learning discriminative dictionaries with partially labeled data. In: 2012 19th IEEE International Conference on Image Processing (ICIP), pp. 3113–3116. IEEE (2012)
18. http://www.iro.umontreal.ca/∼lisa/twiki/bin/view.cgi/Public/DeepVsShallowComparisonICML2007
19. https://www.cs.toronto.edu/∼kriz/cifar.html
20. http://www.cad.zju.edu.cn/home/dengcai/Data/MLData.html
21. Lawson, C.L., Hanson, R.J.: Solving least squares problems, vol. 161. Prentice-hall, Englewood Cliffs (1974)
22. Hinton, G.E., Salakhutdinov, R.R.: Reducing the dimensionality of data with neural networks. Science **313**(5786), 504–507 (2006)
23. http://ceit.aut.ac.ir/∼keyvanrad/DeeBNet%20Toolbox.html
24. Ng, A.: Sparse autoencoder. CS294A Lecture notes 72 (2011)

Sampling-Based Gradient Regularization for Capturing Long-Term Dependencies in Recurrent Neural Networks

Artem Chernodub$^{(\boxtimes)}$ and Dimitri Nowicki

Institute of MMS of NASU, Center for Cybernetics,
42 Glushkova Ave., Kiev 03187, Ukraine
a.chernodub@gmail.com, nowicki@nnteam.org.ua

Abstract. Vanishing (and exploding) gradients effect is a common problem for recurrent neural networks which use backpropagation method for calculation of derivatives. We construct an analytical framework to estimate a contribution of each training example to the norm of the long-term components of the target functions gradient and use it to hold the norm of the gradients in the suitable range. Using this subroutine we can construct mini-batches for the stochastic gradient descent (SGD) training that leads to high performance and accuracy of the trained network even for very complex tasks. To check our framework experimentally we use some special synthetic benchmarks for testing RNNs on ability to capture long-term dependencies. Our network can detect links between events in the (temporal) sequence at the range 100 and longer.

1 Introduction

Recurrent Neural Networks (RNNs) are known as universal approximators of dynamic systems [1]. Since RNNs are able to simulate any open dynamical system, they have a broad spectrum of applications such as time series forecasting [2], control of plants [3], language modeling [4], speech recognition, neural machine translation [5] and other domains. The easiest way to create an RNN is adding the feedback connections to the hidden layer of multilayer perceptron. This architecture is known as Simple Recurrent Network (SRN). Despite of the simplicity, it has rich dynamical approximation capabilities mentioned above. However, in practice training of SRNs using first-order optimization methods is difficult [6]. The main problem is well-known vanishing/exploding gradients effect that prevents capturing of long-term dependencies in data. Vanishing gradients effect is a common problem for recurrent and deep neural networks with sigmoid-like activation functions which uses a backpropagation method for calculation of derivatives. Hochreiter and Schmidhuber designed a set of special synthetic benchmarks for testing RNNs on ability to capture long-term dependencies [7]. They showed that ordinary SRNs are very ineffective to learn correlations in sequential data if distance between the target events is more than 10

© Springer International Publishing AG 2016
A. Hirose et al. (Eds.): ICONIP 2016, Part II, LNCS 9948, pp. 90–97, 2016.
DOI: 10.1007/978-3-319-46672-9_11

time steps. The solution could be using more advanced second-order optimization algorithms such as Extended Kalman Filter, LBFGS, Hessian-Free optimization [8], but they require much more memory and computational resources for state-of-the-art networks. We also mention such an alternative to temporal neural networks as hierarchical sequence processing with auto-associative memories [9]. The mainstream solution for the gradient control problem is based on more complex architectures such as LSTM [7] or GRU [5] networks. However, training the SRN's for catching long-term dependencies is highly desirable at least for better understanding of underlying processes of the training inside the recurrent and deep neural networks. Also, SRNs are more compact and fast working models of RNNs in comparison with LSTMs that is very important for implementation to mobile and embedded devices. Recent research shows the ability to train SRNs for long term dependencies up to 100 time steps and more using several new techniques [8,10]. In this paper we propose a new method to perform the gradient regularization by selection of proper samples in dataset.

2   Backpropagation Mechanism Revisited

Consider a SRN that at each time step k receives an external input $\mathbf{u}(k)$, previous internal state $\mathbf{z}(k-1)$ and produces output $\mathbf{y}(k+1)$:

$$\begin{aligned}
\mathbf{a}(k) &= \mathbf{u}(k)\mathbf{w}_{in} + \mathbf{z}(k-1)\mathbf{w}_{rec} + \mathbf{b}, \\
\mathbf{z}(k) &= f(\mathbf{a}(k)), \\
\mathbf{y}(k+1) &= g(\mathbf{z}(k)\mathbf{w}_{out}),
\end{aligned} \tag{1}$$

where $\mathbf{w}_{in}$ is a matrix of input weights, $\mathbf{w}_{rec}$ is matrix of recurrent weights, $\mathbf{w}_{out}$ is matrix of output weights, $\mathbf{a}(k)$ is known as "presynaptic activations", $\mathbf{z}(k)$ is a network's state, $f(\cdot)$ and $g(\cdot)$ are nonlinear activation functions for hidden and output layer respectively. In this work we always use $tanh$ function for hidden layer and optionally $softmax$ or $linear$ function depending on the target problem (classification or regression) for output layer.

The dynamic error derivative is a sum of immediate derivatives: $\frac{\partial E}{\partial \mathbf{w}} = \sum_{n=1}^{h} \frac{\partial E}{\partial \mathbf{w}(k-n)}$, where $n = 1, ..., h$, where h is BPTT's truncation depth. An intermediate variable $\delta \equiv \frac{\partial E}{\partial \mathbf{a}}$ called a "local gradients" or simply "deltas" is usually introduced for convenience,

$$\delta(k-h) = \delta(k-h+1)\mathbf{w}_{rec}^{T} diag(f'(\mathbf{a}(k-h))). \tag{2}$$

Equation (2) may be rewritten using Jacobian matrix $\mathbf{J}(n) = \frac{\partial \mathbf{z}(n)}{\partial \mathbf{z}(n-1)}$:

$$\delta(k-h) = \delta(k-h+1)\mathbf{J}(k-h). \tag{3}$$

Now we can use an intuitive understanding of exploding/vanishing gradients problem that was deeply investigated in classic [6] and modern papers [10]. As it can be seen from (3), norm of the backpropagated deltas is strongly dependent on norm of the Jacobians. Moreover, they actually are product of Jacobians:

$\delta(k-h) = \delta(n)\mathbf{J}(k)\mathbf{J}(k-1)...\mathbf{J}(k-h+1)$. The "older" deltas are, the more Jacobian matrices were multiplied. If norm of Jacobians are more than 1 if the gradients will grow exponentially in most cases. It refers to the RNN's behavior where long-term components are more important than short-term ones. Vice versa, if norm of Jacobians are less than 1, this leads to vanishing gradients and "forgetting" the long-term events. In [10] a universal "gradient regularization" approach that forces the gradient norm to stay in a stable range via modification of the training objective function proposed. However, they used a complex regularizer to preserve norm in the relevant direction.

3 Differentiation of the Gradient's Norm

Let $\mathbf{d} = \{\mathbf{u}_1; \mathbf{t}_1; ...; \mathbf{u}_N, \mathbf{t}_N\}$ be a minibatch with N_D training examples. We do forward and back propagation in the network for this minibatch and we get the difference (correction) vector $d\mathbf{w}$. Lets check how $d\mathbf{w}$ influences on gradient vanishing or explosion. Let $\mathbf{w}_{rec}^l$ be a weight matrix for recurrent layer of the SRN at the current iteration l of weight update. Suppose we have made the back and forward pass, so $\mathbf{w}_{rec}^l$ is a correction for the recurrent layer such that $\mathbf{w}_{rec}^{l+1} = \mathbf{w}_{rec}^l + d\mathbf{w}_{rec}$.

Consider a function $S(\mathbf{w}_{rec}^{(l)})$ that is equal to squared Euclidian norm of (2) for iteration l:

$$S(\mathbf{w}_{rec}^{(l)}) = \frac{1}{2} \left\| \delta(k-h, \mathbf{w}_{rec}) \right\|_2^2 . \tag{4}$$

Since $\left\| d\mathbf{w}_{rec}^{(l)} \right\|_2^2$ is supposed to be small, we can use Taylor expansion of (4) at the current point of weight matrix space:

$$S(\mathbf{w}_{rec}^{(l+1)}) = S(\mathbf{w}_{rec}^{(l)} + dS + o(\left\| d\mathbf{w}_{rec}^{(l)} \right\|_2^2)). \tag{5}$$

Lemma 1. *Linear term dS in (5) could be expressed as a scalar product of the auxiliary vectors $\mathbf{g}$ and $d\mathbf{g}$,*

$$dS = (\mathbf{g}, d\mathbf{g}), \tag{6}$$

where

$$\mathbf{g} = \left(\prod_{i=h}^{1} diag(f'(\mathbf{a}(k-i+1))) \mathbf{w}_{rec} \right) \delta(k),$$
$$d\mathbf{g} = \sum_{i=1}^{h} \left(\left(\prod_{j=h}^{1} diag\left[f'(\mathbf{a}(k-j+1)) \right] \mathbf{v} \right) \delta(k) \right), \tag{7}$$
$$\mathbf{v} = d\mathbf{w}_{rec}, if\ i = j; \mathbf{v} = \mathbf{w}_{rec}, if\ i \neq j.$$

Proof. Using (2), (3) we get $\delta^{(l)}(k-h)$:

$$\delta(k-h) = \delta(k)\mathbf{w}_{rec}^T diag(f'(\mathbf{a}(k-1)))...\mathbf{w}_{rec}^T diag(f'(\mathbf{a}(k-h+1))). \tag{8}$$

We introduce a $\mathbf{D}_n$ notation as follows:

$$\mathbf{D}_n \equiv diag(f'(\mathbf{a}(n))). \tag{9}$$

Now (8) becomes:

$$\delta^{(l)}(k-h) = \delta^{(l)}(k)\mathbf{w}_{rec}^{T(l)}\mathbf{D}_{k-1}w_{rec}^{T(l)}\mathbf{D}_{k-2}...\mathbf{w}_{rec}^{T(l)}\mathbf{D}_{k-h}. \tag{10}$$

Auxillary vector $\mathbf{g}$ is a transposed (10):

$$\mathbf{g} = \delta^{(l)}(k-h)^T. \tag{11}$$

Since $(\mathbf{AB})^T = \mathbf{B}^T\mathbf{A}^T$, (10) and (11) lead to :

$$\mathbf{g} = \mathbf{D}_{k-1}\mathbf{w}_{rec}^{(l)}\mathbf{D}_2\mathbf{w}_{rec}^{(l)}...\mathbf{D}_{k-h}\mathbf{w}_{rec}^{(l)}\delta^{(l)}(k). \tag{12}$$

Since $\|x\|_2 = \|x^T\|_2$, regarding (11) and (12), the function $S(\mathbf{w}_{rec}^{(l)})$ (4) has an equivalent form:

$$S(\mathbf{w}_{rec}^{(l)}) = \frac{1}{2}\|\mathbf{g}\|_2^2. \tag{13}$$

Now we get the differential dS of (13):

$$dS = (\mathbf{g}, d\mathbf{g}), \tag{14}$$

where the vector $\mathbf{g}$ in (14) is obtained from (12) that is equivalent (7). Also, we get vector $d\mathbf{g}$ in (14) by differentiating vector $\mathbf{g}$ in (12) as follows:

$$d\mathbf{g} = \sum_{i=1}^{h}\delta(k)\mathbf{D}_{k-h}\mathbf{w}_{rec}^{(l)}...\mathbf{D}_i d\mathbf{w}_{rec}^{(l)}...\mathbf{D}_{k-1}\mathbf{w}_{rec}^{(l)}, \tag{15}$$

that is the same as (7) up to usage notation $\mathbf{D}_n$. The lemma is proven. $\square$

We have to figure out the direction of change of the (Euclidean) norm of $\delta(k-h)$ since the gradient is propagated for h steps back at time step k where the correction $d\mathbf{w}_{rec}$ is used.

Theorem 1. *The condition $dS > 0$ is sufficient to increase the norm of $\|\delta^{(l+1)}(k-h)\|_2$ comparing to $\|\delta^{(l)}(k-h)\|_2$ at the next iteration $l+1$ of weight correction then the correction matrix $d\mathbf{w}_{rec}$ is used. dS here is defined by (6) and $d\mathbf{w}_{rec}$ is contained in dS, and $\mathbf{w}_{rec}^{(l+1)} = \mathbf{w}_{rec}^{(l)} + d\mathbf{w}_{rec}$. Similarily, $dS < 0$ is a sufficient condition for decrease of $\|\delta(k-h, \mathbf{w}_{rec}^{(l+1)})\|_2$.*

Proof. Let's compare values $S(\mathbf{w}_{rec}^{(l)})$ and $S(\mathbf{w}_{rec}^{(l+1)})$ that correspond to current iteration l and next one $l+1$ of the weight update. We use Taylor expansion (5) and the Lemma 1. From (5) follows that the sign of (6) defines a direction of change for the Euclidean norm of $\delta(k-h)$ between the iterations l and $l+1$ and absolute value $|dS|$ defines the magnitude of this change. $\square$

Idea of our sampling-based gradient regularization algorithm is selection of "proper" samples of data for training. Using Theorem 1 we can clearly find out an impact of each mini-batch on norm of backpropagated gradients.

Algorithm 1. Algorithm of sampling-based gradient regularization

---
Input: training data $\{\mathbf{U}, \mathbf{T}\}$, $r_0 > 0$.
for each minibatch $\mathbf{u}_i; \mathbf{d}_i$ with N_D vectors **do**
   calculate dS (6), if $|dS| > 0$
   continue
   make forward and backward propagation
   calculate $Q(\delta, h)$ (16)
   if $Q(\delta, h) \in [Q_{min}; Q_{max}]$ **then**
      use current minibatch for training
   else
      if $(Q(\delta, h) < Q_{min}$ **and** $dS > 0)$ **or** $(Q(\delta, h) > Q_{max}$ **and** $dS < 0)$ **then**
         use current minibatch for training
      else
         continue
      end if
   end if
end for

We introduce auxiliary variable called Q-factor that measures how much the norm of the gradient is decreased or increased during the backpropagation. For ideal catching of long-term dependencies Q-factor must be close to 0.

$$Q(\delta, h) = \log_{10}\left(\frac{\|\delta(k)\|}{\|\delta(k - h)\|}\right). \tag{16}$$

Here we use the simplest and the most straightforward method: we watch a norm of the gradients; if the norm becomes too small, we omit mini-batches of data such that decrease this norm. Vice versa, if norm becomes very large, we skip mini-batches increasing this norm even more. Also, note that it is better to skip minibatches with large $|dS|$: they can cause high "leaps" of the gradient norm and therefore its self-oscillations.

4   Experiments

We follow [10] and use the following synthetic problems for catching long-term dependencies: "Adding", "Multiplication", "Temporal order", "Temporal order 3-bit". Two sets containing 10 SRN with 100 hidden units each were initialized by random values and saved. Thus, for different training methods initial weights of neural networks were the same. "Safe" range $[Q_{min}; Q_{max}]$ for (16) was set to $[-1; 1]$.

We use SGD optimization, training speed $\alpha = 10^{-5}...10^{-3}$, momentum $\mu = 0.9$, size of mini-batch is 10. Train/validation/test datasets contains 20,000/1000/10,000 samples respectively. After each epoch, network's performance is tested on validation dataset; network that has the best performance on the validation dataset is tested on the test dataset, this result is recognized as the final result. We trained SRNs during 2000 epochs, each epoch consists 50 iterations, i.e. 100,000 corrections of weights at all.

Weights were initialized by small values from Gaussian distribution with zero mean and standard deviation σ. On Fig. 1 average norms of gradients as function of backpropagation depth (before training, further referred as initial gradients) are graphed for different values σ for the "Temporal order problem". We see that good initialization of weights is very important because vanishing/exploding gradients has monotonous flow in most cases because gradients are propagated through the same matrix of recurrent weights.

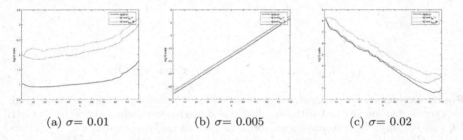

(a) $\sigma = 0.01$ (b) $\sigma = 0.005$ (c) $\sigma = 0.02$

Fig. 1. Average norms of backpropagated (initial) gradients for SRNs, horizon BPTT $h = 100$. (Color figure online)

Each chart at Fig. 1 contains three curves: average norms of local gradients $\delta(k)$ (blue) and average norms of gradients $\Delta\mathbf{w}(k)_{in} \equiv \frac{\partial E}{\partial \mathbf{w}_{in}}$ and $\Delta\mathbf{w}(k)_{rec} \equiv \frac{\partial E}{\partial \mathbf{w}_{rec}}$ (red and green). From the graphs at Fig. 1 one can ensure on practice that to control the norms $\frac{\partial E}{\partial \mathbf{w}}$ which actually make changes to the weights and are under the main scope of our interest it is enough to control the norms of local gradients $\delta(k)$ because they are highly correlated.

Finally we used $\sigma = 0.01$ as in [10]. However, proper initialization doesn't guarantee successful training. Particular case of forward and backward dynamics (norms of the backpropagated gradients are depicted on the top, mean and median activation values) during training of SRN network is shown on Fig. 2. SRN that is depicted on Fig. 2 was initialized with $\sigma = 0.01$ and initial norms of backpropagated gradients were similar to Fig. 1(a). However, after 500 iterations we got norm of gradients less than 10^{-7} for $h = 100$. After that almost all the time neural networks had small gradients in the range $10^{-7}...10^{-8}$. From the graphs on Fig. 2, on the left, we see that area of small gradients is related to area of saturation for neuron's activations. This is a symptom of bad network abilities for successful training and obtaining good generalization properties.

Using our sampling-based gradients regularization allows to refine the quality of training (Table 1). For lengths $T = 100$ and $T = 150$ improvement is 10–20 % in average. Samples rejected by the algorithm during the training not necessarily are lost for using in future training process because they may be used when network is in "safe region" or we may need to change norms of gradients in the opposite direction.

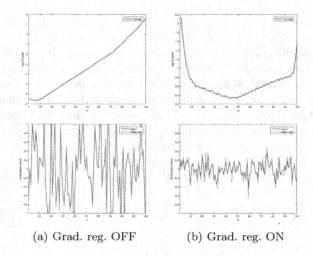

(a) Grad. reg. OFF   (b) Grad. reg. ON

Fig. 2. Evolution of internal dynamics inside the SRN during training without gradients regularization (a) and with sampling-based gradients regularization (b). On both (a) and (b) upper graphs are mean norms in time of backpropagated via BPTT local gradients $\delta(k)$; lower ones are mean and median values of activations $\mathbf{a}(k)$.

Table 1. Accuracies of trained SRNs for synthetic problems which have long-term dependencies without gradient regularization (traditional training method) and with sampling-based gradient regularization (proposed method), for T = 100 and 150.

	Adding		Multiplication		Temporal order		Temporal order 3-bit	
	Best	Mean	Best	Mean	Best	Mean	Best	Mean
T=100, grad. reg. OFF	99 %	68 %	> 99 %	72 %	96 %	44 %	99 %	50 %
T=100, grad. reg. ON	> 99 %	96 %	> 99 %	68 %	> 99 %	60 %	> 99 %	62 %
T=150, grad. reg. OFF	34 %	11 %	N/A	N/A	51 %	30 %	32 %	24 %
T=150, grad. reg. ON	47 %	13 %	N/A	N/A	72 %	42 %	37 %	30 %

5   Conclusion

We provided a novel solution of the problem of exploding and vanishing gradient effects, applied to the Simple Recurrent Networks. We analytically derived sufficient conditions on increase and decrease the Euclidean vector norm of backpropagated gradients for SRNs. Using this theorem we designed the algorithm that controls norm of the gradient operating solely with presence of the minbatches in the training sequence. This framework was tested for long-term prediction on a comprehensive set of appropriate benchmarks. Resulting accuracy outperforms best known SRN learning algorithms by 10–20 %. This paradigm could be generalized to deep and multi-layered recurrent networks, that is a subject of our future research.

References

1. Horne, B.G., Siegelmann, H.T.: Computational capabilities of recurrent narx neural networks. IEEE Trans. Syst. Man Cybern. B **27**(2), 208–215 (1997)
2. Cardot, H., Bone, R.: Advanced Methods for Time Series Prediction Using Recurrent Neural Networks, pp. 15–36. Intech, Croatia (2011)
3. Prokhorov, D.V.: Toyota prius hev neurocontrol and diagnostics. Neural Netw. **21**, 458–465 (2008)
4. Mikolov, T., Karafiát, M., Burget, L., Cernocký, J., Khudanpur, S. Recurrent neural network based language model. In: Interspeech, vol. 2, p. 3 (2010)
5. Bahdanau, D., Cho, K., van Merrienboer, B.: On the properties of neural machine translation: encoder decoder approaches. In: SSST-8, Doha, Qatar (2014)
6. Frasconi, P., Bengio, Y., Simard, P.: Learning long-term dependencies with gradient descent is difficult. IEEE Trans. Neural Netw. **5**(2), 157–166 (1994)
7. Schmidhuber, J., Hochreiter, S.: Long short-term memory. Neural Comput. **9**(8), 1735–1780 (1997)
8. Sutskever, I., Martens, J. Learning recurrent neural networks with hessian-free optimization. In: Proceedings of the ICML (2011)
9. Kussul, E.M., Rachkovskij, D.A.: Multilevel assembly neural architecture and processing of sequences. Neurocomput. Atten. Connectionism Neurocomput. **2**, 577–590 (1991)
10. Bengio, Y., Pascanu, R.: On the difficulty of training recurrent neural networks. Technical report, Universite de Montreal (2012)

Face Hallucination Using Correlative Residue Compensation in a Modified Feature Space

Javaria Ikram[✉], Yao Lu, Jianwu Li, and Nie Hui

Beijing Key Laboratory of Intelligent Information Technology,
School of Computer Science and Technology,
Beijing Institute of Technology, Beijing 100081, China
jikram@bit.edu.cn

Abstract. Local linear embedding (LLE) is a promising manifold learning method in the field of machine learning. Number of face hallucination (FH) methods have been proposed due to its neighborhood preserving nature. However, the projection of low resolution (LR) image to high resolution (HR) is "one-to-multiple" mapping; therefore manifold assumption does not hold well. To solve the above inconsistency problem we proposed a new approach. First, an intermediate HR patch is constructed based on the non linear relationship between LR and HR patches, which is established using partial least square (PLS) method. Secondly, we incorporate the correlative residue compensation to the intermediate HR results by using only the HR residue manifold. We use the same combination coefficient as for the intermediate hallucination of the first phase. Extensive experiments show that the proposed method outperforms some state-of-the-art methods in both reconstruction error and visual quality.

1 Introduction

In recent years, face hallucination (FH) methods have attracted growing attention due to its practical importance in many multimedia applications, such as video surveillance, identity recognition and image retrieval. Different FH methods have been proposed and yielded promising results. However, due to constrained imaging conditions in many scenarios, the captured face images loose detailed facial features to be identified by humans. Therefore, there is a great need of FH algorithms with high level visual results. The existing FH methods can roughly be divided into two classes: reconstruction based and learning based. Learning based technique however have gained more attention and outperform the reconstruction based techniques.

Nie Hui—This work was supported in part by the National Natural Science Foundation of China under Grant Nos. 61273273 and 61271374, and by Research Fund for the Doctoral Program of Higher Education of China under Grant No. 20121101110034.

© Springer International Publishing AG 2016
A. Hirose et al. (Eds.): ICONIP 2016, Part II, LNCS 9948, pp. 98–107, 2016.
DOI: 10.1007/978-3-319-46672-9_12

Baker and Kanade [1] were the first to propose FH method by learning the prior with strong cohesion to face domain. Although its reconstruction is better than that of various interpolation approaches, it still doesn't allow the recovery of detailed high frequency components in the reconstructed HR image. Liu et al. [14] proposed a hybrid approach with the integration of global parametric model and local non parametric model. The hybrid approach [14,18] is basically a two step approach; first step is to use global prior model and second step is to use local prior model to compensate the missing high frequencies in first step using residue compensation. Following the work of [1,14], learning based methods draw enormous attention in the FH research community. Among them, one of the most representative work is image hallucination based on neighbor embedding (NE) [2]. Inspired by the strong structural property of face images, Ma et al. [15] proposed position patch embedding through solving least square regression (LSE) problem which later was implemented in many FH algorithm [8–10]. All of the above mentioned methods used Local linear embedding (LLE) [17]. LLE is a manifold learning method and assumes that the LR and HR manifold share the same geometric structure. However, the researcher in [5,6,9,11,13] show that the manifold assumption doesn't always hold and produce distorted embedding. Local geometry exploited by the reconstruction weights is not well determined. This is because the projection of low resolution (LR) to high resolution (HR) image is "one-to-multiple" mapping. The existing research confirms that guaranteeing the consistency between the LR and corresponding HR manifold can reasonably highlight the effectiveness of FH methods [5,7,9,11,13,18].

1.1 Motivation and Contributions

A good deal of research has been done to guarantee the above consistency. Li et al. [12,13] proposed to project the LR and HR manifold onto a common manifold. NE is then perform using the new projections of LR and HR images and the original LR and HR images. Huang et al. [7] maximizes the correlation between LR and HR images by using Canonical Correlation Analysis (CCA). Recently, Hao et al. [6] proposed to guarantee the consistency relationship between LR and HR manifold by projecting them onto the unified feature space using Easy partial least square (EZ-PLS). Additionally they incorporate smoothness constraints via maximum a posteriori (MAP) formulation. Motivated by the work in [6], we employ EZ-PLS method to project LR and HR image patches onto a unified feature space. Secondly, NE is perform by incorporating locality constraints in the new feature space. Furthermore, in order to make the final HR image having both good reconstruction fidelity and visual quality we incorporate the residue compensation to the preliminary HR results and generate the final result. In the residue compensation phase, we assume that HR residue patch and the preliminary HR patch share the same combination weights and constituent samples. Therefore, instead of finding new combination weights and neighboring patches from residue pair dictionaries we acquire the HR residue patch using the same combination wights and neighboring index in

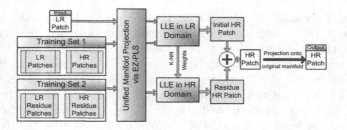

Fig. 1. Flow chart of the proposed method.

first phase. Thus, compared with other two-step approach [14,18], the proposed method bridges the gap between the first and the second phase. Figure 1 gives the graphical illustration of the proposed method.

Specifically, in this work; we improve the performance of face hallucination methods through two measures:

1- In the first step, for each input patch the selection of the $K - NN$ based on the Euclidean distance is introduced. K nearest patches from LR and corresponding HR patch manifold are then mapped onto a unified feature space via learned projection matrices employing EZ-PLS. Finally, refined Z nearest patches are selected from new feature space to find the optimal weights.

2- Secondly, we propose a correlative residue compensation learning method for recovery of lost components. Residue compensation is usually followed by global compensation. However, in this paper we proposed correlative residue compensation for each patch individually and add it to the preliminary halluci-nated patch for the final patch representation. Proposed method greatly reduces the computational complexity without compromising the quality of the facial details as compared to the traditional residue compensation methods.

The rest of the paper is organized as follows. In Sect. 2, we first claim some notations and then present the proposed method. Section 3 presents experiments and evaluation of the proposed framework. Finally concluding remarks are given in Sect. 4.

2   Face Hallucination Using Correlative Residue Compensation in a Modified Feature Space

In this section, we first claim some notations, then present the implementation details of the proposed method.

2.1   Notations

Let $\{T_{L_i}\}_{i=1}^{N}$, $\{T_{H_i}\}_{i=1}^{N}$ be the training set of LR and HR images respectively and $\{R_{L_i}\}_{i=1}^{N}$, $\{R_{H_i}\}_{i=1}^{N}$ are corresponding residual image pairs. Here, N is the number of training images. Input image, LR and HR training images are first divided into M overlapping patches using the same dividing scheme. The patches at q^{th} location from the LR and HR training images are collected to create the patch dictionaries $\{T_L^q\}_{q=1}^M = \{l_1^q, ..., l_N^q\}_{q=1}^M$ and $\{T_H^q\}_{q=1}^M = \{h_1^q, ..., h_N^q\}_{q=1}^M$. The corresponding residue patch LR and HR dictionaries are represented as $\{R_L^q\}_{q=1}^M = \{l_1^{'q}, ..., l_N^{'q}\}_{q=1}^M$ and $\{R_H^q\}_{q=1}^M = \{h_1^{'q}, ..., h_N^{'q}\}_{q=1}^M$. Here each column represents the patch at q^{th} position of the ith training image. The proposed method is composed of training and reconstruction stage. For both training and reconstruction we enlarge the input image and the LR images to the same size as the HR image by using the bi-cubic interpolation to control the number of projection vectors.

2.2   Methadology

The proposed method is a correlative two step method. An intermediate HR image is first constructed by projecting both the LR and HR training images into a unified feature space. Secondly, correlative residue compensation is employed in a new feature space using only the HR coefficients to further improve the details.

Training. For each input patch l^q at q^{th} location, we first find K-NN based on the Euclidean distance between input patch and LR training patches and select the corresponding HR patches with the same neighboring index. Then we trained M pair of projection matrices using EZ-PLS, denoted as $\{P_{TH}^q, P_{TL}^q\}_{q=1}^M$. Here, P_{TH}^q and P_{TL}^q represent the projection matrices for HR and LR patches at q^{th} position. The projection coefficients of LR and HR patches at q^{th} position can be obtained using:

$$v_i^q = (P_{TH}^q)^T(h_i^q - \mu T_H^q), i = 1, ..., N \tag{1}$$

$$u_i^q = (P_{TL}^q)^T(l_i^q - \mu T_L^q), i = 1, ..., N \tag{2}$$

The column vectors μT_H^q and μT_L^q represents the vectors consisting of the mean value of the variable in T_H^q and T_L^q respectively. All the projection coefficients of HR and LR patches at the qth position are collected into the matrices $V^q = \{v_1^q, ..., v_r^q\}$ and $U^q = \{u_1^q, ..., u_r^q\}$.

Reconstruction. In training stage we project the LR and HR images into a modified feature space by using the projection matrices. Similarly, we first convert the input image patch into new feature space using same projection matrices to perform further operations i.e. searching nearest neighbors and finding the optimal weight. Input image patch l^q at q^{th} location is transformed into new feature space as follows.

$$u^q = (P_{TL}^q)^T (l^q - \mu T_L^q) \tag{3}$$

Once the input image is transformed into new feature space, Z nearest training patches are then searched in new feature space based on the Euclidean distance between the low resolution input patch u^q and the patches in the low resolution training images. We can rewrite V^q and U^q as $V^q = \{v_1^q, ..., v_Z^q\}$ and $U^q = \{u_1^q, ..., u_Z^q\}$. The optimal weight $\{\omega_i, i = 1, ..., Z\}$ is obtained by minimizing the local reconstruction error:

$$\omega_i^q = \underset{\omega_i^q}{\arg\min} \left\{ \left\| u^q - \sum_{u_i^q \in Z^q} \omega_i^q u_i^q \right\|^2 + \tau \|d_i^q \omega_i^q\|_2^2 \right\} \tag{4}$$

Here, d_i^q is the locality constraint obtained by finding the Euclidean distance between the intermediate HR input patch and the training patches from actual HR domain in new feature space.

$$d_i^q = \|u^q - u_i^q\|_2, 1 \leq i \leq K \tag{5}$$

The distance matrix d_i^q preserves the locality by penalizing the distance between input patch and training patches in the new feature space. Inducing the locality constraints gives the freedom to adaptively select the most relevant patches from the new feature space and effectively avoids the blurriness. Once we obtain the weights ω^q, we can get the estimation v^q of for the projection of HR patch h^q in latent space as follows

$$v^q = \sum_{u_i^q \in Z^q} \omega_i^q v_i^q \tag{6}$$

The desired HR patch in latent space can be derived as

$$h^q = \left((P_{TH}^q)^T \right)^\dagger v^q + \mu T_H^q \tag{7}$$

where $\dagger$ represents the pseudo inverse, i.e. $\left((P_{TH}^q)^T \right)^\dagger = \left(P_{TH}^q (P_{TH}^q)^T \right)^{-1} P_{TH}^q$

2.3 Residue Compensation

In propose a correlative two-step face hallucination approach to bridge the gap between the preliminary HR and residue construction. The patch pairs of residue images are first used to find the projection matrices using EZ-PLS, denoted as: $\{P_{RH}^q, P_{RL}^q\}_{q=1}^M$. Accordingly, the projection coefficients of HR-LR patches at q_{th} position can be obtained using:

$$v_i^{'q} = (P_{RH}^q)^T \left(h_i^{'q} - \mu R_H^q \right) \tag{8}$$

$$u_i^{'q} = (P_{RL}^q)^T \left(l_i^{'q} - \mu R_L^q \right) \tag{9}$$

Now, Instead of finding the best neighboring patches in LR residue domain and computing new optimal weight vector we propose to use the same index neighboring patches as in first phase from the HR residue patches and combine them using the same optimal weight vector in the first phase. The assumption is reasonable to be applied on image patches because the image patches at the same patches denote the same information of face images in different ways and tends to form a similar local topology structure. The desired HR residue patch can be derived as $\left(\left(P_{RH}^q \right)^T \right)^\dagger v'^q + \mu R_H^q$. The final HR patch is the sum of intermediate HR patch and residue HR patch. Finally integrate all the reconstructed HR patches according to the position of the input image. The HR image is generated by averaging the pixel values in the overlapping region.

3   Experiments and Results

In this section we will describe the details of the extensive experiments performed to evaluate the usefulness of the proposed method. To evaluate the proposed method we make the comparative analysis with some state-of-the-art methods such as LSE [15], LcR [8], LINE [9] and EZ-PLS [6]. The experiments are performed on CAS-PEAL-R1 [4] database and subset of FERET [16] and Stereo-pair database [3]. Hallucination results are demonstrated using two objective metrics, i.e. PSNR and SSIM.

3.1   Database Description and Parameter Setting

For CAS-PEAL-R1 database, we only use the normal lightening and neutral expression face images from the frontal subset. We select 1000 images for training and the rest of 40 images are used for testing. The size of HR image is 112×100. The other database consist of 200 frontal face images. The size of the HR image is 128×96. LR images are formed by smoothing and down sampling (by a factor of 4) from the corresponding HR images. We set HR patch to the size of 12×12 with 4 columns of the patches overlapped. The corresponding LR patch is of size 3×3 with one column of the patches overlapped. The neighborhood size $K = 100$ and the number of latent vectors are taken as 40 for both databases. For the parameter τ we set it at $1e^{-3}$. Some hallucinated results on FERET and Stereo-pair database generated by different state of art methods are shown in Fig. 2. We can see that the proposed method can provide a superior performance than other methods i.e. all the facial feature (eyes, nose, mouth and face contour) are properly inferred and is much sharper than the other methods.

The quantitative results on CAS-PEAL-R1 database and subset of FERET and Stereo-pair database are tabulated in Table 1. The proposed method produces the highest average PSNR and SSIM, which shows the significance of our proposed method.

3.2   Impact of Residue Compensation

Motivated by work in [6] we perform neighbor embedding in feature space of EZ-PLS. However, since the local representation ability is enhanced in EZ-PLS

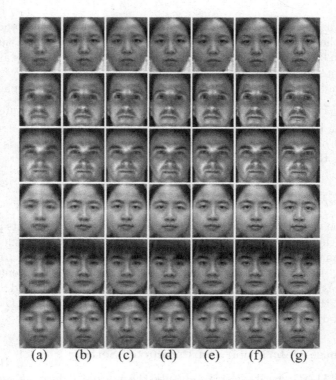

(a)   (b)   (c)   (d)   (e)   (f)   (g)

Fig. 2. The results of visual comparison. (a) 32×24 Input LR faces; (b) LSE [15]; (c) LcR [8]; (d) LINE [9]; (e) EZ-PLS [6]; (f) proposed method; (g) Original HR faces

it bring artifacts in overlapped region. To overcome this they proposed to incorporate global smoothness constraints via MAP formulation. We proposed to deal it with correlative residue compensation. As the face images are highly structured and the position patches denote the same information in different ways to form similar local topological structure. Therefore we suppose that HR patches and HR residue patches at q^{th} position share the same combination coefficients. The results are shown in Fig. 3. We can see that the proposed method efficiently recover the fine details and are much more similar to the ground truth images.

3.3   Results on Real World Images

To further testify our method on real world images, we use subset of FERET, CAS-PEAL and stereo-pair database. From the real world image, face images are extract and align according to the training samples. Several results are shown in Fig. 4.

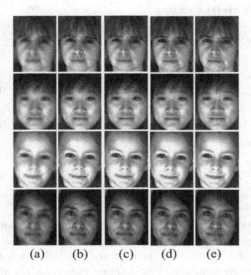

(a)   (b)   (c)   (d)   (e)

Fig. 3. Comparison of Reconstructed HR images in EZ-PLS feature space. (a) Input LR images; (b) HR images without residue compensation; (c) HR images using global smoothness constraints [6]; (d) HR images using correlative residue compensation; (e) Original HR faces

Fig. 4. Results from real life images. (a) Original image (b) LR Images (c) Hallucinated images from proposed method

Table 1. Performance comparison using PSNR and SSIM on FERET and Stereo-pair database and CAS-PEAL-R1 database

DATABASE	FERET and stereo-pair		CAS-PEAL-R1	
Method	PSNR	SSIM	PSNR	SSIM
LcR [8]	30.6685	0.9134	29.2507	0.9104
LINE [9]	30.9021	0.9145	29.398	0.9132
EZ-PLS [6]	31.0755	0.9153	30.8436	0.9155
Proposed method	31.5677	0.9188	31.0455	0.9190
Improvement	0.4922	0.0035	0.2312	0.0035

4   Conclusion

In this paper, we presented a new face hallucination scheme. The proposed method is a correlative two step framework. The distinction of the proposed approach is that it first synthesize an intermediate HR patch by employing the patch embedding in an enhanced consistency subspace rather than the original subspace. NE is then performed by incorporating locality constraints from modified feature space. Then we employ the residue compensation for each patch separately in the new subspace space based on the combination weights obtained in the first step. Moreover, in residue compensation we only use the HR manifold. The proposed method does not only results in minimization of the reconstruction error to produce artifact free results but also produces unbiased and sharp image due to neighbor embedding in a unified feature space. In future we will introduce more advance image representation techniques to cope with difficulties in real world images.

References

1. Baker, S., Kanade, T.: Hallucinating faces. In: FG, pp. 83–88 (2000)
2. Chang, H., Yeung, D., Xiong, Y.: Super-resolution through neighbor embedding. In: CVPR, vol. 1, pp. 275–282 (2004)
3. Fransens, R., Strecha, C., Van Gool, L.: Parametric stereo for multi-pose face recognition and 3D-face modeling. In: Zhao, W., Gong, S., Tang, X. (eds.) AMFG 2005. LNCS, vol. 3723, pp. 109–124. Springer, Heidelberg (2005)
4. Gao, W., Cao, B., Shan, S., Chen, X., Zhou, D., Zhang, X., Zhao, D.: The cas-peal large-scale chinese face database and baseline evaluations. IEEE Trans. Syst. Man Cybern. Part A Syst. Hum. **38**(1), 149–161 (2008)
5. Gao, X., Zhang, K., Tao, D., Li, X.: Joint learning for single-image super-resolution via a coupled constraint. IEEE Trans. Image Proc. **21**(2), 469–480 (2012)
6. Hao, Y., Qi, C.: Face hallucination based on modified neighbor embedding and global smoothness constraint. IEEE Signal Proc. Lett. **21**(10), 1187–1191 (2014)
7. Huang, H., He, H., Fan, X., Zhang, J.: Super-resolution of human face image using canonical correlation analysis. Pattern Recogn. **43**(7), 2532–2543 (2010)

8. Jiang, J., Hu, R., Han, Z., Lu, T., Huang, K.: Position-patch based face hallucina-
    tion via locality-constrained representation. In: IEEE International Conference on
    Multimedia and Expo (ICME), pp. 212–217 (2012)
9. Jiang, J., Hu, R., Han, Z., Wang, Z., Lu, T., Chen, J.: Locality-constraint iterative
    neighbor embedding for face hallucination. In: IEEE International Conference on
    Multimedia and Expo (ICME), pp. 1–6 (2013)
10. Jiang, J., Hu, R., Wang, Z., Han, Z.: Noise robust face hallucination via locality-
    constrained representation. IEEE Trans. Multimedia 16(5), 1268–1281 (2014)
11. Li, B., Chang, H., Shan, S., Chen, X.: Locality preserving constraints for super-
    resolution with neighbor embedding. In: IEEE International Conference on Image
    Processing (ICIP), pp. 1189–1192 (2009)
12. Li, B., Chang, H., Shan, S., Chen, X.: Low-resolution face recognition via coupled
    locality preserving mappings. IEEE Signal Process. Lett. 17(1), 20–23 (2010)
13. Li, B., Chang, H., Shan, S., Chen, X.: Aligning coupled manifolds for face halluci-
    nation. IEEE Signal Process. Lett. 16(11), 957–960 (2009)
14. Liu, C., Shum, H., Zhang, C.: A two-step approach to hallucinating faces: global
    parametric model and local nonparametric model. In: Proceedings of IEEE Con-
    ference on Computer Vision and Pattern Recognition (CVPR), vol. 1, p. I-192
    (2001)
15. Ma, X., Zhang, J., Qi, C.: Hallucinating face by position-patch. Pattern Recogn.
    43(6), 2224–2236 (2010)
16. Phillips, P.J., Moon, H., Rizvi, S.A., Rauss, P.J.: The feret evaluation methodology
    for face-recognition algorithms. IEEE Trans. Pattern Anal. Mach. Intell. 22(10),
    1090–1104 (2000)
17. Roweis, S., Saul, L.: Nonlinear dimensionality reduction by locally linear embed-
    ding. Science 290(5500), 2323–2326 (2000)
18. Zhuang, Y., Zhang, J., Wu, F.: Hallucinating faces: LPH super-resolution and
    neighbor reconstruction for residue compensation. Pattern Recogn. 40(11), 3178–
    3194 (2007)

Modal Regression via Direct Log-Density Derivative Estimation

Hiroaki Sasaki[1]([✉]), Yurina Ono[2], and Masashi Sugiyama[3]

[1] Graduate School of Information Science,
Nara Institute of Science and Technology, Nara, Japan
hsasaki@is.naist.jp
[2] Faculty of Science, The University of Tokyo, Tokyo, Japan
[3] Graduate School of Frontier Science, The University of Tokyo, Tokyo, Japan
sugi@k.u-tokyo.ac.jp

Abstract. Regression is aimed at estimating the conditional expectation of output given input, which is suitable for analyzing functional relation between input and output. On the other hand, when the conditional density with multiple modes is analyzed, *modal regression* comes in handy. *Partial mean shift* (PMS) is a promising method of modal regression, which updates data points toward conditional modes by gradient ascent. In the implementation, PMS first obtains an estimate of the joint density by kernel density estimation and then computes its derivative for gradient ascent. However, this two-step approach can be unreliable because a good density estimator does not necessarily mean a good density derivative estimator. In this paper, we propose a novel method for modal regression based on *direct* estimation of the log-density derivative without density estimation. Experiments show the superiority of our direct method over PMS.

1 Introduction

Modal regression is aimed at estimating the modes of the conditional probability density function of an output variable given input variables [2,5,10]. This allows analysis of data consisting of multiple modes in the conditional density, while ordinary regression methods cannot capture such multimodal structures (see Fig. 1 for illustrative examples). Modal regression would have various real-world applications such as speed-flow analysis in traffic engineering [5] and meteorology [8].

Earlier works on modal regression focused on estimating the *global* conditional modes [10]. However, *local* conditional modes are often informative in practice (see Fig. 1 again). The aim of *mixture regression* [1] is similar to modal regression, which typically models the conditional density by a mixture of Gaussian densities. However, the number of mixing components is a critical tuning parameter, and automatically finding a suitable value is not straightforward in practice.

Partial mean shift (PMS) [2,5] is a non-parametric method that can capture multiple local conditional modes without restrictive parametric assumptions.

© Springer International Publishing AG 2016
A. Hirose et al. (Eds.): ICONIP 2016, Part II, LNCS 9948, pp. 108–116, 2016.
DOI: 10.1007/978-3-319-46672-9_13

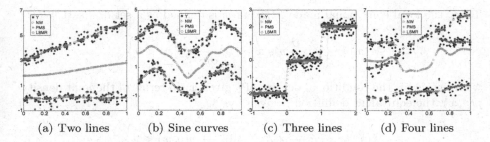

(a) Two lines   (b) Sine curves   (c) Three lines   (d) Four lines

Fig. 1. Four examples of simulated data for modal regression. The conditional mean is estimated by the Nadaraya-Watson estimator (NW), while modal regression methods, PMS and LSMR, perform conditional mode estimation.

PMS can be interpreted as an application of a well-known mode-seeking clustering method called *mean shift* [3, 7] to conditional mode estimation, and it updates data points toward conditional modes by gradient ascent. To do so, PMS first estimates the joint density by kernel density estimation and then computes its derivative for gradient ascent.

However, such a two-step approach can be unreliable because a good density estimator does not necessarily mean a good density derivative estimator. A more reliable way is to avoid density estimation and directly estimate the derivative. Following this idea, a direct log-density derivative estimator has recently been employed for mode-seeking clustering. This clustering method, called *least-squares log-density gradient clustering* (LSLDGC), was experimentally shown to significantly outperform the ordinary mean shift method especially for higher-dimensional data [4, 11].

The purpose of this paper is to propose a novel method for modal regression by applying LSLDGC to conditional mode estimation. Through numerical experiments, we demonstrate that the proposed method can more accurately identify the conditional modes than PMS.

This paper is organized as follows: We formulate the problem of modal regression and review PMS in Sect. 2. Then, we describe our proposed modal regression method in Sect. 3, and its performance is experimentally investigated in Sect. 4. Finally, this paper is concluded in Sect. 5.

2 Review of Partial Mean Shift

In this section, we first formulate the problem of modal regression. Then, partial mean shift is reviewed.

2.1 Problem Formulation

Suppose that n pairs of real output y and d-dimensional real input $\boldsymbol{x}$ independently drawn from a joint distribution with density $p(y, \boldsymbol{x})$ are available:

$$\mathcal{D} = \left\{ (y_i, \boldsymbol{x}_i) \mid y_i \in \mathbb{R}, \boldsymbol{x}_i = (x_i^{(1)}, x_i^{(2)}, \ldots, x_i^{(d)})^\top \in \mathbb{R}^d \right\}_{i=1}^n.$$

Then, the local conditional mode $M(\boldsymbol{x})$ at $\boldsymbol{x}$ is defined as a set of y satisfying

$$M(\boldsymbol{x}) = \left\{ y : \frac{\partial}{\partial y} p(y|\boldsymbol{x}) = 0, \frac{\partial^2}{\partial y^2} p(y|\boldsymbol{x}) < 0 \right\},$$

where $p(y|\boldsymbol{x})$ is the conditional density of y given $\boldsymbol{x}$. In terms of the joint density $p(y, \boldsymbol{x})$, the equivalent definition is given by

$$M(\boldsymbol{x}) = \left\{ y : \frac{\partial}{\partial y} p(y, \boldsymbol{x}) = 0, \frac{\partial^2}{\partial y^2} p(y, \boldsymbol{x}) < 0 \right\}.$$

Here, the goal is to estimate $M(\boldsymbol{x})$ from $\mathcal{D}$.

2.2 Partial Mean Shift

Partial mean shift (PMS) [5] is a nonparametric method to estimate the conditional modes, and is essentially an application of a mode-seeking clustering method called *mean shift* [3,7] to conditional mode estimation. The basic idea is to employ gradient ascent with respect to y given $\boldsymbol{x}$. To estimate the derivative of the joint density, PMS takes a two-step approach: First, the joint density is estimated by kernel density estimation (KDE) as

$$\widehat{p}_{\text{KDE}}(y, \boldsymbol{x}) = \frac{1}{Z_n} \sum_{i=1}^{n} K\left(\frac{|y - y_i|}{h_y} \right) K\left(\frac{\|\boldsymbol{x} - \boldsymbol{x}_i\|}{h_x} \right),$$

where Z_n is the normalization constant, K denotes a kernel function, and h_y and h_x are the kernel bandwidths, respectively. Then its partial derivative is computed with respect to y.

The actual algorithm of PMS for $K = \exp(-t^2/2)$ is given as

$$y \leftarrow \frac{\sum_{i=1}^{n} y_i K\left(\frac{|y-y_i|}{h_y} \right) K\left(\frac{\|\boldsymbol{x}-\boldsymbol{x}_i\|}{h_x} \right)}{\sum_{i=1}^{n} K\left(\frac{|y-y_i|}{h_y} \right) K\left(\frac{\|\boldsymbol{x}-\boldsymbol{x}_i\|}{h_x} \right)}, \tag{1}$$

which is derived from a fixed point method and can be interpreted as gradient ascent with adaptive step size [3]. Thus, repeatedly updating $\{y_i\}_{i=1}^{n}$ by (1) leads them to the conditional modes of the estimated density.

PMS is simple and computationally efficient, but its derivative estimation is not necessarily the best because a good density estimator does not necessarily provide a good density derivative estimator. A more promising way is to avoid density estimation and directly estimate the derivative. The *least-squares log-density gradients* (LSLDG) follows this approach, and directly estimates the derivatives of log-densities [4,11]. In the next section, we review LSLDG in terms of derivative estimation of the logarithmic joint density, and then propose a modal regression method based on LSLDG.

3   Modal Regression via Direct Derivative Estimation

In this section, we first review LSLDG, and then propose a new method for modal regression.

3.1   Review of LSLDG

Here, our tentative goal is to estimate $\frac{\partial}{\partial y} \log p(y, \boldsymbol{x}) = \frac{\frac{\partial}{\partial y} p(y, \boldsymbol{x})}{p(y, \boldsymbol{x})}$ from $\mathcal{D} = \{(y_i, \boldsymbol{x}_i)\}_{i=1}^{n}$. The main idea is to fit a derivative model $g(y, \boldsymbol{x})$ to the true derivative under the squared-loss:

$$
\begin{aligned}
J(g) &= \int \left\{ g(y, \boldsymbol{x}) - \frac{\frac{\partial}{\partial y} p(y, \boldsymbol{x})}{p(y, \boldsymbol{x})} \right\}^2 p(y, \boldsymbol{x}) \mathrm{d}\boldsymbol{x} - C \\
&= \int g(y, \boldsymbol{x})^2 p(y, \boldsymbol{x}) \mathrm{d}\boldsymbol{x} - 2 \int g(y, \boldsymbol{x}) \frac{\partial}{\partial y} p(y, \boldsymbol{x}) \mathrm{d}\boldsymbol{x} \\
&= \int g(y, \boldsymbol{x})^2 p(y, \boldsymbol{x}) \mathrm{d}\boldsymbol{x} + 2 \int \left\{ \frac{\partial}{\partial y} g(y, \boldsymbol{x}) \right\} p(y, \boldsymbol{x}) \mathrm{d}\boldsymbol{x},
\end{aligned}
$$

where $C = \int \left\{ \frac{\frac{\partial}{\partial y} p(y, \boldsymbol{x})}{p(y, \boldsymbol{x})} \right\}^2 p(y, \boldsymbol{x}) \mathrm{d}\boldsymbol{x}$, and the last equation was obtained by applying the *integration by parts* under the mild assumption that $\lim_{|y| \to \infty} g(y, \boldsymbol{x}) p(y, \boldsymbol{x}) = 0$. Then, the empirical version of J is given by

$$
\tilde{J}(g) = \frac{1}{n} \sum_{i=1}^{n} g(y_i, \boldsymbol{x}_i)^2 + 2 \frac{\partial}{\partial y} g(y_i, \boldsymbol{x}_i). \tag{2}
$$

To estimate $g(y, \boldsymbol{x})$, we employ a linear-in-parameter model as

$$
g(y, \boldsymbol{x}) = \sum_{i=1}^{n} \theta_i \psi_i(y, \boldsymbol{x}) = \boldsymbol{\theta}^\top \boldsymbol{\psi}(y, \boldsymbol{x}),
$$

where θ_i are parameters and $\psi_i(y, \boldsymbol{x})$ are basis functions (Our choice will be shown in Sect. 3.2). Substituting the model and adding the ℓ_2 regularizer into (2) yield the following quadratic function:

$$
\tilde{J}(\boldsymbol{\theta}) = \boldsymbol{\theta}^\top \boldsymbol{G} \boldsymbol{\theta} - 2 \boldsymbol{\theta}^\top \boldsymbol{h} + \lambda \boldsymbol{\theta}^\top \boldsymbol{\theta},
$$

where $\lambda \geq 0$ is the regularization parameter,

$$
\boldsymbol{G} = \frac{1}{n} \sum_{i=1}^{n} \boldsymbol{\psi}(y_i, \boldsymbol{x}_i) \boldsymbol{\psi}(y_i, \boldsymbol{x}_i)^\top, \quad \text{and} \quad \boldsymbol{h} = -\frac{1}{n} \sum_{i=1}^{n} \frac{\partial}{\partial y} \boldsymbol{\psi}(y_i, \boldsymbol{x}_i).
$$

All θ_i are further constrained to be non-negative, and the solution in this paper can be obtained as

$$
\widehat{\boldsymbol{\theta}} = \underset{\boldsymbol{\theta}}{\operatorname{argmin}} \left[\boldsymbol{\theta}^\top \left(\boldsymbol{G} + \lambda \mathbf{I}_n \right) \boldsymbol{\theta} - 2 \boldsymbol{\theta}^\top \boldsymbol{h} \right], \quad \text{s.t.} \quad \theta_i \geq 0 \text{ for } i = 1, 2, \ldots, n, \tag{3}
$$

where $\mathbf{I}_n$ is the n by n identity matrix. As will be explained in Sect. 3.2, imposing the non-negativity constraint gives a better theoretical guarantee. However, in practice, omitting the constraint does not yield performance degradation and is computationally more efficient. Indeed, the optimization problem can be solved analytically as

$$\widehat{\boldsymbol{\theta}} = (\boldsymbol{G} + \lambda \mathbf{I}_n)^{-1} \boldsymbol{h}.$$

Finally, a derivative estimator is given by $\widehat{g}(y, \boldsymbol{x}) = \sum_{i=1}^{n} \widehat{\theta}_i \psi_i(y, \boldsymbol{x}) = \widehat{\boldsymbol{\theta}}^{\top} \boldsymbol{\psi}(y, \boldsymbol{x})$. Previously, a mode-seeking clustering called *LSLDG clustering* (LSLDGC) has been proposed based on LSLDG, and LSLDGC was experimentally shown to significantly outperform the ordinary mean shift method [11]. Next, we apply LSLDGC to conditional mode estimation.

3.2 Least-Squares Modal Regression

Here, we proposes a new method of modal regression by applying LSLDGC to conditional mode estimation. As in PMS, the algorithm is derived based on the fixed-point method. To derive the algorithm, we choose the following basis function:

$$\psi_i(y, \boldsymbol{x}) = \frac{\partial}{\partial y} \phi_i(y, \boldsymbol{x}), \tag{4}$$

where $\phi_i(y, \boldsymbol{x}) = \exp\left(-\frac{(y - y_i)^2}{2\sigma_y^2}\right) \exp\left(-\frac{\|\boldsymbol{x} - \boldsymbol{x}_i\|^2}{2\sigma_x^2}\right)$, and σ_y and σ_x are Gaussian bandwidths. From the LSLDG estimator, we have

$$\widehat{g}(y, \boldsymbol{x}) = \sum_{i=1}^{n} \widehat{\theta}_i \frac{y_i - y}{\sigma_y^2} \phi_i(y, \boldsymbol{x}) = \frac{1}{\sigma_y^2} \sum_{i=1}^{n} \widehat{\theta}_i \phi_i(y, \boldsymbol{x}) \left[\frac{\sum_{i=1}^{n} \widehat{\theta}_i y_i \phi_i(y, \boldsymbol{x})}{\sum_{i=1}^{n} \widehat{\theta}_i \phi_i(y, \boldsymbol{x})} - y \right],$$

where we assumed that $\sum_{i=1}^{n} \widehat{\theta}_i y_i \phi_i(y, \boldsymbol{x}) \neq 0$. Setting $\widehat{g}(y, \boldsymbol{x}) = 0$ leads to the following update formula:

$$y \leftarrow \frac{\sum_{i=1}^{n} \widehat{\theta}_i y_i \phi_i(y, \boldsymbol{x})}{\sum_{i=1}^{n} \widehat{\theta}_i \phi_i(y, \boldsymbol{x})}. \tag{5}$$

Compared with the update formula of PMS, (5) can be interpreted as a weighted PMS algorithm.

As in the partial mean shift algorithm, a simple calculation indicates that (5) can be expressed as

$$y \leftarrow y + \frac{\sigma_y^2}{\sum_{i=1}^{n} \widehat{\theta}_i \phi_i(y, \boldsymbol{x})} \widehat{g}(y, \boldsymbol{x}).$$

Thus, when $\sum_{i=1}^{n} \widehat{\theta}_i \phi_i(y, \boldsymbol{x}) > 0$, (5) is gradient ascent. For this reason, we have imposed non-negativity constraints in (4). However, our experiments will show that even without the non-negativity constraints, data points $\{y_i\}_{i=1}^n$ converge to the conditional modes. Under the non-negative constraint on θ_i, the convergence to conditional modes can be proved in a similar way as in mean shift [3,6]. We call this method the *least-squares modal regression* (LSMR).

4   Experiments

In this section, we experimentally investigate the behavior of LSMR, and compare it with PMS on simulated data and speed-flow diagrams.

4.1   Illustration

We generated data according to $y_i = f_k(x_i) + 0.3\epsilon_i$, where $x_i^{(j)}$ was sampled from the uniform density on $[0, 1]$, ϵ_i was sampled from the standard normal density and $f_k(x)$ was specified as $f_1(x) = 3\sum_{j=1}^d x^{(j)}/d + 3$ and $f_2(x) = 0.1\sum_{j=1}^d x^{(j)}/d$ (Fig. 1(a)). For LSMR, we performed five-fold cross-validation with respect to J to optimize the Gaussian bandwidths σ_x, σ_y and regularization parameter λ. σ_x and σ_y were selected from $10^{-0.5}$ to 10 at the regular interval in logarithmic scale, while λ was chosen from 10^{-4} to 10^{-1}. Furthermore, we applied LSMR with the analytic solution in LSLDG, and denote it by LSMR$_a$. For PMS, the Gaussian kernel $K(t) = \exp\left(-t^2/2\right)$ was employed, and the kernel bandwidths h_x and h_y were chosen based on least-squares cross-validation from 10^{-2} to 10^0 at the regular interval in logarithmic scale.[1] The estimation error was defined by $\frac{1}{n}\sum_{i=1}^n \min_k |\hat{f}_i - f_k(x_i)|$, where $\hat{f}_i$ denotes the output of LSMR, LSMR$_a$ or PMS from (y_i, x_i).

LSMR accurately captures the conditional modes, while some points in PMS are deviated from the conditional modes (Fig. 1(a)). Figure 2 quantitatively shows the superior performance of both LSMR and LSMR$_a$ to PMS. The notable point is that LSMR and LSMR$_a$ work significantly better than PMS for relatively high-dimensional data (Fig. 2(b)). Although LSMR and LSMR$_a$ perform similarly, LSMR$_a$ is computationally more efficient than LSMR (Fig. 1(c)).

To further substantiate the superior performance over PMS, we performed more experiments for other three simulated data following the model $y_i = f_k(x_i) + 0.3\epsilon_i$ where f_k was specified as follows:

- (Figure 1(b)): $f_1(x) = \sin(3\pi x^{(1)}) + 3$ and $f_2(x) = 0.1\sin(3\pi x^{(1)})$, where $x^{(1)}$ was sampled from the uniform density on $[0, 1]$.
- (Figure 1(c)): $f_1(x) = 2$, $f_2(x) = 0$ and $f_3(x) = -2$, where $x^{(1)}$ was sampled from the uniform density on $[-1, 0]$, $[0, 1]$ or $[1, 2]$, respectively.
- (Figure 1(d)): $f_1(x) = 5x^{(1)} + 2$, $f_2(x) = x^{(1)}$ and $f_3(x) = 4.5$, where $x^{(1)}$ was sampled from the uniform density on $[0, 1]$.

LSMR and LSMR$_a$ accurately and flexibly find the conditional modes on a wide range of data (Figs. 1(b,c,d) and 3).

---

[1] The intervals of σ_x and h_x (or σ_y and h_y) were further changed by multiplying the median value of $|x_i^{(j)} - x_k^{(j)}|$ with respect to i, j, k (or $|y_i - y_k|$ with respect to i, k).

4.2   Application to Speed-Flow Diagrams

Next, we apply LSMR with the non-negative constraint on θ_i to the speed-flow diagrams used in [5].[2] Here, we employed adaptive Gaussian bandwidths in LSMR, i.e., σ_y and σ_x depend on samples, and set them at the k-nearest neighbor Euclidean distance from y_i and $\boldsymbol{x}_i$ where k was cross-validated. To decrease computational costs, here 500 samples randomly chosen from $\{(y_i, \boldsymbol{x}_i)\}_{i=1}^{n}$ were used as the Gaussian centers in (4). For PMS, we set $h_y = 3.10$ (or $h_y = 3.55$) and $h_x = 108.08$ (or $h_x = 199.56$) for lane 2 (or lane 3) as done in [5]. Furthermore, we applied another variant of PMS by modifying and cross-validating the bandwidth parameters in a similar way as LSMR, and denote it by PMS_m.

Figure 4 indicates that LSMR finds the bimodal structures in data better than PMS_m, and compares favorably with PMS.

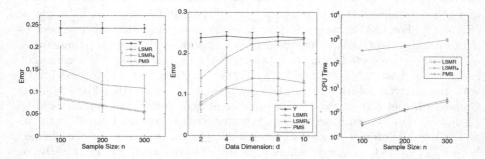

Fig. 2. Comparison of estimation errors between LSMR and PMS against (left) sample size n in $d = 1$ and (middle) input data dimension d in $n = 200$. The right figure is for CPU time when $d = 1$, and the vertical axis is displayed in logarithmic scale. Each point and error bar denote the average and standard deviation over 30 runs, respectively. An example of data in these experiments is illustrated in Fig. 1(a). As a base line, the errors from y_i are also plotted.

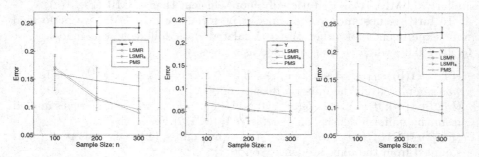

Fig. 3. Comparison of estimation errors between LSMR and PMS over 30 runs on three datasets in $d = 1$, whose examples from the left to right plot are illustrated in Fig. 1(b), (c) and (d), respectively.

---

[2] The datasets were downloaded at http://www.blackwellpublishing.com/rss.

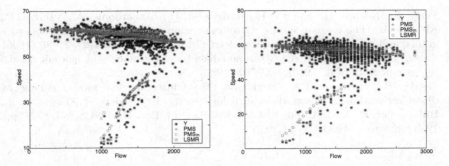

Fig. 4. Speed-flow diagrams for lane 2 (left) and lane 3 (right) of Californian uninterrupted highway collected by [9].

5   Conclusion

In this paper, we proposed a new method for modal regression by applying a mode-seeking clustering method based on a direct log-density derivative estimator. Experimental results showed the superior performance of the proposed method over the existing method. In future, we will apply the proposed method to other real-world datasets.

Acknowledgments. HS was supported by KAKENHI 15H06103 and MS was supported by KAKENHI 25700022.

References

1. Bishop, C.: Pattern Recognition and Machine Learning. Springer, New York (2006)
2. Chen, Y.C., Genovese, C., Tibshirani, R., Wasserman, L.: Nonparametric modal regression. Ann. Stat. **44**(2), 489–514 (2016)
3. Comaniciu, D., Meer, P.: Mean shift: a robust approach toward feature space analysis. IEEE Trans. PAMI **24**(5), 603–619 (2002)
4. Cox, D.D.: A penalty method for nonparametric estimation of the logarithmic derivative of a density function. Annals Inst. Stat. Math. **37**(1), 271–288 (1985)
5. Einbeck, J., Tutz, G.: Modelling beyond regression functions: an application of multimodal regression to speed-flow data. J. Roy. Stat. Soc.: Ser. C (Appl. Stat.) **55**(4), 461–475 (2006)
6. Fashing, M., Tomasi, C.: Mean shift is a bound optimization. IEEE Trans. PAMI **27**(3), 471–474 (2005)
7. Fukunaga, K., Hostetler, L.: The estimation of the gradient of a density function, with applications in pattern recognition. IEEE Trans. IT **21**(1), 32–40 (1975)
8. Hyndman, R., Bashtannyk, D., Grunwald, G.: Estimating and visualizing conditional densities. J. Comput. Graph. Stat. **5**(4), 315–336 (1996)

9. Petty, K., Noeimi, H., Sanwal, K., Rydzewski, D., Skabardonis, A., Varaiya, P., Al-Deek, H.: The freeway service patrol evaluation project: database support programs, and accessibility. Transp. Res. Part C Emerg. Technol. **4**(2), 71–85 (1996)
10. Sager, T.W., Thisted, R.A.: Maximum likelihood estimation of isotonic modal regression. Ann. Stat. **10**(3), 690–707 (1982)
11. Sasaki, H., Hyvärinen, A., Sugiyama, M.: Clustering via mode seeking by direct estimation of the gradient of a log-density. In: Calders, T., Esposito, F., Hüllermeier, E., Meo, R. (eds.) ECML PKDD 2014, Part III. LNCS, vol. 8726, pp. 19–34. Springer, Heidelberg (2014)

Simplicial Nonnegative Matrix Tri-factorization: Fast Guaranteed Parallel Algorithm

Duy-Khuong Nguyen[1,3](✉), Quoc Tran-Dinh[2], and Tu-Bao Ho[1,4]

[1] Japan Advanced Institute of Science and Technology, Nomi, Japan
khuongnd@gmail.com
[2] The University of North Carolina at Chapel Hill, Chapel Hill, USA
[3] University of Engineering and Technology,
Vietnam National University, Hanoi, Vietnam
[4] John von Neumann Institute, Vietnam National University, Ho Chi Minh, Vietnam

Abstract. Nonnegative matrix factorization (NMF) is a linear powerful dimension reduction and has various important applications. However, existing models remain the limitations in the terms of interpretability, guaranteed convergence, computational complexity, and sparse representation. In this paper, we propose to add simplicial constraints to the classical NMF model and to reformulate it into a new model called simplicial nonnegative matrix tri-factorization to have more concise interpretability via these values of factor matrices. Then, we propose an effective algorithm based on a combination of three-block alternating direction and Frank-Wolfe's scheme to attain linear convergence, low iteration complexity, and easily controlled sparsity. The experiments indicate that the proposed model and algorithm outperform the NMF model and its state-of-the-art algorithms.

Keywords: Dimensionality reduction · Nonnegative matrix factorization · Simplicial nonnegative matrix tri-factorization · Frank-wolfe algorithm

1 Introduction

Nonnegative matrix factorization (NMF) has been recognized as a linear powerful dimension reduction, and has a wide range of applications including text mining, image processing, bioinformatics [9]. In this problem, a given nonnegative observed matrix $V \in \mathbb{R}_+^{n \times m}$ consists of m vectors having n dimensions, which is factorized into a product of two nonnegative factor matrices, namely latent components $W^T \in \mathbb{R}_+^{n \times r}$ and new coefficients $F \in \mathbb{R}_+^{r \times m}$. In the classical setting, due to noise and nonnegative constraints, NMF is approximately conducted by minimizing the objective function $D(V \| W^T F) = \| V - W^T F \|_2^2$.

Despite having more than a decade of rapid developments, NMF still has the limitations of interpretability and computation. First, the values of factor matrices in NMF do not concisely represent the roles of latent components over instances and the contributions of attributes over latent components. Simply,

© Springer International Publishing AG 2016
A. Hirose et al. (Eds.): ICONIP 2016, Part II, LNCS 9948, pp. 117–125, 2016.
DOI: 10.1007/978-3-319-46672-9_14

they express the appearances of latent components over instances and attributes over latent components rather than their roles. In other words, it is not reasonable to determine how may percentages each latent component contributes to instances via the values in F. Second, concerning the computation, Wang et al., 2012 [4] proposed one of the most state-of-the algorithms which has sub-linear convergence $O(1/k^2)$, high complexity $\mathcal{O}(r^2)$ at each iteration, and considerable difficulties of parallelability and controlling sparsity because it works on the whole of matrices and all the variable gradients.

To overcome the above mentioned limitations, we introduce a new formulation of NMF as simplicial nonengative matrix tri-factorization (SNMTF) by adding simplicial constraints over factor matrices, and propose a fast guaranteed algorithm which can be conveniently and massively parallelized. In this model, the roles of latent components over instances and the contributions of attributes over latent components are represented via the factor matrices F and W. To the end, this work has the major contributions as follows:

(a) We introduce a new model of NMF which is named by SNMTF using L_2 regularizations on both the latent components W and the new coefficients F. The new model has not only more concise interpretability, but also retains the generalization of NMF.
(b) We propose a fast parallel algorithm based on a combination of three-block alternating direction and Frank-Wolfe algorithm [7] to attain linear convergence, low iteration complexity, and easily controlled sparsity.

2   Problem Formulation

Concerning the interpretability, NMF is a non-convex problem having numerous solutions as stationary points, which has rotational ambiguities [1]. Particularly, if $V \approx W^T F$ is a solution, $V \approx W^T[D_F F'] = [W^T D_F]F'$ are also equivalent solutions; where D_F is a positive diagonal matrix satisfying $F = D_F F'$. Hence, it does not consistently explain the roles of latent components over instances and the contributions of attributes over latent components.

To resolve the limitation, we propose a new formulation as SNMTF, in which the data matrix is factorized into a product of three matrices $V \approx W^T DF$, where D is a positive diagonal matrix, $\sum_{k=1}^{r} W_{ki} = 1 \forall i$, and $\sum_{k=1}^{r} F_{kj} = 1 \forall j$.

However, the scaling of factors via the diagonal matrix D can lead to the inconsistency of interpreting the factor matrices W and F, and the existence of bad stationary points such as $\lambda_k = 0 \Rightarrow W_{ki} = 0, F_{kj} = 0$. Hence, we restrict the formulation by adding the condition $\lambda_1 = ... = \lambda_r = \lambda$. This model can be considered as an extension of the probabilistic latent semantic indexing (PLSI) [5] for scaling data with an additional assumption that the weights of latent factors are the same. Remarkably, their roles of latent components over instances and of attributes over latent components are represented via the values of the factor matrices W and F. As a result, it is more easier to recognize these roles than in the case that the weights of latent factors can be different. Therefore, the objective function of SNMTF with L_2 regularizations is written as follows:

$$\begin{cases} \min_{W,D,F} \left\{ \Phi(W,D,F) := \frac{1}{2}\|V - W^T DF\|_F^2 + \frac{\alpha_1}{2}\|W\|_F^2 + \frac{\alpha_2}{2}\|F\|_F^2 \right\} \\ \text{s.t.} \quad \sum_{k=1}^{r} W_{ki} = 1 \ (i=1,\cdots,n), \quad \sum_{k=1}^{r} F_{kj} = 1 \ (j=1,\cdots,m), \\ \quad W \in \mathbb{R}_+^{r \times n}, \quad F \in \mathbb{R}_+^{r \times m}, \quad D = \mathrm{diag}(\lambda,\cdots,\lambda). \end{cases} \quad (1)$$

There are three significant remarks in this formulation. Firstly, adding simplicial constraints leads to more concise interpretability of the factor matrices and convenience for post processes such as neural network and support vector machine because the sum of attributes is normalized to 1. Secondly, L_1 regularization is ignored because $\|W\|_1$ and $\|F\|_1$ equal to a constant. Finally, the diagonal of $D = \mathrm{diag}(\lambda,...,\lambda)$ has the same value because of two main reasons: First, it is assumed that V_{ij} is generated by W_i, F_j and a scale as $V_{ij} \approx \lambda W_i^T F_j$; Second, it still retains the generalization of SNMTF which every solution of NMF can be equivalently represented by SNMTF, which will be proved in Sect. 4.

3   Proposed Algorithm

We note that problem (1) is nonconvex due to the product $W^T DF$. We first propose to use three-block alternating direction method to decouple there three blocks. Then, we decompose the computation onto column of the matrix variables, which can be conducted in parallel. Finally, we apply Frank-Wolfe's algorithm [6,7] to solve the underlying subproblems.

3.1   Iterative Multiplicative Update for Frobenius Norm

We decouple the tri-product $W^T DF$ by the following alternating direction scheme, which is also called iterative multiplicative update:

$$\begin{cases} F^{t+1} := \arg\min_{F \in \mathbb{R}^{r \times m}} \left\{ \Phi(W^t, D^t, F) : F \in \Delta_r^n \right\}, \\ W^{t+1} := \arg\min_{W \in \mathbb{R}^{r \times n}} \left\{ \Phi(W, D^t, F^{t+1}) : W \in \Delta_r^m \right\}, \\ D^{t+1} := \arg\min_{\lambda \in \mathbb{R}} \left\{ \Phi(W^{t+1}, D, F^{t+1}) : D = \mathrm{diag}(\lambda,\cdots,\lambda) \right\}. \end{cases} \quad (2)$$

Clearly, both F-problem and W-problem in (2) are convex but are still constrained by simplex, while the D-problem is unconstrained. We now can solve the D-problem in the third line of (2). Due to the constraint $D = \mathrm{diag}(\lambda,\cdots,\lambda)$, this problem turns out to be an univariate convex program, which can be solved in a closed form as follows:

$$\lambda_{t+1} = \arg\min_{\lambda \in \mathbb{R}} \|V - W^{t+1^T} DF^{t+1}\|_2^2 = \frac{\langle V, W^{t+1^T} F^{t+1} \rangle}{\langle W^{t+1} W^{t+1^T}, F^{t+1} F^{t+1^T} \rangle}. \quad (3)$$

Algorithm 1. Iterative multiplicative update for Frobenius norm

Input: Data matrix $V = \{V_j\}_{j=1}^{m} \in \mathbb{R}_+^{n \times m}$, and r, $\alpha_1, \alpha_2 \geq 0, \beta \geq 0$.
Output: Coefficients $F \in R_+^{r \times m}$ and latent components $W \in R_+^{r \times n}$
1 **begin**
2   Pick an arbitrary initial point $W \in \mathbb{R}_+^{r \times n}$ (e.g., random);
3   **repeat**
4        $q = -\lambda WV$; $Q = \lambda^2 WW^T + \alpha_1 \mathbb{I}$;
5   /***Inference step**: Fix W and D to find new F;
6   **for** $j = 1$ to m **do**
7          $F_j \approx \underset{x \in \Delta_r}{\mathrm{argmin}} \left\{ \frac{1}{2} x^T Q x + q_j^T x \right\}$/*Call Algorithm 2 in parallel */;
8        $q = -\lambda FV^T$; $Q = \lambda^2 FF^T + \alpha_2 \mathbb{I}$;
9   /***Learning step**: Fix F and D to find new W;
10   **for** $i = 1$ to n **do**
11        $W_i \approx \underset{x \in \Delta_r}{\mathrm{argmin}} \left\{ \frac{1}{2} x^T Q x + q_i^T x \right\}$/*Call Algorithm 2 in parallel */;
12       $\lambda = \underset{\lambda \in \mathbb{R}}{\mathrm{argmin}} \|V - W^T DF\|_2^2 = \frac{\langle V, W^T F \rangle}{\langle WW^T, FF^T \rangle}$;
13   **until** *convergence condition is satisfied*;

Then, we form the matrix D^{t+1} as $D^{t+1} = \mathrm{diag}(\lambda_{t+1}, \cdots, \lambda_{t+1})$.

If we look at the F-problem, then fortunately, it can be separated into n independent subproblems $(j = 1, \cdots, m)$ of the form:

$$F_j^{t+1} = \arg \min_{F_j} \left\{ \frac{1}{2} \|V_j - (W^t)^T D^t F_j\|_2^2 + \frac{\alpha_2}{2} \|F_j\|_2^2 : F_j \in \Delta_r \right\}. \qquad (4)$$

The same trick is applied to the W-problem in the second line of (2). Now, we assume that we apply the well-known Frank-Wolfe algorithm to solve (4), then we can describe the full algorithm for solving (1) into Algorithm 1.

The stopping criterion of Algorithm 1 remains unspecified. Theoretically, we can terminate Algorithm 1 using the optimality condition of (1). However, computing this condition requires a high computational effort. We instead terminate Algorithm 1 if it does not significantly improve the objective value of (1) or the differences $W_{t+1} - W_t$ and $F_{t+1} - F_t$ and the maximum number of iterations.

3.2 Frank-Wolfe's Algorithm for QP over Simplex Constraint

Principally, we can apply any convex optimization method such as interior-point, active-set, projected gradient and fast gradient method to solve QP problems of the form (4). However, this QP problem (4) has special structure and is often sparse. In order to exploit its sparsity, we propose to use a Frank-Wolfe algorithm studied in [7] to solve this QP problem. Clearly, we can write (4) as follows:

$$x \approx \underset{x \in \Delta_r}{\mathrm{argmin}} \frac{1}{2} \|v - A^T x\|_2^2 + \frac{\alpha}{2} \|x\|_2^2 = \underset{x \in \Delta_r}{\mathrm{argmin}} \frac{1}{2} x^T Q x + q^T x \qquad (5)$$

where $v = V_j, A = DW, Q = AA^T + \alpha$, and $q = -Av$. By applying the Frank-Wolfe algorithm form [7] to solve this problem, we obtain Algorithm 2 below.

Algorithm 2. Fast Algorithm for NQP with Simplicial Constraint

---

Input: $Q \in \mathbb{R}^{r \times r}, q \in \mathbb{R}^r$.
Output: New coefficient $x \approx \underset{x \in \Delta_r}{\operatorname{argmin}} f(x) = \frac{1}{2}x^T Q x + q^T x$.

1 **begin**
2   Choose $k = \underset{k}{\operatorname{argmin}} \frac{1}{2}e_k^T Q e_k + q^T e_k$, where e_k is the kth basis vector;
3   Set $x = \mathbf{0}^k$; $x_k = 1$; $Qx = Qe_k$; $qx = q^T x$ and $\nabla f = Qx + q^T$;
4   **repeat**
5   Select $k = \underset{k \in \{1..r\}}{\operatorname{argmin}} \{\langle e_k - x, \nabla f \rangle\}$ or $\{\langle x - e_k, \nabla f \rangle | x_k > 0\}$;
6   Select $\alpha = \underset{\alpha}{\operatorname{argmin}} f(\alpha e_k + (1 - \alpha)x)$;
7          $\alpha = min(1, max(\alpha, -\frac{x_k}{1 - x_k}))$;
8          $Qx = (1 - \alpha)Qx + \alpha Q e_k$; $\nabla f = Qx + q$;
9          $qx = (1 - \alpha)qx + \alpha q e_k$;
10         $x = (1 - \alpha)x$; $x_k = x_k + \alpha$;
11   **until** *converged conditions are staisfied*;

In Algorithm 2, the first derivative of $f(x) = \frac{1}{2}x^T Q x + q^T x$ is computed by $\nabla f = Qx + q$. In addition, the steepest direction in the simplex is selected by this formula: $k = \underset{k \in \{1..r\}}{\operatorname{argmin}} \{\langle e_k - x, \nabla f \rangle\}$ or $\{\langle x - e_k, \nabla f \rangle | x_k > 0\}$.

For seeking the best variable α to minimize $f(\alpha x + (1 - \alpha)e_k)$ where e_k is the kth unit vector. Let consider $f(\alpha x + (1 - \alpha)e_k)$, we have:

$$\begin{aligned} \frac{\partial f}{\partial \alpha}(\alpha = 0) &= (x - \mathbf{e}_k)^T(Qx + q) = x^T(Qx + q) - [Qx]_k - q_k \\ \frac{\partial^2 f}{\partial \alpha^2}(\alpha = 0) &= (x - \mathbf{e}_k)^T Q(x - \mathbf{e}_k) = x^T Q x - 2[Qx]_k + Q_{kk}. \end{aligned} \quad (6)$$

Since f is a quadratic function of α, its optimal solution is $\alpha = \underset{\alpha \in [-\frac{x_k}{1 - x_k}, 1]}{\operatorname{argmin}}$ $f((1 - \alpha)x + \alpha \mathbf{e}_k) = [-\frac{\nabla f_{\alpha=0}}{\nabla^2 f_{\alpha=0}}]_{[-\frac{x_k}{1 - x_k}, 1]}$. The projection of solution over the interval $[-\frac{x_k}{1 - x_k}, 1]$ is to guarantee $x_k \geq 0$ $\forall k$. The updates $x = (1 - \alpha)x$ and $x_k = x_k + \alpha$ are to retain the simplicial constraint $x \in \Delta_r$.

In Algorithm 2, duplicated computation is removed to reduce the iteration complexity into $\mathcal{O}(r)$ by maintaining Qx and $q^T x$. This result is highly competitive with the state-of-the-art algorithm having a sub-linear convergence rate $\mathcal{O}(1/k^2)$ and complexity of $\mathcal{O}(r^2)$ [4].

4   Theoretical Analysis

This section discusses three important aspects of the proposed algorithm as convergence, complexity, and generalization. Concerning the convergence, setting $\alpha_1, \alpha_2 > 0$, based on Theorem 3 in Lacoste-Julien, S., & Jaggi, M. (2013) [7], since $f(x) = \frac{1}{2}x^T Q x + q^T x$ is smoothness and strongly convex, we have:

Theorem 1. *Algorithm 2 linearly converges as* $f(x_{k+1}) - f(x^*) \leq (1 - \rho_f^{FW})^k$ $(f(x_0) - f(x^*))$, *where* $\rho_f^{FW} = \{\frac{1}{2}, \frac{\mu_f^{FW}}{C_f}\}$, C_f *is the curvature constant of the convex and differentiable function* f, *and* μ_f^{FW} *is an affine invariant of strong convex parameter.*

Since Algorithm 2 always linearly converges and the objective function restrictedly decreases, Algorithm 1 always converges stationary points. Regarding the complexity of the proposed algorithm, we have:

Theorem 2. *The complexity of Algorithm 2 is* $\mathcal{O}(r^2 + \bar{t}r)$, *and the complexity of each iteration in Algorithm 1 is* $\mathcal{O}(mnr + (m+n)r^2 + \bar{t}(m+n)r)$.

Proof. The complexity of the initial computation $Qx + q$ is $O(r^2)$, and each iteration in Algorithm 2 is $O(r)$. Hence, the complexity of Algorithm 2 is $\mathcal{O}(r^2 + \bar{t}r)$. The complexity of main operators in Algorithm 1 as WV, FV^T, FF^T, and WW^T is $O(mnr + (m+n)r^2)$. Overall, the complexity of each iteration in Algorithm 1 is $\mathcal{O}(mnr + (m+n)r^2 + \bar{t}(m+n)r)$.

This is highly competitive with the guaranteed algorithm [4] having the complexity of $\mathcal{O}(mnr + (m+n)r^2 + \bar{t}(m+n)r^2)$. Furthermore, adding the simplicial constants into NMF does not reduce the generalization and flexibility of NMF:

Theorem 3. *Each solution of NMF can be equivalently transfered into SNMTF.*

Proof. Assume that $V \approx W^T F$, which leads to the existence of λ large enough to satisfy $W^T F = W'^T D' F'$, where $D' = \mathrm{diag}(\lambda, .., \lambda)$, $\sum\limits_{k=1}^{r} W_{ki} < 1 \ \forall i$, and $\sum\limits_{k=1}^{r} W_{kj} < 1 \ \forall j$. Therefore, $\exists W'', D''$, and F'': $W''^T D'' F'' = W'^T D' F' = W^T F$, where $W''_{ij} = W'_{ij}$, $F''_{ij} = F'_{ij}$, $W''_{r+1,i} = 0, W''_{r+2,i} = 1 - \sum\limits_{k=1}^{r+1} W''_{ki} \ \forall i$, $F''_{r+1,j} = 1 - \sum\limits_{k=1}^{r} F''_{kj}, F''_{r+2,j} = 0 \ \forall j, D'' = \mathrm{diag}(\lambda, .., \lambda)$. $W''^T D'' F''$ is SNMTF of V.

The generalization is crucial to indicate the robustness and high flexibility of the proposed model in comparison with NMF models, although many constraints have been added to enhance the quality and interpretability of the NMF model.

5   Experimental Evaluation

This section investigates the effectiveness of the proposed algorithm via three significant aspects of convergence, classification performance, and sparsity. The proposed algorithm **SNMTF** is compared with the following methods:

- **NeNMF** [4]: It is a guaranteed method, each alternative step of which sublinearly converges at $\mathcal{O}(1/k^2)$ that is highly competitive with the proposed algorithm.
- **LeeNMF** [8]: It is the original gradient algorithm for NMF.
- **PCA:** It is considered as a based-line method in dimensionality reduction, which is compared in classification and sparsity.

Datasets: We compared the selected methods in three typical datasets with different size, namely Faces[1], Digits[2] and Tiny Images [3].

Environment Settings: We develop the proposed algorithm SNMTF in Matlab with embedded code C++ to compare them with other algorithms. We set system parameters to use only six threads in the machine Mac Pro 6-Core Intel Xeon E5 3 GHz 32 GB. The initial

Table 1. Dataset Information

Datasets	n	m	Testing size	r	#class
Faces	361	6,977	24,045	30	2
Digits	784	6.10^4	10^4	60	10
Cifar-10	3,072	5.10^4	10^4	90	10

matrices W^0 and F^0 are set to the same values, the maximum number of iterations is 500. The source code is published on our homepage [4] (Table 1).

5.1   Convergence

We investigate the convergence of the compared algorithms by $\frac{f_1}{f_k}$ because they have the different formulations and objective functions. Figure 1 clearly shows that the proposed algorithm converges much faster than the other algorithms. The most steepest line of the proposed algorithm represents its fast convergence. This result is reasonable because the proposed algorithm has a faster convergence rate and lower complexity than the state-of-the-art algorithm NeNMF.

5.2   Classification

Concerning the classification performance, the training datasets with labels are used to learn gradient boosting classifiers [2,3], one of the robust ensemble methods, to classify the testing datasets. The proposed algorithm outperforms the other algorithms and PCA over all the datasets. For the small and easy dataset

---

[1] http://cbcl.mit.edu/cbcl/software-datasets/FaceData.html.
[2] http://yann.lecun.com/exdb/mnist/.
[3] http://horatio.cs.nyu.edu/mit/tiny/data/index.html.
[4] http://khuongnd.appspot.com/.

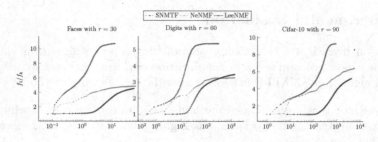

Fig. 1. Convergence of loss information f_1/f_k versus time

Face, the result of the proposed algorithm is close to the results of NeNMF. However, for larger and more complex datasets Digit and Tiny Images, the proposed algorithm has much better accuracy than the other algorithms. Noticeably, the result of Tiny Images is much worse than the result of the other datasets because it is highly complicated and contains backgrounds. This classification result obviously represents the effectiveness of the proposed model and algorithm.

5.3 Sparsity

We investigate the sparsity of factor matrices F, W, and the sparsity average in both F and W. For the dataset Digit, the proposed algorithm outperforms in all these measures. For the other datasets

Table 2. Classification inaccuracy

Dataset	PCA	LeeNMF	NeNMF	SNMTF
Faces	6.7	3.3	2.7	**2.6**
Digits	30.15	12.16	3.9	**3.6**
Tiny Images	59.3	71.4	51.4	**50.1**

Faces and Tiny Images, it has better representation F and more balance between sparsity of F and W. Frankly speaking, in these datasets, achieving more sparse representation F is more meaningful than achieving more sparse model W because F is quite dense but W is highly sparse, which is a reason to explain why SNMTF has the best classification result (Tables 2 and 3).

Table 3. Sparsity of factor matrices (%) of F, W, and (both F and W)

Dataset	PCA	LeeNMF	NeNMF	SNMTF
Faces	(0, 0.26, 0.24)	(0.58, 0, 0.55)	(7.64, **60.33**, 10.23)	(**9.78**, 59.98, **12.24**)
Digits	(1.37, 0, 0.02)	(31.24, 50.47, 31.49)	(41.20, 92.49, 41.86)	(**50.49**, **93.47**, **51.04**)
Tiny Images	(0, 0, 0)	(0.02, 0, 0.02)	(9.97, **86.58**, 14.40)	(**11.71**, 85.29, **15.97**)

6   Conclusion

This paper proposes a new model of NMF as SNMTF with L_2 regularizations, which has more concise interpretability of the role of latent components over instances and attributes over latent components while keeping the generalization in comparison with NMF. We design a fast parallel algorithm with guaranteed convergence, low iteration complexity, and easily controlled sparsity to learn SNMTF, which is derived from Frank-Wolfe algorithm [7]. Furthermore, the proposed algorithm is convenient to massively parallelize, and control sparsity of both new representation F and model W. Based on the experiments, the new model and the proposed algorithm outperform the NMF model and its state-of-the-art algorithms in three significant aspects of convergence, classification, and sparsity. Therefore, we strongly believe that SNMTF is highly potential for many applications and extensible for nonnegative tensor factorization.

Acknowledgments. This work was supported by Asian Office of Aerospace R&D under agreement number FA2386-15-1-4006.

References

1. Cichocki, A., Zdunek, R., Phan, A.H., Amari, S.I.: Nonnegative Matrix and Tensor Factorizations: Applications to Exploratory Multi-way Data Analysis and Blind Source Separation. Wiley, Hoboken (2009)
2. Friedman, J., Hastie, T., Tibshirani, R.: The Elements of Statistical Learning. Springer Series in Statistics, vol. 1. Springer, New York (2001)
3. Friedman, J.H.: Greedy function approximation: a gradient boosting machine. Ann. Stat. **29**(5), 1189–1232 (2001)
4. Guan, N., Tao, D., Luo, Z., Yuan, B.: Nenmf: an optimal gradient method for nonnegative matrix factorization. IEEE Trans. Signal Process. **60**(6), 2882–2898 (2012)
5. Hofmann, T.: Probabilistic latent semantic indexing. In: Proceedings of the 22nd Annual International ACM SIGIR Conference on Research and Development in Information Retrieval, pp. 50–57. ACM (1999)
6. Jaggi, M.: Revisiting frank-wolfe: projection-free sparse convex optimization. In: Proceedings of the 30th International Conference on Machine Learning (ICML 2013), pp. 427–435 (2013)
7. Lacoste-Julien, S., Jaggi, M.: An affine invariant linear convergence analysis for frank-wolfe algorithms (2013). arXiv preprint arXiv:1312.7864
8. Lee, D.D., Seung, H.S.: Algorithms for non-negative matrix factorization. In: Advances in Neural Information Processing Systems, pp. 556–562 (2001)
9. Wang, Y.X., Zhang, Y.J.: Nonnegative matrix factorization: a comprehensive review. IEEE Trans. Knowl. Data Eng. **25**(6), 1336–1353 (2013)

Active Consensus-Based Semi-supervised Growing Neural Gas

Vinícius R. Máximo[1], Mariá C.V. Nascimento[1], Fabricio A. Breve[2], and Marcos G. Quiles[1(✉)]

[1] Federal University of São Paulo (UNIFESP), São José dos Campos, Brazil
{maximo,mcv.nascimento,quiles}@unifesp.br
[2] São Paulo State University (UNESP), Rio Claro, SP, Brazil
fabricio@rc.unesp.br

Abstract. In this paper, we propose a new active semi-supervised growing neural gas (GNG) model, named *Active Consensus-Based Semi-Supervised* GNG, or ACSSGNG. This model extends the former CSS-GNG model by introducing an active mechanism for querying more representative samples in comparison to a random, or passive, selection. Moreover, as a semi-supervised model, the ACSSGNG takes both labelled and unlabelled samples in the training procedure. In comparison to other adaptations of the GNG to semi-supervised classification, the ACSSGNG does not assign a single scalar label value to each neuron. Instead, a vector containing the representativeness level of each class is associated with each neuron. Here, this information is used to select which sample the specialist might label instead of using a random selection of samples. Computer experiments show that our model can deliver, on average, better classification results than state-of-art semi-supervised algorithms, including the CSSGNG.

1 Introduction

Learning in real scenarios demands computer models that can deal with both labelled and unlabelled data. Usually, unlabelled samples are less costly to obtain than labelled data, which are generally more expensive. Thus, semi-supervised learning (SSL) becomes an interesting and cheaper approach in situations where only a small part of the data can be labelled [1–3].

A number of unsupervised problems has been effectively approached by machine learning techniques in the literature. In particular, topological maps approaches, as, for example, the Kohonen Self-Organizing Maps (SOM) [4] and the Growing Neural Gas (GNG) were widely employed to map high-dimensional features into a low dimensional space. For this, the so-called incremental network GNG works by learning the intrinsic topological features of a dataset, being particularly attractive for non-stationary learning settings. The main reason behind is the self-adjustment of the GNG network that happens during its learning stage, when neurons and links can be removed, unlike the SOM networks [5].

© Springer International Publishing AG 2016
A. Hirose et al. (Eds.): ICONIP 2016, Part II, LNCS 9948, pp. 126–135, 2016.
DOI: 10.1007/978-3-319-46672-9_15

Even though the GNG technique has been proposed to approach unsupervised problems, it can be used for supervised and semi-supervised data classification. In these cases, Heiken and Hamker [6] showed that besides being robust in considering parameter changes, its performance in comparison to other benchmark classifiers is superior.

Zaki and Yin [7] suggested the SSGNG (Semi-Supervised Growing Neural Gas) for approaching the semi-supervised classification problem. Beyer and Cimiano [8] proposed the OSSGNG algorithm (On-line Semi-Supervised Growing Neural Gas) that different from SSGNG, it does not keep records of the samples in the learning stage. OSSGNG then works in an online form with the samples. These techniques can achieve a good accuracy on classifying data, however, they can be not indicated in real problems. This happens because of the high time required for the classification by label propagation approaches.

Recently, a new GNG model for semi-supervised learning was proposed in [9], named *Consensus-Based Semi-Supervised* GNG, or CSSGNG. In contrast to the prior approaches, by integrating a consensus mechanism and a pertinence vector to store the class information, the CSSGNG has shown superior results than other GNG models for SSL.

Here, by using the information stored in the pertinence vector, which allow us to observe the classification confidence of each neuron or the level of representativeness of each class, we propose a new model, named Active Consensus-Based Semi-Supervised GNG, or ACSSGNG, which incorporates an active selection process into the learning algorithm. Thus, instead of randomly selecting labelled samples from the dataset, our model can query the specialist for information and label only the most representative samples. Thus, the ACSSGNG can learn using a reduced number of labelled samples in contrast to the CSSGNG, moreover, as it will be demonstrated later, the ACSSGNG can achieve higher accuracy than several state-of-art SSL algorithms, including the CSSGNG model [9].

This paper is organized as follows. Section 2 revisits the CSSGNG and introduces the ACSSGNG. Section 3 describes our experiments and results. Section 4 offers a few concluding remarks.

2   The ACSSGNG Model

The original Growing Neural Gas (GNG) is a competitive self-organizing and incremental neural network proposed by Fritzke [5]. The GNG algorithm can be summarized as follows: the network starts with two neurons randomly inserted into the feature space; iteratively a sample x_i is selected and presented to the network; through a competitive learning process, two neurons, named s_1 and s_2, are selected. These two neurons are, respectively, the first and the second winners. Between this pair of neurons, a connection is (re)established. Afterward, the weights of the winner neuron and its neighbourhood, according to the network, are updated leading to the movement of those neurons towards the position of the input signal in the feature space.

After each λ iterations, a new neuron is added to the network between the linked neurons with the largest error values. Neurons can also be removed whether they are no longer activated by external samples. For a complete overview of the original GNG model, see [5].

The GNG model and its variations have been extended to solve several problems, such as clustering data streams [10], prototype generation [11], semi-supervised learning [7–9], among others [12].

2.1 Revisiting the CSSGNG

The CSSGNG proposed in [9] consists of a GNG network trained with both labeled and unlabeled samples. In addition to the weight vector that represents the position of a neuron into the feature space, each neuron also has a *pertinence vector* which defines its classification ability. It means, the pertinence vector stores the pertinence of neurons regarding all the classes of the problem.

Formally, let $c_j = \{c_{j,1}, c_{j,2}, ..., c_{j,t}\}$ be the pertinence vector of neuron j, in which each scalar $c_{j,l}$ defines the representativeness level of each class l assigned to neuron j. If we normalize the vector c_j in order to make $\sum_{l \in L} c_{j,l} = 1$, for each sample x_i represented by neuron j, $c_{j,l} = p(x_i|l)$ for all class $l \in L$. For assigning a label to a sample x_i, consider its winner neuron j, i.e., $y_i = \arg\max_{l \in L}(c_{j,l})$.

The learning phase of the CSSGNG follows the standard unsupervised GNG algorithm to evolve the network. However, to perform data classification, the authors have introduced two major modifications into the original GNG algorithm: a mechanism to absorb the information from labelled samples and a consensus mechanism responsible for propagating the labels between neighbour neurons.

Specifically, the absorption mechanism is responsible for setting the values into the pertinence vector of the winner neuron. After the label absorption, the consensus mechanism spread the label information through the network in order to reach an equilibrium, or agreement, between neighbour neurons. It means the pertinence vector associated with each neuron is not set only by the label information provided by samples that activate this neuron. It is the result of the combination of the local information provided by the labeled samples with the neighbourhood information, thus, a consensus by the whole of neighbour neurons. A complete description of the CSSGNG can be found in [9].

2.2 Incorporation of the Active Learning

Instead of randomly labelling samples from the pool of data, the active learning allows the learning algorithm to, interactively, choose the samples to be labelled. It aims to improve the learning by using more representative samples from the pool. By taking the active learning into account, we can reduce the number of labelled samples necessary to achieve a good classification accuracy or even reach a better accuracy using the same amount of samples randomly selected from the pool.

In this context, we observe that the information stored in pertinence vector introduced in the CSSGNG model can be used to indicate which neuron should be labelled. By analysing the $c_{j,\hat{l}} = \max(c_{j,l})$ that represents the most representative class of neuron j, we can evaluate how confident a neuron is regarding to the classes it represents. Whether a small value is set into the pertinence vector, it means that this neuron has received low information about the data classes it might represent. Thus, we can infer that or this neuron is located far from labelled samples or it close to a border region. In this last case, a competition between distinct classes can reduce the levels stores in the pertinence vector.

By using the pertinence vector information, we can actively select neurons using a method named *Least confident sampling*. This procedure returns the neurons less confident regarding to the classes they represent. Formally, the less confident neuron is defined as $\hat{j}$, accordingly to the following equation:

$$\mathbf{c}_{\hat{j},\hat{l}} = \min_{j \in V}(\mathbf{c}_{j,\hat{l}}) \tag{1}$$

where V represents the set of all neurons and $\mathbf{c}_{\hat{j},\hat{l}}$ informs, among the most representative class assign to neurons, the smallest value. It means the least confident neuron. Next, using this information, the sample closest to this neuron is selected to be labelled by the specialist.

This selection of neuron $\hat{j}$ might be carried out in order to select representative neurons. Thus, it is necessary to store the frequency in which a neuron $\hat{j}$ is selected to avoid the selection of several samples in the same region. This value is stored as p_j. This behaviour provides a space distribution of the labelled samples allowing a better learning process.

The active selection is implemented as follows. First, we define a set P that represents the possible candidates $\hat{j}$. All neurons $j \in P$ must have the same value associated to p_j, which, in fact, must represent the smallest values p_j of the network, it means, $P = \{j \in V \mid \hat{p}_j = \arg\min(p_j)\}$. This method provides good distributions of the labelled samples. However, it should be stated that other selection methods might be considered.

Thus, the active selection is performed based on Eq. (2), which represents how the neurons $hatj$ are picked.

$$\mathbf{c}_{\hat{j},\hat{l}} = \min_{j \in P}(\mathbf{c}_{j,\hat{l}}) \tag{2}$$

To incorporate the active selection into the CSSGNG, we first evolve the network topology following the unsupervised GNG; next, for each sample to be labelled, the algorithm select the sample x_i closest to neuron $\hat{j}$ and query the specialist regarding its label; finally, the absorption and the consensus mechanism are iterated until the network stabilize (φ steps). The ACSSGNG is summarized in Algorithm 1.

It is worth noting that, after the unsupervised phase, the network topology is already established, however, no information about labels has been provided. It means, all neurons have the same absence of confidence. Thus, the selection of the first neuron if performed according to its centrality measure, computed as follows:

$$\hat{j} = \arg\min_{j \in V}(\max_{i \in V}(d_{ij})) \tag{3}$$

Algorithm 1. The ACSSGNG Algorithm

Input: Set of input data
1 Evolve the network topology;
2 **foreach** *sample to be labeled* **do**
3   Find the neuron $\hat{j}$ accordingly to Eq. 2;
4   Label the sample x_i closest to neuron $\hat{j}$;
5   **for** $it \rightarrow \varphi$ **do**
6   Run the Absorption mechanism [9];
7   Run the Consensus mechanism [9];
8   **end**
9 **end**

---

in which d_{ij} represents the shortest path between neurons i and j. After that, when the network starts absorbing the label information, the model evolves following Algorithm 1.

3   Computer Experiments and Results

In this section, we present a comprehensive evaluation of our experiments and results. We commence by giving a brief overview of the datasets taken into account in our experiments. First, we considered nine versions of UCI datasets using the same methodology used in other semi-supervised studies [13–15]. These datasets are shown in Table 1. Specifically, each dataset was equally divided into two parts, one for training and the other for testing. Each training set contains only ten labeled samples. Second, for the sake of completeness, we also carry out experiments using the benchmark datasets proposed in [1]. The Chapelle's datasets are summarized in Table 2.

Table 1. UCI datasets

Dataset	Class	Dimension	Samples	Type
House	2	16	232	Real
Heart	2	9	270	Real
Vehicle	2	16	435	Real
Wdbc	2	14	569	Real
Isolet	2	51	600	Real
Austra	2	14	690	Real
Optdigits	2	42	1143	Real
Ethn	2	30	2630	Real
Sat	2	36	3041	Real

Table 2. Datasets from [1]

Dataset	Class	Dimension	Samples	Type
g241c	2	241	1500	Synthetic
g241n	2	241	1500	Synthetic
Digit1	2	241	1500	Synthetic
USPS	2	241	1500	Real
COIL	6	241	1500	Real
BCI	2	117	400	Real
TEXT	2	11960	1500	Real

3.1   Inductive Classification

In all the computer experiments, the following parameters were used: $\lambda = (1000)$, $a_{max} = 120$, $\varepsilon_b = 0.04$, $\varepsilon_n = 10^{-3}$, $\alpha = 0.5$, $\beta = 0.995$, $\gamma = 0.85$, $\kappa = 3$, and $\phi = 2$. These parameters were fine-tuned empirically, albeit, based on our fidings and as reported in [6,9,16] for the original GNG parameters, these values are not sensitive and do not need further adjustment for each scenario.

Before presenting the results using the datasets described in Tables 1 and 2, we run our model with a toy dataset shown in Fig. 1(a). This dataset is composed of 500 samples divided into five Gaussian distributions. After the unsupervised phase responsible for learning the neural network structure, the supervised phase takes places. Whether the passive selection is used, all samples have the same probability of being selected. Thus, to obtain a sufficient number of representative samples, we need to label several samples. On the other hand, by using the active selection process (Algorithm 1), the number of labelled samples can be reduced.

To illustrate the advantages of the Active Selection (ACSSGNG) in contrast to its Passive Selection version (CSSGNG), we conduct an experimental evaluation of both models by varying the number of labelled samples. The results are illustrated in Fig. 1(d). When a small number of labelled samples is considered, the ACSSGNG delivers a much higher accuracy in comparison to the CSSGNG, which demonstrates its superiority and advantages in real scenarios.

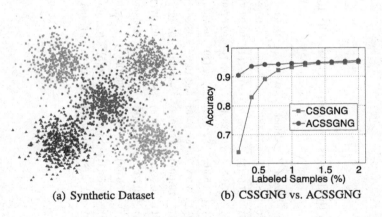

(a) Synthetic Dataset   (b) CSSGNG vs. ACSSGNG

Fig. 1. Illustration of the ACSSGNG versus the CSSGNG (toy scenario).

For each dataset shown in Table 1, the samples were randomly divided into two equal parts using the same approach considered in [13,15]. The first part was used to train the ACSSGNG network, the second part was used to test the accuracy of the network. By selecting only 10 labelled samples (randomly and actively), we have performed 500 realizations with each dataset shown in Table 1.

Table 3. Accuracy rate for the datasets described in Table 1 with 10 labelled samples. The number in parenthesis indicates the algorithm ranking.

Method	House	Heart	Vehicle	Wdbc	Isolet	Austra	Optdigits	Ethn	Sat	Avg.	Avg. rank
LapRLS [14]	87.90(10)	78.11(2)	72.53(7)	89.59(2)	93.92(8)	75.68(2)	98.75(6)	73.51(5)	98.67(8)	85.41	5.56
LapSVM [13]	89.95(6)	77.96(3)	71.38(9)	91.07(1)	93.93(7)	74.38(3)	98.34(7)	74.60(3)	99.12(7)	85.64	5.11
LDS [13]	89.35(8)	77.11(5)	66.28(10)	85.07(7)	92.07(9)	66.00(10)	96.4(8)	67.16(9)	94.20(10)	81.52	8.44
means3vm-iter [15]	91.72(3)	74.56(6)	82.47(4)	79.39(10)	98.75(3)	68.12(8)	98.93(5)	73.21(6)	99.56(3)	85.19	5.33
means3vm-mkl [15]	91.90(2)	73.22(7)	82.15(5)	80.19(9)	98.98(2)	67.59(9)	99.09(3)	73.57(4)	99.56(2)	85.14	4.78
SB-SVM [13]	90.65(5)	79.00(1)	72.29(8)	88.82(4)	95.12(6)	71.36(7)	96.35(9)	67.57(8)	87.71(11)	83.21	6.56
SSCCM [14]	92.58(1)	70.72(9)	83.58(3)	80.44(8)	99.67(1)	72.49(6)	99.34(2)	71.53(7)	99.41(4)	85.53	4.56
SVM [13]	91.16(4)	70.59(10)	78.28(6)	75.74(11)	89.58(11)	65.64(11)	90.31(11)	67.04(10)	99.13(6)	80.83	8.89
TSVM [13]	86.55(11)	77.63(4)	63.62(11)	86.40(6)	90.38(10)	73.38(5)	92.34(10)	54.69(11)	98.26(9)	80.36	8.56
CSSGNG	87.92(9)	67.53(11)	88.15(2)	88.00(5)	95.58(5)	73.99(4)	99.02(4)	82.48(2)	99.31(5)	86.89	5.22
ACSSGNG	89.50(7)	71.01(8)	**91.74(1)**	89.53(3)	97.60(4)	**76.36(1)**	**99.54(1)**	**89.37(1)**	**99.77(1)**	**89.38**	**3.00**

Table 4. Accuracy rate for the datasets described in Table 2 with 10 labelled samples. The number in parenthesis indicates the algorithm ranking.

	G241c	G241d	Digit1	USPS	COIL	BCI	TEXT	Avg.	Avg. rank
1-NN [1]	55.95(10)	56.78(8)	76.53(11)	80.18(9)	34.09(5)	51.26(8)	60.56(11)	59.34	8.86
Cluster-Kernel [1]	51.72(13)	57.95(4)	81.27(9)	80.59(7)	32.68(6)	51.69(4)	57.28(12)	59.03	7.86
LapRLS [1]	56.05(8)	54.32(10)	94.56(1)	81.01(4)	—	51.03(9)	66.32(7)	—	6.50
LapSVM [1]	53.79(11)	54.85(9)	91.03(2)	80.95(5)	—	50.75(11)	62.72(9)	—	7.83
LDS [1]	71.15(5)	49.37(13)	84.37(5)	82.43(3)	38.10(2)	50.73(12)	72.85(1)	64.14	5.86
means3vm-iter [15]	72.22(4)	57.00(5)	82.98(7)	76.34(12)	—	51.88(3)	69.57(4)	—	5.83
means3vm-mkl [15]	65.48(7)	58.94(3)	83.00(6)	77.84(11)	—	52.07(2)	66.91(6)	—	5.83
OCM [17]	55.95(9)	56.78(7)	76.42(12)	80.51(8)	35.50(4)	51.26(7)	60.88(10)	59.61	8.14
SSCCM [14]	66.99(6)	56.91(6)	79.17(10)	80.88(6)	—	54.04(1)	71.05(2)	—	5.17
SVM [1]	52.68(12)	53.34(11)	69.40(13)	79.97(10)	31.64(8)	50.15(13)	54.63(13)	55.97	11.43
TSVM [1]	75.29(2)	49.92(12)	82.23(8)	74.80(13)	32.50(7)	50.85(10)	68.79(5)	62.05	8.14
CSSGNG	75.20(3)	74.79(2)	87.37(4)	84.18(2)	36.59(3)	51.32(6)	65.52(8)	67.85	4.00
ACSSGNG	**81.09(1)**	**85.89(1)**	89.87(3)	**89.27(1)**	**45.19(1)**	51.65(5)	69.68(3)	**73.23**	**2.14**

The obtained results are depicted in Table 3. The results highlight the advantages of the ACSSGNG over several state-of-art models, in which the ACSSGNG achieved the best results in 5 of 9 datasets considered in our experiments. Moreover, the ACSSGNG delivered the best average result and the best average ranking in contrast to all the other models taken into account.

Regarding the experiments using the Chapelle's datasets (Table 2), we considered the same methodology proposed in [1]. Specifically, two scenarios were considered: (a) 10 samples of each dataset were labelled and (b) 100 samples of each dataset were labelled. For each dataset, the same 12 *folds* provided by [1] were used, which enables a straight comparison of our results to those introduced in [1].

Results with the 10−labelled and 100−labelled samples configurations are qualitative similar, thus, we show only the first configuration (10−labelled), which is depicted in Table 4. It can be seen that our model achieved results superior or equivalent to the SSGNG and the OSSGNG models in all datasets, except to the BCI dataset. Again, the ACSSGNG achieved, on average, the best results and rank in contrast to all the competitors.

4 Conclusions

We have proposed a new active semi-supervised model based on the Growing Neural Gas model. In contrast to former GNG-Based SSL models in which a single scalar was used to assign a label to each neuron, here a pertinence vector has been considered to store the representativeness level of each class of the problem. Thus, more information about the classes can be retrieved from each neuron.

By using this information stored in the pertinence vector, we have proposed a novel active selection method, which allows the selection of the most representative samples to be queried to the specialist instead of using a random selection of samples.

Finally, the ACSSGNG has delivered higher accuracy than the CSSGNG in all experiments conducted in this study, which, indeed, demonstrated its robustness and advantages over the CSSGNG and several state-of-art semi-supervised models. Moreover, with the active selection, a small number of labelled samples can be used to train the classifier. As a future work, we plan to adapt the ACSSGNG to work in an on-line fashion.

Acknowledgments. The authors would like to thank the CNPq and FAPESP (2011/18496-7, 2011/17396-9, and 2015/21660-4), for financial support.

References

1. Chapelle, O., Schölkopf, B., Zien, A. (eds.): Semi-supervised Learning. The MIT Press, Cambridge (2006)
2. Zhou, Z.H., Li, M.: Semi-supervised learning by disagreement. Knowl. Inf. Syst. **24**(3), 415–439 (2010)
3. Quiles, M.G., Zhao, L., Breve, F.A., Rocha, A.: Label propagation through neuronal synchrony. In: IJCNN 2010, pp. 2517–2524. IEEE Press, Barcelona (2010)
4. Kohonen, T.: Self-organizing Maps. Springer Series in Information Sciences, 3rd edn. Springer, Heidelberg (2000)
5. Fritzke, B.: A growing neural gas network learns topologies. In: Advances in Neural Information Processing Systems 7, pp. 625–632. MIT Press (1995)
6. Heinke, D., Hamker, F.H.: Comparing neural networks: a benchmark on growing neural gas, growing cell structures, and fuzzy artmap. IEEE TNN **9**(6), 1279–1291 (1998)
7. Mohd Zaki, S., Yin, H.: Semi-supervised growing neural gas for face recognition. In: Fyfe, C., Kim, D., Lee, S.-Y., Yin, H. (eds.) IDEAL 2008. LNCS, vol. 5326, pp. 525–532. Springer, Heidelberg (2008)
8. Beyer, O., Cimiano, P.: Online semi-supervised growing neural gas. Int. J. Neural Syst. **22**(5), 1250023 (2012)
9. Maximo, V., Quiles, M.G., Nascimento, M.: A consensus-based semi-supervised growing neural gas. In: IJCNN 2014, pp. 2019–2026. IEEE Press, Beijing (2014)
10. Ghesmoune, M., Lebbah, M., Azzag, H.: A new growing neural gas for clustering data streams. Neural Netw. **78**, 36–50 (2016)
11. Dias, J., Quiles, M.G., Lorena, A.C.: Using growing neural gas in prototype generation for nearest neighbor classifiers. In: Arik, S., Huang, T., Lai, W.K., Liu, Q. (eds.) ICONIP 2015. LNCS, vol. 9490, pp. 276–283. Springer, Heidelberg (2015). doi:10.1007/978-3-319-26535-3_32
12. Escalante, D.A.C., Taubin, G., Nonato, L.G., Goldenstein, S.K.: Using unsupervised learning for graph construction in semi-supervised learning with graphs. In: SIBGRAPI 2013, pp. 24–30. IEEE Press, Arequipa (2013)
13. Mallapragada, P.K., Jin, R., Jain, A., Liu, Y.: Semiboost: boosting for semi-supervised learning. IEEE TPAMI **31**(11), 2000–2014 (2009)
14. Wang, Y., Chen, S., Zhou, Z.H.: New semi-supervised classification method based on modified cluster assumption. IEEE TNNLS **23**(5), 689–702 (2012)
15. Li, Y.F., Kwok, J.T., Zhou, Z.H.: Semi-supervised learning using label mean. In: 26th ICML, pp. 633–640. ACM, Montreal (2009)
16. Holmström, J.: Growing neural gas. Experiments with GNG, GNG with utility and supervised GNG, Uppsala Master Thesis in Computer Science, Uppsala University Department of Information Technology (2002)
17. Quiles, M.G., Basgalupp, M.P., Barros, R.: An oscillatory correlation model for semi-supervised classification. Learn. Nonlinear Models **11**, 3–10 (2013)

Kernel L1-Minimization: Application to Kernel Sparse Representation Based Classification

Anupriya Gogna and Angshul Majumdar[✉]

Indraprastha Institute of Information Technology, Delhi, India
{anupriyag, angshul}@iiitd.ac.in

Abstract. The sparse representation based classification (SRC) was initially proposed for face recognition problems. However, SRC was found to excel in a variety of classification tasks. There have been many extensions to SRC, of which group SRC, kernel SRC being the prominent ones. Prior methods in kernel SRC used greedy methods like Orthogonal Matching Pursuit (OMP). It is well known that for solving a sparse recovery problem, both in theory and in practice, l_1-minimization is a better approach compared to OMP. The standard l_1-minimization is a solved problem. For the first time in this work, we propose a technique for Kernel l_1-minimization. Through simulation results we show that our proposed method outperforms prior kernelised greedy sparse recovery techniques.

Keywords: L_1-minimization · Kernel machine · Sparse classification

1 Introduction

In sparse recovery, the problem is to find a solution to the linear inverse problem

$$y = Ax + n \tag{1}$$

where, y is the observation, A is the system matrix, x is the solution and n is the noise assumed to be Normally distributed. The solution x is sparse, i.e. it is assumed to be have only 'k' non-zeroes. Such a problem arises in machine learning and signal processing; in fact there is a branch of signal processing called Compressed Sensing that evolves around the solution of such problems.

The exact solution to (1) is NP hard [1] and is expressed as,

$$\min_x \|y - Ax\|_2^2 \text{ such that } \|x\|_0 = k \tag{2}$$

Here the l_0-norm (not exactly a norm in the strictest sense of the term) simply counts the number of non-zeroes in the vector. There are two approaches to solve (2) – the first one is a greedy approach, where the support of x is iteratively detected and the corresponding values are estimated. The orthogonal matching pursuit (OMP) [2] is the most popular greedy technique. There are several extensions to the basic OMP approach like the stagewise orthogonal matching pursuit and the CoSamp.

© Springer International Publishing AG 2016
A. Hirose et al. (Eds.): ICONIP 2016, Part II, LNCS 9948, pp. 136–143, 2016.
DOI: 10.1007/978-3-319-46672-9_16

However OMP is fraught with several limitations. First, the guarantees are only probabilistic; besides several strict assumptions need to be made regarding the nature of the system matrix 'A' in order for OMP to succeed (theoretically). Both in theory and in practice, a much better way to solve the sparse recovery problem is to relax the NP hard l_0-minimization problem by its closest convex surrogate the l_1-norm. This is expressed as,

$$\min_x \|y - Ax\|_2^2 \text{ such that } \|x\|_1 \leq \tau \tag{3}$$

This formulation was first proposed in Tibshirani's paper on LASSO [3]. Although convex, this (3) is a constrained optimization problem and is hard to solve; hence in [3], the unconstrained version was solved instead.

$$\min_x \|y - Ax\|_2^2 + \lambda \|x\|_1 \tag{4}$$

This is a quadratic programming problem and can be solved efficiently using iterative soft thresholding [4].

This formulation (4) is a typical linear regression problem. In this work, we are interested in kernel regression/classification problems. A typical non-linear regression is expressed as,

$$y = \varphi(A)x + n \tag{5}$$

Here the output is expressed as a linear combination of a non-linear system matrix. The Tikhonov regularized solution of (5) has a closed form solution via the kernel trick.

In this work we are interested in solving problems where a non-linear combination of the output can be expressed as linear combination of non-linear inputs, i.e.,

$$\varphi(y) = \varphi(A)x + n \tag{6}$$

Here both the input (A) and the output (y) are of non-linear forms. Such a problem does not arise in regression, where the problem is to predict the output (5) and not a non-linear version of the output (6), but it does arise in kernel sparse representation based classification [5–7]; these studies were based on modifying the OMP algorithm. Issues arising in the linear version of the OMP also persists in the non-linear version. A better approach would be modify the l_1-minimization algorithm to support kernels. This is the topic of this paper.

2   Brief Review on Sparse Representation Based Classification

The SRC assumes that the training samples of a particular class approximately form a linear basis for a new test sample belonging to the same class. One can write the aforesaid assumption formally. If x_{test} is the test sample belonging to the kth class then,

$$x_{test} = \alpha_{c,1}x_{c,1} + \alpha_{c,2}x_{c,2} + \ldots + \alpha_{c,n_k}x_{c,n_k} + n \tag{7}$$

where $x_{c,i}$ are the training samples and η is the approximation error.

In a classification problem, the training samples and their class labels are provided. The task is to assign the given test sample with the correct class label. This requires finding the coefficients $\alpha_{c,i}$ in Eq. (8). Equation (8) expresses the assumption in terms of the training samples of a single class. Alternately, it can be expressed in terms of all the training samples so that

$$x_{test} = X\alpha + n \tag{8}$$

$$whereX = [x_{1,1}|\ldots|x_{n,1}|\ldots|x_{c,1}|\ldots|x_{c,n_c}|\ldots x_{C,1}|\ldots|x_{C,n_c}] \quad \text{and}$$

$\alpha = [\alpha_{1,1}\ldots\alpha_{1,n_1}\ldots\alpha_{c,1}\ldots\alpha_{c,n_c}\ldots\alpha_{C,1}\ldots\alpha_{C,n_c}]^T$.

According to the SRC assumption, only those α's corresponding to the correct class will be non-zeroes. The rest are all zeroes. In other words, α will be sparse. Therefore, one needs to solve the inverse problem (8) with sparsity constraints on the solution. This is formulated as:

$$\min_{\alpha}\|x_{test} - X\alpha\|_2^2 + \lambda\|x\|_1 \tag{9}$$

Once (9) is solved, the representative sample for every class is computed: $x_{rep}(c) = \sum_{j=1}^{n_c}\alpha_{c,j}x_{c,j}$. It is assumed that the test sample will look very similar to the representative sample of the correct class and will look very similar, hence the residual $\varepsilon(c) = \|x_{test} - x_{rep}(c)\|_2^2$, will be the least for the correct class. Therefore once the residual for every class is obtained, the test sample is assigned to the class having the minimum residual.

There are several extensions to the basic SRC; in its pristine form it is an unsupervised approach – it does not utilize information about the class labels. In [8–10] it was argued that α is supposed to be non-zero for all training samples corresponding to the correct class. The SRC assumes that the training samples for the correct class will be automatically selected by imposing the sparsity inducing l_1-norm; it does not explicitly impose the constraint that if one class is selected, all the training samples corresponding to that class should have corresponding non-zero values in α. It was claimed in [2–4] that better recovery can be obtained if selection of all the training samples within the class is enforced. This was achieved by employing a supervised $l_{2,1}$-norm instead of the l_1-norm.

$$\min_{\alpha}\|x_{test} - X\alpha\|_2^2 + \lambda\|\alpha\|_{2,1} \tag{10}$$

where the mixed norm is defined as $\|\alpha\|_{2,1} = \sum_{k=1}^{c}\|\alpha_k\|_2$.

The inner l_2-norm enforces selection of all the training samples within the class, but the sum-of- l_2-norm over the classes acts as an l_1-norm over the selection of classes and selects very few classes. The block sparsity promoting $l_{2,1}$-norm ensures that if a class is selected, ALL the training samples within the class are used to represent the test sample.

A recent addition to the suite of sparse representation based classifiers is the group sparse representation based classifier [11]. This is a generalization of all of the above that can handle multiple kinds of datasets (like multi-modal biometrics) and multiple types of features in a single framework.

Several studies independently proposed the Kernel Sparse Representation based Classification (KSRC) approach [5–7]. KSRC is a simple extension of the SRC using the Kernel trick. The assumption here is that the non-linear function of the test-sample can be represented as a linear combination of the non-linear functions of the training samples, i.e.

$$\phi(x_{test}) = \phi(X)\alpha + n \tag{11}$$

Here $\phi(.)$ represents a non-linear function. As mentioned before, the prior studies solved this problems by modifying the Orthogonal Matching Pursuit.

3 Proposed Approach

3.1 L_1-Minimization

First we will study the vanilla implementation of iterative soft thresholding algorithm. The goal is to solve (9). The derivation can be followed from [12]. The algorithm is given as follows.

Initialize: $x_0 = \min_{x} \|y - Ax\|_2^2$

Continue till convergence

Landweber Iteration – $b = x_{k-1} + \dfrac{1}{a} A^T (y - Ax_{k-1})$

Soft thresholding – $x_k = signum(b) \max\left(0, |b| - \dfrac{\lambda}{2a}\right)$

Here the step-size 'a' is the maximum Eigenvalue of $A^T A$. The iterations converge when the objective function or the value of x does not change significantly over successive iterations.

With a slight modification, one can have an iterative hard thresholding algorithm [13]. The only difference between the hard and soft thresholding algorithm is the thresholding step. In the hard thresholding only those values are kept that are greater than a pre-defined threshold. Such an algorithm is supposed to approximately solve the l0-minimization problem. In practice, it does not yield very good results.

3.2 Kernel L_1-Minimization

Here we are interested in solving (6). We repeat it for the sake of convenience.

$$\varphi(y) = \varphi(A)x + n$$

If we write down the soft thresholding algorithm for the same, we get

---

Initialize: $x_0 = \min_x \|\varphi(y) - \varphi(A)x\|_2^2$

Continue till convergence

Landweber Iteration $- b = x_{k-1} + \dfrac{1}{a}\varphi(A)^T(\varphi(y) - \varphi(A)x_{k-1})$

Soft thresholding $- x_k = signum(b)\max\left(0, |b| - \dfrac{\lambda}{2a}\right)$

First, let us look at the initialization. The normal equations are of the form,

$$(\varphi(A)^T\varphi(A))x_0 = \varphi(A)^T\varphi(y) \tag{12}$$

One can easily identify the kernels: $K(A, A) = \varphi(A)^T\varphi(A)$ and $K(A, y) = \varphi(A)^T\varphi(y)$. With the kernel trick, we can express (12) as,

$$K(A, A)x_0 = K(A, Y) \Rightarrow x_0 = K(A, A)^{-1}K(A, Y) \tag{13}$$

The inversion is guaranteed by the positive definiteness of the kernel.

Now, we look at the Landweber iteration step. One can easily see that, it can be expressed as $b = x_{k-1} + \frac{1}{a}(\varphi(A)^T\varphi(y) - \varphi(A)^T\varphi(A)x_{k-1})$. Identifying the kernels, this is represented as,

$$b = x_{k-1} + \frac{1}{a}(K(A, y) - K(A, A)x_{k-1}) \tag{14}$$

The soft-thresholding step does not require any change.

4   Experimental Evaluation

Once the sparse recovery problem is solved, the residual error needs to be expressed in terms of kernels. This is easily done (keeping the same notation for SRC as before).

$$\begin{aligned}
\varepsilon(c) &= \|\varphi(x_{test}) - \varphi(x_{rep}(c))\|_2^2 \\
&= (\varphi(x_{test}) - \varphi(x_{rep}(c)))^T(\varphi(x_{test}) - \varphi(x_{rep}(c))) \\
&= \varphi(x_{test})^T\varphi(x_{test}) + \varphi(x_{rep}(c))^T\varphi(x_{rep}(c)) - \varphi(x_{test})^T\varphi(x_{rep}(c)) - \varphi(x_{rep}(c))^T\varphi(x_{test}) \\
&= K(x_{test}, x_{test}) + K(x_{rep}(c), x_{rep}(c)) - 2K(x_{rep}(c), x_{test})
\end{aligned}$$

4.1   Results on Benchmark Classification Tasks

In [5] the KSRC was tested on benchmark datasets from the UCI Machine Learning repository. We use the same datasets and follow the same experimental protocol here. No feature extraction or dimensionality reduction was applied on these datasets. We compare the KSRC formulation in [5] with ours; both of them use an RBF kernel. To benchmark, the results from SRC and SVM are also shown (Table 1).

Table 1. Error rate % on benchmark classification tasks

Dataset	SVM	SRC	KSRC [5]	Proposed
Breast	4.09	54.57	5.78	4.09
Glass	30.29	33.77	32.46	30.52
Heart	18.7	22.8	23.8	20.35
Hepatitis	34.51	45.49	38.04	34.51
Ionosphere	5.38	8.21	13.42	6.23
Iris	5.63	20	**4.79**	4.79
Liver	31.05	35.88	32.81	31.05
Musk	6.84	14.75	10	8.29
Pima	24.67	34	30.4	26.8
Sonar	12.9	23.48	12.46	12.09
Soy	4.35	11.65	**3.41**	3.41
Vehicle	18.5	**18.72**	22.96	18.72
Vote	5.59	7.11	7.04	5.92
Wdbc	2.7	6.4	3.44	3.4
Wine	0.86	2.41	1.55	1.55
Wpbd	20.31	26.46	26	22.52

One can see from the table that in most cases SVM outperforms the SRC based methods. However comparison within SRC and its variants show that our method always yields the best results. The prior formulation of KSRC [5] had a naive implementation, therefore even with the kernel trick it was unable to improve upon the SRC which benefitted from more sophisticated optimization algorithm. In this work, our proposed method enjoys the dual benefit of non-linear kernels and better optimization; hence the results always outperform prior techniques.

4.2   Results on Hyperspectral Image Classification

In [6] it was shown that the KSRC (based on KOMP) performed exceptionally well for hyperspectral image classification problems. In this work, we show that our proposed method improves upon the prior work.

We evaluate our proposed Hyperspectral Image Classification on – 1. Indian Pines dataset which has 200 spectral reflectance bands after removing bands covering the region of water absorption and 145*145 pixels of sixteen categories; and, 2. Pavia University dataset which has 103 bands of 340*610 pixels of nine categories.

The background i.e. Class 0 was excluded from the second dataset. For each dataset, we randomly select 10 % of the labelled data as training set and rest as testing set. Input consists of raw data of all the spectral channels pixel-wise.

In [6] a thorough study had been carried out by comparing KOMP based KSRC with SVM, SRC, KSRC etc. In [6] it was claimed that their KOMP based technique outperforms others. Therefore, in this work, we only need to show that our proposed method outperforms [6]. The results can be visualised from Fig. 1. One can see that our proposed method yields better results compared to the prior approach.

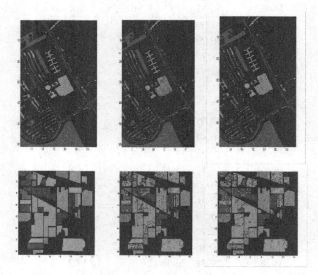

Fig. 1. Top: pavia university; bottom: indian pines. left to right: groundtruth, KOMP based KSRC [6] and Proposed.

5 Conclusion

In this work, we propose a technique for solving a kernel l_1-minimization problem. To the best of our knowledge this is the first work on this topic. We start with the vanilla implementation of the l_1-minimization problem via iterative soft thresholding and show how the kernel trick can be employed on it.

The proposed kernel l_1-minimization problem is employed here to solve the kernel sparse representation based classification problem. We have compared our proposed technique on two implementations of the same – [5, 6]. Experiments on benchmark classification datasets from the UCI machine learning repository show that our method is better than [5]. Evaluation of our proposed technique with [6] for hyperspectral imaging problems show that our method is also better than the kernel OMP based implementation [6].

In the future, we would like to extend this formulation to solve other variants of SRC like group sparse classification, robust sparse classification and robust group sparse representation based classification. We will also compare the proposed methods on a host of other real life problems.

References

1. Natarajan, B.: Sparse approximate solutions to linear systems. SIAM J. Comput. **24**, 227–234 (1995)
2. Tropp, J., Gilbert, A.C., Strauss, M.: Algorithms for simultaneous sparse approximations; Part I: greedy pursuit. Sig. Proc. **86**, 572–588 (2006). Special Issue on Sparse approximations in signal and image processing.
3. Tibshirani, R.: Regression shrinkage and selection via the lasso. J. Royal. Statist. Soc B. **58** (1), 267–288 (1996)
4. Daubechies, I., Defrise, M., De Mol, C.: An iterative thresholding algorithm for linear inverse problems with a sparsity constraint. Commun. Pure Appl. Math. **57**(11), 1413–1457 (2004)
5. Zhang, L., Zhou, W.-D., Chang, P.-C., Liu, J., Yan, Z., Wang, T., Li, F.-Z.: Kernel sparse representation-based classifier. IEEE Trans. Signal Process. **60**(4), 1684–1695 (2012)
6. Chen, Y., Nasrabadi, N., Tran, T.: Hyperspectral image classification via kernel sparse representation. IEEE Trans. Geosci. Remote Sens. **51**(1), 217–231 (2013)
7. Yin, J., Liu, Z., Jin, Z., Yang, W.: Kernel sparse representation based classification. Neurocomputing **77**(1), 120–128 (2012)
8. Majumdar, A., Ward, R.K.: Robust classifiers for data reduced via random projections. IEEE Trans. Syst. Man Cybern. B **40**(5), 1359–1371 (2010)
9. Yuan, X.T., Liu, X., Yan, S.: visual classification with multitask joint sparse representation. IEEE Trans. Image Process. **21**(10), 4349–4360 (2012)
10. Elhamifar, E., Vidal, R.: Robust Classification using Structured Sparse Representation. In: IEEE CVPR (2011)
11. Goswami, G., Mittal, P., Majumdar, A., Singh, R., Vatsa, M.: Group sparse representation based classification for multi-feature multimodal biometrics. Inf. Fus **32**(B), 3–12 (2016)
12. Sparse signal restoration. cnx.org/content/m32168/
13. Bredies, K., Lorenz, D.A.: Iterated hard shrinkage for minimization problems with sparsity constraints. SIAM J. Sci. Comput. **30**(2), 657–683 (2008)
14. Wright, J., Yang, A., Ganesh, A., Sastry, S., Ma, Y.: Robust face recognition via sparse representation. IEEE Trans. Pattern Anal. Mach. Intell. **31**(2), 210–227 (2009)

Nuclear Norm Regularized Randomized Neural Network

Anupriya Gogna and Angshul Majumdar$^{(\boxtimes)}$

Indraprasatha Institute of Information Technology, Delhi, India
{anupriyag,angshul}@iiitd.ac.in

Abstract. Extreme Learning Machine (ELM) or Randomized Neural Network (RNN) is a feedforward neural network where the network weights between the input and the hidden layer are not learned; they are assigned from some probability distribution. The weights between the hidden layer and the output targets are learnt. Neural networks are believed to mimic the human brain; it is well known that the brain is a redundant network. In this work we propose to explicitly model the redundancy of the human brain. We model redundancy as linear dependency of link weights; this leads to a low-rank model of the output (hidden layer to target) network. This is solved by imposing a nuclear norm penalty. The proposed technique is compared with the basic ELM and the Sparse ELM. Results on benchmark datasets, show that our method outperforms both of them.

Keywords: Feedforward neural network · Extreme learning machine · Low-rank · Nuclear norm

1 Introduction

Neural networks are believed to mimic the human brain. The conventional architecture for a neural network is an input layer (for the samples), followed by a hidden layer and at the output is the target or class labels. Traditional neural network learns the link weights between the input and hidden layer nodes as well as the weights between the hidden layer nodes and the target. Randomized neural networks (RNN) or extreme learning machines (ELM) do not learn the weights between the input and the hidden layer; these weights are assigned (fixed) following some random probability distribution.

There are some studies in cognitive sciences supporting the usage of random filters in early vision; also there is mounting mathematical evidence from random matrix theory that points to linear separability [1, 2]. Basically, random projections play the same role as a non-linear kernel, it projects the data to a space such that it is linearly separable. ELM/RNN [3] is based on the same principle. However, using kernels for ELM [4, 5] seems to be an overkill, since the purpose of using a deterministic kernel and random projection is the same – linear separability; thus using the kernel on top of random projection is not likely to improve accuracy significantly.

The usual model of neural network is not sparse, there all the link weights are non-zeroes. This increases model complexity and reduces speed. The seminal work that introduced sparsity into neural network learning is Lecun's Optimal Brain Damage

© Springer International Publishing AG 2016
A. Hirose et al. (Eds.): ICONIP 2016, Part II, LNCS 9948, pp. 144–151, 2016.
DOI: 10.1007/978-3-319-46672-9_17

(OBD) [6]. In this work, the link weights were iteratively pruned by thresholding the saliency of the network. Optimization has evolved significantly since the publication of OBD almost 3 decades back; currently sparsity is introduced by imposing an l_1-norm or l_0-norm on the link weights [7–9]. Sparsity has also been introduced in the ELM framework [10]; however the formulation is slightly different, it introduces sparsity in a manner similar to sparse support vector machines.

The requirement of sparsity arises from the redundancy of the network. Sparsity kills the redundant connections and keeps only the most relevant ones. However, this is not the way human brains operate. There is a large redundancy in our brain, that is why even though thousands of neurons die in our brain regularly after a certain age, we are able to carry forth all our memory and cognitive abilities without any impairment. Even in extreme situations like shock or trauma or ischemic attacks, our brain is able to recover most of its cognitive functions. All this points to the redundancy of our brain. Since modelling the human brain is the holy grail of machine learning, instead of killing the links, we propose to explicitly model the redundancy into the neural network. In principle, we believe, our proposed model will better mimic the human brain compared to existing ones.

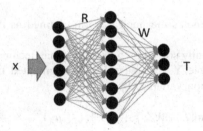

Fig. 1. Neural network

2 Proposed Formulation

The basic architecture for a neural network is shown above (Fig. 1). X is the input training data. In ELMs the network weights (R) is not learnt – it is fixed. Therefore the input to the hidden layer is simply RX. There is an activation function (φ) at the hidden nodes. Therefore the output from the hidden nodes is given by,

$$Z = \varphi(RX) \tag{1}$$

The output network connects the Z to the targets T. This is given by,

$$T = WZ \tag{2}$$

Assuming a Euclidean cost function (2) has a nice closed form solution in the form of a pseudo-inverse, given by,

$$W = (ZZ^T)^{-1}ZT^T \tag{3}$$

This solution (3) does not include any prior regarding the network weights W. The link weights are independent. We propose to incorporate redundancy; the redundancy is modeled in terms of linear dependency of the columns/rows of W. On other words this would lead to a matrix that is rank deficient. Mathematically this can be expressed as,

$$\arg\min_{W} \|T - WZ\|_F^2 \text{ such that W is low-rank} \tag{4}$$

Unfortunately (4) is an NP hard problem; the complexity of solving this problem is doubly exponential. Researchers in machine learning and signal processing have been interested in this problem in the past few years for a variety of applications – Collaborative Filtering [11], Distributed Sensor Network [12, 13], Direction of Arrival estimation [14] etc. What they do is to relax the NP hard rank minimization problem with their closest convex surrogate the nuclear norm, leading to,

$$\arg\min_{W} \|T - WZ\|_F^2 \text{ such that} \|W\|_{NN} \leq \tau \tag{5}$$

Here the subscript denotes the nuclear norm, defined as the sum of the singular values of a matrix.

This (5) can be efficiently solved by a Split Bregman technique proposed in [15]. It introduces a proxy variable Y = Z and solve an augmented Lagrangian by incorporating a Bregman relaxation variable (B).

$$\arg\min_{W,Y} \|T - WZ\|_F^2 + \lambda\|Y\|_{NN} + \mu\|Y - Z - B\|_F^2 \tag{6}$$

The problem (6) can be solved using alternating directions method of multipliers (ADMM) leading to the following two sub-problems. The idea is to have sub-problems that can be solved using stock off-the-shelf algorithms.

$$\arg\min_{W} \|T - WZ\|_F^2 + \mu\|Y - W - B\|_F^2 \tag{7}$$

$$\arg\min_{Y} \lambda\|Y\|_{NN} + \mu\|Y - Z - B\|_F^2 \tag{8}$$

The first sub-problem (7) is a simple least squares problem that is solved using conjugate gradient. The second sub-problem can be efficiently updated using singular value shrinkage [16, 17]; shown as below

$$Y \leftarrow \arg\min_{Y} \lambda\|Y\|_{NN} + \mu\|Y - Z - B\|_F^2$$
$$USV^T = SVD(Z + B)$$
$$\Sigma = diag(\max(0, S - \lambda/2\mu))$$
$$Y = U\Sigma V^T$$

This concludes the derivation of the algorithm. There are two stopping criteria for the Split Bregman algorithm. Iterations continue till the objective function converges (to a local minima). The other stopping criterion is a limit on the maximum number of iterations. We have kept it to be 200.

Our method requires specification of a parameter λ and a hyper-parameter μ. In Split Bregman techniques, usually the hyper-parameter is fixed and the parameter is tuned. We follow the same routine here; we fix $\mu = 1$ and tune λ by the L-curve method.

3 Experimental Evaluation

3.1 Results on Benchmark Classification Datasets

Our experiments were carried out on some well known databases from the UCI Machine Learning repository [18]. Leave-one-out cross validation is used for avoiding variance due to random splits. Also, in order to avoid variations arising out of assignment of link weights for the first layer, the same i.i.d Gaussian random projection matrix (between the input and the hidden layer) is used for all the ELM classifiers.

Here, we compare with the basic ELM [3] and Sparse ELM [10] with linear and rbf kernels. In [10] an empirical analysis showed that for all kinds of ELM, the performance saturates when the number of hidden nodes are about 10 times the dimensionality of the vectors. Therefore we follow the same rule-of-thumb in our experiments (Table 1).

Table 1. Classification accuracy on benhmark datasets

Name	# classes	Basic ELM	Sparse ELM (linear)	Sparse ELM (rbf)	Proposed
Page Block	5	**96.86**	95.32	95.78	96.33
Abalone	29	24.22	26.49	27.39	28.98
Segmentation	7	95.87	96.31	**97.22**	97.22
Yeast	10	54.32	57.71	57.75	59.00
German credit	2	75.88	75.40	76.16	78.43
Tic-Tac-Toe	2	86.72	85.31	85.31	86.88
Vehicle	4	72.97	73.46	74.51	77.88
Australian Cr	2	87.15	86.52	87.14	89.64
Balance scale	3	85.52	93.33	94.33	95.33
Ionosphere	2	91.67	91.67	92.20	94.12
Liver	2	69.04	69.04	69.04	70.21
Ecoli	8	80.26	81.26	81.45	83.86
Glass	7	69.23	69.23	**70.19**	70.19
Wine	3	74.69	**85.51**	85.45	85.45
Iris	3	92.00	96.00	96.67	98.67
Lymphography	4	88.64	86.32	86.32	88.81
Hayes Roth	3	34.85	41.01	43.94	45.38
Satellite	6	89.73	80.30	83.15	86.22
Haberman	2	65.22	63.28	63.20	67.78

Experimental results show that our method yields the best results, except in one dataset (Wine), where the sparse ELM with linear kernel yields the best results. One interesting observation that can be made here is that adding a non-linear kernel to the ELM formulation does not help much; one can see that the difference between the linear and rbf kernel sparse ELM is not much different (less than 1 %). This phenomenon has been explained before – both random projections and non-linear kernel randomize make the data linearly separable, hence adding one to of the other does not change much. It must be noted, this observation is not available in the original paper for sparse ELM since they had not compared with linear kernels.

3.2 Experiments on Face Recognition

We follow the experimental protocol outlined in [19]. The experiments are carried on the Extended Yale B Face Database (Fig. 2). For each subject, we randomly select half of the images for training and the other half for testing. Table 2 contains the results for face recognition. The features are selected using the simple Eigenface method. Although more sophisticated feature extraction techniques exist, our goal is to investigate that given the feature set how different classifiers perform. To compare our results with [19], we select the same number of Eigenfaces as proposed.

Fig. 2. Samples from extended YaleB

We do not compare the results with SVM and ANN, since it has already shown in [19] that the SRC (sparse representation based classification) outperforms them for face recognition problems. We compare our results with basic ELM, and sparse (rbf kernel) ELMs as well. As before, in order to avoid variations due to random assignment of the link weights between the input and the hidden layer, the same random projection matrix (i.i.d Gaussian) is used for all the different types of ELM classifiers (Table 2).

Table 2. Face Recognition

Method	Number of Eigenfaces			
	30	56	120	504
ELM	86.49	91.71	93.87	96.77
Sparse ELM	86.96	92.05	94.26	97.13
SRC	**89.40**	**93.37**	**95.14**	**97.79**
Proposed	87.11	92.56	95.08	97.25

SRC is a lazy learning classifier; it has no training time but a large testing time since it requires solving a sophisticated optimization problem. Our proposed method cannot beat SRC but yields better results than the other ELM variants.

3.3 Experiments on Handritten Digit Recognition

The MNIST digit classification task is composed of 28 × 28 images of the 10 handwritten digits. There are 60,000 training images with 10,000 test images in this benchmark. The images are scaled to [0,1] and we do not perform any other pre-processing. However, we do not carry out experiments on the standard MNIST dataset; experiments are also carried out on the more challenging variations of the MNIST dataset [20]. These were introduced as benchmark deep learning datasets. All these datasets have 10,000 training, 2000 validation and 50,000 test samples. The size of the image as before is 28 × 28 and the number of classes are 10.

Dataset	Description
basic-rot	Smaller subset of MNIST with random rotations.
bg-rand	Smaller subset of MNIST with uniformly distributed random noise background.
bg-img	Smaller subset of MNIST with random image background.
bg-img-rot	Smaller subset of MNIST digits with random background image and rotation.

As before we compare our proposed technique with the basic ELM and the sparse ELMs with linear and rbf kernels. The results are shown in Table 3. We want to eliminate effects arising out of random assignment of link weights in the first layer; in order to do so, we use the same random projection matrix (i.i.d Gaussian) for all the classifiers.

Table 3. Digit classification

Dataset	ELM	Sparse ELM (linear)	Sparse ELM (rbf)	Proposed
basic	92.79	93.16	93.89	**95.08**
basic-rot	86.70	87.47	86.70	**88.21**
bg-rand	86.06	86.70	90.27	**90.94**
bg-img	80.69	80.32	80.69	**82.59**
bg-img-rot	52.61	56.24	52.61	**54.58**

From Table 3, as expected, our proposed technique yields significantly better results than the others (Table 3).

4 Conclusion

In this work we have proposed a variation for randomized neural network/extreme learning machine. Prior variants of the basic technique included kernels and sparsity. In sparsity based techniques the redundant connections are pruned; only the most relevant ones stay. In this work, our goal is to better mimic the human brain. Therefore instead of pruning the connections, we actively promote redundancy in the system. This is achieved by modelling redundancy in terms of linear dependency. In turn, this leads to a low-rank representation of the matrix containing the link weights between the hidden layer and the targets. Following signal processing literature, we formulate a nuclear norm regularized ELM problem. Efficient solutions for this problem already exist.

We carry out thorough experimental validation. We validate on 1. benchmark machine learning datasets from the UCI Machine Learning Repository; 2. Face Recognition (YaleB) and 3. Handrwitten digit recognition (MNIST variations). In all the cases, our method outperforms the basic ELM and the sparse ELM (linear and rbf).

References

1. Paul, S., Boutsidis, C., Magdon-Ismail, M., Drinea, P.: Random projections for linearsupport vector machines. ACM Trans. Knowl. Disc. Data **8**(4) (2014). Article 22
2. Shi, Q., Shen, C., Hill, R., van den Hengel. A.: Is margin preserved after random projection? In: ICML (2012)
3. Huang, G.-B., Zhu, Q.-Y., Siew, C.-K.: Extreme learning machine: theory and applications. Neurocomputing **70**, 489–501 (2006)
4. Scardapane, S., Comminiello, D., Scarpiniti, M., Uncini, A.: Online sequential extreme learning machine with kernels. IEEE Trans. Neural Netw. Learn. Syst. **26**(9), 2214–2220 (2015)
5. Zhou, Y., Peng, J., Chen, C.L.P.: Extreme learning machine with composite kernels for hyperspectral image classification. IEEE J. Sel. Top. Appl. Earth Obs. Remote Sen. **8**(6), 2351–2360 (2015)
6. LeCun, Y.: Optimal Brain Damage. In: NIPS (1990)
7. Thom, M., Palm, G.: Sparse activity and sparse connectivity in supervised learning. J. Mach. Learn. Res. **14**, 1091–1143 (2013)
8. Gripon, V.: Sparse neural networks with large learning diversity. IEEE Trans. Neural Netw. Learn. Syst. **22**(7), 1087–1096 (2011)
9. Glorot, X., Bordes, A., Bengio, Y.: Deep sparse rectifer neural networks. In: AISTATS 2011 (2011)
10. Bai, Z., Huang, G.-B., Wang, D., Wang, H., Westover, M.B.: Sparse extreme learning machine for classification. IEEE Trans. Cybern. **44**(10), 1858–1870 (2014)
11. Gogna, A., Majumdar, A.: Matrix completion incorporating auxiliary information for recommender system design. Expert Syst. Appl. **42**(5), 5789–5799 (2015)
12. Majumdar, A., Ward, R.K.: Increasing energy efficiency in sensor networks: blue noise sampling and non-convex matrix completion. Int. J. Sens. Netw. **9**(3/4), 158–169 (2011)
13. Jindal, A., Psounis, K.: Modeling spatially correlated data in sensor networks. ACM Trans. Sensor Netw. **2**(4), 466–499 (2006)
14. Pal, P., Vaidyanathan, P.P.: A grid-less approach to underdetermined direction of arrival estimation via low rank matrix denoising. IEEE Sign. Proces. Lett. **21**(6), 737–741 (2014)
15. Gogna, A., Shukla, A., Majumdar, A.: Matrix Recovery using Split Bregman. In: International Conference on Pattern Recognition (2014)
16. Majumdar, A., Ward, R.K.: Some empirical advances in matrix completion. Sign. Process. **91**(5), 1334–1338 (2011)

17. Chartrand, R.: Nonconvex splitting for regularized low-rank + sparse decomposition. IEEE Trans. Sig. Process. **60**, 5810–5819 (2012)
18. http://archive.ics.uci.edu/ml/
19. Wright, J., Yang, A., Ganesh, A., Sastry, S., Ma, Y.: Robust face recognition via sparse representation. IEEE Trans. Pattern Anal. Mach. Intell. **31**(2), 210–227 (2009)
20. http://www.iro.umontreal.ca/ ~ lisa/twiki/bin/view.cgi/Public/MnistVariations

Gram-Schmidt Orthonormalization to the Adaptive ICA Function for Fixing the Permutation Ambiguity

Yoshitatsu Matsuda[✉] and Kazunori Yamaguchi

Department of General Systems Studies, Graduate School of Arts and Sciences,
The University of Tokyo, 3-8-1, Komaba, Meguro-ku, Tokyo 153-8902, Japan
{matsuda,yamaguch}@graco.c.u-tokyo.ac.jp

Abstract. Recently, we have proposed a new objective function of ICA called the adaptive ICA function (AIF). AIF is a summation of weighted 4th-order statistics, where the weights are determined by adaptively esti- mated kurtoses. In this paper, the Gram-Schmidt orthonormalization is applied to the optimization of AIF. The proposed method is theoretically guaranteed to extract the independent components in the unique order of the degree of non-Gaussianity. Consequently, it enables us to fix the permutation ambiguity. Experimental results on blind image separation problems show the usefulness of the proposed method.

1 Introduction

Independent component analysis (ICA) is a widely-used method in signal processing [4,5] and feature extraction [7]. It solves blind source separation problems under the assumptions that non-Gaussian source signals are statis- tically independent of each other. The linear model of ICA is given as $x = As$ where x is the observed (known) signals. A and s are the mixing matrix and the sources, respectively. Only x is known and the others are unknown. Though ICA can estimate A by using the independency among the sources, there are still some difficult problems. In this paper, we focus on the permutation ambiguity. Regarding the permutation ambiguity, many ICA methods can not determine the order of the rows of A. In other words, when A is a solution in the usual ICA, any permutation of the rows of A is a solution also. Though some previous works can fix the permutation ambiguity (e.g. [10] for audio signals, [2] using a tracking filter, and [11] if the order of the kurtoses of sources is known), they need prior knowledge of the sources or the mixing matrix.

In this paper, we propose a new method which extracts independent compo- nents in the unique order of the degree of non-Gaussianity. In other words, it can determine the order of components uniquely without the permutation ambiguity. In order to construct this new method, we apply the Gram-Schmidt orthonor- malization to the adaptive ICA function (AIF) which is an objective function of ICA. AIF has been proposed by the authors [9], which is a weighted summation of the 4th-order statistics where the weights depend on the adaptive estimators

© Springer International Publishing AG 2016
A. Hirose et al. (Eds.): ICONIP 2016, Part II, LNCS 9948, pp. 152–159, 2016.
DOI: 10.1007/978-3-319-46672-9_18

of kurtoses. Though the similar deflation approach is also employed in fast ICA [6], it can not avoid the permutation ambiguity. On the other hand, we show that the Gram-Schmidt orthonormalization to AIF is theoretically guaranteed to extract the independent components in the unique order.

This paper is organized as follows. In Sect. 2, AIF is described briefly. Moreover, AIF under the orthonormality constraint is derived. In Sect. 3. a gradient algorithm is proposed for optimizing AIF via the Gram-Schmidt orthonormalization. Section 4 proves that the Gram-Schmidt orthonormalization to AIF can extract all the independent sources in the unique order of the degree of non-Gaussianity. Section 5 shows the experimental results on blind image separation problems. Lastly, this paper is concluded in Sect. 6.

2 Objective Function

We use the adaptive ICA function (AIF), which was originally proposed in [9]. The outline of AIF is described below. The detailed derivation of AIF is described in [9]. Let $\boldsymbol{X} = (x_{im})$ be observed signals. $\boldsymbol{X}$ is an $N \times M$ matrix, where N and M are the number of signals and the sample size, respectively. Let $\boldsymbol{W}$ be the $N \times N$ separating matrix. $\boldsymbol{Y} = \boldsymbol{W}\boldsymbol{X}$ is regarded as the estimated sources. Now, we assume that $\boldsymbol{Y} = (y_{im})$ estimates the independent sources accurately. Then, AIF is derived as the likelihood of $\boldsymbol{X}$ by applying the Gaussian approximation to the distribution of the accurately estimated $\boldsymbol{Y}$ in the second-order feature space. Let $\varphi_2(\boldsymbol{X}, m) = (x_{im}x_{jm})$ and $\varphi_2(\boldsymbol{Y}, m) = (y_{im}y_{jm})$ $(i \leq j)$ be the vectors in the second-order polynomial feature space of $\boldsymbol{X}$ and $\boldsymbol{Y}$ for a sample m, respectively. Consequently, a conditional $\frac{N(N+1)}{2}$-dimensional Gaussian distribution on $\varphi_2(\boldsymbol{X}, m)$ is given as

$$P(\varphi_2(\boldsymbol{X}, m) | \boldsymbol{\alpha}, \boldsymbol{W}) = \prod_{i,j>i} G(y_{im}y_{jm}, 1) \prod_i G(y_{im}^2, \alpha_i) |\boldsymbol{W}|^{N+1} \qquad (1)$$

where $\boldsymbol{\alpha} = (\alpha_i)$ is additional unknown parameters. As shown later, each α_i is related to the estimator of the kurtosis of the i-th source. $G(u, V)$ is the Gaussian distribution on u with the mean of 0 and the variance of V. $|\boldsymbol{W}|$ is the determinant of $\boldsymbol{W}$. Then, AIF (denoted as $\Psi(\boldsymbol{\alpha}, \boldsymbol{W})$) is defined as the following log-likelihood function:

$$\Psi(\boldsymbol{\alpha}, \boldsymbol{W}) = \frac{\sum_m \log P(\varphi_2(\boldsymbol{X}, m) | \boldsymbol{\alpha}, \boldsymbol{W})}{M}$$

$$= -\sum_i \log \alpha_i - \sum_{i,j} \left(\frac{1 - \delta_{ij}}{2} + \frac{\delta_{ij}}{\alpha_i} \right) \frac{\sum_m \left(y_i^m y_j^m - \delta_{ij} \right)^2}{M} + 2(N+1) \log |\boldsymbol{W}| \qquad (2)$$

where it is divided by the constant factor of $\frac{M}{2}$ and some constant terms are removed. Regarding $\Psi(\boldsymbol{\alpha}, \boldsymbol{W})$ of Eq. (2), the following theorem holds (see [9] for the proof):

Theorem 1. *It is assumed that $\boldsymbol{x}$ is given by the linear ICA model $\boldsymbol{x} = \boldsymbol{As}$ whose sources in $\boldsymbol{s}$ do not include more than one Gaussian signal nor any uniform*

Bernoulli variable. In addition, it is assumed that M is sufficiently large. Under the constraint that every estimated signal y_{im} is normalized ($\frac{\sum_m y_{im}^2}{M} = 1$), the true solution of ICA ($W = A^{-1}$) is a stable maximum of $\Psi(\alpha, W)$ of Eq. (2).

This theorem guarantees the local optimality of AIF for estimating the separating matrix W. In addition, the optimal value of α_i is given as

$$\hat{\alpha}_i = \frac{\sum_m y_{im}^4}{M} - 1 \tag{3}$$

by the Karush-Kuhn-Tucker (KKT) conditions. Therefore, it can be regarded as the adaptive estimation of the kurtosis. Note that $\hat{\alpha}_i = 2$ if the i-th source is Gaussian.

In Theorem 1, only the normalization constraint is needed. In this paper, the following orthonormality constraint is added for every i and j:

$$\frac{\sum_m y_{im} y_{jm}}{M} = \delta_{ij} \tag{4}$$

where δ_{ij} is the Kronecker delta. Note that Theorem 1 holds under the orthonormality constraint instead of the normalization one. It is because the orthonormality constraint is strictly stronger than the normalization one and the orthonormality constraint is naturally satisfied when Y estimates the independent sources accurately. Now, it is easily shown that only an arbitrary unitary transformation to W is permitted under the orthonormality condition. Therefore, it is shown that $\sum_i y_{im}^2$ is a constant for any W. In addition, $|W|$ is the constant of 1. Then, the terms $|W|$, $\sum_i (y_{im})^2$, and $\sum_{i,j} (y_{im} y_{jm})^2$ in Eq. (2) are constant irrespective of W. Thus, AIF of Eq. (2) is simplified as follows:

$$\Psi(\alpha, W) = -\sum_i \log \alpha_i + \sum_i \left(\frac{1}{2} - \frac{1}{\alpha_i}\right) \frac{\sum_m \left(y_{im}^4 - 1\right)}{M}. \tag{5}$$

This is equivalent to the objective function derived by the authors in a different way [8]. In the following, Eq. (5) is referred as AIF under the orthonormality constraint.

3   Optimization

Here, we construct a stochastic optimization method maximizing $\Psi(\alpha, W)$ of Eq. (5) by the Gram-Schmidt orthonormalization. $\Psi(\alpha, W)$ for each i (denoted by Ψ_i) is given as

$$\Psi_i(\alpha_i, w_{i1}, \cdots, w_{iN}) = -\log \alpha_i + \left(\frac{1}{2} - \frac{1}{\alpha_i}\right) \frac{\sum_m \left(y_{im}^4 - 1\right)}{M} \tag{6}$$

where $y_{im} = \sum_k w_{ik} x_{km}$ and the factor $\frac{1}{2}$ is removed. In the Gram-Schmidt orthonormalization, Ψ_i is maximized for each i sequentially under the constraints that $\frac{\sum_m y_{im}^2}{M} = 1$ and $\frac{\sum_m y_{im} y_{jm}}{M} = 0$ for every $j < i$. The stochastic gradients of Ψ_i with respect to w_{ik} and α_i for a sample m are given as

$$\frac{\partial \Psi_i}{\partial w_{ik}} = \left(\frac{1}{2} - \frac{1}{\alpha_i} \right) 4 y_{im}^3 x_{km} \tag{7}$$

and

$$\frac{\partial \Psi_i}{\partial \alpha_i} = -\frac{1}{\alpha_i} + \frac{y_{im}^4 - 1}{\alpha_i^2}. \tag{8}$$

In order to reduce the computational costs for satisfying the orthonormality constraint, X is pre-whitened and the Gram-Schmidt orthonormalization is applied to the row vectors of W instead of those of Y at each update. In addition, α is constrained to be non-negative because of the property of kurtosis. Thus, one step of the stochastic optimization of each Ψ_i is given as follows:

Stochastic Optimization
---

1. Pick up a sample m randomly.
2. Calculate each $y_{im} = \sum_k w_{ik} x_{km}$.
3. Update each $w_{ik} := w_{ik} + \rho \frac{\partial \Psi_i}{\partial w_{ik}}$ by Eq. (7) and each $\alpha_i := \alpha_i + \rho \frac{\partial \Psi_i}{\partial \alpha_i}$ by Eq. (8), where ρ is the stepsize which was annealed slowly.
4. $\alpha_i := \epsilon$ if $\alpha_i < \epsilon$ (ϵ is a small positive threshold).
5. Orthonormalize each row of W by the following process: $w_{ik} := w_{ik} - \sum_{j<i} \left(\sum_l w_{il} w_{jl} \right) w_{jk}$ and $w_{ik} := \frac{w_{ik}}{\sqrt{\sum_l w_{il}^2}}$.

In order to avoid the local minima, the maximization of Ψ_i for each i is repeated from different initializations. In the estimation of the value of Ψ_i, we utilize the theoretically optimal $\hat{\alpha}_i$ in Eq. (3) instead of the estimated α_i. Therefore, when a separating row $(w_{i1}, \cdots, w_{iN})$ is given, we use the following estimator $\hat{\Psi}_i$ defined as

$$\hat{\Psi}_i (w_{i1}, \cdots, w_{iN}) = \Psi_i (\hat{\alpha}_i, w_{i1}, \cdots, w_{iN}) = -\log \hat{\alpha}_i + \frac{\hat{\alpha}_i}{2} \tag{9}$$

where some constants are removed. After the L repetitions, the separating row with the largest $\hat{\Psi}_i$ is employed. Consequently, the stochastic gradient algorithm optimizing AIF via the Gram-Schmidt orthonormalization is given as follows:

Complete Algorithm
---

1. *Initialization.* Pre-whiten X and let i be 1 (the first component).
2. *Estimation.* For a given i, repeat the following process L times:
   (a) Set $(w_{i1}, \cdots, w_{iN})$ randomly and normalize it by $w_{ik} := \frac{w_{ik}}{\sqrt{\sum_l w_{il}^2}}$. In addition, set α_i to 2 (corresponding to the Gaussian distribution).
   (b) Maximize Ψ_i by repeating the above stochastic optimization T times (T is the number of iterations).
   (c) Calculate $\hat{\Psi}_i$ by Eq. (9).
3. *Selection.* Select the optimal $(w_{i1}, \cdots, w_{iN})$ with the largest $\hat{\Psi}_i$ in the L trials.
4. *Deflation.* Let i be $i+1$ (the next component) and return to Step 2 (Estimation) until all the components are estimated.

4   Analysis of AIF with the Gram-Schmidt Orthonormalization

Here, it is proved that the global minimum of AIF with the Gram-Schmidt orthonormalization is the ICA solution without any permutation ambiguity under some reasonable conditions. Rigorously, the following theorem is proved:

Theorem 2. *The following four conditions are assumed:*

1. *The linear ICA model $x = As$ holds, where the mean and the variance of each $s_i \in s$ are 0 and 1, respectively.*
2. *The sample size M is sufficiently large. In other words, the average over samples is equivalent to the accurate expectation.*
3. *There is no uniform Bernoulli source (in other words, $\kappa_i > -2$ for every i).*
4. *γ_i for the i-th source is different from each other, where γ_i is defined as*

$$\gamma_i = -\log\left(\kappa_i + 2\right) + \frac{\kappa_i}{2} + \log 2. \tag{10}$$

Here, κ_i is the kurtosis of the i-th source s_i. In addition, $\log 2$ is a useful constant guaranteeing the non-negativity of γ_i.

Then, all the sources of s are extracted in descending order of γ_i when $\Psi(\alpha, W)$ of Eq. (5) is globally minimized under the Gram-Schmidt orthonormalization.

Proof. First, the objective function Ψ_i of Eq. (6) is simplified under the given conditions. We used $\hat{\Psi}_i$ of Eq. (9) instead of Ψ_i of Eq. (6). It is easily shown that W minimizing $\hat{\Psi}_i$ of Eq. (9) is equivalent to W minimizing Ψ_i. Here, a new matrix $B = (b_{ij}) = WA$ is introduced as a useful alternative notation of W. Note that B is constrained to be orthonormal. As the average over samples is equivalent to the accurate expectation (M is large), $\hat{\alpha}_i$ is given as

$$\hat{\alpha}_i = E\left(\left(\sum_k b_{ik}s_k\right)^4 - 1\right) = \left(\sum_k b_{ik}^4 \kappa_k + 2\right) \tag{11}$$

where $E()$ is the expectation operator and κ_i is the kurtosis of s_i. Thus, $\hat{\Psi}_i$ of Eq. (9) is rewritten as the following function $F(z)$:

$$\hat{\Psi}_i(b_{i1}, \cdots, b_{iN}) = F(z) = -\log(z+2) + \frac{z}{2} + 1 \tag{12}$$

where the argument z is given as $z = \sum_k b_{ik}^4 \kappa_k$. In the Gram-Schmidt orthonormalization, $\hat{\Psi}_i$ of Eq. (12) is maximized under the constraints of $\sum_k b_{ik}^2 = 1$ for every i and $\sum_k b_{ik}b_{jk} = 0$ for every $j < i$.

Next, it is proved that the maximization of each $\hat{\Psi}_i$ is equivalent to the maximization or the minimization of $z = \sum_k b_{ik}^4 \kappa_k$ under the given conditions. Since $F(z)$ in Eq. (12) is a convex function for $z > -2$, it is maximized if and only if z is the maximum or the minimum over the feasible region χ (which includes no region under $z = -2$ because $\kappa_k > -2$ for every k). Therefore, the following equation holds:

$$\arg \max_{z \in \chi} F(z) = \begin{cases} z_{\max} & F(z_{\max}) > F(z_{\min}), \\ z_{\min} & \text{otherwise,} \end{cases} \tag{13}$$

where $z_{\max}$ and $z_{\min}$ are the maximum and the minimum of z over χ, respectively.

Lastly, it is proved by the mathematical induction that all the sources are extracted in descending order of γ_k when each $\hat{\Psi}_i$ of Eq. (12) is maximized via the Gram-Schmidt orthonormalization.

Basis: At the first phase in the Gram-Schmidt orthonormalization, $\hat{\Psi}_1$ is maximized under the constraint $\sum_k b_{1k}^2 = 1$. By Eq. (13), it is equivalent to the maximization or the minimization of $\sum_k b_{1k}^4 \kappa_k$ under $\sum_k b_{1k}^2 = 1$. Note that $\sum_k b_{1k}^4 \leq 1$ and $0 \leq b_{1k}^4 \leq 1$ always hold because of $\sum_k b_{1k}^2 = 1$. Therefore, the inequality $\min_k (\kappa_k) \leq \sum_k b_{1k}^4 \kappa_k \leq \max_k (\kappa_k)$ holds, where $\min_k ()$ and $\max_k ()$ are the minimum and the maximum along the index k. The maximum (or the minimum) equality holds if and only if $b_{1l} = \pm 1$ and $b_{1k} = 0$ for every $k \neq l$, where l corresponds to the index of the maximum (or the minimum) of κ_k. Consequently, the source with the largest $F(\kappa_p)$ is extracted, which corresponds to the maximum or the minimum of κ_k (see Eq. (13)). Considering that $F(\kappa_p)$ is equal to γ_p except for a constant, it is proved that the maximization of $\hat{\Psi}_1$ extracts the independent component with the largest γ_p.

Inductive step: Regarding the i-th phase, it is assumed that the sources $l \in \Omega$ are extracted at the previous phases, where Ω denotes the set of $i-1$ sources with larger γ_l. Thus, under the Gram-Schmidt orthonormalization ($\sum_k b_{ik} b_{jk} = 0$ for every $j < i$), b_{il} is bound to 0 for $l \in \Omega$. Therefore, the maximization of $\hat{\Psi}_i$ at the i-th phase is equivalent to the maximization or the minimization of $\sum_{k \in \bar{\Omega}} b_{ik}^4 \kappa_k$ under $\sum_{k \in \bar{\Omega}} b_{1k}^2 = 1$ ($\bar{\Omega}$ denotes the complement of Ω). Similarly as in the basis step, it is easily shown that $b_{ip} = \pm 1$ and $b_{1k} = 0$ for every $k \neq p$ hold at the maximum of $\hat{\Psi}_i$ where the corresponding γ_p is the i-th largest. In other words, the source with the current largest γ_p is extracted.

Conclusion: It is proved by the principle of the mathematical induction that all the sources are extracted in descending order of γ_k. $\square$

This theorem guarantees that the proposed method finds all the sources in the unique order if each Ψ_i is globally maximized under some reasonable conditions. Note that γ_i is regarded as a degree of non-Gaussianity. γ_i is always non-negative and is equal to 0 if and only if κ_i is equal to 0 (corresponding to the Gaussian distribution). In addition, γ_i increases as κ_i is more distant from 0.

5   Results

Here, the proposed method in Sect. 3 was compared with the two widely-used ICA methods: JADE [3] and fast ICA [6] (using the two types of objective function: (A) the kurtosis-based one and (B) the log (cosh)-based one). In the proposed method using AIF, the length of the updates T was set to 300,000. The initial stepsize was set to 0.03. The annealing rate at each update was set to 0.999999. The annealing rate was set to be very slow because we prioritized the accuracy of the solution instead of the efficiency of the calculation. The small threshold ϵ was set to 0.01. L (the number of trials for each i) was set to 10. These methods were compared in the following blind image separation problem. The original dataset consists of 44 images from the USC-SIPI image database (Volume 3: Miscellaneous). They were transformed into grayscale images of 256×256 pixels. Each dataset for the experiments consists of 10 images selected randomly from the original dataset. The source images were sorted by γ_i. Each pixel corresponds to a sample. The size of samples M was set to $256 \times 256 = 65,536$ (the total size of pixels). The values in each pixel are normalized over the samples so that their means and their variances are 0 and 1, respectively. 10 randomly-selected datasets were used. The square mixing matrix A was randomly initialized. The errors were measured by the averages of the separating errors about $B = WA$ over the 10 datasets. The two types of separating errors were employed: E (insensitive to permutation) proposed in [1] and E^* (sensitive to permutation) in the similar way as in [11]. They are defined as

$$E = \sum_{i,j} \frac{b_{ij}}{\max_k (b_{kj})} - 1 + \sum_{i,j} \frac{b_{ij}}{\max_k (b_{ik})} - 1 \tag{14}$$

$$E^* = \sum_{i,j} \text{abs} (b_{ij} - \delta_{ij}), \tag{15}$$

where abs () shows the absolute value.

The experimental results were shown in Table 1. Regarding the usual error E insensitive to permutation, AIF was slightly inferior to the other ICA methods and JADE is the best. Regarding the error E^* sensitive to permutation, however, AIF is definitely superior to the other methods. It verifies that the maximization of AIF via the Gram-Schmidt orthonormalization can extract the independent sources in the unique order.

Table 1. Averaged separating errors of ICA methods in the blind image separation problems over 10 datasets: each value shows the averaged error over 10 datasets with its standard deviation. The upper and lower rows correspond to E (insensitive to permutation) and E^* (sensitive to permutation).

	AIF	JADE	fastICA (kurtosis)	fastICA (cosh)
E	5.4 ($\pm$ 2.4)	**3.6** ($\pm$ 2.3)	4.5 ($\pm$ 1.7)	4.3 ($\pm$ 2.3)
E^*	**4.1** ($\pm$ 2.5)	9.3 ($\pm$ 1.3)	7.9 ($\pm$ 1.8)	7.6 ($\pm$ 3.4)

6   Conclusion

In this paper, we proposed a new algorithm of ICA by applying the Gram-Schmidt orthonormalization to the maximization of the adaptive ICA function (AIF). We theoretically proved that the proposed algorithm finds all the sources in the unique order according to the degree of non-Gaussianity and solves the permutation problem in ICA. The experimental results in blind image separation problems also showed the usefulness of the proposed method. We are now planning to use the proposed method for determining the number of independent components. For this purpose, we are now analyzing the threshold about the non-Gaussianity. Moreover, we are planning to apply the proposed method to many datasets for the feature extraction. We are also planning to improve the annealing rate control in order to achieve the calculation efficiency. This work is partially supported by Grant-in-Aid for Young Scientists (KAKENHI) 26730013.

References

1. Amari, S., Cichocki, A.: A new learning algorithm for blind signal separation. In: Advances in Neural Information Processing Systems 8, pp. 757–763. MIT Press, Cambridge (1996)
2. Amishima, T., Okamura, A., Morita, S., Kirimoto, T.: Permutation method for ICA separated source signal blocks in time domain. IEEE Trans. Aerosp. Electron. Syst. **46**(2), 899–904 (2010)
3. Cardoso, J.F., Souloumiac, A.: Blind beamforming for non Gaussian signals. IEE Proc.-F **140**(6), 362–370 (1993)
4. Cichocki, A., Amari, S.: Adaptive Blind Signal and Image Processing: Learning Algorithms and Applications. Wiley, Hoboken (2002)
5. Comon, P., Jutten, C.: Handbook of Blind Source Separation: Independent component analysis and applications. Academic press, Cambridge (2010)
6. Hyvärinen, A.: Blind source separation by nonstationarity of variance: a cumulant-based approach. IEEE Trans. Neural Netw. **12**(6), 1471–1474 (2001)
7. Hyvärinen, A., Karhunen, J., Oja, E.: Independent Component Analysis. Wiley, Hoboken (2001)
8. Matsuda, Y., Yamaguchi, K.: Objective function of ICA with smooth estimation of Kurtosis. In: Arik, S., Huang, T., Lai, W.K., Liu, Q. (eds.) ICONIP 2015. LNCS, vol. 9491, pp. 164–171. Springer, Heidelberg (2015). doi:10.1007/978-3-319-26555-1_19
9. Matsuda, Y., Yamaguchi, K.: Adaptive objective function of ICA by gaussian approximation in second-order polynomial feature space. In: Proceedings of IJCNN2016, Vancouver, Canada (2016, in press)
10. Sawada, H., Mukai, R., Araki, S., Makino, S.: A robust and precise method for solving the permutation problem of frequency-domain blind source separation. IEEE Trans. Speech Audio Process. **12**(5), 530–538 (2004)
11. Zarzoso, V., Comon, P., Phlypo, R.: A contrast function for independent component analysis without permutation ambiguity. IEEE Trans. Neural Netw. **21**(5), 863–868 (2010)

Data Cleaning Using Complementary Fuzzy Support Vector Machine Technique

Ratchakoon Pruengkarn[✉], Kok Wai Wong, and Chun Che Fung

School of Engineering and Information Technology,
Murdoch University, Perth, Australia
{r.pruengkarn, k.wong, l.fung}@murdoch.edu.au

Abstract. In this paper, a Complementary Fuzzy Support Vector Machine (CMTFSVM) technique is proposed to handle outlier and noise in classification problems. Fuzzy membership values are applied for each input point to reflect the degree of importance of the instances. Datasets from the UCI and KEEL are used for the comparison. In order to confirm the proposed methodology, 40 % random noise is added to the datasets. The experiment results of CMTFSVM are analysed and compared with the Complementary Neural Network (CMTNN). The outcome indicated that the combined CMTFSVM outperformed the CMTNN approach.

Keywords: Data cleaning · Noise removal · CMTNN · CMTFSVM

1 Introduction

Machine learning classification algorithms have been used in many business applications such as credit risk prediction and wait-time prediction for patients in emergency department waiting room [1, 2]. Hence, the data quality has an enormous effect on the accuracy and efficiency of these algorithms [3]. There are two main concerns relating to data quality and they are imbalance and noisy data. With respect to imbalanced data which the instances of one class considered as the majority class or negative class overwhelmed the other class which is the minority class or positive class. Negative class can lead to ignoring the positive class examples as noise and they could be wrongly discarded by the classifier [3, 4]. On the other hand, there are two types of noise which can reduce the accuracy of system performance and they are class noise and attribute noise. Attribute noise is due to errors in the data attributes such as missing values and redundant data, while class noise is due to errors of instances. For example, similar instances are considered as difference classes and instances are classified into wrong class [5].

A study by Garcia, Lorena and Carvalho [6] used consensus and majority voting strategies to identify class mislabeling from ensemble classifiers. The results showed that consensus voting was unable to identify most of noisy datasets whereas majority voting was more successful. Although, the ensemble technique performed well and could handle noisy data, however, it only provides good results for some datasets. Sluban, Gamberger and Lavrac [7] proposed high agreement random forest filter for noise detection. It performed better than other classification filters such as Naïve Bayes

© Springer International Publishing AG 2016
A. Hirose et al. (Eds.): ICONIP 2016, Part II, LNCS 9948, pp. 160–167, 2016.
DOI: 10.1007/978-3-319-46672-9_19

and SVM but not for all datasets. Verbaeten and Van Asshe [8] used ensemble methods such as committees, bagging and boosting to identify noisy and removing outliers from training set. The results showed that the cost of finding new training instance was high on small dataset. Jeatrakul, Wong and Fung [9] presented Complementary Neural Network (CMTNN) techniques to identify noisy data and enhance the performance of a neural network classifier. This technique was implemented to address both binary and multiclass classification problems. It can also handle misclassification problem and improve the prediction accuracy. However, the networks function of a trained network is black boxes whose rules of operation are completely unknown. Lee, Taur and Tao [10] proposed the outlier detection with Fuzzy Support Vector machine (FSVM). The results found that FSVM was more robust against outliers.

In order to handle noise and outlier problem, this study proposed an extension of the Complementary (CMT) technique by combining CMT with FSVM. The structure of this paper is as follows. Section 2 provided background on the complementary technique and fuzzy support vector machine. Section 3 presented the proposed methodology and evaluation method used in this study. The results and conclusion are presented in Sect. 4 and 5 respectively.

2   Background

In this section, the concept of complementary, data cleaning and evaluation techniques are described.

2.1   Complementary Neural Network

Complementary Neural Network (CMTNN) [11] is a misclassification analysis technique used to enhance the quality of training data by comparing the prediction results from both truth and false classified data. Truth Neural Network (Truth NN) and Falsity Neural Network (Falsity NN) form a pair of complementary feed-forward back-propagation neural network as shown in Fig. 1. Training data is trained by Truth NN and Falsity NN in order to predict the degree of the truth memberships and false memberships respectively. The architecture of the Truth NN and Falsity NN is similar except the target outputs of Falsity NN are complementary of the Truth NN target outputs. The differences between the Truth memberships and the False memberships values represent the uncertainty in the classification process.

2.2   Fuzzy Support Vector Machine

Support Vector Machine (SVM) [12] is widely used in machine learning and it works effectively with balanced datasets. It aims to find an optimal separating hyperplane using an expression as shown below:

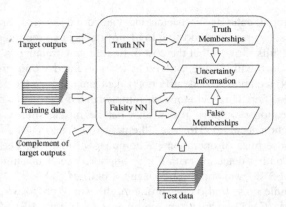

Fig. 1. Complementary Neural Network (Source: Advanced Computational Intelligence and Intelligent Informatics 14 (2010), p. 298)

$$Min\left(\frac{1}{2}\omega \cdot \omega + C\sum\nolimits_{i=1}^{l} \xi_i\right) \tag{1}$$

subject to $y_i(\omega \cdot \Phi(x_i) + b) \geq 1 - \xi_i, \xi_i \geq 0$ $i = 1, \ldots, l$

where y_i is the class label, ω is the weighted normal vector, C is a penalty parameter, ξ is the slack variable for misclassified examples, Φ is a mapping function to transform data into higher dimensional feature space, and b is the bias.

However, SVM is sensitive to outlier and noise [13]. Therefore, Fuzzy Support Vector Machine (FSVM) has been proposed to handle the problems. Fuzzy membership value m_i is applied to FSVM for each input point to represent their importance for their own class [14, 15] as shows in Eq. (2). The lower membership values are assigned to less important examples such as outlier and noise. Also, the detail of m_i computing is described in Sect. 2.3

$$Min\left(\frac{1}{2}\omega \cdot \omega + C\sum\nolimits_{i=1}^{l} m_i\xi_i\right) \tag{2}$$

Subject to $y_i(\omega \cdot \Phi(x_i) + b) \geq 1$

2.3 Complementary Fuzzy Support Vector Machine

Complementary Fuzzy Support Vector Machine (CMTFSVM) applies concepts of complementary (CMT) of Truth target output in CMTNN by using Fuzzy Support Vector Machine (FSVM) as a classifier to identify uncertainty data. The exponential-decaying function based on the distance from the actual hyperplane [13] is used in fuzzy membership value as follows:

$$f(x_i) = \frac{2}{1 + exp\left(\beta d_i^{hyp}\right)} \tag{3}$$

where β is the steepness of the decay which $\beta \in [0, 1]$, d_i^{hyp} is the functional margin for each example x_i which is equivalent to the absolute value of the SVM decision value and it is defined in Eq. (4).

$$d_i^{hyp} = y_i(\omega \cdot \Phi(x_i) + b) \tag{4}$$

2.4   Cleaning Techniques

First the truth and falsity membership values are trained using FSVM. Secondly, the prediction outputs from both truth and falsity membership values are compared with the actual outputs. The misclassification patterns (M_{Truth}, $M_{Falsity}$) are detected. Finally, the misclassification instances are eliminated from the training data (T) and the new training data set is created. There are two kinds of new training data: CMT1 and CMT2 that can be created depend on the elimination techniques used as follows:

$$CMT1 = T - (M_{Truth} \cup M_{Falsity}) \tag{5}$$

$$CMT2 = T - (M_{Truth} \cap M_{Falsity}) \tag{6}$$

CMT1 training data set is constructed by eliminating all misclassification instances by truth and falsity membership respectively. On the other hand, misclassification instances in CMT2 are eliminated if it appeared in both truth and falsity membership.

2.5   Evaluation Method

The confusion matrix [16] as shown in Table 1 is a typical measurement to assess the classification performance in order to record the results of correctly and incorrectly recognised class. The accuracy is defined in Eq. (7).

Table 1. Confusion matrix terminology

	Positive prediction	Negative prediction
Positive class	True Positive (TP)	False Negative (FN)
Negative class	False Positive (FP)	True Negative (TN)

$$Accuracy = \frac{TP + TN}{TP + TN + FP + FN} \tag{7}$$

3   Proposed Methodology

3.1   Datasets

In this paper, the binary classification problem is considered. Two benchmark imbalanced datasets from the UCI machine learning repository [17] and KEEL (Knowledge Extraction based on Evolutionary Learning) [18] are used. They are showed in Table 2.

Table 2. Detail of datasets

Dataset	Source	# Instances	# Training	# Testing
German	UCI	1000	800	200
Ionosphere	UCI	351	281	70
Pima	UCI	768	614	154
Yeast3	KEEL	1484	1187	297

3.2   Experimental Processes

The experimental processes in this work are based on the Complementary technique (CMT) to handle the noise and outlier problems. There are three processes: (1) data cleaning with CMT technique, (2) classification with FSVM, and (3) evaluation and comparing results. The detail steps are presented in Fig. 2.

In the first step of this study, the original datasets are normalised using range transformation method. They are divided randomly into training and testing data sets with a ratio of 80 % and 20 % respectively. In this process, a 10-fold cross validation method is applied for each dataset. Once the samples are divided, the training samples are replicated into falsity and truth data. The complementary technique is then applied to the falsity data by complementing the target output of truth data. After that, both falsity data and truth data are classified by back-propagation neural network (NN) using CMT technique known as CMTNN method and by a CMT fuzzy support vector machine (FSVM) known as CMTFSVM method. CMTFSVM in this experimental uses exponentially decaying membership function and radial basis function kernel. The optimal value of β is chosen from a range of {0.1, 0.2, 0.3, 0.4, 0.5, 0.6, 0.7, 0.8, 0.9, 1} for each falsity and truth data. The results from falsity and truth datasets are compared. Following this process, the new training CMTNN1 and CMTFSVM1 are created by removing the detected misclassification instances from the falsity and truth methods. Also, CMTNN2 and CMTFSVM2 are generated by eliminating misclassification instances which are detected by both falsity and truth methods. The new training datasets are then classified by FSVM. Finally, the testing data are evaluated by comparing the results. In order to further assess the proposed data cleaning techniques, a level of 40 % of the class noise is randomly selected from the original training datasets. The class values are changed to the compliment of the original class before they are injected into the original training datasets. The data cleaning techniques, CMTNN and CMTFSVM are then applied to the noisy datasets and classified by Neural Network and FSVM respectively.

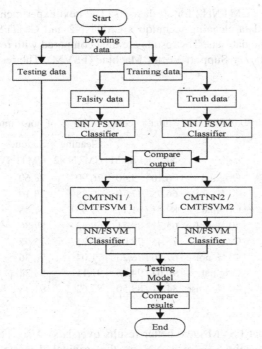

Fig. 2. Comparison of outlier and noise handling with CMTNN and CMTFSVM techniques

4   Experiment Results

Datasets with binary classification problem from the UCI and KEEL are used in the experiment. A comparison of the results obtained by CMTNN and CMTFSVM techniques is shown in Table 3.

Table 3. Comparing the accuracy of results between CMTNN and CMTFSVM techniques.

Datasets	CMTNN1	CMTFSVM1	CMTNN2	CMTFSVM2
German	76.45 %	78.40 %	76.90 %	78.60 %
Ionosphere	89.71 %	90.43 %	90.00 %	90.57 %
Pima	75.52 %	75.91 %	75.71 %	75.97 %
Yeast3	89.02 %	92.39 %	91.11 %	94.34 %

It is observed that CMTFSVM1 technique gave the better results over CMTNN1 by approximately 2 %. Moreover considering on CMTNN2 and CMTFSVM2 techniques, CMTFSVM2 presented a higher accuracy over CMTNN2 for all datasets. Finally, an analysis on CMTFSVM1 and CMTFSVM2 indicated that CMTFSVM2 performed well with approximately 2 %, 1 %, 0.3 % and 2 % for German, Ionosphere, Pima and Yeast3 respectively. Overall, these results indicated that CMTNN2 and CMTFSVM2 gave better results over CMTNN1 and CMTFSVM1. In additional, CMTFSVM2 performed better than CMTNN2 for all datasets. The next experiment is to investigate the performance of data cleaning techniques CMTNN2 and CMTFSVM2 by adding 40 % of noise into the datasets. The results are then compared with results from Neural Network (NN) and Fuzzy Support Vector Machine (FSVM). The results are shown in Table 4.

Table 4. Comparison of accuracy after injection of 40 % of noise into the datasets

Data		NN	FSVM	Cleaning techniques	
				CMTNN2	CMTFSVM2
German	Original	76.10	79.40	76.90	78.60
	40 % noise	66.95	77.70	70.85	79.10
Ionosphere	Original	89.71	91.17	91.43	91.88
	40 % noise	65.71	88.71	68.43	87.86
Pima	Original	75.45	78.05	75.71	75.97
	40 % noise	67.86	76.23	71.04	76.36
Yeast3	Original	90.67	94.58	91.11	94.34
	40 % noise	84.01	89.50	85.62	91.48

It can be seen that FSVM gave better results over NN. Also, FSVM showed the best result approximately 3 % over NN on the original datasets. When applying cleaning techniques into the original datasets, CMTFSVM2 gave approximately 1.5 % higher than CMTNN2 although CMTFSVM2 showed the lower accuracy comparing with FSVM. However after adding 40 % of noise into the original datasets, it was obvious that CMTFSVM2 could handle noisy datasets better than FSVM. Moreover, the results presented that the accuracy of CMTFSVM2 was higher than CMTNN2 by approximately 8 %, 19 %, 5 % and 6 % on German, Ionosphere, Pima and Yeast3 respectively.

5   Conclusion

In this study, the CMTFSVM data cleaning technique is proposed to eliminate outlier and class noise. 40 % of the noise is added to the training dataset. The classification based on FSVM classifier is then compare with NN classifier. Accuracy is used to evaluate the system performance. The four well-known datasets from UCI and KEEL repositories are analysed. The classification accuracy showed that CMTFSVM is robust and perform better with noisy data in the comparison study. Moreover, it can enhance the classification performance in term of accuracy by approximately 10 % for all datasets.

References

1. Twala, B.: Impact of noise on credit risk prediction: does data quality really matter? J. Intell. Data Anal. **17**, 1115–1134 (2013)
2. Ang, E., Kwansnick, S., Bayati, M., Plambeck, E.L., Aratow, M.: Accurate emergency department wait time prediction. J. Manufact. Serv. Oper. Manage. **18**, 141–156 (2015)
3. Sessions, V., Valtorta, M.: The effects of data quality on machine learning algorithms. In: The 11th International Conference on Information Quality (ICIQ-06) (2006)
4. López, V., et al.: An insight into classification with imbalanced data: empirical results and current trends on using data intrinsic characteristics. Inf. Sci. **250**, 113–141 (2013)
5. Zhu, X., Wu, X.: Class noise vs. attribute noise: a quantitative study of their impacts. J. Artif. Intell. Rev. **22**, 177–210 (2004)
6. Garcia, L.P.F., Lorena, A.C., Carvalho, A.C.P.L.F.: A study on class noise detection and elimination. In: The 2012 Brazilian Symposium on Neural Networks (SBRN), pp. 13–18. IEEE, Curitiba (2012)
7. Sluban, B., Gamberger, D., Lavra, N.: Advances in class noise detection. In: 19th European Conference on Artificial Intelligence, pp. 1105–1106. IOS Press, The Netherlands (2010)
8. Verbaeten, S., Van Assche, A.: Ensemble methods for noise elimination in classification problems. In: Windeatt, T., Roli, F. (eds.) Multiple Classifier Systems, vol. 2709, pp. 317–325. Springer, Heidelberg (2003)
9. Jeatrakul, P., Wong, K.W., Fung, C.C.: Using misclassification analysis for data cleaning. In: International Workshop on Advanced Computational Intelligence and Intelligent Informatics (2009)
10. Lee, G.H., Taur, J.S., Tao, C.W.: A robust fuzzy support vector machine for two-class pattern classification. Int. J. Fuzzy Syst. **8**, 76–86 (2006)
11. Jeatrakul, P., Wong, K.W., Fung, C.C.: Data cleaning for classification using misclassification analysis. J. Adv. Comput. Intelligence Intell. Inform. **14**, 297–302 (2010)
12. Jing, R., Zhang, Y.: A view of support vector machines algorithm on classification problems. In: International Conference on Multimedia Communications (MEDIACOM), pp. 13-16, Hong Kong (2010)
13. Batuwita, R., Palade, V.: FSVM-CIL: fuzzy support vector machines for class imbalance learning. IEEE Trans. Fuzzy Syst. **18**, 558–571 (2010)
14. Lin, C.-F., Wang, S.-D.: Fuzzy support vector machines. IEEE Trans. Neural Netw. **13**, 464–471 (2002)
15. Samma, H., Lim, C.P., Ngah, U.K.: A hybrid PSO-FSVM model and its application to imbalanced classification of mammograms. In: Selamat, A., Nguyen, N.T., Haron, H. (eds.) ACIIDS 2013, Part I. LNCS, vol. 7802, pp. 275–284. Springer, Heidelberg (2013)
16. Ramyachitra, D., Manikandan, P.: Imbalanced dataset classification and solutions: a review. Int. J. Comput. Bus. Res. (IJCBR) **5**(4), 1–29 (2014)
17. Lichman, M.: UCI Machine Learning Repository. University of California, Irvine, School of Information and Computer Sciences (2013)
18. Alcalá-Fdez, J., Fernandez, A., Luengo, J., Derrac, J., García, S., Sánchez, L., et al.: KEEL data-mining software tool: data set repository, integration of algorithms and experimental analysis framework. J. Multiple-Valued Logic Soft Comput. **17**, 255–287 (2011)

Fault-Tolerant Incremental Learning
for Extreme Learning Machines

Ho-Chun Leung, Chi-Sing Leung$^{(\boxtimes)}$, and Eric W.M. Wong

Department of Electronic Engineering,
City University of Hong Kong, Hong Kong, Hong Kong
eeleungc@cityu.edu.hk

Abstract. The extreme learning machine (ELM) framework provides
an efficient way for constructing single-hidden-layer feedforward networks
(SLFNs). Its main idea is that the input bias terms and the input weights
of the hidden nodes are selected in a random way. During training, we
only need to adjust the output weights of the hidden nodes. The existing
incremental learning algorithms, called incremental-ELM (I-ELM) and
convex I-ELM (CI-ELM), for extreme learning machines (ELMs) cannot
handle the fault situation. This paper proposes two fault-tolerant incremental ELM algorithms, namely fault-tolerant I-ELM (FTI-ELM) and
fault-tolerant CI-ELM (FTCI-ELM). The FTI-ELM only tunes the output weight of the newly additive node to minimize the training set error
of faulty networks. It keeps all the previous learned weights unchanged.
Its fault-tolerant performance is better than that of I-ELM and CI-ELM.
To further improve the performance, the FTCI-ELM is proposed. It tunes
the output weight of the newly additive node, as well as using a simple
scheme to modify the existing output weights, to maximize the reduction
in the training set error of faulty networks.

Keywords: Weight noise · Fault tolerance · Extreme learning
machines · Single hidden layer network

1 Introduction

Some classical neural network theories [1] showed that SLFNs with sufficient
hidden nodes are universal approximators. In the conventional conception, all
the connection weights of SLFNs should be adjustable. However, training all the
connection weights may create a number of difficulties in the learning process,
such as local minimum. In [2], Huang et al. formally proved that with random
values for the input biases and input weights of the hidden nodes, SLFNs can
work as universal approximators too. Based on the pseudo-inverse technique, a
number of batch mode learning methods for ELM were proposed [2]. Besides,
they proposed two incremental learning methods [2,3], namely incremental ELM
(I-ELM) and convex incremental ELM (CI-ELM). Under the faultless situation,
the performances of these two incremental learning methods work very well.

© Springer International Publishing AG 2016
A. Hirose et al. (Eds.): ICONIP 2016, Part II, LNCS 9948, pp. 168–176, 2016.
DOI: 10.1007/978-3-319-46672-9_20

However, in the development of these two incremental methods, the imperfect conditions in the implementation are ignored.

In the implementation of neural networks, network faults take place unavoidably [4]. For example, when the finite precision technology is used, multiplicative weight/node noise is introduced [5,6]. Also, for traditional neural networks, if special methods are not considered during training, the fault-tolerant ability of the trained neural networks is very poor. For traditional neural network models, some fault-tolerant works related to batch mode learning were reported [6–9]. To the best of our knowledge, there are not many literatures related to the fault-tolerant ability of ELMs.

This paper proposed two fault-tolerant incremental ELM algorithms, namely fault-tolerant I-ELM (FTI-ELM) and fault-tolerant CI-ELM (FTCI-ELM). In the FTI-ELM, the previous learned output weights of the hidden nodes are unchanged. We optimize the output weight of the newly additive node to minimize the training set error of faulty networks. Its fault-tolerant performance is better than that of I-ELM and CI-ELM. To further improve the performance, we propose the FTCI-ELM. It tunes the output weight of the newly additive node, as well as using a simple scheme to modify the existing output weights, to maximize the reduction in the training set error of faulty networks. The fault-tolerant performance of the FTCI-ELM is much better than the performances of I-ELM, CI-ELM, and FTI-ELM. Also, we show that in the FTCI-ELM, the training set error of faulty SLFNs converges.

The rest of this paper is organized as follows. Section 2 provides an introduction to the ELM concept and the multiplicative noise. Section 3 presents the two fault-tolerant incremental ELM algorithms. Simulation results are included in Sect. 4. We then conclude the paper in Sect. 5.

2   ELM and Multiplicative Noise

In this paper, we consider the regression problem. We denote the training set as $\mathbb{D}_t = \{(\boldsymbol{x}_k, y_k) : \boldsymbol{x}_k \in \mathbb{R}^d, y_k \in \mathbb{R}, k = 1, \cdots, N\}$, where $\boldsymbol{x}_k$ and y_k are the input and the target output of the k-th sample, respectively. Similarly, we denote the test set as $\mathbb{D}_f = \{(\boldsymbol{x}'_{k'}, y'_{k'}) : \boldsymbol{x}'_{k'} \in \mathbb{R}^d, y'_{k'} \in \Re, k' = 1, \cdots, N'\}$. In the SLFN approach [1], the network output is given by

$$f_n(\boldsymbol{x}) = \sum_{i=1}^{n} \beta_i g_i(\boldsymbol{x}) \tag{1}$$

where $g_i(\boldsymbol{x})$ denotes the hidden node output function, and β_i is the connection weight between the ith hidden node and the output node. This paper considers the sigmoid hidden nodes. They are defined as

$$g_i(\boldsymbol{x}) = \frac{1}{1 + \exp\{-(\boldsymbol{a}_i^{\mathrm{T}}\boldsymbol{x} + b_i)\}}. \tag{2}$$

where b_i is the bias term of the ith hidden node, and $\boldsymbol{a}_i$ is the input weight vector of the ith hidden node.

For the ELM approach [2,3], the bias terms b_i's and the input weights $\boldsymbol{a}_i$'s are selected randomly. We only need to adjust the output weights β_i's. For a batch mode ELM learning method, we need to minimize the following objective function:

$$\mathcal{E} = \sum_{k=1}^{N}(y_k - \sum_{i=1}^{n}\beta_i g_i(\boldsymbol{x}_k))^2 = \left\| \boldsymbol{y} - \sum_{i=1}^{n}\beta_i \boldsymbol{g}_i \right\|_2^2, \tag{3}$$

where $\boldsymbol{y} = [y_1, \cdots, y_N]^{\mathrm{T}}$, and $\boldsymbol{g}_i = [g_i(\boldsymbol{x}_1), \cdots, g_i(\boldsymbol{x}_N)]^{\mathrm{T}}$.

In the implementation of a network, weight failure or node failure cannot be avoided. For multiplicative noise, the deviation from the original value is proportional to the magnitude of the original value. When we use the digital implementation, finite precision can be modelled as multiplicative noise [5,6]. Also, in the analog implementation, we usually specify the precision in term of percentage of error. For multiplicative weight noise, an implemented weight can be modelled as

$$\tilde{\beta}_i = (1 + \delta_i)\beta_i, \forall i = 1, \cdots, n, \tag{4}$$

where δ_i's are the noise factors that describe the deviation. When the multiplicative node noise is considered, the node output can be modelled in the same way.

In this paper, we assume that the noise factors δ_i's are zero-mean identically independent random variables with variance equal to σ_δ^2.

3   Incremental Learning for Faulty SLFN

3.1   Errors for Faulty SLFN

For a faulty network with weight failure under a particular fault pattern, the training set error can be expressed as

$$\tilde{\mathcal{E}}_{\mathrm{W}} = \sum_{k=1}^{N}(y_k - \sum_{i=1}^{n}\tilde{\beta}_i g_i(\boldsymbol{x}_k))^2 = \sum_{k=1}^{N}(y_k - \sum_{i=1}^{n}(1+\delta_i)\beta_i g_i(\boldsymbol{x}_k))^2. \tag{5}$$

Similarly, the training set error for node failure can be also formulated. The expression would equal to the (5). Hence under either node or weight failure, the training set error can be expressed as

$$\begin{aligned} \tilde{\mathcal{E}} &= \sum_{k=1}^{N}\left(y_k - \sum_{i=1}^{n}(1+\delta_i)\beta_i g_i(\boldsymbol{x}_k) \right)^2 \\ &= \sum_{k=1}^{N}\left(y_k^2 - 2\sum_{i=1}^{n}(1+\delta_i)\beta_i g_i(\boldsymbol{x}_k) + \sum_{i=1}^{n}\sum_{i'=1}^{n}(1+\delta_i)(1+\delta_{i'})\beta_i\beta_{i'}g_i(\boldsymbol{x}_k)g_{i'}(\boldsymbol{x}_k) \right). \end{aligned} \tag{6}$$

Taking the expectation over all possible fault patterns, the faulty training set error is

$$\bar{\mathcal{E}} = \left\| \boldsymbol{y} - \sum_{i=1}^{n}\beta_i \boldsymbol{g}_i \right\|_2^2 + \sigma_\delta^2 \sum_{i=1}^{n}\beta_i^2 \|\boldsymbol{g}_i\|_2^2. \tag{7}$$

It should be noticed that when the network has multiplicative node noise, we get the same equation for the faulty training set error.

3.2 Fault-Tolerant Incremental ELM

This section develops a fault-tolerant incremental ELM (FTI-ELM) algorithm for SLFNs. At the nth step, we add a new node $g_n(\cdot)$ into the network. We need to determine the output weight β_n of the newly additive node, but we keep the previous weights β_i's, for $i = 1, \cdots, n-1$, unchanged. The faulty training set error at the nth step is

$$
\begin{aligned}
\bar{\mathcal{E}}_n &= \left\| \boldsymbol{y} - \sum_{i=1}^{n} \beta_i^2 \boldsymbol{g}_i \right\|_2^2 + \sigma_\delta^2 \sum_{i=1}^{n} \beta_i^2 \|\boldsymbol{g}_i\|_2^2 \\
&= \|\boldsymbol{e}_{n-1} - \beta_n \boldsymbol{g}_n\|_2^2 + \sigma_\delta^2 \sum_{i=1}^{n-1} \beta_i^2 \|\boldsymbol{g}_i\|_2^2 + \sigma_\delta^2 \beta_n^2 \|\boldsymbol{g}_n\|_2^2,
\end{aligned}
\tag{8}
$$

where $\boldsymbol{e}_{n-1} = \boldsymbol{y} - \sum_{i=1}^{n-1} \beta_i \boldsymbol{g}_i$. Hence to minimize $\bar{\mathcal{E}}_n$ (without modifying the previous weights), the output weight β_n is given by

$$
\beta_n = \frac{\boldsymbol{e}_{n-1}^{\mathrm{T}} \boldsymbol{g}_n}{(1 + \sigma_\delta^2) \|\boldsymbol{g}_n\|_2^2}.
\tag{9}
$$

Algorithm 1. FTI-ELM

1: Initialization: Let the number of hidden nodes $n = 0$ and the residual error $\boldsymbol{e}_0 = \boldsymbol{y}$.
2: **while** $n \leq n_{\max}$ and $\bar{\mathcal{E}}_n > \varepsilon$ **do**
3:     $n = n + 1$.
4:     Add a new hidden node $g_n(\cdot)$ to the network, where $(\boldsymbol{a}_n, b_n)$ are randomly generated.
5:     Compute the new weight β_n: $\beta_n = \frac{\boldsymbol{e}_{n-1}^{\mathrm{T}} \boldsymbol{g}_n}{(1+\sigma_\delta^2)\|\boldsymbol{g}_n\|_2^2}$.
6:     $\boldsymbol{e}_n = \boldsymbol{e}_{n-1} - \beta_n \boldsymbol{g}_n$.
7: **end while**

Algorithm 1 summarizes the steps in the FTI-ELM. We can easily observe that the complexity for each iteration is $O(N)$ only. Figure 1 gives us a preview of the performances of the FTI-ELM and the I-ELM. It shows the training set and test set mean square errors (MSEs) under the fault situation for the abalone example. The description of the dataset is in the simulation section. From the figure, it can be seen that around 500 hidden nodes are enough for the problem that the performance of FTI-ELM is better than that of the original I-ELM.

3.3 Fault-Tolerant Convex Incremental ELM

In [3], the CI-ELM algorithm was proposed. Its main idea is that the previous weights are also updated by a simple formula [10]. Its performance is better than that of the I-ELM. One may argue that the original CI-ELM can also handle the multiplicative noise. In fact, unlike the I-ELM and the FTI-ELM, the performance of the original CI-ELM is very poor under the multiplicative noise situation. Therefore, in this section, we will develop the fault-tolerant convex incremental ELM (FTCI-ELM) algorithm. Its performance is much better that the performances of CI-ELM, I-ELM, and FTI-ELM.

At the nth iteration after we determine the new weight β_n, we update the previous weights based on

$$\beta_i = (1 - \beta_n)\beta_i, \text{ for all }, i = 1, \cdots, n - 1. \tag{10}$$

At the nth iteration, we consider the faulty training set error difference between the $(n-1)$th iteration and the nth iteration, given by

$$
\begin{aligned}
\Delta_n &= \bar{\mathcal{E}}_n - \bar{\mathcal{E}}_{n-1} \\
&= \left\| y - (1 - \beta_n)\sum_{i=1}^{n-1}\beta_i g_i - \beta_n g_n \right\|_2^2 - \left\| y - \sum_{i=1}^{n-1}\beta_i g_i \right\|_2^2 \\
&\quad + \sigma_\delta^2(1 - \beta_n)^2\sum_{i=1}^{n-1}\beta_i^2\|g_i\|_2^2 + \sigma_\delta^2\beta_n^2\|g_n\|_2^2 - \sigma_\delta^2\sum_{i=1}^{n-1}\beta_i^2\|g_i\|_2^2.
\end{aligned}
\tag{11}
$$

Hence, (11) can be rewritten as

$$\Delta_n = \beta_n^2\left(\|g_n - f_{n-1}\|_2^2 + \sigma_\delta^2\|g_n\|_2^2 + \xi_{n-1}\right) - 2\beta_n\left(e_{n-1}^{\mathrm{T}}(g_n - f_{n-1}) + \xi_{n-1}\right), \tag{12}$$

where $f_{n-1} = \sum_{i=1}^{n-1}\beta_i g_i$, $e_{n-1} = y - f_{n-1}$ and $\xi_{n-1} = \sigma_\delta^2\sum_{i=1}^{n-1}\beta_i^2\|g_i\|_2^2$. To minimize Δ_n, β_n should be equal to

$$\beta_n = \frac{\left(e_{n-1}^{\mathrm{T}}(g_n - f_{n-1}) + \xi_{n-1}\right)}{\left(\|g_n - f_{n-1}\|_2^2 + \sigma_\delta^2\|g_n\|_2^2 + \xi_{n-1}\right)}. \tag{13}$$

Furthermore, when $\beta_n = \frac{\left(e_{n-1}^{\mathrm{T}}(g_n - f_{n-1}) + \xi_{n-1}\right)}{\left(\|g_n - f_{n-1}\|_2^2 + \sigma_\delta^2\|g_n\|_2^2 + \xi_{n-1}\right)}$, we have

$$\Delta_n = \left(\frac{-\left(e_{n-1}^{\mathrm{T}}(g_n - f_{n-1}) + \xi_{n-1}\right)^2}{\left(\|g_n - f_{n-1}\|_2^2 + \sigma_\delta^2\|g_n\|_2^2 + \xi_{n-1}\right)}\right). \tag{14}$$

Since ξ_{n-1} is positive, $\Delta_n < 0$. That means, $\bar{\mathcal{E}}_n$ is strictly decreasing. Also, $\bar{\mathcal{E}}_n$ is greater than zero. Hence $\bar{\mathcal{E}}_n$ converges.

Algorithm 2 summarizes the steps in the FTCI-ELM. After a simple analysis, one can easily show that the complexity at each update is "$O(n) + O(N)$". Figure 1 shows a preview of the performance of the four incremental learning methods, I-ELM, FTI-ELM, CI-ELM, and FTI-ELM. It can be seen that around 500 hidden nodes are enough for the problem, and that the performance of FTCI-ELM is the best. Also, the original CI-ELM cannot handle multiplicative noise situation.

Algorithm 2. FTCI-ELM

1: Initialization: Let the number of hidden nodes $n = 0$, the residual error $e_0 = y$, $f_0 = 0$, and $\xi_0 = 0$

2: **while** $n \le n_{\max}$ and $\bar{\mathcal{E}}_n > \varepsilon$ **do**

3:     $n = n + 1$.

4:     Add a new hidden node $g_n(\cdot)$ to the network, where (a_n, b_n) are randomly generated.

5:     Compute the new weight β_n: $\beta_n = \frac{(e_{n-1}^{\mathrm{T}}(g_n - f_{n-1}) + \xi_{n-1})}{(\|g_n - f_{n-1}\|_2^2 + \sigma_\delta^2 \|g_n\|_2^2 + \xi_{n-1})}$.

6:     $f_n = (1 - \beta_n)f_{n-1} + \beta_n g_n$.

7:     $e_n = y - f_n$.

8:     $\xi_n = (1 - \beta_n)^2 \xi_{n-1} + \sigma_\delta^2 \beta_n^2 \|g_n\|_2^2$.

9:     $\beta_i = (1 - \beta_n)\beta_i$, for all $i = 1, \cdots, n - 1$.

10: **end while**

4 Simulation

This paper uses three real life data sets, Abalone [11], Housing Price [12], and Compressive Strength [13]. In each dataset, half of samples are used as training set, and the rest of them as test set. The input attitudes of these three datasets are normalized into the range of $[-1, 1]$, while the output attitudes of these three datasets are normalized into the range of $[0, 1]$. The input weights of biases of the hidden nodes are randomly generated from the range of $[-1, 1]$.

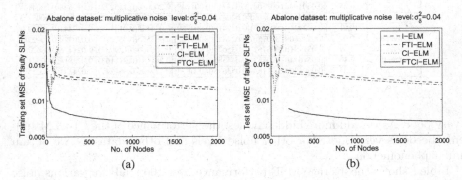

Fig. 1. The performance of different incremental methods versus the number of additive nodes. (a) The training set error of faulty SLFNs. (b) The test set error of faulty SLFNs. The multiplicative noise intensity level is σ_δ^2. The dataset is Abalone.

Figure 1 shows the performance comparison among the I-ELM, CI-ELM, FTI-ELM, and FTCI-ELM under multiplicative noise with noise level equal to $\sigma_\delta^2 = 0.04$ for the abalone dataset. It can be seen that the FTI-ELM is better than the I-ELM. Also, the FTCI-ELM is the best. Unlike the faultless situation presented in [3], the performance of the CI-ELM under the multiplicative noise

Table 1. Average MSE performances of various methods. The average performances are taken over 100 trials. The number of hidden nodes is equal to 500.

	data set	noise level σ_δ^2	Original I-ELM mean(std)	Proposed FTI-ELM mean(std)	Original CI-ELM mean(std)	Proposed FTCI-ELM mean(std)
Training Set	Abalone	0	0.008239(0.000413)	0.008239(0.000413)	0.006552(0.000142)	0.006552(0.000142)
		0.01	0.009304(0.000409)	0.009287(0.000406)	0.014290(0.004354)	0.007159(0.000063)
		0.04	0.012500(0.000397)	0.012201(0.000387)	0.037505(0.017520)	0.007554(0.000065)
		0.09	0.017827(0.000376)	0.016399(0.000359)	0.076195(0.039465)	0.007982(0.000071)
		0.16	0.025285(0.000347)	0.021208(0.000329)	0.130362(0.070189)	0.008385(0.000075)
		0.25	0.034874(0.000310)	0.026074(0.000299)	0.200004(0.109690)	0.008740(0.000079)
	Housing Price	0	0.013586(0.000853)	0.013586(0.000853)	0.011216(0.000403)	0.011216(0.000403)
		0.01	0.015397(0.000844)	0.015389(0.000845)	0.018852(0.004537)	0.012826(0.000206)
		0.04	0.020830(0.000818)	0.020403(0.000826)	0.041760(0.018595)	0.013706(0.000208)
		0.09	0.029885(0.000776)	0.027639(0.000800)	0.079942(0.042040)	0.014630(0.000238)
		0.16	0.042562(0.000716)	0.035952(0.000773)	0.133395(0.074867)	0.015660(0.000273)
		0.25	0.058862(0.000639)	0.044393(0.000749)	0.202121(0.117073)	0.016758(0.000309)
	Concrete Compressive Strength	0	0.019087(0.000597)	0.019087(0.000597)	0.017052(0.000473)	0.017052(0.000473)
		0.01	0.021032(0.000591)	0.021024(0.000591)	0.033574(0.014587)	0.018540(0.000171)
		0.04	0.026868(0.000573)	0.026409(0.000576)	0.083139(0.058617)	0.019291(0.000178)
		0.09	0.036594(0.000543)	0.034181(0.000556)	0.165748(0.132005)	0.020268(0.000213)
		0.16	0.050211(0.000502)	0.043110(0.000536)	0.281401(0.234751)	0.021422(0.000253)
		0.25	0.067718(0.000448)	0.052179(0.000520)	0.430097(0.366852)	0.022670(0.000293)
Test Set	Abalone	0	0.008677(0.000444)	0.008677(0.000444)	0.006788(0.000203)	0.006788(0.000203)
		0.01	0.009740(0.000438)	0.009724(0.000434)	0.014462(0.004309)	0.007430(0.000074)
		0.04	0.012932(0.000422)	0.012634(0.000411)	0.037484(0.017400)	0.007864(0.000074)
		0.09	0.018250(0.000405)	0.016826(0.000384)	0.075854(0.039224)	0.008325(0.000078)
		0.16	0.025696(0.000403)	0.021629(0.000366)	0.129571(0.069777)	0.008752(0.000080)
		0.25	0.035270(0.000436)	0.026487(0.000360)	0.198637(0.109060)	0.009120(0.000083)
	Housing Price	0	0.014695(0.001254)	0.014695(0.001254)	0.011814(0.000529)	0.011814(0.000529)
		0.01	0.016515(0.001252)	0.016511(0.001253)	0.019345(0.004492)	0.013439(0.000290)
		0.04	0.021975(0.001277)	0.021558(0.001274)	0.041937(0.018321)	0.014381(0.000290)
		0.09	0.031074(0.001411)	0.028844(0.001366)	0.079590(0.041402)	0.015374(0.000329)
		0.16	0.043814(0.001743)	0.037213(0.001538)	0.132304(0.073720)	0.016471(0.000372)
		0.25	0.060193(0.002304)	0.045713(0.001749)	0.200080(0.115272)	0.017628(0.000413)
	Concrete Compressive Strength	0	0.020475(0.000802)	0.020475(0.000802)	0.017679(0.000587)	0.017679(0.000587)
		0.01	0.022421(0.000787)	0.022418(0.000789)	0.034148(0.014367)	0.019441(0.000232)
		0.04	0.028260(0.000756)	0.027820(0.000762)	0.083555(0.057720)	0.020329(0.000235)
		0.09	0.037991(0.000762)	0.035621(0.000759)	0.165900(0.129987)	0.021476(0.000273)
		0.16	0.051614(0.000883)	0.044586(0.000800)	0.281184(0.231163)	0.022814(0.000317)
		0.25	0.069130(0.001163)	0.053698(0.000875)	0.429406(0.361247)	0.024231(0.000359)

is very poor. As we add more hidden nodes, the performance of the CI-ELM suddenly becomes very poor. For other noise levels and other datasets, we obtain similar phenomenon.

Table 1 shows the average MSE performance over 100 trials for various noise levels and datasets. In Table 1, the number of hidden nodes is equal to 500. From the table, when there is no multiplicative noise, i.e., $\sigma_\delta^2 = 0$, the performance of the FTI-ELM is identical to that of the I-ELM, and the performance of the FTCI-ELM is identical to that of the CI-ELM. Also, the CI-ELM is the best when there is no multiplicative noise. However, when the multiplicative noise exists, the performance of the CI-ELM becomes very poor.

When multiplicative noise exists, the performance of the FTI-ELM is better than that of the I-ELM. Besides, the performance of the FTCI-ELM is the best. The rationale of the improvement is that our development is based on the training set error of faulty networks, rather than based on the training set error of faultless networks. The performance improvement becomes more significant under high noise levels. For example, for the Abalone dataset with noise level $\sigma_\delta^2 = 0.01$, the average test set MSE of the I-ELM is equal to 0.009740. When we use the FTI-ELM, we slightly reduce the test set MSE to 0.009724. When we use the FTCI-ELM, we can further reduce the test MSE to 0.007430.

For the high noise level, such as $\sigma_\delta^2 = 0.16$, the improvement of the FTI-ELM becomes more significant. The average test set MSE of the I-ELM is equal to 0.025696. When we use the FTI-ELM, we can reduce the test set MSE to 0.021629. When we use the FTCI-ELM, we further reduce the test MSE to 0.008752. Clearly, the improvements of using FTI-ELM and FTCI-ELM are more significant for high noise levels.

5   Conclusion

This paper proposes two incremental algorithms, namely FTI-ELM and FTCI-ELM, for ELM under the multiplicative noise situation. The FTI-ELM tunes the output weight of the newly additive node without modifying the previous learned weights. Its fault-tolerant performances are better than those of the original I-ELM and CI-ELM. To further improve the performance, we propose the FTCI-ELM. It tunes the output weight of the newly additive node, as well as modifying the existing output weights, to maximize the reduction in the training set error of faulty networks. Simulation results show that the fault-tolerant performance of the FTCI-ELM is much better than the performances of I-ELM, CI-ELM, and FTI-ELM. Also, we show that in the FTCI-ELM, the training set error of faulty SLFNs converges.

References

1. Hornik, K.: Approximation capabilities of multilayer feedforward networks. Neural Netw. **4**(2), 251–257 (1991)
2. Huang, G.B., Chen, L., Siew, C.K.: Universal approximation using incremental constructive feedforward networks with random hidden nodes. IEEE Trans. Neural Netw. **17**(4), 879–892 (2006)
3. Huang, G.B., Chen, L.: Convex incremental extreme learning machine Guang-Bin. Neurocomputing **70**, 3056–3062 (2007)
4. Burr, J.: Digital neural network implementations, in Neural Networks, Concepts, Applications, and Implementations, pp. 237–285. Prentice Hall, Englewood Cliffs (1995)
5. Liu, B., Kaneko, K.: Error analysis of digital filter realized with floating-point arithmetic. Proc. IEEE **57**(10), 1735–1747 (1969)
6. Bernier, J.L., Ortega, J., Rojas, I., Ros, E., Prieto, A.: Obtaining fault tolerant multilayer perceptrons using an explicit regularization. Neural Process. Lett. **12**(2), 107–113 (2000)
7. Leung, C.-S., Hongjiang, W., John, S.: On the selection of weight decay parameter for faulty networks. IEEE Trans. Neural Netw. **21**(8), 1232–1244 (2010)
8. Leung, C.-S., Sum, J.: RBF networks under the concurrent fault situation. IEEE Trans. Neural Netw. Learn. Syst. **23**(7), 1148–1155 (2012)

9. Leung, C.-S., Wan, W.Y., Feng, R.: A regularizer approach for RBF networks under the concurrent weight failure situation. IEEE Trans. Neural Netw. Learn. Syst. (Accepted)
10. Kwok, T.Y., Yeung, D.Y.: Objective functions for training new hidden units in constructive neural networks. IEEE Trans. Neural Netw. **8**(5), 11311148 (1997)
11. Sugiyama, M., Ogawa, H.: Optimal design of regularization term and regularization parameter by subspace information criterion. Neural Netw. **15**(3), 349–361 (2002)
12. Lichman, M.: UCI machine learning repository. http://archive.ics.uci.edu/ml (2013)
13. I-Cheng, Y.: Analysis of strength of concrete using design of experiments and neural networks. J. Mater. Civ. Eng. **18**(4), 597–604 (2006)

Character-Aware Convolutional Neural Networks for Paraphrase Identification

Jiangping Huang[1], Donghong Ji[1(✉)], Shuxin Yao[2], and Wenzhi Huang[1]

[1] Computer School, Wuhan University, Wuhan 430072, China
{hjp,dhji,hwz208}@whu.edu.cn
[2] Language Technologies Institute, Carnegie Mellon University,
Pittsburgh, PA 15213, USA
shuxiny@cs.cmu.edu

Abstract. Convolutional Neural Network (CNN) have been successfully used for many natural language processing applications. In this paper, we propose a novel CNN model for sentence-level paraphrase identification. We learn the sentence representations using character-aware convolutional neural network that relies on character-level input and gives sentence-level representation. Our model adopts both random and one-hot initialized methods for character representation and trained with two paraphrase identification corpora including news and social media sentences. A comparison between the results of our approach and the typical systems participating in challenge on the news sentence, suggest that our model obtains a comparative performance with these baselines. The experimental result with tweets corpus shows that the proposed model has a significant performance than baselines. The results also suggest that character inputs are effective for modeling sentences.

Keywords: CNN · Sentence model · Paraphrase identification · Twitter

1 Introduction

Convolutional neural networks (CNN) have achieved good success in vast fields, such as image recognition [12] and natural language processing (NLP) [4]. In recent years, CNN have been shown to be effective for various NLP tasks and have achieved excellent results in semantic parsing [18], sentence modeling [11] and other traditional NLP problems [20]. The target of CNN applied for NLP tasks is to analyze and represent the semantic content of texts or sentences. The sentence modeling problem is at the core of many tasks referring to natural language comprehension. The aim of a sentence model involves a feature function that defines the process by which the features of the sentence are extracted from the feature of words or characters.

Compared to some traditional models, it has advantage to adopt CNN to accurately represent sentences because CNN can be directly applied to distribution embedding of words, without any knowledge on the syntactic or semantic structures of a language [13]. Word-level CNN models mainly depend on

© Springer International Publishing AG 2016
A. Hirose et al. (Eds.): ICONIP 2016, Part II, LNCS 9948, pp. 177–184, 2016.
DOI: 10.1007/978-3-319-46672-9_21

the input of the distribution representation of words because semantically close words are likewise close in the induced vector space. However, embeddings of rare words are poorly estimated because sub-word or morphemes information are shielded to neural language models, which will lead to high perplexities for rare words [19]. This is especially problematic in morphologically rich languages with long-tailed frequency distributions or domains with dynamic vocabularies (e.g. Twitter). Some language models that leverages sub-word information or morphemes have proposed through a character-level convolutional neural networks. These include using character representations for part-of-speech (POS) tagging [15] and incorporating character-level features to CNN for text classification [21].

In this paper, we propose a novel convolutional neural network model that depends on character-level input to perform sentence semantic representation. We show that when the proposed model trained on datasets, the model does not require the knowledge of words, in addition to conclusion from previous research that CNN do not require the knowledge about syntactic or semantic structure of a language. Working on only characters also has the advantage that abnormal character combinations such as misspelling, ungrammatical expression and emotion icons may be naturally learned. We evaluate performance of the proposed model on two sentence-level paraphrase identification tasks, the model gains significant performance compared to existing word-level input baselines.

2   Preliminary

Firstly, we introduce the typical structure of convolutional neural network sentence model. Then, we describe sentence-level paraphrase identification tasks.

2.1   Convolutional Sentence Model

The input to sentence model is a tokenized sentence $\mathbf{s}$ treated as a sequence of words $\mathbf{s} = w_1, \cdots, w_{|s|}$, where each word is drawn from a vocabulary V [9]. Words are represented by distributional vectors $\mathbf{w} \in \mathbb{R}^d$ looked up in a word embeddings matrix $\mathbf{W} \in \mathbb{R}^{d \times |V|}$ which is formed by concatenating embeddings of all words in V. For each input sentence $\mathbf{s}$ we build a sentence matrix $\mathbf{S} \in \mathbb{R}^{d \times |s|}$, where each column i represents a word embedding w_i at the corresponding position i in a sentence. The sentence matrix is shown in Fig. 1.

To learn to capture and compose features of individual words in a given sentence from low-level word embeddings into higher level semantic concepts, the neural network applies a series of transformations to the input sentence matrix $\mathbf{S}$ using convolution, non-linearity and pooling operations. As shown in Fig. 1, the convolution in layer-1 operates on sliding windows of words (width $m_1 = 3$), which are used to extracted features from the input feature maps via different convolution kernels. The convolutions in deeper layers are defined in a

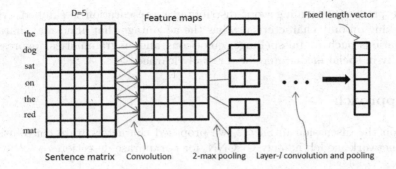

Fig. 1. Illustration of a CNN architecture for modeling sentence.

similar way. Generally, with input sentence $\mathbf{s}$, the convolution unit for feature map of type-f (among f_l of them) on layer-l is

$$c_i^{(l,f)} \overset{def}{=} c_i^{(l,f)}(\mathbf{s}) = \alpha(\mathbf{w}^{(l,f)}\hat{\mathbf{c}}_i^{(l-1)}), f = 1, 2, \cdots, f_l \tag{1}$$

and its matrix form is $\mathbf{c}_i^{(l)} \overset{def}{=} \mathbf{c}_i^{(l)}(\mathbf{s}) = \alpha(\mathbf{W}^{(l)}\hat{\mathbf{c}}_i^{(l-1)} + \mathbf{b}^l)$, where

- $c_i^{(l,f)}(\mathbf{s})$ gives the output of feature map of type-f for location i in layer-l;
- $\mathbf{w}^{(l,f)}$ is the parameters for f on layer-l, with $\mathbf{W}^{(l)} \overset{def}{=} [\mathbf{w}^{(l,1)}, \cdots, \mathbf{w}^{(l,f_l)}]$;
- $\alpha()$ is the non-linear activation function, sigmoid (or logistic), hyperbolic tangent $tanh$, and a rectified linear (ReLU) function are the most common choices of activation functions;
- $\hat{\mathbf{c}}_i^{(l-1)}$ denotes the segment of layer-l-1 for the convolution at location i, while

$$\hat{\mathbf{c}}_i^{(0)} = \mathbf{s}_{i:i+m_1-1} \overset{def}{=} [\mathbf{s}_i^\top, \mathbf{s}_{i+1}^\top, \cdots, \mathbf{s}_{i+m_1-1}^\top]^\top$$

concatenates the vectors for m_1 (width of sliding window) words from input sentence $\mathbf{s}$.

2.2 Paraphrase Identification

Sentence-level paraphrase identification (PI) is the task of detecting whether two sentences are semantically equivalent [16]. It is usually formalized as a binary classification: for two sentences (S_a, S_b), deciding whether they roughly have the same meaning.

Convolution network method [8] and recursive neural model [3] have been applied for paraphrase detection, however, these methods depend on word-level embedding input, which need pre-trained word distributed representation on large-scale corpus. In addition, the methods are evaluated on news corpus, the performance of these models on other paraphrase dataset (e.g. Twitter) is unknown. In this work, we propose a character-aware convolutional neural network model for paraphrase identification, which depends on character-level input and the process of training word distributed representation is skipped. Above all, working on only characters also has the advantage that abnormal character combinations such as misspellings, emoticons and ungrammatical expression, especially in social media, may be naturally learned.

3   Approach

Based on the discussion in Sect. 2, we proposed character-aware convolutional neural network model, namely CwCNN, for paraphrase detection.

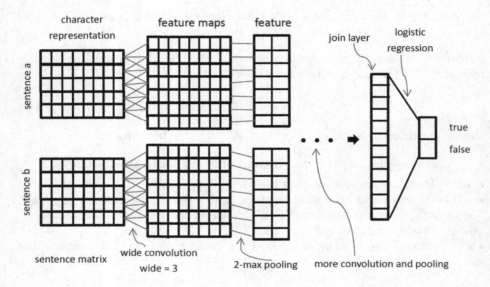

Fig. 2. Illustration of the CwCNN architecture for paraphrase identification.

CwCNN, as illustrated in Fig. 2, takes a convolutional approach. It first finds the representation of each sentence, and then applies convolutional layer to extract features. The pooling operation is applied for these mapped feature matrices, fixed length vector is obtained for each sentence in the convolutional output. After that, a joint layer is used to concatenate the vectors of sentence pair as a dense vector s_o. Finally, a logistic regression implements the binary classification. In the following we describe the details. Our model accepts a sequence of encoded characters as input. The encoding is done with two ways. The first is prescribing an alphabet of size n for the input language, and then quantize each character using 1-of-n encoding (or "one-hot"). The other is to initialize a 50-dimensional vector in $[-0.25, 0.25]$ randomly for each character. The input in all of our model consists of 69 characters, including 26 english letters, 10 digits, 33 other characters. The upper case in sentence is converted to lowercase.

We employ wide one-dimensional convolution [11]. Denoting the number of characters of S_i as $|S_i|$, we convolve weight matrix $\mathbf{M} \in \mathbb{R}^{d \times m}$ over sentence representation matrix $\mathbf{S} \in \mathbb{R}^{d \times |S_i|}$ and generate a matrix $\mathbf{C} \in \mathbb{R}^{d \times (|S_i|+m-1)}$ each column of which is the representation of an m-character. d is the dimension of character embeddings and m is filter width. Our aim for using convolution is that after training, a convolutional filter corresponds to a feature detector that learns to recognize a class of m-grams that is useful for paraphrase identification. Applying the weights $\mathbf{M}$ in a wide convolution has some advantages over applying them in a narrow one. A wide convolution ensures that all weights in the filter reach the entire sentence, including the words at margins.

The k-max pooling operator is applied in the network after each convolutional layer. This guarantees that the input to the fully connected layers is independent of the length of the input sentence. Given a value k and a sequence $\mathbf{p} \in \mathbb{R}^p$ of length $p \geq k$, k-max pooling chooses the subsequence $\mathbf{p}^k_{max}$ of the k highest values of $\mathbf{p}$. The order of the values in $\mathbf{p}^k_{max}$ corresponds to their original order in $\mathbf{p}$. The k-max pooling operations makes it possible to pool the k most active features in $\mathbf{p}$ that may be a number of positions apart. It preserves the order of the features, but is insensitive to their specific positions. The dynamic k-max pooling [11] operation is a k-max pooling operations where let k be a function of the length of the sentence and the depth of the network. Although many functions are possible, we simply model the pooling parameter as follows:

$$k_l = max(k_{tl}, \lceil \frac{L-l}{L} s \rceil) \tag{2}$$

where l is the number of the current convolutional layer to which the pooling is applied and L is the total number of convolutional layers in the network. k_{tl} is the fixed pooling parameter for the topmost convolutional layer.

The model uses logistic regression to compute the probability distribution over the labels, the sigmoid function is following:

$$P(y = t|\mathbf{s_o}) = h_\theta(\mathbf{s_o}) = \frac{1}{1 + e^{\theta^\top \mathbf{s_o}}} \tag{3}$$

Our goal is to find a value of θ so that the probability $P(y{=}t|s_o){=}h_\theta(s_o)$ is large when s_o belongs to the "$true(t)$" class, which indicates paraphrase, and small when s_o belongs to the "$false(f)$" class, which means non-paraphrase.

4   Experiments

We perform a set of comparative experiments on the proposed and several conventional paraphrase detecting methods with Microsoft Research Paraphrase (MSRP) corpus and SemEval-2015 Task 1 Paraphrases in Twitter (PIT) corpus.

4.1   Setup

In MSRP [6], the training set contains 2753 paraphrase pairs and 1323 non-paraphrase pairs, the test set contains 1147 true and 578 false paraphrase pairs,

Table 1. Statistics of PIT-2015 Twitter Paraphrase Corpus. Debatable cases are those received a medium-score from annotators, which are excluded in our experiments.

Dataset	Unique sent	Sent pair	Paraphrase	Non-paraphrase	Debatable
Train	13231	13063	3996 (30.6 %)	7534 (57.7 %)	1533 (11.7 %)
Dev	4772	4727	1470 (31.1 %)	2672 (56.5 %)	585 (12.4 %)
Test	1295	972	175 (18.0 %)	663 (68.2 %)	134 (13.8 %)

respectively. The PIT corpus uses a training and development set of 17,790 sentence pairs and test set of 972 sentence pairs with paraphrase annotations. Table 1 shows the basic statistics of the corpus. For the test set, 972 sentence pairs are selected for evaluating paraphrase recognizing systems. The dataset is more realistic and balanced, containing about 70 % non-paraphrase vs. the 34 % non-paraphrases in the MSRP dervied from new articles.

For MSRP corpus, we use the following four methods as baselines. The first is a supervised model, which uses semantic heuristic features to detect paraphrase [17]. A vector-based similarity was proposed for paraphrase identification that used additive composition of vectors and cosine distance [14]. In order to address the problem of modeling compositional meaning for phrases and sentences, a simple distribution semantic space model is proposed for the distributed representation of sentences [2]. A semantic similarity of short texts method was also used to paraphrase detection [10]. For the PIT corpus, the methods compared to the proposed model are recursive auto-encoders (RAEs) [16], a supervised logistic regression (LR) [5], a weighted matrix factorization (WTMF) [7] and referential machine translation (RTM) [1]. We use accuracy (Acc) and F-measure (F_1) to evaluate experimental results on MSRP corpus. For PIT, precision ($Prec$), recall (Rec) and F_1 are used for experimental evaluation.

4.2 Results

We report the experimental results on both MSRP and PIT corpus using the proposed model, Table 2 shows the evaluated results on the proposed model for the first corpus with four benchmark systems.

From the Table 2, the performance of CwCNN with random initialization surpasses one-hot input, which indicates that 1-of-n encoding pattern will result in sparse representation. And our model has significant performance than semantic similarity of texts and vector-base similarity methods although the Acc is lower than the semantic heuristic features and the F_1 of distribution semantic space method is better than our model.

Table 3 shows the comparison of experimental results between our model and some other paraphrase detection models, we can find that CwCNN with random initialization performs better than other methods and the one-hot input method also obtains a competitive result. Both tables indicate that the proposed method can be used for modeling sentence, especially the social media texts, which also indicate the proposed model has efficient on paraphrase identification tasks.

Table 2. The experimental results for MSRP with our model and other four methods.

Models	$Acc(\%)$	$F_1(\%)$
Semantic similarity of texts [10]	72.6	81.3
Semantic heuristic features [17]	74.7	81.8
Vector-based similarity [14]	73.0	82.0
Distribution semantic space [2]	73.0	82.3
CwCNN with one-hot	68.9	75.0
CwCNN with random	73.3	82.1

Table 3. The experimental results for PIT with our model and baselines.

Models	$Acc(\%)$	$Rec(\%)$	$F_1(\%)$
RAEs [16]	54.3	39.4	45.7
LR [5]	67.9	52.0	58.9
WTMF [7]	45.0	66.3	53.6
RTM [1]	85.9	41.7	56.2
CwCNN with one-hot	51.4	64.0	57.0
CwCNN with random	67.2	54.9	60.4

5   Conclusions

In this paper, we research on the task of sentence-level paraphrase identification by proposing a character-aware convolutional neural network. Our contributions are threefold. Firstly, a novel convolutional neural network is proposed to model sentence for representing the semantic content. The model successfully avoids the hand-craft feature extraction through character-level input. Secondly, the input of model depends only on characters not words, which can be used for more languages and capturing fine-grained semantic and structure information. Thirdly, the proposed model is evaluated in MSRP and PIT, which obtained obvious performance improvement better than the baseline system. In future, we are interested in exploring more effective CNN models, as well as evaluating the proposed model in various of paraphrase identification corpus.

Acknowledgements. We thank the reviewers for their valuable comments and suggestions. This work is supported by grants: State Key Program of National Natural Science Foundation of China (61133012), National Natural Science Foundation of China (61373108 and 61170148).

References

1. Bicici, E.: RTM-DCU: predicting semantic similarity with referential translation machines. In: SemEval 2015, pp. 56–63. Association for Computational Linguistics, June 2015

2. Blacoe, W., Lapata, M.: A comparison of vector-based representations for semantic composition. In: EMNLP, pp. 546–556 (2012)
3. Cheng, J., Kartsaklis, D.: Syntax-aware multi-sense word embeddings for deep compositional models of meaning. In: EMNLP, pp. 1531–1542 (2015)
4. Collobert, R., Weston, J., Bottou, L., Karlen, M., Kavukcuoglu, K., Kuksa, P.P.: Natural language processing (almost) from scratch. JMLR **12**, 2493–2537 (2011)
5. Das, D., Smith, N.A.: Paraphrase identification as probabilistic quasi-synchronous recognition. In: ACL, pp. 468–476. Association for Computational Linguistics, August 2009
6. Dolan, B., Quirk, C., Brockett, C.: Unsupervised construction of large paraphrase corpora: exploiting massively parallel news sources. In: COLING, pp. 350–356, August 23– August 27 2004
7. Guo, W., Diab, M.: Modeling sentences in the latent space. In: ACL, pp. 864–872. Association for Computational Linguistics, July 2012
8. He, H., Gimpel, K., Lin, J.: Multi-perspective sentence similarity modeling with convolutional neural networks. In: EMNLP, pp. 1576–1586, September 2015
9. Hu, B., Lu, Z., Li, H., Chen, Q.: Convolutional neural network architectures for matching natural language sentences. In: Advances in Neural Information Processing Systems, vol. 27, pp. 2042–2050. Curran Associates, Inc. (2014)
10. Islam, A., Inkpen, D.: Semantic similarity of short texts. In: RANLP, pp. 291–297 (2007)
11. Kalchbrenner, N., Grefenstette, E., Blunsom, P.: A convolutional neural network for modelling sentences. In: ACL, pp. 655–665, June 2014
12. Kiela, D., Bottou, L.: Learning image embeddings using convolutional neural networks for improved multi-modal semantics. In: EMNLP, pp. 36–45, October 2014
13. Kim, Y.: Convolutional neural networks for sentence classification. In: EMNLP, pp. 1746–1751, October 2014
14. Milajevs, D., Kartsaklis, D., Sadrzadeh, M., Purver, M.: Evaluating neural word representations in tensor-based compositional settings. In: EMNLP, pp. 708–719 (2014)
15. dos Santos, C.N., Zadrozny, B.: Learning character-level representations for part-of-speech tagging. In: ICML, pp. 1818–1826 (2014)
16. Socher, R., Huang, E.H., Pennin, J., Manning, C.D., Ng, A.Y.: Dynamic pooling and unfolding recursive autoencoders for paraphrase detection. In: NIPS, pp. 801–809 (2011)
17. Ul-Qayyum, Z., Altaf, W.: Paraphrase identification using semantic heuristic features. Res. J. Appl. Sci. Eng. Technol. **4**(22), 4894–4904 (2012)
18. Yih, W.t., He, X., Meek, C.: Semantic parsing for single-relation question answering. In: ACL, pp. 643–648, June 2014
19. Yoon, K., Yacine, J., David, S., Alexander, M.R.: Character-aware neural language models. In: AAAI (2016)
20. Zeng, D., Liu, K., Lai, S., Zhou, G., Zhao, J.: Relation classification via convolutional deep neural network. In: COLING, pp. 2335–2344, August 2014
21. Zhang, X., Zhao, J., LeCun, Y.: Character-level convolutional networks for text classification. In: NIPS, pp. 649–657 (2015)

Learning a Discriminative Dictionary with CNN for Image Classification

Shuai Yu[1], Tao Zhang[1], Chao Ma[1], Lei Zhou[1], Jie Yang[1(✉)],
and Xiangjian He[2]

[1] Institute of Image Processing and Pattern Recognition,
Shanghai Jiao Tong University, Shanghai, China
{yushuai9471,zhb827,sjtu_machao,311821-8.15,jieyang}@sjtu.edu.cn
[2] University of Technology, Sydney, Australia
sean@it.uts.edu.au

Abstract. In this paper, we propose a novel framework for image recognition based on an extended sparse model. First, inspired by the impressive results of CNN over different tasks in computer vision, we use the CNN models pre-trained on large datasets to generate features. Then we propose an extended sparse model which learns a dictionary from the CNN features by incorporating the reconstruction residual term and the coefficients adjustment term. Minimizing the reconstruction residual term guarantees that the class-specific sub-dictionary has good representation power for the samples from the corresponding class and minimizing the coefficients adjustment term encourages samples from different classes to be reconstructed by different class-specific sub-dictionaries. With this learned dictionary, not only the representation residual but also the representation coefficients will be discriminative. Finally, a metric involving these discriminative information is introduced for image classification. Experiments on Caltech101 and PASCAL VOC 2012 datasets show the effectiveness of the proposed method on image classification.

Keywords: Image classification · Convolutional Neural Networks · Sparse model · Unsupervised dictionary learning

1 Introduction

As one of the most active research areas in computer vision, image classification has been widely studied. Conventional approaches for image classification use carefully designed hand-crafted features, e.g., SIFT [13]. Recently, in contrast to the hand-crafted features, features learnt by deep network architectures, represented by deep convolutional neural networks (CNN) [9] have got impressive results over different areas, especially in image classification. Specifically, deep learning attempts to model the visual data of high level abstract structural composites using multivariate nonlinear transformations. Several works [16,19] show that the pre-trained CNN models on colossal datasets with data diversity can be transferred to extract discriminative features for other tasks.

© Springer International Publishing AG 2016
A. Hirose et al. (Eds.): ICONIP 2016, Part II, LNCS 9948, pp. 185–194, 2016.
DOI: 10.1007/978-3-319-46672-9_22

After image representation by CNN features, we adopt sparsity-based classification (SRC) to achieve the image classification task, due to the applicability of SRC with nonlinear features. For SRC, Wright et al. [20] proposed a general classification scheme based on sparse representation and applied it on robust face recognition (FR). Obviously, SRC is impressive that a test sample can be represented by a weighted combination of those training samples belonging to the same class. Since the SRC scheme achieves competitive performance in FR, it triggers the researchers' interest in sparsity-based pattern classification. How to learn a discriminative dictionary for both sparse data representation and classification is still an open problem.

According to predefined relationship between dictionary atoms and class labels, we can divide current supervised dictionary learning (DL) into three categories: shared DL, class-specific DL and hybrid DL. In the shared DL, a dictionary shared by all classes is learned but also the discriminative power of the representation coefficients is mined. It is popular to learn a shared dictionary and simultaneously train a classifier using the representation coefficients. In [12], Marial et al. proposed a scheme which learned discriminative dictionaries while training a linear classifier over coding coefficients. Inspired by KSVD [1], Zhang and Li [24] proposed discriminative KSVD learning algorithm on FR. Following the work in [24], Jiang et al. [7] proposed to enhance the discriminative power via adding a label consistent term. Recently, Mairal et al. [10] proposed to minimize different risk functions over the coding coefficients for different tasks, called a task-driven DL. Generally, in this scheme, a shared dictionary and a classifier over the representation coefficients are learned together. However, there is no relationship between the dictionary atoms and the class labels, and thus no class-specific representation residuals are introduced to perform classification task.

In the class-specific DL, a dictionary whose atoms are predefined to correspond to subject class labels is learned and thus the class-specific reconstruction error could be used to perform classification. Via adding a discriminative reconstruction penalty term in the KSVD model [1], Mairal et al. [11] proposed a DL algorithm for texture segmentation and scene analysis. Yang et al. [21] proposed to learn a structural dictionary and impose the Fisher discrimination criterion on both the sparse coding coefficients to enhance class discrimination power. In [3], via adding non-negative penalty on both dictionary atoms and representation coefficients, Castrodad and Sapiro proposed to learn a set of action-specific dictionaries. In [15], Ramirez et al. introduced an incoherence promoting term to the DL model for ensuring the dictionaries representing different classes to be as independent as possible. H. Wang et al. [18] learned a dictionary with similarity constrained term and the dictionary incoherence term and applied it to human action recognition. Based on each atom in the learned dictionary fixed to a single class label, the representation residual associated with each class-specific dictionary could be used to perform classification.

Very recently, the hybrid dictionary models which combine shared dictionary atoms and class-specific dictionary atoms have been proposed. Using a Fisher-like penalty term on the coding coefficients, Zhou et al. [25] learned a hybrid dictionary, while introducing a coherence penalty term on different sub-dictionaries,

Kong et al. [8] learned a hybrid dictionary. Although the shared dictionary atoms could encourage learned hybrid dictionary compact to some extent, how to balance the shared part and class-specific part in the hybrid dictionary is not a trivial task.

Based on the previous works, we propose an extended sparse framework to learn a class-specific dictionary with input features extracted from CNN, i.e., the dictionary atoms corresponding to the class labels. In this proposed framework, two terms named the reconstruction residual term and the coefficients adjustment term, are introduced to ensure the learned dictionary with the powerful discriminative ability. The reconstruction residual term is utilized to enforce that class-specific sub-dictionary has good reconstruction capability for the training samples from the same class. The coefficients adjustment term is utilized to enforce that class-specific sub-dictionaries have poor reconstruction capability for training samples from different classes. Therefore, both the representation residual and the representation coefficients of a query sample will be discriminative, and a corresponding classification scheme is proposed to exploit such information. Then we test our classification scheme on Caltech-101 [6] and VOC 2012 [5].

The remainder of this paper is organized as follows. In Sect. 2, we introduce the proposed extended sparse framework and a supervised class-specific DL method for classification. In Sect. 3, we demonstrate experimental results. In Sect. 4, we make conclude our method.

2 Methodology

Since the previous works show that pre-trained CNN models on colossal datasets with data diversity can be transferred to extract CNN features for other image datasets, we use the VGG-Net [17] model pre-trained on ImageNet [4] to extract features for sparse representation-based DL. As for the selection of the features from CNN, those belonging to the shallow layers contain too many dimensions and they are too sparse to get effective results for classification. Meanwhile, the features of the deepest layer is totally corresponding to the original dataset, which is limited in transferring to other tasks. Therefore we select some middle layers of CNN to get features for classification. In our experiments, we adopt the fc_7 layer of 4096 dimensions as the feature for sparse representation-based DL.

2.1 Sparse Representation and Dictionary Learning

Due to the diversity of CNN features, we adopt the class-specific dictionary. In the class-specific DL, each dictionary atom in the learned dictionary $D = [D_1, D_2, \ldots, D_K]$ have class label corresponding to the subject classes, where D_i is the sub-dictionary corresponding to class i. By representing a test sample over the learned dictionary D, the representation residual associated with each class can be naturally employed to classify it, as in the SRC method.

Given $a_{i,j}$, $i = 1, \ldots K$, $j = 1, \ldots, n_i$ denotes a training sample described by CNN features in class i, where K is the sum of classes, and n_i is the number of samples in class i. We form $A_i = [a_{i,1}, a_{i,2}, \ldots, a_{i,n_i}]$. The dictionary D can be learned by the following extended sparse model:

$$< D, Z >= \arg\min_{D,Z} \sum_{i=1}^{K} \{ \|A_i - DZ_i\|_F^2 + \lambda_1 \|Z_i\|_1 + \lambda_2 \|A_i - D_i Z_i^i\|_F^2$$

$$+ \kappa \sum_{j \neq i} \|\tilde{Z}_j^T Z_i\|_F^2 \} \qquad\qquad s.t. \|d_l\|_2 = 1, \forall l \qquad (1)$$

where Z_i is the sub-matrix containing the coding coefficients of A_i over D. Z_i can be written as $Z_i = [Z_i^1; \ldots; Z_i^j; \ldots; Z_i^K]$, where Z_i^j represents the coefficients of A_i over D_j; and $\tilde{Z}_j$ is $\tilde{Z}_j = [\tilde{Z}_{j,1}, \tilde{Z}_{j,2}, \ldots, \tilde{Z}_{j,n_j}]$, where $\tilde{Z}_{j,i} = Z_{j,i}/\|Z_{j,i}\|$ is normalized coefficients of the i-th sample in A_i over D. And $\lambda_1, \lambda_2, \kappa$ are weight coefficients.

Different from the conventional sparse model SRC in [20], the reconstruction residual term $\|A_i - D_i Z_i^i\|_F^2$ and coefficients adjustment term $\sum_{j \neq i} \|\tilde{Z}_j^T Z_i\|_F^2$ are introduced in Eq.(1).

Reconstruction Residual Term: For A_i, it should be well represented by the dictionary D, hence there is $A_i \approx DZ_i$. Since A_i is associated with the class i, it is expected that A_i could be represented further well by D_i. This implies that Z_i should have some significant coefficients Z_i^i such that $\|A_i - D_i Z_i^i\|_F^2$ is small.

Coefficients Adjustment Term: In the SRC scheme proposed by Wright et al. [20], given a test sample, the accurate classification can be conducted based on that the largest coefficients are associated with the training samples that belong to the same class as the test sample. It implies that the reconstruction error is minimized when test sample are sparsely represented by its own training samples. Likewise, in the class-specific DL, it is expected that the largest coefficients of A_i are associated with the sub-dictionary D_i. In Eq. (1), minimizing the coefficients adjustment term $\sum_{j \neq i} \|\tilde{Z}_j^T Z_i\|_F^2$ encourages that for the A_i and A_j, the largest coefficients are associated with the corresponding different sub-dictionary D_i and D_j as illustrated in Fig. 1. This means that similar samples over dictionary D have similar coefficients and samples belonging to different classes over dictionary D have absolutely different coefficients. Therefore, the value of the object function Eq.(1) is minimized when samples are sparsely represented by dictionary atoms in their own sub-dictionaries.

Overall, minimizing the reconstruction residual term $\|A_i - D_i Z_i^i\|_F^2$ guarantees that the class-specific sub-dictionary has good representation power for the samples from the corresponding class and minimizing the coefficients adjustment term $\sum_{j \neq i} \|\tilde{Z}_j^T Z_i\|_F^2$ encourages samples from different classes are reconstructed by different class-specific sub-dictionaries. By incorporating the reconstruction residual term and coefficients adjustment term, our proposed sparse representation algorithm is more effective for classification.

The optimization: Although the objective function in Eq.(1) is not jointly convex to (D, Z). Like the work done by other authors [18,22] when trying to

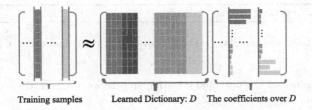

Training samples   Learned Dictionary: D The coefficients over D

Fig. 1. Sparse representation of training samples using the learned dictionary D. The green and yellow training samples belong to class i and j; the green and yellow atoms in D have class labels corresponding to class i and j. The sparse coefficients of green and yellow training samples recovered are plotted in the coefficients matrix with the green and yellow largest vales associated with the green and yellow atoms in D which have class labels corresponding to class i and j. (Color figure online)

solve similar optimization problems, here we divide the objective function in Eq.(1) into two sub-problems by optimizing D and Z alternatively: updating coefficient matrix Z while fixing the dictionary D, and updating dictionary D while fixing the coefficient matrix Z.

Update of Z: When we fix the dictionary D, the objective function in Eq.(1) is reduced to a sparse representation problem to compute $Z = [Z_1, Z_2, \ldots, Z_K]$. We can compute Z_i class by class by fixing $Z_j, j \neq i$. The objective function in Eq.(1) is further reduced to:

$$\min_{Z_i}\{\|A_i - DZ_i\|_F^2 + \lambda_1\|Z_i\|_1 + \lambda_2\|A_i - D_iZ_i^i\|_F^2 + \kappa\sum_{j\neq i}\|\tilde{Z}_j^T Z_i\|_F^2\} \quad (2)$$

It can be proved that $\varphi_i(Z_i) = \|A_i - DZ_i\|_F^2 + \lambda_2\|A_i - D_iZ_i^i\|_F^2 + \kappa\sum_{j\neq i}$ $\|\tilde{Z}_j^T Z_i\|_F^2$ is convex with Lipschitz continuous gradient. Hence, in this work we adopt a new fast iterative shrinkage-thresholding algorithm (FISTA) [2] to solve Eq.(2), as described in Algorithm 1.

Update of D: Then we describe how to update $D = [D_1, D_2, \ldots, D_K]$, while fixing the coefficient matrix Z. When updating D_i, all D_j, $j \neq i$, are fixed and $D_i = [d_1, d_2, \ldots, d_{p_i}]$ is updated class by class. We can reduce objective function in Eq.(1) as:

$$\min_{D_i}\{\|\bar{A} - D_iZ^i\|_F^2 + \lambda_2\|A_i - D_iZ_i^i\|_F^2\} \quad s.t.\|d_l\|_2 = 1, l = 1, \ldots, p_i \quad (3)$$

where $\bar{A} = A - \sum_{j=1,j\neq i}^{K}$; Z^i represents the coefficient matrix of A over D_i. Equation (3) can be further reduced to:

$$\min_{D_i}\|\Lambda_i - D_i\Gamma_i\|_F^2 \quad s.t.\|d_l\|_2 = 1, l = 1, \ldots, p_i \quad (4)$$

where $\Lambda_i = [\bar{A}A_i]$, $\Gamma_i = [Z^iZ_i^i]$. Equation(4) can be efficiently solved by updating each dictionary atom one by one via the algorithm like [22], as presented in Algorithm 2.

Algorithm 1. Learning sparse code Z_i.

Input:
   A training subset A_i from class i; the dictionary D; the parameters $\rho, \tau > 0$.
Initialize:
   $\hat{Z}_i^{(1)} \leftarrow 0$ and $t \leftarrow 1$;
   while convergence or the maximal iteration step is not reached **do**
      $t \leftarrow t + 1; u^{t-1} \leftarrow \hat{Z}_i^{(t-1)} - 1/2\rho \nabla \varphi_i(\hat{Z}_i^{(t-1)})$,
      where $\nabla \varphi_i(\hat{Z}_i^{(t-1)})$ is the derivative of $\varphi_i(\hat{Z}_i^{(t-1)})$ w.r.t. $\hat{Z}_i^{(t-1)}$;
      $\hat{Z}_i^{(t)} \leftarrow soft(u^{(t-1)}, \tau/\rho)$, where $soft(u, \tau/\rho)$ is defined by :
      $soft(u, \tau/\rho) = 0$, if $\|u_j\| \leq \tau/\rho; soft(u, \tau/\rho) = u_j - sign(u_j)\tau/\rho$, otherwise
   end while
Output:
   $\hat{Z}_i = \hat{Z}_i^{(t)}$.

Algorithm 2. Learning dictionary D_i.

Input:
   A training subset A_h from class i; the coefficients Z_i; the dictionary D_i^o.
   Let $Z_i = [z_1; z_2; \ldots; z_{p_i}]$ and $D_i^o = [d_1; d_2; \ldots; d_{p_i}]$, where $z_j, j = 1, 2, \ldots, p_i$,
   is the row vector of Z_i and d_j is the jth column vector of D_i^o;
   for $j = 1$ to p_i **do**
      Fix all $d_l, l \neq j$ and update d_j. Let $X = \Lambda_i - \sum_{l \neq j} d_l z_l$. The minimization of
      Eq.(4) becomes: $\min_{d_j} \|X - d_j z_j\|_F^2$ s.t. $\|d_j\|_2 = 1$
      By solving this objective function, we could get the solution
      $d_j = X z_j^T / \|X z_j^T\|_2$.
   end for
Output:
   The updated version of $D_i^o : D_i$.

Complete dictionary D learning algorithm: The complete algorithm is summarized in Algorithm 3. The algorithm converges since the cost function in Eq.(1) is lower bounded and can only decrease in the two alternative minimization stages (i.e., updating Z and updating D).

The classification scheme: Once the dictionary D have been trained, it could be adopted to represent a query sample y and do a classification task. According to different schemes for learning the dictionary D, different information can be utilized to perform the classification task.

   In our proposed sparse representation model, not only the desired dictionary D is learned from the training dataset A, but also the normalized representation matrix $\tilde{Z}_i$ of each class A_i is computed. Considering both the representation residual and the representation coefficients are discriminative, we can make use of both of them to achieve more accurate classification results. Hence, we propose the following representation model:

$$\hat{\alpha} = \arg \min_{\alpha} \{\|y - D\alpha\|_2^2 + \gamma \|\alpha\|_1\} \tag{5}$$

where, γ constant.

Algorithm 3. The complete algorithm of dictionary D learning.

Initialize D.
   We initialize the atoms of Di as the eigenvectors of A.
Update coeffcients Z.
   Fix D and solve Z_i, $i = 1, 2, \ldots, K$, one by one by solving Eq.(2) with Algorithm 1.
Update coeffcients D.
   Fix Z and update each D_i, $i = 1, 2, \ldots, K$, by solving Eq.(3) with Algorithm 2.
   return
   Update D and Z when the objective function values between adjacent
   iterations are not close enough or the maximum number of iterations is not reached.
Output:
   Z and D

Let $\hat{\alpha} = [\hat{\alpha}^1, \hat{\alpha}^2, \ldots, \hat{\alpha}^K]$, where $\hat{\alpha}^i$ is the coefficient sub-vector associated with sub-dictionary D_i. In the training stage, we have enforced the class-specific representation residual to be discriminative. Therefore, if y is from class i, the residual $\|y - D_i\hat{\alpha}^i\|_2^2$ should be small while $\|y - D_j\hat{\alpha}^j\|_2^2$, $j \neq i$, should be large. In addition, the representation sub-vector $\hat{\alpha}^j$ should be far different from the representation vector of other classes. By considering the discrimination capability of both representation residual and representation vector, we could define the following metric for classification:

$$e_i = \|y - D_i\hat{\alpha}^i\|_2 + w\sum_{j \neq i} \|\tilde{Z}_j^T\hat{\alpha}\|/n_j \qquad (6)$$

where w is preset weight to balance the contribution of the two terms for classification. The classification rule is simply set as $identity(y) = \arg\min_i\{e_i\}$.

3   Experimental Results

3.1   Datasets and Experiments Settings

For image classification task, test our classification scheme on the widely used datasets, Caltech-101 and VOC 2012. In our proposed sparse representation model, there are two stages: DL stage and classification stage. In DL stage we set weight coefficients by experiment: $\lambda_1 = 0.005$, $\lambda_2 = 1$, $\kappa = 0.01$; in classification stage we set $\gamma = 1$, $w = 0.05$. In the proposed model, the number of atoms in D_i, denoted by p_i, is important and it is set as the number of training samples by default. All of the experiments are executed on a workstation with 2.8GHz CPU and 16GB RAM.

3.2   Experiments on Caltech-101

To verify the effectiveness of our proposed sparse model for image classification, we make comparisons with other classifiers. We use the same features exacted from CNN as the input of SRC [20], SVM and our sparse model incorporating reconstruction residual and coefficients adjustment terms (IRRCA).

We evaluate our algorithm on Caltech101 dataset with cross-validation: 5-30 random images are used for training, and the remaining for testing. For each size of training images, we process 10 times with our method and the results are averaged. The SVM parameters are estimated by 5-fold cross-validation on each training data. The result is shows in Table 1.

The accuracy of SVM is higher than that of SRC which only uses the original training samples as dictionary, and our IRRCA achieves the highest accuracy. It proves that the proposed sparse model with supervised DL method is discriminative for image classification, and by incorporating the reconstruction residual term and coefficients adjustment term, our proposed sparse representation model is more effective for classification.

3.3 Experiments on VOC 2012

To further verify the effectiveness of our method, which combined the CNN features and sparse representation-based DL, we make experiments on VOC 2012 for classification task and compare with the results of state-of-art methods using CNN models pre-trained on ILSVRC-2012 (1000 classes)[4]. The experiment result that our method compared with other methods is shown in Table 2. The overall mAP we obtain on this dataset is 82.9 %(AP). For most of the categories, the results reach more than 80 %(AP). The classification result of our method surpass others with similar CNN features in most of the categories. However,

Table 1. The IRRCA performance comparison on Caltech-101(Accuracy)

Training images	5	10	15	20	25	30
SRC	62.38	69.44	73.13	77.25	79.04	82.56
SVM	62.07	70.01	74.52	77.49	81.97	83.98
IRRCA	64.38	72.21	76.17	81.34	83.36	85.88

Table 2. The performance comparison on VOC 2012(AP)

Category	Aero	Bike	Bird	Boat	Bottle	Bus	Car	Cat	Chair	Cow	
Wei et al. [19]	97.7	83.0	93.2	87.2	59.6	88.2	81.9	94.7	66.9	81.6	
Oquab et al. [14]	93.5	78.4	87.7	80.9	57.3	85.0	81.6	89.4	66.9	73.8	
Zeiler et al. [23]	96.0	77.1	88.4	85.5	55.8	85.8	78.6	91.2	65.0	74.4	
Ours	97.6	83.4	90.2	88.7	62.3	89.1	80.9	91.5	68.6	77.5	
Category	Table	Dog	Horse	Motor	Person	Plant	Sheep	Sofa	Train	TV	mAP
Wei et al. [19]	68.0	93.0	88.2	87.7	92.7	59.0	85.1	55.4	93.0	77.2	81.7
Oquab et al. [14]	62.0	89.5	83.2	87.6	95.8	61.4	79.0	54.3	88.0	78.3	78.7
Zeiler et al. [23]	67.7	87.8	86.0	85.1	90.9	52.2	83.6	61.1	91.8	76.1	79.0
Ours	71.7	92.8	90.1	91.9	94.5	59.2	84.2	65.4	94.4	83.5	82.9

some categories like "Chair" and "Table" got lower AP, one possible explanation is that the CNN features of these categories are not discriminative enough.

4   Conclusion

In this paper, we have propose an extended sparse model to learn a discriminative dictionary for classification. We have adopted a pre-trained CNN model on large datasets to exact input features. In the proposed sparse model, the reconstruction residual term and the coefficients adjustment term are introduced to ensure the learned dictionary to obtain powerful discriminative ability. With this learned dictionary, both the representation residual and the representation coefficients are discriminative. Finally, we have present a corresponding classification scheme by exploiting such information. The experiments show the proposed method are effective for classification.

Acknowledgments. This research is partly supported by 973 Plan, China (No. 2015CB856004) and NSFC, China (No: 61572315).

References

1. Aharon, M., Elad, M., Bruckstein, A.: K-svd: An algorithm for designing overcomplete dictionaries for sparse representation. IEEE Trans. Signal Process. **54**(11), 4311–4322 (2006)
2. Beck, A., Teboulle, M.: A fast iterative shrinkage-thresholding algorithm for linear inverse problems. SIAM J. Imaging Sci. **2**(1), 183–202 (2009)
3. Castrodad, A., Sapiro, G.: Sparse modeling of human actions from motion imagery. Int. J. Comput. Vis. **100**(1), 1–15 (2012)
4. Deng, J., Dong, W., Socher, R., Li, L.J., Li, K., Fei-Fei, L.: Imagenet: a large-scale hierarchical image database. In: 2009 IEEE Conference on Computer Vision and Pattern Recognition, CVPR 2009, pp. 248–255. IEEE (2009)
5. Everingham, M., Van Gool, L., Williams, C.K.I., Winn, J., Zisserman, A.: The PASCAL visual object classes challenge (VOC 2012) results (2012). http://www.pascal-network.org/challenges/VOC/voc2012/workshop/index.html
6. Fei-Fei, L., Fergus, R., Perona, P.: Learning generative visual models from few training examples: an incremental bayesian approach tested on 101 object categories. Comput. Vis. Image Underst. **106**(1), 59–70 (2007)
7. Jiang, Z., Lin, Z., Davis, L.S.: Label consistent K-SVD: Learning a discriminative dictionary for recognition. IEEE Trans. Pattern Anal. Mach. Intell. **35**(11), 2651–2664 (2013)
8. Kong, S., Wang, D.: A dictionary learning approach for classification: separating the particularity and the commonality. In: Fitzgibbon, A., Lazebnik, S., Perona, P., Sato, Y., Schmid, C. (eds.) ECCV 2012, Part I. LNCS, vol. 7572, pp. 186–199. Springer, Heidelberg (2012)
9. Le Cun, B.B., Denker, J.S., Henderson, D., Howard, R.E., Hubbard, W., Jackel, L.D.: Handwritten digit recognition with a back-propagation network. In: Advances in neural information processing systems. Citeseer (1990)
10. Mairal, J., Bach, F., Ponce, J.: Task-driven dictionary learning. IEEE Trans. Pattern Anal. Mach. Intell. **34**(4), 791–804 (2012)

11. Mairal, J., Bach, F., Ponce, J., Sapiro, G., Zisserman, A.: Discriminative learned dictionaries for local image analysis. In: 2008 IEEE Conference on Computer Vision and Pattern Recognition, CVPR 2008, pp. 1–8. IEEE (2008)

12. Mairal, J., Ponce, J., Sapiro, G., Zisserman, A., Bach, F.R.: Supervised dictionary learning. In: Advances in neural information processing systems, pp. 1033–1040 (2009)

13. Ng, P.C., Henikoff, S.: Sift: predicting amino acid changes that affect protein function. Nucleic Acids Res. **31**(13), 3812–3814 (2003)

14. Oquab, M., Bottou, L., Laptev, I., Sivic, J.: Learning and transferring mid-level image representations using convolutional neural networks. In: Proceedings of the IEEE Conference on Computer Vision and Pattern Recognition, pp. 1717–1724 (2014)

15. Ramirez, I., Sprechmann, P., Sapiro, G.: Classification and clustering via dictionary learning with structured incoherence and shared features. In: 2010 IEEE Conference on Computer Vision and Pattern Recognition, CVPR 2010, pp. 3501–3508. IEEE (2010)

16. Razavian, A., Azizpour, H., Sullivan, J., Carlsson, S.: CNN features off-the-shelf: an astounding baseline for recognition. In: Proceedings of the IEEE Conference on Computer Vision and Pattern Recognition Workshops. pp. 806–813 (2014)

17. Szegedy, C., Liu, W., Jia, Y., Sermanet, P., Reed, S., Anguelov, D., Erhan, D., Vanhoucke, V., Rabinovich, A.: Going deeper with convolutions. In: Proceedings of the IEEE Conference on Computer Vision and Pattern Recognition, pp. 1–9 (2015)

18. Wang, H., Yuan, C., Hu, W., Sun, C.: Supervised class-specific dictionary learning for sparse modeling in action recognition. Pattern Recogn. **45**(11), 3902–3911 (2012)

19. Wei, Y., Xia, W., Huang, J., Ni, B., Dong, J., Zhao, Y., Yan, S.: CNN: single-label to multi-label (2014). arXiv preprint arXiv:1406.5726

20. Wright, J., Yang, A.Y., Ganesh, A., Sastry, S.S., Ma, Y.: Robust face recognition via sparse representation. IEEE Trans. Pattern Anal. Mach. Intell. **31**(2), 210–227 (2009)

21. Yang, M., Zhang, L., Feng, X., Zhang, D.: Fisher discrimination dictionary learning for sparse representation. In: 2011 IEEE International Conference on Computer Vision (ICCV), pp. 543–550. IEEE (2011)

22. Yang, M., Zhang, L., Feng, X., Zhang, D.: Sparse representation based fisher discrimination dictionary learning for image classification. Int. J. Comput. Vis. **109**(3), 209–232 (2014)

23. Zeiler, M.D., Fergus, R.: Visualizing and understanding convolutional networks. In: Fleet, D., Pajdla, T., Schiele, B., Tuytelaars, T. (eds.) ECCV 2014, Part I. LNCS, vol. 8689, pp. 818–833. Springer, Heidelberg (2014)

24. Zhang, Q., Li, B.: Discriminative K-SVD for dictionary learning in face recognition. In: Computer Vision and Pattern Recognition, CVPR 2010, pp. 2691–2698. IEEE (2010)

25. Zhou, N., Shen, Y., Peng, J., Fan, J.: Learning inter-related visual dictionary for object recognition. In: 2012 IEEE Conference on Computer Vision and Pattern Recognition, CVPR 2012, pp. 3490–3497. IEEE (2012)

Online Weighted Multi-task Feature Selection

Wei Xue[1] and Wensheng Zhang[1,2(✉)]

[1] School of Computer Science and Engineering,
Nanjing University of Science and Technology, Nanjing 210094, China
`mailweixue@163.com`
[2] Institute of Automation, Chinese Academy of Sciences, Beijing 100190, China
`wensheng.zhang@ia.ac.cn`

Abstract. The goal of multi-task feature selection is to learn explanatory features across multiple related tasks. In this paper, we develop a weighted feature selection model to enhance the sparsity of the learning variables and propose an online algorithm to solve this model. The worst-case bounds of the time complexity and the memory cost of this algorithm at each iteration are both in $\mathcal{O}(N \times Q)$, where N is the number of feature dimensions and Q is the number of tasks. At each iteration, the learning variables can be solved analytically based on a memory of the previous (sub)gradients and the whole weighted regularization, and the weight coefficients used for the next iteration are updated by the current learned solution. A theoretical analysis for the regret bound of the proposed algorithm is presented, along with experiments on public data demonstrating that it can yield better performance, e.g., in terms of convergence speed and sparsity.

1 Introduction

Multi-task learning, other than traditional single-task learning that treats each task separately and independently, is a machine learning method that is aimed to improve the performance of multiple related tasks by exploiting the intrinsic relationships among them [5]. Learning multiple related tasks simultaneously by utilizing shared information across tasks has demonstrated advantages over those models learned through individual tasks, see [2,14,15,19] for example. Recently, multi-task learning has received increasing attention and has been successfully employed in various applications including medical diagnosis [1], human action recognition [9], image classification [11], fine grained visual recognition [20], etc. A key problem in multi-task learning problem is to find the explanatory features across these multiple related tasks. Many methods have been proposed to solve this problem by utilizing some variants of l_1 norm, particularly matrix norms, such as $l_{2,1}$ and $l_{\infty,1}$ norms. The $l_{2,1}$ norm is the sum of the l_2 norm of the rows, and the $l_{\infty,1}$ norm is the sum of the l_∞ norm of the rows. A major advantage of the two matrix norms is that they can not only encourage row sparsity, but also achieve group sparsity among features, e.g., group lasso [18]. Despite their success in many applications, the previous *multi-task feature selection* (MTFS) methods still have defects that need to be addressed. (1) These methods are implemented

© Springer International Publishing AG 2016
A. Hirose et al. (Eds.): ICONIP 2016, Part II, LNCS 9948, pp. 195–203, 2016.
DOI: 10.1007/978-3-319-46672-9_23

in a batch training model. However, the data in real-world applications may appear sequentially. In this situation, these methods cannot be used. (2) The real-world data can be in a large volume, over millions both in the sample size and the feature space. These algorithms will suffer inefficiency in this case due to their inefficiency or poor scalability, especially when the data cannot be loaded into memory simultaneously. (3) Most previous MTFS methods only select features in individual tasks or across the tasks, while how to enhance the sparsity of the learned variables is not taken into consideration.

To address the above problems, we first develop a novel MTFS model, *weighted multi-task feature selection* (wMTFS), which is aimed to enhance the sparsity of the learned variables during the learning process. Furthermore, we propose an online learning algorithm to solve the wMTFS model. Unlike traditional batch learning methods, online learning represents a promising family of efficient and scalable machine learning algorithms for large scale applications. To the best of our knowledge, we are the first to study the (online) weighted multi-task feature selection model. The main contributions of this paper include the following.

- We develop a weighted multi-task learning model that can not only select important features across all tasks but also achieve more sparse solutions.
- We propose an online learning algorithm based on a recently developed first-order method to solve the wMTFS model.
- We give a regret bound of the proposed algorithm, which also theoretically guarantees its convergence rate.

The rest of this paper is organized as follows. We introduce the proposed wMTFS formulation in Sect. 2. In Sect. 3, we present the online optimization algorithm together with its theoretical analysis. Experimental results are presented in Sect. 4. Finally, we conclude this paper in Sect. 5.

2 Weighted Multi-task Learning Formulation

In this section, we first introduce the problem setup of multi-task learning and then give the formulation of the weighted multitask feature selection.

2.1 Problem Setup

Assume that we are given Q tasks with all data coming from the same space $\mathcal{X} \times \mathcal{Y}$, where $\mathcal{X} \subset \mathbb{R}^N$ and $\mathcal{Y} \subset \mathbb{R}$. Each task has M_q samples. Hence, it consists of a dataset of $\mathfrak{S} = \cup_{q=1}^{Q} \mathfrak{S}_q$, where $\mathfrak{S}_q = \{\mathbf{z}_i^q = (\mathbf{x}_i^q, y_i^q)\}_{i=1}^{M_q}$ are sampled from a distribution $\mathcal{P}_q$ on $\mathcal{X} \times \mathcal{Y}$. The goal of multi-task learning is to learn Q decision functions $f^q : \mathbb{R}^N \to \mathbb{R}$, $q = 1, \ldots, Q$ such that $f^q(\mathbf{x}_i^q)$ approximates y_i^q.

2.2 Formulation

In typical multi-task learning models, the decision function f^q for the q-th task is often parameterized by a weight vector $\mathbf{w}^q$, which means, $f^q(\mathbf{x}) = \mathbf{w}^{q\top}\mathbf{x}$, $q = 1, \ldots, Q$. Hence, all weight vectors learned consist of a matrix in the size of $N \times Q$. For convenience, we express the learned weight matrix into rowwise and columnwise vectors, i.e., $\mathbf{W} = (\mathbf{w}^1, \ldots, \mathbf{w}^Q) = (\mathbf{W}_{1\bullet}; \ldots; \mathbf{W}_{N\bullet}) = (\mathbf{W}_{\bullet 1}, \ldots, \mathbf{W}_{\bullet Q})$. The objective of MTFS models is to learn a weight matrix by solving the *regularized empirical risk minimization* problem: $\min_{\mathbf{W}} \sum_{q=1}^{Q} \frac{1}{M_q} \sum_{i=1}^{M_q} l^q(\mathbf{W}_{\bullet q}, \mathbf{z}_i^q) + \lambda R(\mathbf{W})$, where $\lambda > 0$ is a regularization parameter to balance the loss and the regularization term, $l^q(\mathbf{W}_{\bullet q}, \mathbf{z}_i^q)$ is the loss on the sample $\mathbf{z}_i^q$ for the q-th task, and $R(\mathbf{W})$ is a regularization (matrix norm). We assume that both the loss function and the regularization are convex. Typical loss functions include least square loss, logistic loss and hinge loss. Examples of the regularization include:

- $l_{1,1}$ norm, $R(\mathbf{W}) = \sum_{q=1}^{Q} \|\mathbf{W}_{\bullet q}\|_1$, which simply sums the l_1 norm on the weights of all tasks together to yield sparse solutions, but it does not find the information across the tasks [14].
- $l_{2,1}$ norm, $R(\mathbf{W}) = \sum_{j=1}^{N} \|\mathbf{W}_{j\bullet}\|_2$, which penalizes the l_1 norm on the l_2 norm of the weight vectors across all tasks. As argued by Liu et al. [10], this norm tends to select features based on the strength of the input variables of all tasks jointly.
- $l_{p,1}$ norm, $R(\mathbf{W}) = \sum_{j=1}^{N} \|\mathbf{W}_{j\bullet}\|_p$, $1 \leq p \leq \infty$, which contains other matrix norms in addition to $l_{1,1}$ and $l_{2,1}$, and the choice of p usually depends on how much feature sharing that we wish to impose among learning problems, from none ($p = 1$) to full sharing ($p = \infty$) [14].

In this paper, we consider the $l_{2,1}$ norm, and to promote additional sparsity, we propose a weighted $l_{2,1}$ norm[1], $R(\mathbf{W}) = \sum_{j=1}^{N} \sigma_j \|\mathbf{W}_{j\bullet}\|_2$, where $\sigma_1, \ldots, \sigma_j$ are positive weight coefficients. Then the weighted multi-task feature selection (wMTFS) model we are going to optimize is in the following form

$$\min_{\mathbf{W} \in \mathbb{R}^{N \times Q}} \sum_{q=1}^{Q} \frac{1}{M^q} \sum_{i=1}^{M^q} l^q(\mathbf{W}_{\bullet q}, \mathbf{z}_i^q) + \lambda \sum_{j=1}^{N} \sigma_j \|\mathbf{W}_{j\bullet}\|_2, \tag{1}$$

Next, we are to show how to solve this model and give some theoretical analysis.

3 Algorithm and Theoretical Analysis

In this section, we present an online learning framework for the weighted multi-task feature selection model (1), as well as its theoretical analysis.

Inspired by the recently developed first-order method for optimizing convex composite functions [13], we propose an online learning framework to solve (1), which is outlined in Algorithm 1. For this framework, we have some remarks.

---

[1] Our motivation comes from the recently developed reweighted l_1 minimization model in compressive sensing [4,6]. Due to space limitations, we do not elaborate here.

Algorithm 1. Online Learning Framework for Solving Eq. (1)

1: **Input:** $\mathbf{W}_0 \in \mathbb{R}^{N \times Q}$, λ, γ, ϵ, a strongly convex function $h(\mathbf{W})$, a nonnegative and nondecreasing sequence $\{\beta_t\}$ for $h(\mathbf{W})$, $\mathbf{W}_1 = \mathbf{W}_0$, $\bar{\mathbf{G}}_0 = \mathbf{O}_{N \times Q}$, $\sigma_1 = \mathbf{1}_{N \times 1}$.

2: **for** $t = 1, 2, \ldots$ **do**

3:     Given l_t, compute a (sub)gradient on $\mathbf{W}_t$, $\mathbf{G}_t \in \partial l_t$, for the coming Q samples with each for one task, $(\mathbf{z}_t^1, \ldots, \mathbf{z}_t^Q)$.

4:     Update the average (sub)gradient $\bar{\mathbf{G}}_t$: $\bar{\mathbf{G}}_t = \frac{t-1}{t}\bar{\mathbf{G}}_{t-1} + \frac{1}{t}\mathbf{G}_t$.

5:     Calculate the next iteration $\mathbf{W}_{t+1}$ by

$$\mathbf{W}_{t+1} = \underset{\mathbf{W} \in \mathbb{R}^{N \times Q}}{\arg\min} \left\{ \bar{\mathbf{G}}_t^\top \mathbf{W} + \lambda R(\mathbf{W}) + \beta_t h(\mathbf{W}) \right\}. \tag{2}$$

6:     Update the weight coefficients: $(\sigma_j)_{t+1} = 1/(\|(\mathbf{W}_{j\bullet})_t\|_2 + \epsilon)$, $j = 1, 2, \ldots, N$.

7: **end for**

Remarks:

(1) In Algorithm 1, the samples for each task arrive at each iteration. This is consistent with the online scheme for multi-task learning in [7].

(2) This algorithm only needs $\mathcal{O}(N \times Q)$ space to store the required information. In addition, the worst-case time complexity for updating $\mathbf{W}$ is also $\mathcal{O}(N \times Q)$.

(3) This learning framework for the wMTFS model is motivated by the regularized dual averaging method for lasso [16]. We can also design an accelerated algorithm, for example, using the Nesterov's accelerated technique [12].

(4) In the step 6, $\epsilon > 0$ is a parameter to ensure the algorithm to be well-defined.

(5) In this learning framework, the key to solving the online algorithm efficiently lies in the simplicity of updating the weight matrix $\mathbf{W}_{t+1}$ in Eq. (2). Here, we have the following theorem to update $\mathbf{W}_{t+1}$.

Theorem 1. *Given* $h(\boldsymbol{W}) = \frac{1}{2}\|\boldsymbol{W}\|_F^2$, $\beta_t = \gamma/\sqrt{t}$ *and the average (sub)gradient* $\bar{\boldsymbol{G}}_t$, *the optimal solution of Eq. (2) can be updated by*

$$(\boldsymbol{W}_{j\bullet})_{t+1} = -\frac{\sqrt{t}}{\gamma}\left(1 - \frac{\lambda\sigma_j}{\|(\bar{\boldsymbol{G}}_{j\bullet})_t\|_2}\right)_+ \cdot (\bar{\boldsymbol{G}}_{j\bullet})_t, \; j = 1, \ldots, N. \tag{3}$$

The proof of Theorem 1 can follow [17], so we omit it here.

A main issue to guarantee an online algorithm is to analyze its regret[2]. Following [17], we define the regret as follows:

$$\Phi_T(\mathbf{W}) \triangleq \frac{1}{Q}\sum_{q=1}^{Q}\left(\sum_{t=1}^{T}[l_t(\mathbf{W}_t) + R((\mathbf{W}_{\bullet q})_t)] - \sum_{t=1}^{T}[l_t(\mathbf{W}_T) + R(\mathbf{W}_T)]\right).$$

Here we use l_t to simplify the expression of the loss suffered by the t-th coming samples for all tasks. Directly from [16, Theorem 1], we have the following theorem that provides a bound of the regret $\Phi_T(\mathbf{W})$ for Algorithm 1.

---

[2] Regret is the difference of the objective function value up to the T-th step and the smallest objective function value.

Theorem 2. *Suppose that W^* is the optimal solution of (1) satisfying $h(W^*) \leq \theta$ for some $\theta > 0$, and there exists a nonnegative constant α such that $\|(\bar{G}_{\bullet q})_T\|_* \leq \alpha$ for all $T \geq 1$ and $q = 1, \ldots, Q$. Then it holds that $\Phi_T(W^*) \leq (\gamma\theta + \alpha^2/\gamma)\sqrt{T}$.*

Theorem 2 means that Algorithm 1 has an $\mathcal{O}(\sqrt{T})$ regret bound. Moreover, it converges to the optimal solution of (1) with the optimal rate $\mathcal{O}(1/\sqrt{T})$.

4   Experiments

In the following, we provide experiments for the online learning algorithm to solve the multi-task feature selection problem by using the school dataset[3]. This dataset consists of examination scores of 15362 students from 139 secondary schools in London during the years of 1985, 1986 and 1987. Each school is viewed as one task, therefore, there are 139 tasks, corresponding to predicting student

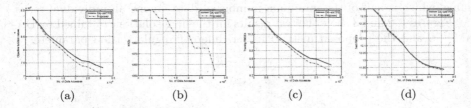

    (a)   (b)   (c)   (d)

Fig. 1. Convergence results of each method when $\lambda = 5$, $\gamma = 20$ and $\epsilon = 1e - 2$.

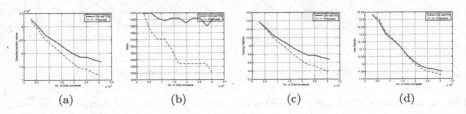

    (a)   (b)   (c)   (d)

Fig. 2. Convergence results of each method when $\lambda = 10$, $\gamma = 20$ and $\epsilon = 1e - 2$.

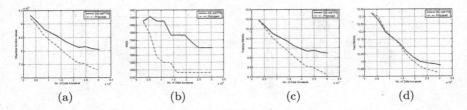

    (a)   (b)   (c)   (d)

Fig. 3. Convergence results of each method when $\lambda = 15$, $\gamma = 20$ and $\epsilon = 1e - 2$.

---

[3] http://cvn.ecp.fr/personnel/andreas/code/mtl/index.html.

performance in each school. The input consists of the year of the exam (YR), 4 school-specific and 3 student-specific attributes. Attributes that are constant in each school in a certain year are: percentage of students eligible for free school meals, percentage of students in VR band one (highest band in a verbal reasoning test), school gender (SG) and school denomination (SD). Student-specific attributes are: gender (GE), VR band (can take the values 1, 2 or 3) and ethnic group (EG). Following [3,8], we replaced categorical attributes (that is, all attributes that are not percentages) with one binary variable for each possible attribute value. In total, we obtained 27 attributes.

Since the task is a regression problem to predict the students' exam scores, we employ the least square as the loss function and use the objective function value, the number of nonzero elements (NNZs) and the root mean square errors (RMSEs) to measure the model performance. In the experiments, we compare the proposed method with a recently developed method, DA-aMTFS [17], which is also designed to solve the online multi-task learning problem. Following the setting in [17], in the training process, we randomly generate 20 sets of training data and use the rest as the test set. The number of training data is set to the same, which is half of the minimum number of data among all individual tasks.

Figures 1, 2 and 3 show the test results for the final solutions obtained by DA-aMTFS and the proposed method under different regularization parameter λ.

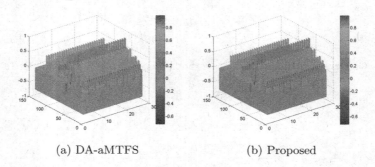

(a) DA-aMTFS   (b) Proposed

Fig. 4. 3D plots of the learned weight matrices when $\lambda = 10$, $\gamma = 50$ and $\epsilon = 1e - 2$.

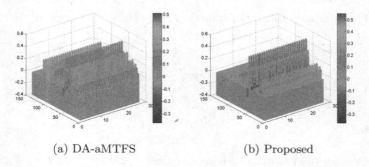

(a) DA-aMTFS   (b) Proposed

Fig. 5. 3D plots of the learned weight matrices when $\lambda = 10$, $\gamma = 100$ and $\epsilon = 1e - 2$.

Table 1. Sensitivity analysis results for the proposed method on the school dataset.

γ, ϵ	Training RMSEs	Test RMSEs	NNZs
$\gamma = 20, \epsilon = 1$	9.3514±0.16885	11.8626±0.10760	1394.65±16.4294
$\gamma = 20, \epsilon = 1e-1$	9.4520±0.16744	11.8625±0.10631	1303.70±31.6080
$\gamma = 20, \epsilon = 1e-2$	9.4598±0.16575	11.8622±0.10631	1298.15±35.5058
$\gamma = 20, \epsilon = 1e-3$	9.4604±0.16578	11.8622±0.10629	1298.15±35.5058
$\gamma = 20, \epsilon = 1e-4$	9.4605±0.16579	11.8622±0.10629	1298.15±35.5058
$\gamma = 20, \epsilon = 1e-5$	9.4605±0.16579	11.8622±0.10629	1298.15±35.5058
$\epsilon = 1e-2, \gamma = 5$	7.8002±0.20521	12.3499±0.18509	1384.00±22.3159
$\epsilon = 1e-2, \gamma = 10$	8.3581±0.16987	11.7727±0.12426	1357.10±26.4931
$\epsilon = 1e-2, \gamma = 15$	9.0195±0.16510	11.7957±0.11208	1330.70±25.9414
$\epsilon = 1e-2, \gamma = 20$	9.4598±0.16575	11.8622±0.10631	1298.15±35.5058
$\epsilon = 1e-2, \gamma = 25$	9.7525±0.16769	11.9236±0.10308	1280.05±30.4674
$\epsilon = 1e-2, \gamma = 30$	9.9597±0.17234	11.9740±0.10048	1272.90±25.0744

We observe that besides yielding comparable values for the objective function, the solution obtained by the proposed method is sparser than that obtained by DA-aMTFS, which verifies the results of the learned weight matrices shown in Figs. 4 and 5. In terms of error, we see that our method achieves lower root mean square errors than DA-aMTFS method. Additionally, we observe that NNZs decrease as γ increases and become stable within a certain range.

Sensitivity Analysis. Finally, we try to answer what is the effect of the online algorithm parameters γ and ϵ in our learning framework with respect to the regularization parameter λ? In the test, we set $\lambda = 15$, and then we fix one parameter and vary the other. ϵ is searched in $\{1e-5, 1e-4, 1e-3, 1e-2, 1e-1, 1\}$, and γ is searched in $\{5, 10, 15, 20, 25, 30\}$. Table 1 reports the trade-off between λ, γ and ϵ. From Table 1, we have the following observations.

- When we fix λ and γ, the training RMSE increases as ϵ decreases and becomes stable when $\epsilon < 0.01$. Contrarily, the test RMSE and the NNZs decrease as ϵ decreases and becomes stable when $\epsilon \leq 0.01$.
- When we fix λ and ϵ and vary γ, we find that the training RMSE increases as ϵ increases, while the NNZs decrease as ϵ increases. It is worth noting that the Test RMSE decreases at first and then becomes larger gradually. In terms of the performance, for the regression task on the school dataset, maybe the best result is obtained when $\gamma = 10$ and $\epsilon = 0.01$ on condition that $\lambda = 15$.

The sensitivity analysis indicates that for a specific dataset, the best λ, γ and ϵ are often data-dependent and should be tuned correspondingly.

5   Conclusion

In this paper, we proposed a weighted multi-task learning model to capture the shared features among multiple related tasks and simultaneously enhance the sparsity of the learned weight matrices. We presented an online learning algorithm to solve this model and also theoretically analyzed its regret bound, which guarantees that the proposed algorithm can work efficiently both in time cost and memory cost. Experimental results presented here highlight the efficiency and effectiveness of our approaches.

Acknowledgments. This work was supported by the National Natural Science Foundation of China under Grants 61305018, 61432008, 61472423 and 61532006.

References

1. Altmann, A., Ng, B.: Joint feature extraction from functional connectivity graphs with multi-task feature learning. In: Proceedings of the International Workshop on Pattern Recognition in NeuroImaging, pp. 29–32 (2015)
2. Argyriou, A., Evgeniou, T., Pontil, M.: Multi-task feature learning. In: Advances in Neural Information Processing Systems, pp. 41–48 (2006)
3. Argyriou, A., Evgeniou, T., Pontil, M.: Convex multi-task feature learning. Mach. Learn. **73**(3), 243–272 (2008)
4. Candès, E.J., Wakin, M.B., Boyd, S.P.: Enhancing sparsity by reweighted l_1 minimization. J. Fourier Anal. Appl. **14**(5), 877–905 (2008)
5. Caruana, R.: Multitask learning. Mach. Learn. **28**(1), 41–75 (1997)
6. Chartrand, R., Yin, W.: Iteratively reweighted algorithms for compressive sensing. In: Proceedings of the IEEE International Conference on Acoustics, Speech and Signal Processing, pp. 3869–3872 (2008)
7. Dekel, O., Long, P.M., Singer, Y.: Online multitask learning. In: Lugosi, G., Simon, H.U. (eds.) COLT 2006. LNCS (LNAI), vol. 4005, pp. 453–467. Springer, Heidelberg (2006)
8. Evgeniou, T., Micchelli, C.A., Pontil, M.: Learning multiple tasks with kernel methods. J. Mach. Learn. Res. **6**, 615–637 (2005)
9. Liu, A.-A., Su, Y.-T., Nie, W.-Z., Kankanhalli, M.: Hierarchical clustering multi-task learning for joint human action grouping and recognition. IEEE Trans. Pattern Anal. Mach. Intell. DOI 10.1109/TPAMI.2016.2537337
10. Liu, J., Ji, S., Ye, J.: Multi-task feature learning via efficient $l_{2,1}$ norm minimization. In: Proceedings of the 25th Conference on Uncertainty in Artificial Intelligence, pp. 339–348 (2009)
11. Luo, Y., Wen, Y., Tao, D., Gui, J., Xu, C.: Large margin multi-modal multi-task feature extraction for image classification. IEEE Trans. Image Process. **25**(1), 414–427 (2016)
12. Nesterov, Y.: A method of solving a convex programming problem with convergence rate $O(1/k^2)$. Sov. Math. Dokl. **27**(2), 372–376 (1983)
13. Nesterov, Y.: Primal-dual subgradient methods for convex problems. Math. Program. **120**(1), 221–259 (2009)
14. Obozinski, G., Taskar, B., Jordan, M.I.: Joint covariate selection and joint subspace selection for multiple classification problems. Statist. Comput. **20**(2), 231–252 (2010)

15. Quattoni, A., Carreras, X., Collins, M., Darrell, T.: An efficient projection for $l_{1,\infty}$ regularization. In: Proceedings of the 26th International Conference on Machine Learning, pp. 857–864 (2009)
16. Xiao, L.: Dual averaging methods for regularized stochastic learning and online optimization. J. Mach. Learn. Res. **11**, 2543–2596 (2010)
17. Yang, H., Lyu, M.R., King, I.: Efficient online learning for multitask feature selection. ACM Trans. Knowl. Discov. Data **7**(2), 1–27 (2013)
18. Yuan, M., Lin, Y.: Model selection and estimation in regression with grouped variables. J. Roy. Statist. Soc. B **68**(1), 49–67 (2006)
19. Zhang, Y., Yeung, D.-Y., Xu, Q.: Probabilistic multi-task feature selection. In: Advances in Neural Information Processing Systems, pp. 2559–2567 (2010)
20. Zhou, Q., Zhao, Q.: Flexible clustered multi-task learning by learning representative tasks. IEEE Trans. Pattern Anal. Mach. Intell. **38**(2), 266–278 (2016)

Multithreading Incremental Learning Scheme for Embedded System to Realize a High-Throughput

Daisuke Nishio[✉] and Koichiro Yamauchi

Department of Computer Science, Chubu University,
1200 Matsumoto, Kasugai, Aichi, Japan
tp15015-0215@sti.chubu.ac.jp, yamauchi@cs.chubu.ac.jp
http://sakura.cs.chubu.ac.jp/

Abstract. Recent improvement of the microcomputer enables it to execute complex intelligent algorithms on embedded systems. However, when using conventional incremental learning methods, its resources are often increased with learning, and continuing the execution of the incremental learning becomes difficult on small embedded systems. Moreover, for real applications, the response time should be reduced. This paper proposes a technique for implementing incremental learning methods on a budget. Normally, they proceed online learning by alternating recognition and learning, so that they cannot respond to the next new instance until the previous learning is finished. Unfortunately, their computational learining complexities are extremely high to realize a quick response to new input. Therefore, this paper introduces a multithreading technique for such learning schemes. The recognition and learning threads are executed in parallel so that the system can respond to a new instance even when it is in the progress of learning. Moreover, this paper shows that such multithreading learning schemes sometime need a "sleep-period" to complete the learning similar to a biological brain. During the "sleep-period," the leaning system prohibits the receival of any sensory inputs and yielding outputs.

Keywords: Incremental learning · Learning on a budget · Limited general regression neural network (LGRNN) · Embedded systems · Response time · Real-time operating system (RTOS) · Sleep

1 Introduction

Incremental learning systems realize one-pass learning. Such systems are assumed to be used in an online manner. To realize such one-pass learning on a small embedded system, we must overcome two major problems. The first problem is how to bound the resources for learning. In the embedded systems, the learning algorithm must not waste the storage space, which is larger than the device capacity. Fortunately, several low-cost learning methods have been

© Springer International Publishing AG 2016
A. Hirose et al. (Eds.): ICONIP 2016, Part II, LNCS 9948, pp. 204–213, 2016.
DOI: 10.1007/978-3-319-46672-9_24

proposed [1–5]. We may be able to overcome the limitation in storage capacity by using these models. The second problem is the reduction of computational complexity for the learning tasks. To realize high throughput, we must reduce the computational complexity for learning rather than for calculation of its outputs. In particular, approximation techniques for deriving inverse matricies can be used to speed up the learning [4]. However, in many cases, such light-weighted learning is still heavier than that used for output calculation. This is not suitable for the real use of embedded systems. In many cases, machine learning systems in embedded systems are usually implemented in FPGA chips to speed up their learning speed (e.g., [6,7]). However, there is a possibility of embedding a software-learning machine into a microcomputer directly. We expect that such software-learning machines are cost-effective in providing adaptive properties to the embedded systems. Therefore, we propose an implementation scheme to parallelly execute learning and recognition methods by using a multithreading technique. The learning thread is initiated when the recognition thread for each instance is completed. The learning thread executes the learning of the last recognized sample. However, under such a learning scheme, the learning machine must ignore part of the instances presented during the learning process. To catch up the learning, the learning scheme is extended to periodically introduce a sleep period. During the sleep period, the learning machine prohibits receiving sensory inputs and yielding outputs. The remainder of this paper is organized as follows. Section 2 describes the general scheme of multithreading learning and Sect. 3 illustrates a multithreading learning machine. In Sect. 4, we introduce a sleep period to extend multithread learning. Section 5 presents the comparison of the multithreading and normal learning. Finally, Sect. 6 concludes this study.

2   General Scheme of Multithreading Learning

Normally, cost-effective online learning methods, such as those described in [1–4] execute the learning of new current input followed by yielding the output of the predicted label. Therefore, a new input is obtained just after the previous learning process is completed. However, in embedded systems, sensory inputs are

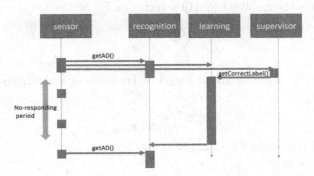

Fig. 1. Sequence of original learning

provided at an arbitrary timing when an external interruption is occurs. Moreover, the time interval for learning is much longer than calculating the output $y(\boldsymbol{x}_t)$; thus, the learner is sometimes unable to respond to new sensory inputs (Fig. 1).

Therefore, the original learning methods sometimes cannot respond to the next new sensory input. To overcome this, we divided the learning algorithm into two parts: recognition and learning threads. These two threads must be executed in parallel (Fig. 2).

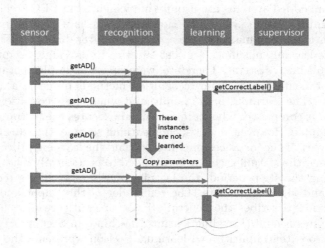

Fig. 2. Sequence diagram of multithreading learning

2.1　Recognition Thread

The recognition algorithms are based on kernel perceptron-based algorithm. Therefore, in [1, 2, 4],

$$f[\boldsymbol{x}] \equiv \langle f_t, K(\boldsymbol{x}, \cdot) \rangle, \tag{1}$$

where f_t denotes the resultant function vector on Hirbert space at the t-th round and

$$f_t \equiv \sum_{i=1}^{B} w_i K(\boldsymbol{x}_i, \cdot). \tag{2}$$

Some algorithms are based on kernel regresion-based algorithm [3]. Therefore,

$$f[\boldsymbol{x}] \equiv \frac{\langle f_t, K(\boldsymbol{x}, \cdot) \rangle}{\langle g_t, K(\boldsymbol{x}, \cdot) \rangle}, \tag{3}$$

where f_t is equivalent to Eq. (2). g_t is

$$g_t \equiv \sum_{i=1}^{B} K(\boldsymbol{x}_i, \cdot). \tag{4}$$

As suggested by Webb [8], the network using Eq. (3) is optimal for maximizing the robustness of noisy inputs.

2.2   Learning Thread

Learning threads are divided into the following two phases: resource-allocation and learning-under-fixed-budget phases. The resource-allocation phase allocates new kernels to record new instances, whereas the learning-under-fixed-budget phase realizes one of the following two processes: replacement of one of ineffective kernel with a new kernel, or the learning of the new instance by changing parameters of existing kernels. The next section describes an example of the multithreading learning scheme.

3   Multithreading Limited General Regression Neural Network

We have already proposed a kernel regression model: "limited general regression neural network (LGRNN)," which realizes incremental learning with a bounded number of kernels [3,5]. The behavior of LGRNN is similar to that of the k-nearest neighbors. Although there are many nearest neighbor learning methods that reduce prototypes [9], almost all of them consider clustering. However, LGRNN considers regression, and reduces redundant kernel considering approximated linear dependency. In this section, we briefly explain the learning algorithm as an example of learning on a budget. Under multithreading environments, the learning and recognition methods are executed in parallel. However, if the LGRNN parameters are changed during the recognition, the output of LGRNN might be abnormal. To avoid this, we must prepare two sets of LGRNN parameters. The first parameter set defines the output calculation for new inputs, whereas the second parameter defines learning. The parameters for the output calculation were modified after each learning by copying the newest parameter generated by the learning. It is necessary to perform exclusive control semaphore operations to prevent modification of the parameter during recognition.[1] During thread learning, LGRNN proceeds as follows.

1. Determine the most ineffective kernel to be replaced with a new one.
2. Calculate expected loss of the four learning options.
3. Execute the learning option with the least expected loss.

3.1   Find the Most Ineffective Kernel

For clustering, condensation or editing techniques are usually used to reduce the number of prototypes. However, for regression, it is desirable that the output function is not changed because of prototype selection. To realize this, the most

---

[1] Semaphore is the most commonly used method of performing an exclusive control.

ineffective kernel is determined by using approximated linear dependency. If the i-th kernel is represented by a linear combination of the other kernels, the output of kernel regression is represented by the kernels without the i-th kernel.[2]

To measure the linear dependency, the following estimator is used.

$$\delta_j = \min_{a_j} \left\| K(\boldsymbol{x}_j, \cdot) - \sum_{k \neq j} a_{jk} K(\boldsymbol{x}_k, \cdot) \right\|^2 \tag{5}$$

The most ineffective kernel is detected by $i = \arg\min_j \delta_j$.

3.2 Learning Options

To proceed with the incremental learning by keeping the memory constant, the following four learning options must be executed. LGRNN selects one option with the least expected loss.

– Modification: Current input is simply projected to the existing kernels.

$$f_t = f_{t-1} + y_t P_{t-1} K(\boldsymbol{x}_t, \cdot), \quad g_t = g_{t-1} + P_{t-1} K(\boldsymbol{x}_t, \cdot), \tag{6}$$

where $P_{t-1} K(\boldsymbol{x}_t, \cdot)$ denotes the projected vector of $K(\boldsymbol{x}_t, \cdot)$ to the space spanned by the existing kernels.
– Replacement and substitution: The most ineffective kernel is replaced with a new kernel, recording the current instance.

$$f_t = f_{t-1-i} + w_i P_{t-1-i} K(\boldsymbol{x}_i, \cdot) + y_t K(\boldsymbol{x}_t, \cdot)$$
$$g_t = g_{t-1-i} + w_i P_{t-1-i} K(\boldsymbol{x}_i, \cdot) + K(\boldsymbol{x}_t, \cdot), \tag{7}$$

where $P_{t-1-i} K(\boldsymbol{x}_t, \cdot)$ denotes the projected vector of $K(\boldsymbol{x}_i, \cdot)$ to the space spanned by the kernels except for the i-th kernel.
– Replacement: The most ineffective kernel is simply replaced with the new kernel.

$$f_t = f_{t-1-i} + y_t K(\boldsymbol{x}_t, \cdot), \quad g_t = g_{t-1-i} + K(\boldsymbol{x}_t, \cdot) \tag{8}$$

– Ignore: Do nothing. $f_t = f_{t-1}$, $g_t = g_{t-1}$.

3.3 Expected Loss of Each Learning Option

One of the four learning options, explained in Sect. 3.2, is selected according to the expected loss of each learning option. The expected loss is the sum of the predicted magnitude of forgetting learned patterns and the predicted remained error of the new sample. Normally, the expected loss varies depending on the

---

[2] Although the LGRNN output for the i-th kernel center is recovered through the linear combination of the other kernels, there are no guarantees that the outputs for the other inputs are not changed.

distribution of inputs. However, the learning machine cannot know the precise distribution of inputs during learning. To approximate the distribution, the LGRNN counts the activation time of each kernel. If the output value of the i-th kernel is the highest, the number of activation for the kernel N_i is incremented. Furthermore, N_i is increased when a kernel is projected to it as follows:

$$N_i+ = N_{new} \frac{|a_i|}{\sum_j |a_j|}, \tag{9}$$

where N_i and N_{new} are the number of activation of the i-th kernel and an importance weight for the new sample, respectively. This equation is used when the "modification" option is applied. If the "replacement with substitution" option is applied, N_{new} should be replaced with the number of activations of the most ineffective kernel. If the LGRNN applies the "replacement" option, $N_{nearest(i)}+ = N_i$ where $N_{nearest(i)}$ denotes the number of activations of the nearest kernel to those of the i-th most ineffective kernel.

For example, the expected loss for the "replacement with substitution" option is

$$e_{substitute} \equiv \sum_{j \neq i} N_j \left(\frac{a_{ij}(w_j^* - w_i^*)}{R_j + a_{ij}R_i} \right)^2 + (w_{nearest(i)}^* - w_i^*)^2 \delta_i, \tag{10}$$

where $w_i^* \equiv w_i/R_i$ and δ_i is given by Eq. (5). The first term of Eq. (10) is the predicted magnitude of forgetting, and the second term denotes the predicted constant error for the most ineffective kernel center. The expected loss in the other learning options is described in our previous study [5].

4   Beyond Multithreading Learning

Although the multithreading learning realizes quick response to new input, it misses learning for some inputs because the previously activated learning thread is still running (Fig. 2). Such a period varies according to the computational complexity of learning. For example, if we employ LGRNN Light [5] as the learning thread, the computational complexity is no more than $O(B^2 + k^2)$, where B is the maximum number of kernels and k denotes the number of neighbors used for the approximated calculation of LGRNN (see [5] for detailed explanations). On the other hand, the computational complexity for recognition is $O(k)$. This increases the learning time more than the time required for recognition. Therefore, LGRNN misses the learning of a part of samples. To prevent these missing, a sleep-learning period, which realizes the learning for such inputs, is introduced. When the input data is assigned to the system when a previously activated learning thread is running, the system stores the input and corresponding label data into a buffer, namely a short-term memory. If the short-term memory becomes full, the system moves into a sleep phase, in which the system prohibits any input assignment or output yield. During the sleep phase, the system learns the remained samples by using the same algorithm used for the learning thread (Fig. 3). Note that even if the system introduce such sleep phase, the total learning period is eventually the same as that of original LGRNN. However, when input data interval is long for example, the sleep phase completes within the interval. Therefore, users have to set the capacity for short-term memory properly according to the property of the environment.

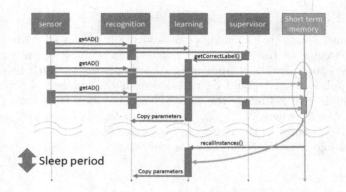

Fig. 3. Multithreading learning with a sleep period

5   Experiments

We implemented the multithreading learning method on an operating system (OS): T-kernel Engine http://www.tron.org/specifications/, which is a derivative of TRON. The multithreading LGRNN was tested on the "T-board" CPU: ARM7TDMI, Clock: 33 MHz in which the OS is installed http://www.t-engine4u.com/archive/teaboard_arm7-at91_e.pdf. A sensor substrate is attached to the T-board, and from the learning data, the sensor data of up to two inputs is obtained from five-input data. For the sake of simplicity, the recognition task was performed by using a periodic handler. The learning task is initiated immediately after the completion of the recognition task. However, the round-robin scheduling is not conducted in the T-kernel. Instead, an absolute priority scheduling is performed. Therefore, the period handler is used every time to activate the recognition task on returning to the state waiting for a learning task to be activated. Of the light sensor data, update must be performed from the recognition task. Figure 2 shows the sequence diagram. Further, as it is difficult to conduct experiments with the same data by being periodically activated by the periodic handler, the sensory inputs were stored and used for the experiment.

5.1   Comparison with a Singlethreading LGRNN

The multithreading LGRNN was compared with a singlethreading LGRNN. First, we checked the CPU time for executing multithreading LGRNN by varying the number of input dimensions. The tested sizes of the input dimension were two, three, four, and five. Both the multi and singlethreading LGRNN with both recognition and learning processes were run 50 times. The input prepared was light data for an AD convertor by $x = [x(t - n\tau), x(t - (n - 1)\tau)), \cdots, x(t - \tau), x(t)]$, and the corresponding desired output was $gx(t - 1)x(t)$. Here, g is the coefficient and was set at g = 1.8. We compared the difference between the response times calculated using our proposed method and the original calculation method. In addition, we investigated their behaviors when the number of input dimensions is increased. Table 1 shows the results of the average response time when the number of input dimensions is increased. The recognition task is executed periodically, with the execution interval as 150 ms and the upper limit of a storage volume kernel as 10. When multitasking is performed, which is shorter than ever by our computation, the result obtained by increasing the number of input dimensions is remarkable.

Table 1. The difference between the average response time and the number of input dimensions

Input dimentions	Original [ms]	Multitask [ms]
Two-inputs	3.72	1.94
Three-inputs	4.16	2.7
Four-inputs	4.66	2.88
Five-inputs	27.64	3.6

Figure 4 shows the cumulative square error results at the time of LGRNN and multithreading. An expression of cumulative square error is shown as follows. In this case the result is $y(x_t)$, which is actually determined by the learning of LGRNN, y_t is the original desirable result.

$$CumulativeSquareError+ = (y_t - y(x_t))^2 \qquad (11)$$

Figure 4 shows the error results and the original LGRNN when the execution cycle time in a state that was applied to the light source is 50, 60, and 70 [ms]. The cumulative square error value was the largest when the period is 50 [ms]. The proposed methods cumulative square error value was closest to that of origitnal LGRNN when the cycle time was 70[ms].

Next multithreading LGRNN with sleep was compared to the

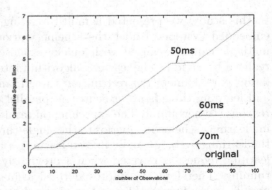

Fig. 4. Comparison of cumulative errors under various cycle-times.

original LGRNN. The cumulative square error results in the widening of the interval of the execution of recognition were examined. Further, this responce time was compared with that of the multithreading learning without sleep by using the same data; however, the same data cannot be obtained to run in real time in the actual environment.

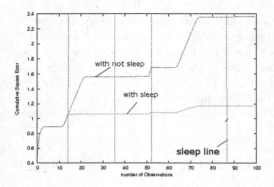

Fig. 5. Cumulative errors of multithreading learning with and without sleep periods

Figure 5 shows a comparison of the cumulative square error results in multithreading and sleep learning. The time of recognition was 60 [ms]. The size of the short-time memory, which is the data-storage for sleep-learning, sleep-learning was set to 10.

Table 1 compared the response time of the mutithreading and original LGRNN with various input dimension. Although this experiment was conducted to compare the input data of multidimensional samples, the difference in computation time was considered to increase by further increasing the number of input dimensions. However, it is necessary to note that this method is not intended for shortening the entire program time. This technique is for merely shortening the response time as much as possible. Therefore, the system skips the learning of new input during the previous learning thread is running. Thus, the cumulative error is greater than in the original LGRNN. This is the stage where the inevitable omission of learned part occurs. Thus, the users adjustment is required for sleep-learning.

6   Conclusion

In this study, we presented a multithreading incremental learning scheme for embedded systems. Under the scheme, various low-cost incremental learning methods can be realized with quick responses to new inputs. This ability is realized by dividing the learning algorithm into the two parts: the output calculation part, namely the recognition thread, and learning part. The recognition and learning threads are executed in parallel so that the system realizes quick response to new input. The experimental results suggested that the multithreading learning scheme is superior to the single-threading learning scheme. However, the learning system usually missed learning a few parts of inputs because a new learning request is rejected when a previously activated learning thread is still running. Therefore, we improved the learning scheme by introducing a sleep period. In the future, the capacity of the short term memory should be adjusted according to the allowed error set by user.

References

1. Dekel, O., Shalev-Shwartz, S., Singer, Y.: The forgetron: a kernel-based perceptron on a budget. SIAM J. Comput. (SICOMP) **37**(5), 1342–1372 (2008)
2. Orabona, F., Keshet, J., Caputo, B.: The projectron: a bounded kernel-based perceptron. In: ICML, pp. 720–727 (2008)
3. Yamauchi, K.: Pruning with replacement and automatic distance metric detection in limited general regression neural networks. In: Proceedings of International Joint Conference on Neural Networks, San Jose, California, USA, July 31 - August 5, 2011, pp. 899–906. The Institute of Electrical and Electronics Engineers, Inc., New York, July 2011
4. He, W., Si, W.: A kernel-based perceptron with dynamic memory. Neural Netw. **25**, 105–113 (2011)
5. Yamauchi, K.: Incremental learning on a budget and its application to quick maximum power point tracking of photovoltaic systems. J. Adv. Comput. Intell. Intell. Inf. **18**(4), 682–696 (2014)
6. Genov, R., Cauwenberghs, G.: Kerneltron: support vector "machine" in silicon. IEEE Trans. Neural Netw. **14**(5), 1426–1434 (2003)
7. Hikawa, H., Kaida, K.: Novel FPGA implementation of hand sign recognition system with SOM-Hebb classifier. IEEE Trans. Circ. Syst. Video Technol. **25**(1), 153–166 (2015)
8. Webb, A.R.: Functional approximation by feed-forward networks: a least-squares approach to generalization. IEEE Trans. Neural Netw. **5**(3), 363–371 (1994)
9. Garcìa, S., Derrac, J., Cano, J., Herrera, F.: Prototype selection for nearest neighbor classification: taxonomy and empirical study. IEEE Trans. Pattern Anal. Mach. Intell. **34**(3), 417–435 (2012)

Hyper-Parameter Tuning for Graph Kernels via Multiple Kernel Learning

Carlo M. Massimo, Nicolò Navarin$^{(\boxtimes)}$, and Alessandro Sperduti

Department of Mathematics, University of Padova, Padua, Italy
cmassimo@studenti.math.unipd.it, {nnavarin,sperduti}@math.unipd.it

Abstract. Kernelized learning algorithms have seen a steady growth in popularity during the last decades. The procedure to estimate the performances of these kernels in real applications is typical computationally demanding due to the process of hyper-parameter selection. This is especially true for graph kernels, which are computationally quite expensive. In this paper, we study an approach that substitutes the commonly adopted procedure for kernel hyper-parameter selection by a multiple kernel learning procedure that learns a linear combination of kernel matrices obtained by the same kernel with different values for the hyper-parameters. Empirical results on real-world graph datasets show that the proposed methodology is faster than the baseline method when the number of parameter configurations is large, while always maintaining comparable and in some cases superior performances.

1 Introduction

The behavior of a learning algorithm usually depends on some external parameters that are given by the user. For example, in kernel methods [12], learning is cast as a regularized constrained optimization problem involving a regularization cost parameter, and one or more kernel parameters. Of course, in general these parameters have to be tuned in order to achieve satisfying predictive performances on real-world applications. Such tuning problem is referred to as *hyper-parameter optimization*, and it constitutes a key step for almost any machine learning method. The problem of hyper-parameter optimization is in general posed as a search problem. In some cases, it is possibile to efficiently find the optimal value on a validation set for the regularization cost parameter, e.g., in the context of kernel methods, an algorithm for fitting the entire path of SVM solutions for every value of the cost parameter C, incurring in a computational cost similar to the one payed to train a single SVM model, has been proposed [9]. The usual approach, however, consists in performing a grid-search, i.e. a range of possible values for each parameter is fixed beforehand, and then for each combination of parameter values (each point in the grid) learning is performed. The hyper-parameter values that produce the classifier with the best performance on the validation set are the ones selected for the final model learning. Depending on the number of hyper-parameters, and on the granularity of the grid, this procedure may become time consuming (e.g. 2 hyper-parameters

© Springer International Publishing AG 2016
A. Hirose et al. (Eds.): ICONIP 2016, Part II, LNCS 9948, pp. 214–223, 2016.
DOI: 10.1007/978-3-319-46672-9_25

with 10 values each generate a grid with 100 points, so 100 training procedures have to be executed). A possible approach for speeding-up the computation is to randomize the search, but in this case there is no guarantee to find a good parameter configuration.

When considering graph kernels [6] parameter selection is computationally very expensive due to the typically high computational cost involved in their computation. In this paper, we propose an alternative approach to parameter selection that casts the problem of hyper-parameter tuning as a multiple kernel learning problem [8] where: *(i)* the involved kernel matrices are obtained by using different kernel hyper-parameter values for the same kernel; *(ii)* an approach for multiple kernel learning that scales linearly with the number of involved kernel matrices is adopted [2]. In this way, we aim to reduce the computational burden of hyper-parameter tuning on large grids. Moreover, we do not select the point in the grid that minimizes the error on the validation set, but we find the best (linear) combination among all the (models resulting from) different points in the grid. Finally, since it is well known that multiple kernel learning is more effective when combining orthogonalized kernel matrices, we also explore this venue. We study the classification and computational performances obtained by this approach using four different graph kernels on five commonly adopted real world bio-chemical datasets. Experimental results show that when the number of involved kernel matrices is large, the proposed approach is computationally less demanding than the traditional approach. Moreover, in some cases better generalization performances can be obtained.

2   Background

In this paper, we deal with graphs with labeled nodes. Kernels for graphs are used to deal with learning tasks. An example of learning task is to learn a function defined on the molecular graph[1] of chemical compounds that tells whether the compound is toxic or not. In the following, we introduce the state-of-the-art graph kernels studied in this paper as well as a multiple kernel learning approach that scales linearly with the number of involved kernel matrices. Finally, we recall the commonly adopted hyper-parameter selection procedure.

Graph Kernels. Among the different graph kernels available in literature, in this paper we will focus on the Ordered Decomposition DAG (ODD) Kernels [4,5] and on the Weisfeiler-Lehman (WL) kernels [13]. The ODD kernel framework [4] considers as non-zero features in the RKHS the trees that appear as subtrees of the input graphs. It exploits the shortest-path (up to length h) DAG decompositions starting from each node in the graph to generate DAG structures, and then extracts tree features from them. Tree features are then weighted by the λ parameter at the power of the size of the tree. The kernel is referred to as ODD_{ST}.

---

[1] A graph where vertices are atoms and edges are chemical bonds; the label attached to each vertex reports the atom type.

The WL kernel framework [13] is based on the recursive WL color refinement procedure. The principal member of this family, the Fast Subtree WL kernel, in an efficient way (i.e., linearly in the number of edges), maps a graph in a RKHS where each feature represents a subtree-walk pattern (subtrees where vertices can appear multiple times). The value associated to a feature is the frequency of the particular subtree-walk in the input graph. WL kernel is computed in an iterative fashion, that stops after h (user-specified parameter) iterations.

The ODD_{ST} and WL kernels decompose a graph in a set of simple (local) features. A recent work [11] aims at improving local feature expressiveness by enriching the feature space with contextual information. The Tree Context Kernel (TCK_{ST}) is the extension of ODD_{ST} considering contexts. In this kernel, the contexts are defined as feature themselves. Thus, two identical tree features f of height l from two different graphs will match only if they appear in the same context, i.e. the tree feature of height $l + 1$ that has f as proper subtree. In [14] also the WL kernel is extended considering contexts. Basically, we augment the information of a local feature by adding the structural information of its neighborhood. This process is repeated up to a user-defined parameter h. During the last iteration we cannot further determine the structural information around each node, therefore the contextualized features created in the last iteration will have an empty context. The resulting kernel is referred to as WLC.

Multiple Kernel Learning. In Multiple Kernel Learning (MKL), fixed a pool of kernels, the learning algorithm learns a combination of them that hopefully performs better than the single kernels. The MKL problem can be formulated as $K_{MKL} = \sum_{r=1}^{R} \eta_r K_r$, $\eta_r \geq 0$, where the goal is to learn a vector of coefficients $\eta = \{\eta_r\}_{1 \leq r \leq R}$ that minimize some cost function. Among the different methods present in literature [8], in this paper, we focus on the method proposed in [2], referred to as EasyMKL. The proposed algorithm requires the solution of an optimization problem, but is very efficient (linear complexity in the number of kernels). A recent work [3] has managed to integrate EasyMKL with the ODD kernels. The proposed technique defines orthogonalized kernels starting from a single feature space, by dividing the features into independent sets. In the case of ODD_{ST} kernel, the features are divided according to their height; in this way, no dependent features ends up in the same set. Every set composed in this way is then used as a separate feature space to compute a separate kernel and the whole pool becomes the input for the EasyMKL algorithm.

Hyper-Parameter Selection. One of most widely adopted and unbiased approaches for evaluating a learning algorithm on a dataset is the nested K-fold cross validation, schematized in Fig. 1. The dataset is iteratively splitted into a test set and a validation set (outer K-fold). Then another K-fold cross validation is performed on the training set (that is iteratively splitted in the actual training set and the validation set). This inner K-fold is used for selecting the algorithm parameters (kernel and learning machine). Finally, just the selected values are adopted to classify the test set. Usually, step (3) in the figure (the selection of the best model/parameters) is performed using a grid search, i.e. a different classifier is trained for each possible parameter combination and

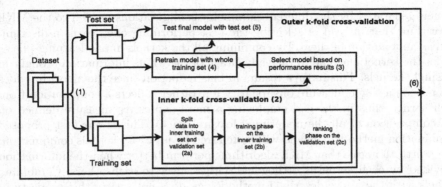

Fig. 1. A schematization of nested k-fold cross validation.

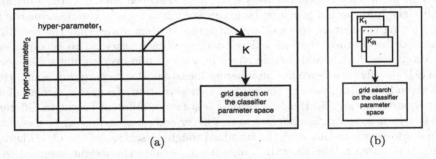

Fig. 2. Comparison between standard grid-search (a) and our proposed approach (b).

tested on the validation set. The best performing one is the model that will be selected. However, when many parameters have to be validated, this step may become very expensive. Our proposed approach constitutes an alternative to this step.

3 Proposed Methodology

As previously discussed, MKL methods generally provide a consistent way of using different, possibly weak, kernels to build a model which is the result of their composition. While these methods are usually employed to boost the performance of single kernels, we approach the combination technique from a different perspective. The proposed methodology consists in avoiding the kernel hyperparameter selection process entirely by considering all the kernels according to a finite subset of the kernel parameter space and composing them together with EasyMKL. Figure 2(a) shows the classic grid-search approach, where for each kernel parameter configuration (each point in the grid), a grid search (in k-fold cross validation) has to be performed on the classifier parameter space (e.g. the SVM C parameter). Figure 2(b) shows our proposed approach: exploiting MKL, a single grid search on the classifier parameter space has to be performed, since all the kernels generated by the different parameter configurations are fed to the MKL algorithm, that instead of selecting the best performing one, learns a discriminative combination of them. By combining all the kernels in one learning phase, we let the kernel machine to take advantage of the whole information brought in by single kernels. This in turn results in bias reduction since the algorithm does not rely on a single measure of similarity but rather collects a contribution from each kernel, thus in the end employing a more expressive measure defined on the composition of all the underlying feature spaces. This is our first proposed combination methodology, and will be referred to as "c" for kernels combination. It is worth to notice that MKL algorithms perform better when the information provided by each kernel is orthogonal with respect to each others. Combining kernels with the above described methodology does not satisfy this property in general, because the kernels are computed by the same function, although with different parameter values. Recently some work has been done to further orthogonalize the information provided by graph kernels [3]. The technique consists in partitioning the feature space of a given kernel in such a way that the resulting subsets define features that are independent from each others. This is equivalent to decompose the original feature space in a set of non-overlapping sub-spaces. The ODD_{ST} feature space has an inherent hierarchical structure that is, if a tree of a given height is present in the explicit representation of a graph, also all of its proper sub-trees are [3]. Hence, grouping features of different height to different sub-spaces makes two dependent features never end up contributing to the same kernel. Moreover, in this way, the standard weighing scheme of the ODD kernel can be delegated to the EasyMKL algorithm, because the weight assigned to a particular kernel by the algorithm would directly correlate with the underlying feature sub-space thus rendering the λ parameter of the ODD kernel superfluous. The WL kernel has an already orthogonally structured feature space since each feature extracted at iteration $i \in \{0, \ldots, h\}$ is independent from all the feature extracted at iteration $j \in \{0, \ldots, h\} \setminus \{i\}$, hence the partition technique consists in considering each iteration as a different feature space. The adoption of this techniques results in the possibility of removing some hyper-parameter thus typically reducing the overall number of generated kernels. Note that the orthogonalization of the ODD kernel has already been presented in [3], while the one of WL is a minor contribution of this paper. We will refer to this second approach as "oc", for orthogonalized kernels combination.

4   Experiments

The experiments were conducted on five different datasets, namely: AIDS [16], CAS[2], CPDB [10], NCI1 [15] and GDD [7]. The first four datasets contain chemical particles encoded in graph form, while the latter encodes proteins. All the nodes are labelled and none have self loops. The AIDS dataset contains 1503 chemical compounds, labelled according to their antiviral activity; CAS and CPDB store 4337 and 684 mutagenic compounds respectively; NCI1 consists of

---

[2] http://cheminformatics.org/datasets/bursi/.

Table 1. AUROC results (± standard deviation) relative to the experiments conducted with the kernels ODD_{ST} and TCK_{ST}. hs: standard hyper-parameter selection; c: kernels combination with MKL; oc: orthogonalized kernels combination with MKL.

(kernels)^{method}	CAS	NCI1	AIDS	CPDB	GDD
$(ODD_{ST})^{hs}$	89.82 ± 0.17	90.69 ± 0.10	82.62 ± 0.52	84.42 ± 0.67	84.73 ± 0.38
$(TCK_{ST})^{hs}$	90.06 ± 0.13	91.50 ± 0.11	82.25 ± 0.67	84.22 ± 0.80	$\mathbf{86.74 \pm 0.26}$
$(ODD_{ST} + TCK_{ST})^{hs}$	90.10 ± 0.11	91.10 ± 0.11	83.23 ± 0.65	84.97 ± 0.72	86.27 ± 0.18
$(ODD\text{-}MKL)^{hs}$	$\mathbf{90.49 \pm 0.08}$	91.44 ± 0.08	85.15 ± 0.31	85.64 ± 0.56	84.98 ± 0.26
$(ODD_{ST})^{c}$	88.99 ± 0.13	89.76 ± 0.10	83.88 ± 0.44	84.01 ± 0.33	80.13 ± 0.19
$(TCK_{ST})^{c}$	89.54 ± 0.13	90.95 ± 0.06	84.87 ± 0.40	85.25 ± 0.30	86.17 ± 0.22
$(ODD_{ST}, TCK_{ST})^{c}$	89.60 ± 0.12	90.76 ± 0.07	84.68 ± 0.42	85.17 ± 0.34	86.12 ± 0.18
$(ODD_{ST})^{oc}$	90.33 ± 0.10	91.95 ± 0.06	86.27 ± 0.35	86.32 ± 0.29	85.43 ± 0.24
$(TCK_{ST})^{oc}$	90.36 ± 0.10	91.89 ± 0.06	$\mathbf{86.34 \pm 0.34}$	86.25 ± 0.32	84.58 ± 0.21
$(ODD_{ST}, TCK_{ST})^{oc}$	90.38 ± 0.10	$\mathbf{91.96 \pm 0.06}$	86.32 ± 0.34	$\mathbf{86.32 \pm 0.38}$	85.28 ± 0.22

4110 chemical compounds screened for activity against non-small lung cancer cells; GDD is composed of 1178 X-ray crystal structures of proteins.

Description. The kernels tested in our experiments are: ODD_{ST}, TCK_{ST}, and the kernel resulting from their sum; WL, WLC and their sum; we furthermore tested the two kernel families together. In our experiments, we compared three hyper-parameter selection methods. The first method serves as a baseline method: we employed hyper-parameter selection performed through a grid search. Each kernel function is used to train a SVM classifier whose C parameter was validated on the set $\{10^{-4}, 10^{-3}, \dots, 10^{3}\}$. The kernel functions parameters have been validated on the following sets: $h = \{1, \dots, 10\}$ and $\lambda = \{0.1, 0.5, 0.8, 0.9, \dots, 1.5, 1.8\}$, for the kernels derived from the ODD framework, $h = \{1, \dots, 10\}$ for the kernels derived from the WL framework. We refer to this set of experiments as hs. In the second method, we replace the kernel parameter selection procedure with EasyMKL as detailed in Sect. 3. The parameters of the graph kernels are the same as detailed before. The kernel machine employed is KOMD [1], and its hyper-parameter β has been validated in the set $\beta = \{0.0, 0.1, \dots, 0.7\}$. We refer to this methodology as c. Moreover, for the set of experiments concerning ODD_{ST} and TCK_{ST}, we compare our results against those of the experiments in [3] i.e.: $i)$ the combination of the $ODD - MKL$ kernel, an orthogonalized version of the ODD_{ST} kernel, with EasyMKL where the parameters have been selected with the hs technique; $ii)$ the ODD_{ST} kernel with the SVM. Finally, we tested the proposed methodology with orthogonalized kernels. The set of kernel matrices is pre-computed for each of the possible combinations of the following parameter values: $h = \{1, \dots, 10\}$ and λ fixed to 1 for the kernels derived from the ODD framework. For the fast subtree kernels, the parameter h was set to 10 (since with the orthogonalization, all the matrices for heights smaller than 10 are computed). Due to the complexity of the structures contained in the GDD dataset the h parameter of the ODD kernels has been validated on the set $h = \{1, 2, 3\}$. We refer to this methodology as oc.

Table 2. AUROC results ($\pm$ standard deviation) relative to the experiments conducted with the kernels WL and WLC.

(kernels)^{method}	CAS	NCI1	AIDS	CPDB	GDD
$(WL)^{hs}$	90.15 ± 0.20	91.53 ± 0.07	82.78 ± 0.45	85.48 ± 0.63	85.97 ± 0.23
$(WLC)^{hs}$	89.45 ± 0.28	91.55 ± 0.08	82.74 ± 0.56	85.91 ± 0.47	72.44 ± 0.39
$(WL+WLC)^{hs}$	89.95 ± 0.20	91.60 ± 0.08	82.95 ± 0.63	85.18 ± 0.56	$\mathbf{86.85 \pm 0.19}$
$(WL)^{c}$	89.84 ± 0.16	91.41 ± 0.07	85.83 ± 0.32	85.66 ± 0.36	85.51 ± 0.19
$(WLC)^{c}$	90.02 ± 0.15	91.74 ± 0.08	85.96 ± 0.32	85.89 ± 0.33	72.63 ± 0.36
$(WL,WLC)^{c}$	$\mathbf{90.10 \pm 0.15}$	$\mathbf{91.76 \pm 0.07}$	85.94 ± 0.31	$\mathbf{85.97 \pm 0.32}$	85.83 ± 0.15
$(WL)^{oc}$	88.79 ± 0.18	91.37 ± 0.08	85.99 ± 0.43	85.10 ± 0.51	81.75 ± 0.15
$(WLC)^{oc}$	88.75 ± 0.19	91.36 ± 0.08	$\mathbf{86.03 \pm 0.44}$	84.99 ± 0.46	72.66 ± 0.35
$(WL,WLC)^{oc}$	88.76 ± 0.19	91.36 ± 0.08	86.02 ± 0.43	85.08 ± 0.52	81.59 ± 0.16

Table 3. AUROC results ($\pm$ standard deviation) relative to the experiments conducted combining the kernels ODD_{ST}, TCK_{ST}, WL and WLC.

(kernels)^{method}	CAS	NCI1	AIDS	CPDB	GDD
$(ODD_{ST},WL)^{c}$	90.08 ± 0.13	90.80 ± 0.08	84.39 ± 0.47	86.05 ± 0.31	85.68 ± 0.21
$(TCK_{ST},WLC)^{c}$	90.15 ± 0.14	$\mathbf{91.63 \pm 0.06}$	$\mathbf{85.15 \pm 0.39}$	86.26 ± 0.40	85.86 ± 0.20
$(ODD_{ST},TCK_{ST},$ $WL,WLC)^{c}$	$\mathbf{90.24 \pm 0.14}$	91.49 ± 0.06	84.98 ± 0.40	$\mathbf{86.43 \pm 0.32}$	$\mathbf{86.20 \pm 0.20}$

Discussion. The following tables report the AUROC measure with the standard deviation obtained with a nested 10-fold cross validation. Experiment names that feature a list of kernels between brackets separated by a comma refer to the combination of the kernels with MKL, while names that contains a plus sign refer to the sum of the involved kernels. Table 1 shows the results of the three methods when using ODD_{ST} and TCK_{ST} kernels. The results achieved by the c and oc methods are always comparable when not better with respect to hs except when considering the GDD dataset, due probably to the high similarity between each kernel given the smaller parameter grid. $(ODD_{ST},TCK_{ST})^{oc}$, $(TCK_{ST})^{oc}$, and $(ODD_{ST})^{oc}$ are the best performers on the NCI1, AIDS, and CPDB datasets respectively. On the CAS dataset, both c and oc achieve performances comparable to the baseline presented in [3]. Table 2 shows the situation with WL and WLC kernels. The performances of the c and oc methods remain significantly above the baseline on the AIDS dataset, while remaining comparable on the CAS, CPDB and NCI1 datasets. The employed kernels seem not to gain any advantage from being orthogonalized in general: in the case of the GDD dataset we even register an outright performance loss. In Table 3 we present the results obtained while combining the ODD_{ST}, TCK_{ST}, WL and WLC kernel functions together with the c method. These results show no appreciable gain with respect the previously exposed data, but indicate that our proposed method can be used with multiple kernel functions.

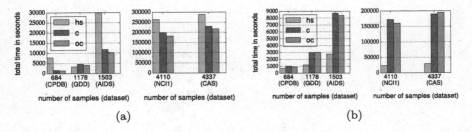

Fig. 3. Time required by each method to compute a full nested 10-fold cross-validation on the benchmark datasets using: (a) $(TCK)_{ST}$ and (b) WLC.

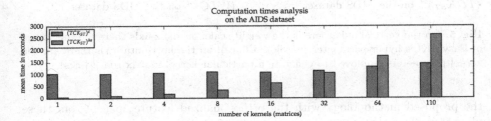

Fig. 4. Comparison of mean times for the $(TCK_{ST})^c$ and $(TCK_{ST})^{hs}$ experiments, employing an increasing number of kernels on AIDS dataset.

We analysed the computational performances of each method. Figure 3a shows that, with the TCK_{ST} kernel, the proposed methodology is more efficient with respect to the baseline method. $(TCK_{ST})^{oc}$ registers a 36.4 % average decrement in the overall time required to perform a full nested 10-fold cross validation with respect to the standard method, i.e. $(TCK_{ST})^{hs}$. The only case where the proposed methodology performs worse is with the GDD dataset. Figure 3b shows the data obtained from the experiments concerning the WLC kernel. The baseline is consistently faster than the proposed methodology; moreover the gap between the two methods is directly proportional to the size of the dataset. Figure 4 shows how the computational time required from hyper-parameter selection hs and our proposed method c varies as a function of the number of kernels that are considered. The figure shows that, when the number of kernels is small, the hs approach is the more convenient, while over a certain treshold the proposed method is faster. Moreover, it is clear that the complexity of the proposed method grows linearly in the number of kernels.

Kernels Contribution Analysis. Weight coefficients calculated by EasyMKL represent the contribution that each kernel gives to the overall learning process; they range in the interval $[0, 1]$ and their sum is equal to 1. Figure 5 shows some examples of the obtained weight distributions and highlights two peculiar patterns that we encountered. The experiments are selected among those employing

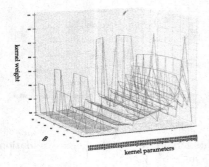 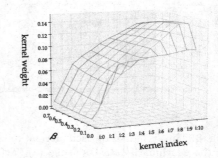

(a)   Kernel weights distribution for $(TCK_{ST})^{oc}$ on the AIDS dataset

(b)   Kernel weights distribution for $(WLC)^{oc}$ on the AIDS dataset

Fig. 5. On the vertical axis there is the weight scale; on the z-axis there are the values of EasyMKL's hyper-parameter β (colored lines); on the horizontal axis there are the kernel parameters employed to generate a particular kernel matrix (dark lines).

the proposed methodology with the orthogonalized feature space. From these plots it is clear that there is a common pattern occurring in the data: all the orthogonalized kernels show a tendency of contributing more the more complex they are, i.e. the larger the height of the features they represent. Moreover, in most cases the MKL algorithm is able to exploit the contribution of more than one kernel at once, mitigating the bias otherwise deriving from the single model selection.

5   Conclusions

In this paper, we proposed an alternative method to the classic grid-search for hyper-parameter tuning of graph kernels, based on the EasyMKL algorithm. We showed two different approaches: the first one is to directly combine the kernels generated from the different parameter configurations, and the second, more principled one, is to divide the feature space induced by the graph kernels in such a way that the generated kernels are orthogonal one from each other. The proposed approach is faster than grid-search when the number of considered parameter configurations is large. Moreover, on some datasets the predictive performance are slightly higher with respect to the grid-search approach.

Acknowledgments. This work was supported by the University of Padova under the strategic project BIOINFOGEN.

References

1. Aiolli, F., Da San Martino, G., Sperduti, A.: A kernel method for the optimization of the margin distribution. In: Kůrková, V., Neruda, R., Koutník, J. (eds.) ICANN 2008, Part I. LNCS, vol. 5163, pp. 305–314. Springer, Heidelberg (2008)
2. Aiolli, F., Donini, M.: EasyMKL: a scalable multiple kernel learning algorithm. Neurocomputing **169**, 215–224 (2015)
3. Aiolli, F., Donini, M., Navarin, N., Sperduti, A.: Multiple graph-kernel learning. In: IEEE SSCI, Cape Town, pp. 1607–1614. IEEE (2015)
4. Da San Martino, G., Navarin, N., Sperduti, A.: A tree-based kernel for graphs. In: SDM, pp. 975–986 (2012)
5. Da San Martino, G., Navarin, N., Sperduti, A.: Ordered decompositional, DAG kernel enhancements. Neurocomputing **192**, 92–103 (2016)
6. Da San Martino, G., Sperduti, A.: Mining structured data. IEEE Comput. Intell. Mag. **5**(1), 42–49 (2010)
7. Dobson, P.D., Doig, A.J.: Distinguishing enzyme structures from non-enzymes without alignments. J. Mol. Biol. **330**(4), 771–783 (2003)
8. Gnen, M., Alpaydin, E.: Multiple kernel learning algorithms. JMLR **12**, 2211–2268 (2011)
9. Hastie, T., Rosset, S., Tibshirani, R., Zhu, J.: The entire regularization path for the support vector machine. JMLR **5**(2), 1391–1415 (2004)
10. Helma, C., Cramer, T., Kramer, S., Raedt, L.D.: Data mining and machine learning techniques for the identification of mutagenicity inducing substructures and structure activity relationships of noncongeneric compounds. J. Chem. Inf. Model. **44**(4), 1402–1411 (2004)
11. Navarin, N., Sperduti, A., Tesselli, R.: Extending local features with contextual information in graph kernels. In: Arik, S., Huang, T., Lai, W.K., Liu, Q. (eds.) ICONIP 2015. LNCS, vol. 9492, pp. 271–279. Springer, Heidelberg (2015). doi:10.1007/978-3-319-26561-2_33
12. Shawe-Taylor, J., Cristianini, N.: Kernel Methods for Pattern Analysis. Cambridge University Press, New York (2004)
13. Shervashidze, N., Borgwardt, K.M.: Fast subtree kernels on graphs. In: Bengio, Y., Schuurmans, D., Lafferty, J.D., Williams, C.K.I., Culotta, A. (eds.) NIPS, pp. 1660–1668. Curran Associates Inc., Red Hook (2009)
14. Tesselli, R.: Adding contextual information to graph kernels. Master's thesis, Università di Padova (2015)
15. Wale, N., Watson, I.A., Karypis, G.: Comparison of descriptor spaces for chemical compound retrieval and classification. Knowl. Inf. Syst. **14**(3), 347–375 (2008)
16. Weislow, O.S., Kiser, R., Fine, D.L., Bader, J., Shoemaker, R.H., Boyd, M.R.: New soluble-formazan assay for HIV-1 cytopathic effects: application to high-flux screening of synthetic and natural products for AIDS-antiviral activity. J. Natl. Cancer Inst. **81**(8), 577–586 (1989)

A Corrector for the Sample Mahalanobis Distance Free from Estimating the Population Eigenvalues of Covariance Matrix

Yasuyuki Kobayashi[✉]

Teikyo University, Faculty of Science and Engineering, Utsunomiya, Japan
ykoba@ics.teikyo-u.ac.jp

Abstract. To correct the effect deteriorating the recognition performance of the sample Mahalanobis distance by a small number of learning sample, a new corrector for the sample Mahalanobis distance toward the corresponding population Mahalanobis distance is proposed without the population eigenvalues estimated from the sample covariance matrix defining the sample Mahalanobis distance. To omit computing the population eigenvalues difficult to estimate, the corrector uses the Stein's estimator of covariance matrix. And the corrector also uses accurate expectation of the principal component of the sample Mahalanobis distance by the delta method in statistics. Numerical experiments show that the proposed corrector improves the probability distribution and the recognition performance in comparison with the sample Mahalanobis distance.

Keywords: Mahalanobis distance · Stein's estimator · Delta method

1 Introduction

The Mahalanobis distance (MD) is frequently used as a discriminator for statistical machine learning or pattern recognition because the MD can be defined easily and fast by covariance matrix from learning sample. The MD has applied to many applications [1]. However, for the population MD (PMD), it is difficult to estimate the unknown population covariance matrix Σ from a limited number of learning sample, thus the sample MD (SMD), Hotelling's T^2 statistic, is calculated for a p-variate test vector y by

$$T^2 = (y - \bar{x})' S^{-1} (y - \bar{x}), \tag{1}$$

where the sample covariance matrix S and the mean vector $\bar{x}$ are estimated by p-variate learning sample vectors, $x_1, \cdots, x_n$ following a p-variate normal distribution $N_p(\mu, \Sigma)$ with mean μ and population covariance matrix Σ. If y follows the same $N_p(\mu, \Sigma)$, then the SMD is known to follow a central F-distribution $F(p, n - p)$ with p and $n - p$ degrees of freedom. In contrast, the PMD follows a chi-squared distribution $\chi^2(p)$ with p degrees of freedom. Therefore as the number n of the learning sample is smaller, the estimated SMD follows more different distribution from the distribution followed by PMD.

© Springer International Publishing AG 2016
A. Hirose et al. (Eds.): ICONIP 2016, Part II, LNCS 9948, pp. 224–232, 2016.
DOI: 10.1007/978-3-319-46672-9_26

There have been many studies to correct the SMD. One of the correctors is regularizing the sample covariance matrix S of the SMD with the appropriate constant [2, 3]. Regularizing S performs well, however the probability distribution of the regularized SMD is unknown, and regularizing S needs numerical experiments to determine the regularizing constant. Another of the correctors is estimating the population covariance matrix Σ. Estimating the population eigenvalues λ_i of Σ and substituting the estimated λ_i into the sample eigenvalues l_i of S also performs well [4]. Furthermore, both estimating λ_i and considering estimation error on sample eigenvectors of S by Monte Carlo (MC) simulation performs better than only estimating λ_i [5]. However, MC simulation needs large computing time. To overcome this weakness, estimating the expectation of each principal component of S by the delta method in statistics without MC simulation performs well with much less time [6]. The expectation is composed of the ratio $\lambda_i/E[l_i]$, where $E[l_i]$ is the expectation of l_i [6], thus excellent estimation of λ_i is required to improve the correcting performance.

In decision-theoretic estimation, an estimation of sample covariance matrix by minimizing quadratic loss (the Stein's estimator) was proposed [7]. The Stein's estimator gives an improved eigenvalue from l_i by a function of n and p. A corrector by substituting the ratio calculated by the Stein's estimator without estimating λ_i into $\lambda_i/E[l_i]$ was proposed [8]. However, Ref. [8] did not show that the proposed SMD improved the recognition performance.

This paper shows that the SMD corrector without estimating the population eigenvalues by using the Stein's estimator can improve the recognition performance of the corrected SMD. Section 2 shows the theory of the corrector. Section 3 shows that numerical experiments give the results of the improved probability distribution and the recognition performance of the corrected SMD. The last section summarizes the proposal and the remained problems to be solved.

2 Theory

Let $x_1, \cdots, x_n$ and y be n learning sample vectors and a test sample vector independently following a p-variate normal distribution $N_p(\mu, \Sigma)$ with mean μ and population covariance matrix Σ, where Σ has the population eigenvalues $\lambda_1 \geq \lambda_2 \geq \cdots \geq \lambda_p$ and the corresponding population eigenvectors $\phi_1, \cdots, \phi_p$. Let $\bar{x}$ be the mean vector of the $x_1, \cdots, x_n$ and X be the data matrix, $X = (x_1 - \bar{x}, \cdots, x_n - \bar{x})'$. And let S be the sample covariance matrix, $S = X'X/(n-1)$, where S is supposed to have the sample eigenvalues $l_1 \geq l_2 \geq \cdots \geq l_p$ and the corresponding eigenvectors $f_1, \cdots, f_p$.

The PMD for the test sample vector y is defined by

$$D^2 = (y - \mu)' \Sigma^{-1} (y - \mu) = \sum_{i=1}^{p} \frac{\{(y - \mu) \cdot \phi_i\}^2}{\lambda_i}. \tag{2}$$

And the corresponding SMD is decomposed by the principal components of S into

$$T^2 = (y - \bar{x})' S^{-1} (y - \bar{x}) = \sum_{i=1}^{p} \frac{\{(y - \bar{x}) \cdot f_i\}^2}{l_i}. \tag{3}$$

Assuming that f_i can be approximated by ϕ_i, T^2 is approximated from (3) by

$$T^2 \cong \sum_{i=1}^{p} \frac{\{(y - \bar{x}) \cdot \phi_i\}^2}{l_i} = \sum_{i=1}^{p} \frac{\lambda_i h_i}{l_i} = \sum_{i=1}^{p} g_i(l_i, h_i), \tag{4}$$

where $\{(y - \bar{x}) \cdot \phi_i\}^2$ follows a $\chi^2(1)$ distribution so that $\{(y - \bar{x}) \cdot \phi_i\}^2 \sim \lambda_i \chi^2(1)$, and let h_i be a random variable following the $\chi^2(1)$ distribution. Furthermore, let $g_i(l_i, h_i) = \lambda_i h_i / l_i$, where the random variables l_i and h_i in g_i are independent.

The expectation of (4) is given by the sum of the expectations of $g_i(l_i, h_i)$, i.e. $E[g_i(l_i, h_i)]$ that can be approximated by the delta method. The delta method is an intuitive technique for approximating the moments of functions of random variables by using the Taylor's expansion [9]. Let $\mu_l = E[l_i]$ and $\mu_h = E[h_i]$ and $E[g_i(l_i, h_i)] \cong \sum_{m,k=0} \partial^{m+k} g_i(\mu_l, \mu_h) / \partial l_i^m \partial h_i^k \cdot E\left[(l_i - \mu_l)^m (h_i - \mu_h)^k \right]$. As $\partial^{m+k} g_i / \partial l_i^m \partial h_i^k = 0 (k \geq 2)$, and $E\left[(l_i - \mu_l)^k (h_i - \mu_h) \right] = 0 (k \geq 0)$, approximating $E[g_i(l_i, h_i)]$ to the 4th order by the delta method is given by

$$\widehat{E}[g_i(l_i, h_i)] = g_i(\mu_l, \mu_h) + \frac{1}{2} \frac{\partial^2 g_i(\mu_l, \mu_h)}{\partial l_i^2} \cdot E\left[(l_i - \mu_l)^2 \right]$$

$$+ \frac{1}{6} \frac{\partial^3 g_i(\mu_l, \mu_h)}{\partial l_i^3} \cdot E\left[(l_i - \mu_l)^3 \right] + \frac{1}{24} \frac{\partial^4 g_i(\mu_l, \mu_h)}{\partial l_i^4} \cdot E\left[(l_i - \mu_l)^4 \right]. \tag{5}$$

The variance, skewness, and kurtosis of l_i (the moments of l_i) are approximately given by $2\lambda_i^2 / (n-1)$, $8\lambda_i^3 / (n-1)^2$, and $48\lambda_i^4 / (n-1)^3$, respectively [10]. Substituting the partial derivatives of g_i, $E[h_i] = 1$, and the moments of l_i into (5), $\widehat{E}[g_i]$ is given by

$$\widehat{E}[g_i(l_i, h_i)] = \frac{\lambda_i}{E[l_i]} \left\{ 1 + \frac{2}{n-1} \left(\frac{\lambda_i}{E[l_i]} \right)^2 - \frac{8}{(n-1)^2} \left(\frac{\lambda_i}{E[l_i]} \right)^3 + \frac{48}{(n-1)^3} \left(\frac{\lambda_i}{E[l_i]} \right)^4 \right\}. \tag{6}$$

Each $g_i(l_i, h_i)$ has a bias given by (6), therefore dividing $g_i(l_i, h_i)$ by (6) gives correction for the bias included in $g_i(l_i, h_i)$ in the SMD. If $n \to \infty$, i.e. the SMD converges to the PMD, then $\widehat{E}[g_i] \to 1$ and dividing $g_i(l_i, h_i)$ by (6) has no effect of correction. Hence a corrected SMD is defined by

$$\widehat{T}^2 = \sum\nolimits_{i=1}^{p} \frac{\{(y - \bar{x}) \cdot f_i\}^2}{l_i} \cdot \widehat{E}[g_i(l_i, h_i)]^{-1}. \tag{7}$$

Here, supposing that l_i in (7) is $E[l_i]$ and substituting (6) into (7), $\widehat{T}^2$ is given by

$$\widehat{T}^2 = \sum\nolimits_{i=1}^{p} \frac{\{(y - \bar{x}) \cdot f_i\}^2}{\lambda_i} \cdot \left[1 + \frac{2}{n-1}\left(\frac{\lambda_i}{E[l_i]}\right)^2 - \frac{8}{(n-1)^2}\left(\frac{\lambda_i}{E[l_i]}\right)^3 + \frac{48}{(n-1)^3}\left(\frac{\lambda_i}{E[l_i]}\right)^4\right]^{-1}. \tag{8}$$

(8) shows that an SMD corrector proposed in Ref. [4] by only substituting λ_i into l_i in (3), i.e., $\sum_{i=1}^{p}\{(y - \bar{x}) \cdot f_i\}^2/\lambda_i$ is not enough, and Ref. [6] shows that the 2nd order or less of $\lambda_i/E[l_i]$ of (8) is superior to $\sum_{i=1}^{p}\{(y - \bar{x}) \cdot f_i\}^2/\lambda_i$ by experiment. However, it is difficult to estimate λ_i correctly if the Σ is unknown. Here, an approximate λ_i is supposed to be given by the Stein's estimator. The Stein's estimator [7] gives the improved eigenvalue l_i^* from l_i of S such that $l_i^* = l_i \cdot (n-1)/(n+p-2i)$. Here, supposing that $\lambda_i \cong l_i^*$ and substituting $E[l_i]$ into l_i,

$$\frac{\lambda_i}{E[l_i]} = \frac{n-1}{n+p-2i}. \tag{9}$$

Applying (9) to (6), $\widehat{E}[g_i]$ can be approximated without estimating λ_i by

$$\widehat{E}[g_i(l_i, h_i)] = \frac{n-1}{n+p-2i}\left\{1 + \frac{2}{n-1}\left(\frac{n-1}{n+p-2i}\right)^2 - \frac{8}{(n-1)^2}\left(\frac{n-1}{n+p-2i}\right)^3 + \frac{48}{(n-1)^3}\left(\frac{n-1}{n+p-2i}\right)^4\right\}. \tag{10}$$

Therefore, this paper proposes an SMD corrector without estimating λ_i such that

$$\widehat{T}^2 = \sum\nolimits_{i=1}^{p} \frac{\{(y - \bar{x}) \cdot f_i\}^2}{l_i}\left[\left(\frac{n-1}{n+p-2i}\right)\left\{1 + \frac{2}{n-1}\left(\frac{n-1}{n+p-2i}\right)^2 - \frac{8}{(n-1)^2}\left(\frac{n-1}{n+p-2i}\right)^3\right.\right.$$
$$\left.\left. + \frac{48}{(n-1)^3}\left(\frac{n-1}{n+p-2i}\right)^4\right\}\right]^{-1}. \tag{11}$$

Ref. [8] showed the case up to the 2nd order approximation of $\widehat{E}[g_i]$ of (11) corrected the SMD visually on graph. However, expanding the higher order approximation up to the 4th order in (11), can be expected for more accuracy than the 2nd order of Ref. [8].

3 Numerical Experiments

3.1 Procedure

To confirm the performance of the proposed SMD corrector (11), numerical experiments are conducted on Microsoft Excel 2010 or 2013 with a source code written in Excel VBA. In the numerical experiments, test sample of the SMD are generated by using Monte Carlo simulation, and the SMD sample corrected by the SMD correctors are investigated about their probability distribution and recognition performance.

The algorithm of the source code is described below. In short, k of the p-variate population covariance matrix Σ are randomly generated except the predefined population eigenvalues $\lambda_1 \geq \lambda_2 \geq \cdots \geq \lambda_p$. The corresponding population eigenvectors $\phi_1, \cdots, \phi_p$ are also defined by Σ. For each Σ, both learning sample $x_1, \cdots, x_n$ and the test sample $y_1, \cdots, y_m$ are randomly generated for central sample by $N_p(0, \Sigma)$, and the sample eigenvalues $l_1 \geq l_2 \geq \cdots \geq l_p$ and the corresponding eigenvectors $f_1, \cdots, f_p$ are calculated by the sample covariance matrix S of x_i. In a similar way, sample eigenvalues $\tilde{l}_1 \geq \cdots \geq \tilde{l}_p$, and the corresponding eigenvectors $\tilde{f}_1, \cdots, \tilde{f}_p$ are calculated for non-central sample, however test sample $\tilde{y}_1, \cdots, \tilde{y}_m$ are generated by $N_p(\Delta, \Sigma)$ where $\Delta = (\delta_1, \cdots, \delta_p)$ and δ_i is randomly generated by predefined non-centrality parameter δ $(\delta > 0)$ to satisfy $\delta = \delta_1^2 + \cdots + \delta_p^2$. For y as a sample of a central distribution, the PMD, SMD, and the following corrected SMDs are calculated by the formulae in Table 1, and all the MDs are normalized by dimensionality p. And for $\tilde{y}$ as a sample of a non-central distribution, the corresponding MDs are calculated in a similar way with the above tilde-marked variables $\tilde{y}$, $\tilde{l}$, and $\tilde{f}$. The expectation of l_i ($E[l_i]$) is numerically estimated before the experiments. In Table 1, the delta terms mean the square brackets in (8) and (11), respectively.

In 3.2 Results, the curves of p, pΔ2, S, and SΔ2 are especially shown in Figs. 2 through 5, therefore the corresponding formulae are shown in (12)–(15).

Table 1. Calculated MDs in numerical experiments.

Abbr.	Equation with condition	Abbr.	Equation with condition
PMD	(2) with $\mu = 0$	SMD	(3)
p	(12)	S	(14)
pΔ2	(8) changed by the delta terms up to the 2nd order, i.e. (13)	SΔ2	(11) changed by the delta terms up to the 2nd order, i.e. (15)
pΔ3	(8) changed by the delta terms up to the 3rd order	SΔ3	(11) changed by the delta terms up to the 3rd order
pΔ4	(8) changed by the delta terms up to the 4th order	SΔ4	(11) changed by the delta terms up to the 4th order

p:

$$\widehat{T}^2 = \sum_{i=1}^p \frac{\{(y-\bar{x})\cdot f_i\}^2}{\lambda_i}. \tag{12}$$

pΔ2:

$$\widehat{T}^2 = \sum_{i=1}^p \frac{\{(y-\bar{x})\cdot f_i\}^2}{\lambda_i}\left\{1 + \frac{2}{n-1}\left(\frac{\lambda_i}{E[l_i]}\right)^2\right\}^{-1}. \tag{13}$$

S:

$$\widehat{T}^2 = \sum_{i=1}^p \frac{\{(y-\bar{x})\cdot f_i\}^2}{l_i}\left(\frac{n-1}{n+p-2i}\right)^{-1}. \tag{14}$$

SΔ2:

$$\widehat{T}^2 = \sum_{i=1}^p \frac{\{(y-\bar{x})\cdot f_i\}^2}{l_i}\left[\left(\frac{n-1}{n+p-2i}\right)\left\{1 + \frac{2}{n-1}\left(\frac{n-1}{n+p-2i}\right)^2\right\}\right]^{-1}. \tag{15}$$

After all $k \cdot m$ sample of y are generated, the empirical probability density functions are calculated. To evaluate the performance of the correctors, the following indicators of the above MDs with depending on non-centrality δ are calculated:

(a) expectation, (b) variance, (c) the Hellinger distance (HL) to the $\chi^2(p)$ distribution that the PMD follows with the corresponding δ, where the distance is given by

$$HL = 2\int_0^\infty \left(\sqrt{f(x)} - \sqrt{g(x)}\right)^2 dx, \tag{16}$$

where $f(x)$ and $g(x)$ are the probability density functions of the empirical and the $\chi^2(p)$ distributions, respectively, (d) the minimum misclassification rate $(\alpha+\beta)_{min}$, and (e) the threshold value Θ of MD to $(\alpha+\beta)_{min}$.

$(\alpha+\beta)_{min}$ is defined by the sum of type I error rate α of the central distribution and type II error rate β of the non-central distribution at the MD value Θ at which the probability density curves of both the distributions are crossing [11]. The experiments assume that both the central sample and the non-central sample follow the same population covariance matrix Σ, which linear discriminant analysis also assumes [2].

3.2 Results

The experiment condition is as follows: The dimensionality $p = 30$, the number of learning sample $n = 40$, the number of test sample $m = 10000$, and the number of generated population covariance matrix $k = 1000$. The population eigenvalues $\lambda_i = 10^{-0.3(i-1)}(i = 1,\cdots,p)$, where the maximum $\lambda_1 = 1$ and the minimum $\lambda_{30} = 2 \times 10^{-9}$. The non-centrality $\delta = 0, 5, \cdots, 50$.

In comparison of the Hellinger distance among the SMD correctors in Table 1, Fig. 1 shows that the SMD correctors with λ_i, i.e. p - pΔ4 are better than those with the Stein's estimators, i.e. S - SΔ4. And Fig. 1 also shows that the SMD corrector (13), pΔ2, with the 2nd order approximation is the best of p - pΔ4, and that the SMD corrector (15), SΔ2, with the 2nd order approximation is the best of S - SΔ4. Figure 2 shows that the SMD corrected by the SMD correctors are closer to the PMD than the

SMD, and that p and pΔ2 are better than S and SΔ2. Hence, the performances of the SMD correctors of pΔ2 and SΔ2 are investigated for comparison.

Figure 3 shows that both probability density curves corrected by pΔ2 and SΔ2 are closer to that of PMD than that of SMD when $\delta = 0$ and $\delta = 30$.

Figure 4 shows that the SMD corrected by p, pΔ2, S and SΔ2 have much smaller expectation and variance than the SMD. Applying the delta terms in pΔ2 makes the expectation and variance of pΔ2 smaller than those of p, and similarly in case of SΔ2 and S.

Figure 5 shows that the SMD corrected by p, pΔ2, S and SΔ2 have smaller $(\alpha + \beta)_{min}$ and smaller threshold Θ than the SMD. Although the Θ of p and pΔ2 are better than those of S and SΔ2, the $(\alpha + \beta)_{min}$ of p, pΔ2, S and SΔ2 are almost equal, respectively.

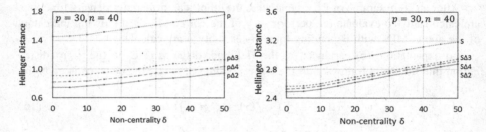

Fig. 1. Hellinger distance depending on SMD correctors

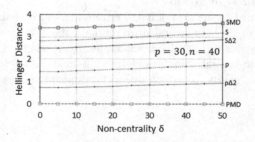

Fig. 2. Hellinger distance from PMD

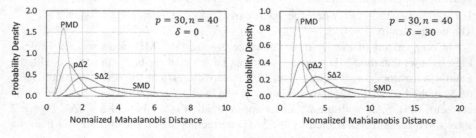

Fig. 3. Probability density distribution

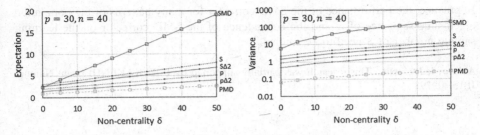

Fig. 4. Expectation and variance of normalized Mahalanobis distances

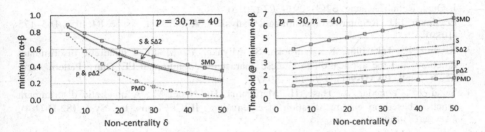

Fig. 5. Minimum misclassification rate $(\alpha + \beta)_{min}$ and corresponding threshold normalized by dimension

As a result, if λ_i can be obtained, the SMD corrector (13), pΔ2 can correct the SMD toward the PMD more accurately than SΔ2. However, the proposed SMD corrector (15), SΔ2 can moderately correct the SMD without estimating λ_i.

4   Conclusion

Many studies have proposed correctors for the sample Mahalanobis distance (SMD) toward the population Mahalanobis distance (PMD). They need to estimate the population eigenvalues λ_i of the population covariance matrix Σ, however λ_i are difficult to estimate correctly from the sample covariance matrix S if Σ is unknown.

This paper has proposed an SMD corrector that does not need λ_i by using the Stain's estimator of covariance matrix and the delta method in statistics. Numerical experiments show that the SMD corrected by the proposed corrector without λ_i is closer to the PMD and improves the probability distribution and the recognition performance in comparison with the SMD.

However, the proposed SMD corrector is still inferior to the SMD corrector with λ_i. Hence in order to improve the proposed SMD corrector, there are following problems to be solved: The Stain's estimator may be a rough approximation of λ_i, therefore more refined approximation needs to be studied. The proposed SMD corrector supposes that the sample eigenvector f_i of S is equal to the population eigenvector ϕ_i of Σ, however the SMD corrector needs to consider the fact that f_i is a random vector variable.

In practice, the performance of the SMD corrector for many patterns of the population eigenvalue sequence of λ_i and the dependence of the SMD corrector on the learning sample number of S need to be studied.

Acknowledgement. This research was partially supported by JSPS KAKENHI Grant Number JP15H02798.

References

1. Ghasemi, E., Aaghaie, A., Cudney, E.A.: Mahalanobis Taguchi system: a review. Int. J. Qual. Reliab. Manag. **32**(3), 291–307 (2015)
2. Friedman, J.H.: Regularized discriminant analysis. J. Am. Statist. Assoc. **84**(405), 165–175 (1989)
3. Kimura, F., Takashina, K., Tsuruoka, S., Miyake, Y.: Modified quadratic discriminant functions and the application to chinese character recognition. IEEE Trans. Pattern Anal. Mach. Intell. **PAMI-9**(1), 149–153 (1987)
4. Sakai, M., Yoneda, M., Hase, H., Maruyama, H., Naoe, M.: A quadratic discriminant function based on bias rectification of eigenvalues. IEICE Trans. Inf. Syst. **J82-D-II**(4), 631–640 (1999). (in Japanese)
5. Iwamura, M., Omachi, S., Aso, H.: Estimation of true Mahalanobis distance from eigenvectors of sample covariance matrix. IEICE Trans. Inf. Syst. **J86-D-II**(1), 22–31 (2003). (in Japanese)
6. Kobayashi, Y.: A proposal of simple correcting scheme for sample Mahalanobis distances using delta method. IEICE Trans. Inf. Syst. **J97-D**(8), 1228–1236 (2014). (in Japanese)
7. James, W., Stein, C.: Estimation with quadratic loss. In: Proceedings of the 4th Berkeley Symposium on Mathematical Statistics and Probability, vol. 1, pp. 361–379 (1961)
8. Kobayashi, Y.: A simplified corrector for sample Mahalanobis distance. In: Proceedings of the 12th International Conference on Ubiquitous Healthcare, p. 109 (2015)
9. Oehlert, G.W.: A note on the delta method. Am. Statist. **46**(1), 27–29 (1992)
10. Lawley, D.N.: Tests of significance for the latent roots of covariance and correlation matrix. Biometrika **43**, 128–136 (1956)
11. Fukunaga, K.: Introduction to Statistical Pattern Recognition, 2nd edn. Academic Press, New York (1990)

Online Learning Neural Network
for Adaptively Weighted Hybrid Modeling

Shao-Ming Yang, Ya-Lin Wang[✉], Yong-fei Xue, Bei Sun,
and Bu-song Yang

School of Information Science and Engineering,
Central South University, Changsha, China
ylwang@csu.edu.cn

Abstract. The soft sensor models constructed based on historical data have poor generalization due to the characters of strong non-linearity and time-varying dynamics. Moving window and recursively sample updating online modeling methods can not achieve a balance between accuracy and training speed. Aiming at these problems, a novel online learning neural network (LNN) selects high-quality samples with just-in-time learning (JITL) for modeling. And the local samples could be further determined by principal component analysis (PCA). The LNN model shows better performance but poor stability. Weighted multiple sub models, the hybrid model improves accuracy by covering deficiencies. Additionally, the weights could be developed with mean square error (MSE) of each sub model. And the detailed simulation results verify the superiority of adaptive weighted hybrid model.

Keywords: Hybrid modeling · Just-in-time learning · Learning neural network

1 Introduction

The product quality or some key parameters of complex industrial processes are difficult to detect online. Generally, some soft models based on historical data could be used for prediction, such as neural networks (NN). And the back propagation (BP) neural network has been widely used in various fields due to its simple structure, strong parallel processing ability, good nonlinear approximation ability, etc. And yet, for low learning efficiency and local minimum problems, many researchers have done much work on learning rate [1], initial weights [2], structure [3] and parameters [4]. On the other hand, functional networks [5] are generalized networks based on stand neural networks, but the functions of neurons are multivariate and alterable. And that there are not weights between neurons. Sometimes the performance of functional networks is

This work is supported by the National Natural Science Foundation of China (Grant No. 61273187); the Major Program of the National Natural Science Foundation of China (Grant No. 61590921); the Fundamental Research Funds for the Central Universities of Central South University (Grant No. 502210008).

© Springer International Publishing AG 2016
A. Hirose et al. (Eds.): ICONIP 2016, Part II, LNCS 9948, pp. 233–240, 2016.
DOI: 10.1007/978-3-319-46672-9_27

better than general neural networks. The problem is that the optimal initial structures are difficult to determine.

Based on the general neural networks and functional networks, a novel active function learning neural network (LNN) [6] makes multidimensional output of neurons map to one-dimensional for next layer neurons. The LNN inherits advantages of the two networks. And its structure could be easily online adjusted. Moreover, the LNN is trained by extreme learning machine (ELM) [7] without iterative learning, which greatly improves the learning speed.

When the process changes abruptly, the global model would functions worse. Then a local modeling method, just-in-time learning (JITL) [8], has been proposed, which selects local modeling samples by considering neighborhood of the query sample. For example, the k-Nearest Neighbors (k-NN) method [9] calculated the distance between query sample and database, yet it can not commendably describe the similarity. The k-Vector Nearest Neighbor (k-VNN) method [10] was fused both distance and angular, but the weight allocation affects the similarity and even the prediction accuracy. Later, a novel index is proposed by incorporating the two indexes without weights. To some extent, it improves reliability of modeling samples.

The hybrid modeling approach [11] could make up the deficiencies of sub model and show better stability. In this article, we build multiple LNN models with JITL for hybrid model, and that the weights of sub model are achieved with their mean square errors (MSE) [12]. Finally, function approximate and an industrial application verify the superiority of the adaptive weighted hybrid model based on LNN model.

2 Preliminaries

2.1 Just-in-Time Learning (JITL)

Generally, the global model may function worse as the strong non-linearity and time-varying in the process. Then a method divides the process into multiple small regions and builds local model in each small region. It is just-in-time learning (JITL) model, also called "lazing learning". It is that similar outputs corresponding to similar inputs. Additionally, the similar samples are fulfilled by similarity with query sample.

The current input sample is $\mathbf{X}_q$, $\mathbf{X}_q = (x_{q1}, x_{q2}, \cdots, x_{qn})$. The database sample is $\mathbf{X}_i$, $\mathbf{X}_i = (x_{i1}, x_{i2}, \cdots, x_{in})$. Then the distance and angle between the two samples are

$$d_{qi} = \sqrt{\sum_{j=1}^{n} (x_{qj} - x_{ij})^2} \tag{1}$$

$$cos(\theta_{qi}) = \frac{\mathbf{X}_q \mathbf{X}_i^{\mathrm{T}}}{\|\mathbf{X}_q\|_2 \|\mathbf{X}_i\|_2} \tag{2}$$

Define the similarity as

$$C_{qi} = \cos(\theta_{qi}) \sqrt{e^{-d_{qi}^2}} \tag{3}$$

According to rank of the similarity, the early k samples can be well selected for local modeling. The samples are as follows.

$$\Omega_k = \{(\mathbf{X}_1, y_1), (\mathbf{X}_2, y_2), \cdots, (\mathbf{X}_k, y_k) | C_{q1} > C_{q2} > \cdots > C_{qk}\} \tag{4}$$

Further, the modeling samples could be optimized in PCA model. From the viewpoint of the correlation among variables, Q is described as distance and T^2 is to guarantee the samples in local region. Then the two indexes are fused to a novel index, whose minimum corresponding to k is the optimal size for local samples [10].

For JITL model, there are multiple linear regression (MLR), artificial neural network (ANN), partial least squares (PLS) and so on. And main steps are the same as follows:

Step1: Search the relevant samples to construct a set Ω_k from the database;
Step2: Build JITL model $f_{JITL}(x)$ with local sampling set Ω_k;
Step3: Predict the output $\hat{y}_q$ corresponding to the query sample $\mathbf{X}_q$ with $f_{JITL}(x)$, then the JITL model would be discarded.
Step4: If the output y_q of $\mathbf{X}_q$ is detected, the data $(\mathbf{X}_q, y_q)$ should be stored in the database. When the next query sample comes, it will back the step1.

2.2 Learning Neural Networks (LNN)

The active function leaning neural network (LNN) is proposed on base of neural network and functional network. Different from other networks, the number and kind of the hidden neurons of LNN are not fixed. For a single neuron, all input variables of sample are weighted as Σ_1 and Σ_2 could be reached by linear combination of basic functions as its output. And the internal structure is shown in Fig. 1.

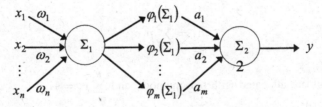

Fig. 1. The internal structure of the neuron

2.3 Extreme Learning Machine (ELM)

For SLFN, Huang [6] proposed extreme learning machine (ELM), that the input layer weights ω and hidden basis $\mathbf{b}$ are randomly determined. Then the weights of output layer are extremely solved without iteration. Then the training speed has been greatly improved. When the samples are given, the structures of input and output are known. And the algorithm can be summarized as follows:

Step 1: Decide l hidden neurons and infinite differential active functions;
Step 2: Randomly assign input weight ω and bias $\mathbf{b}$;
Step 3: Calculate the output matrix of hidden layer $\mathbf{H}$ and weights matrix $\mathbf{A}$.

3   Adaptive Hybrid Modeling

The LNN shows high prediction accuracy with fast speed but poor stability. This paper [11] developed a hybrid model based on multiple excellent ELM models, which fuse advantages of each sub model. Considering the performance of each sub model, in this paper, we propose the adaptive weighted hybrid model based on LNN, and the weights are calculated with prediction errors. The final output could be got as follows (Fig. 2).

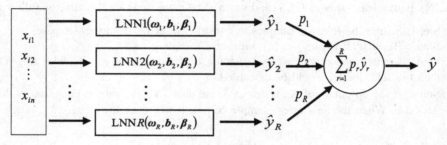

Fig. 2. The structure of hybrid model

For example, there are R sub models for hybrid modeling, and the $\hat{y}$ and total variance σ^2 are

$$\hat{y} = \sum_{r=1}^{R} p_r \hat{y}_r \tag{5}$$

$$\sigma^2 = \sum_{r=1}^{R} p_r^2 \sigma_r^2 \tag{6}$$

The weights are allocated with two methods, and $\sum_{r=1}^{R} p_r = 1$.

(1) $p_r = \frac{1}{R}$, the output of hybrid model and variance are $\hat{y}1$ and σ^2,

$$\hat{y}1 = \sum_{r=1}^{R} \frac{1}{R} \hat{y}_r = \frac{1}{R} \sum_{r=1}^{R} \hat{y}_r \tag{7}$$

$$\sigma^2 = \sum_{r=1}^{R} \frac{1}{R^2} \sigma_r^2 = \frac{1}{R^2} \sum_{r=1}^{R} \sigma_r^2 \tag{8}$$

The weights of the hybrid model are the same, and the variance is least among these sub models. It is obvious that σ^2 is multivariate function with two-dimensional about the weights of sub models. Based on the theory [12], minimal variance σ^2 will be reached with the higher effective weights.

(2) $p_r = \dfrac{1}{\sigma_r^2 \sum\limits_{r=1}^{R} \frac{1}{\sigma_r^2}}$, the output of hybrid model and variance are $\hat{y}2$ and $\sigma_{\min}^2$,

$$\hat{y}2 = \sum_{r=1}^{R} \frac{1}{\sigma_r^2 \sum\limits_{r=1}^{R} \frac{1}{\sigma_r^2}} \hat{y}_r = \frac{1}{\sum\limits_{r=1}^{R} \frac{1}{\sigma_r^2}} \sum_{r=1}^{R} \frac{1}{\sigma_r^2} \hat{y}_r \tag{9}$$

$$\sigma_{\min}^2 = \frac{1}{\sum\limits_{r=1}^{R} \frac{1}{\sigma_r^2}} \tag{10}$$

Judging from that, we choose the second method and the weights are calculated with variances of sub models for hybrid modeling. Meanwhile, the three indexes of root mean square error (RMSE), mean relative error (MRE) and model coefficients (R2) are shown the performance,

$$R^2 = \frac{\left(M \sum\limits_{i=1}^{M} \hat{y}_i y_i - \sum\limits_{i=1}^{M} \hat{y}_i \sum\limits_{i=1}^{M} y_i \right)^2}{\left(M \sum\limits_{i=1}^{M} \hat{y}_i^2 - \left(\sum\limits_{i=1}^{M} \hat{y}_i \right)^2 \right) \left(M \sum\limits_{i=1}^{M} y_i^2 - \left(\sum\limits_{i=1}^{M} y_i \right)^2 \right)} \tag{11}$$

As above, $\hat{y}_i$ is prediction and y_i is true. When RMSE and MRE are closer to 0, the performance is better. R^2 is closer to 1, the better. $i \in [1, M]$, $R^2 \in [0, 1]$.

4   Case Study

4.1   Function Approximation

$$F(x_1, x_2) = 0.5x_1 x_2 + x_1 e^{-x_1 x_2}, x_1 \in [0, 1], x_2 \in [0, 1] \tag{12}$$

To verify the hybrid model, we can use the rand function, embedded in MATLAB, to generate 1000 groups of two dimensional data as input, $x_{ij} \in [0, 1]$. The outputs are achieved from (14). The 990 groups of data are optionally selected as train set, and the remaining data are test set. And the simulations results are as follows.

In Fig. 3 (a), the approximation of BP, ELM and LNN are all good and the LNN is better than the others from the above three indexes. It is because the different neurons of LNN play a key role on model training. As for BP and ELM, the same neurons have relatively weak effect on the model learning. In Fig. 3(b), the LNN-PCA shows better performance with LNN and PCA models. That because the local modeling samples are further optimized. In addition, the adaptive hybrid model can approach the true by closing to nil, yet the hybrid model of simple linear superposition shows relatively poor effect. That illustrates the validity of the weights for sub models, calculated by prediction errors. For the adaptive hybrid model, the strategy promotes the contributions of good sub models and reduces the effect of worse sub model. The detailed comparisons of experiment results are obviously seen in Table 1. From the point of the three indexes, the LNN-PCA is with best performance among these models. But the stability of that is lower according to multiple experiments. In general, we had better choose adaptive hybrid model for online prediction.

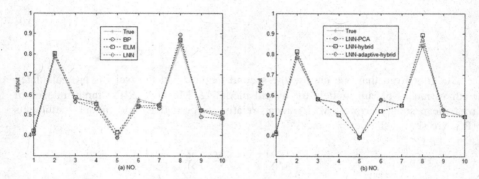

Fig. 3. Results of different models in function approximation

Table 1. Performance comparison of different models in function approximation

	BP	ELM	LNN	LNN PCA	LNN PCA hybrid	LNN PCA adaptive hybrid
MRE	0.0388	0.0279	0.0160	0.0011	0.0385	0.0077
RMSE	0.0283	0.0190	0.0114	0.0007	0.0344	0.0112
R2	0.9810	0.9840	0.9971	1.0000	0.9647	0.9983

4.2 Predicting Gasoline Octane

Octane is used to evaluate the quality of gasoline. In recent years, a method based on near infrared (NIR) spectra is commonly used as the low cost and pollution-free. Firstly, we collect 60 groups of gasoline samples and record 401 wavelength points for each sample. Then the octane of all samples could be detected with traditional laboratory methods. Using the spectra data and corresponding octane, build models for online prediction. To certify the performance of the hybrid model, we also choose several other models for comparison. The results are as follows.

Table 2. Performance comparison of different models in predicting octane

	BP	ELM	LNN	LNN PCA	LNN PCA hybrid	LNN PCA adaptive hybrid
MRE	0.0082	0.0110	0.0094	0.0061	0.0022	0.0014
RMSE	0.9321	1.1295	1.0715	0.6416	0.2348	0.1393
R2	0.8035	0.5516	0.4536	0.7986	0.9812	0.9926

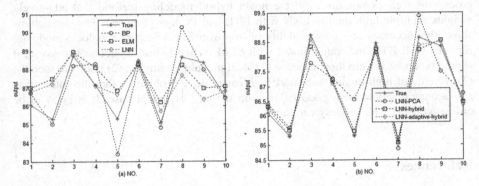

Fig. 4. Results of different models in predicting octane

Similarly, in Fig. 4(a), compared with BP, the ELM and LNN models both show good prediction error but the model coefficient. Additionally, the LNN model shows better performance with JITL and PCA models as Fig. 4(b). The two hybrid models express low error, meanwhile the adaptive weighted hybrid model approaches the true by lower errors. In the application, the improved hybrid model proves that the new weights for hybrid modeling are well making up mutual deficiencies of these sub models. And the detailed comparisons are as in Table 2. Therefore, before modeling, we may as well use JITL and PCA model to optimize local modeling samples, and the adaptively weighted hybrid LNN model would be better for online prediction.

To illustrate the good stability, we have done fifty groups of simulations. By the mean μ and standard deviation σ of the three indexes, the models can be evaluated again.

Table 3. Simulation results of different models

	BP	ELM	LNN	LNN PCA	LNN PCA hybrid	LNN PCA adaptive hybrid
MRE μ	0.01142	0.00972	0.01085	0.0109	0.01020	0.00340
MRE σ	0.00398	0.00278	0.00317	0.0034	0.00139	0.00114
RMSE μ	1.34932	1.06894	1.22588	1.12581	0.36881	0.29748
RMSE σ	0.57464	0.31734	0.36342	0.35428	0.14650	0.11587
R2 μ	0.61959	0.62006	0.56316	0.58369	0.95420	0.96747
R2 σ	0.19452	0.19362	0.16450	0.17007	0.04277	0.02845

In Table 3, it is obvious that μ and σ are both declining along with BP, ELM, LNN, LNN-PCA, LNN-PCA hybrid and LNN-PCA adaptive hybrid as well as RMSE and R2. By the two measures, the adaptive hybrid model shows better performance and stability. Thus adaptive hybrid LNN model is more worthy to be online prediction.

5 Conclusion

In the present work, to develop a general hybrid model that cope with changes in process as well as non-linearity, the novel hybrid modeling method that adaptively weights multiple high-quality LNN with JITL and PCA is proposed. The simulations show high accuracy and good stability of improvement, which selects local modeling samples with JITL and trains model with ELM. When hybrid model works worse, it should be updated from three aspects: (1) local modeling samples; (2) poor sub models; (3) weights of high-quality sub models. Meanwhile, a problem is that the calculation and accuracy should be a good balance on the size of model base. It is good to be online adjusted with process variations.

References

1. Zweiri, Y.H., Seneviratne, L.D., Althoefer, K.: Stability analysis of a three-term back-propagation algorithm. Neural Netw. **18**(10), 1341–1347 (2005)
2. Yam, Y.F., Leung, C.T., Tam, P.K.S., et al.: An independent component analysis based weight initialization method for multilayer perceptrons. Neurocomputing **48**(1), 807–818 (2002)
3. Leung, F.H.F., Lam, H.K., Ling, S.H., et al.: Tuning of the structure and parameters of a neural network using an improved genetic algorithm. IEEE Trans. Neural Netw. **14**(1), 79–88 (2003)
4. Valian, E., Mohanna, S., Tavakoli, S.: Improved cuckoo search algorithm for feedforward neural network training. Int. J. Artif. Intell. Appl. **2**(3), 36–43 (2011)
5. Castillo, E., Cobo, A., Gómez-Nesterkin, R., et al.: A general framework for functional networks. Networks **1**, 70–82 (2000)
6. Yang, S.M., Wang, Y.L., Wang, M.Y.: Active functions learning neural network. J. Jiangnan Univ. (Nat. Sci. Ed.) **14**(6), 689–694 (2015)
7. Huang, G.B., Zhu, Q.Y., Siew, C.K.: Extreme learning machine: theory and applications. Neurocomputing **70**(1), 489–501 (2006)
8. Bontempi, G., Birattari, M., Bersini, H.: Lazy learning for local modeling and control design. Int. J. Control **72**(7–8), 643–658 (1999)
9. Cheng, C., Chiu, M.S.: A new data-based methodology for nonlinear process modeling. Chem. Eng. Sci. **59**(13), 2801–2810 (2004)
10. Fujiwara, K., Kano, M., Hasebe, S., et al.: Soft - sensor development using correlation - based just – in - time modeling. AIChE J. **55**(7), 1754–1765 (2009)
11. Yang, S., Wang, Y., Sun, B., et al.: ELM weighted hybrid modeling and its online modification. In: 28th Chinese Control and Decision Conference, May 2016
12. Zeng, X.: The algorithm of CFNN image data fusion in multi-sensor data fusion. Sens. Transducers **166**(3), 197 (2014)

Semi-supervised Support Vector Machines - A Genetic Algorithm Approach

Gergana Lazarova[✉]

Sofia University "St. Kliment Ohridski", Sofia, Bulgaria
gergana1@fmi.uni-sofia.bg

Abstract. Semi-supervised learning combines both labeled and unlabeled examples in order to find better future predictions. Semi-supervised support vector machines (SSSVM) present a non-convex optimization problem. In this paper a genetic algorithm is used to optimize the non-convex error - GSSSVM. It is experimented with multiple datasets and the performance of the genetic algorithm is compared to its supervised equivalent and shows very good results. A tailor-made modification of the genetic algorithm is also proposed which uses less unlabeled examples – the closest neighbors of the labeled instances.

Keywords: Semi-supervised learning · Semi-supervised support vector machines · Genetic algorithms

1 Introduction

Semi-supervised learning [1, 9] is a state-of-the-art discipline of artificial intelligence, which is responsible for training an agent when only few labeled examples exist. There is not enough past history the agent can learn from. In this field, unlabeled instances are also used for training. The aim is to extract relevant information from the unlabeled instances, so that a good prediction algorithm can be learned.

Semi-supervised learning has been used for web-page classification [2], co-tracking [3], semi-supervised image segmentation [15] and sentiment analysis [16].

Semi-supervised support vector machines (SSSVM) are a modern semi-supervised equivalent of the supervised support vector machines (SVM). They punish unlabeled examples, which are very close to the separation hyperplane.

Since 1999 when SVM were implemented by Joachims [17], a variety of improvements have been applied to solve this non-convex optimization problem. For instance, gradient descent [18], continuation techniques [1], deterministic annealing [20]. Chapelle et al. [21] apply branch and bound techniques for obtaining exact, globally optimal solutions. But, global optimization can be computationally very expensive. The authors themselves describe in the paper that the approach is impractical for large data sets.

The proposed in this paper approach to non-convex function optimization uses a genetic algorithm and the global optimum is not guaranteed. Although, the global optimum is not guaranteed, it achieves good results. In this paper, a modified genetic algorithm is also proposed, which does not use all the unlabeled examples, but only the most significant ones (the ones that are close to labeled examples).

© Springer International Publishing AG 2016
A. Hirose et al. (Eds.): ICONIP 2016, Part II, LNCS 9948, pp. 241–249, 2016.
DOI: 10.1007/978-3-319-46672-9_28

2 Semi-supervised Learning

Semi-supervised learning uses both labeled and unlabeled examples. A teacher, an expert in the field, has already labeled some of the instances - D_1. In semi-supervised learning unlabeled instances D_2 are also used and added to the pool of training examples. The final training data contains both the examples of D_1 and D_2 ($D = D_1 \cup D_2$). Let the number of labeled examples be l and the number of the unlabeled examples-u.

$$D_1 = \{(x_i, y_i)\}_{i=1}^{l}, \quad D_2 = \{x_j\}_{j=1}^{u}$$

2.1 Support Vector Machines

A support vector machine (SVM) constructs a hyperplane in a high dimensional space, which can be used for classification or regression. A good separation is achieved by the hyperplane that has the largest distance to the nearest training data point of any class.

Let $y \in \{1, -1\}$ and $f(x) = w^T x + b$.

We are interested in the decision boundary that minimizes:

$$\min_{w,b} \sum_{i=1}^{l} \max(1 - y_i(w^T x_i + b), 0) + \lambda \|w\|^2$$

2.2 Semi-supervised Support Vector Machines

Since labeled examples do not have labels, we do not know on which side of the boundary they lie. Let's define the function $\acute{y}$ as: $y' = sign(f(x))$.

Hat   *loss*   *function*:
$$c(x, y', f(x)) = \max(1 - y'(w^T x_i + b), 0) = \max(1 - |w^T x_i + b|, 0))$$

Regularizer: $\Omega(f) = \lambda_1 \|w\|^2 + \lambda_2 \sum_{j=l+1}^{l+u} \max(1 - |w^T x_j + b|, 0)$

Semi-supervised support vector machines search for such a hyperplane so that the following error is minimized:

$$\min_{w,b} \sum_{i=1}^{l} \max(1 - y_i(w^T x_i + b), 0) + \lambda_1 \|w\|^2 + \lambda_2 \sum_{j=l+1}^{l+u} \max(1 - |w^T x_j + b|, 0)$$

$$(1)$$

$$subject\ to\ \frac{1}{u} \sum_{j=l+1}^{l+u} w^T x_j + b = \frac{1}{l} \sum_{i=1}^{l} y_i$$

An unlabeled instance x is always on the correct side of the decision boundary. The examples for which $f(x) \geq 1$ $or f(x) \leq -1$ are far from the boundary.

When the optimization is over, the algorithm has learned the parameter b and the vector w. It corresponds to the decision boundary. Based on this boundary we predict the labels of future examples.

3 GSSSVM - A Genetic Algorithm for SSSVM

Darwin's theory of natural selection revolutionized nineteenth century natural science by revealing that all plants and animals had slowly evolved from earlier live forms. Biology has been the impetus for the development of a highly efficient method for computer optimization – genetic algorithms [11].

3.1 Non-convex Optimization

The proposed approach is suitable for both convex and non-convex functions. Approaches based on gradient descend require a convex function. When a function is not convex, it is a hard optimization problem. As described in Sect. 2.2 semi-supervised support vector machines incorporate the non-convex hat function.

3.2 Individuals, Fitness Function, Mutation and Crossover

All living organisms (individuals) consist of cells, and each cell contains the same set of one or more chromosomes - strings of DNA. The individuals used in the semi-supervised learning genetic algorithm consist of the weights of the features.

Individual: [$w_1 \ldots w_s$, b]

A better fitted individual has a higher fitness function and will contribute the most offspring to the next generation. Since the goal is to optimize (1), a better fitted individual will have a smaller error. Therefore, the fitness function can be defined as:

$$fitness(p) = -(\sum_{i=1}^{l} \max(1 - y_i(w^T x_i + b), 0) + \lambda_1 \|w\|^2 + \lambda_2 \sum_{j=l+1}^{l+u} \max(1 - |w^T x_j + b|, 0))$$

Individuals, whose fitness function is high are selected for reproduction with higher probability. Uniform crossover was chosen and used in the genetic algorithm. For mutation – a small part of the chromosome is altered, newly generated weights replace some of the genes of the individuals.

Because the algorithm is an iterative procedure, which terminates when a maximum number of iterations has been reached - MAX_ITER, the complexity of the genetic algorithm is: O(MAX_ITER*GENETIC_S). GENETIC_S is the sum of:

- **crossover steps**: $s*m + 2*s*(l + l+u)$: m is the number of attributes, s – number of parent pairs. Each parent pair produces two children. So, for the newly generated 2*s children the fitness function should be calculated - $2*s*(l + l+u)$;
- **mutation steps**: k – number of bits mutated;

- **sort** function which orders the individuals based on their fitness function ($N*\log N$): N – number of individuals in the population.

3.3   Optimized Modification of the Algorithm

At each iteration step (generation) $2*s*(l + l+u)$ steps are performed. For example, if we terminate the algorithm when 10000 iterations are performed (MAX_ITER = 10000) and the number of unlabeled examples is 10000, it results in $2*s*100000000$ steps only for fitness function calculation of the newly-generated children (regarding the unlabeled examples). In semi-supervised learning usually the number of labeled examples is small and a genetic algorithm is applicable. On the contrary, unlabeled examples are easy to find and obtain and their number can be numerous. In order to find a way to reduce the execution time a modification of the algorithm concerning the unlabeled examples is required. Not all unlabeled examples should be used. In this paper the following approach is used: only the closest neighbors of labeled examples in the training set are selected. This idea was inspired by graph-based semi-supervised learning, where label propagation is applied to neighboring unlabeled examples.

Therefore, the fitness function can be defined as:

$$fitness(p) = -(\sum_{i=1}^{l} \max(1 - y_i(w^T x_i + b), 0) + \lambda_1 \|w\|^2 + \lambda_2 \sum_{j=l+1}^{l+v} \max(1 - |w^T x_j + b|, 0))$$

v ($v < u$) is the number of the closest unlabeled neighbors of the labeled examples.

4   Experiments

An experiment is held comparing the following three genetic algorithms:

- *supervised genetic algorithm* (SVM). No unlabeled examples are used.
- *semi-supervised genetic algorithm* (GSSSVM) – the proposed algorithm, which also relies on unlabeled examples.
- *semi-supervised genetic algorithm – optimized* (GSSSVM optimized): the optimized modification of the proposed algorithm.

The performance of the algorithms is compared based on the classification accuracy – percentage of correctly classified instances. The aim of the comparison is to see whether using unlabeled examples improves the accuracy of the algorithm.

An N-folded cross-validation is used. Each fold corresponds to the amount of labeled examples. "One fold" number of examples is used as labeled instances and (N-1) folds are used as unlabeled instances. The classification accuracy is evaluated on the unlabeled examples, which makes the algorithm a transductive one.

Because genetic algorithms depend on the randomly generated initial weights, the genetic algorithm was restarted multiple times and averaged results are provided after the cross-validation steps.

4.1 Datasets

It was experimented with two datasets:

- *Diabetes:* "Diabetes" can be obtained from the UCI Machine Learning Repository [13]. The attributes of *diabetes* are real-valued, there are no missing values. It consists of 768 examples, 8 attributes and 2 classes (tested negative and tested positive).
- *Coil20:* "Coil20" contains images distributed in 20 classes. The size of each image is 128 × 128 pixels. Therefore, the number of attributes is 16384 (the number of pixels). The preprocessing step included a dimensionality reduction procedure. Principal component analysis was performed and 171 features were selected. The number of examples in the dataset is 1440 (72 examples of each class).

4.2 Experimental Results

-*Diabetes* Table 1 contains results (on *diabetes*) based on the percentage of labeled examples −5 %, 10 % and 20 %. Respectively, the percentage of unlabeled examples used is 95 %, 90 % and 80 %. The parameters of the algorithms are: MAX_ITER = 50000, regularization parameters: $\lambda_1 = 0.5$, $\lambda_2 = 0.5$, mutation rate: 5 %, N = 100.

GSSSVM improves the classification accuracy and outperforms the supervised equivalent, the improvement is at most 0.38 %.

Table 1. Classification accuracy based on the percentage of labeled examples

% labeled	GSSSVM	SVM
20 %	66.63 %	66.27 %
10 %	65.30 %	64.92 %
5 %	62.95 %	62.64 %

Table 2 shows the influence of the number of iteration steps. The percentage of labeled examples used is 20 % (80 % unlabeled instances). The bigger the number of iterations, the bigger the classification accuracy.

Table 2. GSSSVM - performance based on the number of iterations

Iterations	GSSSVM (20 %)
200000	67.22 %
100000	66.80 %
50000	66.63 %
10000	64.61 %
5000	64.18 %
1000	63.18 %

For comparison, the performance of the algorithm SSSVM^{light} [17] was also evaluated. Implementation is provided by the author and was applied on the dataset diabetes. The results follow in Table 3. Both algorithms use the same number of labeled examples (20 %).

Table 3. Comparison between GSSSVM and SSSVM^{light}

Algorithm	Accuracy
GSSSVM	67.22 %
SSSVM^{light}	66.80 %

In part Sect. 3.3 a modification of the genetic algorithm was proposed, which uses only the closest unlabeled neighbors of the labeled examples. The aim of the following experiment was to see if we can achieve similar improvement with less unlabeled examples. In Table 4 the performance of the semi-supervised algorithm is compared to the one that uses all available unlabeled examples.

What is interesting is that when the percentage of labeled examples is 5 % using only 10 % unlabeled examples (the closest) leads to better results −63.22 % compared to 62.95 % (when using 95 % unlabeled). Therefore, when using all the unlabeled examples, not all of them are informative and some even manage to worsen the final performance.

Results are also provided for *a semi-supervised genetic algorithm – optimized*, which uses not the closest unlabeled neighbors, but instead those that are farthest away from the labeled ones. It can be seen in the table that the performance (62.63 %) when using 5 % labeled examples is equivalent to the supervised one and there is no point in using them.

Table 4. Performance based on the amount of unlabeled examples

% labeled	% unlabeled	Accuracy
20 %	80 %	66.63 %
20 %	20 %	66.52 %
20 %	0 %	66.27 %
5 %	95 %	62.95 %
5 %	10 % (closest)	63.22 %
5 %	10 % (farthest)	62.63 %
5 %	0 %	62.64 %

-*Coil20* Coil20 contains 20 groups of images (classes). For the experimental results the problem was reduced to a binary classification one – the examples of a definite class versus the rest of the examples. Therefore, the accuracy of an algorithm, which draws an example randomly in accordance to the distribution of the examples is 90.50 % (72 positive examples versus 72*19 negative examples). This procedure was repeated for each of the 20 classes and averaged results after the process of cross-validation are provided in Table 5. for each class.

The parameters of the algorithms are: 5 % labeled examples, 20 %, 40 % and 95 % unlabeled examples – the closest ones, MAX_ITER = 10000, regularization parameters: $\lambda_1 = 0.5$, $\lambda_2 = 0.5$, mutation rate: 5 %, N = 100.

Table 5. Classification accuracy (Coil20) based on the number of unlabeled examples

Class	1	2	3	4	5
GSSSVM, 20 %	95.22 %	94.90 %	94.94 %	94.85 %	94.88 %
GSSSVM, 40 %	95.21 %	94.98 %	95.00 %	95.01 %	94.96 %
GSSSVM, 95 %	95.17 %	95.06 %	95.07 %	95.06 %	95.02 %
SVM	95.08 %	94.74 %	94.86 %	94.82 %	94.67 %
Class	6	7	8	9	10
GSSSVM, 20 %	94.95 %	94.85 %	94.86 %	94.78 %	94.81 %
GSSSVM, 40 %	94.98 %	94.91 %	94.90 %	94.92 %	94.90 %
GSSSVM, 95 %	95.03 %	95.02 %	95.04 %	95.04 %	95.03 %
SVM	94.78 %	94.75 %	94.75 %	94.75 %	94.65 %
Class	11	12	13	14	15
GSSSVM, 20 %	94.81 %	94.91 %	94.78 %	94.80 %	94.88 %
GSSSVM, 40 %	94.92 %	94.93 %	94.94 %	94.87 %	94.92 %
GSSSVM, 95 %	95.05 %	95.04 %	95.04 %	94.99 %	95.00 %
SVM	94.74 %	94.78 %	94.73 %	94.70 %	94.77 %
Class	16	17	18	19	20
GSSSVM, 20 %	94.85 %	94.84 %	94.76 %	94.87 %	94.88 %
GSSSVM, 40 %	94.89 %	94.92 %	94.95 %	94.93 %	94.92 %
GSSSVM, 95 %	94.98 %	95.00 %	95.03 %	95.02 %	95.01 %
SVM	94.70 %	94.67 %	94.58 %	94.70 %	94.79 %

Table 6 contains the performance of the algorithms after averaging the results for the 20 classes.

Table 6. Average classification accuracy (Coil20)

% labeled	% unlabeled	Accuracy
5 %	95 %	95.04 %
5 %	40 %	94.95 %
5 %	20 %	94.87 %
5 %	0 %	94.75 %

5   Conclusions

Genetic algorithms are based on natural evolution and use heuristic search techniques to find decent solutions when an optimal one does not exist or takes a lot of time to complete. Such a problem is presented in this paper in the field of machine learning.

For non-convex function optimization, a genetic algorithm was proposed and multiple experiments were held. Because genetic algorithms are iterative procedures, taking into consideration the large number of features and amount of examples, an optimized modification was also put forward, which uses the closest unlabeled neighbors of the labeled examples. How many unlabeled examples we are going to use depends on the execution time we can afford and when there is a time limit, we can use the heuristic proposed in this publication. Furthermore, it is not reasonable to use the unlabeled examples, which are farther from those, which are already labeled.

The experiments show that using unlabeled examples in conjunction with labeled ones improves the classification accuracy and similar improvement can be achieved with less unlabeled examples.

For future work, a distributed architecture can be implemented (based on Hadoop and Spark) so that the iterative algorithm can make use of the map-reduce paradigm.

References

1. Chapelle, O., Schölkopf, B., Zien, A.: Semi-supervised Learning. MIT Press, Cambridge (2006)
2. Blum, A., Mitchell, T.: Combining labeled and unlabeled data with co-training. In: Proceedings of the eleventh Annual Conference on Computational Learning Theory, pp. 92–100 (1998)
3. Tang, F., Brennan, S., Zhao, Q., Tao, H.: Co-tracking using semi-supervised support vector machines. In: IEEE 11th International Conference on Computer Vision ICCV 2007, vol. 14, pp. 1–8 (2007)
4. Chapelle, O., Vapnik, V., Bousquet, O., Mukherjee, S.: Choosing multiple parameters for support vector machines. Mach. Learn. **46**, 131–159 (2002)
5. Brefeld, U., Scheffer, T.: Co-EM support vector learning. In: Proceedings of the Twenty-First International Conference on Machine learning, p. 16 (2004)
6. Bennett, K., Demiriz, A.: Semi-supervised support vector machines. Adv. Neural Inf. Process. Syst. 368–374 (1999)
7. Fung, G., Mangasarian, O.: Semi-supervised support vector machines for unlabeled data classification. Optim. Methods Softw. **15**, 29–44 (2001)
8. Chapelle, O., Sindhwani, V., Keerthi, S.S.: Optimization techniques for semi-supervised support vector machines. J. Mach. Learn. Res. **9**, 203–233 (2008)
9. Zhu, X., Goldberg, A.: Introduction to Semi-supervised Learning: Synthesis Lectures on Artificial Intelligence and Machine Learning. Morgan and Claypool Publishers, San Rafael (2009)
10. Mitchell, M.: An Introduction to Genetic Algorithms. MIT Press, Cambridge (1998)
11. Golberg, D.: Genetic algorithms in search, optimization, and machine learning. Addison wesley, Boston (1989)
12. Whiteley, D.: Applying genetic algorithms to neural network problems. Neural Netw. **1**, 230 (1988)
13. Bache, K., Lichman, M.: UCI Machine Learning Repository (2013)
14. Farquhar, J., Hardoon, D., Meng, H., Shawe-taylor, J., Szedmak, S.: Two view learning: SVM-2K, theory and practice. In: Advances in neural information processing systems, pp. 355–362 (2005)

15. Lazarova, G.: Semi-supervised image segmentation. In: Agre, G., Hitzler, P., Krisnadhi, A. A., Kuznetsov, S.O. (eds.) Artificial Intelligence: Methodology, Systems, and Applications. Lecture Notes in Computer Science, vol. 8722, pp. 59–68. Springer, Heidelberg (2014)
16. Lazarova, G.: Semi-supervised Multi-view Sentiment Analysis. In: Núñez, M., Nguyen, N. T., Camacho, D., Trawiński, B. (eds.) Computational Collective Intelligence 2015. Lecture Notes in Computer Science, vol. 9329, pp. 181–190. Springer, Heidelberg (2014)
17. Joachims, T.: Transductive inference for text classification using support vector machines. In: ICML, pp. 200–209 (1999)
18. Chapelle, O., Zien, A.: Semi-supervised classification by low density separation. In: AISTATS, pp. 57–64 (2005)
19. Chapelle, O., Chi, M., Zien A.: A continuation method for semi-supervised SVMs. In: Proceedings of the 23rd International Conference on Machine Learning, pp. 185–192 (2006)
20. Sindhwani, V., Keerthi, S.S., Chapelle, O.: Deterministic annealing for semi-supervised kernel machines. In Proceedings of the 23rd International Conference on Machine Learning, pp. 841–848 (2006)
21. Chapelle, O., Sindhwani, V., Keerthi, S.: Branch and bound for semi-supervised support vector machines. In: Advances in Neural Information Processing Systems, pp. 217–224 (2006)

Hinge Loss Projection for Classification

Syukron Abu Ishaq Alfarozi[1][(✉)], Kuntpong Woraratpanya[1],
Kitsuchart Pasupa[1], and Masanori Sugimoto[2]

[1] Faculty of Information Technology,
King Mongkut's Institute of Technology Ladkrabang, Bangkok 10520, Thailand
syukron@outlook.com, {kuntpong,kitsuchart}@it.kmitl.ac.th
[2] Department of Computer Science, Hokkaido University, Sapporo 060-814, Japan
sugi@ist.hokudai.ac.jp

Abstract. Hinge loss is one-sided function which gives optimal solution
than that of squared error (SE) loss function in case of classification. It
allows data points which have a value greater than 1 and less than -1 for
positive and negative classes, respectively. These have zero contribution
to hinge function. However, in the most classification tasks, least square
(LS) method such as ridge regression uses SE instead of hinge function.
In this paper, a simple projection method is used to minimize hinge loss
function through LS methods. We modify the ridge regression and its
kernel based version i.e. kernel ridge regression so that it can adopt to
hinge function instead of using SE in case of classification problem. The
results show the effectiveness of hinge loss projection method especially
on imbalanced data sets in terms of geometric mean (GM).

Keywords: Hinge loss · Least square · Squared error · Projection ·
Ridge regression · Kernel ridge regression

1 Introduction

Different loss function will lead to different performance with respect to the
problem and type of data. One of the popular loss functions is squared error
(SE), which assumes the distribution of errors are Gaussian. Mainly, it is used
for estimating function in either regression or classification problems using LS
method. Ridge regression (RR) [4], which is the regularized LS technique, is
widely used in many applications on machine learning field. Although this technique is very useful and well known, RR with its loss function i.e. SE, only work
optimally if the errors are Gaussian [10]. However, in the real world application,
the distribution might not be Gaussian. Meanwhile, hinge loss was designed only
for classification problem and focus on the separation of data. It was reported
that using hinge loss gives a better solution than that of SE [6]. It is used in
SVM [9] solved by QP, which is sophisticated convex optimization method. However, there is least square version, LS-SVM [8], which turns the problem into LS
method in place of QP and reduces the learning complexity.

© Springer International Publishing AG 2016
A. Hirose et al. (Eds.): ICONIP 2016, Part II, LNCS 9948, pp. 250–258, 2016.
DOI: 10.1007/978-3-319-46672-9_29

In classification problem, we are given a set of training data $(x_1, y_1), (x_2, y_2), \ldots, (x_n, y_n)$, where $x_i \in \mathbb{R}^m$ is m dimension input features and $y_i \in \{-1, 1\}$ is the output. Let us denote our decision boundary function as $f(x)$ and decide the class for a given input with $sign[f(x)]$. Then, the SE loss function, $L_{SE} = \sum_{i=1}^{n}(y_i - f(x_i))^2$, tends to make all points have value 1 and -1 for positive and negative classes respectively, otherwise it is counted as error. In other words, for the correctly classified points i.e. $y_i f(x_i) > 1$ (included outliers) will quadratically contribute to the SE loss function [3]. It may face a problem when the data is not balanced in which the majority class takes more concern where the outliers inside the majority class margin i.e. $y_i f(x_i) \gg 1$ will move the separation hyperplane. In the contrary, hinge loss function, $L_{hinge} = \sum_{i=1}^{n} \max\{0, 1 - y_i f(x_i)\}$, is more flexible which allows points have value greater than 1 for positive class and less than -1 for negative class. Intuitively, it tells us the points that satisfy $y_i f(x_i) > 1$ are far enough from the decision boundary which are easy to classify and have less important role to affect the hyperplane's position. Moreover, the outliers inside class margin, $y_i f(x_i) \gg 1$, will have no contribution to the hinge function. So that, hinge function is more robust to outliers inside the margin.

In this work, we provide a method to minimize hinge loss function through the projection (Sect. 3). We also evaluate the implementation of hinge loss function in kernel ridge regression (KRR) also known as LS-SVM so that it uses hinge loss function in place of SE. In addition, we briefly explain RR and KRR in Sect. 2. Through the projection method, the hinge loss is reduced for each iteration proofed by our developed theory. The results (Sect. 4) show the effectiveness of our method in a number of cases imbalanced data sets which usually have the one sided outliers on the majority class.

2 RR and KRR for Classification

2.1 Ridge Regression [4]

Ridge regression is similar to ordinary LS but it minimizes also the norm of weight parameter. With the given a set of classification training data then RR minimizes the following problem,

$$\arg\min_{w} \sum_{i=1}^{n}(y_i - f(x_i))^2 + \lambda \sum_{i=1}^{m} w_i^2 \qquad (1)$$

The optimal solution of this problem is

$$w^* = (X^{T}X + \lambda I)^{-1}X^{T}y, \quad \text{where } X = [x_1, ..., x_n]^{T}. \qquad (2)$$

Now, from the solution given by (2), we have our decision boundary function $f(x) = x^{T}w^*$. The regularization parameter, $\lambda \geq 0$, plays important role to define the trade off between minimizing the error function and the norm of weight parameter, $\|w\|$. If $\lambda = 0$, we will have non-regularized solution which refers to ordinary LS solution.

Figure 1(a) gives the geometrical intuition how LS and RR do the projection and obtain the solutions. For visualization purpose, Fig. 1 only shows 2 features. Let us denote the principal components of X as $c_1, ..., c_m$ which span the column space of X and are also orthogonal each other, i.e. $c_i \perp c_j$ for $i \neq j$. Then, the ordinary LS solution of this problem is, $\beta_i = \frac{y^T c_i}{\|c_i\|}$ and $y_i = c_i \beta_i$ is the projection of y to i-th principal component. This solves the linear system, $C\beta = y$, where $C = [c_1, ..., c_m]$, and gives the smallest residual error, e_r, i.e. orthogonal to column space of X. Meanwhile, the RR solution uses shrinkage factor that shrinks the parameter with the given λ in direction of its principal component such that

$$\beta_i^{ridge} = \frac{\|c_i\|^2}{\|c_i\|^2 + \lambda} \beta_i \tag{3}$$

Note that, the residual error obtained by RR is greater or equal than that of ordinary LS solution because RR gives non orthogonal error to the column space of X for non zero λ.

The relation between w^* and β^{ridge} is $w^* = P\beta^{ridge}$ where P is transformation matrix of X to its principal component C. In practice, we directly find the solution w^* using (2) instead of finding C, β, and P.

2.2   Kernel Ridge Regression [7,8]

KRR is a kernel based version of RR method, also known as LS-SVM. This method utilizes kernel function that transforms the input feature of training data into another feature space. With the given training data set, instead of using $x \in \mathbb{R}^m$, KRR transforms it into $\phi(x) \in \mathbb{R}^d$ where $\phi(.)$ is a mapping function to a particular feature space. Hence, the decision function is $f(x) = \phi(x)^T w + b$ where b is its bias parameter. As well explained in [8], KRR solves the following problem:

$$\arg\min_{w,b} \frac{1}{2} \sum_{i=1}^{m} w_i^2 + \frac{\gamma}{2} \sum_{i=1}^{n} e_i^2 \tag{4}$$

$$\text{subject to } y_i = \phi(x_i)^T w + b + e_i, \quad i = 1, ..., n$$

where e_i is a squared error of point x_i. The optimal condition of dual problem of (4) can be formed in a linear system, such that

$$\begin{bmatrix} 0 & 1^T \\ 1 & \Omega + \lambda I \end{bmatrix} \begin{bmatrix} b \\ \alpha \end{bmatrix} = \begin{bmatrix} 0 \\ y \end{bmatrix} \tag{5}$$

where $\lambda = \gamma^{-1}$ is the regularization parameter, $\Omega_{ij} = \phi(x_i)^T \phi(x_j)$ is Gramian matrix and our decision function becomes $f(x) = \sum_{i=1}^{n} \alpha_i K(x, x_i) + b$, which $K(.,.)$ is a kernel function that satisfies Mercer condition. One can simply solve (5) using LS method to obtain the optimal parameters α and b.

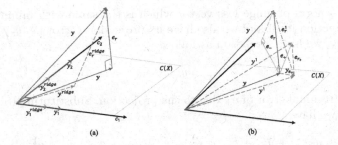

Fig. 1. Geometrical illustration of; (a) The solution of LS projection $\hat{y}$ and ridge regression projection $\hat{y}^{ridge}$ which shrinks each component y_i based on its principal components c_i; (b) hinge loss projection method decomposes the residual error e_r and also projects the hinge error/loss e_+ to the $C(X)$ so that the new solution is $\hat{y}^1$ which is the resultant of the previous solution $\hat{y}$ and projection of e_+ i.e. $\hat{y}_{e_+}$, this process continues until considerable e_+.

3   Hinge Loss Projection

Inspired by RR method (Sect. 2), hinge loss projection (HLP) takes advantages from RR's geometrical properties. Basically, RR does projection to the output vector, y, into the column space of X (Fig. 1(a)) shrinkage by parameter λ. First, let us write RR's solution (2) as a projection function to column space of X that returns the optimal weight parameter w and the output of training data y acts as input of the function, such that

$$w = \text{Proj}_{X,\lambda}(y) \tag{6}$$

where $(X^{\mathrm{T}}X + \lambda I)^{-1}X^{\mathrm{T}}$ is its projection operation matrix and $\hat{y} = Xw$ is its projection vector. Then, we have

$$y = \hat{y} + e_r \tag{7}$$

where e_r is the residual error vector. Second, let us divide e_r into two partitions, denoted as e_+ and e_- refer to error of points that contribute to hinge loss and otherwise respectively (Fig. 1(b)), such that

$$e_r = e_+ + e_-. \tag{8}$$

The element of e_+ in e_r is indicated by points that satisfy $y_i\hat{y}_i \leq 1$ and otherwise indicated as e_-'s element. Because of the partition, it yields the following properties: (i) Orthogonal, $e_+ \perp e_-$; (ii) Complementary, which means if i-th element of e_+ is non zero, then i-th element of e_+ must be zero, vice versa; and (iii) $\|e_+\| \leq \|e_r\|$ and $\|e_-\| \leq \|e_r\|$.

As obtained e_+, then we have the hinge loss such that $L_{hinge} = \sum_1^n \|e_{+_i}\|$. One can minimize this hinge loss using the projection function (6). So that

$$w_{e_+} = \text{Proj}_{X,\lambda}(e_+) \tag{9}$$

minimizes L_2 norm of hinge loss vector which is the same with minimizing the hinge loss function itself and we also have its projection vector i.e. $\hat{y}_{e_+} = Xw_{e_+}$. Consequently, we have a relation as follows

$$e_+ = \hat{y}_{e_+} + e_{r_+} \tag{10}$$

where e_{r_+} is residual error of e_+ due to the projection. Substituting (8) and (10) to (7), then we have

$$y = \hat{y} + \hat{y}_{e_+} + e_{r_+} + e_- = \hat{y}^1 + e_{r_+} + e_- = \hat{y}^1 + e_r^1 \tag{11}$$

where $\hat{y}^1$ is the optimal solution after doing hinge loss projection in column space of X and e_r^1 is its next residual error. Another way, we can obtain

$$\hat{y}^1 = X\left[\mathrm{Proj}_{X,\lambda}(\hat{y} + e_+)\right]. \tag{12}$$

Note that, the solution $\hat{y}^1$ obtained by (11) and (12) is the same when $\lambda = 0$, otherwise it is different. In our case, we used (12) approach which project in terms of modified target value, $y^1 = \hat{y} + e_+$, rather than project its individual hinge loss vector, e_+. Furthermore, one can also repeatedly divide e_r^k into e_+^k and e_-^k until it reaches a certain value of hinge loss. In our case, the iteration is stopped when the relative error is less than 10^{-4} or exceeded the allowed maximum iteration. Figure 1(b) gives the full illustration for hinge loss projection process. Obviously, if we use the (6) solution without any further projection process, then the solution will be the same as RR solution.

Based on these explanations, we need to show the hinge loss $\|e_+\|$ has convergence property. Hence, we have

Theorem 1. *The hinge loss in k-th iteration, $\|e_+^k\|$, monotonously decreases with an increasing the number of iterations, $\|e_+^{k+1}\| \leq \|e_+^k\|$.*

Proof. Let us consider the solution in k-th iteration,

$$y = \hat{y}^k + e_r^k \tag{13}$$

For pleasurable notation, we reorder the element of e_r^k such that the upper elements are element that contributes to hinge loss, and the lower elements are element that does not contribute to the hinge loss. Hence, from (8) and (13), the solution can be reformed as

$$y = \hat{y}^k + e_r^k = \hat{y}^k + \begin{bmatrix} e_+^k \\ e_-^k \end{bmatrix} = \hat{y}^k + \begin{bmatrix} e_+^k \\ 0 \end{bmatrix} + \begin{bmatrix} 0 \\ e_-^k \end{bmatrix} = \hat{y}^k + e_+^k + e_-^k \tag{14}$$

After doing the projection of e_+^k then

$$\hat{y}^k + \hat{y}_{e_+^k} + e_{r_+}^k + e_-^k = \hat{y}^{k+1} + e_{r_+}^k + e_-^k$$
$$= \hat{y}^{k+1} + \begin{bmatrix} e_{r_+U}^k \\ e_{r_+L}^k \end{bmatrix} + \begin{bmatrix} 0 \\ e_-^k \end{bmatrix} = \hat{y}^{k+1} + e_r^{k+1} = y. \tag{15}$$

Because of the projection, the norm of error residual hinge loss, $e_{r_+}^k$, is less than or equal to that of the previous hinge loss, e_+^k, i.e. $\|e_{r_+}^k\| \leq \|e_+^k\|$. Now, let us assume that hinge loss is monotonously increasing, $\|e_+^{k+1}\| \geq \|e_+^k\|$, in which some or all of points that previously do not contribute to hinge loss make contributions in the next iteration and assume that all points contribute to hinge function in the next iteration. Then,

$$e_+^{k+1} = e_r^{k+1} = \begin{bmatrix} e_{r+U}^k \\ e_-^k + e_{r+L}^k \end{bmatrix}. \tag{16}$$

Obviously, if i-th point previously does not contribute to hinge loss i.e. in e_-^k group then makes contribution in the next iteration that means $e_{r_i}^k \cdot e_{r_i}^{k+1} \leq 0$ i.e. different sign property. Thus, the i-th element of e_-^k and $e_{r_+}^k$ also have different sign property. Subsequently, we have $\|e_{-i}^k + e_{r+i}^k\| \leq \|e_{r+i}^k\|$, then, it implies that $\|e_-^k + e_{r+L}^k\| \leq \|e_{r+L}^k\|$. Therefore, we can conclude that $\|e_+^{k+1}\| \leq \|e_{r+}^k\| \leq \|e_+^k\|$ which is contradiction. This is the complete proof of Theorem 1.

For a kernel based model, one can simply update the output value, y, for each iteration in (5), such that

$$y^k = \hat{y}^{k-1} + e_+^{k-1}. \tag{17}$$

Finally, for the solutions i.e. α and b from the projection process can be obtained as follows

$$\begin{bmatrix} b^k \\ \alpha^k \end{bmatrix} = \begin{bmatrix} 0 & 1^{\mathrm{T}} \\ 1 & \Omega + \lambda I \end{bmatrix}^{-1} \begin{bmatrix} 0 \\ y^k \end{bmatrix} = \mathrm{Proj}_{\Omega,\lambda}\left(\begin{bmatrix} 0 \\ y^k \end{bmatrix}\right). \tag{18}$$

In addition, we relabel the points that satisfy $\hat{y}_i^k y_i \leq -1$ to its opposite class i.e. $sign(\hat{y}_i^k)$ which indicate as outliers outside the margin on each iteration to reduce its negative influences. So that, it is not only robust to the outliers inside the margin but also outliers outside the margin[1].

4   Experimental Results

To evaluate the performance of HLP, a well-known two spiral classification benchmark [5] is chosen with using radial basis function kernel i.e. $K(x_i, x_j) = \exp(-\gamma\|x_i - x_j\|^2)$. The data set contains 194 samples in two dimensional features so that it is easy to visualize. The result of classification boundary between HLP and RR method in kernel RBF is quite similar (Fig. 2). This implies that HLP has a good performance like RR method in standard and balanced data.

Furthermore, we also evaluate HLP on 16 pre-split five-fold imbalanced benchmark data sets[2] [1] with Imbalanced Ratio (IR) between majority and minority classes greater than 9, i.e. varying in range of 9 to 40, to perform five-fold cross validation. For each data set, we consider the average performance of

---

[1] The Matlab code implementation of HLP is available in the author's repository.
[2] Available in http://sci2s.ugr.es/keel/imbalanced.php#sub20.

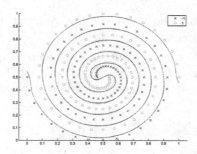

Fig. 2. Two spiral class boundary i.e. ridge regression SE (dashed line) and hinge loss projection (solid line) with $\lambda = 2^{-10}$ and $\gamma = 64$.

those five partitions. The data sets are normalized such that $x \in [0, 1]$. Because of the imbalance characteristic, geometric mean (GM) of sensitivity and specificity metric

$$GM = \sqrt{\frac{TP}{TP + FN} \cdot \frac{TN}{TN + FP}} \tag{19}$$

Table 1. Performance evaluation of hinge loss projection method and ridge regression in linear space and its kernel space i.e. RBF kernel.

Data set (IR)	Linear				RBF			
	H-RR		RR		H-KRR		KRR	
	GM_{tr}	GM_{ts}	GM_{tr}	GM_{ts}	GM_{tr}	GM_{ts}	GM_{tr}	GM_{ts}
ecoli0137vs26 (39.14)	99.64	**74.01**	95.80	68.15	**99.95**	79.82	99.86	79.82
ecoli4 (15.8)	99.53	**99.22**	98.31	78.92	99.69	**99.53**	99.84	99.38
glass016vs2 (10.29)	**29.17**	00.00	00.00	00.00	100.0	**68.69**	96.88	48.48
glass016vs5 (19.44)	96.95	**73.94**	78.27	19.72	100.0	79.16	99.93	**89.99**
glass2 (11.59)	**35.10**	00.00	00.00	00.00	92.20	**68.15**	96.43	46.91
glass4 (15.47)	87.84	**44.97**	97.37	39.04	100.0	95.64	100.0	95.64
glass5 (22.78)	99.76	**87.87**	78.52	00.00	99.82	**93.90**	100.0	90.23
yeast05679vs4 (9.35)	88.99	**90.38**	95.52	76.66	89.58	92.23	88.07	**93.82**
yeast1289vs7 (30.57)	**39.39**	00.00	00.00	00.00	**99.35**	67.42	99.22	67.42
yeast1458vs7 (22.1)	00.00	00.00	00.00	00.00	99.72	**44.57**	99.33	39.74
yeast1vs7 (14.3)	97.05	**58.32**	00.00	00.00	**98.02**	97.53	97.86	97.53
yeast2vs4 (9.08)	95.62	95.76	96.72	**96.54**	97.11	**97.13**	97.23	97.03
yeast2vs8 (23.1)	98.01	98.94	**98.09**	98.94	98.94	98.94	98.94	98.94
yeast4 (28.1)	87.25	**91.21**	00.00	00.00	87.25	**92.69**	91.80	91.71
yeast5 (32.73)	81.92	**80.67**	00.00	00.00	98.41	**92.02**	99.72	92.01
yeast6 (41.4)	87.98	**83.81**	00.00	00.00	93.48	87.72	96.06	**89.76**
Mean	76.51	**61.19**	46.16	29.87	97.09	**84.70**	97.57	82.40

is used as suggested in [2]. Generally speaking, if we use accuracy metric, a classifier that obtains 90 % in data set that has IR equal to 9 might not be accurate if all of minority class points are incorrectly classified. For non-linear cases, the RBF kernel is used and the parameters λ and γ are chosen from 2^{-12} to 2^6 in multiple of 2.

Table 1 shows the performance differences between HLP method named as hinge loss RR (H-RR) and the original RR in either linear space or kernel space i.e. KRR and H-KRR. The boldface value is the best GM_{ts} performance classifier on testing set for each data sets, if the both GM_{ts} are the same then their training performance, GM_{tr}, are compared. It can be seen in the linear and kernel space, HLP approach gives better performance in many data sets than that of RR. If we look at the data sets that give $GM_{ts} = 0$, i.e. glass016vs2, glass2 and yeast1289vs7 data sets in linear space, which means the minority class points are all misclassified in the testing set, but in the training set HLP method can perform better to recover the minority classes. However, in few data sets, the original RR performs better than HLP which might have too few points (i.e. small margin) to describe the hyperplane, because in our case we stopped the iteration when the relative error of hinge loss is less than 10^{-4} or exceeds allowed iteration number. As we know, the solution of HLP (6) is the same as original RR in the first iteration and marks the error points inside class boundary as zero which do not contribute to hinge loss on each iteration.

5   Conclusion

In this paper, we introduce the hinge loss projection method (HLP) which uses the hinge loss function instead of squared error (SE) to solve classification problems. It can describe the separation hyperplane better than ridge regression with SE loss function in the case imbalanced data sets which are usually have many outliers inside the margin. HLP might have too small margin and degrades its performance, thus it needs to define proper stopping criteria method in the future.

References

1. Alcalá, J., Fernández, A., Luengo, J., Derrac, J., García, S., Sánchez, L., Herrera, F.: Keel data-mining software tool: data set repository, integration of algorithms and experimental analysis framework. J. Multiple-Valued Log. Soft Comput. **17**(11), 255–287 (2011)
2. Barandela, R., Sánchez, J.S., Garcıa, V., Rangel, E.: Strategies for learning in class imbalance problems. Pattern Recogn. **36**(3), 849–851 (2003)
3. Friedman, J., Hastie, T., Tibshirani, R.: The Elements of Statistical Learning. Springer Series in Statistics, vol. 1. Springer, Berlin (2001)
4. Hoerl, A.E., Kennard, R.W.: Ridge regression: biased estimation for nonorthogonal problems. Technometrics **12**(1), 55–67 (1970)
5. Lang, K.J., Witbrock, M.J.: Learning to tell two spirals apart. In: Proceeding of 1988 Connectionist Models Summer School (1988)

6. Rosasco, L., De Vito, E., Caponnetto, A., Piana, M., Verri, A.: Are loss functions all the same? Neural Comput. **16**(5), 1063–1076 (2004)
7. Saunders, C., Gammerman, A., Vovk, V.: Ridge regression learning algorithm in dual variables. In: (ICML-1998) Proceedings of the 15th International Conference on Machine Learning, pp. 515–521. Morgan Kaufmann (1998)
8. Suykens, J.A., Vandewalle, J.: Least squares support vector machine classifiers. Neural Process. Lett. **9**(3), 293–300 (1999)
9. Vapnik, V.: The Nature of Statistical Learning Theory. Springer, New York (1995)
10. Zhang, S., Hu, Q., Xie, Z., Mi, J.: Kernel ridge regression for general noise model with its application. Neurocomputing **149**, 836–846 (2015)

Analytical Incremental Learning: Fast Constructive Learning Method for Neural Network

Syukron Abu Ishaq Alfarozi[1,2(✉)], Noor Akhmad Setiawan[1],
Teguh Bharata Adji[1], Kuntpong Woraratpanya[2], Kitsuchart Pasupa[2],
and Masanori Sugimoto[3]

[1] Department of Electrical Engineering and Information Technology,
Universitas Gadjah Mada, Yogyakarta 55281, Indonesia
syukron@outlook.com, {noorwewe,adji}@ugm.ac.id
[2] Faculty of Information Technology, King Mongkut's Institute of Technology
Ladkrabang, Bangkok 10520, Thailand
kuntpong@gmail.com, kitsuchart@it.kmitl.ac.th
[3] Department of Computer Science, Hokkaido University, Sapporo 060-814, Japan
sugi@ist.hokudai.ac.jp

Abstract. Extreme learning machine (ELM) is a fast learning algorithm for single hidden layer feed-forward neural network (SLFN) based on random input weights which usually requires large number of hidden nodes. Recently, novel constructive and destructive parsimonious (CP and DP)-ELM which provide the effectiveness generalization and compact hidden nodes have been proposed. However, the performance might be unstable due to the randomization either in ordinary ELM or CP and DP-ELM. In this study, analytical incremental learning (AIL) algorithm is proposed in which all weights of neural network are calculated analytically without any randomization. The hidden nodes of AIL are incrementally generated based on residual error using least square (LS) method. The results show the effectiveness of AIL which has not only smallest number of hidden nodes and more stable but also good generalization than those of ELM, CP and DP-ELM based on seven benchmark data sets evaluation.

Keywords: Analytical incremental learning · Extreme learning machine · Parsimonious model · Recursive generalized inversion · Orthogonality

1 Introduction

From the last decade, a popular learning algorithm for single hidden layer feed-forward neural network (SLFN) which concerns about learning speed is known as extreme learning machine (ELM) [5] that tends to solve the lack of conventional gradient based learning such as back propagation (BP) [8]. Huang et al. [5]

© Springer International Publishing AG 2016
A. Hirose et al. (Eds.): ICONIP 2016, Part II, LNCS 9948, pp. 259–268, 2016.
DOI: 10.1007/978-3-319-46672-9_30

have already compared the performance of ELM and BP. The reported results show that ELM speeds up the training time for about 400 times faster and has good generalization in some cases compared to BP. ELM is developed based on Moore-Penrose generalized inverse. First, ELM randomizes the weights between input and hidden layers. Second, it calculates weights between hidden layer and output layer using generalized inverse. However, randomizing the weights can cause the performance of the algorithm unstable. In addition, ELM requires a large number of hidden nodes to reach its best performance which causes slow response time to test new examples (testing time).

Several studies try to find the optimal subset model of original ELM using constructive and destructive techniques. Incremental ELM [4] is proposed based on constructive approach which is starting with empty hidden node and add a random hidden node for each iteration. In addition, another version of I-ELM i.e. enhanced I-ELM [3], provides more compact nodes, has better generalization and reduces the redundant nodes using random search. However, those algorithms cannot realize the optimal structure and solution automatically. Recently, Wang et al. [10], proposed constructive and destructive parsimonious (CP and DP)-ELM which based on recursive orthogonal least square (ROLS) method [1]. Moreover, CP and DP ELM can find the optimal subset of hidden nodes regressors using constructive and destructive approaches recursively. Although from the aforementioned algorithms give a compact and optimal solution for a given set of random hidden nodes, ELM and its variants are still based on ELM randomization which has different solution for different randomization.

In this paper, we introduce a novel algorithm that tends to be an alternative for solving those problems, namely analytical incremental learning (AIL) based on recursive generalized inverse. Different from ELM and its variants, AIL calculates not only the output weights but also the input weights analytically based on residual error at each iteration and adds the hidden node incrementally. For a given a set of training examples and same learning parameters, AIL will give the same solution since there is no randomization and weight initiation. The results show AIL has very small number of hidden nodes and has good generalization compared to state of the art algorithms.

This paper is organized as follows. Section 2 introduces the brief explanation of ELM. Sections 3 and 4 explain the preliminaries and the proposed algorithm AIL, respectively. The experimental results and the evaluation of developed algorithm are discussed in Sect. 5. Finally, we conclude our work in Sect. 6.

2   Extreme Learning Machine

Given a SLFN model with N training examples $\{(\boldsymbol{x}_i, t_i)\}_{i=1}^{N}$ where $\boldsymbol{x}_i \in \mathbb{R}^m$ is the input vector and $t_i \in \mathbb{R}$ is its output. Then, the SLFN with L hidden nodes is modeled as

$$f(\boldsymbol{x}_i) = \sum_{i=1}^{L} \beta_i g(\boldsymbol{x}_j, b_i, \boldsymbol{w}_i)) = t_i, \quad j = 1, ..., N \tag{1}$$

where β_i is the output weight parameter that connects i-th hidden node to the output node, b_i is the bias parameter, w_i is input weight vector for each hidden nodes and $g(.)$ is the activation function. Equation (1) can be written in compact matrix form such that

$$H\beta = t \quad \text{where,} \quad H = \begin{bmatrix} g(x_1, b_1, w_1) & \cdots & g(x_{\hat{1}}, b_L, w_L) \\ \vdots & \ddots & \vdots \\ g(x_N, b_1, w_1) & \cdots & g(x_N, b_L, w_L) \end{bmatrix}_{N \times L} . \tag{2}$$

Here, w_i and b_i are generated randomly and H is called hidden output matrix. Then, by using LS method, ELM simply solves the following optimization,

$$\arg\min_{\beta} \| t - H\beta \|_2^2. \tag{3}$$

The optimum solution of (3) can be analytically determined by generalize inverse such that $\beta^* = H^\dagger t$ where $H^\dagger = (H^T H)^{-1} H^T$ is Moore-Penrose generalized inverse of matrix H. According to ELM learning theory, H is full column rank based on probability one randomization. Subsequently, if $L = N$, the ELM model can learn all of training data with zero error. In real world application, the number of hidden nodes are always less than the number of training examples, $L < N$. However, the number of hidden nodes L learned by ELM is still quite large. Therefore, to solve the large and redundant number of hidden nodes problem, one can use the extended version of ELM such as [2–4,7] etc. which give the subset solution of hidden nodes following the ELM randomization.

3 Preliminaries

3.1 Recursive Block Inversion

Given a $n \times n$ matrix X partitioned as follows

$$X = \begin{bmatrix} A & B \\ C & D \end{bmatrix}, \quad A \in \mathbb{R}^{k \times k} \tag{4}$$

then the inverse matrix $M = X^{-1}$ can be obtained from well known block matrix inversion such that

$$M = \begin{bmatrix} A^{-1} + A^{-1}B(D - CA^{-1}B)^{-1}CA^{-1} & -A^{-1}B(D - CA^{-1}B)^{-1} \\ -(D - CA^{-1}B)^{-1}CA^{-1} & (D - CA^{-1}B)^{-1} \end{bmatrix}. \tag{5}$$

The complexity of (5) is $O(n^{2+\epsilon})$ with the full recursion and $k = \lfloor \frac{n}{2} \rfloor$ [6,9]. Furthermore, in case of symmetric matrices, the complexity can be $O(n^2)$.

3.2   AIL Incremental Hidden Node

For a given standard SLFN structure with the bias output node parameter, β_0. Then, SLFN of AIL at k-th iteration is represented as

$$f(\boldsymbol{x}_i) = \sum_{i=1}^{k} \beta_i g(\boldsymbol{x}_j, b_i, \boldsymbol{w}_i) + \beta_0 = t_i, \quad j = 1, ..., N \tag{6}$$

So that, Eq. (2) at k-iteration is reformed such that

$$[\mathbf{1}_{N \times 1} \quad \boldsymbol{H}_{N \times k}] \begin{bmatrix} \beta_0 \\ \beta \end{bmatrix} = \boldsymbol{H}^{(k)} \boldsymbol{\beta}^{(k)} = t. \tag{7}$$

Following the least square optimization (3), the optimum weight parameter at the current iteration is

$$\boldsymbol{\beta}^{*(k)} = (\boldsymbol{H}^{(k)^{\mathrm{T}}} \boldsymbol{H}^{(k)})^{-1} \boldsymbol{H}^{(k)^{\mathrm{T}}} t = \boldsymbol{S}^{(k)^{-1}} \boldsymbol{u}^{(k)} \tag{8}$$

where $\boldsymbol{S}^{(k)} = \boldsymbol{H}^{(k)^{\mathrm{T}}} \boldsymbol{H}^{(k)}$ and $\boldsymbol{u}^{(k)} = \boldsymbol{H}^{(k)^{\mathrm{T}}} t$. Note that, $\boldsymbol{S}^{(k)}$ is Hermitian matrix so that also its inverse $\boldsymbol{S}^{(k)^{-1}}$. Hence, the next optimum weight parameter $\boldsymbol{\beta}^{*(k+1)}$ with a additional hidden node vector $\boldsymbol{h}_{k+1}$ can be obtained recursively such that

$$\begin{aligned} \boldsymbol{\beta}^{*(k+1)} &= \left(\begin{bmatrix} \boldsymbol{H}^{(k)^{\mathrm{T}}} \\ \boldsymbol{h}_{k+1}^{\mathrm{T}} \end{bmatrix} \begin{bmatrix} \boldsymbol{H}^{(k)} & \boldsymbol{h}_{k+1} \end{bmatrix} \right)^{-1} \begin{bmatrix} \boldsymbol{H}^{(k)^{\mathrm{T}}} \\ \boldsymbol{h}_{k+1}^{\mathrm{T}} \end{bmatrix} t \\ &= \begin{bmatrix} \boldsymbol{H}^{(k)^{\mathrm{T}}} \boldsymbol{H}^{(k)} & \boldsymbol{H}^{(k)^{\mathrm{T}}} \boldsymbol{h}_{k+1} \\ \boldsymbol{h}_{k+1}^{\mathrm{T}} \boldsymbol{H}^{(k)} & \boldsymbol{h}_{k+1}^{\mathrm{T}} \boldsymbol{h}_{k+1} \end{bmatrix}^{-1} \begin{bmatrix} \boldsymbol{H}^{(k)^{\mathrm{T}}} t \\ \boldsymbol{h}_{k+1}^{\mathrm{T}} t \end{bmatrix} \\ &= \begin{bmatrix} \boldsymbol{S}^{(k)} & \boldsymbol{H}^{(k)^{\mathrm{T}}} \boldsymbol{h}_{k+1} \\ \boldsymbol{h}_{k+1}^{\mathrm{T}} \boldsymbol{H}^{(k)} & \boldsymbol{h}_{k+1}^{\mathrm{T}} \boldsymbol{h}_{k+1} \end{bmatrix}^{-1} \begin{bmatrix} \boldsymbol{u}^{(k)} \\ \boldsymbol{h}_{k+1}^{\mathrm{T}} t \end{bmatrix} = \boldsymbol{S}^{(k+1)^{-1}} \boldsymbol{u}^{(k+1)}. \end{aligned} \tag{9}$$

Applying (5) to symetric matrix $\boldsymbol{S}^{(k+1)^{-1}}$ in (9) and denoting $v = \boldsymbol{H}^{(k)^{\mathrm{T}}} \boldsymbol{h}_{k+1}$ and $d = \boldsymbol{h}_{k+1}^{\mathrm{T}} \boldsymbol{h}_{k+1}$, we have

$$\boldsymbol{S}^{(k+1)^{-1}} = \begin{bmatrix} \boldsymbol{S}^{(k)^{-1}} + \alpha^{-1} \boldsymbol{\theta} \boldsymbol{\theta}^{\mathrm{T}} & -\alpha^{-1} \boldsymbol{\theta} \\ -\alpha^{-1} \boldsymbol{\theta}^{\mathrm{T}} & \alpha^{-1} \end{bmatrix} \tag{10}$$

where $\boldsymbol{\theta} = \boldsymbol{S}^{(k)^{-1}} v$ and $\alpha = d - v^{\mathrm{T}} \boldsymbol{\theta}$ which is a scalar value. Here, α tells us about linear independent property of new node to the existing nodes. If alpha is zero then the new additional node $\boldsymbol{h}_{k+1}$ is not linear independent to $\boldsymbol{H}^{(k)}$.

The residual error of current iteration can be obtained such that

$$e_r^{(k)} = t - \boldsymbol{H}^{(k)} \boldsymbol{\beta}^*. \tag{11}$$

Thus, we have

Theorem 1. *For a given new additional node h_{k+1}, the norm of residual error $\|e_r^{(k)}\|$ will always decrease if and only if $e_r^{(k)} \not\perp h_{k+1}$.*

Proof. It is easy to show that the residual error will decrease at least because of the projection of $e_r^{(k)}$ to h_{k+1} with the projection vector

$$e_{h_{k+1}}^{(k)} = e_r^{(k)} - \frac{e_r^{(k)^T} h_{k+1}}{h_{k+1}^T h_{k+1}} h_{k+1} \tag{12}$$

where $\|e_{r_{h(k+1)}}^k\| < \|e_r^{(k)}\|$ according to projection theory if $e_r^{(k)} \not\perp h_{k+1}$. Hence, we have $\|e_r^{(k+1)}\| \leq \|e_{r_{h(k+1)}}^k\| < \|e_r^{(k)}\|$ and the equality $\|e_r^{(k+1)}\| = \|e_{h_{k+1}}^{(k)}\| < \|e_r^{(k)}\|$ is satisfied when $e_{h_{k+1}}^{(k)^T} H^{(k)} = 0$ which means the error reduced by the new node, $e_{h_{k+1}}^{(k)}$, is perpendicular to column space of $H^{(k)}$, otherwise the inequality is held because it is able to project into column space of $H^{(k)}$ which reduces $e_{h_{k+1}}^{(k)}$. However, if $e_r^{(k)} \perp h_{k+1}$ in (12) then $\|e_r^{(k+1)}\| = \|e_{h_{k+1}}^{(k)}\| = \|e_r^{(k)}\|$. This is a complete proof of Theorem 1.

4 Analytical Incremental Learning Algorithm

The main focus of AIL is to reduce the residual error $e_r^{(k-1)}$ as much as possible on each iteration in (12). Obviously, if $h_k = ce_r^{(k-1)}$ for any scalar c then according to (12) we can obtain zero error. Thus, AIL does the optimization to find h_k that minimize the residual error.

Let us consider the hidden node of SLFN uses additive node with activation function $g(.)$, then we have relation on k-th hidden node such that

$$\begin{bmatrix} g(x_1^T w_k + b_k) \\ \vdots \\ g(x_N^T w_k + b_k) \end{bmatrix} = ce_r^{(k-1)}. \tag{13}$$

Then (13) can be formed in compact matrix notation

$$G\left(\begin{bmatrix} 1_{N \times 1} & X_{N \times n} \end{bmatrix} \begin{bmatrix} b_k \\ w_k \end{bmatrix} \right) = G(\hat{X}\hat{w}_k) = ce_r^{(k-1)} \tag{14}$$

where $G(.)$ is the element wise activation function of $g(.)$ and $X = [x_1, ..., x_N]^T$ is the input matrix with n attributes, w_k and b_k are the input weight and bias parameters of k-node respectively. Thus, from (14), one can simplify the problem in linear system form such that

$$\hat{X}\hat{w}_k = G^{-1}(ce_r^{(k-1)}) \tag{15}$$

in which $G^{-1}(.)$ is the inverse of element wise function $G(.)$.

Furthermore, Applying least square method to (15), we have the optimal input weights parameter for the k-th node

$$\hat{w}_k = \hat{X}^\dagger G^{-1}(ce_r^{(k-1)}). \tag{16}$$

To avoid the ill-condition and obtain better generalization, Tikhonov reguarization method is used which modified the optimization of least square problem (3) not only minimizes the norm of error but also minimizes the norm of weight parameters such that

$$\arg\min_{\hat{w}_k} \|G^{-1}(ce_r^{(k-1)}) - \hat{X}\hat{w}_k\|_2^2 + \lambda\|\hat{w}_k\|_2^2 \tag{17}$$

where λ is the regularization parameter. Hence, the regularized Moore-Penrose generalized inverse of X is calculated as

$$\hat{X}^\dagger = (\hat{X}^T\hat{X} + \lambda I)^{-1}\hat{X}^T. \tag{18}$$

Then, the approximation of k-th hidden node for a given residual error $e_r^{(k-1)}$ is

$$h_k = G(\hat{X}\hat{w}_k). \tag{19}$$

Considering the inverse of activation function $g^{-1}(.)$, AIL uses the non-linear sigmoid activation function $g(x) = \tanh(x)$ which has function range $\in [-1, 1]$. Obviously, for the element of $e_r^{(k-1)}$ outside that range, there is no way to do the inversion except using complex number. To solve this problem, AIL bounds the element of $e_r^{(k-1)}$ such that it satisfies the range of function. This is done by setting the scalar

$$c = \frac{1 - \epsilon}{\max\{\text{abs}(\max\{e_{r_i}^{(k-1)}\}), \text{abs}(\min\{e_{r_i}^{(k-1)}\})\}} \tag{20}$$

which ensures the value of element in $e_r^{(k-1)}$ in the range $[-1 + \epsilon, 1 - \epsilon]$ where ϵ is a small positive value to avoid the infinity value. In our case, we use $\epsilon = 0.01$. Moreover, the inverse function of $g(x) = \tanh(x)$ is defined as

$$g^{-1}(x) = \frac{1}{2}\ln\left(\frac{1 + x}{1 - x}\right). \tag{21}$$

Remark 1. For each a new additional node, AIL considers the linear independent property (idp) in (10) of the new hidden node to the existing nodes and also based on its relative error of its residual. So that, one can stop the addition of node when idp less than a particular value $0 < \eta \ll 1$ such that

$$idp = \text{abs}\left(\frac{\alpha}{d}\right) \leq \eta \tag{22}$$

and also when the relative error of its residual

$$\xi^{(k)} = \text{abs}\left(\frac{\|e_r^{(k)}\| - \|e_r^{(k-1)}\|}{\|e_r^{(k)}\|}\right) \leq \eta. \tag{23}$$

Given a set of N training examples with input in matrix form $\boldsymbol{X} = [\boldsymbol{x}_1, ..., \boldsymbol{x}_N]^{\mathrm{T}} \in \mathbb{R}^{N \times n}$, the output vector $\boldsymbol{t} \in \mathbb{R}^{N \times 1}$, the number of hidden node L, the activation function $g(x) = \tanh(x)$ and regularization parameter λ, then the AIL algorithm is defined as follows:

Step 1. Calculate the input inverse $\hat{\boldsymbol{X}}^{\dagger}$ using (18).

Step 2. Initiate $\boldsymbol{H}^{(0)} = \boldsymbol{h}_0 = \boldsymbol{1}_{N \times 1}$ and calculate $\boldsymbol{S}^{(0)^{-1}}$, $\boldsymbol{u}^{(0)}$ and $\boldsymbol{\beta}^{*(0)}$ by using (8). Then, calculate the initial residual error $\boldsymbol{e}_r^{(0)}$ by using (11). Finally, set $k = 1$.

Step 3. Calculate c by using (20) and the input weight $\hat{\boldsymbol{w}}_k$ by using (16). Then, calculate $\boldsymbol{h}_k$ with a given $\boldsymbol{e}_r^{(k-1)}$ by using (19).

Step 4. Calculate d and α in (10). If (22) is satisfied then set $k = k - 1$ and go to Step 7, otherwise update $\boldsymbol{S}^{(k)^{-1}}$, $\boldsymbol{u}^{(k)}$ $\boldsymbol{\beta}^{*(k)}$ by using (9) and (10).

Step 5. Calculate the residual error $\boldsymbol{e}_r^{(k)}$ and check the relative error, if (23) is satisfied then update $\boldsymbol{\beta}^{*(k)} = \boldsymbol{\beta}^{*(k-1)}$ and set $k = k - 1$ then go to Step 7.

Step 6. If $k < L$ then update $k = k + 1$ and go to Step 3.

Step 7. Return the optimal parameter $\{\hat{\boldsymbol{w}}_1, ..., \hat{\boldsymbol{w}}_k\}$ and $\boldsymbol{\beta}^{*(k)}$.

5   Experimental Results

The AIL algorithm is evaluated on 7 benchmark data sets[1] with a single output task because AIL is currently only supports for a single output structure implemented on Octave 3.8.1. All input attributes and output are normalized into the range $[-1, 1]$ and $[0, 1]$, respectively. Following [10] experimental setup, for each data sets was randomly split into training and testing sets on each trials. Fifty trials have been conducted and the average of testing root mean square error ($RMSE_{test}$) and its standard deviation were recorded. The data set specification is shown in Table 1. In addition, the optimal regularization parameter

Table 1. Data sets description of real world benchmark problems.

Data set	#Attributes	#Training	#testing
Autompg	7	198	200
Bank	8	4501	3692
Boston	13	250	256
California	8	8000	12640
Ailerons	5	3000	4129
Elevators	6	4000	5517
Servo	4	80	87

[1] Available in http://www.dcc.fc.up.pt/~ltorgo/Regression/DataSets.html.

Table 2. Performance evaluation AIL with state of the art algorithms in [10].

Data set	Algorithms	RMSE$_{test}$		L_{max}	L_{final}	
		mean	dev.		mean	dev.
Autompg	ELM [5]	0.0737	0.0037	30	30	-
	OP-ELM [7]	0.1126	0.0131	30	16.5	4.0066
	PCA-ELM [2]	0.0890	-	7	**7**	-
	CP-ELM [10]	**0.0735**	0.0030	30	19.45	2.5644
	DP-ELM [10]	0.0743	0.0040	30	16.2	2.5874
	AIL	0.0832	0.0033	8	*7.98*	0.1414
Bank	ELM [5]	0.0472	0.0004	100	100	-
	OP-ELM [7]	0.1564	0.0016	100	81.5	5.2967
	PCA-ELM [2]	0.1553	-	8	7	-
	CP-ELM [10]	0.0473	0.0005	100	66.2	5.4324
	DP-ELM [10]	0.0473	0.0004	100	58.6	4.0332
	AIL	**0.0441**	0.0004	2	**2**	0
Boston	ELM [5]	0.1084	0.0077	50	50	-
	OP-ELM [7]	0.1568	0.1100	50	28.25	4.6665
	PCA-ELM [2]	0.1274	-	13	9	-
	CP-ELM [10]	0.1074	0.0047	50	30.15	2.9069
	DP-ELM [10]	0.1076	0.0059	50	25.3	2.6378
	AIL	**0.0960**	0.0078	4	**4**	0
California	ELM [5]	0.1332	0.0012	80	80	-
	OP-ELM [7]	0.1984	0.0105	80	59	5.6765
	PCA-ELM [2]	0.1578	-	8	5	-
	CP-ELM [10]	**0.1331**	0.0007	80	44.9	3.9567
	DP-ELM [10]	0.1336	0.0007	80	37.3	2.4518
	AIL	0.1434	0.0018	4	**3.9**	0.3030
Ailerons	ELM [5]	0.0395	0.0002	70	70	-
	OP-ELM [7]	0.0585	0.0023	70	41.5	7.8351
	PCA-ELM [2]	0.0624	-	5	**5**	-
	CP-ELM [10]	**0.0390**	0.0001	70	34.3	4.3729
	DP-ELM [10]	0.0392	0.0001	70	32.5	3.8658
	AIL	0.0401	0.0008	6	**5**	0
Elevators	ELM [5]	**0.0524**	0.0001	70	70	-
	OP-ELM [7]	0.0650	0.0029	70	44.5	3.6893
	PCA-ELM [2]	0.0572	-	6	6	-
	CP-ELM [10]	0.0525	0.0001	70	36	4.8990
	DP-ELM [10]	0.0525	0.0001	70	30.6	3.0258
	AIL	0.0538	0.0007	5	**3.56**	0.5014
Servo	ELM [5]	0.1339	0.0185	20	20	-
	OP-ELM [7]	0.1641	0.0141	20	10.6	3.4464
	PCA-ELM [2]	0.2278	-	4	**4**	-
	CP-ELM [10]	**0.1328**	0.0169	20	13.48	1.2493
	DP-ELM [10]	0.1338	0.0160	20	15.16	1.2835
	AIL	0.1395	0.0165	4	**4**	0

λ and the number of hidden nodes L_{max} were obtained by cross validation and the stopping criteria is set to $\eta = 10e - 05$.

From those data sets, the evaluation performance of AIL and state of the art algorithms reported in [10] is directly compared and shown in Table 2. Generally speaking, the AIL has great performance in terms of the number of hidden nodes which is the smallest number of node L_{final} in the most cases and also has good generalization performance i.e. $RMSE_{test}$ especially on bank and boston data sets.

Furthermore, because of stopping criteria (22) and (23), AIL can stop before the maximum number of nodes L_{max} obtained by cross validation which has very small unit of hidden nodes when compared to the other algorithms. For autompg, california and elevators data sets, AIL shows that the stopping criteria is satisfied although the number of hidden nodes is very small without losing the generalization performance. In addition, for the OP-ELM [7] and PCA-ELM [2] which are generally the most compact hidden node when compared to three other algorithms i.e. ELM [5], CP and DP-ELM [10], AIL in the all cases is still outperformed in terms of $RMSE_{test}$ to OP-ELM and PCA-ELM. Moreover, the testing time depends on the number of hidden nodes where AIL is the fastest.

6   Conclusion

In this paper, the analytical incremental learning (AIL) is proposed based on recursive generalized inverse without any randomization which incrementally adds a new hidden node based on the residual error. Theoretically, AIL will give the same trained model of neural network with the same parameters because there is no randomization and weight initiation. The results show that AIL has great performance with very small number of hidden nodes and also more stable due to its deterministic method compared to state of the art algorithms in [10].

References

1. Bobrow, J.E., Murray, W.: An algorithm for rls identification parameters that vary quickly with time. IEEE Trans. Autom. Control **38**(2), 351–354 (1993)
2. Castaño, A., Fernández-Navarro, F., Hervás-Martínez, C.: PCA-ELM: a robust and pruned extreme learning machine approach based on principal component analysis. Neural Process. Lett. **37**, 1–16 (2013)
3. Huang, G.B., Chen, L.: Enhanced random search based incremental extreme learning machine. Neurocomputing **71**(16), 3460–3468 (2008)
4. Huang, G.B., Chen, L., Siew, C.K.: Universal approximation using incremental constructive feedforward networks with random hidden nodes. IEEE Trans. Neural Netw. **17**(4), 879–892 (2006)
5. Huang, G.B., Zhu, Q.Y., Siew, C.K.: Extreme learning machine: theory and applications. Neurocomputing **70**(1), 489–501 (2006)
6. Petković, M.D., Stanimirović, P.S.: Generalized matrix inversion is not harder than matrix multiplication. J. Comput. Appl. Math. **230**(1), 270–282 (2009)

7. Rong, H.J., Ong, Y.S., Tan, A.H., Zhu, Z.: A fast pruned-extreme learning machine for classification problem. Neurocomputing **72**(1), 359–366 (2008)

8. Rumelhart, D.E., Hinton, G.E., Williams, R.J.: Learning internal representations by error propagation. In: Readings in Cognitive Science: A Perspective from Psychology and Artificial Intelligence, pp. 399–421. Kaufmann, San Mateo (1988)

9. Strassen, V.: Gaussian elimination is not optimal. Numer. Math. **13**(4), 354–356 (1969)

10. Wang, N., Er, M.J., Han, M.: Parsimonious extreme learning machine using recursive orthogonal least squares. IEEE Trans. Neural Netw. Learn. Syst. **25**(10), 1828–1841 (2014)

Acceleration of Word2vec Using GPUs

Seulki Bae and Youngmin Yi[✉]

School of Electrical and Computer Engineering, University of Seoul,
Seoul, Republic of Korea
{zpero422,ymyi}@uos.ac.kr

Abstract. Word2vec is a widely used word embedding toolkit which generates word vectors by training input corpus. Since word vector can represent an exponential number of word cluster and enables reasoning of words with simple algebraic operations, it has become a widely used representation for the subsequent NLP tasks. In this paper, we present an efficient parallelization of word2vec using GPUs that preserves the accuracy. With two K20 GPUs, the proposed acceleration technique achieves 1.7M words/sec, which corresponds to about 20× of speedup compared to a single-threaded CPU execution.

Keywords: Machine learning · Natural language processing · Neural network · Word2vec · Word embedding · CUDA

1   Introduction

Word embedding is mapping of words onto a continuous vector space. It is a distributed representation of words in a real value vector space, where the similar words in terms of semantics or syntax are close together. For example, 'learn' and 'studied' will be located nearby 'study'. Word embedding has become very successful since word embedding can capture various latent features of a word with a low-dimensional vector and can represent an exponential number of word clusters, which is not possible with the traditional approaches such as N-gram where words are treated as atomic units. Also, as similar words are located closely, it is natural and easy to reason words or phrases with this representation. For example, vector("smallest") can be obtained by simple algebraic operation of vector("biggest") – vector("big") + vector("small"), and it is also true with semantic reasoning: vector("France") – vector("Paris") + vector ("Korea") = vector("Seoul"). It can also reason the word in a sentence since, for example, Score(I, am, a, researcher) would be higher than Score(I, am, a, eat) or Score (I, am, a, him). Hence, the word vectors obtained by word embedding are widely used as a representational basis for subsequent NLP tasks such as named entity recognition, part-of-speech tagging, syntactic parsing, semantic role labeling, and so forth.

Bengio et al. first used a neural network to project words onto continuous vector space [3] since neural networks can learn better ways to represent the data

This work was supported by ICT R&D program of MSIP/IITP. [R0101-15-0054, WiseKB: Big data based self-evolving knowledge base and reasoning platform].

© Springer International Publishing AG 2016
A. Hirose et al. (Eds.): ICONIP 2016, Part II, LNCS 9948, pp. 269–279, 2016.
DOI: 10.1007/978-3-319-46672-9_31

automatically. Mikolov et al. proposed efficient neural network model architectures called Continuous Bag-of-Words (CBoW) and Skip-gram [1, 2] for faster training, which do not have a hidden layer but have only two layers. Many efforts have continued to improve word embedding [7–9].

The quality of word vectors depends on many factors such as the amount of the training data, size of the vectors, and the training algorithm. In particular, it is much affected by the amount of training data. Thus, faster training implies larger amount of data in a given time, which in turn could result in higher accuracy. In fact, *word2vec*, a widely used word embedding toolkit provided by Mikolov et al. exploits multiple CPUs to process the training algorithm in parallel. Still, the training speed is 660 kWords/sec.

In this paper, we present an efficient parallelization of word2vec using a GPU, achieving an order of magnitude speedup compared to the sequential C implementation. Moreover, we present multi-GPU acceleration of word2vec, which scales well with an efficient model synchronization. This multi-GPU implementation also achieves an order of magnitude speedup against the multiple-CPU (i.e., multi-threaded) implementation of the original word2vec. Since CBoW and Skip-gram, the neural network model architectures in word2vec, do not have any hidden layer, they do not have the time-consuming matrix multiplication, which other DNNs usually have and are accelerated with cuDNN included in many popular deep learning toolkits. Thus, the proposed parallelization and acceleration strategy of word2vec using GPUs involves an elaborate design from scratch with comprehensive understanding of the algorithm. In contrast to other approaches which also attempted GPU acceleration of word2vec, we carefully considered the algorithmic dependency when parallelizing the model architectures, and was able to maintain the accuracy without any loss.

The rest of the paper is organized as follows: In Sect. 2, we review the related work to accelerate word2vec either using a GPU or using a cluster with multiple nodes. In Sect. 3, we presents the model architecture and the training algorithm in word2vec, and the proposed parallelization strategy is explained in detail in Sect. 4. Section 5 presents the experimental results, followed by a conclusion in Sect. 6.

2 Related Work

cuDNN is de facto standard GPU accelerated Deep Learning library, where optimized implementation of forward and backward convolution, pooling, normalization, and neuron activation are provided. As mentioned in the previous section, the neural network models in word2vec is not actually deep, even without any hidden layer. Thus, utilizing cuDNN for accelerating word2vec using GPUs would not work well. The recent seemingly ongoing work [5] leverages cuDNN in Keras and Theano to accelerate the word2vec implemented in Gensim, but shows even slower training time than the CPU version.

Another open source project named word2vec_cbow [4] accelerates CBoW model in word2vec using GPUs. However, it employs word-level parallelization without any synchronization, which results in significant accuracy drops. The original word2vec toolkit provides two input corpora, one with new line delimiter and the other without it.

It also provides a test set with about 20,000 questions and the answers to measure the accuracy obtained from the training. The accuracy obtained by the original word2vec with a corpus that distinguishes each sentence is 60 %, whereas the one with the other corpus that does not distinguish sentences is 44 %. Although word2vec_cbow achieves 20 times speedup against the original word2vec sequential version, the accuracy by word2vec_cbow drops to 26 % for the corpus without new line delimiter. For the corpus with new line delimiter, as their kernel is launched only per sentence, it is even slower than the CPU sequential version by 1.8 times, also showing huge accuracy drop to 7.2 %.

As word2vec is so popular that it has been implemented in MLlib for Spark [6], which is a widely used distributed framework. Word2vec in MLlib scales well as the node in the Spark increases but we found that the accuracy drops significantly when the training iteration, or epoch, increases in a distributed environment. More detailed results are shown in Sect. 5.

As a short summary, the proposed GPU-accelerated word2vec in this paper, to the best of our knowledge, is the first implementation that achieves an order of magnitude speedup without any loss of accuracy.

3 Background

Word2vec has is a word embedding toolkit that has released in 2013, which takes text corpus as input and generates word vectors. Word2vec provides two model architectures: CBoW and Skip-gram. The former predicts the current word based on the context, or the neighboring words, and the latter predicts the neighboring words based on the current word [1, 2].

Figure 1 illustrates CBoW model architecture and the pseudocode below presents the training algorithm for CBoW with negative sampling. As shown in Fig. 1 the input layer consists of the neighboring words in a sentence. Instead of having hidden layers, word2vec introduces a projection layer where vector values of each word in the context are averaged, which is denoted as *neu1* in line 8 in pseudocode. Then, forward propagation is done in line 14 obtaining *f*. The current word is used as the positive sample (line 11) while randomly chosen words are used as negative samples (line 12–13). Gradient, *g*, is calculated with the supervised label and *f*, which are back propagated to the projection layer (line 15–16), updating the weight (line 17) for each

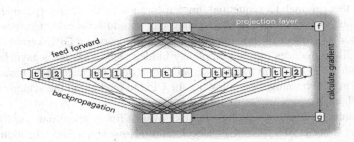

Fig. 1. The forward projection and backward projection in word2vec when w = 2.

```
Input: train_data /* the input train data */
       window /* window size */
       vDim /* dimensionality of the word vector */
       nSample /* samples that selected randomly */
Output: updated syn0 /* word vectors */
01 repeat until train_data run out
02     // read a sentence from train data
03     sen[] = ReadWords(train_data);
04     foreach tWord ∈ sen
05         // feed forward
06         foreach wWord ∈ sideWords(tWord, window)
07             for d=0 to vDim
08                 neu1[d] = neu1[d] + syn0[wWord][d];
09         // negative sampling
10         foreach sample ∈ (nSample || tWord)
11             if sample == tWord then label = true
12             else // sample ∈ negative samples
13                 label = false
14             for d=0 to vDim do f += neu1[d] * syn1[sample][d];
15             g = getGradient(f, label);
16             for d=0 to vDim do neu1e[d] += g * syn1[sample][d];
17             for d=0 to vDim do syn1[sample][d] += g * neu1[d];
18         // backpropagation
19         foreach wWord ∈ sideWords(tWord, window)
20             for d=0 to vDim do syn0[wWord][d] += neu1e[d];
```

Fig. 2. Pseudocode for word2vec with CBOW & negative sampling

sample. Once back propagations to the projection layer for all samples have been completed, the back propagation to the input layer is done by updating *syn0* using *neu1e* (line 19–20).

4   The Proposed Parallelization

Word2vec mainly consists of two parts: reading sentences and training. As for CBoW and Skip-gram models, the training part takes more than 90 % of the total execution time when the original C sequential implementation with default parameters was executed. Thus, it is obvious that the training part should be accelerated using GPUs. Also, the good side-effect of implementing the training part in a GPU kernel is that the time for reading sentences can be completely hidden since a kernel is asynchronously executed with a CPU and the training kernel output does not need to be transferred to the CPU during the training loop, allowing the CPU to read ahead the sentences for the next kernel launch.

Since training algorithm exhibits massive level of fine grained data-parallelism as shown in Fig. 2, it fits well to the GPU acceleration. However, a parallelization strategy without meeting the required dependency would only result in huge accuracy drop. Thus, the dependency in the algorithm is explained first, followed by the proposed mapping and synchronization in the GPU implementation, which incurs no accuracy drop. Then, further issue on the synchronization is discussed to accelerate the training algorithm using multiple GPUs.

4.1   Dependency in the Training Algorithm

Word2vec reads a sentence from the input text and trains the vectors for each word in the sentence *sequentially*. The context of a word is defined as $2w$ neighboring words, which are the left w and the right w words from the word. For the current word t, the vectors of the neighboring words from word $t - w$ to word $t + w$ are updated. Likewise, when the next word $t + 1$ is trained, the vectors of the words from word $t + 1 - w$ to word $t + 1 + w$ is updated. In other words, the window of which the size is $2w + 1$ including the current word slides from the left to the right in the sentence. Since the window of the previous word and the one of the current word overlap, the window must move sequentially (i.e., train the words sequentially) in order not to discard the training result of the previous words. The vector of a word in the sentence can be updated up to $2w$ times since a word can belong to a window, or a context, as the rightmost word, and also to another window as the leftmost word. If we parallelize the training at the word-level, as many updates are done independently, failing to capture the latent information in the sentence: it can only capture the information in the context, or the window. Such parallelization would result in a huge accuracy drop (Fig. 3).

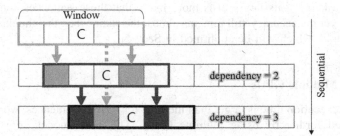

Fig. 3. The window whose size is 5 (w = 2) updates the vector of the words in the window except the current word (represented as bold rectangle), and slides to the next word. The node represents the vector of a word, and it gets darker as it is updated. The arrow indicates the dependency.

4.2   The Sentence-Wise Mapping

As discussed in the previous subsection, word-wise parallelization is not acceptable, and words in a sentence must be executed sequentially. However, the vector values in a word is independent of one another as shown in Fig. 1 (line 7, 16, 17, 20) and they are usually hundreds, typical value being 200. Thus, we map each dimension in the vector to each CUDA thread, and a sentence to a CUDA block.

A sentence that is mapped on a CUDA block runs each word in the sentence sequentially, but updating the vector of the word in parallel, as each dimension in the vector is mapped to a CUDA thread in the block. This corresponds to executing line 7, 14, 16, 17, and 20 in parallel. Note that, however, sum reduction is required among the threads for line 14. Three different well known reduction methods among threads in the block is used and compared to find the most efficient one in our context: atomic operation, parallel reduction using shared memory, and using shuffle instructions.

Mapping a sentence to a block means that as many sentences as the number of CUDA blocks in the kernel run concurrently. Since a typical input corpus usually has tens of millions of sentences, there are enough number of sentences to fully utilize the GPU in the kernel. In fact, a single kernel may not train all the sentences at once as the number of CUDA blocks in a kernel is limited to about 2 billion. In addition, the sentence buffer size would be several GBs, which could be a limiting factor, as most GPUs has less than 10 GBs of memory. Therefore, we divide the input corpus to the chunks of sentences, of which the size is determined empirically to achieve the highest throughput of the device.

The number of words in each sentence can vary drastically, which results in a different completion time of a block. However, as a CUDA block can run independently of one another, the variation in completion time does not matter as long as sufficient number of sentences are provided in the kernel. Since each sentence size is different, indexes that mark the end of each sentence are maintained.

Although the sentence-wise mapping has the same issue as in the word-wise mapping the same words in different sentences are much less dependent than the ones in different contexts in the same sentence. As the number of sentences running concurrently in a kernel gets larger, it is more likely that those sentences contain many same words, yet the accuracy still remains almost the same when the chunk size is less than a threshold, which will be confirmed in Sect. 5.

4.3 Using Multiple GPUs

If the kernel execution time is larger than the time for reading sentences, word2vec can be accelerated further by using multiple GPUs. The same model, or vectors, are duplicated to each GPU memory, and each GPU device needs to train only half of the corpus. However, the trained result, or the updated weights in the vectors, must be used in training the other words in order to maintain the accuracy. This means the models that reside in different GPU memories should be synchronized. As the total number of words in the model is in the order of hundreds of millions and the dimension size is in the order of hundreds, the model size is in the order of hundreds of MBs. Thus, the synchronization time is costly, especially if we synchronize the model at every kernel launch. In order to reduce the synchronization overhead, the model should be synchronized at less frequency yet can prevent the accuracy drop.

Note that CUDA provides a peer-to-peer data transfer API which does not go through host memory. Each GPU sends its own model to the other using this API, while at the same time receiving the model of the other to the temporary buffer.

5   Experimental Result

Table 1 shows the target platform information and word2vec parameter values used in the experiments. Gcc 4.4.6 was used for the CPU implementation and CUDA 7.0 for the GPU implementation. As an input corpus, we used Google News dataset, which contains 1.5 million sentences with 300 million words, with each sentence separated by a new line character. The accuracy in Figs. 4 and 6 was measured with about 12,000 questions in the test set, which cover 30k words appeared frequently. But Tables 3 and 4 was measured with every questions in the test set.

Table 1. Specification of the target platform and parameters

CPU	Intel Xeon E5-2630@2.30 GHz × 2	Vector dimension	200
GPU	Tesla K20 × 2	Window size	8
System memory	64 GB 1333 MHz	Negative sampling	25
		Iterations	15

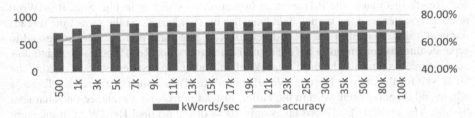

Fig. 4. Training throughput and accuracy when number of sentences per kernel varies.

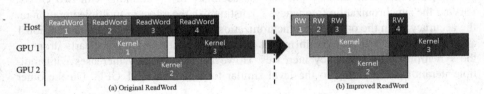

Fig. 5. (a) 2-GPU execution with the original ReadWord and (b) with the improved ReadWord

As discussed in Sect. 4, the number of sentences that are executed in a single kernel affects both the throughput and the accuracy. Figure 4 shows the overall throughput of word2vec when varying the number of sentences per kernel. The throughput saturates at 5k, reaching the training speed of about 880kWords/sec for the entire corpus. Since the accuracy saturates at 15k, we chose the number to be 15k.

We also applied multiple streams, expecting that the throughput may increase in case a GPU is not fully utilized. However, only 1.2 % of performance gain was achieved, which implies that a GPU is fully utilized. In fact, the variation in the completion time of CUDA blocks due to different sentence size mapped to each block

Table 2. Accelerated throughput using GPUs and multiple streams for the CBoW model with negative sampling

Version	kWords/sec
1-GPU CUDA (1-stream)	885.5
1-GPU CUDA (2-stream)	896.1
2-GPU CUDA (2-stream)	1502.9
2-GPU CUDA (2-stream + improved ReadWord)	1704.7

can result in load imbalance. However, since 15k sentences are provided to the kernel, there is no SM that is idle, executing the next sentence as soon as it completes the current one. It is only at the final phase of the kernel execution when some SMs become idle since there is no more input sentences to train. With two streams, this idleness can be avoided but it is already too little fraction.

Then, we accelerated the training further with two GPUs, achieving 1.68 time of speedup compared to 1-GPU execution (896kWords/sec → 1,503kWords/sec). The reason that 1.68× of speedup is achieved instead of near 2.0 times of speedup was that it was ReadWord that is the performance bottleneck. As shown in Fig. 5(a), ReadWord reads 15k sentences for a kernel, then launches the first kernel. At the same time it reads the next 15k sentence for the second kernel. Note that ReadWord can be processed in separate threads concurrently but the result is the same since ReadWord is bound to disk access time. As a result, the third ReadWord ends after the first kernel ends, making the GPU wait for the input to be read. We refactored ReadWord such that it reads a line as a whole and tokenize each word in the buffer, instead of reading a sentence one character by one. The revised ReadWord takes only 40 % of the original ReadWord implementation, removing the idle time in the GPU, as illustrated in Fig. 5(b). As a result, it achieves the Table 2 summaries the throughput of GPU each version.

Figure 6 shows the throughput and the accuracy of the training with two GPUs, varying the synchronization frequency. To distinguish the effect of multiple iteration on the accuracy from the one of the synchronization frequency, the results with iteration 1 and 5 are shown. We could confirm that the accuracy with iteration 1 clearly drops as the synchronization frequency increases. However, the accuracy increases with multiple iterations, recovering to the level similar to the one with 1 GPU. On the other

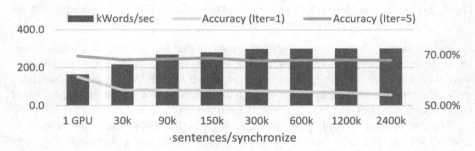

Fig. 6. Training throughput and accuracy when the frequency of the synchronization for 2 GPUs varies

hand, the throughput saturates at 300k. Thus, we perform synchronization between two GPUs at 300k sentences (i.e., at every 20 kernel launches).

Tables 3 and 4 summaries the increased throughput of the four different models by the proposed acceleration technique, as well as the Top-1 accuracy. The speedups ranging from 19.8× to 38.2× were achieved, while accuracy drops only by 0.2 %p, except for Skip-gram with hierarchical softmax algorithm where 2.1 %p of accuracy drop was observed. Note that, for some GPU implementations, the accuracy increases on the contrary.

Table 3. Acceleration results for CBoW model with negative sampling and softmax

Negative sampling	Throughtput [kWords/sec]	Acceleration over 1-thread CPU	Top1 accuracy [%]
1-thread CPU	86.2	1.0×	58.2 %
12-thread CPU	660.2	7.7×	60.5 %
1-GPU	897.7	10.4×	56.7 %
2-GPU	1704.7	19.8×	60.3 %
Hierarchical softmax	Throughtput [kWords/sec]	Acceleration over 1-thread CPU	Top1 accuracy [%]
1-thread CPU	112.4	1.0×	43.8 %
12-thread CPU	825.2	7.3×	45.4 %
1-GPU	2117.1	18.8×	41.8 %
2-GPU	3169.3	28.2×	45.3 %

Table 4. Acceleration results for Skip-gram model with negative sampling and softmax

Negative sampling	Throughtput [kWords/sec]	Acceleration over 1-thread CPU	Top1 accuracy [%]
1-thread CPU	16.4	1.0×	61.4 %
12-thread CPU	108.8	6.6×	61.8 %
1-GPU	177.5	10.8×	61.8 %
2-GPU	349.6	21.3×	59.9 %
Hierarchical softmax	Throughtput [kWords/sec]	Acceleration over 1-thread CPU	Top1 accuracy [%]
1-thread CPU	26.8	1.0×	55.0 %
12-thread CPU	174.2	6.5×	55.6 %
1-GPU	535.3	20.0×	54.1 %
2-GPU	1021.7	38.2×	54.4 %

Finally, we compared our results with the distributed word2vec version in MLlib of Spark. The train data and parameters were set identically as in Table 1, but the iteration count was set 1 since word2vec in MLlib seems to have a critical limitation of significant accuracy drop when the iteration is set larger.

Each node in the cluster has 12 CPU cores as specified in Table 1, thus 12 partitions are assigned to a node. And, the nodes are connected to one another through ConnectX-3 InfiniBand. Figure 7 shows the reduced throughput and the accuracy when multiple nodes are used in Spark. As the number of nodes increases, the throughput increases almost proportionally, but the accuracy also drops significantly (29 %p). Using four nodes (48 partitions) achieves the throughput similar to one with one K20 GPU.

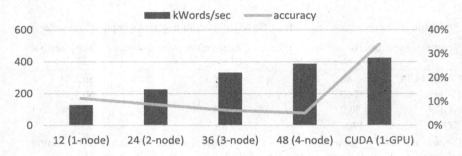

Fig. 7. Throughput and accuracy of word2vec in MLlib of Spark, and the comparison with ours (1-GPU)

6   Conclusion

In this paper, we proposed an efficient parallelization of word2vec using GPUs, which preserves almost the same accuracy. By carefully considering the algorithm dependency when mapping the algorithm concurrency to the GPU threads, we could maintain the accuracy while accelerating word2vec by an order of magnitude. Compared to a single threaded CPU implementation, the proposed technique achieved up to 38.2 times of speedup using two GPUs, while showing only 0.2 %p decrease in the accuracy except one model that shows 1.2 %p drop. This contrasts sharply with the currently available GPU implementations or a Spark implementation, where the accuracy drops drastically (29 %p – 59 %p).

We leave it as a future work to utilize more GPUs in a machine, and to utilize both GPUs and Spark on a GPU-cluster, for further acceleration of word2vec.

References

1. Mikolov, T., Sutskever, I., Chen, K., Corrado, G., Dean, J.: Distributed representations of words and phrases and their compositionality. NIPS (2013)
2. Mikolov, T., Chen, K., Corrado, G., Dean, J.: Efficient estimation of word representations in vector space. In: ICLR Workshop (2013)
3. Bengio, Y., Ducharme, R., Vincent, P., Jauvin, C.: A neural probabilistic language model. J. Mach. Learn. Res. **3**, 1137–1155 (2003)

4. word2vec_cbow. https://github.com/ChenglongChen/word2vec_cbow
5. word2vec-keras-in-gensim. https://github.com/niitsuma/word2vec-keras-in-gensim
6. Meng, X., Bradley, J., Yavuz, B., Sparks, E., Venkataraman, S., Liu, D., Freeman, J., Tsai, D., Amde, M., Owen, S., et al.: MLlib: machine learning in apache spark. arXiv preprint arXiv:1505.06807 (2015)
7. Huang, E., Socher, R., Manning, C., Ng, A.: Improving word representations via global context and multiple word prototypes. In: Association for Computational Linguistics, pp. 873–882 (2012)
8. Collobert, R., Weston, J.: A unified architecture for natural language processing: deep neural networks with multitask learning. In: International Conference on Machine Learning (2008)
9. Mnih, A., Hinton, G.: Three new graphical models for statistical language modelling. In: International Conference on Machine Learning (2007)

Automatic Design of Neural Network Structures Using AiS

Toshisada Mariyama[1(✉)], Kunihiko Fukushima[1,2],
and Wataru Matsumoto[1]

[1] Mitsubishi Electric Corporation, Information Technology R&D Center,
Kamakura, Kanagawa, Japan
Mariyama.Toshisada@ab.MitsubishiElectric.co.jp,
fukushima@m.ieice.org,
Matsumoto.Wataru@ai.MitsubishiElectric.co.jp
[2] Fuzzy Logic System Institute, Fukuoka, Japan

Abstract. Structures of neural networks are usually designed by experts to fit target problems. This study proposes a method to automate small network design for a regression problem based on the Add-if-Silent (AiS) function used in the neocognitron. Because the original AiS is designed for image pattern recognition, this study modifies the intermediate function to be Radial Basis Function (RBF). This study shows that the proposed method can determine an optimized network structure using the Bike Sharing Dataset as one case study. The generalization performance is also shown.

Keywords: Neural networks · Automatic design · Add-if-Silent · Neocognitron · Regression

1 Introduction

The structure of a neural network is usually designed by experts to fit a target task. This is because a neural network has many parameters, and a person who is not familiar with neural network design cannot use it well. Recently, deep neural networks have become regarded as a powerful solution for image recognition [1, 2], and many engineers in industry are eager to apply it to many practical fields. However, most of them are not familiar with a neural network and find it too difficult to use because of the above parameter-setting problem. If these parameters can be automatically designed, neural networks could be applied in more industrial fields and this technology would become more popular.

The computation complexity of a neural network is also a key issue for various fields in industry. Network structures have a strong tendency to become deeper and the number of parameter becomes correspondingly huge [3]. This may not be a big problem for cloud computing, but it is critical in an environment that cannot use powerful computer resources for reasons such as information security.

This study proposes a method to automate small network design for regression problems using the Add-if-Silent (AiS) function in the neocognitron [4, 5]. AiS is a kind of unsupervised learning technology and determines the number of neurons in the

© Springer International Publishing AG 2016
A. Hirose et al. (Eds.): ICONIP 2016, Part II, LNCS 9948, pp. 280–287, 2016.
DOI: 10.1007/978-3-319-46672-9_32

intermediate layers and parameters by training on a data set once. This means that the network structure determined by AiS is based on training data and not an expert's guesses or experiments. Parameters regarding the network structure are fixed once after they are determined by unsupervised learning and are not changed during supervised learning. However, the results of AiS depend on several parameters, and the way to choose these parameters is not fixed. This study proposes a method to determine the best set of parameters for a given training data set.

Regression problems are targeted in this study. A deep neural network is successful at image pattern recognition but performs comparatively less well on non-image regression problems. However, there are many regression problems in industry, and solutions to them are needed. The original AiS of the neocognitron was designed for image pattern recognition and is modified for regression problems in this study. The target regression problem is to estimate the number of bike sharing rentals corresponding to the years 2011 and 2012 from the Capital Bikeshare System in Washington D.C. [6]. Our suggested model in this study predicts the total number of rental bikes given weather conditions, holidays, and other factors.

Section 2 describes our suggested neural network model and the target data sets of this study. Section 3 presents the performance of AiS. Section 4 reviews the results of Sect. 3.

2 Procedure

2.1 Model

This study's neural network model consists of three layers. The learning data set used in this study consists of the number of rental bikes per day and 11 input parameters (e.g., weather or day of the week). Its details are described in Sect. 2.2. The number of neurons in an input layer is 11 because the data consist of 11 input parameters, and the number of outputs is one. The number of neurons in an intermediate layer is determined by AiS, which was proposed for use in the neocognitron.

AiS is a kind of unsupervised learning, and inputting all input data into the network creates neurons in the intermediate layers. In the neocognitron, the intermediate layers are determined by AiS. Once the network parameters of the intermediate layer are determined, the intermediate parameters are fixed and not changed by supervised learning.

An intermediate neuron receives excitatory connections from an input layer whose output is $x = (x_1, \cdots, x_i, \cdots, x_I)$, where i is the index for the input neurons and I is the total number of input neurons. The output y_j of the j-th neuron in an intermediate layer is given by:

$$y_j = exp\left(-\frac{\sum_{i=1}^{I}(x_i - \mu_{i,j})^2}{\sigma_j^2}\right) \tag{1}$$

where j is the index of the intermediate layer neurons, $\boldsymbol{\mu}_j = (\mu_{1,j}, \cdots \mu_{i,j} \cdots \mu_{I,j})$ is the central coordinate of the j-th intermediate neuron, and σ_j is the standard deviation. The original function of an intermediate neuron of AiS is based on a rectified linear function (presented in Sect. 4) to represent the activities of biological neurons in the brain for visual pattern recognition. And various types of function will be necessary to adopt various regression problems. Thus, this paper tries to expand the variation for regression problems, hence the Radial Basis Function (RBF), which is a time-tested function for this problem, is used.

The basic idea of AiS is that if all postsynaptic intermediate neurons are silent for a training signal, a new postsynaptic neuron is generated and added to the layer. If any neuron of an intermediate layer is less than a threshold a, a new intermediate neuron (of index $J + 1$) is created, whose μ_{J+1} is the same as an input signal x, so that $\mu_{i,J+1} = x_i$.

The number of neurons in the intermediate layer depends on a and the starting value of σ. Because AiS creates a new neuron based on the output of Eq. (1) for a new training input signal, which spreads from the trained input signal, both a and the starting value of σ need to be determined based on the resolution of the input space. The resolution of input space δ is determined by $\sqrt{-\ln a}$, then by substituting δ for x in Eq. (1). The above parameters are determined by the desirable resolution.

Supervised learning starts after the above network structure is built by AiS. Weights between the intermediate layer and the output layer represented by W_j and standard deviation of RBF σ_j are learned during this supervised learning process. The update rule is

$$Y = \sum_{j=1}^{J} W_j \cdot y_j \tag{2}$$

$$E = \frac{1}{2}(\hat{Y} - Y)^2 \tag{3}$$

$$W_j^{k+1} = W_j^k + \alpha \cdot \beta^l \cdot (1+\gamma) \cdot \frac{\partial E}{\partial W_j} \tag{4}$$

$$\sigma_j^{k+1} = \sigma_j^k + \alpha \cdot \beta^l \cdot (1+\gamma) \cdot \frac{\partial E}{\partial \sigma_j} \tag{5}$$

where $\hat{Y}$ is a training output signal of the training data set, Y is the output of this model, α, β, γ are learning ratios that vary from 0 to 1, k is the number of training iterations, and l is the number of epochs, where one epoch starts from the initial data set and ends after the last data set.

2.2   Data Set

This study uses bike sharing data sets [6] as training data. This data set is taken from the two-year historical log corresponding to years 2011 and 2012 from Capital

Bikeshare System, Washington D.C., USA. It provides daily data of the number of rentals. The dimension of this input data is 11;

- Season (1: spring, 2: summer, 3: fall, 4: winter)
- Year (0: 2011, 1: 2012)
- Month (1 to 12)
- Holiday: holiday or not
- Day of the week
- Working day: 1 if the day is neither a weekend day nor holiday and 0 otherwise.
- Weather:
  - 1: Clear, Few Clouds, Partly Cloudy, Cloudy
  - 2: Mist + Cloudy, Mist + Broken clouds, Mist + Few clouds, Mist
  - 3: Light Snow, Light Rain + Thunderstorm, and so on
  - 4: Heavy Rain + Ice Pellets + Thunderstorm + Mist, Snow + Fog
- Temperature in Celsius
- Wind chill temperature in Celsius
- Humidity
- Wind speed

This study normalizes all input parameters by dividing each parameter by the maximum value for that parameter to that all the values range from 0 to 1.

3 Results

3.1 Experiment 1: Generalization Performance

Experiments were conducted under four conditions in which the ratio r of training data to total number of data (720) was varied in order to determine the generalization performance (i.e., $r = 1/2, 1/4, 1/8, 1/10, 1/30$). For example, a ratio of $r = 1/10$ means 72 data were used as training data. The parameters were set as follows: $\alpha = 0.03, \beta = 0.99995, \gamma = 0.1$, initial $\sigma = 1.0$, and $\delta = 0.5$. An epoch, in this study, starts when the initial data set are trained and ends after all training data set have been trained once. For each experimental condition, 200,000 epochs were performed.

Table 1. MSE and the number of intermediate neurons with respect to r. The total number of training and non-training data equals the number of all data. For example, if the number of training data is 72, the number of non-training data is 648 for $r = 1/10$.

r		1/2	1/4	1/8	1/10	1/30
MSE	Training data	0.0057	0.0053	0.0046	0.0039	0.0041
	Non-training data	0.0068	0.0080	0.0075	0.0078	0.0095
	All data	0.0063	0.0073	0.0071	0.0075	0.0093
Number of intermediate neurons		30	24	18	19	14

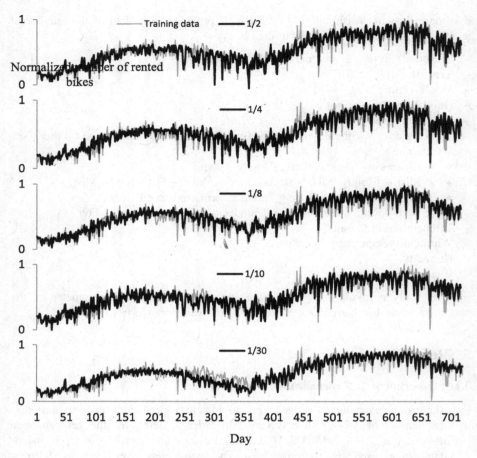

Fig. 1. The x-axis shows the date from 2011 January 1st to 2012 December 31st. The y-axis shows the number of rented bikes, normalized from 0 to 1 by dividing the number of rented bikes by the maximum number rented during this period. The legend values (1/2, 1/4, 1/8, 1/10, 1/30) show the ratio of training data to total data (720 data). The gray line indicates all the training data and the black lines shows the prediction results of the suggested models.

Table 1 shows the mean squared error (MSE) at $r = 1/2$, 1/4, 1/8, 1/10, and 1/30. Three MSEs are measured in this study. The MSEs of the training data and non-training data are measured to check the generalization performance. The MSE of all data is also measured.

The MSE of the non-training data is the most important factor because the most important metric of a regression problem is its performance against unknown data. The lowest MSE score for the non-training data is obtained when $r = 1/2$, and the score is stable until $r = 1/10$. Table 1 also shows the number of intermediate neurons.

Figure 1 shows the actual data (gray line) and the predicted results (black line) of the suggested model. These results also show no significant difference between

$r = 1/2$ and $1/8$. However the results of $r = 1/30$ cannot estimate well the data from days 251 to 351 and from days 601 to 650.

3.2   Experiment 2: Optimization of Resolution Parameters

Experiments for optimizing the resolution parameters were performed by changing the resolution parameter a while keeping $r = 1/8$ fixed. The other parameters are the same as in Experiment 1 and a was varied around the value of 0.8, which was hypothesized to have the least MSE of the non-training data. Table 2 shows the experimental results and implies that trials nos. 5 and 6 obtained the lowest MSE of the eight data.

Table 2. Results of Experiments.

Trial	a	δ	Intermediate neurons	MSE training	MSE non-training	Evaluation index
1	0.8288	0.433	29	0.0036	0.00792	-161.4
2	0.8188	0.447	26	0.0032	0.00727	-173.0
3	0.8088	0.461	23	0.0037	0.00715	-173.1
4	0.8038	0.467	23	0.0037	0.00713	-172.9
5	0.7988	0.474	21	0.0041	0.00712	-172.9
6	0.7958	0.478	21	0.0041	0.00712	-172.9
7	0.7938	0.481	21	0.0046	0.00737	-168.6
8	0.7888	0.487	20	0.0048	0.00745	-168.5

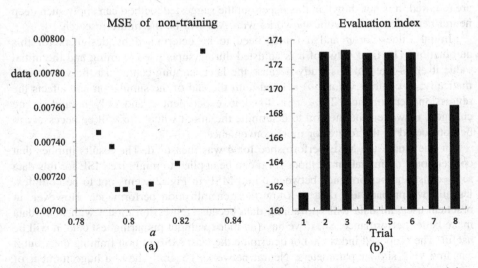

Fig. 2. (a) The x-axis shows the value of a and the y-axis shows the MSE of the non-training data. (b) The x-axis shows the trial number (listed in Table 2) and the y-axis shows the evaluation index.

This study uses Akaike's Information Criterion (AIC) [7] to evaluate each model and determine the best set of parameters. Because the MSE of the non-training data usually cannot be known, the method is used to determine the best parameters without knowing the MSE of the non-training data.

AIC is one of the evaluation indexes of model. The model with the lowest AIC is determined to be the best model. AIC can be calculated as $AIC = N \cdot \log(\Sigma^2) + 2 \cdot K$, where N is the number of training data sets ($N = 90$), Σ is the standard deviation of error between the training output and output of this model, and K is the number of parameters in the original AIC. However, this study uses K as the number of inter-mediate neurons J, as discussed in Sect. 4. Thus, the evaluation index in Table 2 and Fig. 2(b) is calculated as $N \cdot \log(\Sigma^2) + 2J$. Trial 3 obtains the least evaluation index, but not the lowest MSE of the non-training data. However, the difference of the index between Trial 5 (or 6) and Trial 3 is 0.003 and it is very slight.

4   Discussion

This study proposed an automatic design method for the number of intermediate neurons and their RBF central coordinates. Such parameters for RBF are usually determined by designers, and prior parameter settings are necessary. Generally speaking, the central coordinates can be determined by back propagation in supervised learning. However, the initial central coordinates sometimes affect the performance, and some engineers analyze the input data and determine the coordinates themselves. Dirichlet process Gaussian mixture model is a method to determine the parameters [8, 9]. However, this can apply three layers neural networks. Recently, neural networks with more layers have better performance and design methods for deep neural networks are required. It is not shown in this paper but the suggested method can apply such deep neural networks. Deep neural networks with AiS will be the future research.

Initial values for σ and a (or δ) need to be determined by designers for this automation. The final value of σ is adjusted during supervised learning and the initial value itself is fixed in this study because the later learning can adjust the value auto-matically. In contrast, a (and δ) is fixed until the end of the simulation and affects the regression performance. Thus, σ is fixed in experiment 2 and only a (and δ) are changed. However, the method to determine the initial value of σ will be necessary to be considered in the future for more convenience.

In this study, the MSE performance for a was measured. The results implies that conventional optimization methods for a can be applied to minimize MSE for this data set because the performance between a and MSE in Fig. 2 seems not to be complex. Usually we prepare test data to know the generalization performance. However, in practical uses, all data will be training data because regression model with more data make better performance. So, if we have an index without preparing test data, it will be useful. The proposed index cannot determine the least MSE of non-training data, but it can find very similar parameters. Neural networks generally have a huge number of parameters, and it has been pointed out that AIC does not work well for neural net-works [10]. This study originally used the original AIC, but confirmed that it does not work well (the results are not shown in this paper). Hence, a modified AIC was used to search for the neighbor parameters to achieve the lowest MSE. More research will be necessary to determine this in the future.

The suggested RBF is different from the AiS in the neocognitron. The intermediate function of neocognitron is defined as $y_j = \varphi \left[\sum_{i=1}^{I} V_{i,j} \cdot x_i - \theta \cdot \|x\| \right] / 1 - \theta$, where $V_{i,j}$ are the weight parameters between the input layer, the i-th neuron, and an intermediate layer j-th neuron; θ is the inhibitory connection of strength ($0 \le \theta < 1$); and $\varphi[\,]$ is a rectified linear function defined by $\varphi[x] = \max(x, 0)$. Here, $\mu_{i,j}$ of (1) is the same as $V_{i,j}$ because $V_{i,j}$ is set in the same manner as $\mu_{i,j}$ in Sect. 2. This means that the original method outputs a kind of similarity among $V_{i,j}$ and further that the input signal is vector x and θ is a threshold of the similarity. Because σ_j of (1) is also a kind of threshold, the original intermediate neuron can be considered to be similar with (1). The original function is not used in this study, but there is a possibility that similar results should be expected if the backpropagation of θ is introduced. The determination method of θ is not fixed and it is instead determined by a designer's experiments. The value of θ in unsupervised learning is different from that in recognition in the neocognitron. In the next step of this research, the method to determine θ for the original function will be investigated.

References

1. Fukushima, K.: Neocognitron: a self-organizing neural network model for a mechanism of pattern recognition unaffected by shift in position. Biol. Cybern. **36**(4), 193–202 (1980)
2. Hinton, G.E., Osindero, S., The, Y.: A fast learning algorithm for deep belief nets. Neural Comput. **18**, 1527–1554 (2006)
3. He, K., Zhang, X., Ren, S., Sun, J.: Deep residual learning for image recognition. arXiv preprint arXiv:1512.03385 (2015)
4. Fukushima, K.: Artificial vision by multi-layered neural networks: neocognitron and its advances. Neural Netw. **37**, 103–109 (2013)
5. Fukushima, K., Shouno, H.: Deep convolutional network neocognitron: improved interpolating-vector. In: International Joint Conference on Neural Networks 2015, Killarney, Ireland, pp. 1603–1610 (2015)
6. Hadi, F., Joao, G.: Event labeling combining ensemble detectors and background knowledge. Prog. Artif. Intell. **2**, 1–15 (2013). Springer, Heidelberg
7. Akaike, H.: Information theory and an extension of the maximum likelihood principle. In: Proceedings of the 2nd Informational Symposium on Information Theory, Akadimiai Kiado, Budapest, pp. 267–281 (1973)
8. Kim, S., Tadesse, M.G., Vannucci, M.: Variable selection in clustering via Dirichlet mixture models. Biometrika **93**(4), 877–893 (2006)
9. Kurihara, K., Welling, M., Vlassis, N.: Accelerated variational dirichlet process mixtures. In: NIPS (2006)
10. Hagiwara, K., Toda, N., Usui, S.: On the problem of applying AIC to determine the structure of a layered feedforward neural network. In: Proceedings of 1993 International Joint Conference on Neural Networks, Nagoya, pp. 2263–2266 (1993)

Sequential Collaborative Ranking Using (No-)Click Implicit Feedback

Frédéric Guillou[1], Romaric Gaudel[2(✉)], and Philippe Preux[2]

[1] Inria, Univ. Lille, CNRS, Lille, France
frederic.guillou@inria.fr
[2] Univ. Lille, CNRS, Centrale Lille, Inria, UMR 9189 - CRIStAL, Lille, France
{romaric.gaudel,philippe.preux}@univ-lille3.fr

Abstract. We study Recommender Systems in the context where they suggest a list of items to users. Several crucial issues are raised in such a setting: first, identify the relevant items to recommend; second, account for the feedback given by the user after he clicked and rated an item; third, since new feedback arrive into the system at any moment, incorporate such information to improve future recommendations. In this paper, we take these three aspects into consideration and present an approach handling click/no-click feedback information. Experiments on real-world datasets show that our approach outperforms state of the art algorithms.

Keywords: Recommender systems · Sequential recommendation · Learning to rank · Implicit feedback

1 Introduction

Recommender Systems (RS) seek to suggest to users items they might like or need. The recommendation builds upon the feedback given by users at previous recommendations. In this paper, we focus on Collaborative Filtering (CF) approaches, when they aim to recommend a list of items. Standard approaches in that setting discuss which loss is the most adapted and how to optimize that loss [1,10,12]. In their attempts, they omit one effect deriving from recommending a list of items: we could expect more feedback than a rating on a single item. For example, we have access to the list of items items which have been seen or clicked by the user. In short, we are confronted to a mix of explicit and implicit feedback. The feedback gathered depends on the model of interaction for users. The first contribution of the paper is to propose two approaches to handle the feedback involved by the interaction model discussed in [3]. These two approaches lead to relevant recommendation lists for three real-life datasets.

The second contribution of our paper relates to the experimental setting. A RS works in a sequential context: it suggests items, receives feedback, suggests another set of items, receives another feedback, etc. Thus, the data used by a CF based RS depend on previous recommendations. However, the evaluation of

A. Hirose et al. (Eds.): ICONIP 2016, Part II, LNCS 9948, pp. 288–296, 2016.
DOI: 10.1007/978-3-319-46672-9_33

CF is usually done in a fixed, batch setting. In this paper, we propose a proper experimental setting which mimics the interaction between the RS and the users.

The paper is organized as follows. We first go through the related work in Sect. 2. Then, in Sect. 3, we specify the setting of the recommendation process and we present our two approaches. Section 4 provides an experimental study on real datasets. Finally, we conclude and draw some future lines of work in Sect. 5.

2   Related Work

In this section, we briefly review state of the art in CF. Among approaches recommending a list of items, we omit the ones only handling implicit data due to space limitations. Approaches recommending a list of items are looking at the best loss function. These approaches replace the squared loss on ratings by a loss measuring the ranking ability of the learned model. Examples of such models include [12] and [1] which both optimize a smooth approximation of the *Normalized Discounted Cumulative Gain* (NDCG), or [6,8,10] which respectively optimize approximations of the *Expected Reciprocal Rank* (ERR), of a pairwise learning to rank loss, and of a structured output loss. [11] also optimizes a pairwise learning to rank loss, but it starts with a feature construction step.

These approaches only handle either explicit or implicit feedback. However, when recommending a list of items, a mix of both kinds of feedback is often to be expected: the rating of selected items (explicit), the list of items which have not been clicked despite having been shown to the user (implicit), the list of items which have been selected but not purchased, etc.

SVD++ [5] and Co-Rating [7] mix explicit and implicit data when learning. On one hand, SVD++ integrates implicit feedback in the representation of a user and limits itself to a restrictive kind of implicit feedback: a list of clicked items. On the other hand, Co-Rating considers implicit feedback as complementary values to fit. As such, Co-Rating manages any kind of implicit data. Still, it assumes that explicit and implicit feedback equal themselves as soon as they are scaled to $[0, 1]$. SVD++ and Co-Rating both learn by optimizing a squared loss.

In the remainder of the paper, we develop an approach which properly handles any kind of feedback, and therefore leads to better experimental results.

3   Ranking Recommender System Using Click Feedback

We now propose two approaches to handle both explicit and implicit feedback. Before that presentation, we have to express and define the model of interaction we will use in what follows. This model governs the feedback which are gathered, and is used to build our second approach (Sect. 3.2).

We consider the recommendation setting described in [3]. Let us consider N users and M items and the unknown matrix $\mathbf{R}^*$ of size $N \times M$ such that $r_{i,j}^*$ is the rating of user i with regard to item j. We assume the ratings range from 0 to R. From $\mathbf{R}^*$, we derive a relevance probability $p(i,j)$ which represents how much a user i is eager to select an item j:

$$p(i,j) = \frac{2^{r^*_{i,j}} - 1}{2^R}. \tag{1}$$

Then we set up the interaction between the RS, the users and the items. At each time-step t, a user i_t requests ℓ items. The RS provides the ranked list $j_1^{(t)},\ldots,j_\ell^{(t)}$, one item at a time, starting with the top-ranked item. While observing the s-th item, the user has a chance to pick it with probability $p(i_t, j_s^{(t)})$. Once an item is picked (denoted j_t), the user stops observing the list, and reveals the rating $r^*_{i_t,j_t}$. Thus, note that the user does not observe following items.

This setting implies two types of feedback are received at every recommendation list displayed: the first one is the list of skipped items and the clicked item (aka. implicit feedback), and the second one is the rating of the clicked item, if there is one (aka. explicit feedback). Note that the rank (in the recommendation list) of the clicked item brings only information about items ranked above in the list: we know these items were skipped. On the contrary, no information is gathered about items placed below the clicked one, as the user did not observe them according to our setting. The following parts describe two different approaches attempting to use these types of feedback to improve recommendations.

In the following, we denote $\mathbf{A}$ and $\mathbf{S}$ the matrices of size $N \times M$ which respectively stores for each user i and item j the number of clicks received by item j when presented to i, and the number of times j was skipped when shown to i on a recommendation list. We also note $\mathcal{S}$ the set of known entries of $\mathbf{R}^*$.

3.1 Feature Engineering

The first method denoted SVD+- assumes we can embed the implicit feedback into features of the model. Following the data representation used in Factorization Machine (FM) [9], we associate to any user-item couple (i,j) a representation $\phi(i,j)$ and we look at a function f s.t. $f(\phi(i,j)) = r^*(i,j)$. We take

$$\phi(i,j) = (\underbrace{0,\ldots,1,0,\ldots}_{N},\underbrace{0,\ldots,1,0,\ldots}_{M},\underbrace{\mathbf{C}_{i1},\ldots,\mathbf{C}_{is},\ldots,\mathbf{C}_{iM}}_{M}), \tag{2}$$

where the first section indexes i, the second section indexes j, and $\mathbf{C}_{is} = (\mathbf{A}_{is} - \mathbf{S}_{is})/(\sum_{s'=1}^{M} \mathbf{A}_{is'} + \mathbf{S}_{is'})$. Value $\mathbf{C}_{is}$ represents how many times item s was clicked or skipped out of all interactions with user i. This value ranges from -1 to 1.

From that feature model, the Factorization Machine learns the function

$$\hat{f}(\phi(i,j)) = w_0 + w_i + w_j + v_u.v_j^T + \left(\sum_{s=1}^{M} \mathbf{C}_{is}v_s\right) \cdot v_j^T$$

$$+v_u \cdot \left(\sum_{s=1}^{M} \mathbf{C}_{is}v_s\right)^T + \sum_{s=1}^{M} \mathbf{C}_{is}w_s + \left(\sum_{s=1}^{M} \mathbf{C}_{is}v_s\right) \cdot \left(\sum_{s=1}^{M} \mathbf{C}_{is}v_s\right)^T, \tag{3}$$

where w_0, w_i, w_j and $(w_s)_{1 \leqslant s \leqslant M}$ are real values, and v_0, v_i, v_j and $(v_s)_{1 \leqslant s \leqslant M}$ are feature vectors of size k. These parameters are chosen to optimize a trade-off between (i) the average square loss with respect to known values in $\mathbf{R}^*$ and (ii) the L_2 norm of the parameters.

3.2 Dual Matrix Factorization

From another perspective, we design a second approach called DualMF, which considers both types of feedback as values to fit. More specifically, we look at a low rank approximation $\widehat{\mathbf{R}} = \mathbf{U}.\mathbf{V}^T$ of $\mathbf{R}^*$, where $\mathbf{U}$ and $\mathbf{V}$ are matrices of respective sizes $N \times k$ and $M \times k$. Obviously, we want $\widehat{\mathbf{R}}$ to fit known values in $\mathbf{R}^*$ (aka. explicit feedback).

But we also make use of implicit feedback. Based on the model of the probability of click $p(i, j)$, we can build a biased estimator $\hat{r}_{ij}^{imp}$ of r_{ij}^* for any user-item couple (i, j) s.t. item j has been clicked or skipped at least one time by user i:

$$\hat{r}_{ij}^{imp} = \log_2 \left(1 + 2^R \frac{\mathbf{A}_{ij} + 0.5}{\mathbf{A}_{ij} + \mathbf{S}_{ij} + 1} \right). \tag{4}$$

The 0.5 added at the numerator and the 1 added at the denominator act similarly to a prior and help the model to not penalize too much items for which only little information has yet been collected.

Overall, the approach DualMF looks for a matrix $\widehat{\mathbf{R}}$ fitting both known values in $\mathbf{R}^*$, and $\hat{r}_{ij}^{imp}$ values. It solves the minimization problem:

$$\min_{\widehat{\mathbf{R}}=\mathbf{U}.\mathbf{V}^T} \mu \sum_{(i,j) \in \mathcal{S}} (\hat{r}_{ij} - r_{ij}^*)^2 + \sum_{(i,j): \mathbf{A}_{ij}+\mathbf{S}_{ij} \neq 0} (\hat{r}_{ij} - \hat{r}_{ij}^{imp})^2 + \lambda(\|\mathbf{U}\|^2 + \|\mathbf{V}\|^2), \tag{5}$$

where μ and λ are non-negative real values. μ controls the impact of explicit data compared to implicit one, and λ weights the regularization term.

4 Experimental Investigation

We empirically evaluate the algorithms in a sequential setting on real-world datasets. For each dataset, we start with an empty matrix $\mathbf{R}$ to simulate an extreme cold-start scenario where no information is available at all. Then, a list of items is recommended for each user according to the following procedure:

1. we select a user i_t uniformly at random among possible users (the ones to which no recommendation list was displayed yet),
2. the algorithm chooses a list of 5 items to recommend,
3. the user observes the list, clicks or not on an item j_t according to the setting described in Sect. 3. The value of r_{i_t,j_t} is revealed if there was a click. The user is then discarded from possible users.

When all users have been shown a recommendation list once, we reintegrate all of them in the list of possible users, and loop on this procedure again, while keeping all feedback gathered until now. These steps are done up to 50 recommendations shown for every user. As the ground truth is unknown for every item, we restrict the possible choices for a user at each time-step to the items with a known rating in the dataset. Note that it is allowed to include an item in the list of recommendations for a user even if it has already been rated in the past by him. Metrics used for the evaluation are abandonment, ERR@5 and NDCG@5. Let us consider a user i, to which the list of items $j_1, \ldots, j_\ell$ has been recommended. The abandonment metric represents the probability $\prod_{r=1}^{\ell}(1 - p(i, j_r))$ for a user not to be satisfied by any item in the list (no click), while the ERR and NDCG depict how adequate is the list of items suggested by the RS regarding the user's tastes. The ERR is based on the click probability defined by Eq. (1):

$$ERR@\ell(i, (j_1, \ldots, j_\ell)) = \sum_{r=1}^{\ell} \frac{1}{r} \prod_{s=1}^{r-1}(1 - p(i, j_s))p(i, j_r). \qquad (6)$$

The formula used for the Discounted Cumulative Gain at ℓ [2] is:

$$DCG@\ell(i, (j_1, \ldots, j_\ell)) = \sum_{r=1}^{\ell} \frac{2^{r^*_{ij_r}} - 1}{\log_2(r)}. \qquad (7)$$

The NDCG is the DCG divided by the best possible DCG. As we have no access to the ground-truth matrix $\mathbf{R}^*$, the NDCG is computed $w.r.t.$ the known values.

We consider three real-world datasets for our experiments: Movielens1M, Yahoo! Music ratings for User Selected and Randomly Selected songs, and a subset of Yahoo! Music user ratings of musical artists. To build this subset, we first select the 10,000 most rated artists, and then the 50,000 users who rated the highest number of artists. We also normalize the ratings from the range 0–100 to 1–5 and put the ratings with 255 (meaning "do not recommend this ever again") to 0.5. Characteristics of these datasets are reported in Table 1.

Table 1. Dataset characteristics

	Movielens1M	Yahoo! songs	Yahoo! artists
Number of users	6,040	15,400	50,000
Number of items	3,706	1,000	10,000
Number of ratings	1,000,209	365,703	32,997,016

Movielens1M is a standard dataset to compare CF approaches, but it contains mostly high ratings. On the other hand, the Yahoo! datasets contain a lot more low ratings, which allows for more realistic experiments (since items to recommend are chosen only among the ones for which we have ratings).

Our two approaches are compared to the following baselines:

- Oracle: this strategy knows the ground truth matrix and makes recommendation accordingly, with the list of best possible items for each user.
- Random: at each iteration, a random list of items is recommended to the user.
- Popular: we assume we know the most popular items based on the ground-truth matrix, and at each iteration, the 5 most popular items (restricted to the items rated by the user on the dataset) are recommended.
- xCliMF [10]: this model is built by optimizing the ERR on explicit ratings. It does not use implicit features but targets an appropriate ranking of items.
- CoFiRank [12]: CoFiRank optimizes a ranking measure using explicit features. In our experiments, we use the version optimizing the ordinal regression.
- ALS: we use a Factorization Machine implementation of Alternating Least Square method, taking into account only the explicit ratings given by users.
- SVD++ [5]: this approach also consists in a FM with data arranged to make the model mimic SVD++ (while adding pairwise interactions between features), as described in [9]. It also incorporates both explicit data and implicit data, but only implicit data about the item clicked.
- Co-Rating [7]: this approach tries to unify explicit and implicit feedback. To do so, it first normalizes both explicit and implicit scores between 0 and 1 and combines them when solving the minimization problem, using ALS. The main difference with DualMF is that it does not assume any model behind the observed clicks. In order to compare fairly with DualMF, we also use a FM to learn the model, and choose the implicit score to be $\hat{r}^{imp}(ij) = \frac{\mathbf{A}_{i,j}+0.5}{\mathbf{A}_{i,j}+\mathbf{S}_{i,j}+1}$.

Since most state of the art algorithms are not designed to handle the sequential aspect which implies frequent updates, we update the model of each approach each time 50 % of the users have been recommended a list of items (overall, each algorithm will have 100 updates spread along the 50 recommendations). Results of all algorithms on the datasets are shown on Fig. 1. We use existing implementation called fastFM for models using Factorization Machines (ALS, SVD++, SVD+-, Co-Rating, DualMF). For Co-Rating and DualMF, we give different weight to explicit and implicit feedback when training the model, and best results are obtained by giving twice more weight to explicit feedback for DualMF and five times more weight to explicit feedback for Co-Rating. Parameters for regularization or number of steps were chosen as the ones giving the best results.

From the results of experiments, we can draw two main conclusions: firstly, the approaches learning from both explicit and implicit feedback (DualMF, Co-Rating and SVD+-) perform significantly better than all other approaches on all datasets. Our approach DualMF in particular reach the best performance on Movielens1M and Yahoo! songs, and is on par with Co-Rating on Yahoo! artists except on the NDCG@5 metric. Note that the implicit score given during the learning phase by DualMF comes directly from the click model we defined. In practice, it would not be possible to access such knowledge and it would have to

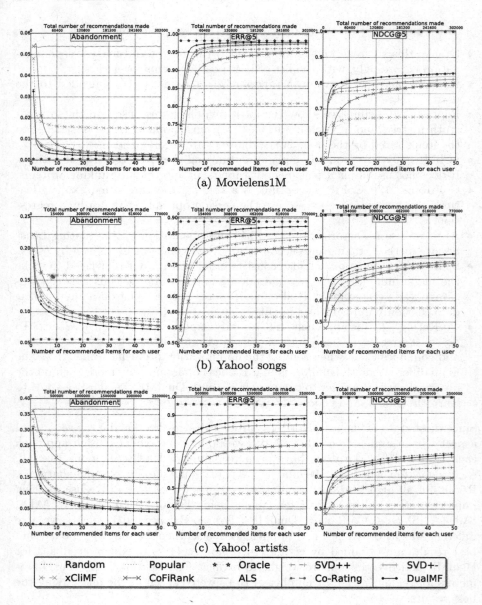

Fig. 1. Evaluation of algorithms on three datasets and three metrics (from the left column to the right one: Abandonment, ERR@5, NDCG@5). Plotted values correspond to the average score obtained while recommending a list of items to each user. For all algorithms learning matrices of features to represent users and items, we fix the number of columns of these matrices to 15. All algorithms using FM use a L2 penalty weight of $\lambda = 5.0$ for both pairwise coefficients and linear coefficients, and do one more step of learning at each update. CoFiRank and xCliMF do from 20 to 50 steps of learning at each update depending on the dataset, but relearn from scratch due to implementation. (a) Movielens1M. (b) Yahoo! songs. (c) Yahoo! artists

be inferred. Results of these three methods emphasizes the importance of using any feedback given by the user. Secondly, approaches targeting specifically a good ranking by optimizing ranking loss (xCliMF and CoFiRank) surprisingly performs the worst, even compared to those using only explicit feedback like ALS. Aiming at the solving the learning to rank aspect does not seem to be a priority to reach a good performance.

5   Conclusion and Future Work

We study Recommender Systems from a novel point of view, where at every step a list of recommendation is provided to the user, and the system receives both explicit and implicit feedback. This model of interaction impacts the learning algorithm but also the experimental setting. We provide a case study for a specific interaction model for which we propose two approaches, tackling the problem from two distinct perspectives. We evaluate various state of the art methods on several metrics, and results on experiments display how considering both implicit and explicit feedback can significantly improve the performance.

Several aspects could be targeted as future work. First, would it be possible to infer the click model directly as new feedback is received? Second, a sequential setting also raises concerns about finding good balance between gathering information about the user and providing good recommendations, a.k.a. the exploration-exploitation dilemma [4]. We leave the attempt to incorporate a solution about this dilemma in the proposed approaches as future work.

Acknowledgments. The authors would like to acknowledge the stimulating environment provided by SequeL research group, Inria and CRIStAL. This work was supported by French Ministry of Higher Education and Research, by CPER Nord-Pas de Calais/FEDER DATA Advanced data science and technologies 2015–2020, and by FUI Herms. Experiments presented in this paper were carried out using the Grid'5000 testbed, hosted by Inria and supported by CNRS, RENATER and several Universities as well as other organizations.

References

1. Balakrishnan, S., Chopra, S.: Collaborative ranking. In: Proceedings of the Fifth ACM International Conference on Web Search and Data Mining (WSDM 2012), pp. 143–152. ACM (2012)
2. Burges, C., Shaked, T., Renshaw, E., Lazier, A., Deeds, M., Hamilton, N., Hullender, G.: Learning to rank using gradient descent. In: Proceedings of the 22nd International Conference on Machine Learning (ICML 2005), pp. 89–96. ACM (2005)
3. Chapelle, O., Metlzer, D., Zhang, Y., Grinspan, P.: Expected reciprocal rank for graded relevance. In: Proceedings of CIKM 2009, pp. 621–630. ACM (2009)
4. Kawale, J., Bui, H., Kveton, B., Thanh, L.T., Chawla, S.: Efficient thompson sampling for online matrix-factorization recommendation. In: NIPS 2015 (2015)
5. Koren, Y.: Factorization meets the neighborhood: a multifaceted collaborative filtering model. In: Proceedings of SIGKDD 2008, pp. 426–434. ACM (2008)

6. Lee, J., Bengio, S., Kim, S., Lebanon, G., Singer, Y.: Local collaborative ranking. In: Proceedings of the 23rd International Conference on World Wide Web (WWW 2014), pp. 85–96 (2014)

7. Liu, N.N., Xiang, E.W., Zhao, M., Yang, Q.: Unifying explicit and implicit feedback for collaborative filtering. In: Proceedings of the 19th ACM International Conference on Information and Knowledge Management, pp. 1445–1448. ACM (2010)

8. Liu, X., Aberer, K.: Towards a dynamic top-N recommendation framework. In: Proceedings of the 8th conference on Recommender Systems (RecSys 2014), pp. 217–224 (2014)

9. Rendle, S.: Factorization machines. In: 2010 IEEE 10th International Conference on Data Mining (ICDM), pp. 995–1000. IEEE (2010)

10. Shi, Y., Karatzoglou, A., Baltrunas, L., Larson, M., Hanjalic, A.: xCLiMF: optimizing expected reciprocal rank for data with multiple levels of relevance. In: Proceedings of the 7th ACM Conference on Recommender Systems, pp. 431–434. ACM (2013)

11. Volkovs, M., Zemel, R.S.: Collaborative ranking with 17 parameters. In: Advances in Neural Information Processing Systems, pp. 2294–2302 (2012)

12. Weimer, M., Karatzoglou, A., Le, Q.V., Smola, A.J.: Cofi rank - maximum margin matrix factorization for collaborative ranking. Adv. Neural Inf. Process. Syst. **20**, 1593–1600 (2008)

Group Information-Based Dimensionality Reduction via Canonical Correlation Analysis

Haiping Zhu[1], Hongming Shan[1], Youngjoo Lee[2], Yiwei He[1], Qi Zhou[1], and Junping Zhang[1(✉)]

[1] School of Computer Science,
Shanghai Key Lab of Intelligent Information Processing, Fudan University,
Shanghai, China
{hpzhu14,hmshan,ywhe15,qizhou15,jpzhang}@fudan.edu.cn
[2] Manufacturing Technology Center, Samsung Electronics, Suwon, South Korea
yj6.lee@samsung.com

Abstract. As an effective way of avoiding the curse of dimensionality and leveraging the predictive performance in high-dimensional regression analysis, dimension reduction suffers from small sample size. We proposed to utilize group information generated from pairwise data, to learn a low-dimensional representation highly correlated with target value. Experimental results on four public datasets imply that the proposed method can reduce regression error by effective dimension reduction.

Keywords: Dimension reduction · Canonical correlation analysis · Group information

1 Introduction

The high-dimensional regression analysis is a challenging task because of the curse of dimensionality [1]. One way to address this issue is to perform dimension reduction, either linear or nonlinear. Linear dimension reduction methods such as principal component analysis (PCA) [7,9] can minimize the residuals between data and principal components, and maximize the variance of data along the principal components. Assuming the data resided in high-dimensional space lie in a low-dimensional manifold, manifold learning algorithms such as locally linear embedding (LLE) [4,12], maximum variance unfolding (MVU) [15], and local tangent space alignment (LTSA) [16] can discover nonlinear geometrical structure compared with PCA. However, the performance of those methods may degenerate when the low-dimensional representation is less correlated to the target value.

Note that grouped data are usually available in the case of designed experiments. If the grouped information can be used to guide the reduction of high-dimensional data, the predictive performance may be improved. Focusing on regression analysis, this paper thus introduces a group information based dimension reduction method to obtain a low-dimensional representation which is highly

© Springer International Publishing AG 2016
A. Hirose et al. (Eds.): ICONIP 2016, Part II, LNCS 9948, pp. 297–305, 2016.
DOI: 10.1007/978-3-319-46672-9_34

correlated to the target value. Specifically, we utilize group information from unlabeled data to perform canonical correlation analysis (CCA) [6], named as grouped CCA (**GCCA**) so that a reduced low-dimensional space will provide a good predictive performance in regression. Experiments demonstrate that the proposed algorithm is effective and practically works well even under imperfect group information. To our knowledge, it is the first research to utilize group information for high-dimensional regression analysis.

The rest of paper is organized as follows. Section 2 offers a brief review on related works. Section 3 describes our proposed **GCCA** algorithm in details. Experiments on four public datasets are performed in Sect. 4. We conclude this paper in Sect. 5.

2 Related Works

The dimensionality reduction plays an important role to extract key information from high-dimensional data and to get a better model for inference. PCA is an exemplar in extracting useful information from data. It aims to find an orthogonal transformation to convert observed data into a set of value of linearly uncorrelated variables which are called principal components. When data are assumed to lie in a low-dimensional manifold, many manifold learning algorithms are devoted to uncover such manifold structure from data. Among them, LLE [12] is to preserve locally linear relationship of each data point and its neighbors when projecting data into a low-dimensional space. Isomap [14] attempts to approach geodesic distances among data. However, these linear and nonlinear dimension reduction algorithms are unsupervised learning, which can not utilize the group information when data are grouped.

Another widely used dimension reduction technique is CCA where two or more views of observation are available [8]. CCA tries to find two projection vectors for each view that maximize the correlation between the projected data by mapping them into a common space. Shen *et al.* [13] proposed a semi-supervised canonical correlation method which utilizes partly given label information in multi-view semi-supervised scenarios. Chen *et al.* [3] proposed a sparse CCA to genome-wide association study problem, which could exploit unknown group structure information. However, most of CCA based dimension reduction methods are designed for multi-view problems which have an underlying assumption that the pairs of feature were obtained from an identical object. Different from the multi-view learning, McWilliams *et al.* [10] proposed a method that generated two views consisting of random features with CCA, but it does not consider the information when groups are available. In this paper, we propose a novel dimension reduction method which utilizes CCA on two disjoint groups with same features.

3   The Proposed Method

In this section, we give a detailed description of the proposed **GCCA** algorithms. Besides, a toy example is given to facilitate understanding of the proposed method.

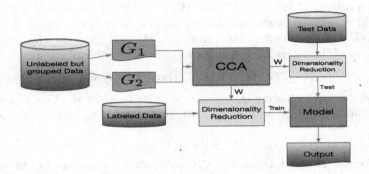

Fig. 1. The workflow of the proposed **GCCA**, where W is a linear projection matrix from CCA

3.1   Group Canonical Correlation Analysis

Noticed that although unlabeled data cannot provide predictive information directly, the relationship between unlabeled but grouped data indirectly uncover a changing tendency of predictive parameters. If the principal tendency of such relationship can be captured based on a large amount of unlabeled grouped data, then prediction model would attain high prediction performance. In literature, CCA focuses on mining the relationship between two or more sets of heterogeneous data. Therefore, we attempt to utilize CCA to achieve this goal. The difference between ours and classical CCA is that we try to capture the principal tendency of correlation of two homogeneous data to improve the performance.

Assume we have labeled data $\mathcal{L} = \{(\boldsymbol{x}_i, y_i)\}_{i=1}^{n_l}$ where $\boldsymbol{x}_i$ resides in a d-dimensional space and y_i is the corresponding target value, and unlabeled data $\mathcal{U}$, which consists of $\boldsymbol{G}_1 = [\boldsymbol{x}_{11}, \ldots, \boldsymbol{x}_{1n_G}] \in \mathbb{R}^{d \times n_G}$ and $\boldsymbol{G}_2 = [\boldsymbol{x}_{21}, \ldots, \boldsymbol{x}_{2n_G}] \in \mathbb{R}^{d \times n_G}$.

CCA finds projection vectors $\boldsymbol{w}_{g1}$ and $\boldsymbol{w}_{g2}$ such that the correlation coefficient between $\boldsymbol{w}_{g1}^T \boldsymbol{G}_1$ and $\boldsymbol{w}_{g2}^T \boldsymbol{G}_2$ is maximized. Each pair of the data from two groups has the same dimensionality, which means $\boldsymbol{w}_{g1} \in \mathbb{R}^d$ and $\boldsymbol{w}_{g2} \in \mathbb{R}^d$. The correlation coefficient ρ is calculated by,

$$\rho = \frac{\boldsymbol{w}_{g1}^T \boldsymbol{G}_1 \boldsymbol{G}_2^T \boldsymbol{w}_{g2}}{\sqrt{(\boldsymbol{w}_{g1}^T \boldsymbol{G}_1 \boldsymbol{G}_1^T \boldsymbol{w}_{g1})(\boldsymbol{w}_{g2}^T \boldsymbol{G}_2 \boldsymbol{G}_2^T \boldsymbol{w}_{g2})}}. \tag{1}$$

Multiple projections of CCA can be computed simultaneously by solving the following problem:

$$\max_{\boldsymbol{W}_{g1}, \boldsymbol{W}_{g2}} \operatorname{Tr}\left[\boldsymbol{W}_{g1}^T \boldsymbol{G}_1 \boldsymbol{G}_2^T \boldsymbol{W}_{g2}\right]$$
$$\text{s.t.} \quad \boldsymbol{W}_{g1}^T \boldsymbol{G}_1 \boldsymbol{G}_1^T \boldsymbol{W}_{g1} = \boldsymbol{I}, \quad \boldsymbol{W}_{g2}^T \boldsymbol{G}_2 \boldsymbol{G}_2^T \boldsymbol{W}_{g2} = \boldsymbol{I}. \tag{2}$$

where each column of $\boldsymbol{W}_{g1} \in \mathbb{R}^{d \times r}$ and $\boldsymbol{W}_{g2} \in \mathbb{R}^{d \times r}$ corresponds to a projection vector and r ($r \leq d$) is the number of projection vectors computed. Assume that $\boldsymbol{G}_{g2} \boldsymbol{G}_{g2}^T$ is nonsingular, the projection $\boldsymbol{W}_{g1}$ is given by the r principal eigenvectors of the following generalized eigenvalue problem:

$$\boldsymbol{G}_{g1} \boldsymbol{G}_{g2}^T (\boldsymbol{G}_{g2} \boldsymbol{G}_{g2}^T)^{-1} \boldsymbol{G}_{g2} \boldsymbol{G}_{g1}^T \boldsymbol{w}_{g1} = \eta \boldsymbol{G}_{g1} \boldsymbol{G}_{g1}^T \boldsymbol{w}_{g1}, \tag{3}$$

where η is the corresponding eigenvalue. In order to make good use of the information from both $\boldsymbol{W}_{g1}$ and $\boldsymbol{W}_{g2}$, final projection $\boldsymbol{W} \in \mathbb{R}^{d \times 2r}$ is calculated by,

$$\boldsymbol{W} = [\boldsymbol{W}_{g1}, \boldsymbol{W}_{g2}]. \tag{4}$$

To leverage the performance of **GCCA**, labeled data can be merged into grouped unlabeled data by (1) sorting by its target value, (2) splitting into two groups, and (3) combining two labeled data groups with existing two unlabeled groups. In this manner, **GCCA** can exploit the information from both of labeled and unlabeled grouped data. If all data in each group are ordered according to its underlying target value, the performance of **GCCA** can be maximized and we denote it as Best-match GCCA (**BGCCA**).

3.2   A Toy Example

Given two grouped data, the projection directions of PCA, **GCCA**, and **BGCCA** on five representative cases are plotted in Fig. 2. We assume the color of samples changes according to target value which is x-axis in the plots. Five cases are as follows: (a) two groups are separated with small margin; (b) two groups are separated with large margin; (c) two groups have different distributions; (d) two groups are partly overlapped; (e) two groups are separated by irrelevant feature.

From Fig. 2 we can see that **BGCCA** always finds the correct projection direction if we know exactly the match between two groups of data. **GCCA** approximately obtains the correct direction since we only have two group information without the match information. The widely used PCA almost cannot find the correct direction because it only consider the data variance information without considering group information. From this toy example, with group information, our proposed **GCCA** can achieve better performance than PCA.

4   Experiments

4.1   Experimental Settings

The performance of **GCCA** was evaluated in comparing with three prevalent dimensionality reduction techniques (PCA, locality preserving projection (LPP)

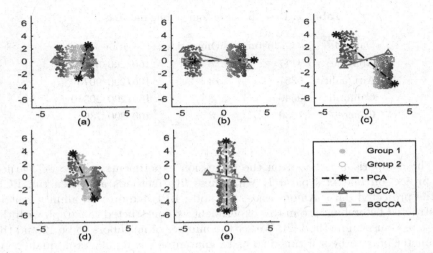

Fig. 2. A toy example on five distribution situation of two groups data, each groups includes 200 points

[11], and factor analysis (FA) [5]) on four public regression problems[1]. Since it is hard to directly evaluate the performance of dimensionality reduction, principal component regression (PCR) was combined to dimensionality reduction methods and regression error of each dimensionality reduction method were compared. For the regression error measure, mean absolute error (MAE) is employed in this work. Unfortunately, there is no public dataset which group information is explicitly available so that an artificial imperfect group information was employed in **GCCA** for experiments as following:

1. Take a specific ratio of train data $\{(x_i, y_i)\}_{i=1}^{2n_G}$ and add noise on its target values.

$$y_i^{noise} = y_i + \epsilon * \text{mean}(y) \tag{5}$$

where noise level $\epsilon \sim \mathcal{U}[-.5, .5]$ in the experiments.

2 Divide data into two groups by noisy target value

$$\begin{cases} G_1 = \{x_i \,|\, y_i^{noise} \leq \text{median}(y^{noise})\}, \\ G_2 = \{x_i \,|\, y_i^{noise} > \text{median}(y^{noise})\}. \end{cases} \tag{6}$$

3. Remove target value to make them unlabeled.
4. Randomize the inner order for each group.

The statistic information about the datasets is shown in Table 1. For each dataset, we normalized its input features to zero mean, unit variance. All experiments were repeated for 100 random trials. In each trial we split the data into three parts: (1) l labeled points; (2) u unlabeled points; (3) t test points.

---

[1] Available in http://archive.ics.uci.edu/ml/datasets.html.

Table 1. Description of four public datasets

Datasets	# Of instances	Dimensions	Partition $l/u/t$
Abalone	4177	8	100/200/3877
AirQuality	9357	14	100/200/9075
California	20640	8	100/200/20340
Concrete	1030	8	100/200/730

The comparison is based on the regression experiments, where we learned the projection matrix separately with these five methods and then run PCR on the projected data. Among these methods, LPP requires an affinity matrix constructed beforehand, so in this experiment we constructed the graph with the self-tune Gaussian method [2]. We set the number of neighbors to be 12 and the parameter of σ to be self-tuned so as to guarantee the input graph quality. As for PCA, FA, **GCCA** and **BGCCA**, there are no parameter need to be tune.

4.2 Experimental Results

The regression error over dimensionality change shows in Fig. 3. It can be seen that the MAE of the proposed **CGGA** and **BGCCA** are always smaller than that of others such as PCA, LPP, and FA in every dataset regardless of dimensionality. As expected, **BGCCA** performs better than **GCCA** because **BGCCA** is the ideal case of **GCCA**. The MAE of the competing five methods in four datasets are summarized in Table 2 at the fixed dimensionality 3. In a comparison to FA which is superior to PCA and LPP in three out of four datasets, the MAEs of the proposed **GCCA** are reduced by 24.9 % in average and 1.2 %, 72.5 %, 15.9 %, and 10.1 % in four datasets, respectively.

Table 2. Mean absolute error: all differences are statistically significant

DataSet	PCA [7]	LPP [11]	FA [5]	GCCA	BGCCA
Abalone	1.84 ± 0.08	1.91 ± 0.08	1.70 ± 0.07	$\mathbf{1.68 \pm 0.06}$	$\mathbf{1.68 \pm 0.06}$
AirQuality	7.23 ± 1.80	8.71 ± 8.43	4.43 ± 0.69	$\mathbf{1.22 \pm 1.02}$	1.35 ± 0.96
California	73138 ± 5218	82838 ± 9879	72091 ± 5169	60654 ± 6608	$\mathbf{56847 \pm 5472}$
Concrete	11.28 ± 0.43	12.28 ± 0.40	11.53 ± 0.77	10.14 ± 0.91	$\mathbf{9.86 \pm 0.93}$

To investigate the influence of the number of labeled data, MAE was evaluated with changing of the number of labeled data from 10 to 150 while the number of unlabeled data is fixed to 200. Figure 4 shows that **GCCA** and **BGCCA** are outperform than other methods by increasing of labeled data on "AirQuality", "California", and "Concrete" datasets. Exceptionally, FA performs better than the proposed method on "Abalone" dataset when the number of labeled data is smaller than 70.

Additionally, the influence of the number of unlabeled data is analyzed by evaluating MAE over the number of unlabeled data changing from 30 to 300 with the fixed number of labeled data 100. Figure 5 shows influence of the number of unlabeled data that the proposed method is quite robust to the number of unlabeled data and it only requires small amount of unlabeled data.

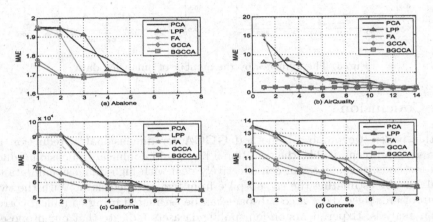

Fig. 3. Empirical performance estimation of PCA, LPP, FA, and the proposed **GCCA** and **BGCCA**

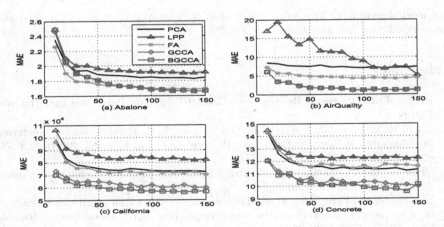

Fig. 4. The influence by the number of labeled data

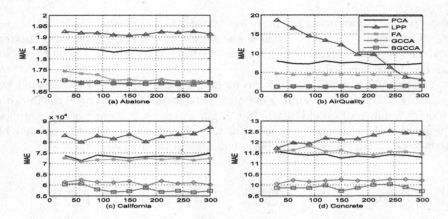

Fig. 5. The influence by the number of unlabeled data

5   Conclusion

In this paper, we propose a method **GCCA** for dimensionality reduction in regression problem. Unlike the traditional linear and nonlinear dimension reduction methods, this algorithm can effectively deal with high-dimensional data of small size by estimating the co-subspace from two grouped data. With the two group of unlabeled data, a co-subspace can be estimated using canonical correlation analysis. Experiments on four public datasets indicate that our proposed method **GCCA** can attain better predictive performance than other published dimension reduction methods.

Acknowledgements. This work was funded by Samsung Electronics Co., Ltd and the NSFC (Grant No. 61273299).

References

1. Burges, C.J.: Dimension Reduction: A Guided Tour. Now Publishers Inc., Hanover (2010)
2. Chen, W.Y., Song, Y., Bai, H., Lin, C.J., Chang, E.Y.: Parallel spectral clustering in distributed systems. IEEE Trans. Pattern Anal. Mach. Intell. **33**(3), 568–586 (2011)
3. Chen, X., Liu, H., Carbonell, J.G.: Structured sparse canonical correlation analysis. AISTATS **12**, 199–207 (2012)
4. Ebtehaj, A.M., Bras, R.L., Foufoula-Georgiou, E.: Shrunken locally linear embedding for passive microwave retrieval of precipitation. IEEE Trans. Geosci. Remote Sens. **53**(7), 3720–3736 (2015)
5. Fruchter, B.: Introduction to factor analysis (1954)
6. Hardoon, D.R., Szedmak, S., Shawe-Taylor, J.: Canonical correlation analysis: an overview with application to learning methods. Neural Comput. **16**(12), 2639–2664 (2004)

7. Hotelling, H.: Analysis of a complex of statistical variables into principal components. J. Educ. Psychol. **24**(6), 417 (1933)
8. Hotelling, H.: Relations between two sets of variates. Biometrika **28**(3/4), 321–377 (1936)
9. Jolliffe, I.: Principal Component Analysis. Wiley, Hoboken (2002)
10. McWilliams, B., Balduzzi, D., Buhmann, J.M.: Correlated random features for fast semi-supervised learning. In: Advances in Neural Information Processing Systems, pp. 440–448 (2013)
11. Niyogi, X.: Locality preserving projections. In: Neural Information Processing Systems, vol. 16, p. 153. MIT (2004)
12. Roweis, S.T., Saul, L.K.: Nonlinear dimensionality reduction by locally linear embedding. Science **290**(5500), 2323–2326 (2000)
13. Shen, X., Sun, Q.: A novel semi-supervised canonical correlation analysis and extensions for multi-view dimensionality reduction. J. Vis. Commun. Image Represent. **25**(8), 1894–1904 (2014)
14. Tenenbaum, J.B., De Silva, V., Langford, J.C.: A global geometric framework for nonlinear dimensionality reduction. Science **290**(5500), 2319–2323 (2000)
15. Dymowa, L.: Introduction. In: Dymowa, L. (ed.) Soft Computing in Economics and Finance. ISRL, vol. 6, pp. 1–5. Springer, Heidelberg (2011)
16. Zhang, Z., Zha, H.: Nonlinear dimension reduction via local tangent space alignment. In: Liu, J., Cheung, Y., Yin, H. (eds.) IDEAL 2003. LNCS, vol. 2690, pp. 477–481. Springer, Heidelberg (2003)

Neuromorphic Hardware

Simplification of Processing Elements in Cellular Neural Networks

Working Confirmation Using Circuit Simulation

Mutsumi Kimura[1,2(✉)], Nao Nakamura[2], Tomoharu Yokoyama[2],
Tokiyoshi Matsuda[2], Tomoya Kameda[1], and Yasuhiko Nakashima[1]

[1] Graduate School of Information Science,
Nara Institute of Science and Technology, Ikoma, Japan
mutsu@rins.ryukoku.ac.jp
[2] Department of Electronics and Informatics, Ryukoku University, Shiga, Japan

Abstract. Simplification of processing elements is greatly desired in cellular neural networks to realize ultra-large scale integration. First, we propose reducing a neuron to two-inverter two-switch circuit, two-inverter one-switch circuit, or two-inverter circuit. Next, we propose reducing a synapse only to one variable resistor or one variable capacitor. Finally, we confirm the correct workings of the cellular neural networks using circuit simulation. These results will be one of the theoretical bases to apply cellular neural networks to brain-type integrated circuits.

Keywords: Simplification · Processing element · Cellular neural network · Circuit simulation · Neuron · Synapse · Ultra-large scale integration · Brain-type integrated circuit

1 Introduction

Cellular neural networks are neural networks where a neuron is connected to only neighboring neurons [1], which are extremely suitable for integration of electron devices and promising for image processing [2], pattern recognition [3], etc. Until now, fundamental theory, operation principle, and potential applications have been actively investigated using formal models and logical simulation. However, there are few reports on actual hardware of cellular neural networks [4], although they are suitable for integration of electron devices. We imagine that this is because the conventional circuits of the processing elements are still complicated, even though the network structure is simple.

We are investigating artificial neural networks using actual devices and hardware [5–8]. Based on our discussions until now, we concluded that simplification of processing elements is greatly desired in cellular neural networks to realize ultra-large scale integration. In this study, first, we propose reducing a neuron circuit to "two-inverter two-switch circuit", "two-inverter one-switch circuit", or "two-inverter circuit". Next, we propose reducing a synapse device only to "one variable resistor" or "one variable capacitor". Although these simplifications have been individually tried

© Springer International Publishing AG 2016
A. Hirose et al. (Eds.): ICONIP 2016, Part II, LNCS 9948, pp. 309–317, 2016.
DOI: 10.1007/978-3-319-46672-9_35

[9], they have not been systematically combined especially in cellular neural networks. Finally, we confirm the correct workings of the cellular neural networks having 3 × 3 neurons using circuit simulation by its learning of arbitrary two-input one output logics, such as AND and OR. Although the current results are very fundamental because the number of the processing elements is not so many, it can be expected that the cellular neural network acquires various abilities if a lot of processing elements are provided, which is possible because we succeeded in simplification of processing elements. In summary, our results will be one of the theoretical bases to apply cellular neural networks to brain-type integrated circuits.

2 Neuron

We consider the working of the neuron and reach an idea that the requisite minimum functions are (1) generating a binary state and (2) alternating the binary state according to the input signal. We propose three-type neuron circuits, 2-inverter 2-switch, 2-inverter 1-switch, and 2-inverter circuits, as shown in Fig. 1. An inverter consists of a pair of n-type and p-type transistors, and a switch also consists of a pair of the transistors.

The 2-inverter 2-switch circuit is a circuit where the two inverters and two switches are circularly connected. The inverters generate a binary state, which is maintained when the switches are on, whereas it is alternated when the switches are off and some input signal is received. The two terminals are bi-directional, namely, work as both input and output terminals. One is for positive logic, whereas the other is for negative logic. This neuron circuit consists of eight transistors. The number of the synapse per connected neighboring neuron is two. Because the switching pulses are periodically applied to the switches, some switching noise exists.

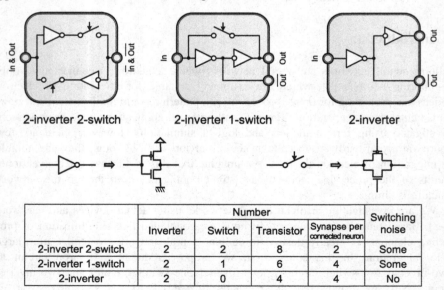

	Number				Switching noise
	Inverter	Switch	Transistor	Synapse per connected neuron	
2-inverter 2-switch	2	2	8	2	Some
2-inverter 1-switch	2	1	6	4	Some
2-inverter	2	0	4	4	No

Fig. 1. Neuron circuits.

The 2-inverter 1-switch circuit is a circuit where the two inverters and one switch are circularly connected. In contrast to the former, the three terminals are uni-directional, namely, work as either input or output terminal. One is an input terminal, another is an output terminal for positive logic, whereas the other is an output terminal for negative logic. This neuron circuit consists of six transistors. The number of the synapse per connected neighboring neuron is four, twice of that for the former, because a synapse sending the signal from a neuron to the neighboring neuron and another synapse vice versa are necessary. Some switching noise exists, which is similar to the former.

The 2-inverter circuit is a circuit where the two inverters are connected in series. The inverters generate a binary state, which is alternated whenever some input signal is received. The three terminals are uni-directional. This neuron circuit consists of four transistors. The number of the synapse per connected neighboring neuron is four. Because the switching pulses are not applied, no switching noise exists.

3 Synapse

We consider the working of the synapse and reach an idea that the requisite minimum functions are (1) sending the signal from a neuron to the neighboring neuron, (2) merging the signals from the multiple neurons, and (3) controlling the synaptic connection strength. We propose two-type synapse devices, variable register and variable capacitor, as shown in Fig. 2.

The variable resistor sends the signal as an electric current, where the conductance corresponds to the synaptic connection strength. The electric currents are added by bundling them in parallel, which corresponds to merging the signals. The advantage for the variable resistors is the operation stability because the constant currents settle all the conditions in the network circuit.

The variable capacitor sends the signal as a voltage shift through capacitive coupling, where the capacitance corresponds to the synaptic connection strength. The voltage shifts are added by bundling them in parallel. The advantage for the variable capacitors is the low power consumption because there is no constant current.

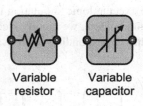

Variable   Variable
resistor   capacitor

Fig. 2. Synapse devices.

4 Neural Network

We form the network structures for the 2-inverter 2-switch, 2-inverter 1-switch, and 2-inverter circuits, as shown in Fig. 3. We propose two-type synapse connections, concordant connection and discordant connection. The concordant connection connects the same logics of the two neurons and tends to make the states of the two neurons the same, whereas, the discordant connection connects the different logics and make the states different. The reason why we prepare two-type synapse connections is to obtain the same effect that the synaptic connection strength becomes both stronger and weaker

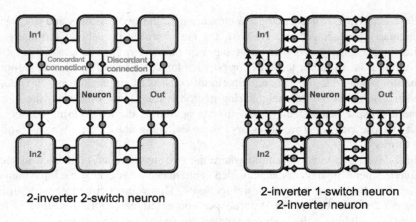

Fig. 3. Neural networks.

even if the actual strength becomes either one. Here, the cellular neural networks have 3 × 3 neurons, and a neuron is connected to the four neighboring neurons. Two input neurons and one output neuron are assigned for its learning of arbitrary two-input one output logics, such as AND and OR.

The structure for the 2-inverter 2-switch circuit has concordant and discordant connections between a pair of the neighboring neurons. The structure for the 2-inverter 1-switch neuron and 2-inverter neuron has forward and reverse connections in addition to the concordant and discordant connections.

5   Synaptic Connection Strength

We execute logic simulation to determine the entire pattern of the synaptic connection strength. Here, we modify Hebbian learning rule [10]. First, we apply In1, In2, and Out to the corresponding neurons. Next, we calculate a steady pattern for the states of the

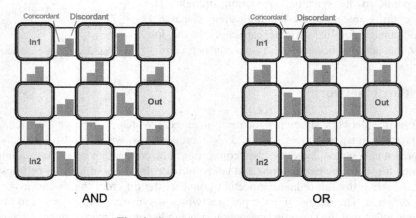

Fig. 4. Synaptic connection strength.

neurons based on the normal theory of the dynamics of the neural network. After that, the synaptic connection strength of the concordant connection is kept the same when both neurons connected to the synapse are in the same states, and is impaired otherwise, whereas that of the discordant connection vice versa. Finally, we obtain the entire pattern of the synaptic connection strength for AND and OR, as shown in Fig. 4.

6 Simulation and Results

We compose the neuron circuits, synapse devices, and network structures in a circuit simulator, OrCAD PSpice, as shown in Fig. 5, circuit schematic example for the 2-inverter 2-switch circuit and variable resistor. It should be noted that the synapse device should be undoubtedly materialized by some actual electron devices such as memristors and ferroelectric capacitors to realize ultra-large scale integration. The entire patterns of the resistances and capacitances are set to the synaptic connection strength determined using the modified Hebbian leaning rule.

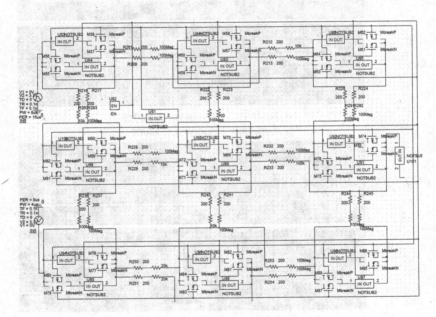

Fig. 5. Circuit schematic.

We observe the input and output waveforms, as shown in Fig. 6. In1 and In2 are periodical rectangle pulses, which make all four combinations of high voltage (H) and low voltage (L), and applied to In1 and In2, whereas Out is measured at the output terminal. It is confirmed that the correct workings are achieved for all logics, AND and OR, all neuron circuits, 2-inverter 2-switch, 2-inverter 1-switch, and 2-inverter circuits, and all synapse devices, variable resistor and capacitor.

314   M. Kimura et al.

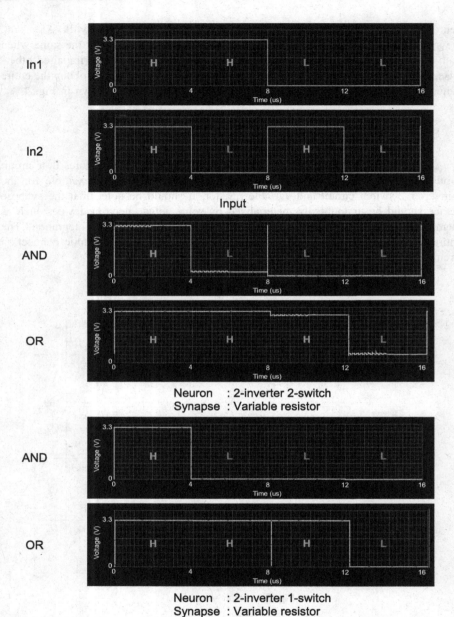

Fig. 6. Input and output waveforms.

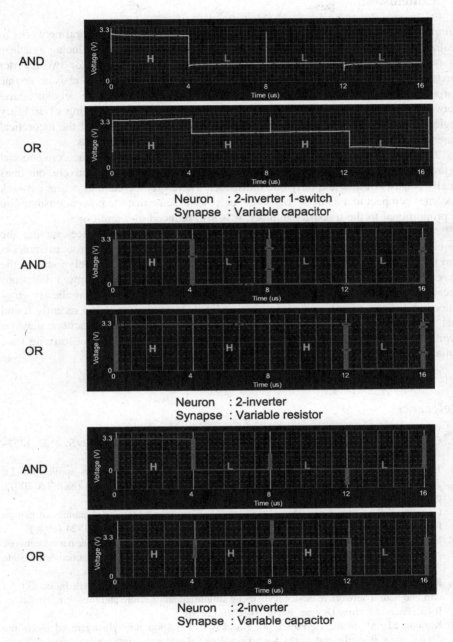

Fig. 6. *(Continued)*

7   Conclusion

Simplification of processing elements is greatly desired in cellular neural networks to realize ultra-large scale integration. In this study, first, we proposed reducing a neuron to two-inverter two-switch circuit, two-inverter one-switch circuit, or two-inverter circuit. Next, we proposed reducing a synapse only to one variable resistor or one variable capacitor. Finally, we confirmed the correct workings of the cellular neural networks having 3×3 neurons using circuit simulation by its learning of arbitrary logics, such as AND and OR. In summary, our results will be one of the theoretical bases to apply cellular neural networks to brain-type integrated circuits.

The quantitative merits of the simplification are: First, whereas the conventional neuron and synapse circuits consist of more than ten transistors, respectively, our ones finally consists of four and only one transistors. Therefore, the cellular neural network becomes compact to a tenth or so. Moreover, if we assume that the power consumption is proportional to the transistor number, it is also reduced to a tenth or so.

Although we need great efforts to develop actual electron devices having the requisite minimum functions especially for the synapse devices, such as memristors and ferroelectric capacitors, our results indicates that the researchers solely focus on the development of the circuits and devices having the requisite minimum functions because the correct workings of the cellular neural networks have been already guaranteed once such circuits and devices are realized. Moreover, we have recently found that the dynamic features of the memristors and ferroelectric capacitors may be available for the unsupervised learning based on the modified Hebbian learning rule, which we would like to develop in the near future.

References

1. Chua, L.O., Yang, L.: Cellular neural networks: theory. IEEE Trans. Circuits Syst. **32**, 1257–1272 (1988)
2. Koeppl, H., Chua, L.O.: An adaptive cellular non-linear network and its application. In: 2007 International Symposium on Nonlinear Theory and its Applications (NOLTA 2007), pp. 15–18, 2007
3. Crounse, K.R., Chua, L.O., Thiran, P., Setti, G.: Characterization and dynamics of pattern formation in cellular neural networks. Int. J. Bifurcation Chaos **6**, 1703–1724 (1996)
4. Morie, T., Miyake, M., Nagata, M., Iwata, A.: A 1-D CMOS PWM cellular neural network circuit and resistive-fuse network operation. In: 2001 International Conference Solid State Devices and Materials (SSDM 2001), pp. 90–91 (2001)
5. Kasakawa, T., Tabata, H., Onodera, R., Kojima, H., Kimura, M., Hara, H., Inoue, S.: An artificial neural network at device level using simplified architecture and thin-film transistors. IEEE Trans. Electron Devices **57**, 2744–2750 (2010)
6. Kimura, M., Miyatani, T., Fujita, Y., Kasakawa, T.: Apoptotic self-organized electronic device using thin-film transistors for artificial neural networks with unsupervised learning functions. Jpn. J. Appl. Phys. **54**, 03CB02 (2015)

7. Kimura, M., Fujita, Y., Kasakawa, T., Matsuda, T.: Novel architecture for cellular neural network suitable for high-density integration of electron devices - learning of multiple logics. In: Arik, S., Huang, T., Lai, W.K., Liu, Q. (eds.) ICONIP 2015 Part I. LNCS, vol. 9489, pp. 12–20. Springer, Heidelberg (2015)
8. Kimura, M., Morita, R., Koga, Y., Nakanishi, H., Nakamura, N., Matsuda, T.: Simplified architecture for cellular neural network suitable for high-density integration of electron devices. In: 2015 International Symposium on Nonlinear Theory and its Applications (NOLTA 2015), pp. 499–502 (2015)
9. Tanaka, M., Saito, T.: Neural Nets and Circuits. Corona Publishing, Tokyo (1999). (in Japanese)
10. Hebb, D.O.: The Organization of Behavior. Wiley, Hoboken (1949)

Pattern and Frequency Generation Using an Opto-Electronic Reservoir Computer with Output Feedback

Piotr Antonik[1]([✉]), Michiel Hermans[1], Marc Haelterman[2], and Serge Massar[1]

[1] Laboratoire d'Information Quantique, Université Libre de Bruxelles,
50 Avenue F.D. Roosevelt, CP 225, 1050 Brussels, Belgium
pantonik@ulb.ac.be
[2] Service OPERA-Photonique, Université Libre de Bruxelles,
50 Avenue F.D. Roosevelt, CP 194/5, 1050 Brussels, Belgium

Abstract. Reservoir Computing is a bio-inspired computing paradigm for processing time dependent signals. The performance of its analogue implementations matches other digital algorithms on a series of benchmark tasks. Their potential can be further increased by feeding the output signal back into the reservoir, which would allow to apply the algorithm to time series generation. This requires, in principle, implementing a sufficiently fast readout layer for real-time output computation. Here we achieve this with a digital output layer driven by an FPGA chip. We demonstrate the first opto-electronic reservoir computer with output feedback and test it on two examples of time series generation tasks: pattern and frequency generation. The good results we obtain open new possible applications for analogue Reservoir Computing.

1 Introduction

Reservoir Computing (RC) is a set of methods for designing and training artificial recurrent neural networks [7,10]. A typical reservoir consists of randomly connected fixed network with random input coupling coefficients. Only the output weights are optimised, which reduces the training process to solving a system of linear equations [9]. The RC algorithm has been successfully applied to channel equalisation and chaotic series forecasting [7], phoneme recognition [16] and won an international competition on prediction of future evolution of financial time series [1]. Its simplicity makes it well suited for analogue implementations. To cite a few examples: electronic [4], opto-electronic [8,11] and all-optical [17,18] implementations have been reported (for a review see [13]).

The list of possible applications of these systems can be significantly broadened by looping the output signal back into the reservoir. It has been shown that this additional feedback allows the algorithm to solve long horizon prediction tasks, such as forecasting chaotic time series, which are impossible to solve otherwise [7]. It also allows the RC to run autonomously, that is, produce an output signal without receiving any input signal, and thus makes it capable of generating time series.

A. Hirose et al. (Eds.): ICONIP 2016, Part II, LNCS 9948, pp. 318–325, 2016.
DOI: 10.1007/978-3-319-46672-9_36

In this work, we demonstrate the first experimental opto-electronic reservoir computer with output feedback. The setup is successfully applied to pattern and frequency generation (thus confirming previously reported numerical results [3]). The tasks investigated here have various applications in motion generation and robot control [14]. Solving them in hardware can allow opto-electronic reservoir computers to be applied to fast control applications, for example high-speed robot control [6]. This work thus opens new perspectives for analogue Reservoir Computing.

In the present experiment, an FPGA chip is used to compute the output signal and feed it back into the reservoir in real time. Our previous RC implementation with an FPGA chip has shown that programmable electronics is well suited for this task, offering high speed and versatility [2]. In perspective, the digital readout layer could be replaced by an analogue one [5,12] in order to benefit from the advantages of a fully analogue opto-electronic implementation, such as high speed and low power consumption.

2   Reservoir Computing with Output Feedback

A general reservoir computer is described in [9]. In our experimental implementation, schematised in Fig. 1, we use a sine function $f = \sin(x)$ and a ring topology, so that only the first neighbour nodes are connected [8,11]. The evolution equations are given by

$$x_0(n+1) = \sin\left(\alpha x_{N-1}(n-1) + \beta M_0 I(n)\right), \tag{1a}$$
$$x_i(n+1) = \sin\left(\alpha x_{i-1}(n) + \beta M_i I(n)\right), \tag{1b}$$

where $x_i(n)$, $i = 0, \ldots, N-1$ are the internal variables, evolving in discrete time $n \in \mathbb{Z}$, α and β parameters are used to adjust the feedback and the input signals, respectively, and M_i is the input mask, drawn from a uniform distribution over the interval $[-1, +1]$ [11]. $I(n)$ is a time multiplexed input signal, which is a teacher sequence $I(n) = u(n)$ during training and the reservoir computer output signal $I(n) = y(n)$ during autonomous run, with

$$y(n) = \sum_{i=0}^{N} w_i x_i(n), \tag{2}$$

where $x_N = 1$ is a constant neuron used to adjust the bias of the output signal and w_i are the readout weights, trained offline, using ridge regression algorithm [15] in order to minimise the Normalised Mean Square Error (NMSE) between the output signal $y(n)$ and the target signal.

At training stage the reservoir receives a periodic sequence $u(n)$ as input and is taught to predict the next value of the sequence from the current one. Then the reservoir input $I(n)$ is switched from $u(n)$ to $y(n)$ and the system is left running autonomously.

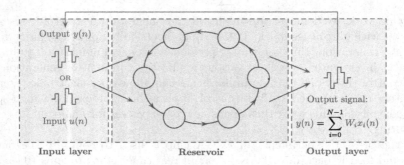

Fig. 1. Schematic representation of our reservoir computer with output feedback. The recurrent neural network with N nodes denoted $x_i(n)$ in ring-like topology (in brown) is driven by either a time multiplexed input signal $u(n)$, or its own output signal $y(n)$, given by a linear combination of the readout weights w_i with the reservoir states $x_i(n)$.

3   Time Series Generation Tasks

Pattern and frequency generation are two relatively simple examples of time series generation tasks.

Pattern Generation. A pattern is a short sequence of randomly chosen real numbers (here within the interval $[-0.5, 0.5]$) that is repeated periodically to form an infinite time series [3]. The aim is to obtain a stable pattern generator, that reproduces precisely the pattern and doesn't deviate to another periodic behaviour. To evaluate the performance of the generator, we compute the NMSE between the reservoir output signal and the target pattern signal during the training phase and the autonomous run.

Frequency Generation. The system is trained to generate a sine wave given by

$$u(n) = \sin(\nu n), \tag{3}$$

where ν is a relative frequency and n is the discrete time. This task allows to measure the bandwidth of the system and investigate different timescales within the neural network.

   As the generated frequency is never exactly the same as desired, the resulting phase accumulation shifts the output signal from the target, thus rendering point-by-point error estimation meaningless. For this reason we used the FFT algorithm to compute the frequency of the reservoir output signal and compare it to the frequency of the target signal [3].

4   Experimental Setup

Figure 2 depicts the experimental setup. The opto-electronic reservoir, a replica of previously reported works [2,11], is driven by a Xilinx ML605 evaluation

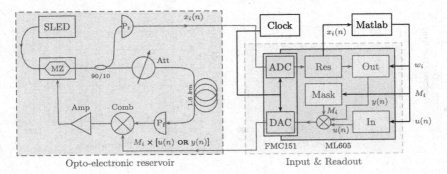

Fig. 2. (a) Schematic representation of the simulated setup, based on the experimental system [2,11]. Optical and electronic components of the opto-electronic reservoir are shown in red and green, respectively. It contains an incoherent light source (SLED), a Mach-Zehnder intensity modulator (MZ), a 90/10 beam splitter, an optical attenuator (Att), a 1.6 km fibre spool, two photodiodes (P_r and P_f), a resistive combiner (Comb) and an amplifier (Amp). The FPGA board acquires the reservoir states $x_i(n)$ and generates analogue input and output signals to the reservoir. A personal computer, running Matlab, computes the readout weights w_i.

board, powered by a Virtex 6 FPGA chip and paired with a 4DSP FMC-151 daughter card, used for signal acquisition and generation. The FPGA is programmed to record the reservoir states $x_i(n)$ and send them to the personal computer, running Matlab, through an Ethernet connection. The readout weights w_i are uploaded on the chip for real-time computation of the reservoir output signal $y(n)$ during the autonomous run.

The experiment roundtrip time is defined by the length of the delay loop. Here we use a relatively short loop of approximately 1.6 km fibre in order to obtain a delay of $T = 7.93\,\mu s$. This allows to create a reservoir containing $N = 100$ neurons.

The experiments are conducted as follows. The reservoir is first driven by a teacher signal for 1k–2k timesteps, depending on the task parameters. The reservoir states $x_i(n)$ are recorded by the FPGA and sent to Matlab, which computes optimal readout weights w_i and uploads them to the FPGA. The system then runs autonomously, with the masked output signal sent into the reservoir as input. Because of a relatively long pause between the training and the autonomous run, we initiate the reservoir by driving it with the desired signal for 128 timesteps prior to coupling the output back, hence letting it run autonomously.

As the neurons are processed sequentially, due to propagation delay between the intensity modulator (MZ) and the ADC, the output signal $y(n)$ can only be computed in time for the 24-th neuron $x_{23}(n)$. For this reason, we set the first 23 elements of the input mask M_i to zero. That way, all 100 neurons contribute to solving the task, but only 77 of them "see" the input signal.

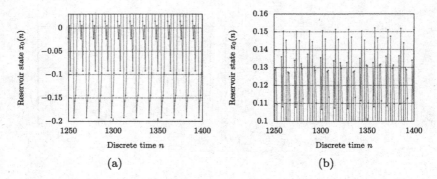

Fig. 3. Examples of neuron behaviour in a **(a)** noiseless numerical and a **(b)** noisy experimental reservoir. For clarity, the range of the Y axis is limited to the area of interest. In the former case $x_0(n)$ cycles between several identical values, while in the latter it takes many similar, but not identical values.

5   Results

5.1   Noisy Reservoir

Previously reported numerical simulations [3] considered a noiseless reservoir. However, our experimental implementation is noisy, which can deteriorate the performance. In order to compare our experimental results to more realistic simulations, we estimated the quantity of noise present in the experimental system.

Figure 3(a) depicts the behaviour of one neuron in a noiseless reservoir, driven by a periodic teacher signal. The reservoir state cycles between several identical values. Figure 3(b) shows what happens in our experimental system. While the noisy neuron exhibits periodic behaviour, it cycles between many similar, but not identical values. We can estimate the noise level by computing the standard deviation of these values, which gives 1.5×10^{-3}. We ran the same simulations as in [3], now with noise, and compare the numerical results with our experimental results, as will be discussed in the following sections.

5.2   Pattern Generation

Figure 4(a) shows an example of the output signal during the autonomous run. The system was trained for 1000 timesteps to generate a pattern of length 10.

We have shown that a noiseless 51-neuron reservoir is capable of generating patterns up to 51-element long [3]. Numerical simulations of a noisy 100-neuron reservoir, with 23 input mask elements zeroed, similar to the experimental setup, show that the maximum pattern length is reduced down to 13 elements. We obtain similar results in the experiments.

Figure 4(b) shows the evolution of the NMSE measured during first 1k timesteps of 10k-timestep autonomous runs with different pattern lengths. Plotted curves are averaged over 100 runs of the experiment, with 5 random input masks and 20 random patterns for each length. The initial minimum (at $n = 128$)

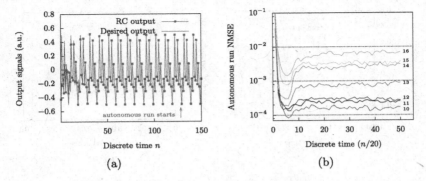

Fig. 4. (a) Example of an autonomous run output signal for pattern generation task. The reservoir is first driven by the desired signal for 128 timesteps (see Sect. 4), and then the input is connected to the output. Note that in this example the reservoir output requires about 50 timesteps to match the driver signal. The autonomous run continues beyond the scope of the figure. (b) Determination of the maximum pattern length: patterns shorter than 13 are reproduced with low NMSE $< 10^{-3}$, while patterns longer than 14 are not generated correctly with NMSE $> 10^{-3}$.

corresponds to the initialisation of the reservoir (see Sect. 4), then the output is coupled back and the system runs autonomously. Patterns with 12 elements or less are generated very well and the error stays low. Patterns of length 13 show an increase in NMSE, but they are still generated reasonably well. For longer patterns, the system deviates to a different periodic behaviour, and the error grows above 10^{-3}.

We also tested the stability of the generator by running it for several hours ($\sim 10^{9}$ timesteps) with patterns of lengths 10, 11 and 12. The output signal was visualised on a scope and remained stable through the whole test.

5.3   Frequency Generation

Figure 5(a) shows an example of the output signal during the autonomous run. The system was trained for 1000 timesteps to generate a frequency of $\nu = 0.1$.

With this task we found that noise inside the reservoir doesn't affect its capacity to generate sine waves. In previous simulations of a noiseless 100-neuron reservoir we obtained a bandwidth of $\nu \in [0.06, \pi]$, the upper limit being a signal oscillating between -1 and 1. These results are confirmed experimentally here.

We tested our setup on frequencies ranging from 0.01 to π, and more precisely within the $[0.01, 0.1]$ interval, where the lower limit lies. For each frequency, we ran the experiment 10 times for 10k timesteps with different random input masks and counted the number of times the reservoir produced a sine wave with the desired frequency and constant amplitude of 1. The results are shown in Fig. 5(b). Frequencies below 0.05 are not generated correctly with most input masks. At $\nu = 0.7$ the output is correct most of the times, and for $\nu = 0.08$ and above the output sine wave is correct with any input mask. The bandwidth of this experimental RC is thus $\nu \in [0.08, \pi]$.

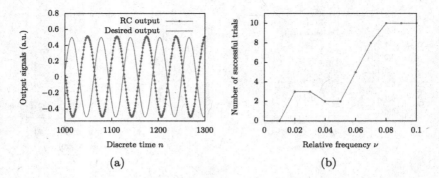

Fig. 5. (a) Example of an autonomous run output signal for frequency generation task with $\nu = 0.1$. The experiment continues beyond the scope of the figure. (b) Determination of the minimal frequency: frequencies below 0.05 fail with most input masks, while those above 0.08 are generated very well with any input mask.

The physical frequency f of the sine wave is $f = \nu/2\pi T$. Given the roundtrip time $T = 7.93\,\mu\mathrm{s}$ it results in a physical bandwidth of 1.5 – 63 kHz.

6   Conclusion

We reported here the first photonic reservoir computer with output feedback, produced by a digital FPGA-based output layer. The setup was able to successfully solve two examples of time series generation tasks. This opens a new area of potential applications for Reservoir Computing, such as robot control and motion generation [6]. With some minor improvements, this experiment could tackle more complex tasks, such as forecasting chaotic time series [3], which will be the target of our future investigations. Upgraded with an analog output layer, this setup could be used as a fast analog control system or prediction machine for e.g. high-frequency trading. This work is thus the first step towards numerous additional applications of photonic Reservoir Computing.

Acknowledgements. We acknowledge financial support by Interuniversity Attraction Poles program of the Belgian Science Policy Office under grant IAP P7-35 photonics@be, by the Fonds de la Recherche Scientifique FRS-FNRS and by the Action de Recherche Concertée of the Académie Universitaire Wallonie-Bruxelles under grant AUWB-2012-12/17-ULB9.

References

1. The 2006/07 forecasting competition for neural networks & computational intelligence (2006). http://www.neural-forecasting-competition.com/NN3/. Accessed 21 Feb 2014

2. Antonik, P., Duport, F., Smerieri, A., Hermans, M., Haelterman, M., Massar, S.: Online training of an opto-electronic reservoir computer. In: Arik, S., et al. (eds.) ICONIP 2015. LNCS, vol. 9490, pp. 233–240. Springer, Heidelberg (2015). doi:10. 1007/978-3-319-26535-3_27

3. Antonik, P., Hermans, M., Duport, F., Haelterman, M., Massar, S.: Towards pattern generation and chaotic series prediction with photonic reservoir computers. In: SPIE's 2016 Laser Technology and Industrial Laser Conference, vol. 9732 (2016)

4. Appeltant, L., Soriano, M.C., Van der Sande, G., Danckaert, J., Massar, S., Dambre, J., Schrauwen, B., Mirasso, C.R., Fischer, I.: Information processing using a single dynamical node as complex system. Nat. Commun. **2**, 468 (2011)

5. Duport, F., Smerieri, A., Akrout, A., Haelterman, M., Massar, S.: Fully analogue photonic reservoir computer. Sci. Rep. **6** (2016). Article number: 22381

6. Ijspeert, A.J.: Central pattern generators for locomotion control in animals and robots: a review. Neural Netw. **21**(4), 642–653 (2008)

7. Jaeger, H., Haas, H.: Harnessing nonlinearity: predicting chaotic systems and saving energy in wireless communication. Science **304**, 78–80 (2004)

8. Larger, L., Soriano, M., Brunner, D., Appeltant, L., Gutiérrez, J.M., Pesquera, L., Mirasso, C.R., Fischer, I.: Photonic information processing beyond turing: an optoelectronic implementation of reservoir computing. Opt. Express **20**, 3241–3249 (2012)

9. Lukoševičius, M., Jaeger, H.: Reservoir computing approaches to recurrent neural network training. Comput. Sci. Rev. **3**, 127–149 (2009)

10. Maass, W., Natschläger, T., Markram, H.: Real-time computing without stable states: a new framework for neural computation based on perturbations. Neural Comput. **14**, 2531–2560 (2002)

11. Paquot, Y., Duport, F., Smerieri, A., Dambre, J., Schrauwen, B., Haelterman, M., Massar, S.: Optoelectronic reservoir computing. Sci. Rep. **2**, 287 (2012)

12. Smerieri, A., Duport, F., Paquot, Y., Schrauwen, B., Haelterman, M., Massar, S.: Analog readout for optical reservoir computers. In: Advances in Neural Information Processing Systems, pp. 944–952 (2012)

13. Soriano, M.C., Brunner, D., Escalona-Morán, M., Mirasso, C.R., Fischer, I.: Minimal approach to neuro-inspired information processing. Front. Comput. Neurosci. **9**, 68 (2015)

14. Sussillo, D., Abbott, L.: Generating coherent patterns of activity from chaotic neural networks. Neuron **63**(4), 544–557 (2009)

15. Tikhonov, A.N., Goncharsky, A., Stepanov, V., Yagola, A.G.: Numerical Methods for the Solution of Ill-Posed Problems, vol. 328. Springer, Netherlands (1995)

16. Triefenbach, F., Jalalvand, A., Schrauwen, B., Martens, J.P.: Phoneme recognition with large hierarchical reservoirs. Adv. Neural Inf. Process. Syst. **23**, 2307–2315 (2010)

17. Vandoorne, K., Mechet, P., Van Vaerenbergh, T., Fiers, M., Morthier, G., Verstraeten, D., Schrauwen, B., Dambre, J., Bienstman, P.: Experimental demonstration of reservoir computing on a silicon photonics chip. Nat. Commun. **5** (2014). Article number: 3541

18. Vinckier, Q., Duport, F., Smerieri, A., Vandoorne, K., Bienstman, P., Haelterman, M., Massar, S.: High-performance photonic reservoir computer based on a coherently driven passive cavity. Optica **2**(5), 438–446 (2015)

A Retino-Morphic Hardware System Simulating the Graded and Action Potentials in Retinal Neuronal Layers

Yuka Kudo, Yuki Hayashida, Ryoya Ishida, Hirotsugu Okuno,
and Tetsuya Yagi[✉]

Graduate Engineering, Osaka University, Suita, Japan
{kudo,hayashida,ishida,h-okuno,yagi}@neuron.eei.eng.osaka-u.ac.jp

Abstract. We recently developed a retino-morphic hardware system operating at a frame interval of 5 ms, that was short enough for simulating the graded voltage responses of neurons in the retinal circuit in a quasi-continuous manner. In the present, we made a further progress, by implementing the Izhikevich model so that spatial spike distributions in a ganglion-cell layer can be simulated with millisecond-order timing precision. This system is useful for examining the retinal spike encoding of natural visual scenes.

Keywords: Neuromorphic hardware · Retina · Real-time simulation · Spike timing precision · Izhikevich model · Analog silicon retina · FPGA

1 Introduction

In vertebrate retinas, incoming visual images are continuously transduced into graded voltage changes in the photoreceptor-cell array, processed by the following neuronal circuitries in a parallel manner, and then encoded into spike trains in the ganglion-cell array. Basic synaptic connections among major classes of neurons and fundamental properties of neuronal responses in the retinas were revealed by previous anatomical and physiological studies [1,2]. However, little was understood about the functions of the retinal neurons in response to natural visual scenes, since most of conventional physiological experiments used simplified or unrealistic images as visual stimuli, and considered less about the influence of eye and/or head movements on the stimulus images. Although recent studies have suggested new insights on information encoding by certain ganglion-cell types by using more feasible visual stimuli [3–5], spatio-temporal responses of the ganglion-cell array in an animal *in vivo*, in a natural environment are still largely unknown. Simulating neuronal responses by using a retinal model *in silico* is one of the powerful approaches to estimate and predict the functional computations and the information encoding in the retinal circuit [6,7]. Recently, we implemented a retinal model with combining the analog silicon retina [8] and digital circuits in a field-programmable gate array (FPGA) so that the model

© Springer International Publishing AG 2016
A. Hirose et al. (Eds.): ICONIP 2016, Part II, LNCS 9948, pp. 326–333, 2016.
DOI: 10.1007/978-3-319-46672-9_37

structures and parameters can be modified, and yet the spatial response patterns at each neuronal layer are visualized at a frame interval of 5 ms [9]. In the present study, we have made a further progress in this system by implementing the Izhikevich model [10] as a spike generation mechanism of the ganglion cells. This model enables us to visualize spatial spike distributions in the ganglion-cell array with millisecond-order timing precision without being affected by the clock timing of the above-mentioned frame interval. We demonstrate a real-time simulation of the spike responses of cat X-type ganglion-cell array under a natural visual condition.

2   Retino-Morphic Hardware System with Izhikevich Spiking Modulus

2.1   Neural Circuit Structure

Figure 1 shows a schematic diagram of the retino-morphic hardware system we developed in the present study. This system consists of the analog silicon retina [8] (left box in the figure) and the digital circuits in FPGA (right box). The figure depicts a single pixel of the retinal circuit and almost all the components are the same as those described in the previous study [9], except for a spiking module based on the Izhikevich model [10] ('Izh1' and 'Izh2'). This model is capable of reproducing various spiking patterns of neurons, and yet is compact in computational cost enough for being implemented with the digital circuits in FPGA. And also this enabled us to reproduce the adaptive nature of spiking frequency in retinal ganglion cells [11], as in the real time emulation.

In the silicon retina, an active pixel photo-sensor ('APS') connecting to a resistive network ('1st RN layer'), and another resistive network ('2nd RN layer'),

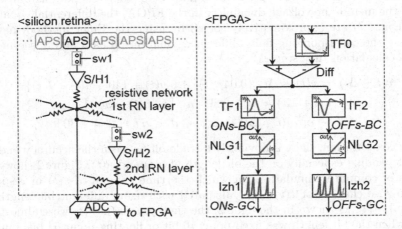

Fig. 1. Schematic diagram of the retino-morphic hardware system. The system is composed of the analog silicon retina (left box), the interface (i.e. A/D converter, left bottom), and the digital circuit modules in the FPGA (right box).

reproduce spatial response profiles (i.e. spatial receptive fields) of a photorecep-
tor cell, and of a horizontal cell, respectively, that are coupled electrically with
their neighbor cells. In the FPGA, a temporal filter, 'TF0' represents the sig-
naling delay due to the synaptic transmission from a horizontal cell to a bipolar
cell. The difference between the inputs from the 1st and 2nd RN layers ('Diff')
reproduces a spatial, center-surround antagonistic receptive field of a bipolar
cell. And subsequent temporal filters, 'TF1' and 'TF2' account for temporal
response profiles of on-type and off-type bipolar cells, respectively ('ONs-BC'
and 'OFFs-BC'). A nonlinear gain function ('NLG1' or 'NLG2') is relevant to a
nonlinear synaptic transmission from a bipolar cell to a ganglion cell. And the
output of this block is fed to the Izhikevich spiking module ('Izh1' or 'Izh2') for
generating spikes as responses of an sustained on-type or off-type ganglion cell
('ONs-GC' or 'OFFs-GC').

2.2   Approximation of the Izhikevich Model

The original Izhikevich model is governed by two differential equations and a
variables resetting via conditional branching, as follows [10]:

$$\frac{dv(t)}{dt} = 0.04\, v(t)^2 + 5\, v(t) + 140 - u(t) + I(t), \tag{1}$$

$$\frac{du(t)}{dt} = \boldsymbol{a} \cdot \{\boldsymbol{b} \cdot v(t) - u(t)\}, \tag{2}$$

$$v(t) \leftarrow \boldsymbol{c}, \quad u(t) \leftarrow u(t) + \boldsymbol{d}, \quad if\ v(t) = +30\,[mV]. \tag{3}$$

Here, $v(t)$ represents the membrane potential; $u(t)$ represents the recovery variable which controls the post-spike refractoriness; $I(t)$ represents the input current determined by the graded voltage response of a bipolar cell (i.e. the output
of 'NLG1' or 'NLG2' in Fig. 1); $\boldsymbol{a}$, $\boldsymbol{b}$, $\boldsymbol{c}$ and $\boldsymbol{d}$ are constant parameters which
define the membrane voltage dynamics. In the FPGA, the differential equations
(Eqs. 1 and 2) are approximatively solved by using the Euler method, namely
by using the corresponding finite-difference equations (Eqs. 4 and 5, see below)
with a calculation time step Δt:

$$v(t + \Delta t) = v(t) + \Delta t \cdot \{0.04\, v(t)^2 + 5\, v(t) + 140 - u(t) + I(t)\}, \tag{4}$$

$$u(t + \Delta t) = u(t) + \Delta t \cdot \boldsymbol{a} \cdot \{\boldsymbol{b} \cdot v(t) - u(t)\}, \tag{5}$$

$$(t + \Delta t) \leftarrow \boldsymbol{c}, \quad u(t + \Delta t) \leftarrow u(t) + \boldsymbol{d}, \quad if\ v(t) = +30\,[mV]. \tag{6}$$

In this computation, there is obviously a trade-off between the accuracy and the
cost, depending especially on the bit length of $v(t)$ and $u(t)$. Figure 2 shows the
results of computer simulations in which the time courses of $v(t)$ in response
to step-wise changes of $I(t)$ were calculated with using different bit lengths of
$v(t)$ and $u(t)$. As shown in the figures, the time courses of $v(t)$ were almost the
same when the bit length was fixed-point 16 bit or floating-point 64 bit, but not
fixed-point 8 or 12 bit. Thus in the present, the bit length of $v(t)$ and $u(t)$ was
set to fixed-point 16 bit for the spiking modules in the FPGA (i.e. 'Izh1' and
'Izh2' in Fig. 1).

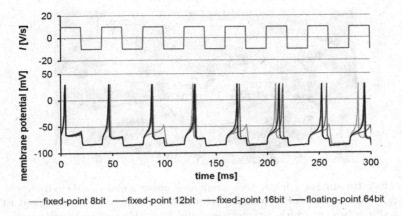

-fixed-point 8bit   -fixed-point 12bit   -fixed-point 16bit   -floating-point 64bit

Fig. 2. Effects of the bit length of $v(t)$ and $u(t)$ on the accuracy of solving the finite-difference equations derived from the Izhikevich model. I is input current to the Izhikevich model. The input start time to the membrane potential of a stable state was plotted as 0 ms.

2.3 Operation of the Hardware System

As similar to our previous study [9], the silicon retina operates at 200 fps for the 128×128-pixel frame and the outputs in parallel are fed to the FPGA through analog-to-digital conversion ('ADC' in Fig. 1) in the 5-ms-frame-based manner. The FPGA operates at 40 MHz of clock rate, and the graded voltage response of each neuronal module is computed every 5 ms. The timing of spiking in each ganglion cell module is determined by the time stamp of the conditional branching followed by the variables resetting in Eq. 6. And the final spike outputs have a timing tag of a 0.5-ms resolution(i.e. Δt in Eqs. 4–6). The data set of the voltage distributions in the RN layers in the silicon retina, and of the voltage responses of the neuronal arrays in the FPGA (e.g. ONs-/OFFs-BC and ONs-.OFFs-GC), can be sent to a computer via the Universal Serial Bus (USB), to be visualized as image streams in quasi-continuous, real-time manner. Figure 3 shows a photograph of the actual hardware equipment of our system. The components of this hardware system are one printed circuit board (PCB) with silicon retina and input-output (I/O) modules including the A/D converter, two PCBs with interface connectors and buffer memories, and one PCB with the FPGA and I/O modules including the USB terminal. As seen in the photograph, the whole system is compact enough for being packed in a box so that a camera lens and a tripod are attached directly to the box, and can be used as if an animal's eye gazes toward an object in a natural visual environment.

Fig. 3. The retino-morphic hardware system used under a real visual environment. The hardware components, i.e. PCBs with silicon retina, interface connectors, and FPGA, are packed in a box, to which a camera lens and a tripod are directly attached.

3   Hardware Simulation of the Neuronal Arrays in Response to a Natural Scene

Our hardware system enables us to visualize the spatial response patterns in each of the retinal neuronal arrays in a real-time manner. For the present simulations, the values of a, b, c and d in Eqs. 5–6, as well as other circuit parameters in the silicon retina and in the FPGA, were tuned for reproducing the intrinsic spiking properties of the cat beta-type ganglion cell [12] as well as the physiological light responses of the cat X-type ganglion cell [13] as an example (data not shown: $a = 0.02$, $b = 0.25$, $c = -58$, $d = 3$). Figure 4 A-C show the images of the voltage distributions in the 1st RN layer (A), of the graded voltage responses in the ONs-BC array (B), and of the spike distributions in the ONs-GC array (C) while a white horizontal bar is vertically moved downward in front of a static object (a stuffed doll) in a situation similar to one shown in Fig. 3. In Fig. 4A, the image represents a blurred image of the visual scene captured by the APS array, since no temporal filtering is applied in the 1st RN layer (Fig. 1). In the images in Fig. 4B, spatial contrast edges in the visual scene were enhanced due to the center-surround antagonistic receptive filed of bipolar cells (i.e. 'ONs-BC' in Fig. 1). Moreover, the moving bar was blurred in shape, and shifted in position to behind from the original position. And in turn, the static object appeared not to be completely occluded by the bar. These were due to the temporal filtering properties in the outer retinal circuit (i.e. 'TF0' and 'TF1' in Fig. 1). In Fig. 4C, the red dots represent positions of the ganglion cells that fired a spike during the last 5 ms. Since the spiking simulated here was the sustained on-type response of the X(beta)-type ganglion cells, basically the spikes were induced in relatively bright regions in the visual scene. In addition, spike firings were highly synchronized among the cells near the leading edge of the moving bar. Figure 4D superimposes the images in the panel A and C so that the location and timing of the spike firings can be examined in combination with the incoming images. For example, the leading edge of the moving bar (arrows in D) can be blurred

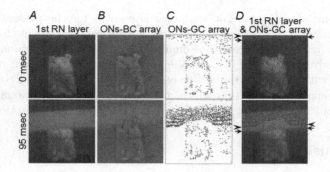

Fig. 4. A real-time simulation of the spatio-temporal responses of the retinal neuronal arrays in a real visual environment. (Color figure online)

at the stage of the bipolar cells (*B*), but can be sharpened at the stage of the ganglion cells' spikes with a certain temporal delay (arrowheads in *D*).

4   Discussion

In the present study, we adopted Izhikevich model as a spiking mechanism of the retinal ganglion cell. Other than this model, for instance, conventional Integrate-and-Fire(I-F) model has been widely used for computer simulations of neuronal spiking. However with this model, when the membrane voltage approaches to the pre-defined threshold value, the timing when a spike is fired can be largely fluctuated dependently on the bit accuracy, and thus would be determined in rather stochastic manner. Such a stochastic spiking behavior of the simple I-F model is less suitable for modeling the ganglion cells spiking in which the spike timing precision is thought to be in milliseconds order [14]. In contrast, the Izhikevich model was suited better for realizing the millisecond-order precision of spike timing (data not shown). Meanwhile, the generalized linear I-F model, which can reproduce a variety of spiking behaviors, has been proposed [15]. However, compared with this I-F models, number of state variables in the Izhikevich equations is less, and thus less of the memory in FPGA is required for the implementation. The adaptive exponential I-F model [16] can be another candidate for emulating the spiking of retinal ganglion cells. And, although we selected the Izhikevich model in the present study, our hardware system was designed to have a capability of testing those other models, as long as the resource in FPGA allows. This would be another benefit from our system configuration.

The visualization of ganglion cells' spiking patterns at a frame interval of 5 ms should be possible also by other technologies, like GPU, multi-core CPUs, or ASIC. On the other hand, the FPGA offers a program-based configurability of the model circuit, as similar to a simulation with GPU or multi-core CPUs. In addition, its compact size enabled us to build the whole system in a relatively small box attached with an optical lens (Fig. 3). Hence, only this system in the box and a laptop/pocket computer is enough for the present visualization. Besides, one can combine this system with a robotics device for the gaze control, thereby enabling a real-time emulation of the binocular vision under eye and head/body movements in a mobile environment. In addition, our system architecture with FPGA is usable for one to create a distributed computation platform of sensory information which may reduce the computational load in a host GPU/multi-core CPUs, and in turn, lower the power consumption as total.

5   Conclusion

The present neuromorphic hardware system makes it possible to visualize and analyze the relation between natural images of the real world and the corresponding neuronal response images which encode significant information for the perception of the real world. The virtual *in-vivo* experiment with using our system (Fig. 4) facilitates the understanding of visual functions and essential circuit structures of the retina.

Acknowledgments. This research was partly supported by JSPS KAKENHI, Grant-in-Aid for Scientific Research (C), 16K01354 to T.Y.

References

1. Kolb, H., Nelson, R., Fernandez, E., Jones, B.: WEBVERSION. http://webvision. med.utah.edu/
2. Demb, J.B., Singer, J.H.: Functional circuitry of the retina. Annu. Rev. Vis. Sci. **1**, 263–289 (2015)
3. Roska, B., Werblin, F.: Rapid global shifts in natural scenes block spiking in specific ganglion cell types. Nat. Neurosci. **6**, 600–608 (2003)
4. Gollisch, T., Meister, M.: Rapid neural coding in the retina with relative spike latencies. Science **319**, 1108–1111 (2008)
5. Zhang, Y., Kim, I.J., Sanes, J.R., Meister, M.: PNAS Plus: the most numerous ganglion cell type of the mouse retina is a selective feature detector. Proc. Natl. Acad. Sci. **109**, E2391–E2398 (2012)
6. Mead, C.A., Mahowald, M.A.: A silicon model of early visual processing. Neural Netw. **1**, 91–97 (1988)
7. Zaghloul, K.A., Boahen, K.: A silicon retina that reproduces signals in the optic nerve. J. Neural Eng. **3**, 257–267 (2006)
8. Kameda, S., Yagi, T.: An analog VLSI chip emulating sustained and transient response channels of the vertebrate retina. IEEE Trans. Neural Netw. **14**, 1405–1412 (2003)
9. Okuno, H., Hasegawa, J., Sanada, T., Yagi, T.: Real-time emulator for reproducing graded potentials in vertebrate retina. IEEE Trans. Biomed. Circ. Syst. **9**, 284–295 (2015)
10. Izhikevich, E.M.: Simple model of spiking neurons. IEEE Trans. Neural Netw. **14**, 1569–1572 (2003)
11. van Rossum, M.C., O'Brien, B.J., Smith, R.G.: Effects of noise on the spike timing precision of retinal ganglion cells. J. Neurophysiol. **89**, 2406–2419 (2003)

12. O'Brien, B.J., Isayama, T., Richardson, R., Berson, D.M.: Intrinsic physiological properties of cat retinal ganglion cells. J. Physiol. **538**, 787–802 (2002)
13. Enroth-Cugell, C., Robson, J.G.: Functional characteristics and diversity of cat retinal ganglion cells. Invest Ophthal. Vis. Sci. **25**, 250–267 (1984)
14. Berry, M.J., Warland, D.K., Meister, M.: The structure and precision of retinal spiketrains. Proc. Natl. Acad. Sci. U.S.A. **94**, 5411–5416 (1997)
15. Mihala, S., Niebur, E.: A generalized linear integrate-and-fire neural model produces diverse spiking behaviors. Neural Comput. **21**, 704–718 (2009)
16. Brette, R., Gerstner, W.: Adaptive exponential integrate-and-fire model as an effective description of neuronal activity. J. Neurophysiol. **94**, 3637–3642 (2005)

Stability Analysis of Periodic Orbits in Digital Spiking Neurons

Tomoki Hamaguchi, Kei Yamaoka, and Toshimichi Saito[✉]

Hosei University, Koganei, Tokyo 184-8584, Japan
tsaito@hosei.ac.jp

Abstract. This paper considers stability of various periodic spike-trains from digital spiking neuron constructed by two coupled shift registers. The dynamics is integrated into a digital spike map defined on a set of points. In order to analyze the stability, we introduce two simple feature quantities that characterize plentifulness and superstability of the periodic spike-trains. Using the feature quantities, stability of typical examples is investigated.

Keywords: Spiking neurons · Neuromorphic hardware · Stability

1 Introduction

This paper studies stability of spike-trains in digital spiking neurons (DSNs). The DSN is constructed by two shift registers connected by a wiring [1,2]. Depending on writing patterns and initial states, the DSN can output various periodic spike-trains (PST). The stability of the PST is based on transient spike-trains. The DSN is well suited for hardware implementation by digital circuits and can be regarded as a digital version of simple analog spiking neurons [3,4].

The dynamics of the DSN can be integrated into the digital spike map (DSM) defined on a set of lattice points [5,6]. The DSM is well suited for computer-aided precise analysis and can be regarded as a digital version of analog one-dimensional map [7]. The DSM is related to various digital dynamical systems including cellular automata [8–10], dynamic binary neural networks [11,12]. The engineering applications include image/signal processing, switching power converters, and ultra-wide-band communications [2,13]. However, systematic analysis is hard because the dynamics of DSN is very complicated.

This paper presents two simple feature quantities for stability analysis of the DSM. The first quantity represents ratio between PSTs and transient spike-trains. The second quantity characterize stability of PSTs. Especially, as the second quantity decreases, the PST approaches to be superstable such that all initial states fall instantaneously into the PST. Calculating the feature quantities of typical DSMs from DSNs, superstability and variety of PSTs are investigated. Analysis of the DSN and DSM can contribute not only to fundamental study of nonlinear dynamics but also to engineering applications.

T. Saito—This work is supported in part by JSPS KAKENHI#15K00350.

A. Hirose et al. (Eds.): ICONIP 2016, Part II, LNCS 9948, pp. 334–341, 2016.
DOI: 10.1007/978-3-319-46672-9_38

2   Digital Spiking Neuron and Digital Spike Map

In this section, we introduce outline of the DSN and DSM presented in [1,2]. The
DSN consists of two shift registers connected by a wiring as shown in Fig. 1. The
left and right shift registers are referred to as P-cells and X-cells, respectively.
The P-cells consist of M elements and operate as a pacemaker with period N. Let
$P(\tau) \equiv (P_1(\tau), \cdots, P_M(\tau))$ denote the P-cells at discrete time $\tau \in \{1, 2, 3, \cdots\}$
and let $P_i(\tau) \in \{0, 1\}$ be the i-th element. The dynamics is described by

$$P_i(\tau) = \begin{cases} 1 & \text{if } \tau = i + nN \\ 0 & \text{otherwise} \end{cases} \tag{1}$$

where $i \in \{1, \cdots, M\}$ and n denotes integers. The X-cells consist of M elements and are state variables corresponding to an action potential. Let $X(\tau) \equiv$
$(X_1(\tau), \cdots, X_M(\tau))$ denote the X-cells and let $X_j(\tau) \in \{0, 1\}$ be the j-th

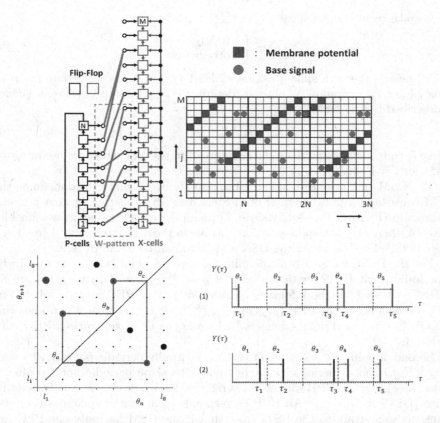

Fig. 1. Digital spiking neuron. (a) Coupled shift registers. (b) DSM (red: PEO, blue:
DEPP, black: EPP). (c) Periodic spike train with period 3 ($\theta_1 = \theta_a$, $\theta_2 = \theta_b$, $\theta_3 = \theta_c$).
(d) Transient spike train. (Color figure online)

element. If $X_j = 1$ then X_j is said to be an active element. An element of the P-cells is connected to an element of the X-cells in one-way. The wiring is characterized by the connection vector

$$W = (w_1, w_2, \cdots, w_j, \cdots, w_N), \quad w_i = j \text{ if } P_i \text{ is connected to } X_j. \tag{2}$$

The connection vector determines the base signel

$$B(\tau) = (B_1(\tau), B_2(\tau), \cdots, B_j(\tau), \cdots, B_N(\tau)) \tag{3}$$

For example, a connection vector $W = (2, 5, 9, 4, 6, 10, 13, 13)$ gives the base signal in Fig. 1. In the X-cells, position of the active element moves upward. If the top cell is active at time τ then the DSN outputs an spike $Y(\tau) = 1$ and the active element is jumps to the base signal. The time evolution is described by

$$X_{k+1}(\tau + 1) = \begin{cases} 1 \text{ if } X_k(\tau) = 1 \\ 1 \text{ if } X_M(\tau) = 1 \text{ and } B_k(\tau) = 1 \\ 0 \text{ otherwise} \end{cases} \tag{4}$$

A spike-train is defined by

$$Y(\tau) = \begin{cases} 1 \text{ if } X_M(\tau) = 1 \\ 0 \text{ otherwise} \end{cases} \tag{5}$$

Let τ_n denote the n-th spike position where $Y(\tau_n) = 1$. Let θ_n denote the n-th spike-phase: $\theta_n = \tau_n \bmod N$. Since θ_n determines θ_{n+1}, the spike-phase sequence is described by the DSM

$$\theta_{n+1} = f(\theta_n), \quad \theta_n \in \{l_1, l_2, \cdots, l_N\} \equiv L_N \tag{6}$$

where l_1 to l_N denotes points in basic interval. In the case where we use integers, $l_i \equiv i$ for $i = 1 \sim N$.

The DSM and its spike-train are illustrated in Fig. 1. Since the domain on the DSM consists of a finite number of integers, the DSN must generate a periodic spike-train (PST) in the steady state. For simplicity, in this paper, we consider the case where the n-th spike-position appears in the n-th interval $\tau_n \in [n-1, n)$. More detailed definition on the DSN and DSM can be found in [1,2].

For the DSM, we give basic definitions. First, a point $p \in L_N$ is said to be a periodic point (PEP) with period k if $p = f^k(p)$ and $f(p)$ to $f^k(p)$ are all different where f^k is the k-fold composition of f. One PEP corresponds to one PST. A sequence of the PEP $\{p, f(p), \cdots, f^{k-1}(p)\}$ is said to be a periodic orbit (PEO). In Fig. 1, red points denotes PEPs and the DSM generates a PEO with period 3.

Second, a point $q \in L_N$ is said to be an eventually periodic point (EPP) with step k if q is not a periodic point but falls into some periodic point p after k steps: $f^k(q) = p$. An EPP with step 1 is referred to as a direct eventually periodic point (DEPP): $f(q) = p$. An EPP corresponds to an initial spike-position of a transient spike-train to the PST. Note that if the DSM has only one PEO, all the other points are EPPs that fall into the PEO. If all the EPPs are DEPP, the PEO is said to be super-stable. In Fig. 1, black points are EPP and blue points are DEPP.

3   Feature Quantities

In order to analyze the dynamics of DSN, we define two simple feature quantities. The first feature quantity is the rate of PEPs that characterizes plentifulness of the steady states (or transient phenomena):

$$\alpha = \frac{\text{the number of PEPs}}{N}, \quad \frac{1}{N} \leq \alpha \leq 1 \tag{7}$$

The second feature quantity is related to the maximum step to PEPs that characterizes falling speed into a PEP:

$$S_{max} = \frac{\text{The maximum step of EPPs}}{N}, \quad \frac{1}{N} \leq S_{max} \leq \frac{N-1}{N} \tag{8}$$

where we assume that at least one PEO exists. If the DSM has one PEO and $S_{max} = 1/N$ then the PEO is superstable.

Using the two feature quantities α and S_{max}, we construct a feature plane as shown in Fig. 2. This plane is useful in visualization and classification of the DSN dynamics.

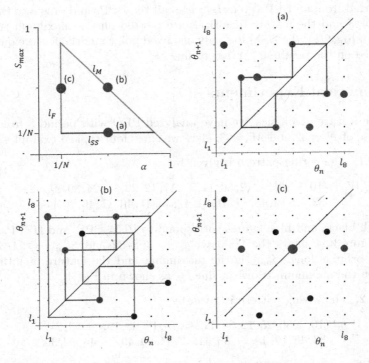

Fig. 2. Feature plane and examples of DSMs. (a) $\alpha = 4/8$, $S_{max} = 1/8$. (b) $\alpha = 4/8$, $S_{max} = 4/8$. (c) $\alpha = 1/8$, $S_{max} = 4/8$.

In the feature plane, we define three line segments that can be criteria of the analysis. The first one is the superstable line on which $S_{max} = 1$ and all the PEOs are superstable:

$$l_{SS}: \ \alpha = \frac{1}{N}, \ \frac{1}{N} \leq S_{max} \leq \frac{N-1}{N} \tag{9}$$

The second one is the maximum transient line on which the length of the transient steps is the maximum, i.e., there exists some orbit that visits all the EPPs and falls into a PEP.

$$l_M: \ S_{max} = -\alpha + 1, \ \frac{1}{N} \leq \alpha \leq 1 \tag{10}$$

The third one is the fixed point line on which the DSM has only one fixed point.

$$l_F: \ S_{max} = \frac{1}{N}, \ \frac{1}{N} \leq \alpha \leq \frac{N-1}{N} \tag{11}$$

Figure 2 shows typical examples of DSMs for $N = 8$. In Fig. 2(a), the DSM has one PEO with period 4 and $\alpha = 4/8$. All the EPPs are DEPPs and the PEO is superstable. $S_{max} = 1/8$ and the feature quantities are plotted on the superstable line l_{SS}. In Fig. 2(b), the DSM has one PEO with period 4. An orbit started from an EPP (point l_8) visits all the EPPs and falls into the PSO. $S_{max} = 4/8$ and the feature quantities are plotted on the maximum transient line l_M. In Fig. 2(c), the DSM has only one fixed point and the feature quantities are plotted on the maximum transient line l_F.

4   Numerical Experiments

Using the feature quantities, we have analyzed PEO with period 8 from DSNs in the case of $N = 32$ and $M = 63$. Here we show four typical examples.

Example 1. The wiring pattern is given by

$$\begin{aligned} W = (&10, 3, 30, 25, 12, 36, 11, 37, 17, 12, 23, 19, 24, 39, 21, 42, \\ &33, 42, 19, 48, 24, 31, 31, 47, 45, 41, 49, 47, 46, 29, 36, 42) \end{aligned} \tag{12}$$

where red, black, and blue figures correspond to PEP, EPP, and DEPP, respectively. Figure 3(a) shows the DSM where $\alpha = 8/32$ and $S_{max} = 24/32$. The transient orbit is long, S_{max} is the maximum, and the feature quantities are plotted on the maximum transient line l_M as shown in Fig. 4.

Example 2. The wiring pattern is given by

$$\begin{aligned} W = (&10, 3, 30, 25, 12, 36, 11, 37, 10, 12, 23, 19, 24, 39, 21, 45, \\ &33, 42, 19, 48, 24, 36, 41, 47, 45, 29, 49, 47, 46, 29, 36, 42) \end{aligned} \tag{13}$$

Figure 3(b) shows the DSM where $\alpha = 8/32$ and $S_{max} = 14/32$. The feature quantities are plotted between maximum transient line and super-stable line as shown in Fig. 4.

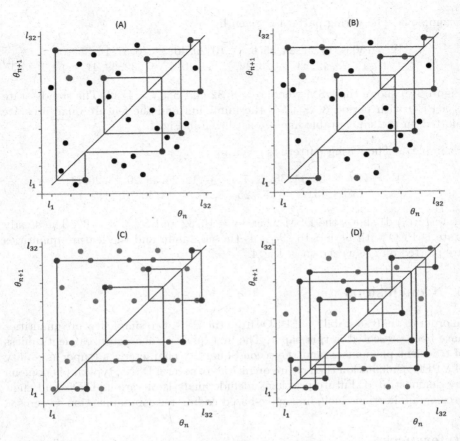

Fig. 3. Typical DSMs (A) $\alpha = 8/32$, $S_{max} = 24/32$. (B)$\alpha = 8/32$, $S_{max} = 14/32$. (C)$\alpha = 8/32$, $S_{max} = 1/32$. (D)$\alpha = 16/32$, $S_{max} = 1/32$.

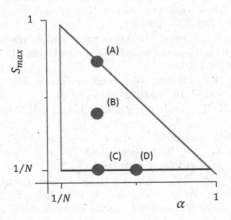

Fig. 4. Feature plane.

Example 3. The wiring pattern is given by

$$W = (30, 3, 10, 27, 24, 35, 6, 37, 10, 20, 40, 19, 20, 37, 14, 35,$$
$$18, 28, 42, 23, 24, 23, 22, 47, 44, 49, 47, 47, 32, 29, 41, 42) \tag{14}$$

Figure 3(c) shows the DSM where $\alpha = 8/32$ and $S_{max} = 1/32$. The steady state is a PEO with period 8. S_{max} is the minimum and the feature quantities are plotted on the super-stable line l_{SS} as shown in Fig. 4.

Example 4. The wiring pattern is given by

$$W = (15, 26, 29, 11, 35, 11, 7, 19, 37, 12, 25, 38, 20, 23, 26, 21,$$
$$43, 18, 43, 27, 39, 52, 25, 28, 39, 56, 54, 40, 53, 31, 36) \tag{15}$$

Figure 3(d) shows the DSM where $\alpha = 16/32$ and $S_{max} = 1/32$. The steady state is PEO with period 16. S_{max} is the minimum and the feature quantities are plotted on l_{SS} as shown in Fig. 4.

5   Conclusions

In order to analyze stability of PEOs from the DSN, two simple feature quantities have been presented in this paper. The first quantity characterizes plentifulness of transient phenomena and the second quantity characterizes (super) stability of a PEO. Calculation the feature quantities of several DSNs, typical phenomena are demonstrated. Future problems include analysis of various DSNs and engineering applications including spike-based coding for communication systems.

References

1. Torikai, H., Hamanaka, H., Saito, T.: Reconfigurable spiking neuron and its pulse-coupled networks: basic characteristics and potential applications. IEEE Trans. Circ. Syst. II **53**(8), 734–738 (2006)
2. Iguchi, T., Hirata, A., Torikai, H.: Theoretical and heuristic synthesis of digital spiking neurons for spike-pattern-division multiplexing. IEICE Trans. Fundam. **E93–A**(8), 1486–1496 (2010)
3. Torikai, H., Saito, T., Schwarz, W.: Synchronization via multiplex pulse-train. IEEE Trans. Circ. Syst. I **46**(9), 1072–1085 (1999)
4. Lee, G., Farhat, N.H.: The bifurcating neuron network 1. Neural Netw. **14**, 115–131 (2001)
5. Horimoto, N., Saito, T.: Digital dynamical systems of spike-trains. In: Lee, M., Hirose, A., Hou, Z.-G., Kil, R.M. (eds.) ICONIP 2013, Part II. LNCS, vol. 8227, pp. 188–195. Springer, Heidelberg (2013)
6. Yamaoka, H., Horimoto, N., Saito, T.: Basic feature quantities of digital spike maps. In: Wermter, S., Weber, C., Duch, W., Honkela, T., Koprinkova-Hristova, P., Magg, S., Palm, G., Villa, A.E.P. (eds.) ICANN 2014. LNCS, vol. 8681, pp. 73–80. Springer, Heidelberg (2014)
7. Ott, E.: Chaos in dynamical systems, Cambridge (1993)

8. Chua, L.O.: A Nonlinear Dynamics Perspective of Wolfram's New Kind of Science, I, II. World Scientific, Singapore (2005)
9. Wada, W., Kuroiwa, J., Nara, S.: Completely reproducible description of digital sound data with cellular automata. Phys. Lett. A **306**, 110–115 (2002)
10. Rosin, P.L.: Training cellular automata for image processing. IEEE Trans. Image Process. **15**(7), 2076–2087 (2006)
11. Kouzuki, R., Saito, T.: Learning of simple dynamic binary neural networks. IEICE Trans. Fundam. **E96–A**(8), 1775–1782 (2013)
12. Moriyasu, J., Saito, T.: A deep dynamic binary neural network and its application to matrix converters. In: Wermter, S., Weber, C., Duch, W., Honkela, T., Koprinkova-Hristova, P., Magg, S., Palm, G., Villa, A.E.P. (eds.) ICANN 2014. LNCS, vol. 8681, pp. 611–618. Springer, Heidelberg (2014)
13. Rulkov, N.F., Sushchik, M.M., Tsimring, L.S., Volkovskii, A.R.: Digital communication using chaotic-pulse-position modulation. IEEE Trans. Circ. Syst. I **48**(12), 1436–1444 (2001)

Letter Reproduction Simulator for Hardware Design of Cellular Neural Network Using Thin-Film Synapses

Crosspoint-Type Synapses and Simulation Algorithm

Tomoya Kameda[1(✉)], Mutsumi Kimura[1,2],
and Yasuhiko Nakashima[1]

[1] Nara Institute of Science and Technology, 8916–5, Takayama-Cho,
Ikoma, Nara 630–0192, Japan
kameda.tomoya.kg0@is.naist.jp
[2] Ryukoku University, 1-5 Yokotani, Seta Oe-Cho, Otsu 520-2194, Japan

Abstract. Recently, neural networks have been developed for variable purposes including image and voice recognitions. However, those based on only software implementation require huge amount of calculation and energy. Therefore, we are now designing a hardware with cellular neural network (CNN) that features low power, high-density, and high-functionality. In this study, we developed a CNN simulator for evaluating some letter reproduction algorithm. In this simulator, each of the neurons is just connected to neighboring neurons with surrounding synapses. Learning process is executed by modifying the strength of each connection. Particularly, we assumed to employ a-IGZO films for crosspoint-type synapses that utilize a phenomenon that the conductance changes when an electric current flows. We modeled this phenomenon and implemented it into the simulator to determine the network architecture and device parameters. In this paper, the structure, allocation method of a-IGZO and the algorithm are described. Finally, we confirmed that our cellular neural network can learn two letters. Furthermore, it was found that the estimated time for learning is around 100 h based on the current characteristic change model of a-IGZO film, and some conditions to enhance the deterioration speed of a-IGZO film should be explored.

Keywords: Cellular neural network · Letter reproduction · Low power · High-density integration · Modeling · Thin-film synapse

1 Introduction

Neural networks imitate biological neural circuit in living brains [1] and hence are promising for image processing, pattern recognition, etc. Neural networks can process many and complicated data by using lots of neurons and synapses. In human brains, there are more than 10^{10} neurons and 10^{13} synapses generally [2]. Neural networks also need astronomical number of neurons and synapses to perform required operations fast and efficiently.

© Springer International Publishing AG 2016
A. Hirose et al. (Eds.): ICONIP 2016, Part II, LNCS 9948, pp. 342–350, 2016.
DOI: 10.1007/978-3-319-46672-9_39

Today, neural networks by software have been able to recognize handwritten character [3]. Since ImageNet Large Scale Visual Recognition Challenge (ILSVRC) in 2012, neural networks are best ways for image recognition [4]. However, neural networks using software need further improvement of calculation efficiency and reduction of large amount of energy consumption. To solve these problems, neural networks are developed at large-scale integration (LSI) level or by using memristors [5, 6]. However, each neuron and synapse is made from more than ten transistors. Moreover, a neuron is connected with many synapses intricately. Therefore, such neural networks are not suitable to be integrated.

Among the neural networks, cellular neural networks (CNN) are suitable for integration of electron devices, because each neuron is connected to only neighboring neurons [7]. Further technologies on CNN preferable for higher integration have been also developed from the viewpoint of device hardware [8]. To make CNN at the circuit level, we are evaluating CNN by using a simulator. In this paper, an architecture of our CNN and characteristic of a-IGZO, which used as synapses, are presented in Sect. 2. In Sect. 3, I explain a letter reproduction simulator. In Sect. 4, after we confirm that this cellular neural network can reproduce two letters and that the simulator can model the deterioration of the synapse of the a-IGZO film correctly, we point out that the learning takes about 100 h when we use this a-IGZO and therefore we have to find some conditions to enhance the deterioration speed of the a-IGZO film. Finally, we conclude this result in Sect. 5.

2 Cellular Neural Network

2.1 Network Architecture

Figure 1 shows the network architecture of the CNN used for our simulator. In this architecture, we arrayed multiple neurons and connected each neuron to eight neurons through synapses. The neurons are in a state of either firing or non-firing according to the states of the adjacent neurons and the connection strength of the synapses placed between the neurons. When both neurons connected to the synapse are in firing states, the synaptic connection strength doesn't change. On the contrary, when the neurons connected to the synapse are in different states, the synaptic connection strength decreases. Such deterioration of the synapses corresponds to the learning of the CNN. The mathematical expressions are written in the prior articles [9].

2.2 Hardware System

Figure 2 shows the hardware system of our CNN. In this time, a field-programmable gate array (FPGA) works as neurons and the a-IGZO films work as synapses. When this CNN is made learn some functions, the states of firing or non-firing are inputted to the neurons, and some synapses of a-IGZO films deteriorate as mentioned later. After the learning, we input some states to the neurons, and if the a-IGZO films are sufficiently deteriorated, the correct function is realized by the CNN.

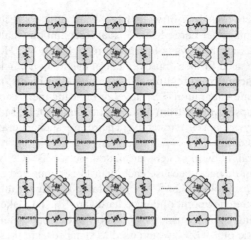

Fig. 1. Network architecture of the CNN used for our simulator

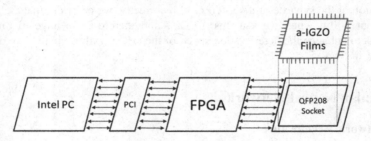

Fig. 2. Hardware system for our CNN

2.3 Synapse Modeling

We assumed to use a-IGZO films as synapses. Nowadays, a-IGZO films are used for channel layers of field-effect transisters [10]. The conductance of the a-IGZO films changes when an electric current flows [11], if they are made using some fabrication process. Figure 3(a) shows the actual device of our crosspoint-type synapses made from a-IGZO films. Eighty horizontal electrodes and eighty vertical electrodes are integrated to form 80 × 80 crossbars, and a-IGZO films are sandwitched between the electrodes at each crosspoint. These a-IGZO thin films are deposited using RF magnetron sputtering in working gas of Ar on a quartz substrate, and the a-IGZO film thickness is 70 nm. Figure 3(b) shows the structure of the crosspoint-type synapses. The red collums are a-IGZO films forming the synapses. The green and blue lines are the horizontal and vertical electrodes, respectively. The width of both electrodes is 150 um.

Each electrode corresponds to each neuron in the FPGA. When the neuron is in the firing or non-firing states, the voltage of 3.3[V] or 0[V] is applied to the electrode, respectively. We evaluated the deterioration speed of the crosspoint-type a-IGZO film

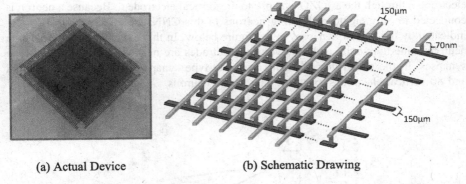

(a) Actual Device   (b) Schematic Drawing

Fig. 3. Crosspoint-type synapses made form a-IGZO films (Color figure online)

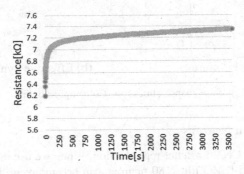

Fig. 4. Deteritoration phenomenon of the crosspoint-type a-IGZO Films

when a voltage of 3.3 [V] is applied. Figure 4 is the result of the deterioration phenomenon. It takes about two minutes for the resistance to increase 1 kΩ. The resistance increases curvely, and increasing rate decrease. After 250 s, the resistance increases almost linearly. We thought that this device is suitable for synapses because synapses of connection strength should be changed continuously. This is because a-IGZO films consist of four atomic elements and the electric characteristics can be relatively freely controlled.

2.4 Synapse Allocation

Figure 5 shows how neurons and synapses are allocated to the crosspoint-type synapses. Here, the CNN has the size of 3 × 3 as an example. We allocated each neuron to each horizontal and vertical electrodes. For example, if the neuron 3 is in the firing state, a voltage of 3.3 V is applied to the horizontal and vertical electrodes 3 from FPGA. If the neuron 6 is in the non-firing state, a ground voltage is applied to the horizontal and vertical electrodes 6. Then, an electric current flows from the horizontal electrode 3 through the a-IGZO synapse to the vertical electrode 6. Because a neuron is connected to only eight neighboring neurons in this CNN, the effective synapses are indicated by the yellow synapses in the figure below. In this example, to connect nine neurons, nine horizontal and nine vertical electrodes are needed in the crospoint-type synapses. In the actual device of the crospoint-type synapses, because 80 horizontal and 80 vertical electrodes exist, we can use 80 neurons.

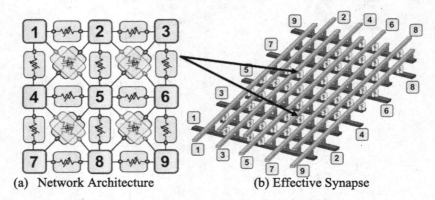

(a) Network Architecture   (b) Effective Synapse

Fig. 5. Synapse allocation (Color figure online)

2.5 I/O and Hidden Neurons

In this time, we use a CNN for letter reproduction. When we use these crosspoint-type synapses made from a-IGZO films, 80 neurons can be employed. Figure 6 shows the I/O and hidden neurons. We put a hidden neuron between each pair of the adjacent I/O neurons. Therefore, we can input number images converted to the binary format of 4×4 pixels. All neurons are indicated by 80 squares, 4×4 matrix neurons surrounded by bold lines are I/O neurons, and other neurons are hidden neurons. In the

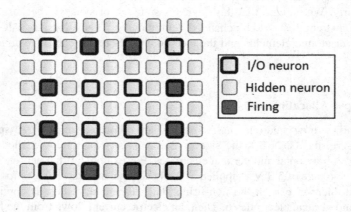

Fig. 6. I/O and hidden neurons (Color figure online)

learning stage, a pixel pattern corresponding to an input letter is inputted to I/O neurons as a firing states. For example, when an input letter of "0" is inputted, a pixel pattern indicated by the red square is inputted.

3 Letter Reproduction Simulator

Figure 7 shows the simulation algorithm of the letter reproduction simulator. At first, this simulator choose a letter and input it to a CNN. Next, this simulator calculates the states of either firing or non-firing of all the neurons. The hidden neurons come to be in the firing or non-firing states obeying the majority rule of the states of the neurons surrounding the hidden neurons with consideration of the connection strength of the synapses between the hidden neuron and the surrounding neuron. After that, synapses are deteriorated, namely, the resistances of some synapses increase. Next, this simulator repeat the learning of other letters and finish the first learning. After the learning, this simulator try to reproduce the learned letters from input letters slightly distorted from the learned letters. In the same way as learning, this simulator chooses a distorted letter and input it in the CNN having deteriorated synapses. When the all letters can be reproduced, this simulation finishes.

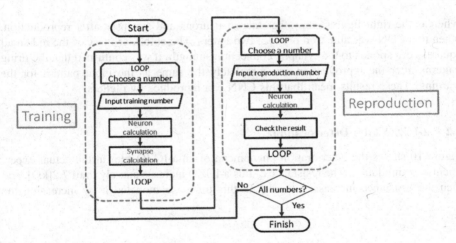

Fig. 7. Simulation algorithm of the letter reproduction simulator

4 Simulation Results

4.1 Letter Reproduction

Figure 8 shows the input pattern for the learning and that for the reproduction slightly distorted form the former. Figure 9 shows the simulation results of the letter reproduction. The left figures show the states of neurons and synapses after learning,

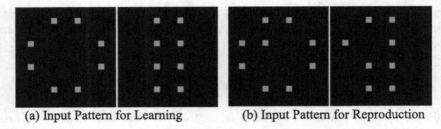

(a) Input Pattern for Learning   (b) Input Pattern for Reproduction

Fig. 8. Input pattern for learning and reproduction (Color figure online)

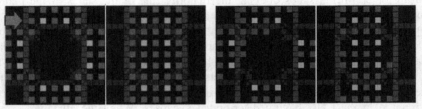

(a) Neurons and synapses after Learning (b) Neurons and synapses after reproduction

Fig. 9. Simulation results of the letter reproduction (Color figure online)

whereas the right figures show the state of neurons and synapses after reproduction, when this CNN was able to reproduce two letters. The color deepness of the red small squares corresponds to the synaptic connection strength. It was confirmed that the firing patterns after the reproduction are completely the same as the input pattern for the learning. These results mean that this CNN can reproduce two letters.

4.2   a-IGZO Film Deterioration

Figure 10 shows the deterioration phenomena of an a-IGZO film in the actual experiment and simulation. The resistance of this a-IGZO increase curvely until 7.2[kΩ], and then the resistance increases linearly. In this time, we modeled it by increasing the

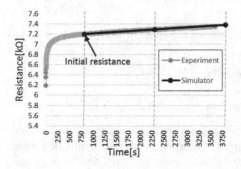

Fig. 10. Deterioration of an a-IGZO Film in the actual experiment and simulation.

resistance linearly. Hence, in this simulator, initial resistance is 7.2[kΩ], and the resistance increases about 0.1[kΩ] in 1500 s. In any case, it is confirmed that the simulator can model the deterioration of the synapse of the a-IGZO film correctly.

4.3   Evaluation Result

In this simulation, we recorded the resistance of the synapses that increases most, which is pointed by the arrow in Fig. 9. The deterioration of a synapse is shown in Fig. 11. When this resistance increased from 7.2 kΩ to 17.9 kΩ, this CNN was able to reproduce two letters. This result indicates that it takes about 100 h, when we use this a-IGZO film. Therefore, we have to find some conditions to enhance the deterioration speed of the a-IGZO film.

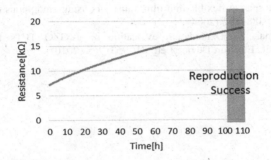

Fig. 11. Deterioration of a synapse that increases most

5   Conclusion

We developed a CNN system using FPGA as neurons and a-IGZO films as synapses and the letter reproduction simulator. We modeled the deterioration phenomenon of the a-IGZO films and implemented it into the simulator. It is confirmed using the simulator that this CNN can reproduce two characters. Moreover, we evaluated the time necessary for the learning two characters and we came to a conclusion that we have to enhance the deterioration speed of the a-IGZO films, because long time is needed based on the present deterioration phenomena of the a-IGZO films. In the future, because a-IGZO films can be fabricated using a printing process, stacked integrated CNN systems will be used as three-dimensional brain-like computers.

References

1. Dayhoff, J.E.: Neural Network Architectures, An Introduction. Van Nostrand Reinhold, New York (1990)
2. Mountcastle, V.B.: The Cerebral Cortex. Harvard University Press, Cambridge and London (1998)

3. LeCun, Y., Bottou, L., Bengio, Y., Haffner, P.: Gradient-based learning applied to document recognition. Proc. IEEE **86**, 2278–2324 (1998)
4. Krizhevsky, A., Sutskever, I., Hinton, G.E.: ImageNet classification with deep convolutional neural networks. In: NIPS 2012 (2012)
5. Mead, C.: Analog VLSI and Neural Systems. Addison-Wesley Publishing, Boston (1989)
6. Prezioso, M.: Training and operation of an integrated neuromorphic network based on metal-oxide memristors. Nature **521**, 61–64 (2015)
7. Chua, L.O.: Cellular neural networks: theory. IEEE Trans. Circuites Syst. **32**, 1257–1272 (1988)
8. Kimura, M.: Neural network using thin-film transistors -working confirmation of asymmetric circuit-. IEICE Technical report, SDM1012-123, 47-52 (2012)
9. Kimura, M., Miyatani, T., Fujita, Y., Kasakawa, T.: Apoptotic self-organized electronic device using thin-film transistors for artificial neural networks with unsupervised learning functions. Jpn. J. Appl. Phys. **54**, 03CB02 (2015)
10. Nomura, K., Ohta, H., Takagi, A., Kamiya, T., Hirano, M., Hosono, H.: Room-temperature fabrication of transparent flexible thin-film transistors using amorphous oxide semiconductors. Nature **432**, 488–492 (2004)
11. Kimura, M., Imai, S.: Degradation evaluation of a-IGZO TFTs for application to AM-OLEDs. IEEE Electron Device Lett. **31**, 963–965 (2010)

Sensory Perception

An Analysis of Current Source Density Profiles Activated by Local Stimulation in the Mouse Auditory Cortex in Vitro

Daiki Yamamura, Sano Ayaka, and Takashi Tateno[✉]

Bioengineering and Bioinformatics, Graduate School of Information Science and Technology, Hokkaido University, Kita 14, Nishi 9, Kita-ku, Sapporo 060-0814, Japan
{Yamamura_Daiki,Sano_Ayaka,tateno}@ist.hokudai.ac.jp

Abstract. To examine microcircuit properties of the mouse auditory cortex (AC) in vitro, we extracellularly recorded spatiotemporal laminar profiles driven by short electric microstimulation on a planar multielectrode array (MEA) substrate. The recorded local field potentials (LFPs) were subsequently evaluated using current source density (CSD) analysis to identify sources and sinks. Current sinks are thought to be an indicator of net synaptic current in a small volume of cortex surrounding the recording site. Thus, CSD analysis combined with MEAs enabled us to compare mean synaptic activity in response to current stimuli on a layer-by-layer basis. Here, we used senescence-accelerated mice (SAM), some strains of which show age-related hearing loss, to examine characteristic spatiotemporal CSD patterns stimulated by electrodes in specific cortical layers. Thus, the CSD patterns were classified into several clusters based on the stimulation sites in the cortical layers. We also found, in a reduced space obtained by principle component analysis, some CSD pattern differences between the two SAM strains in terms of aging and stimulation layers. Finally, on the basis of these results, we discuss the effects of aging on AC microcircuit properties.

Keywords: Auditory cortex · Current source density · Hierarchical clustering · Multielectrode array · Principal component analysis · Senescence-accelerated mice

1 Introduction

Senescence-accelerated mice (SAM) are a model of accelerated senescence and are utilized worldwide as a biogerontological resource in aging research [1]. SAM consist of senescence-accelerated-prone (SAMP) and -resistant mice (SAMR). In animal models of presbycusis, SAMP1 are suitable due to earlier onset of progressive hearing loss compared to SAMR1 [2]. In particular, as assessed by auditory brainstem responses (ABRs) and compared to SAMR1, SAMP1 showed age-related hearing loss manifested as an elevated hearing threshold, prolongation of interpeak intervals of waves I-III and I-IV of ABRs, and decreased amplitude of wave I [3]. Therefore, SAMP1 shows more rapid and severe auditory loss with age than SAMR1. However, although changes and differences in the auditory periphery have been

© Springer International Publishing AG 2016
A. Hirose et al. (Eds.): ICONIP 2016, Part II, LNCS 9948, pp. 353–362, 2016.
DOI: 10.1007/978-3-319-46672-9_40

reported between the two SAM strains with aging [4], little is known about these factors in the auditory central nervous system (CNS) of SAM.

Cortical local field potential (LFP) is an important and observable measure of CNS neural activity. The LFP captures key integrative synaptic processes that cannot be measured by observing the spike activity of a few neurons alone. That is, LFPs are generated by electrical currents flowing across cell membranes together with population spikes, and they provide complementary measures of ensemble neural dynamics at both the input and output levels [5]. Several studies have used LFPs to examine cortical network mechanisms involved in visual [6] and auditory [7] information processing. In addition, because LFPs are more stably obtained in chronic recordings than spikes, they are promising signals for steering neuroprosthetic devices and monitoring neural activity over longer periods [8].

In this study, to increase experimental controllability and recording stability, we recorded LFP responses to short-current local stimuli in the mouse auditory cortex (AC) in vitro. LFPs were recorded with a planar multielectrode array (MEA) substrate [9], which allowed us to sample laminar and columnar structures simultaneously at 64 sites at 150-μm intervals. LFPs were subjected to a current source density (CSD) analysis to identify sources and sinks [10]. Current sinks are thought to be an indicator of net synaptic current in a small volume of cortex surrounding the recording site. Thus, CSD analysis combined with MEAs enabled us to compare mean synaptic activity in response to current stimuli on a layer-by-layer basis. In particular, we examined the following questions: (i) does simple local stimulation at a specific cortical layer induce characteristic spatiotemporal CSD patterns; (ii) can evoked spatiotemporal patterns be classified in a manner that depends on stimulus-driven layers; and (iii) are there any response changes and differences among the two SAM strains with aging? Finally, on the basis of the results, we discuss AC microcircuit properties in the context of aging.

2   Materials and Methods

2.1   Experimental Procedures

All animal experiments described below were carried out in accordance with the National Institutes of Health Guidelines for the Care and Use of Laboratory Animals and with approval of the Institutional Animal Care and Use Committee of Hokkaido University. In this study, we used C57BL/6 J mice aged 10 to 13 weeks (Japan SLC Inc., Japan) and SAM aged over 10 weeks (Japan SLC Inc., Japan). SAM strains contained both SAMR1 and SAMP1. Previous studies that assessed mouse hearing impairment through ABR recording reported that the SAMP1 strain showed age-associated hearing loss at ages over 12 months (52 weeks) [11]. In this study, we used two younger groups in SAM (SAM group A, 12.5 ± 1.2 weeks; SAM group B, 16.4 ± 0.8 weeks) and one older group (SAM group C, 51.6 ± 0.6 weeks), which including both SAMR1 and SAMP1 mice, as well as younger C57BL/6 J mice (C56BL/6 J group A; 12.2 ± 0.5 weeks).

To record LFPs from mouse brain slices in vitro, we prepared 400-μm-thick coronal slices that included the primary AC (A1). During the slicing process, the distance of each coronal section from bregma along the rostral/caudal axis was defined in the following way. A digitized atlas [12] provided drawings of coronal sections through the mouse brain. For coronal slices, we chose the coronal section that best matched the drawing at 2.70–2.80 mm caudal to bregma (Figs. 53 and 54 in the atlas). In slicing a brain block that included the cortex, chilled artificial cerebrospinal fluid (ACSF) saturated with 95 % O_2 and 5 % CO_2 mixed gas was prepared. The ACSF contained (in mM) 119 NaCl, 26.2 $NaHCO_3$, 2.5 KCl, 2.5 $CaCl_2$, 1.3 $MgSO_4$, 1.0 NaH_2PO_4, and 11.0 D-glucose (pH = 7.4). Each mouse was deeply anesthetized with halothane and decapitated. Slices were then cut with a tissue slicer (Linear Slicer Pro7, D.S.K., Japan) in the chilled ACSF. The slices were recovered in a submerged-type holding chamber at 28 °C in a water bath for at least 2.5 h before electrophysiological recordings.

All electrophysiological recordings in brain slices were performed with ACSF perfusion, and the mixed gas was supplied from the top of the recording chamber in an incubator (APC-30, Asteck Co., Japan) maintained at 28.0 °C. During the recording, slices on an MEA substrate (MED-P515A, Alpha MED Scientific, Japan) were covered with a nylon mesh and a stainless steel slice anchor. Each of the 64 recording sites in the array covered an area of 50×50 μm^2, and the distance between the centers of adjacent sites was 150 μm. The submerged recording chamber was continuously circulated with aerated ACSF at 28 °C at a rate of 2 mL/min. In a typical recording, the recording electrode sites were located at layer 2/3 (L2/3), L4, or L5 in the AC. In localizing the AC, the hippocampus and the rhinal fissure were used as landmarks [12].

For electrical stimulation, a single stimulation electrode site at L2/3, L4 or L5 was selected, unless specifically mentioned. Throughout the experiments described here, we used fixed, small stimulus current intensities ranging from 15 to 20 μA. The intensity was usually 20–30 % of the saturation level in the input-output (IO) relationship between input current intensities and output evoked LFPs. In controls, after selecting a single stimulation site in one of the three layers (L2/3, L4, and L5) and confirming the stability of the responses, we applied a test stimulus (a single 200-μs-width bipolar pulse; 100 μs at positive current, followed by 100 μs at negative current) every 20 to 30 s for 10 to 20 min. Evoked LFP responses in each slice were recorded simultaneously at a sampling rate of 20 kHz, and the signals were filtered in a range of frequencies between 100 Hz and 10 kHz. The locations of all recording positions on the multielectrode substrate were digitally imaged before and after recording. After each experiment, the laminar positions were histologically determined by standard Nissl staining with cresyl violet.

2.2 Data Analysis

To detect characteristic spatiotemporal patterns in neural activity through laminarly organized networks in the AC, standard CSD analysis was used because evoked current sinks and sources can be easily identified [13]. In calculating CSDs from LFP data recorded at multiple sites, we used the standard CSD method; i.e., the second spatial derivatives of LFPs were replaced with the corresponding spatial differences, and estimates of the CSDs were calculated (see below). Here, we assumed constant extracellular tissue conductivity and simultaneous nervous tissue activation along the measurement axis. In addition, before the CSD analysis, we used a cubic spline data interpolation technique to estimate three more points between the two original data points obtained experimentally. Thus, the spatial resolution at successive points became finer, and the interval between the points was 37.5 μm. Briefly, simple one-dimensional CSD analysis was performed by LFP values at 29 points along the vertical axis, which is perpendicular to the cortical surface and layers of the A1. Along the vertical "on-line" stimulation sites, we applied the one-dimensional CSD analysis to the estimated LFPs at each of the points and obtained the CSD using the formula:

$$\text{CSD}(z, t) \approx -\sigma[\varphi(z - \Delta z, t) - 2\varphi(z, t) + \varphi(z + \Delta z, t)] \big/ (\Delta z)^2 \tag{1}$$

where z is the spatial coordinate in the laminar direction, t is time, σ is the conductivity of the extracellular space, φ is the LFP at z, and Δz is the inter-point interval (i.e., $\Delta z = 37.5$ μm). Notice that a two-dimensional CSD analysis and a more advanced CSD analysis [14] can provide similar CSD values at the corresponding points as the one-dimensional analysis, if the LFP values horizontally adjacent to the target point at z are similar as that at the target point. This was in fact most often the case for the data in the present study, so we used the simpler conventional one-dimensional CSD analysis to reduce the calculation time and cost. Furthermore, for simplicity and regardless of recording sites, we assumed that σ was uniform over the space. In the experiments, we repeated the same trials 10 to 20 times in the identical condition, so that the CSD values represent the average over the trials.

To classify all spatiotemporal CSD patterns evoked at stimulation sites in the three cortical layers (L2/3, L4, and L5) into several clusters, we considered the individual patterns to be the corresponding vectors located at a multidimensional space. To achieve this end, we first constructed a CSD matrix for each pattern, as follows. When electrical stimulation was applied at an electrode site in layer k (k = L2/3, L4, or L5), we denoted a CSD matrix by $D_k(m, n)$, defined as $\text{CSD}(z = m\Delta z, t = n\Delta t)$, where the variable z denotes the recording position, t time, Δz the inter-point interval, Δt a sampling-time interval, and both m and n integers ($m = 1, \ldots, 29; n = n_0, \ldots, N$). Then, we denoted a labeled CSD vector by V_k, which can be obtained as a vector of all the elements of $D_k(m, n)$ with the label k, which represents the specific layer in which the stimulation site is located.

Once we obtained a spatiotemporal CSD pattern as a vector variable V_k, the next step was to apply a pattern classification method. Here, we used a hierarchical clustering method [15]. To apply the clustering method, we constructed a set of unlabeled sample vectors from all vectors V_k obtained in our experiments. To perform the clustering method, we followed the next three steps [15]. (i) To find the similarity between every pair of vectors, the Euclidean distance was calculated. (ii) Vector pairs that were in close proximity were linked to each other, and a binary, hierarchical cluster tree (dendrogram) was constructed. (iii) Finally, a branch point was determined in the tree, so that all the vectors bellow the branch point in the direction of singleton clusters were grouped into individually identical clusters. In this study, we always classified all vectors into three groups, because the stimulation sites comprised three cortical layer groups (L2/3, L4, and L5). However, after the classification, there is arbitrariness regarding the correspondence of each classified group to one of the three stimulation layers (L2/3, L4, or L5). Therefore, to associate the classified groups with the stimulation layers, we referred the original stimulation sites.

To represent the pattern of similarity of the vectors V_k and to compare the obtained groups with the originally labeled groups by the hierarchical clustering method, we computed a three-dimensional subspace using principal component analysis (PCA) and constructed the corresponding vectors in the reduced dimensional space by projecting all the original vectors onto the three principal components in the PCA space [16]. As usual in the PCA analysis, we calculated the contribution ratios, which represent the importance of each component in the data set. In addition, on the basis of the clustering results, we calculated the accuracy rate (AR), which was defined as the ratio between the number of correct answers and that of all the vectors V_k.

The present study is based on data from experiments in 82 AC slices from 40 animals: six C57BL/6 J, 17 SAM-P1, and 17 SAM-R1 mice. Data are given as mean ± standard errors of the mean (SEM), and measures such as means were compared using a t-test, with P < 0.05 assumed to be significant. In addition, Bonferroni's multiple comparison test was performed, if necessary, using Matlab (MathWorks, USA), and values of P less than 0.05 were considered statistically significant.

3   Results

3.1   Spatiotemporal LFP and CSD Patterns Activated by Local Stimulation

For all AC slices, large negative LFP responses in L2/3 with short latency were evoked by stimulation at single sites in all layers (L2/3, L4 and L5). In all the strains, the maximum amplitudes of the LFPs driven in L4 were always observed at the recording sites in L2/3 (Fig. 1), and the peak amplitudes of the younger mouse strain groups were not significantly different ($P > 0.10$ by Bonferroni's multiple comparison test): 0.710 ± 0.043 mV in C57BL/6 J, 0.830 ± 0.052 mV in SAMR1, and 0.920 ± 0.058 mV

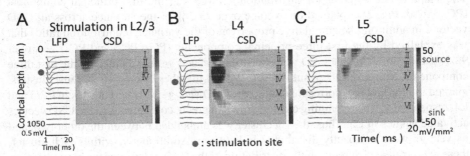

Fig. 1. For SAMP1, typical examples of LFP and CSD profiles driven at different sites in the three layers: L2/3 in A, L4 in B, and L5 in C. A coronal AC slice including the A1 overlaid on the MEA substrate of 64 recording sites, and 8 electrodes of "on-line" sites are selected to show the spatiotemporal profiles. Two more points between the recording sites are interpolated to obtain a higher spatial resolution. In the profiles, the stimulation sites are indicated by red circles. (Color figure online)

in SAMP1. In addition, the corresponding latencies were also not significantly different ($P > 0.10$ by Bonferroni's multiple comparison test): 4.20 ± 0.10 ms in C57BL/6 J, 3.90 ± 0.08 ms in SAMR1, and 3.80 ± 0.10 ms in SAMP1. In contrast, the same stimulation in L4 evoked small positive field responses in L1 and L5/6 with longer latencies.

In response to stimulation in L2/3, we obtained CSD patterns in which a sink and source existed in the supragranular and granular/infragranular layers, respectively (Fig. 1A), with current flow from the source to the sink. Based on the color map patterns (Fig. 1), we could not find any qualitative differences among the three mouse strains, so we performed a quantitative analysis of this point as described in the next subsection. In contrast, after the onset of the stimulation in L4, during a period from 2 to 5 ms (referred to as "earlier phase" hereafter), we could see source-sink-source triplet CSD profiles, which are reminiscent of the profiles observed in vivo recordings of the A1 in response to acoustic sound stimulation, both in mice [17] and rats [18]. After the triplet appeared in the CSD patterns in all mouse strains, a pair of small sink and source was observed over a period of 5 ms (called "later phase") from the stimulus onset. Moreover, the CSD patterns observed in response to stimulation in L5 during the early phase were similar to those in response to stimulation in L4, and extended current sources from L4 to L5/6 could be seen (Fig. 1). In the case of stimulation in L5, however, no sink and source pair was clearly observed in the later phase. Thus, in the AC slices, changing the stimulation layers caused the different CSD patterns described above, indicating that laminar-dependent stimulation can provide a particular combination of current sources and sinks. The results were reflected by the inhomogeneous laminar structure of neural networks in the AC slices. However, no differences among the mouse strains were prominent when only qualitatively comparing the CSD patterns.

3.2   CSD Pattern Classification

Next, we quantitatively analyzed the CSD patterns, and the clustering results of the younger mouse groups (categorized as group A) in the three strains are shown in Fig. 2. The shape of the CSD vector distribution showed that the first principal component (PC_1 axis in Fig. 2) explained on average over 73.2 % of the variance across all CSD vectors. In addition, the cumulative proportion of the variance (contribution ratio) that was explained by the first three principal components (PC_1, PC_2, and PC_3) was over 92.5 %, indicating that the CSD vector distribution was well explained by the first three components. Moreover, the results in Fig. 2 show that there were two extreme clusters, marked in blue and red, corresponding to the CSD vectors activated at sites in L2/3 and L4, respectively. The remaining cluster, drawn in green, corresponds to the CSD vectors activated at sites in L5. Thus, the L5 cluster was embedded between the other two, and its vectors were compactly distributed in between. Additionally, within each cluster, there was no clear difference in the distribution of the CSD patterns of the mouse strains. However, some SAMR1 CSD vectors stimulated in L2/3 and L4 were isolated from the centroids of the highly dense vectors.

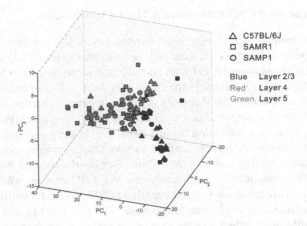

Fig. 2. Classification of CSD patterns. Each sample point corresponds to a CSD vector reformed from one CSD pattern matrix. After the hierarchical clustering and the PCA analysis, all CSD vectors in group A are represented in the reduced 3D space of the first three principal components (PC$_1$, PC$_2$, and PC$_3$). The CSD samples of C57BL/6 J, SAMPR1, and SAMP1 are respectively illustrated by triangles (△), squares (□), and circles (○). The three colors represent the CSD patterns activated at sites in the three layers. (Color figure online)

When the estimated stimulation layers of the clustered CSD patterns were compared with the corresponding original stimulation layers, the overall AR of correctly estimating the stimulation layers from the CSD pattern classification was 84.5 %. It was more difficult to estimate the correct stimulation layer based on the CSD patterns from stimulation of layer 2/3 in the C57BL/6 J mice (AR = 66.7 %) than from stimulation of the two other layers (AR = 93.3 % in L4 and 90.0 % in L5); this was also the case with the two other strains (AR = 75.0 % and 100 % in L2/3 of SAMR1 and SAMP1, respectively).

To identify differences between the younger and older groups, we analyzed the distribution of new CSD vectors (categorized as groups B and C) in the reduced PCA space, vectors that were not included in the PCA analysis mentioned above (see Materials). We specifically focused on only the CSD patterns stimulated at sites in L4. The results showed that in the reduced PCA space, the CSD vectors of the older group C were more compactly concentrated in the high density area than those of the younger group B (data not shown here). Moreover, the variance of older SAMP1 CSD vectors was smaller than that of older SAMR1 CSD vectors ($P < 0.05$ by t-test), suggesting that hearing loss may reduce the variability of the cortical CSD patterns activated in L4.

4   Discussion

In vivo, the A1 receives its major ascending projection from the medial geniculate body (MGB) projecting to A1 L4 [19]. Functionally, the A1 contains a tonotopic map of the cochlea, and specific regions of the A1 may be specialized for processing each frequency and its combinations. Inputs from the contralateral AC and non-auditory inputs impinge on L2 and L6 in the A1 with descending and intracortical outputs from A1 L5 [20, 21].

4.1   Characteristics of CSD Patterns and Their Classification

In this study, to examine stimulation-layer dependency to CSD response profiles, we applied short bipolar-pulse stimuli to L2/3, L4, or L5 in the mouse AC slices. Our results of the L4 stimulation were consistent with cortical laminar responses to MGB stimulation in mouse AC slices [17], although the slices in this study did not include cell bodies in the MGB. With respect to neural activity patterns, in general, current sinks are interpreted to indicate depolarizing events such as active excitatory synaptic populations and axonal depolarization, i.e., indicating inward, usually excitatory, synaptic current flow [22, 23]. Current sources, in contrast, reflect passive return currents most of the time, indicating outward, usually passive, current flow. Our results showed that in the early phase, L2/3 stimulation induced the source-sink pair CSD profile, while L4 and L5 stimulation resulted in the source-sink-source triplet CSD profile. In addition, in the late phase, only L4 stimulation evoked another source-sink pair CSD profile. The characteristic CSD profiles may reflect a serial loop of the granular, supragranular, and infragranular layers, as a reverberating circuit, which allows the AC to control its own excitability.

As stated above, layer-dependent stimulation of the AC slices induced characteristic CSD laminar profiles, and the CSD patterns were characterized by specific combinations of current sources and sinks. Therefore, we used the simple pattern classification method to classify the CSD patterns and estimate the stimulation layers based on unlabeled (unknown) CSD patterns with high accuracy. In the future, to develop neural prostheses applied to the AC, understanding the spatiotemporal LFP patterns driven by electrical stimulation will be important in directly encoding sound information. Therefore, similarities and differences in activated responses between natural sound and electrical stimuli must also be examined to realize such devices in practical ways. To achieve this end, our results are a very first step in exploring evoked cortical laminar profiles in the AC through local electrical stimulation in vitro in a layer-dependent manner, before applying the similar stimulation techniques in vivo.

4.2   Response Differences Among SAM Strains

The SAMP1 strain manifested a neurobiological phenotype of hearing impairment, an age-associated pathology [1–3]. Moreover, morphological studies showed an age-related decrease in both cell density and the size of spiral ganglion neurons in SAMP1 and SAMR1 [3]. An age-related loss of the inner and outer hair cells was also observed in both SAMP1 and SAMR1. In addition, the changes in the spiral ganglion neurons and hair cells appeared earlier and progressed more rapidly in SAMP1 than in SAMR1. Furthermore, SAMP1 showed greater age-related atrophy of the stria vascularis than did SAMR1. These results suggest that hearing impairment in SAM is due to a combination of sensory and strial (metabolic) presbycusis [4] as well as to neural presbycusis [3]. However, age-related properties of the SAM auditory CNS are still unclear.

Several studies reported that in the AC, electrophysiological changes seen in aging were associated with specific neurochemical changes related to GABA neurotransmission [24]. In particular, the largest age-related changes (around 40 % decrease relative to control) in glutamic acid decarboxylase message were found in A1 L2 [25]. In aging, the modulation of GABAergic neurotransmission would be expected to significantly alter spontaneous and stimulus-driven neural activity. Therefore, a decrease of GABAergic neurotransmission in aging may influence the variability of activity in AC laminar profiles. In this study, the reduction in CSD pattern variability in the older SAMP1 mice may reflect age-related disruption of GABA neurotransmission reported in the A1, and is probably associated with the mechanism whereby neurons in the A1 code sensory signals, including frequency and amplitude coding of sound input in an age-dependent manner.

Acknowledgements. In this work, T.T. was supported by the Suzuken Memorial Foundation (Japan), the Nakatani Foundation (Japan), and a Grant-in-Aid for Scientific Research (B) (No. 15H02772) and Exploratory Research (No. 15K12091) (Japan).

References

1. Takeda, T., Hosokawa, M., Takeshita, S., Irino, M., Higuchi, K., Matsushita, T., Tomita, Y., Yasuhira, K., Hamamoto, H., Shimizu, K., Ishii, M., Yamamuro, T.: A new murine model of accelerated senescence. Mech. Ageing Dev. **17**, 183–194 (1981)
2. Hosokawa, M., Ashida, Y., Saito, Y., Shoji, M.: Aging of eyes and ears in SAM mice. In: Takeda, K. (ed.) The Senescence-Accelerated Mouse (SAM): Achievements and Future Directions, pp. 221–227. Elsevier, Amsterdam (2013)
3. Saitoh, Y., Hosokawa, M., Shimada, A., Watanabe, Y., Yasuda, N., Takeda, T., Murakami, Y.: Age-related hearing impairment in senescence-accelerated mouse (SAM). Hear. Res. **75**, 27–37 (1994)
4. Iwai, H., Baba, S., Omae, M., Lee, S., Yamashita, T., Ikehara, S.: Maintenance of systemic immune functions prevents accelerated presbycusis. Brain Res. **1208**, 8–16 (2008)
5. Einevoll, G.T., Kayser, C., Logothetis, N.K., Panzeri, S.: Modelling and analysis of local field potentials for studying the function of cortical circuits. Nat. Rev. Neurosci. **14**, 770–785 (2013)
6. Womelsdorf, T., Fries, P., Mitra, P.P., Desimone, R.: Gamma-band synchronization in visual cortex predicts speed of change detection. Nature **439**, 733–736 (2006)
7. Szymanski, F.D., Garcia-Lazaro, J.A., Schnupp, J.W.: Current source density profiles of stimulus-specific adaptation in rat auditory cortex. J. Neurophysiol. **102**, 1483–1490 (2009)
8. Mehring, C., Rickert, J., Vaadia, E., Cardosa de Oliveira, S., Aertsen, A., Rotter, S.: Inference of hand movements from local field potentials in monkey motor cortex. Nat. Neurosci. **6**, 1253–1254 (2003)
9. Tateno, T., Jimbo, Y., Robinson, H.P.: Spatio-temporal cholinergic modulation in cultured networks of rat cortical neurons: spontaneous activity. Neuroscience **134**, 425–437 (2005)
10. Kaur, S., Rose, H.J., Lazar, R., Liang, K., Metherate, R.: Spectral integration in primary auditory cortex: laminar processing of afferent input, in vivo and in vitro. Neuroscience **134**, 1033–1045 (2005)
11. Takeda, T.: Senescence-accelerated mouse (SAM) with special references to neurodegeneration models, SAMP8 and SAMP10 mice. Neurochem. Res. **34**, 639–659 (2009)
12. Paxinos, G., Franklin, K.B.J.: The Mouse Brain in Stereotaxic Coordinates. Academic Press, San Diego (2008)

13. Haberly, L.B., Shepherd, G.M.: Current-density analysis of summed evoked potentials in opossum prepyriform cortex. J. Neurophysiol. **36**, 789–802 (1973)

14. Pettersen, K.H., Devor, A., Ulbert, I., Dale, A.M., Einevoll, G.T.: Current-source density estimation based on inversion of electrostatic forward solution: effects of finite extent of neuronal activity and conductivity discontinuities. J. Neurosci. Methods **154**, 116–133 (2006)

15. Duda, R.O., Hart, P.E., Stork, D.G.: Pattern Classification, p. 654. Wiley, New York (2001)

16. Szymanski, F.D., Rabinowitz, N.C., Magri, C., Panzeri, S., Schnupp, J.W.: The laminar and temporal structure of stimulus information in the phase of field potentials of auditory cortex. J. Neurosci. **31**, 15787–15801 (2011)

17. Cruikshank, S.J., Rose, H.J., Metherate, R.: Auditory thalamocortical synaptic transmission in vitro. J. Neurophysiol. **87**, 361–384 (2002)

18. Sakata, S., Harris, K.D.: Laminar structure of spontaneous and sensory-evoked population activity in auditory cortex. Neuron **64**, 404–418 (2009)

19. Winer, J.A., Lee, C.C.: The distributed auditory cortex. Hear. Res. **229**, 3–13 (2007)

20. Winer, J.A., Miller, L.M., Lee, C.C., Schreiner, C.E.: Auditory thalamocortical transformation: structure and function. Trends Neurosci. **28**, 255–263 (2005)

21. Winer, J.A.: Decoding the auditory corticofugal systems. Hear. Res. **212**, 1–8 (2006)

22. Telfeian, A.E., Connors, B.W.: Epileptiform propagation patterns mediated by NMDA and non-NMDA receptors in rat neocortex. Epilepsia **40**, 1499–1506 (1999)

23. Mitzdorf, U.: Current source-density method and application in cat cerebral cortex: investigation of evoked potentials and EEG phenomena. Physiol. Rev. **65**, 37–100 (1985)

24. Mendelson, J.R., Ricketts, C.: Age-related temporal processing speed deterioration in auditory cortex. Hear. Res. **158**, 84–94 (2001)

25. Ling, L.L., Hughes, L.F., Caspary, D.M.: Age-related loss of the GABA synthetic enzyme glutamic acid decarboxylase in rat primary auditory cortex. Neuroscience **132**, 1103–1113 (2005)

Differential Effect of Two Types of Anesthesia on Sound-Driven Oscillations in the Rat Primary Auditory Cortex

Hisayuki Osanai and Takashi Tateno[✉]

Bioengineering and Bioinformatics, Graduate School of Information Science and Technology, Hokkaido University, Kita 14, Nishi 9, Kita-ku, Sapporo, Hokkaido 060-0814, Japan
{h-osanai,tateno}@ist.hokudai.ac.jp

Abstract. Neural oscillations are considered to reflect the activity of neural populations, and are thus closely associated with brain function. However, the extent to which different anesthetic agents exert unique effects on such oscillations is unclear. A mixture of three anesthetics (medetomidine, midazolam, and butorphanol) was recently developed as an alternative to ketamine, which has potential addictive effects. Yet, little is known about the effects of this combination of anesthetics on neural oscillations. In this study, we used multi-channel electrophysiological recording and flavoprotein endogenous imaging to compare sound-driven oscillations in primary auditory cortical neurons after administration of ketamine vs. a medetomidine, midazolam, and butorphanol mixture. We observed differences in high gamma activities (over 120 Hz) between these two anesthetics, independent of cortical layers, but found no differences in activities including lower frequency components (<120 Hz). Our results provide new information about how specific anesthetics influence sound-driven neural oscillations.

Keywords: Anesthesia · Flavoprotein fluorescent imaging · Ketamine · Medetomidine-midazolam-butorphanol · Oscillation

1 Introduction

Oscillations in electrical brain activity have been used to explore brain function since the early days of neurophysiology. Neural oscillations are considered to reflect the dynamics of neural circuits and neural populations, and thus contain a wealth of information about sensory processing [1] and sensory perception [2].

Although it is known to have a profound effect on neural oscillations, anesthesia is widely used in investigations of in vivo sensory processing in the brain [3]. Several studies have reported that different anesthetics have different effects on neural oscillations in the olfactory bulb [4]. However, the extent to which the choice of anesthetic agents differently affects neural oscillations in the auditory cortex (AC) is still unclear.

Ketamine, a known N-methyl-D-aspartate (NMDA) antagonist, is one of the most common anesthetics employed in laboratory animals, including those used in auditory research. Furthermore, ketamine is usually administered in combination with an alpha-2

© Springer International Publishing AG 2016
A. Hirose et al. (Eds.): ICONIP 2016, Part II, LNCS 9948, pp. 363–371, 2016.
DOI: 10.1007/978-3-319-46672-9_41

adrenoceptor agonist such as xylazine or medetomidine. However, ketamine has dissociative side-effects, and its abuse as a recreational drug is widely recognized as a social problem [5].

Recently, to avoid the potentially addictive effects of ketamine, a mixture of medetomidine, midazolam, and butorphanol (MMB) was introduced as an alternative for use in rodents [6]. Midazolam is a water-soluble benzodiazepine that potentiates neural inhibition mediated by γ-aminobutyric acidA (GABA$_A$) receptors, while butorphanol acts as a synthetic opioid agonist-antagonist. Neither compound acts as an NMDA antagonist. Therefore, in recent years, MMB has been increasingly used in laboratory animal anesthesia. However, to the best of our knowledge, no studies have reported on the effect of MMB on response characteristics in the AC or compared the effects of MMB with those of other anesthetics, including ketamine.

As each auditory subfield in the AC can play a different role, in this study, we focused on the primary AC (A1). We compared MMB with a ketamine/xylazine mixture (KX) in terms of their effects on neural oscillations in the A1 in vivo. To achieve this, we developed a unique method in which we combined multisite recording in vivo with flavoprotein auto-fluorescence imaging to precisely identify the A1 area and its cortical layers. Thus, in this report, we compare the two types of anesthesia, with the goal of assisting researchers with the selection of anesthetic agents for their future experiments.

2 Materials and Methods

2.1 Experimental Procedure

All animal experiments described below were carried out in accordance with the National Institutes of Health Guidelines for the Care and Use of Laboratory Animals and with approval of the Institutional Animal Care and Use Committee of Hokkaido University. We used three 8- to 15-week-old adult male Wistar/ST rats (Japan SLC, Japan) in this study. Anesthesia was induced with an initial intraperitoneal dose of a solution containing ketamine (80 mg/kg; Daiichi Sankyo, Tokyo, Japan) and xylazine (8 mg/kg; Bayer Japan, Japan). Rectal temperature was monitored and kept at 33 °C via a heating pad. We intentionally maintained a lower-than-normal body temperature because the lower temperature facilitated a more stable condition for recording [7]. The skull and the dura were carefully removed to expose the left AC, and the cortical surface was covered with saline to prevent desiccation. During the experiments, the adequacy of anesthesia was confirmed by the absence of toe-pinch reflexes, and supplemental doses of anesthetics were administered with one-quarter of the initial dose to maintain anesthesia when reflexes were observed.

To minimize the effect of differences between individual animals, each rat received both KX and MMB. We first administered KX before the surgical procedure, as mentioned, and then conducted a round of electrophysiological recordings. We then administered MMB when reflexes were observed, and once 2 h had elapsed from the administration of MMB, we repeated the recordings. Specifically, we conducted the anesthetic conversion by administering a mixture of midazolam (0.5 mg/kg; Astellas Pharma, Japan), medetomidine (0.04 mg/kg; Kobayashi Kako, Japan), and butorphanol (0.62 mg/kg; Meiji Seika Pharma Co., Ltd., Tokyo, Japan), at a dose of one-quarter the amount used for general anesthesia [6]. This procedure was meant to minimize the effect of the KX on the second round of electrophysiological recordings (Fig. 1C).

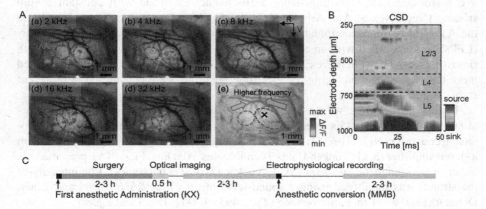

Fig. 1. Determination of an electrode site using flavoprotein fluorescence imaging, and of cortical layers using CSD analysis. A(a)–(d), areas (covering a region of 10.5 mm²) activated in response to pure tone pips (2, 4, 8, 16, and 32 kHz, respectively) were captured and superimposed onto the surface view of the blood vessel pattern in the exposed portion of the rat AC. In all figures, anterior (rostral) is toward the left and ventral is downward, as illustrated by the arrows in the upper right. The activated areas located at the AAF (blue) and A1 (orange) are indicated by two dashed ellipses. The superimposition of all frequency results in (a–d) is shown in (e). The "X" symbol indicates the insertion site of an electrode array. B, a CSD pattern derived using the second spatial derivatives of recorded LFPs. The border between layers 4 and 5 was identified on the basis of the sink (blue) and source (red) in the CSD profile. C, schematic representation of the experimental procedure. We conducted anesthesia conditions successively, with a conversion period from KX to MMB. (Color figure online)

2.2 Flavoprotein Imaging

We conducted flavoprotein endogenous optical imaging according to the previously reported method [8]. Briefly, a CMOS camera system (MiCAM02; BrainVision, Japan) was mounted on a tandem-lens upright microscope (THT, BrainVision). Excitation light was provided by a 465-nm blue LED (LEX2-B; BrainVision) passed through a blue band-pass filter (466 ± 20 nm). Endogenous flavoprotein green fluorescence (500–550 nm) was detected by the CMOS camera system through a dichroic mirror and a green band-pass filter (525 ± 22.5 nm). Cortical images of fluorescent signals were recorded at 20 frames/s before and after pure-tone sound stimulation (60-dB SPL, 100-ms duration with 5-ms linear rise and fall ramps) using an acrostic stimulus system, described below. Trials were repeated eight times at random 5–10 s intervals, and the averaged images were displayed using acquisition and analysis software (BV-Ana; BrainVision). The A1 was recognized by tonotopic organization and its position in relation to brain surface blood vessels (Fig. 1A; see Results).

2.3 LFP Recording and Data Analysis

We performed LFP recording using a 16-channel electrode with constant 50-μm spacing (NeuroNexus, Ann Arbor, MI, USA). Each probe was slowly advanced into the A1 using a micromanipulator (Narishige, Tokyo, Japan). Local field potentials (LFPs) were recorded with an amplification gain of 250 and a low-pass Butterworth filter (0.1–300 Hz) (OmniPlex; Plexon Inc., Dallas, TX, USA). Only data recorded more than 0.5 h after changing the electrode positions were considered to represent stable neural activity.

During the recording, we presented a click pulse sound stimulus, delivered 100 times at 80 dB SPL at 0.3–0.45 s intervals. Sound stimuli were generated with a digital-to-analog converter (NI USB-6341; National Instruments, Austin, USA) and amplified with a stereo amplifier (SA1; Tucker-Davis Technologies, Alachua, FL, USA), presented via either of two magnetic speakers (MF1; Tucker-Davis Technologies). The intensity of the stimuli was monitored using a sound-level meter (Type 2636, Brüel and Kjaer, Denmark) and a 1/4-inch microphone (Type 4939-L-002; Brüel and Kjaer).

We conducted a CSD analysis to estimate electrode positions in relation to cortical layers. Specifically, we calculated the second spatial derivative of the averaged LFP along the z-axis [9]. We obtained the average LFP in response to the click trains by averaging more than five identical trials. We used the presence of a clear sink-source pair to identify putative layer 4, which was consistent with the electrode depth and thickness of the cortical layers [10] (Fig. 1B).

We performed a time-frequency analysis of the LFP oscillations using the matching pursuit algorithm [11], which is an iterative procedure for decomposing a signal as a linear combination of waveforms that are localized in terms of time and frequency. This algorithm uses an overcomplete dictionary of Gabor functions. At each iteration, a Gabor function that best describes the LFP waveform is chosen from the dictionary, and its projection on the LFP waveform is subtracted to create a residual signal waveform. This iteration is repeated until the linear combination of the chosen Gabor functions reconstructs the original LFP waveform. The time-frequency energy distribution is obtained from the Wigner–Ville distribution of the summation of the individual Gabor functions that were chosen during the iterations. The mathematical details are described in previous literature [11].

In the present study, we pre-defined eight frequency bands: delta (0.1–4 Hz), theta (4–8 Hz), alpha (8–12 Hz), beta (12–30 Hz), gamma1 (30–60 Hz), gamma2 (60–120 Hz), gamma3 (120–200 Hz), and gamma4 (200–300 Hz). Data are given as mean ± standard errors of the mean (SEM), and means were compared using a one-sample t-test conducted via Matlab (MathWorks, USA), with $p < 0.05$ assumed to be statistically significant.

3 Results

3.1 Identification of the A1 and Cortical Layers

We located A1 according to the tonotopic organization of this region. At the beginning of each experiment, the A1 and other auditory areas were activated in response to pure-tone pip stimulation (Fig. 1A). To reconstruct tonotopic maps, we clipped the response areas of different frequencies (2–32 kHz) at 60 % of their peak responses (Figs. 1Aa–d). We observed at least two major activation areas (Figs. 1Aa–c) in response to tones with lower frequencies (i.e., 2, 4, and 8 kHz). With increasing tone frequency, two individual centers of the activation areas shifted towards one another. On the basis of the response patterns, the activation areas were superimposed onto one another from low to high frequencies (Fig. 1Ae). As the A1 is located at the posterior area of the AAF [7], it was easy to recognize the A1 and the anterior auditory field (AAF) by their mirror-imaged tonotopic organization (Fig. 1Ae). Also, in the A1 and AAF, the response area of each frequency had a roughly dorsoventral bandlike appearance. In the A1, therefore, the lower frequencies were located posteriorly and the higher frequencies anteriorly. Thus, we were able to identify the A1 location via flavoprotein autofluorescence imaging.

Next, we recorded LFPs in the A1 using a multichannel electrode array that contacted tissue across different layers (Fig. 1B). We inserted the electrode into the cortical area that responded to 8–16 kHz tones during flavoprotein imaging. To minimize the effect of differences between individual animals, we recorded LFPs in each rat under successive KX and MMB conditions ($n = 3$) (Fig. 1C). We determined the position of the putative layer 4 from the surface according to a CSD analysis in response to a click sound (Fig. 1A, B). After the onset of the sound stimulation, we observed an early and strong sink of around 600–750 μm in the CSD profiles during the period from 20 to 25 ms after click tone onset (Fig. 1B). This is classically interpreted as an excitatory event, e.g. axonal depolarization or excitatory/inhibitory synaptic activation. Previous studies have described similar patterns in the CSD in vivo recordings in the A1 in response to acoustic sound stimulation [12]. In addition, considering the anatomical size of the cortical layers [10], we defined the lower and upper boundaries of putative layer 4 as the cortical depths between observed sink-source pairs and the point 150 μm above the lower boundaries, respectively. We then determined the depths of layer 2/3 and 5 accordingly (Fig. 1B).

3.2 Effect of Different Anesthetics on Neural Oscillations

To determine whether the two anesthetics modulated neural oscillations differently, we investigated the frequency structure of the LFPs elicited in response to the click sound stimuli (Fig. 2A). We performed a time-frequency analysis of the LFPs using the matching-pursuit algorism [11] (Fig. 2B). Additionally, we investigated the total power of click acoustic-stimuli-driven oscillations by summation of the oscillatory power during a period from 0 ms to 100 ms after tone onset (Fig. 2C). At first glance, differences among individual recordings and animals were relatively large. Thus, it was difficult to discern any clear relationships between activity levels and anesthesia type. For example, in the KX compared with the MMB condition, one rat showed higher power in the 12–120 Hz range, which includes beta, gamma1 and gamma2 bands, and in the 160–300 Hz range, which includes gamma3 and gamma4 bands (Fig. 2C). Power in the rest of the bands was unchanged in this animal. When data for all three animals was averaged, however, the power of the gamma3 band in the KX condition was significantly higher than that in the MMB condition in layers 4 ($\log_2 (P_{KX}/P_{MMB}) = 1.1 \pm 0.2$, $p < 0.05$; P: average power in the frequency band) and 5 (0.9 ± 1.5, $p < 0.05$) (Figs. 2Db, c). Similarly, in layer 2/3, the power of the higher-frequency bands in the KX condition tended to be greater than that in the MMB condition ($\log_2 (P_{KX}/P_{MMB})$ 0.9 ± 0.1 in gamma2 band, 1.5 ± 0.3 in gamma4 band, $p < 0.07$) (Fig. 2Da). On the other hand, no systematic differences were observed with respect to the lower frequency bands (Fig. 2D).

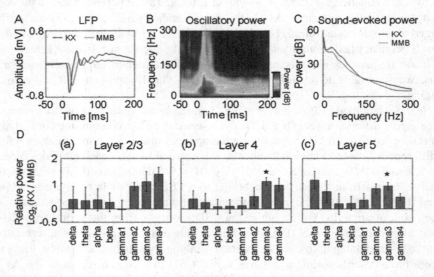

Fig. 2. Effect of different anesthetics on oscillations in the AC. A-C, neural oscillations in one of three animals. A, LFP waveforms obtained from layer 5 after click tone onset ($t = 0$ ms), averaged across 100 trials. Pink: power distribution in the KX condition, blue: that in the MMB condition. B, example of averaged time-frequency structure of the LFP in the KX condition shown in A. C, power distribution of the LFPs in layer 5 from 0 ms to 100 ms after tone onset. Pink: power distribution in the KX condition, blue: that in the MMB condition. D, relative oscillatory power in the KX condition compared with that in the MMB condition across all three animals in each cortical layer, in the pre-defined frequency bands: delta (0.1–4 Hz), theta (4–8 Hz), alpha (8–12 Hz), beta (12–30 Hz), gamma1 (30–60 Hz), gamma2 (60–120 Hz), gamma3 (120–200 Hz), and gamma4 (200–300 Hz). All plots in D show averages across animals. Error bars indicate standard error of mean. *: $p < 0.05$ by one-sample t-test for mean equal to zero. (Color figure online)

4   Discussion

In this study, we identified two core fields (A1 and AAF) in the rat AC according to the criteria for tonotopic organization. Our results regarding rat A1 tonotopic organization are consistent with those of previous studies using optical imaging [7]. The identification of the A1 is critically important for distinguishing the oscillatory characteristics of the A1 and other subfield. Usually, inserting electrodes into multiple sites takes much time and effort, and long-term recordings can cause tissue damage. In addition, owing to the toxicity of several voltage-sensitive dyes, optical imaging techniques using such dyes might not always be suitable for long-term neuronal recording. However, because flavoprotein is auto-fluorescent and the preparation time for the imaging takes less than half an hour, our method enables efficient electrophysiological recording for identification of the A1 and other subfields.

After identifying the A1 and its laminar structure via multi-electrode recording of neural activity, we showed that KX and MMB can have different effects on sound-driven oscillations in the A1. Generally, the low-frequency bands (~30 Hz) are considered to be involved in memory, attention, and interaction with other cortical areas, while the low-gamma band (~60 Hz) appears to be related to sound perception [13]. Compared with awake conditions, anesthesia-induced changes in these low-frequency bands have been reported [3]. However, we observed no systematic differences in the low-frequency power of the LFPs obtained in both the KX and MMB conditions. In contrast, with respect to the higher-frequency bands (over 120 Hz), we found decreased power across cortical layers under the MMB condition compared with the KX condition. These results suggest that the effects of the anesthesia on the oscillations were uniform among the cortical layers between the KX and MMB conditions. However, we cannot exclude the possibility of a residual effect of KX, which was delivered as the initial agent. In addition, we conducted a preliminary experiment and found that, when the two anesthetics were administered in the opposite order (i.e., MMB first and KX second), the power of the gamma3 band in the MMB condition was lower than that in the KX condition in three out of four animals (data not shown), which is consistent with the data described in the present report.

Ketamine is a pharmacological antagonist of NMDA glutamate receptors. Previous studies have reported that the administration of NMDA antagonists increases oscillatory activity in high frequency bands (140–180 Hz) in the nucleus accumbens [14], and increases such activity in different frequencies (40–60, 68–102, and 128–192 Hz) in the motor cortex and other brain areas. High-frequency (50–200 Hz) power in the barrel cortex and thalamus has also been attenuated by propofol and isoflurane [15] anesthesia, which enhance $GABA_A$ receptor activity, as well as by midazolam [16], although no previous studies have reported on the effect of altered synaptic activity on high frequency oscillations in the AC. Thus, our finding of a differential effect of KX and MMB on high frequency oscillations in A1 could be attributable to pharmacological differences between the studied anesthetics. Our results provide new information about how sound-driven neural oscillations in the A1 are affected by specific anesthetics.

Acknowledgements. In this work, T.T. was supported by the Suzuken Memorial Foundation (Japan) and a Grant-in-Aid for Scientific Research (B) (No. 15H02772) and Exploratory Research (No. 15K12091) (Japan). H.O. is supported by the Tateisi Science and Technology Foundation (Japan).

References

1. Belitski, A., Gretton, A., Magri, C., Murayama, Y., Montemurro, M.A., Logothetis, N.K., Panzeri, S.: Low-frequency local field potentials and spikes in primary visual cortex convey independent visual information. J. Neurosci. **28**, 5696–5709 (2008)
2. van der Loo, E., Gais, S., Congedo, M., Vanneste, S., Plazier, M., Menovsky, T., Van de Heyning, P., De Ridder, D.: Tinnitus intensity dependent gamma oscillations of the contralateral auditory cortex. PLoS ONE **4**, e7396 (2009)
3. Lazarewicz, M.T., Ehrlichman, R.S., Maxwell, C.R., Gandal, M.J., Finkel, L.H., Siegel, S.J.: Ketamine modulates theta and gamma oscillations. J. Cogn. Neurosci. **22**, 1452–1464 (2010)
4. Li, A., Zhang, L., Liu, M., Gong, L., Liu, Q., Xu, F.: Effects of different anesthetics on oscillations in the rat olfactory bulb. J. Am. Assoc. Lab. Anim. Sci. **51**, 458–463 (2012)
5. Newport, D.J., Carpenter, L.L., McDonald, W.M., Potash, J.B., Tohen, M., Nemeroff, C.B.: Ketamine and Other NMDA Antagonists: early clinical trials and possible mechanisms in depression. Am. J. Psychiatry **172**, 950–966 (2015)
6. Kirihara, Y., Takechi, M., Kurosaki, K., Kobayashi, Y., Saito, Y., Takeuchi, T.: Effects of an anesthetic mixture of medetomidine, midazolam, and butorphanol in rats-strain difference and antagonism by atipamezole. Exp. Anim. **65**, 27–36 (2016)
7. Noto, M., Nishikawa, J., Tateno, T.: An analysis of nonlinear dynamics underlying neural activity related to auditory induction in the rat auditory cortex. Neuroscience **318**, 58–83 (2016)
8. Shibuki, K., Hishida, R., Murakami, H., Kudoh, M., Kawaguchi, T., Watanabe, M., Watanabe, S., Kouuchi, T., Tanaka, R.: Dynamic imaging of somatosensory cortical activity in the rat visualized by flavoprotein autofluorescence. J. Physiol. **549**, 919–927 (2003)
9. Kaur, S., Rose, H.J., Lazar, R., Liang, K., Metherate, R.: Spectral integration in primary auditory cortex: laminar processing of afferent input, in vivo and in vitro. Neuroscience **134**, 1033–1045 (2005)
10. DeFelipe, J., Alonso-Nanclares, L., Arellano, J.I.: Microstructure of the neocortex: comparative aspects. J. Neurocytol. **31**, 299–316 (2002)
11. Chandran Ks, S., Mishra, A., Shirhatti, V., Ray, S.: Comparison of matching pursuit algorithm with other signal processing techniques for computation of the time-frequency power spectrum of brain signals. J. Neurosci. **36**, 3399–3408 (2016)
12. Sakata, S., Harris, K.D.: Laminar structure of spontaneous and sensory-evoked population activity in auditory cortex. Neuron **64**, 404–418 (2009)
13. Stolzberg, D., Chrostowski, M., Salvi, R.J., Allman, B.L.: Intracortical circuits amplify sound-evoked activity in primary auditory cortex following systemic injection of salicylate in the rat. J. Neurophysiol. **108**, 200–214 (2012)
14. Hunt, M.J., Olszewski, M., Piasecka, J., Whittington, M.A., Kasicki, S.: Effects of NMDA receptor antagonists and antipsychotics on high frequency oscillations recorded in the nucleus accumbens of freely moving mice. Psychopharmacol. (Berl) **232**, 4525–4535 (2015)

15. Plourde, G., Reed, S.J., Chapman, C.A.: Attenuation of High-Frequency (50–200 Hz) Thalamocortical Electroencephalographic Rhythms by Isoflurane in Rats Is More Pronounced for the Thalamus than for the Cortex. Anesth. Analg. (2016)
16. Saari, T.I., Uusi-Oukari, M., Ahonen, J., Olkkola, K.T.: Enhancement of GABAergic activity: neuropharmacological effects of benzodiazepines and therapeutic use in anesthesiology. Pharmacol. Rev. **63**, 243–267 (2011)

Developing an Implantable Micro Magnetic Stimulation System to Induce Neural Activity *in Vivo*

Shunsuke Minusa and Takashi Tateno[✉]

Bioengineering and Bioinformatics, Graduate School of Information Science and Technology,
Hokkaido University, Kita 14, Nishi 9, Kita-ku, Sapporo, Hokkaido 060-0814, Japan
{s-minusa,tateno}@ist.hokudai.ac.jp

Abstract. Although electromagnetic stimulation is widely used in neurological studies and clinical applications, conventional electromagnetic stimulation methods have several limitations. Recent studies have reported that micro magnetic stimulation (μMS), which can directly activate neural tissue and cells via sub-millimeter solenoids, has the possibility to overcome such limitations. However, the development and application of μMS using implantable sub-millimeter solenoids has not yet been reported. Here, we proposed a new implantable μMS system and evaluated its validity. In particular, using flavoprotein fluorescence imaging with a high spatial resolution, we evaluated if the stimuli delivered by our system were large enough to activate the mouse auditory cortex *in vivo*. The results indicated that our system successfully activated neural tissue, and the activity propagation was observed on the brain surface. Thus, this study is the first step to applying μMS implantable devices in investigating basic neuroscience and clinical application tools.

Keywords: Auditory cortex · Flavoprotein fluorescence imaging · Implantable device · Magnetic stimulation · Solenoid

1 Introduction

To activate the periphery and the central nervous system, electromagnetic stimulation is widely used [1]. To clinically treat neurological disorders including depression, transcranial magnetic stimulation (TMS) is applied via brain activation with lower side effects [2]. With respect to electrical stimulation, in addition, deep brain stimulation (DBS) has achieved great success in the treatment of Parkinson's disease [3]. Despite such successes, conventional electromagnetic stimulation methods have several limitations. In a long-term DBS system, for example, local stimulation via implanted electrodes to the targeted brain area is applied during a long period. However, the impedance between the electrodes and brain tissues around them changes over time because of tissue and chemical reactions on the electrode surface [4]. Consequently, the DBS electrodes are gradually required to apply a larger stimulus to activate neurons effectively, so it is more difficult to avoid inflammatory reactions in tissue surrounding the implanted electrodes.

© Springer International Publishing AG 2016
A. Hirose et al. (Eds.): ICONIP 2016, Part II, LNCS 9948, pp. 372–380, 2016.
DOI: 10.1007/978-3-319-46672-9_42

Conversely, magnetic stimulation uses eddy currents induced by time-varying magnetic fields, and it can noninvasively evoke and modulate neural activities [5]. However, TMS also has several drawbacks to overcome. First, because magnetic fields decrease as the square of the distance, magnetic stimulation of targets deeper than the brain surface is rather difficult [6]. Second, compared with the cellular interfaces of electric stimulators, in general, TMS coils are so large (size: 10–20 cm) to generate a high magnetic field that they cannot be compactly implanted [7]. Third, because conventional TMS devices have lower spatial resolutions and experimental controllability, accurate stimulation control to small neuronal circuits is more difficult than in an electric stimulation system.

Recent studies have reported that micro magnetic stimulation (μMS), which uses a sub-millimeter solenoid and is smaller than conventional TMS coils, can be used to evoke action potentials and modulate activity [8]. The μMS has several advantages over conventional methods; (i) the μMS interface can be isolated from contiguous tissues using bio-inert polymers, thus avoiding inflammatory reactions owing to implantation [9], (ii) small μMS solenoids are expected to be implanted near targeted areas and stimulate a deeper target with higher spatial accuracy in the brain.

In this study, we propose an implantable μMS interface and evaluate how well the μMS system functions as an effective magnetic stimulator. In particular, we demonstrate that our μMS system can evoke and modulate neural activities in the auditory central nervous system. Using flavoprotein fluorescence imaging (FFI) with a high spatial resolution, we observed μMS-driven neural activities of the mouse auditory cortex *in vivo*. Additionally, based on the obtained results, we discuss the future improvement of our μMS system.

2   Materials and Methods

2.1   Design and Testing a μMS Interface

For the μMS interface, we used a commercially available sub-millimeter inductor (ELJ-RFR10JFB; Panasonic Electric Devices Corporation of America, Knoxville, TN, USA) and a thin circuit board. To support and fix the inductor (solenoid), a 0.1-mm flame retardant type 4 circuit board (FR-4) was used. The shank length of the μMS interface was determined on the basis of the distance from the surface of the mouse brain to the ventral subnucleus of the medial geniculate body (vMGB) because neurons in the vMGB project to those in the auditory cortex (AC) [10]. To reduce damage to the brain, the long axis of the solenoid was designed to be parallel to the penetrating direction of the interface. After soldering the interface to copper wires, it was coated with a biocompatible polymer, parylene-C, to eliminate current leakage from the circuit. Our magnetic stimulators shown in Fig. 1A were so small that we could not generally measure magnetic flux density through commercially available Gauss meters. Therefore, we produced several custom-made search coils shown in Fig. 1B to measure induced electromotive force generated by the time-varying magnetic field around the μMS interface [6]. Because the μMS interface needed to deeply penetrate into brain tissue, we also tested its mechanical stiffness.

2.2 Animal Preparation

All experiments were carried out in accordance with the National Institutes of Health Guidelines for the Care and Use of Laboratory Animals, and with the approval of the Institutional Animal Care and Use Committee of Hokkaido University. In this study, 6- to 10-week male C57BL/6 J mice were used. The mice were anesthetized using intra-peritoneal injection of urethane (1.3 g/kg bodyweight; Wako, Japan). Atropine (0.005 %; Mitsubishi Tanabe Pharma, Japan) was injected before anesthesia to reduce salivary and bronchial secretions. Dexamethasone (0.5 mg/kg; Kobayashi Kako, Japan) was admin-istered to prevent brain edema. Rectal temperature was monitored and kept at 36 °C using a heat pad. After application of xylocaine, the scalp was transected. The skull and the dura were carefully removed to expose the left AC, and the surface was covered with physiological saline solution to prevent drying out.

2.3 Flavoprotein Fluorescence Imaging

The principles of optical recording with FFI have been described in the literature [11]. Briefly, cortical images of endogenous green fluorescence were recorded using an imaging system with a complementary metal oxide semiconductor (CMOS) camera (MiCAM02; Brainvision Inc., Japan). The CMOS camera was mounted on a tandem

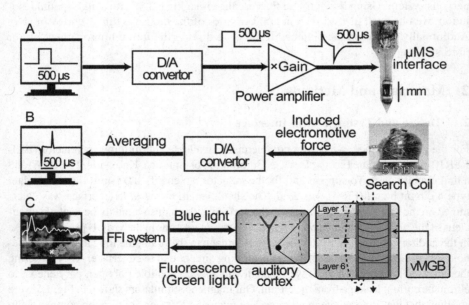

Fig. 1. An implantable µMS interface and its evaluation system. (A) A stimulus was generated by a D/A convertor, amplified by a power amplifier to increase its maximum amplitude three-fold (Gain = 3.0), and finally applied to the µMS interface. (B) Induced electromotive force was measured by a custom-made search coil to confirm the effectiveness of the µMS. (C) The µMS interface was inserted in the mouse A1 or AAF. Neural activities evoked by the µMS were recorded using flavoprotein fluorescence imaging.

lens fluorescence microscope (THT; Brainvision Inc.). Light from a blue light-emitting diode (LEX2-B; Brainvision Inc.) was projected through a blue band-pass filter (wavelength $\lambda = 466 \pm 20$ nm) and reflected by a dichroic mirror ($\lambda = 509$ nm) to activate flavoprotein at the cortical surface. Fluorescence signals were then collected through the dichroic mirror, projected through a green band-pass filter ($\lambda = 525 \pm 22.5$ nm), and detected with the CMOS camera at 20 frames/s. The signals from the cortical surface were acquired by the camera with 188×160 pixels in each frame. Optical signals recorded in response to μMS were averaged over 20 trials to reduce random noise per trial. Spatial averaging in 5×5 pixels and temporal averaging of five consecutive frames were used to improve the quality of the images. Images were normalized pixel by pixel with respect to the fluorescence intensity of a reference image (F_0), which was obtained by averaging eight images taken immediately before μMS. The normalized images were shown in a pseudocolor scale in terms of the relative fluorescence changes ($\Delta F/F_0$), where ΔF is the change from the reference F_0.

2.4 μMS and Neural Responses

Before applying μMS, we identified the primary auditory cortex (A1) and the anterior auditory field (AAF) in the mouse AC, using 5- or 10-kHz tone-burst stimuli through a sound speaker (T250D; FOSTEX, Chesterfield, MO, USA). Sound stimuli were produced by a personal computer using a custom-made Python program, and analog signals were obtained through a D/A convertor (USB-6341; National Instruments, Austin, TX, USA) at a sampling rate of 50 or 100 kHz. The stimuli were passed through a band-pass filter (Multifunction Filter 3611; NF Electronic Instruments, Japan) with a frequency range of 2 Hz to 100 kHz, and amplified with a stereo amplifier (SA1; Tucker-Davis Technologies, Alachua, FL, USA). The duration of the sound stimuli was 100 ms, with a rise and fall time of 2.5 ms.

In response to the sound stimuli, the A1 and the AAF were first identified using FFI. Then, the μMS interface was inserted into the A1 or the AAF by a manipulator. μMS stimuli were generated by a D/A convertor (USB-6341; National Instruments). Each stimulus had a pulsatile shape, and its duration was 0.5 ms. Amplitudes of the stimuli ranged from 0.5 to 5.0 V in 0.5-V steps. Each stimulus was randomly presented 20 times with inter-stimulus intervals of 6.5–8.5 s. The output signals were connected to a power amplifier (PB717; Pyramid Inc., Brooklyn, NY, USA). When output from the power amplifier, the maximum amplitude of each stimulus increased three-fold from the maximum amplitude of the corresponding input signal, and the output waveform was changed from a step-like shape to a biphasic one. To confirm that there were no breaks and no current leakage, the μMS interface was tested before and after each experiment. The μMS coil will be destroyed unless the impedance ranges from 3.5 to 4.5 Ω. In addition, when the μMS interface was submerged in physiological solution, the impedance between one of its terminals and an electrode immersed in the physiological saline solution also needs to be above 2 MΩ.

3   Results

3.1   Evaluation of the μMS Interface

Initially, we examined the output properties of the μMS interface by applying several voltage signals. To measure the induced electromotive force generated by the μMS interface, we produced some custom-made search coils, and selected one with the following parameters: inner radius = 0.75 mm, outer radius = 2.5 mm, and 100 turns. The result of the measurement using the search coil is shown in Fig. 1B. The voltage changes (a 0.5-ms-width biphasic waveform) of the search coil exactly followed the changes of the applied voltage signals (a 0.5-ms-width mono-polar step-pulse with positive voltage). In a few experiments, however, when intensities (>5.0 V) and durations (>0.5 ms) of the applied signals to the interface were respectively large and long, the interfaces were destroyed. Moreover, the stiffness of the μMS interface was also tested, and the interface was able to penetrate the mouse brain smoothly (Fig. 2B).

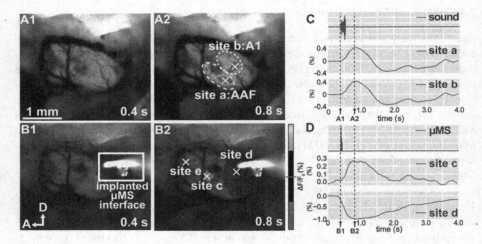

Fig. 2. Sound- and μMS-driven responses obtained by FFI. (A) Neural responses to a 5-kHz tone stimulus (right panel). The identified locations of the A1 and the AAF are indicated by the two dashed ellipses. The time from the stimulus onset is indicated in the bottom right corner. (B) Neural responses to 5.0 V μMS (right panel) are shown. The location of the μMS interface inserted in the brain is indicated by a white box. (C) In response to a 5-kHz tone stimulus (upper panel), fluorescence intensity changes at sites a and b are shown. The imaging process was triggered 400 ms before the stimulus onset. From the onset, the time points A1 and A2 were at 0.4 and 0.8 s, respectively. (D) Fluorescence intensity changes to μMS. The location at site d (cross mark) near the inserted interface showed negative fluorescence changes as compared with other sites (a, b, and c). From the onset, the time points B1 and B2 were at 0.4 and 0.8 s, respectively.

3.2 Comparing Sound- and μMS-Driven Neural Responses

In response to the pure-tone stimuli, response peaks of FFI changes were locally observed in the AC after a period of 0.87 ± 0.14 s (mean $\pm$ SEM) from the sound onset (Fig. 2A). After identifying the A1 and the AAF (Fig. 2A), we slowly inserted and gradually advanced the interface into the A1 (approximately in a range of 0.5–1.4 mm from the brain surface to the tip of the interface). Around site c in Fig. 2B, the time courses of μMS-driven responses were similar to those of the sound-driven responses (Fig. 2C). In contrast, compared with the sound-driven responses, μMS-driven responses at site d (Fig. 2B) showed negative changes (decreases in the bottom panel of Fig. 2D), and these changes rapidly spread over the surface. In this study, latency was defined as the time interval when the FFI change reached the response peak from the stimulation onset. The average latency of the negative peaks at multiple sites in the A1 (e.g., site d) was 0.67 ± 0.08 s. Once these negative peaks in the A1 passed by, subsequent positive peaks of the responses were observed in the AAF. In a few experiments, neural activity was also observed in the rostral area far from the AAF (e.g., site e) after the negative peaks appeared.

Next, from the obtained FFI data, we characterized the relationships between the applied voltage intensities vs. response peak properties such as response intensity, response latency, and duration. The result showed that the response intensities monotonically increased with applied voltage intensity (Fig. 3A). When the applied voltage was not more than 1.0 V, the average latency was 1.33 ± 0.11 s (Fig. 3B). In contrast, when the applied voltage was not less than 1.5 V, the average latency was 0.42 ± 0.08 s. The result suggests that an intensity of 1.5 V is the threshold for the activation of the neural responses to propagate. With respect to duration, the results indicated a similar tendency to that of the response latency, i.e., in a manner that depended on the voltage intensity.

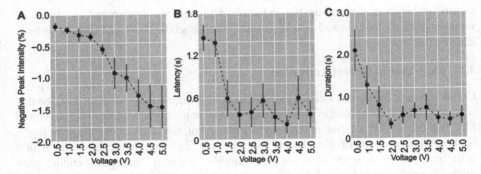

Fig. 3. Relationships between the applied voltage intensity vs. response peak properties such as response intensity (A), response latency (B), and duration (C). Data are represented by mean $\pm$ SEM.

To confirm that the results mentioned above truly corresponded to neural responses induced by μMS, we performed an additional experiment. In the additional experiment, we compared FFI responses before and after the animal died as the previous study demonstrated [12]. The result showed that, negative peaks of μMS driven responses were obtained under the both conditions, the negative transients in the former condition were smaller than those in the latter condition. The result also implied that positive neural signals were superimposed into the transient signals in the former conditions.

4   Discussion

4.1   Validity of the μMS Interface

In this study, we demonstrated the validity of the μMS interface to evoke neural activity. The custom-made search coil measured the electromotive force induced by the μMS interface. Thus, the μMS interface is able to drive magnetic stimulation because the induced electromotive force is derived from the time-varying magnetic field around the μMS interface. Additionally, we showed that the μMS interface had enough stiffness to penetrate into the mouse brain. Because of these results, we confirmed that the μMS interface could work as an implantable magnetic stimulator.

In addition, we examined the intensity of μMS that was adequate to evoke neural activity. In the analysis of the obtained imaging data, we found a long latency in the response to a stimulus with an intensity ranging from 0.5 to 1.0 V. In the voltage range, furthermore, we could not observe clear negative peaks, indicating that such stimuli did not induce detectable neural activity through FFI.

However, in the present μMS interface, several problems need to be solved. First, in some experiments, the interface was burned and destroyed when applying large stimuli. Therefore, it is important to know the exact degree of heating because this heating could result in an inflammatory reaction of the contiguous tissues. Second, we also need to know a practically applicable range of parameters of stimulus waveforms such as amplitude, duration, and whether they are mono/bipolar, to decide on adequate and effective parameters without destroying the inductor. Although we determined the range of voltage intensity that evoked neural activity, further studies will be required to fully address these problems.

4.2   μMS-Driven Neural Activation

Compared with the neural activity induced by sound stimulation, we demonstrated that our μMS interface was able to evoke neural activity at a similar intensity level to that of FFI. We also confirmed that, in the control experiment, the μMS-driven negative peaks were not just artifacts of stimulation, but were the responses associated with neural activity. In addition, when increasing the stimulus intensity, the corresponding peak intensity was increased (Fig. 3A). Thus, these results indicated that our μMS interface could practically activate neural tissues. After responses in the A1 around the interface, we observed delayed neural activity in the AAF. Therefore, these results showed that A1 neural activity induced by μMS was propagated to the AAF. Subsequently, we also observed neural activity in the rostral area far from the AAF with very long latency $(1.03 \pm 0.07 \text{ s})$. In a mouse, the insular auditory field (IAF), one of the belt fields in the AC, is located in the rostral area to the AAF [13], and the IAF is close to the MGB [10].

Therefore, our result implies that, via the MGB, µMS-driven neural activity was indirectly back propagated from the A1 to the IAF.

In the A1 around the interface, µMS-driven fluorescence-intensity changes rapidly decreased and slowly recovered to baseline. Previous studies have reported that, using extracellular and cell-attached voltage recording, in response to µMS, action potentials were superimposed with a very slow transient of a stimulation artifact [12]. In addition, a study using voltage-sensitive dye imaging also reported that neural activity was decreased below the baseline during a period from 20 to 300 ms after a single TMS pulse onset to the cat visual cortex [14]. Our observation regarding the slow intensity changes seems to be related to the slow voltage transients reported by these previous studies. To clarify the mechanism of the slow transient phenomenon, further studies will be required.

4.3 Improvement of the Present Interface

Although we showed that the µMS interface could evoke neural activity, there are several ways in which this interface could be improved. The first point is on the arrangement of the inductor direction. Previous studies have reported that spikes are more easily evoked when the long axis of the solenoid was not perpendicular but parallel to the brain surface [9]. In the present experiments, however, the µMS interface was inserted into the brain perpendicular to the long axis. The reasons are as follows: (i) the thickness of its long axis was much larger than that of standard penetrate-type electrodes and (ii) when the interface is inserted in the parallel direction, the damage to the brain would be greater than in the perpendicular direction. Second, when the µMS device is clinically applied, these devices will be implanted for a long period. Although we used a biocompatible polymer to coat the devices, we have not examined its efficacy yet. Therefore, long-term experiments regarding µMS-drive implantation are required. Third, to obtain better results, the size of the inductor needs to be smaller. Additionally, we have not directly observed the neural activity that is propagated from the vMGB to some subfields in the AC. The shank length of the µMS interface was decided on the basis of the distance from the brain surface to the vMGB. The size of the inductor and the shank length needs to be reconsidered to produce a better activation effect in the AC.

Acknowledgments. In this work, T.T. was supported by the Magnetic Health Science Foundation (Japan), the Suzuken Memorial Foundation (Japan), and a Grant-in-Aid for Scientific Research (B) (No. 15H02772) and Exploratory Research (No. 15K12091) (Japan). The authors also wish to thank Dr. Kaori Kuribayashi-Shigetomi for her help on the parylene-C coating.

References

1. Luan, S., Williams, I., Nikolic, K., Constandinou, T.G.: Neuromodulation: present and emerging methods. Front. Neuroeng. **7**, 1–9 (2014)
2. O'Reardon, J.P., Solvason, H.B., Janicak, P.G., Sampson, S., Isenberg, K.E., Nahas, Z., McDonald, W.M., Avery, D., Fitzgerald, P.B., Loo, C., Demitrack, M.A., George, M.S., Sackeim, H.A.: Efficacy and safety of transcranial magnetic stimulation in the acute treatment of major depression: a multisite randomized controlled trial. Biol. Psychiatry **62**, 1208–1216 (2007)

3. Deuschl, G., Schade-Brittinger, C., Krack, P., Volkmann, J., Schäfer, H., Bötzel, K., Daniels, C., Deutschländer, A., Dillmann, U., Eisner, W., Gruber, D., Hamel, W., Herzog, J., Hilker, R., Klebe, S., Kloß, M., Koy, J., Krause, M., Kupsch, A., Lorenz, D., Lorenzl, S., Mehdorn, H.M., Moringlane, J.R., Oertel, W., Pinsker, M.O., Reichmann, H., Reuß, A., Schneider, G.-H., Schnitzler, A., Steude, U., Sturm, V., Timmermann, L., Tronnier, V., Trottenberg, T., Wojtecki, L., Wolf, E., Poewe, W., Voges, J.: A randomized trial of deep-brain stimulation for parkinson's disease. N. Engl. J. Med. **355**, 896–908 (2006)
4. Kringelbach, M.L., Jenkinson, N., Owen, S.L.F., Aziz, T.Z.: Translational principles of deep brain stimulation. Nat. Rev. Neurosci. **8**, 623–635 (2007)
5. Rossi, S., Hallett, M., Rossini, P.M., Pascual-Leone, A.: Safety, ethical considerations, and application guidelines for the use of transcranial magnetic stimulation in clinical practice and research. Clin. Neurophysiol. **120**, 2008–2039 (2009)
6. Basu, S., Pany, S.S., Bannerjee, P., Mitra, S.: Pulsed magnetic field measurement outside finite length solenoid: experimental results & mathematical verification. J. Electromagn. Anal. Appl. **05**, 371–378 (2013)
7. Ridding, M.C., Rothwell, J.C.: Is there a future for therapeutic use of transcranial magnetic stimulation? Nat. Rev. Neurosci. **8**, 559–567 (2007)
8. Bonmassar, G., Lee, S.W., Freeman, D.K., Polasek, M., Fried, S.I., Gale, J.T.: Microscopic magnetic stimulation of neural tissue. Nat. Commun. **3**, 921 (2012)
9. Lee, S.W., Fried, S.I.: Suppression of subthalamic nucleus activity by micromagnetic stimulation. IEEE Trans. Neural Syst. Rehabil. Eng. **23**, 116–127 (2015)
10. Takemoto, M., Hasegawa, K., Nishimura, M., Song, W.J.: The insular auditory field receives input from the lemniscal subdivision of the auditory thalamus in mice. J. Comp. Neurol. **522**, 1373–1389 (2014)
11. Shibuki, K., Hishida, R., Murakami, H., Kudoh, M., Kawaguchi, T., Watanabe, M., Watanabe, S., Kouuchi, T., Tanaka, R.: Dynamic imaging of somatosensory cortical activity in the rat visualized by flavoprotein autofluorescence. J. Physiol. **549**, 919–927 (2003)
12. Park, H., Bonmassar, G., Kaltenbach, J.A., Machado, A.G., Manzoor, N.F., Gale, J.T.: Activation of the central nervous system induced by micro-magnetic stimulation. Nat. Commun. **4**, 2463 (2013)
13. Sawatari, H., Tanaka, Y., Takemoto, M., Nishimura, M., Hasegawa, K., Saitoh, K., Song, W.-J.J.: Identification and characterization of an insular auditory field in mice. Eur. J. Neurosci. **34**, 1944–1952 (2011)
14. Kozyrev, V., Eysel, U.T., Jancke, D.: Voltage-sensitive dye imaging of transcranial magnetic stimulation-induced intracortical dynamics. Proc. Natl. Acad. Sci. **111**, 13553–13558 (2014)

"Figure" Salience as a Meta-Rule for Rule Dynamics in Visual Perception

Kazuhiro Sakamoto[✉]

Department of Neuroscience, Faculty of Medicine,
Tohoku Medical and Pharmaceutical University,
4-4-1 Komatsushima, Aoba-ku, Sendai 981-8558, Japan
sakamoto@tohoku-mpu.ac.jp

Abstract. The brain faces many ill-posed problems whose solutions cannot be achieved solely on the basis of external conditions. To solve such problems, certain constraints or rules are required. Using a fixed set of rules, however, is not necessarily advantageous in ever-changing environments. Here, I revisit two of our previous psychophysical experiments. One pertains to visual depth perception based on spatial frequency cues, and the other involves apparent group motion. Results from these experiments demonstrate that perceptual rules change dynamically depending on the experimental conditions. They also suggest the existence of a meta-rule governing the dynamics of perceptual rules, which I refer to as a meta-rule of "figure" salience.

Keywords: Rule dynamics · Spatial frequency · Depth perception · Apparent motion

1 Introduction

One of the central issues for information processing in the brain is how to solve ill-posed problems. These are defined as problems for which unique solutions cannot be achieved under the given external conditions [1]. To solve these problems, appropriate constraints or internal rules are needed. For example, the famous random dot stereogram requires observers to establish, for each dot within an image projected onto one eye, a correspondence with a dot in the stimulus observed in the other eye [2]. However, the relationships between the dots perceived in each eye are arbitrary. Thus, some constraints or rules, such as a smoothness rule or a one-to-one correspondence rule, are required to achieve a single stereo percept [2,3]. In most computational models related to particular perceptual or behavioural problems, fixed sets of rules are employed to solve ill-posed problems. However, the use of a stable rule set results in a lack of adaptability to novel situations in ever-changing environments. In the discussion below, I introduce two domains related to psychophysical clues that suggest that

K. Sakamoto—This paper was supported by the grant from MEXT (22120504, 22500283 and 15H05879).

A. Hirose et al. (Eds.): ICONIP 2016, Part II, LNCS 9948, pp. 381–390, 2016.
DOI: 10.1007/978-3-319-46672-9_43

the brain applies rules in a dynamic fashion. Stereo perception arises not only from binocular vision, but also from monocular cues. For example, images consisting of spatial frequency components (as shown in Fig. 1b–e) induce a sense of depth [4]. However, whether the lower- or higher-frequency portion of the image is perceived as being in front changes depending on the stimuli used in the experiments [4–6]. The factors within images that may affect such judgements have yet to be elucidated.

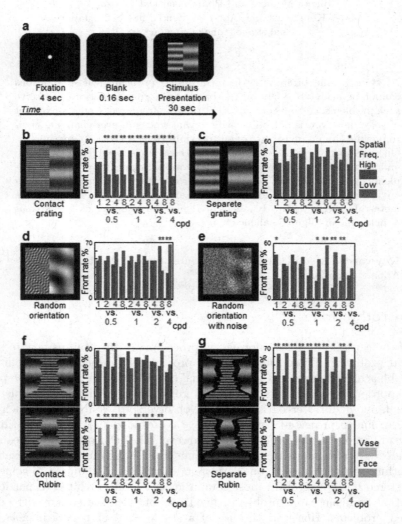

Fig. 1. Depth perception in spatial frequency images. (a) The time course of a trial. (b–g) An example of images and the front rate for the area of interest in each condition. Single and double asterisks correspond to $p < 0.05$ and $p < 0.01$ significance levels in the binomial test, respectively.

Motion perception also presents ill-posed problems. Apparent motion is a well-known phenomenon that arises when spots on a screen are followed by other spots at certain spatiotemporal intervals [7]. Motion percepts cannot be uniquely determined, due to ambiguity in the correspondence between spots in one frame and those in the following frame [2,8]. Apparent group motion, as illustrated in Fig. 2b, is characterised by dynamic changes in motion percepts [9,10]. When the stimulus onset asynchrony (SOA) is long enough, spots are seen to move as a whole (group motion), whereas under short SOA conditions, spots move to the nearest other spots irrespective of the entire configuration (exhibiting local motion). To explain such changes in apparent motion, Dawson has proposed a relative velocity constraint [11,12]. Nevertheless, this does not guarantee the preservation of dot configurations. Thus, more systematic surveys are needed.

In this article, I overview two psychophysical experiments that we previously conducted, involving depth perception of spatial frequency stimuli [13,14] and apparent group motion [15,16]. Our results reproduced previous findings of the dynamic nature of the rules influencing judgements in each of these tasks. Furthermore, they suggest that the strength or salience of the "figure" exerted a strong influence on rule dynamics in these tasks. Based on these results, I hypothesis a meta-rule of "figure" salience that allows for the flexible operation of perceptual rules.

2   Depth Perception in Monocular Spatial Frequency Images

The following is a summary of [13,14].

2.1   Methods

Stimuli were presented on a CRT monitor (Nanao FlexScan 88F) in the 1280 × 1024 pixels mode after gamma correction. The size of a single stimulus was 4° × 4°. The mean luminance was 22 cd/m². Stimulus presentation was controlled by a computer with a Pentium 133 MHz processor and a VSG2/3 board (Cambridge Research Systems). The frame rate was 75 Hz. All experimental sessions were conducted in a dark room.

We used five spatial frequencies (0.5, 1, 2, 4 and 8 cycles/degree), each of which occupied one of two areas in each stimulus. Six types of stimuli were prepared: contact grating, separate grating, random orientation, random orientation with noise, contact Rubin and separate Rubin (Fig. 1). Each stimulus was viewed monocularly.

Four male subjects with normal vision participated in this experiment. After a fixation spot was presented for 4 s, a stimulus was presented for 30 s following a brief blank frame (Fig. 1a). Subjects had to keep pressing the key assigned to an area as long as they perceived that area to be in front of the other one. The front rate for an area was defined as the ratio of the total time that the key was pressed to the total presentation time (30 s). The inter-stimulus interval (ISI) was 20 s. Each subject performed 20 trials in each stimulus condition. The head of each subject was fixed, with a viewing distance from the display of 1.28 m. The performance of subjects was assessed using a binomial test.

2.2   Results

In the contact grating condition, in which the border between the two areas was clear, the higher-frequency area was perceived as being in the front in most cases (Fig. 1b). This tendency was degraded in the separate grading condition (Fig. 1c) and in the random orientation condition (Fig. 1d). In the separate grading condition, the two areas in each image were allocated their own borders, so that competition for a single border between the two areas was unlikely. Images used in the random orientation condition appeared to yield less obvious impressions of a border than those in the contact grating condition. When the border became even more ambiguous in the random orientation with noise condition, results contrasted sharply with those for the contact grating condition (Fig. 1e); i.e. the lower-frequency area was perceived as the nearer one in multiple cases.

Results from trials using Rubin's vase were even more impressive. In the contact Rubin condition, the higher-frequency areas were perceived as being in front much less frequently (Fig. 1f, upper graph) than in the corresponding cases of the contact grating condition. Instead, when subject performance was sorted based on the shape of the area, vase or face, it was apparent that face areas were preferentially interpreted as being in front (Fig. 2f, lower graph). The separate Rubin condition, in which both vase and face areas had borders and could pop out as figures simultaneously, yielded very different results. It did not matter anymore whether the area was a vase or a face (Fig. 1g, lower graph); lower-frequency areas were seen as being in front in all cases (Fig. 1g, upper graph).

3   Apparent Group Motion

The results described in this section are from [15, 16].

3.1   Methods

Apparent motion stimuli were presented on a CRT monitor (Nanao FlexScan 88F) with a black background using a VSG2/3 board (Cambridge Research Systems). The diameter of each spot within a dot configuration was 0.2° and its luminance was 26 cd/m². All experiments were conducted in a dark room. A fixation cross was also presented in the centre of the display throughout each trial.

Each dot configuration was determined as follows (Fig. 2a). First, the length of the configuration, L, was determined. Second, the basic gap between adjacent dots, L, was calculated by dividing L by ($dot\ number$ − 1). Third, virtual lines vertical to the direction of L and with the spacing of G were presumed. Each dot was placed at a pseudo random position on the corresponding virtual line (in practice, the configuration obtained as a result of the aforementioned steps obeyed a certain restriction, to be described later.) Fourth, the gross angle of the configuration, θ, was determined. θ had five steps: $25°$, $35°$, $45°$, $55°$, and $65°$. Each dot configuration and its rotated image were presented in alternation. In general, even when group motion is perceived, the correspondence of dots between alternating displays depends on the configuration angle θ. This is shown in Fig. 2b using a simple example of a straight dot configuration (Fig. 2b). When θ is $0°$ (Fig. 2b, left), dot 1 in the first frame (white square 1) is perceived to move to dot 1 in the second frame (red circle 1). As a whole, the configuration slides horizontally (green arrow). In contrast, when θ is $90°$ (Fig. 2b, right), white square 1 moves to red circle 4. The entire configuration also moves in the same direction (red arrow). These two different dot correspondences should change over when θ is approximately $45°$ (Fig. 2b, middle), which was confirmed by our preliminary experiment using straight configurations (data not shown). Therefore, two types of group motion are possible. One is overhand motion (Fig. 2c, left) and the other is underhand motion (Fig. 2c, middle). Hence, we have two changes for each configuration (Fig. 2d). The one that preserves its configuration when overhand motion is perceived is defined as the overhand alteration (Fig. 2d, left) and the opposite one is referred to as the underhand alteration (Fig. 2d, right).

The restriction on dot configurations alluded to above was as follows. We calculated each configuration so that overhand and underhand motions were perceived with equal probability in the $\theta = 45°$ condition, even when the configuration was presented as an overhand or underhand alteration, as long as motion was perceived in accordance with Dawson's three constraints: one-to-one correspondence, nearest neighbour and minimum relative velocity [11,12]. Four configurations were used within each experiment.

Four male subjects with normal vision performed the task, in which each subject's head was fixed. Viewing distance was set at 1 m. Subjects were first required to fixate the central cross. Then a dot configuration was presented with four alterations. Each stimulus duration was 75 or 125 msec, and ISIs were 25, 50, 75, 100 or 125 msec. Thus, each experiment included $400 (= 2 \times 2 \times 4 \times 5 \times 5)$ trials. Subjects were required to select an answer from among four options: overhand, underhand, nearest neighbour and no motion (Fig. 2c). Note that nearest neighbour motion does not represent group motion.

3.2 Results

We conducted five experimental sessions using different configuration parameters, as indicated in Fig. 2e. The response rates for the four options in each condition, averaged over all subjects in Exp. 1, are plotted in Fig. 2f, as an example. Note that, when $\theta = 45°$, the response rate for underhand motion in the underhand alteration condition (Fig. 2f, lower row, middle, red area) was higher than that in the overhand alteration condition (Fig. 2f, upper row, middle, red area).

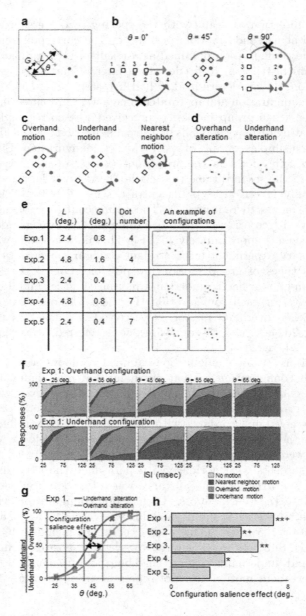

Fig. 2. Apparent motion experiments. (a) Definition of dot configuration parameters. (b) The schematics for the two possible group motion percepts and the transition between them depend on the parameter θ. (c) Perceivable motion. (d) Two alterations for dot configurations. (e) Conditions within each experiment. (f) An example of responses in Exp. 1 averaged across all subjects. (g) Changes in the ratio for underhand motion along with θ in Exp. 1. (h) Comparison of configuration salience effects among experiments. The stimulus duration of the experiments shown in (f–h) was 75 msec. Asterisks and crosses correspond to $p < 0.01$ and $p < 0.05$ significance levels of the binominal test for each of four subjects, respectively. (Color figure online)

This discrepancy could not occur if motion was perceived in accordance with Dawson's constraints, particularly the relative velocity principle. Such differences can be quantified by calculating the ratio of underhand motion curves, as illustrated in Fig. 2g. The ratio of underhand motion for each θ is defined as the average response for underhand motion across all ISI conditions divided by the corresponding averaged response for overall group motion, including both underhand and overhand motion. In this way, two curves representing the ratios for underhand motion were obtained: one for the underhand alteration (red curve in Fig. 2g) and the other for the overhand alteration (green curve in Fig. 2g). The configuration salience effect within an experiment (black arrow in Fig. 2g) was defined as the angle difference between the two curves at the 50 % level of the ratio of underhand motion (dotted line in Fig. 2g). I emphasise here again that the configuration salience effect is an index of the degree of preservation of the configuration of interest. This should be zero if motion perception obeys Dawson's principles, even though the relative velocity principle was intended to minimise local changes in the configuration.

Figure 2h shows the configuration salience effect for each experiment obtained by averaging response rates across four subjects. Statistical significance was evaluated for each subject, as indicated by asterisks or crosses. Comparisons among experiments suggest that the smaller configuration length L resulted in a stronger tendency to preserve the configuration, with Exp. 1 yielding a stronger configuration salience effect than Exp. 2. This was caused by L and not by G, because a small value for the configuration salience effect was observed in Exp. 4, where L was identical to that in Exp. 2, whereas G was the same as that in Exp. 1. This is further supported by the comparison between Exps. 1 and 3.

However, the situation here appears to be somewhat more complex. Let us compare the results of Exps. 3 and 5. In these experiments, the configuration parameters were the same except that the dots in Exp. 5 were placed pseudo-randomly (following the restriction described in 3.1) whereas in Exp. 3, the second, fourth and sixth dots were placed in the middle of their neighbouring dots. The configuration salience effect in Exp. 3 was strong and comparable to that of Exp. 1. In contrast, the results in Exp. 5 did not differ significantly between the underhand and overhand alterations. The configurations in Exp. 3 were apparently recognisable in comparison with those in Exp. 5, which should have affected these results.

4   Discussion

In this article, I have reviewed two psychophysical experiments that we previously conducted, involving stereo perception based on spatial frequency cues and apparent visual motion. Both experiments demonstrate that rules governing perceptual decisions appear to change dynamically.

In the experiment involving depth perception, we reproduced both of the results of prior studies. That is, in the contact grating condition, in which two gratings of a single orientation contacted each other, higher-frequency areas were perceived as being in front. This result is consistent with that of [6]. In contrast, the random orientation with noise condition resulted in the perception that lower-frequency areas were in front, a result that is compatible with [4,5]. Based on these results, it is obvious that depth perception based on spatial frequency cues does not follow a fixed perceptual rule, such as a high-front or low-front rule. We therefore hypothesised that the sharpness of figure-ground segregation influenced these apparent changes in the application of perceptual rules for the following reasons. Images in the contact grating condition contained clear borders that tended to belong to each area individually, rather than to both areas. Border-ownership is well-known to be a critical factor in figure-ground segregation. On the other hand, borders became vague in the random orientation with noise condition, where the law of perspective seemed dominant. This idea is consistent with the results of the separate grating and random orientation conditions, in which the ownership of borders was less relevant. The results of experiments using Rubin's vase stimuli strongly supported this hypothesis. In the contact Rubin condition, the difference between the spatial frequencies of the two areas, face and vase, became irrelevant. Instead, face areas popped out in almost all cases. In contrast, in the separate Rubin condition, both face and vase areas were given borders, which eliminated competition over a single border. Consequently, the law of perspective prevailed, with low-frequency areas preferentially perceived as being in front.

Perceptual constraints seem to change in visual motion perception as well. The Ternus configuration of apparent motion is a good example. In simple cases of apparent motion, involving a small number of dots, the correspondence between dots at different time frames is governed by the one-to-one correspondence and the nearest neighbour constraints [10]. This is also true for Ternus configurations under short SOA conditions. In contrast, under longer SOA conditions, motion percepts that appear to preserve dot configurations are obtained, and these cannot be explained solely on the basis of these two constraints. To explain this group motion phenomena, Dawson invoked another constraint, the so-called relative velocity principle, which minimises the differences among motion correspondences or motion vectors [11,12]. Our experimental results support Dawson's intent that dot configurations should be preserved in group motion, but not his specific implementation using the relative velocity constraint. This constraint prefers configuration preservation but does not guarantee it. In fact, the overhead configuration in Fig. 2d, left, for instance, is not preserved completely when an underhand motion is perceived, even though we prepared all configurations in such a manner as to yield the same response rates for overhand and underhand motion under Dawson's constraints. In lieu of Dawson's constraint, a rule capturing the degree of configuration salience or "figure" salience should be formulated to account for dynamic changes in the perception of group motion. This idea is empirically supported by the configuration salience effect, or the discrepancies in performance between overhand and underhand motion, as exemplified in Fig. 2g, in our experiments. In particular, the distinct results of Exps. 3 versus 5 should be noted. Dot configurations in these two experiments had the same parameters for length (L), gap (G) and dot number. However, in the configurations of Exp. 3, the second, fourth and sixth dots were placed in the median centres between their neighbouring dots. Shapes involving this type of configuration appear to be more easily recognised as "figures" than those in Exp. 5, which must have accounted for the differences in the configuration salience effects shown in Fig. 2h.

Taking the results of these two experiments together, I conclude that rules governing perceptual decisions are not always fixed, but can change dynamically and that a meta-rule for "figure" salience should be one of the factors governing these dynamic changes in perceptual rules. Influences of "figure" salience on psychophysical judgements have also been reported by another group [17].

Other meta-rules that exert effects on perceptual dynamics should also be explored. One promising candidate might be the existence of a rule related to the degree of simplicity, as has been discussed in relation to the law of Prägnanz in Gestalt psychology [18]. The problem of occluded contour completion presents a good example. Superficially, the rules guiding the completion of an occluded contour flip between the law of local continuity of the contour coverture and the law of global symmetry of the contour. However, some studies have pointed out that this phenomenon is governed by a meta-rule for simple representations or compressible codings of contour shape [19,20].

In conclusion, I must emphasise the differences between conventional rules or constraints and the meta-rules discussed here. To develop computational theories of certain cognitive or behavioural processes, one must calculate or operate certain concrete physical parameters, such as spatial frequency or moving spots, obtained from front-end sensors like eyes or cameras. However, if constraints or rules are specific to a limited number of concrete parameters, then computational processing will lose the flexibility to deal with unexpected changes in situations. In contrast, the meta-rules discussed above are somewhat ambiguous, but they do not depend on particular parameters or dimensions. For example, figure-ground separation can be achieved through a wide variety of cues. I have pointed out in this article that the degree of "figure" salience can exert an effect on other perceptual judgements such as those involving depth or motion. The ambiguous but flexible nature of meta-rules is one of key the factors that enable us to adapt to unpredictable environmental changes.

References

1. Marr, D.: Vision. MIT Press, Cambridge (1982)
2. Julesz, B.: Binocular depth perception without familiarity cues. Science **145**, 356–361 (1964)
3. Sato, N., Yano, M.: A model of binocular stereopsis including a global consistency constraint. Biol. Cybern. **82**, 357–371 (2000)
4. Klymenko, V., Weisstein, N.: Spatial frequency differences can determine figure-ground organization. J. Exp. Psychol. Hum. Percept. Perform. **12**, 324–330 (1986)
5. Brown, J.M., Weisstein, N.: A spatial frequency effect on perceived depth. Percept. Psychophys. **44**, 157–166 (1988)

6. Sakai, K., Finkel, L.H.: Characterization of the spatial-frequency spectrum in the perception of shape from texture. J. Opt. Soc. Am. A. **12**, 1208–1224 (1995)
7. Korte, A.: Kinematoskopische Untersuchungen. Zeitschrift fur Psycholgie **72**, 193–296 (1915)
8. Sakamoto, K., Sugiura, T., Kaku, T., Onizawa, T., Yano, M.: Spatiotemporal balance in competing apparent motion is not predicted from the strength of the single-motion percept. Perception **35**, 947–957 (2006)
9. Ternus, J.: Experimentelle Untersuchungen uber phanomenale Identitat. Psychologische Foschung **7**, 81–136 (1926)
10. Ullman, S.: The Interpretation of Visual Motion. MIT Press, Cambridge (1979)
11. Dawson, M.R.W.: The how and why of what went where in apparent motion: modeling solution to the motion correspondence problem. Psychol. Rev. **98**, 569–603 (1991)
12. Dawson, M.R.W., Wright, R.D.: Simultaneity in the Ternus configuration: psychophysical data and computer model. Vis. Res. **34**, 397–407 (1994)
13. Yano, M., Suzuki, T., Sakamoto, K.: Interaction between spatial frequency components on apparent depth: effects of contour shape. Soc. Neurosci. Abstr. 873.3 (1999)
14. Suzuki, T., Sakamoto, K., Yano, M.: Changes in apparent depth through interaction between spatial frequency. In: Proceedings of IEICE General Conference, D-2-28 (1999)
15. Kaku, T., Sakamoto, K., Yano, M.: Grouping in visual apparent motion. In: Proceedings of IEICE General Conference, D-2-29 (1999)
16. Sakamoto, K., Kaku, T., Yano, M.: Effects of global features on grouping in visual apparent motion. In: Annual Conference Japanese Neural Network Society, P3–19 (1999)
17. Leonards, U., Singer, W., Fahle, M.: The influence of temporal phase differences on texture segmentation. Vis. Res. **36**, 2689–2697 (1996)
18. Koffka, K.: Principles of Gestalt Psychology. Kegan Paul, Trench (1935)
19. van Lier, R., van der Helm, P., Leeuwenberg, E.: Integrating global and local aspects of visual occlusion. Perception **23**, 883–903 (1994)
20. Sakamoto, K., Kumada, T., Yano, M.: A computational model that enables global amodal completion based on V4 neurons. In: Wong, K.W., Mendis, B.S.U., Bouzerdoum, A. (eds.) ICONIP 2010, Part I. LNCS, vol. 6443, pp. 9–16. Springer, Heidelberg (2010)

A Neural Network Model for Retaining Object Information Required in a Categorization Task

Yuki Abe[1], Kazuhisa Fujita[1,2], and Yoshiki Kashimori[1(✉)]

[1] Department of Engineering Science, University of Electro-Communications,
Chofu, Tokyo 182-8585, Japan
yuuki.abe@uec.ac.jp, kazu@spikingneuron.net,
kashi@pc.uec.ac.jp
[2] Tsuyama National College of Technology,
654-1 Numa, Tsuyama, Okayama 708-8506, Japan

Abstract. Categorization is our ability to generalize properties of object, and clearly fundamental cognitive capacity. A delayed match-to-categorization task requires working memory of category information, shaped by the interaction between prefrontal cortex (PFC) and inferior temporal (IT) cortex. In the present study, we present the neural mechanism by which working memory is shaped and retained in PFC and how top-down signals from PFC to IT affect the categorization ability.

Keywords: Categorization · Neural model · Working memory · Top-down influence

1 Introduction

In daily life, we can rapidly and effortlessly recognize huge objects. Object information is known to process along two visual pathways; ventral and dorsal pathways. Although, in the early visual areas, features of object's shapes are represented as simple features such as lines and orientations, they are gradually represented as complex features as object information is transmitted to higher visual areas [1, 2]. On the other hand, the representation of complex features is closely related with visual recognition [3]. The ability to categorize objects is a fundamental function of visual recognition. Our raw perception would be useless without our classification of visual items. Visual categorization has a crucial role in giving object's meaning. Although a great deal is known about the neural analysis of visual features, little is known about the neural basis of categorization.

Much effort has been devoted to the studies of how category information is processed in the brain circuits. Sigala and Logothetis [4] reported that responses of inferior temporal (IT) cortex neurons were enhanced by stimulus features relevant to a category task. We demonstrated that the enhanced responses of IT neurons were caused by top-down signals from prefrontal cortex (PFC) [5]. Freedman et al. [6] also trained monkeys to categorize the computer-generated images of dogs and cats, and demonstrated the involvement of the interaction between the IT cortex and the PFC in the categorization task. Moreover, they showed that neurons in the IT were more sensitive

A. Hirose et al. (Eds.): ICONIP 2016, Part II, LNCS 9948, pp. 391–398, 2016.
DOI: 10.1007/978-3-319-46672-9_44

to object features, whereas PFC neurons had a larger sensitivity to object categories than that of IT neurons. They also pointed out that the PFC retains category information as working memory. However, it still remains unclear how working memory of category information is formed in the PFC and what a role the top-down from PFC to IT plays in the categorization task.

To address these issues, we develop a model of visual system that consists of a PFC and an IT networks. Using the model, we show that PFC neurons with lower firing thresholds play a dominant role in formation and retention of working memory, whereas PFC neurons with higher firing thresholds are highly sensitive to category information of dogs/cats-images. We also show that top-down from PFC to IT enables PFC neurons to improve the categorization ability of the dogs/cats-images close to the boundary in discrimination between dog and cat categories. The improvement arises from the mechanism by which top-down signals increase the subthreshold activity of IT neurons even in the absence of stimulus. The increase leads to the increase in the sensitivity of IT neurons to forthcoming stimulus. These results suggest that different firing thresholds of PFC neurons and top-down from PFC may contribute to an efficient performance of categorization task.

2   Model

We constructed a network model for dog/cat-categorization task, which consists of network models of IT and PFC. The PFC network represents category information of dog and cat as dynamical attractors, as shown in Fig. 1a. Similarly, the IT network has attractors of three dog objects and three cat ones. Both networks are reciprocally connected to each other.

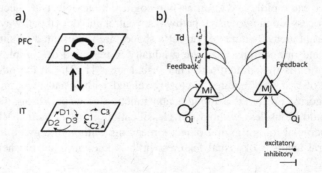

Fig. 1. (a) The models of IT and PFC. Information about objects and categories are represented by dynamical attractors in both networks. The attractors of dog and cat categories are denoted by D and C, and those of three dog and cat images are done by $D1 \sim D3$ and $C1 \sim C3$, respectively. (b) Neural units in the IT network. The single unit consists of a main neuron (M) and an inhibitory interneuron (Q). Synaptic connections have time delays of $t_d^1 - t_d^M$ in the recurrent connections and that of T_d in the feedback one, respectively. The PFC network has the similar units.

2.1   The IT Model

The IT network has two-dimensional array of neuronal units, with the size of the raw N_{IT} x the column M_{IT}. The neural unit consists of a pair of a main neuron and an inhibitory interneuron, as shown in Fig. 1b. The main neuron provides the paired interneuron with an excitatory output and receives an inhibitory input from the interneuron. It also receives a feedforward input from V4 and a feedback input from PFC. Then the membrane potentials of the (i, j)th IT neuron and the paired interneuron, V_{ij}^{IT} and u_{ij}^{IT}, are determined by

$$\tau_{IT}^{exc} \frac{dV_{ij}^{IT}}{dt} = -V_{ij}^{IT} + \sum_{t_d^m = t_d^1}^{t_d^M} \sum_{kl} w_{ij,kl}^{IT}(t, t_d^m) X_{kl}(t - t_d^m) + \sum_{mn} w_{ij,mn}^{IT-PFC,e} X_{mn}^{PFC}(t - T_d)$$
$$+ w^I U_{ij}^{IT} + I_{ij}^{FF}, \tag{1}$$

$$\tau_{IT}^{inh} \frac{du_{ij}^{IT}}{dt} = -u_{ij}^{IT} + w^E X_{ij}^{IT} + \sum_{mn} w_{ij,mn}^{IT-PFC,i} X_{mn}^{PFC}(t - T_d), \tag{2}$$

$$\mathrm{Pr}\,ob(X_{ij}^{IT} = 1) = \frac{1}{1 + \exp(-(V_{ij}^{IT} - V_\theta^{IT})/\varepsilon_{IT}^{exc})}, \tag{3}$$

$$U_{ij}^{IT} = \frac{1}{1 + \exp(-(u_{ij}^{IT} - u_\theta^{IT})/\varepsilon_{IT}^{inh})}, \tag{4}$$

where $w_{ij,kl}^{IT}(t, t_d^m)$ is the weight of the synaptic connection from the (k, l)th IT neuron to the (i, j)th one. $w_{ij,mn}^{IT-PFC,e}$ is the weight of the connection from the (m, n)th PFC neuron to the (i, j)th IT neuron and $w_{ij,mn}^{IT-PFC,i}$ is that of the connection from the (m, n)th PFC neuron to the (i, j)th interneurons. τ_{IT}^{exc} and τ_{IT}^{inh} are the time constants of the membrane potentials, V_{ij}^{IT} and u_{ij}^{IT}, respectively. w^E is the excitatory synaptic weigh to an interneuron and w^I is the inhibitory weight to the paired main neuron. IT neurons are connected through the recurrent connections, accompanied with synaptic transmission with multiple time delays, $t_d^m (m = 1 - M)$. A fixed time delay, T_d, was assumed for spike propagation between IT and PFC. The output of the (i, j)th IT neuron, X_{ij}^{IT}, has the value of 0 or 1, determined by the probability of Eq. (3), and the output of the (i, j)th interneuron, U_{ij}^{IT}, has the value of $0 \sim 1$, given by the sigmoidal function of Eq. (4). X_{mn}^{PFC} is the output of the (m, n)th PFC neuron. I_{ij}^{FF} is the feedforward input from V4.

2.2   The PFC Model

The PFC network has two-dimensional array of neuronal units, with the size of the row N_{PFC} x the column M_{PFC}. The unit has the same neuron pair as the IT unit. PFC neurons receive the output of IT neurons and are sending feedback to IT neurons. A PFC neuron is fully connected to other PFC neurons and has the feedback connection to the entire population of the IT network. Then the membrane potentials of the (i, j)th PFC neuron and the paired interneuron, V_{ij}^{PFC} and u_{ij}^{PFC}, are determined by

$$\tau_{PFC}^{exc}\frac{dV_{ij}^{PFC}}{dt} = -V_{ij}^{PFC} + \sum_{t_d^m=t_d^1}^{t_d^M}\sum_{kl} w_{ij,kl}^{PFC}(t, t_d^m)X_{kl}^{PFC}(t - t_d^m)$$
$$+ \sum_{mn} w_{ij,mn}^{PFC-IT}X_{mn}^{IT}(t - T_d) + w^I U_{ij}^{PFC}, \tag{5}$$

$$\tau_{PFC}^{inh}\frac{du_{ij}^{PFC}}{dt} = -u_{ij}^{PFC} + w^E X_{ij}^{PFC} + \sum_{q\in Q} w_q X_q^{PFC}, \tag{6}$$

where $w_{ij,kl}^{PFC}(t, t_d^m)$ and $w_{ij,mn}^{PFC-IT}$ are, respectively, the weights of the synaptic connection from the (k, l)th PFC neuron to the (i, j)th one and that from the (m, n)th IT neuron to the (i, j)th PFC neuron. The outputs of PFC neuron and interneuron are determined in the similar way to Eqs. (3) and (4). The last term in the right hand side of Eq. (6) indicates the mutual inhibition from PFC neurons responsive to opposite category Q.

2.3   Learning of Visual Objects and Category Information

According to the experiment by Freedman et al. [6], three dog images and cat ones were learned in the IT network. These images were denoted by D1, D2, and D3 for dog images and C1, C2, and C3 for cat ones. They are represented by distributed activities of IT neurons. The activity pattern of D1 is shown in Fig. 2a. Each of other images also has the activity pattern similar to that of D1. Similarly, in the PFC network, the activity patterns of dog and cat categories are represented by the activities shown in Fig. 2b.

The learning was made in a way that two stimuli in the same category were separated by a brief delay. The learning of the visual images and category information was made based on Hebbian learning, given by

$$\tau_w^X\frac{dw_{ij,kl}^X(t, t_d^m)}{dt} = -w_{ij,kl}^X(t, t_d^m) + \lambda_X X_{ij}^X(t)X_{kl}^X(t - t_d^m), \quad (X = IT, \quad PFC), \tag{7}$$

$$\tau_w^{X-Y}\frac{dw_{ij,kl}^{X-Y}(t, T_d)}{dt} = -w_{ij,kl}^{X-Y}(t, T_d) + \lambda_{X-Y}X_{ij}^X(t)X_{kl}^Y(t - T_d),$$
$$(X - Y = IT - PFC, PFC - IT), \tag{8}$$

where $w_{ij,kl}^X(t, t_d^m)$ (X = IT, PFC) are the weights of the synaptic connection from the (k, l)th X neuron to the (i, j)th one, and $w_{ij,kl}^{X-Y}(t, T_d)$ (X–Y = IT−PFC, PFC−IT) are the weight of the connection from the (k, l)th Y neuron to the (i, j)th X neuron. τ_w^X and τ_w^{X-Y} are the time constants of the weights, $w_{ij,kl}^X(t, t_d^m)$ and $w_{ij,kl}^{X-Y}(t, T_d)$, respectively, and λ_X and λ_{X-Y} are the learning rates of these weights.

a)   b)

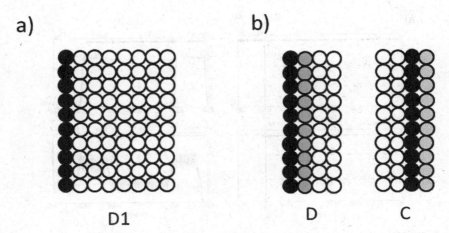

D1   D   C

Fig. 2. Memory patterns in IT and PFC networks. Activity patterns for object information of dog1 (D1) and those for category information of the dog (D) and cat (C) are shown in the panels (a) and (b), respectively. Activated neurons are depicted by filled circles and silent neurons are done by unfilled ones. The activities of neurons with higher and lower firing thresholds in the PFC network are shown by the gray and black colors, respectively.

3   Results

3.1   Response Properties of IT and PFC Neurons in Dog-Cat Category Task

Figure 3a shows the temporal variations of PFC and IT network states during the learning of a delayed match-to-category (DMC) task. In the DMC task used by Freedman et al. [6], the monkeys viewed two stimuli that were separated by a brief delay, and were trained to indicate whether a second (test) stimulus was from the same category as a previously seen (sample) stimulus. The three dog stimuli (D1 - D3) and three cat ones (C1 - C3) were used in the learning, and each test image was chosen from the same category as sample image. Figure 3a also shows the temporal variations of the network state for two DMC tasks after the learning. One task is a category-matching (CM) task, in which a test image is within the same category as a sample image, and another is a category-unmatching (CU) task, in which a test image and a sample image are, respectively, within different categories. In both tasks, the IT network settled down to the attractor relevant to each image, whereas the PFC network fell in the attractor of the category relevant to each image. Thus, the IT and PFC networks, respectively, are capable for representing the object images and categories as the relevant attractors.

Figure 3b shows the time courses of spike counts of PFC and IT neurons during the CM task shown in Fig. 3a. PFC neurons lasted the firing in the delay period, indicating the retention of working memory of the sample image D1. On the other hand, IT neurons showed a rapid disappearance of firing in the delay period and were silenced,

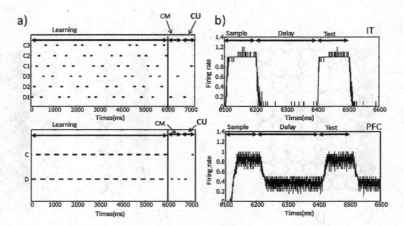

Fig. 3. (a) Temporal variations of PFC and IT network states during the learning and two tasks. Upper: IT network, Lower: PFC network. A category-matching task and a category-unmatching task are denoted by CM and CU, respectively. (b) Temporal courses of firing rates of IT and PFC neurons in the CM task.

indicating that IT neurons respond only in the presence of stimulus. The result exhibits the firing properties similar to the experimental result by Freedman et al. [6].

3.2 Role of Different Firing Thresholds of PFC Neurons in Retention of Working Memory

To understand the mechanism underlying the retention of working memory in the PFC network, we investigated the firing patterns of PFC neurons. The raster plot of PFC firing is shown in Fig. 4a. PFC neurons with lower firing thresholds lasted the firing even in the stimulus offset, leading to the retention of working memory. In contrast, neurons with higher firing thresholds fired only in the presence of stimulus, providing them with a high selectivity to category information. Moreover, a uniform firing threshold was unable to retain the working memory, as shown in Fig. 4b.

3.3 Feedback Influence on Categorization Performance

After the learning, test images were created by linear combinations of a dog and a cat image. When the mixing ratios of two images are close to 50 %: 50 %, e.g. D1:C1 = 40 %:60 % and D1:C1 = 60 %:40 %, one will expect that the monkeys fail to determine the categories of ambiguous images. However, the monkeys clearly discriminated between categories of mixed images [6]. The discrimination ability of ambiguous images is due to the top-down influence on IT responses. Figure 5a shows the temporal courses of the activities of IT and PFC neurons for the DMC task in which the sample and test stimuli were, respectively, D1 and a mixed image (D1:C1 = 60 %:40 %). Top-down signals from PFC elevated the activity of IT neurons in the delay period, enhancing the IT responses to the forthcoming

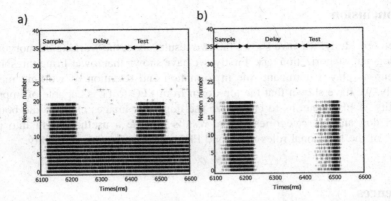

Fig. 4. Raster plots of PFC neurons in a CM task. (a) PFC neurons with different firing thresholds. Of PFC neurons responsive to dog category, the neuron for $i = 1 - 10$ have lower thresholds, while those for $i = 11 - 20$ have higher thresholds. (b) PFC neurons with a uniform firing threshold.

mixed images, even if they are ambiguous images for categorization. This enhancement enables the PFC network to clearly represent the category information relevant to the mixed stimulus. On the other hand, the lack of top-down signals did not elicit the enhanced responses of IT neurons, leading to no improvement of the categorization ability of PFC neurons, as shown in the PFC response in Fig. 5b.

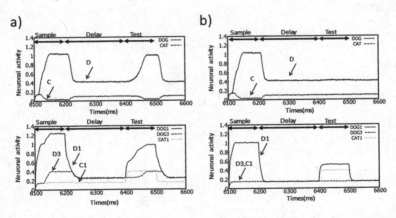

Fig. 5. Temporal variations of neural activities of PFC and IT neurons. Upper: PFC neurons responsive to the categories, D and C. Lower: IT neurons responsive to the images, D1, D3, and C1. The activities of PFC and IT neurons are shown in the presence (a) and absence (b) of top-down signals from PFC to IT.

4 Conclusion

In the present study, we have shown the two results about the working memory in PFC in the dog-cat categorization task. Firstly, we have shown that lower firing thresholds of PFC neurons play a dominant role in formation and retention of working memory. Secondly, we have shown that the top-down from PFC to IT is capable of improving the ability of PFC neurons to categorize the dogs-cats images close to the boundary between dog and cat categories. These results provide a useful insight into understanding of the functional roles of IT and PFC in categorization capability.

References

1. Tanaka, K.: Inferotemporal cortex and object vision. Annu. Rev. Neurosci. **19**, 109–139 (1996)
2. Logothetis, N.K., Sheinberg, D.L.: Visual object recognition. Annu. Rev. Neurosci. **19**, 577–621 (1996)
3. Palmeri, T.J., Gauthier, I.: Visual object understanding. Nat. Rev. Neurosci. **5**, 291–303 (2004)
4. Sigala, N., Logothetis, N.K.: Visual categorization shapes feature selectivity in the primate temporal cortex. Nature **415**, 318–320 (2002)
5. Soga, M., Kashimori, Y.: Functional connections between visual areas in extracting object features critical for a visual categorization task. Vis. Res. **49**, 337–347 (2009)
6. Freedman, D.J., Riesenhuber, M., Poggio, T., Miller, E.K.: A comparison of primate prefrontal and inferior temporal cortices during visual categorization. J. Neurosci. **23**, 5235–5246 (2003)

Pattern Recognition

Weighted Discriminant Analysis and Kernel Ridge Regression Metric Learning for Face Verification

Siew-Chin Chong[1]([✉]), Andrew Beng Jin Teoh[2], and Thian-Song Ong[1]

[1] Faculty of Information Science and Technology,
Multimedia University, Melaka, Malaysia
{chong.siew.chin,tsong}@mmu.edu.my
[2] School of Electrical and Electronic Engineering,
Yonsei University, Seoul, South Korea
bjteoh@yonsei.ac.kr

Abstract. A new formulation of metric learning is introduced by assimilating the kernel ridge regression (KRR) and weighted side-information linear discriminant analysis (WSILD) to enjoy the best of both worlds for unconstrained face verification task. To be specific, we formulate a doublet constrained metric learning problem by means of a second degree polynomial kernel function. The said metric learning problem can be solved analytically for Mahalanobis distance metric due to simplistic nature of KRR in which we named KRRML. In addition, the WSILD further enhances the learned Mahalanobis distance metric by leveraging the within-class and between-class scatter matrix of doublets. We evaluate the proposed method with Labeled Faces in the Wild database, a large benchmark dataset targeted for unconstrained face verification. The promising result attests the robustness and feasibility of the proposed method.

Keywords: Kernel ridge regression · Metric learning · Face verification · Unconstrained

1 Introduction

Face verification has been an active research area for decades and sparked enormous interest among the researchers. Applications of face verification continue blooming in personal security, law-enforcement and military sectors. The objective of the face verification is to determine whether a given pair of face images matches to the same person or different person. The significant variations of a face image caused by varying aging, lighting, expression, pose and others need to be managed carefully in order to fulfill the real world scenarios.

Learning a good distance metric is always a crucial task in machine learning such as computer vision, biometric recognition, activity recognition, and image retrieval. Metric learning techniques plays a vital role in face verification in the

© Springer International Publishing AG 2016
A. Hirose et al. (Eds.): ICONIP 2016, Part II, LNCS 9948, pp. 401–410, 2016.
DOI: 10.1007/978-3-319-46672-9_45

past decade [1,2]. The problem of metric learning relates to learning a distance function adjusted to a specific task. A new distance metric is learned from the training samples to measure the similarity between the samples by enlarging the similarity of similar pairs and reducing the similarity of the dissimilar pairs. Different metric learning algorithms have different objective functions, thus different types of metrics can be learned [3–5].

In spite of the fact that many metric learning algorithms have been proposed and proved to be useful, there are still problems to be further investigated. Some of the metric learning methods that involve additional information in training process might become impractical in certain scenarios such as verifying an intruder who tries to abuse the system repeatedly or an unknown whose identity information is not available in the data bank [6,7]. Furthermore, the flexibility in integrating the metric learning methods in the real world applications remains as an interesting topic to be further investigated.

Wang et al. [8] outlined a new perspective on designing metric learning method via support vector machine, named SVMML. Their approach utilizes doublet-based or triplet based constraints. With inspiration from Wang et al. [8] work and consider the above mentioned problems, a simplistic metric learning is proposed by leveraging Weighted Side-Information based Linear Discriminant Analysis (WSILD) [9] and Kernel Ridge Regression (KRR) [10] for learning the Mahalanobis distance metric in unconstrained face verification task. Specifically, doublets are constructed from the training pairs and are represented in a means of second order polynomial kernel function. KRR provides the closed-form solution for metric learning in which we named KRRML, while WSILD exploits the within-class and between-class scatter matrix by using only the side-information. Furthermore, the samples which are difficult to be classified are stressed through a weighting function of WSILD.

The restricted face verification protocol is adopted to report the experimental results with Labeled Faces in the Wild (LFW) [11] dataset. This dataset is more feasible to demonstrate the real-life scenario, where only labels of same person or not same person are involved in training. No other extra information about the person is provided.

The paper is organized as follows. The related works are reviewed in Sect. 2. In Sect. 3, the contribution of our proposed method in face verification is presented. The concept and flow the proposed method is described in Sect. 4. Section 5 demonstrates the experimental results and analysis. Lastly, Sect. 6 concludes the paper.

2   Related Works

Kernel ridge regression (KRR) [10] is one of the classical algorithms for classification and regression. KRR is a regularized least square method, which nonlinearly mapped the observed predictor variables into a high-dimensional space. Support Vector Machine (SVM) [12] is similar to KRR, except that different constraint objective function is being optimized. SVM relies on a subset of support vectors instead of all the training examples.

Most of the metric learning algorithms seek for an optimized cost function in learning a distance metric from the training data. One of the famous metric learning algorithms was introduced by Weinberger et al. [3] to learn a transformation matrix in order to improve the k nearest neighbor (kNN) classification through maintaining a large margin at the boundaries of different categories. An extended version of large margin nearest neighbor metric learning was presented by Kumar et al. [13] for transformation invariant classification. Davis et al. [14] deals with general pair-wise constraints by proposing Information Theoretic Metric Learning (ITML). A multivariate Gaussian's differential entropy is maximized subject to the constraints on the Mahalanobis distance. A modified ITML is then presented by Saenko et al. [15] for visual category domain adaptation. A logistic discriminant approach known as Logistic Discriminant Metric Learning (LDML) [16] is introduced to learn the metric on sets of image pairs. Besides, a fast gradient-based algorithm is developed by Hieu et al. [17] to study the cosine similarities. In addition, Pairwise-constrained Multiple Metric Learning (PMML) [18] learns a discriminative distance on multiple metrics. Wang et al. [8] proposed to utilize doublet-based constraints in metric learning through support vector machine. On the other hand, Linear Discriminant Analysis (LDA) [19], which is widely use in face recognition, can be viewed as metric learning technique over the intra-personal subspace, due to its objective in minimizing the average distance between similar pairs. Kan et al. [9] proposed a variant of LDA by utilizing the side-information to estimate the between-class and within-class scatter matrices.

Two criteria should be imposed when designing a good metric learning algorithm. First criterion is that the availability of training sample labels should be as minimal as possible. It is difficult to obtain the full label of the training samples in real life scenarios. Compared to the class label, data pair labels are more preferable and more practical in metric learning applications. However, data pair labels are weaker in the sense that only the information of similar or dissimilar is known. Second criterion is that the algorithm should be computationally efficient. In other words, the metric learning techniques should be scalable with respect to the size of the training samples.

3   Contribution

Based on the issues of the existing metric learning techniques and the two criteria of designing a good metric learning algorithm as discussed in the previous sections, we come out with the solution by introducing a new formulation of metric learning, namely KRRML, and coupling with WSILD for unconstrained face verification task. The proposed method offers several merits as follows:

- Does not rely on full label information of samples.
- Emphasize on the pair of samples that are hard to be classified.
- Closed-form solution of KRR promises speedy training process.

4   Overview of the Proposed Solution

In this section, the overview of the proposed method, KRRML with WSILD for face verification is given. Firstly, the cropped face images is applied with Difference of Gaussians (DoG) [20] filter to remove the noises and ameliorate the image quality. After that each face image is divided into several local regions. Over-Complete Local Binary Pattern (OCLBP) [21] is used to extract the face descriptors from each region independently. The face descriptors are remained as a set of blocks to be processed in the proposed framework. Basically the proposed approach involves WSILD in prior and followed by KRRML. Below we show the inherent relationship between these two metric learning methods.

Considering a dataset of subjects $\{x_i\}_{i=1}^{N} \in \mathbb{R}^n$. $S = \{(x_i, x_j) : l(x_i) = l(x_j)\}$ is the set of OCLBP feature vectors belongs to the similar image pairs and $D = \{(x_i, x_j) : l(x_i) \neq l(x_j)\}$ is the set of OCLBP feature vectors belongs to the dissimilar image pairs, with the class label of OCLBP feature vectors x, $l(x)$. As the first process, WSILD is applied to calculate the within-class and between class scatter matrices. Let K blocks of descriptors from each feature vector $x_i \in \mathbb{R}^{K \times n}$ for $i = 1, \cdots, N$, where n is the size of a block, the matrices for within-class and between class can be defined as below:

$$S_{W_k}^{block} = \sum_{(x_{ki}, x_{kj} \in S)} \left(x_i^k - x_j^k\right) \left(x_i^k - x_j^k\right)^T . \tag{1}$$

$$S_{B_k}^{block} = \sum_{(x_{ki}, x_{kj} \in D)} \left(x_i^k - x_j^k\right) \left(x_i^k - x_j^k\right)^T . \tag{2}$$

The k-th block of descriptors of the similar pair images, $(x_i^k, x_j^k) \in S$, and the dissimilar pair images, $(x_i^k, x_j^k) \in D$, are used to calculate the within-class and between class scatter matrices in Eqs. (1) and (2) respectively.

Consider the situation that a dissimilar pair images are near to each other, so the distance will be small and it contributes less to the between-class scatter matrix. Thus, a cosine similarity measure is used as weight function, $c(x_i^k, x_j^k)$ to emphasize the samples that are hard to classify:

$$S_{B_k}^{block} = \sum_{(x_{ki}, x_{kj} \in D)} c(x_i^k, x_j^k) \left(x_i^k - x_j^k\right) \left(x_i^k - x_j^k\right)^T . \tag{3}$$

$$c(x_i^k, x_j^k) = \frac{x_i^k \cdot x_j^k}{\| x_i^k \| \| x_j^k \|} . \tag{4}$$

The projection matrix for block-based WSILD, W_k^{block} can be redesigned as follows:

$$W_k^{block} = \arg\max \frac{| W^T S_{B_k}^{block} W |}{| W^T S_{W_k}^{block} W |} . \tag{5}$$

A projection matrix $W_k^{block} \in \mathbb{R}^{m \times n}$ with $m < n$ largest eigenvalues will be learned from each specific block among all the feature vectors in training. The k-th block face descriptor with reduced dimension can be generated as follows:

$$z_i^k = W_k x_i^k, \quad k = 1, \cdots, K .\tag{6}$$

Thus, K compact face descriptors, z_i^k for $i = 1, \cdots, N$ will be produced. An explicit function of the projection matrix W is maximized by WSILD in the form $Tr\left(\left(WS_W W^T\right)^{-1}\left(WS_B W^T\right)\right)$. When the matrix W consists of the first m eigenvectors of $S_W^{-1}S_B$ corresponding to the largest eigenvalues, the maximum is said to be attained.

KRRML operates on sets of doublets. A doublet (z_i, z_j) is formed by any two samples extracted from the training set. A label e is given to this doublet where $e = -1$ for similar image pairs from S and $e = 1$ for dissimilar image pairs from D. By combining all the doublets built from the training samples, a doublet set is formed by $\{q_i, \cdots, q_{N_d}\}$, where $q_l = (z_{l,1}, z_{l,2})$, $l = 1, \cdots, N_d$. The label of the doublet of q_l is denoted by e_l.

By integrating the degree-2 polynomial kernel, the distance metric learning can be readily created as a classification problem. To apply the kernel function to a pair of doublets, the degree-2 polynomial kernel can be extended as [6]:

$$K_D\left(q_i, q_j\right) = tr\left((z_{i,1} - z_{i,2})(z_{i,1} - z_{i,2})^T (z_{j,1} - z_{j,2})(z_{j,1} - z_{j,2})^T\right)\tag{7}$$

where $q_i = (z_{i,1}, z_{i,2})$ and $q_j = (z_{j,1}, z_{j,2})$ are a pair of doublets.

KRR learns a non-linear regression function in the space induced by the respective kernel functions. KRR uses squared error loss with l_2 regularization. In contrast to support vector machine, fitting KRR can be done in closed-form manner. An advantage of KRR is that the optimization solution has an analytic solution, which can be solved efficiently [33]. With the degree-2 polynomial kernels defined in Eq. (7), we can form the following:

$$g_d\left(q\right) = \sum_l \alpha_l K_D\left(q_l, q\right) + b .\tag{8}$$

where $q = (t_i, t_j)$ is the test doublet, b is the bias and α_l is the weight.

By substituting Eq. (7) into Eq. (8) for doublets, we have

$$\sum_l \alpha_l tr\left((z_{l,1} - z_{l,2})(z_{l,1} - z_{l,2})^T (t_i - t_j)(t_i - t_j)^T\right) + b$$

$$= \left((t_i - t_j) M(t_i - t_j)^T\right) + b .\tag{9}$$

where M is the projection matrix of Mahalanobis distance metric,

$$M = \sum_l \alpha_l (z_{l,1} - z_{l,2})(z_{l,1} - z_{l,2})^T .\tag{10}$$

Given that K is the kernel matrix, λ is the regularization term, I is the identity matrix and $e_l = \{1, -1\}$ is the label of the doublets, we can find:

$$\alpha_l = (K + \lambda I)^{-1} e_l .\tag{11}$$

Therefore, the regression function $g_d(q)$ with a thresholding function can be used to determine whether t_i and t_j are similar or dissimilar to each other. Formally, we need to maximize over λ and it can be chosen by performing the leave-one-out cross validation protocol. Given the doublet training set, the primal form of the KRRML can be formulated as:

$$\min_{M,e} \frac{1}{2} M^T \cdot M + \frac{\lambda}{2} \sum_{l=1}^{N} e_l^2 . \tag{12}$$

$$s.t \quad y_l = (z_{l,1} - z_{l,2})^T M (z_{l,1} - z_{l,2}) + e_l, \quad \forall l . \tag{13}$$

For the testing phase, the k-th block of the reduced feature vectors among the testing sets are projected onto the Mahalanobis matrix, W_k^{block}. A set of K compact vectors can be produced and are combined into a single feature vector. Then the KRR with RBF kernel is carried out to perform binary classification (match or not match). The experimental results in Sect. 4 proves that the proposed method is able to enhance the performance of learning significantly.

Ideally, the mentioned WSILD and KRR in the proposed method are able to integrate in the learning process smoothly. Consider a pair of feature vectors (x_i, x_j) with K blocks of face descriptors produced by OCLBP, the induced Mahalanobis distance d_A over points can be written as:

$$d_A \left(x_i^k, x_j^k \right) = \left(x_i^k - x_j^k \right)^T A \left(x_i^k - x_j^k \right) . \tag{14}$$

where $A = WW^T$ is a positive semidefinite matrix. The problem of metric learning is closely related to WSILD in view of the distance of between-class $W^T S_B^{block} W$ is maximized and the within-class $W^T S_W^{block} W$ is minimized. For the overall amalgamation of the KRRML approach, the Mahalanobis distance can be reformulated as:

$$d_M \left(x_i^k, x_j^k \right) = \left(x_i^k - x_j^k \right) W_k M W_k^T \left(x_i^k - x_j^k \right)^T . \tag{15}$$

where W is the projection matrix generated from blocked-based WSILD in Eq. (5), and M is the projection matrix learned by using KRR as in Eq. (10).

5   Experimental Result

Experiments are conducted using the LFW dataset based on the standard face verification "Restricted, no outside data" protocol to work on the unconstrained environment. There are a total of 13,233 face images in the LFW dataset with 5749 individuals. In the restricted protocol, 6000 pairs of different face images are randomly arranged from the sets to build 5,400 training image pairs (2,700 similar pairs and 2,700 dissimilar pairs)and 600 testing image pairs (300 similar and 300 dissimilar) applied with 10-fold cross validation splits.

In the experiments, the original 250×250 pixels face images are cropped into the size of 73×61 pixels. For a fair comparison, our method does not rely on any external training data and none of the pose estimators or 3D modeling methods is involved in the experiments. The experiment platform is using 64 bit Operating System with 8 GB RAM on fourth Generation Intel Core i7 processor.

Experiments are conducted using different dimensions and different hyper-parameters such as width of the RBF kernel, σ, the bias in the polynomial kernel, β, the degree in polynomial kernel, δ and the ridge, λ to evaluate the performance based on verification rate (VR) in percentage. Several configurations have decided after several trials. The parameters of the DoG filter (the std of smaller Gaussian, ∂_1 and the std of the larger Gaussian, ∂_2)are set to $\partial_1 = 0$ and $\partial_2 = 2$. For OCLBP, the best parameter settings chosen: total number of blocks = 7×7, horizontal and vertical overlap = 0.5 respectively, radius = $\{1, 2\}$ and number of points = 8. Each descriptor produced by OCLBP consists of dimension 256.

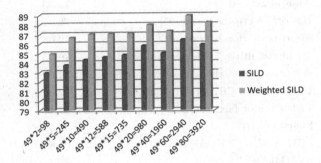

Fig. 1. Comparisons of KRRML with SILD and Weighted SILD using different dimensions (total number of blocks * reduced dimension)

Figure 1 illustrates the accuracies of KRRML with SILD and weighted SILD using different feature dimensions (reduced dimension = 2, 5, 10, 12, 15, 20, 40, 60, 80; number of blocks = 49). The highest accuracy rate can be achieved at 88.96 % with the dimension of 2940. Additionally, we can observe that the weighted SILD achieves better than SILD by effectively exploiting the samples that are hard to classify.

In Table 1, we compare our proposed method with the state-of-the-art algorithms in restricted face verification protocol. To be fair in comparison, we minimized the dependency on the preprocessing technique and followed the standard "Restricted, no outside data" protocol as benchmarked by the state-of-the art methods in their result reporting. By incorporating the classification power in metric learning, our method with 88.96 % is able to outperform the list of the state-of-the-art methods under the restricted protocol except for MRF-Fusion-CSKDA [32]. However, MRF-Fusion-CSKDA is not comparable with our method as it focusses mostly on multiple descriptors, but not the metric learning algorithm. The work introduces the component-based representation coupled with two image descriptors to reduce the sensitivity of misalignments and pose variations. On the other hand, we re-run the SVMML [8] under the face verification protocol and obtain 82.64 %. Utilizing the concept of kernel-classification based metric learning and taking consideration of weighted side-information of samples, our proposed method favorably achieves much better performance than SVMML.

Table 1. Performance comparison for the state-of-the-art methods under restricted protocol

Learning method	VR (%)
Eigenfaces [22]	60.02
Nowak [23]	72.45
3 × 3 multi region histograms [24]	72.95
Joint alignment [23]	73.93
Hybrid descriptor-based [25]	78.47
Multi-resolution framework [26]	79.08
V1-like/multiple kernel learning [27]	79.35
Adaptive probabilistic elastic matching [28]	84.08
Hierarchical-PEP [29] (single layer)	87.20
Fisher vector faces [30]	87.47
Spartans [31]	87.55
MRF-Fusion-CSKDA [32]	95.89
SVMML [8]	82.64
Our proposed method	88.96

6   Conclusion

A simplistic metric learning method is introduced by amalgamating KRR and WSILD to exploit the strengths of both for unconstrained face verification task. KRR provides the closed-form solution for metric learning, namely KRRML. To be precise, a doublet constrained metric learning problem is designed by means of a second degree polynomial kernel function, which can be solved analytically for Mahalanobis distance metric. Moreover, the WSILD further aggrandizes the learned Mahalanobis distance metric by leveraging the within-class scatter matrix and between-class scatter matrix of doublets. Experiments are run on the LFW dataset based on the "Restricted" protocol using ten fold cross validations. The results show that KRRML has reasonable accuracy performance, which is superior to the state-of-the-art methods in terms of verification rate.

Acknowledgments. The authors would like to thank Malaysia's Fundamental Research Grant Scheme for supporting the research under grants MMUE/140026.

References

1. Ying, Y., Li, P.: Distance metric learning with eigenvalue optimization. J. Mach. Learn. Res. **13**, 1–26 (2012)
2. Huang, C., Zhu, S., Yu, K.: Large scale strongly supervised ensemble metric learning with applications to face verification and retrieval. NEC TR115 (2011)
3. Weinberger, K.Q., Blitzer, J., Saul, L.K.: Distance metric learning for large margin nearest neighbor classification. In: NIPS, pp. 1473–1480 (2005)
4. Balcan, M.F., Blum, A., Srebro, N.: A theory of learning with similarity functions. Mach. Learn. **72**, 89–112 (2008)
5. Kedem, D., Tyree, S., Weinberger, K., Sha, F., Lanckriet, G.: Non-linear metric learning. In: Advances in Neural Information Processing Systems, pp. 2582–2590 (2012)
6. Cao, Z., Yin, Q., Tang, X., Sun, J.: Face recognition with learning-based descriptor. In: Computer Vision and Pattern Recognition (CVPR) (2010)
7. Lei, Z., Pietikainen, M., Li, S.Z.: Learning discriminant face descriptor. IEEE Trans. Pattern Anal. Mach. Intell. (PAMI) **36**(2), 289–302 (2013)
8. Wang, F., Zuo, W., Zhang, L., Meng, D., Zhang, D.: A kernel classification framework for metric learning (2013) arXiv: 1309.5823
9. Kan, M., Shan, S., Xu, D., Chen, X.: Side-information based linear discriminant analysis for face recognition. In: British Machine Vision Conference (2011)
10. Hastie, T., Tibshirani, R., Friedman, J.H.: The Elements of Statistical Learning. Springer, Heidelberg (2001)
11. Huang, G.B., Ramesh, M., Berg, T., Learned-Miller, E.: Labeledfaces in the wild: a database for studying face recognition in unconstrained environments. University of Massachusetts, Amherst, Technical report 7–49 (2007)
12. Cristianini, N., Shawe-Taylor, J.: An Introduction to Support Vector Machines. Cambridge University Press, Cambridge (2000)
13. Kumar, M.P., Torr, P.H.S., Zisserman, A.: An invariant large margin nearest neighbor classifier. In: ICCV, pp. 1–8 (2007)
14. Davis, J., Kulis, B., Jain, P., Sra, S., Dhilon, I.: Information-theoretic metric learning. In: ICML (2007)
15. Saenko, K., Kulis, B., Fritz, M., Darrell, T.: Adapting visual category models to new domains. In: Daniilidis, K., Maragos, P., Paragios, N. (eds.) ECCV 2010, Part IV. LNCS, vol. 6314, pp. 213–226. Springer, Heidelberg (2010)
16. Guillaumin, M., Verbeek, J., Schmid, C.: Is that you? Metric learning approaches for face identification. In: ICCV (2009)
17. Nguyen, H.V., Bai, L.: Cosine similarity metric learning for face verification. In: Kimmel, R., Klette, R., Sugimoto, A. (eds.) ACCV 2010, Part II. LNCS, vol. 6493, pp. 709–720. Springer, Heidelberg (2011)
18. Cui, Z., Li, W., Xu, D., Shan, S., Chen, X.: Fusing robust face region descriptors via multiple metric learning for face recognition in the wild. In: CVPR (2013)
19. Etemad, K., Chellappa, R.: Discriminant analysis for recognition of human face images. J. Opt. Soc. Am **14**, 1724–1733 (1997)
20. Anila, S., Devarajan, N.: Preprocessing technique for face recognition applications under varying illumination conditions. Glob. J. Comput. Sci. Technol. Graph. Vis. 12(11) (2012)
21. Barkan, O., Weill, J., Wolf, L., Aronowitz, H.: Fast high dimensional vector multiplication face recognition. In: ICCV, pp. 1960–1967 (2013)
22. Turk, M.A., Pentland, A.P.: Face recognition using eigenfaces. In: CVPR (1991)

23. Nowak, E., Jurie, F.: Learning visual similarity measures for comparing never seen objects. In: CVPR (2007)

24. Sanderson, C., Lovell, B.C.: Multi-region probabilistic histograms for robust and scalable identity inference. In: Tistarelli, M., Nixon, M.S. (eds.) ICB 2009. LNCS, vol. 5558, pp. 199–208. Springer, Heidelberg (2009)

25. Wolf, L., Hassner, T., Taigman, Y.: Descriptor based methods in the wild. In: ECCV (2008)

26. Arashloo, S.R., Kittler, J.: Efficient processing of MRFs for unconstrained-pose face recognition. In: Biometrics: Theory, Applications and Systems (2013)

27. Pinto, N., DiCarlo, J.J., Cox, D.D.: How far can you get with a modern face recognition test set using only simple features? In: CVPR (2009)

28. Li, H., Hua, G., Lin, Z., Brandt, J., Yang, J.: Probabilistic elastic matching for pose variant face verification. In: CVPR (2013)

29. Li, H., Hua, G.: Hierarchical-PEP model for real-world face recognition. In: CVPR (2015)

30. Simonyan, K., Parkhi, O.M., Vedaldi, A., Zisserman, A.: Fisher vector faces in the wild. In: BMVC (2013)

31. Felix, X., Luu, K., Savvides, M.: Spartans: single-sample periocular-based alignment-robust recognition technique applied to non-frontal scenarios. IP **24**(12), 4780–4795 (2015)

32. Arashloo, S.R., Kittler, J.: Class-specific kernel fusion of multiple descriptors for face verification using multiscale binarised statistical image features. IEEE Trans. Inf. Forensics Secur. **9**(12), 2100–2109 (2014)

33. Lee, W.H., Lee, R.B.: Implicit authentication for smartphone security. In: Camp, O., Weippl, E., Bidan, C., Aïmeur, E. (eds.) Information Systems Security and Privacy. CCIS, vol. 576, pp. 160–176. Springer, Heidelberg (2015)

An Incremental One Class Learning Framework for Large Scale Data

Qilin Deng[1], Yi Yang[1], Furao Shen[1(✉)], Chaomin Luo[2], and Jinxi Zhao[1]

[1] National Key Laboratory for Novel Software Technology,
Department of Computer Science and Technology,
Nanjing University, Nanjing, China
qldeng@qq.com, yangyi868@gmail.com,
{frshen,jxzhao}@nju.edu.cn
[2] Department of Electrical and Computer Engineering,
University of Detroit Mercy, Detroit, MI, USA
chaominluo@yahoo.com

Abstract. In this paper, we propose a novel one class learning method for the large scale data. In the context of one class learning, the proposed method could automatically learn the appropriate number of prototypes needed to represent the original target examples, and acquire the essential topology structure of target distribution. Then based on the learned topology structure, a neighbors analysis technique is utilized to separate the target examples from outlier examples. Experimental results show that our method can accommodate the large scale data environment, and achieve comparable or preferable performance than other contemporary methods on both artificial and real word data sets.

Keywords: One class learning · Incremental learning · Topology learning · Neighbors analysis

1 Introduction

One class classification (OCC) [1] addresses such learning problems where only specific examples are available for training the classifier. In this case, the available examples are called the target class, and examples not in this target class the outlier class. The objective of OCC is to separate target examples from examples not belonging to the target class. For the OCC problems, existing methods may have different characteristics concerning the description ability, the robustness against noises, the space and time complexity, etc. However, most of them including the state-of-the-art SVDD [2] or OCSVM [3] could not directly solve large scale learning problems very well. e.g., Nearest Neighbor(s) (NN) and Kernel Density Estimation (KDE) have to maintain the whole training data for testing, while SVDD or OCSVM need to solve a quadratic programming problem which will suffer from the large training data. Consequently, it is practically infeasible or very costly to build description models for OCC problems on large scale

© Springer International Publishing AG 2016
A. Hirose et al. (Eds.): ICONIP 2016, Part II, LNCS 9948, pp. 411–418, 2016.
DOI: 10.1007/978-3-319-46672-9_46

data by traditional methods. Meanwhile, large scale data may have very complex manifold structure, arbitrarily shaped distribution, multiple distributions with different shapes and different density levels. Some OCC methods make a Gaussian or spherical distribution assumption [1], thus these methods may result in a high false positive rate if the assumption is inconsistent with the fact. Moreover, It is common in practice that the training data may be corrupted by noises, and some traditional methods like nearest neighbor are very sensitive to these noisy data. Therefore, an important characteristic of the one class classifiers is their robustness against a few noises or outliers in the target training data.

In this paper, we address the problem of one class learning on large scale data, and propose an incremental one class learning framework. In comparison with other OCC methods, the main contribution of our work can be summarized as follows: (1) It can realize fast incremental one class learning in the situation of large scale data; (2) It could remove the noises and outliers automatically during learning process, thus is more robust to noises or outliers in the target examples; (3) It could detect complex manifold structure or multiple distributions with arbitrarily shapes, and recognize targets and outliers effectively and efficiently.

2   Proposed Method

To deal with those issues stated above, we propose an incremental learning framework named Topology Learning and Neighbors Analysis (TLNA). TLNA first performs incremental topology learning on the target training data to build a graph topological structure which captures the essential knowledge about the target class. Then based on the new graph representation, an effective nearest neighbors analysis technique is utilized to construct the one class classifier.

2.1   Incremental Topology Learning

In order to acquire a concise internal representation for the large scale training data, we adopt topology learning to learn the essential topological relations underlying the training data. The competitive Hebbian learning rule [5], which is an combination of competitive learning and Hebbian theory, is an effective method to construct such topological structure of input data. The principle of this method can be stated as: given some nodes in the input space, for each input pattern x, update the closest node and connect the two closest nodes by an edge. It is proved that the resulting graph is a subgraph of the Delaunay triangulation corresponding to the given set of nodes, and that the resulting graph structure is optimally topology preserving in a very general sense.

In the circumstances of large scale training data, we want to realize incremental topology learning, i.e., successively insert topological nodes and edges connecting these nodes to construct the graph structure $G = (V, E)$, where V is the learned node set, and E is the edge set. Some incremental models such as Growing Neural Gas (GNG) [6] adopt a periodic splitting scheme to insert new nodes nearby the node with maximum accumulated error. However, we employ an adaptive threshold scheme, i.e., associate a similarity threshold T_i with each node i. If node i has topological neighbors, its similarity threshold is defined by the maximum Euclidean distance between node i and all of its topological neighbors $N(i)$. Thus, we have:

$$T_i = \max_{j \in N(i)} \|w_i - w_j\| \tag{1}$$

Otherwise, its similarity threshold is defined as positive infinity [7]. For an input x, the two nearest nodes of x is denoted by n_1 and n_2, respectively. We have:

$$n_1 = \arg\min_{i \in V} \|x - w_i\|, \ n_2 = \arg\min_{i \in V \setminus \{n_1\}} \|x - w_i\| \tag{2}$$

Depending on the similarity (measured by Euclidean distance) between input x and these two nodes, i.e., $d_1 = \|x - w_{n_1}\|$ and $d_2 = \|x - w_{n_1}\|$, we take corresponding update actions as follows:

- If $d_1 > T_1$ (Fig. 1(a)), insert x as a new node, $V = V \cup \{x\}$.
- If $\alpha \cdot T_1 < d_1 \leq T_1$ (Fig. 1(b)), also insert x as a new node, $V = V \cup \{x\}$. Here, parameter α is a real number between 0 and 1.
- If $d_1 \leq \alpha \cdot T_1$ (Fig. 1(c)), update the weight of winner node. Here, η is the learning rate.

$$w_{n_1} = w_{n_1} + \eta \cdot (x - w_{n_1}) \tag{3}$$

- Meanwhile, if $d_1 \leq T_1$ and $d_2 \leq T_2$ (Fig. 1(d)), add an edge between the two winner nodes n_1 and n_2, $E = E \cup \{(n_1, n_2)\}$.

The graphical illustration is shown in Fig. 1. However, as the learning process proceeds, connected nodes might not be neighbors any more because of the motion of nodes. Thus we apply the similar edge aging scheme used by GNG model to deal with these obsolete edges. In this case, we associate each edge with an age value. When the age of some edge exceeds a predefined maximal age parameter τ, then remove that edge. When the edge is activated then reset its age. Let t_i denote the winning times of node i, i.e. the times that node i has become the most nearest node so far during competitive learning process, and we set the learning rate η to $1/t_i$, then the weight update Eq. (3) could also be considered as the mean value of input patterns around node i. Furthermore, the winning times of topological nodes also captures the local density information around these nodes to some extent.

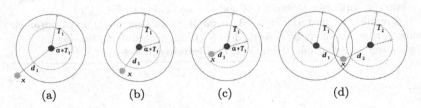

Fig. 1. (a) $d_1 > T_1$ (b) $\alpha \cdot T_1 < d_1 \leq T_1$ (c) $d_1 \leq \alpha \cdot T_1$ (d) $d_1 \leq T_1$ and $d_2 \leq T_2$

2.2   Periodic Refinement

During the incremental topology learning, we adopt the refinement process to eliminate the superfluous nodes and noisy nodes to maintain the main topology. With the assumption that noisy or outlier data is mostly distributed over the regions with lower probability density, and is sparse enough so that the main structure of target density distribution could still be recovered, most nodes created for them will have no or very few neighbors around. Let $N_k(i)$ denote the k nearest neighbors of node i, where k is determined by the maximal neighbors for all the nodes in topology, then the average winning times of its neighbors is defined as:

$$t(N_k(i)) = \frac{1}{|N_k(i)|} \sum_{j \in N_k(i)} t_j \tag{4}$$

Every λ learning iterations, i.e. λ input examples, we find the nodes which have no neighbor or only one neighbor. Based on the presumption that such nodes lie in the low-density area, we eliminate those nodes whose winning times is lower than the averaged winning times of its neighbors to a certain degree $\beta \cdot t(N_k(i))$. Here, λ is a positive integer, and β is a real number between 0 and 1. They are both user defined parameters. By this periodic refinement, we make the learned topology more concise and more robust. As a summary, we give the complete incremental topology learning algorithm in Algorithm 1.

Algorithm 1. Incremental Topology Learning

Input: density factor α, maximal age τ, refinement parameters λ and β.
Output: graph topological structure $G = (V, E)$, where V is the learned node set and
     E is the edge set between nodes.
 1: Initialize node set $V = \phi$, edge set $E = \phi$.
 2: Input two patterns, add them to the node set V.
 3: Input a new pattern x.
 4: Find the nearest nodes n_1 and n_2 of x according to Eq. (2).
 5: Calculate the distance between x and n_1, n_2, take corresponding actions described
     in Sect. 2.1, and update weight of n_1 according to Eq. (3) when necessary.
 6: If n_1 and n_2 are connected by an edge, set the age of this edge to zero. If such an
     edge does not exist, create it and add it to edge set E when necessary.
 7: Increase the age of all edges emanating from n_1, remove edges with an age larger
     than τ. If this results in nodes having no connecting edges, remove them as well.
 8: Update the similarity threshold of node n_1 according to Eq. (1).
 9: If the number of patterns inputed so far is an integer multiple of parameter λ,
     refine current topology as described in Sect. 2.2 according to Eq. (4).
10: Go to step 3 to continue the incremental learning process.

2.3   Weighted Neighbors Analysis

When the topology learning finished, we obtain a concise graph representation $G = (V, E)$ of the original training data. We then construct the k-nearest neighbor (KNN) graph based on the learned topology. The choice of k can be made as the maximal number of neighbors of learned nodes in the topology. The original one class nearest neighbor method takes local density into account [1]. It can be described as: the distance from object x to its nearest neighbor in the training set $NN(x)$ is compared with the distance from this nearest neighbor $NN(x)$ to its nearest neighbor $NN(NN(x))$, and the dissimilarity measure between test object x and the target class X is defined as:

$$d(x, X) = \frac{\|x - NN(x)\|}{\|NN(x) - NN(NN(x))\|} \tag{5}$$

Inspired by the nearest neighbor classifier, for each node i in the learned topology, we define the average distance to its topological neighbors $N(i)$ as:

$$d(i, N(i)) = \frac{1}{|N(i)|} \sum_{j \in N(i)} \|w_i - w_j\| \tag{6}$$

Note that we have constructed the KNN graph, then $|N(i)|$ is equal to k for every node i. Given test object x, the local dissimilarity score $d(x, X)$ to the target class is defined as the average distance ratio for all its k-nearest neighbors $N_k(x)$. Here, k is also the maximal number of neighbors of learned nodes in the topology. Since some nodes are near the distribution boundary, while the others are deep within the region with high distribution density. To indicate different contribution of nodes for our neighbors analysis, we incorporate a weight scheme to get a more meaningful score. Thus we have:

$$\alpha_i = \frac{t_i}{\sum_{j \in N_k(x)} t_j}, \ i \in N_k(x) \tag{7}$$

$$d(x, X) = \frac{1}{|N_k(x)|} \sum_{i \in N_k(x)} \alpha_i \cdot \frac{\|x - w_i\|}{d(i, N(i))} \tag{8}$$

where t_i is winning times of node i, and in some sense it captures the local density information around node i as we stated before. The dissimilarity score $d(x, X)$ is local in the sense that only a restricted neighborhood of each object is taken into account.

This neighbors analysis scheme is comparable but not identical to the Local Outlier Factor(LOF) method [9]. Finally, the decision whether x belongs to the target or outlier class is based on the score compared with threshold θ which is adjusted to accept or reject a user-defined fraction of the target class [1]. In summary, to apply our method to the OCC problems, firstly we perform incremental topology learning for training on the large scale target data or the target data in stream-oriented applications. When testing is performed, we constructed the

KNN graph based on the learned topology. After that, the weighted neighbors analysis technique is utilized to construct the one class classifier. Although it seems that our method is not a pure incremental learning method, it is effective enough for the real large scale OCC problems.

3   Experiments

In this section, we conduct experiments on an artificial data set and some widely used real world data sets, also present detailed experimental results for TLNA in comparison with contemporary one class classifiers. In our comparative experiments, the value of k for the KNN method and value of width parameter for the KDE method is optimized. The Gaussian kernel is used as kernel function in the SVDD methods. For OCC problems, usually Receiver Operating Characteristic (ROC) curve is used as the performance measurement of classifier. In our experiment results, the Area Under the Curve (AUC) for the ROC is also calculated and compared. The parameters of proposed method in all experiments are set as follows: $\alpha = 0.5$, $\tau = 50$, $\lambda = 50$, $\beta = 0.5$.

3.1   Artificial Data

For the artificial dataset, the training data consists of three different distributions: Gaussian distribution, banana-shaped distribution and random noisy data following the uniform distribution. Both Gauss and banana distributions are considered as target class, and others outlier class. Figure 2 shows the plots of the generated dataset, the learned topology structure and the experiment result. We compare the performance of TLNA with contemporary one class classifiers including NN(Neatest Neighbor), KNN(k-Neatest Neighbor), KDE(Kernel Density Estimation), SVDD(Support Vector Data Description), ISVDD(Incremental SVDD) [4] and MST(Minimum Spanning Tree) [8]. The result shows that TLNA can obtain comparable performance with the state-of-the-art methods SVDD and KDE, and preferable performance than others for this synthetic data set.

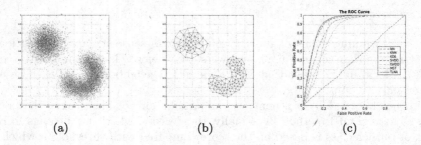

(a)   (b)   (c)

Fig. 2. Artificial dataset and experiment result (a) target training data with noise (b) learned topology structure (c) ROC curve compared with contemporary OCC methods

3.2 Real World Data

For the case of real world data sets, since there is no special data set for one class classifiers, we use the conventional classification data sets. Most of them were originally collected from the UCI machine learning repository [10], and some data sets were collected in the LIBSVM data sets [11]. We divide each data set into two separate class group. One class group is used as target class and the other outlier class. Half of randomly selected data from the target class is used as the training target data, and the rest testing target data. To demonstrate the robustness to noises or outliers of OCC methods, we also add 5 % examples from outlier class to the training target data to construct the noisy target training data. Equal amount of targets and outliers are put together to construct the testing data. To make sure that the achieved results are not coincidental, this scheme is repeated for 20 times to build 20 different training and testing sets. And the final result is achieved by averaging over the best 10 models for each classifier. The result is shown in Table 1.

Table 1. Average AUC of each method for the real world data sets (best method in **underline bold**, second best **bold**)

Data set	TLNA	NN	KNN	KDE	SVDD	ISVDD	MST
Skin	**98.94 ± 0.27**	66.81 ± 0.60	74.70 ± 0.35	95.70 ± 0.36	**98.34 ± 0.28**	81.84 ± 0.21	72.71 ± 0.32
Svmguide1	**96.91 ± 0.57**	58.36 ± 1.07	95.83 ± 0.16	**97.58 ± 0.05**	96.63 ± 0.51	94.11 ± 0.37	93.96 ± 0.23
Cod-rna	**85.95 ± 1.32**	52.94 ± 0.25	79.75 ± 0.14	82.44 ± 0.26	**87.10 ± 0.26**	79.50 ± 0.31	77.25 ± 0.22
Shuttle	**98.79 ± 0.28**	58.75 ± 1.03	68.29 ± 0.30	78.91 ± 0.45	**98.32 ± 0.20**	97.07 ± 0.22	65.29 ± 0.39
Pendigits	**96.08 ± 0.43**	53.23 ± 0.25	94.20 ± 0.47	94.89 ± 0.31	**95.01 ± 0.40**	94.95 ± 0.42	92.23 ± 0.37
Connect4	**77.69 ± 0.29**	58.62 ± 0.11	60.24 ± 0.12	**88.74 ± 0.18**	61.47 ± 0.25	59.64 ± 0.24	66.90 ± 0.20

The experimental results on the real world datasets show that our approach TLNA can obtain comparable or preferable performance compared with the best method in most datasets. Particularly, for the *pendigits* dataset, we use several digits as target class and others as outliers, and TLNA achieves better performance than all the other methods. It verifys that our approach can deal with complex data distribution and is robust to noises or outliers in the target training data.

4 Conclusion

In this paper, we propose an incremental learning framework for OCC problems in large scale data environment. It can automatically acquire a concise graph representation of the target class by incremental topology learning. Based on the graph representation, we build a description model for the target class. As a result of the incremental characteristic, our method need not to store any historical data but the learned graph model. Therefore, the time and space complexity of our method are much smaller than the traditional batch OCC methods. Furthermore, The periodic refinement operation in the topology learning process makes it robust to noises and outliers. Experimental results on artificial and real world data sets confirmed the validity of proposed learning method. The future works may include extension to other one class learning models based on the learned graph representation of original input data.

Acknowledgement. The authors would like to thank the anonymous reviewers for their time and valuable suggestions. This work is supported in part by the National Science Foundation of China under Grant Nos. (61375064, 61373001) and Jiangsu NSF grant (BK20131279).

References

1. Tax, D.M.J.: One-class classification: concept-learning in the absence of counterexamples. Ph.D. thesis, Delft University of Technology (2001)
2. Tax, D.M.J., Duin, R.P.W.: Support vector data description. Mach. Learn. **54**(1), 45–66 (2004)
3. Scholkopf, B., Platt, J.C., Shawe-Taylor, J.C., Smola, A.J., Williamson, R.C.: Estimating the support of a high-dimensional distribution. Neural Comput. **13**(7), 1443–1471 (2001)
4. Tax, D.M.J., Laskov, P.: Online SVM learning: from classification to data description and back. In: Proceedings of the 13th IEEE NNSP Workshop. IEEE Computer Society Press, Los Alamitos (2003)
5. Martinetz, T.M.: Competitive Hebbian learning rule forms perfectly topology preserving maps. In: International Conference on Artificial Neural Networks, pp. 427–434 (1993)
6. Fritzke, B.: A growing neural gas network learns topologies. In: Proceedings of the Advances in Neural Information Processing Systems, pp. 625–632 (1995)
7. Shen, F., Hasegawa, O.: An incremental network for on-line unsupervised classification and topology learning. Neural Netw. **19**(1), 90–106 (2006)
8. Juszczak, P., Tax, D.M.J., Duin, R.P.W.: Minimum spanning tree based one-class classifier. Neurocomputing **72**(7), 1859–1869 (2009)
9. Breunig, M.M., Kriegel, H.P., Ng, R.T., Sander, J.: LOF: identifying density-based local outliers. In: Proceedings of the 2000 ACM SIGMOD International Conference on Management of Data, pp. 93–104 (2000)
10. Asuncion, A., Newman, D.: UCI Machine Learning Repository. University of California, Irvine (2007)
11. Chang, C., Lin, C.: LIBSVM: a library for support vector machines. ACM Trans. Intell. Syst. Technol. **2**(3), 389–396 (2011). http://www.csie.ntu.edu.tw/cjlin/libsvm

Gesture Spotting by Using Vector Distance of Self-organizing Map

Yuta Ichikawa, Shuji Tashiro, Hidetaka Ito, and Hiroomi Hikawa[✉]

Department of Electrical and Electronic Engineering, Kansai University,
Suita, Japan
hikawa@kansai-u.ac.jp

Abstract. This paper proposes a dynamic hand gesture recognition algorithm with a function of gesture spotting. The algorithm consists of two self-organizing maps (SOMs) and a Hebb learning network. Feature vectors are extracted from input images, and these are fed to one of the SOMs and a vector that represents the sequence of postures in the given frame is generated. Using this vector, gesture classification is performed using another SOM. In the SOM, the vector distance between the input vector and the winner neuron's weight vector is used for the gesture spotting. The following Hebb network identifies the gesture class. The experimental results show that the system recognizes eight gestures with the accuracy of 95.8 %.

Keywords: Self organizing map (SOM) · Hebb learning · Hand sign recognition · Gesture spotting

1 Introduction

Hand gestures are used frequently in human daily communication. This is a great intention transmission method that can be used in the same way as words, and sometimes it can communicate sentiment and feeling of human more than words. The hand gesture can be a new effective human-computer interaction (HCI). Many methods for hand gesture recognition have been proposed [1]. Hand gestures are generally either hand postures or dynamic hand gestures. Hand postures are static hand poses without any movements [2,3], and hand gestures are defined as dynamic movement, which is a sequence of hand postures [4,5]. This paper focuses on the dynamic hand gesture recognition.

For the dynamic gesture recognition, it is important to detect the start and the end of the gesture sequence in video frames. This function is called gesture spotting. This paper proposes a new dynamic gesture recognition method with the gesture spotting function. The proposed system is composed of two self-organizing maps (SOMs) [6] and a Hebb learning network. In the proposed method, feature vectors are extracted from RGB video image. The feature vector is fed to one of the two SOMs, which generates a vector that represents the sequence of given postures. The result of the gesture recognition is computed by

A. Hirose et al. (Eds.): ICONIP 2016, Part II, LNCS 9948, pp. 419–426, 2016.
DOI: 10.1007/978-3-319-46672-9_47

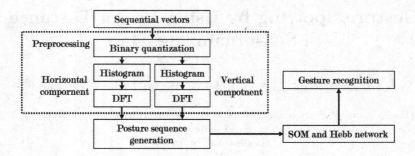

Fig. 1. Flow of gesture recognition.

feeding this vector to a SOM-Hebb network [7]. The gesture spotting is carried out by using the vector distance between the input vector and that of the winner neuron. This vector distance decreases as gesture sequence progresses. Therefore, the end of the gesture can be detected by observing the transition of the vector distance. In the proposed method, the bottom in the vector distance transition is detected to find the end of the gesture. Feasibility of the proposed method is verified by simulations.

2   Gesture Recognition System

Figure 1 shows the flow of the proposed gesture recognition system. The input is a sequence of frames extracted from video footage containing a gesture. In this paper, a single gesture is defined by F video frames. Every video frame contains different postures, and the dynamic gesture can be classified by using a sequence of postures in the video frames. Feature vector is extracted from each video frame, which is fed to the posture sequence generation module that consists of a SOM. Following SOM-Hebb network identifies the gesture class. The gesture spotting function is embedded in this network.

2.1   Preprocessing

Each video frame is $P \times P$ pixels in RGB color format. The system requires users to wear a glove in Fig. 2, to retrieve only a finger segment without depending on

Fig. 2. Input image.

the change of background image and arm position. The preprocessing consists of a binary quantization, two horizontal and vertical projection histogram calculations, and two discrete Fourier transforms (DFTs) that generate a D dimensional feature vector $\overrightarrow{X}$. Details of the feature extraction can be found in [7].

2.2 Posture Sequence Generation

Figure 3 shows the posture sequence generation with SOM. SOM is used to quantize the feature vectors, and a posture sequence vector is generated by using the quantized information.

The operation of SOM is divided into two phases, the learning phase and the recall phase. In the learning phase, the distance between the input vector and the weight vector is calculated and the winner neuron that has the weight vector which is the shortest distance is determined. Then weight vectors of the winner neuron and its neighborhood neurons are adapted toward the input vector. In the recall phase, the winner neuron that has the shortest distance for the input vector is searched without updating the weight of neuron. The winner neuron represents the cluster to which the input vector belongs.

The SOM includes $M \times M$ neurons. The feature vectors $\overrightarrow{X}$, generated from each frame image by the preprocessing is fed into the SOM, and the coordinates of winner neuron is saved sequentially in the shift register. Contents of the shift register represent the posture sequence vector $\overrightarrow{G}$ that is a $2F$-dimensional vector and given by

$$\overrightarrow{G} = \{\xi_1, \xi_2, \cdots \xi_{2F}\} \tag{1}$$

$$\xi_i = \begin{cases} W_x(n) & (0 \le n < F) \\ W_y(n - F) & (F \le n < 2F) \end{cases} \tag{2}$$

where F is the number of frames, which defines a gesture. $W_x(n)$ and $W_y(n)$ are X and Y coordinates of the winner neuron at time n. When all F postures belonging to a gesture are processed, the posture sequence vector $\overrightarrow{G}$ in the shift register represents the gesture. Since the number of video frames for one gesture is defined by 10 frames ($F = 10$), the dimension of the time series vector $\overrightarrow{G}$ is 20.

2.3 SOM-Hebb Network for Gesture Classification

Figure 4 shows a block diagram of the SOM-Hebb network that performs the gesture identification by using the posture sequence vector as its input. This SOM is trained in the same way as the SOM used in the previous module. The class of gesture is identified from the winner neuron of the SOM. This identification is carried out by the following Hebbian learning network. Detail operation of the SOM-Hebb network is discussed in [7]. In this paper, the gesture spotting function is added to the network. The gesture output port after the Hebbian learning network is activated at the end of gesture to give the recognition result. Gesture spotting module finds the posture end, and activates the winner find output port.

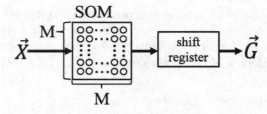

Fig. 3. Posture sequence generation.

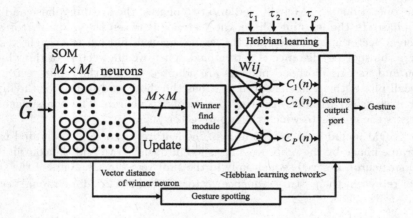

Fig. 4. SOM-Hebb network with gesture spotting.

The gesture spotting is carried out using the vector distance between the input vector and the winner neuron's weight vector. Figure 5 shows the frame-by-frame transition of the posture sequence vector, $\vec{G}$ in the shift register of the posture sequence generation module. $W_X(n)$ is the X coordinate of the winner neuron where f is the frame number, and the question marks indicate the information of the previous gesture. If the training of the SOM in the SOM-Hebb network were properly done, a neuron, weight vector of which is the closest to the input vector, is chosen as winner. Since gesture is defined by 10 frames, the posture sequence vector is completed at $f = 9$. As the figure shows the vector distance between the posture sequence vector and the winner's weight vector gradually decreases until $f = 9$, at which the register contains a complete set of vector elements of the current gesture. Then it increases after $f = 9$ because postures belonging to the next gesture are loaded to the register. Therefore the end of gesture can be detected by searching the bottom of the distance transition. However, actual distance transition of real video is not as smooth as the one shown in Fig. 5. It locally goes up and down frequently, which makes it difficult to find the bottom. In order to eliminate such local minima, moving

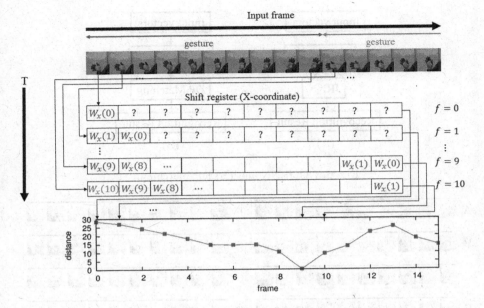

Fig. 5. The number of input frames and vector distance.

average of the vector distance is employed. The moving average $V_{ma}(n)$ used in this study is given by

$$V_{ma}(n) = \frac{1}{N_r} \sum_{i=1}^{N_r} V(n - i + 1) \tag{3}$$

where, $V(n)$ is the vector distance between the input vector (posture sequence vector) and the winner neuron's weight vector at time n. N_r is the number of samples used to obtain the average. The posture spotting module detects the bottom of transition when the transition of $V_{ma}(n)$ is "dip $\to$ dip $\to$ dip $\to$ rise $\to$ rise $\to$" rise. When the end of posture, i.e., the bottom of the distance transition is detected, the gesture spotting module activates the gesture output port to output the recognition result.

3 Simulation

The performance of the proposed algorithm was tested by simulations. In order to compare the performances, Jordan recurrent neural network (JRNN) and dynamic programing (DP) matching were also tested. As shown in Fig. 6, these two classifiers used the same feature vectors for the recognition.

Eight gestures used in the simulation are shown in Fig. 7. Note that A and B, C and D, E and F, G and H are gestures of reverse order, respectively. Input image sets for the simulation were extracted from pre-recorded video. Two data

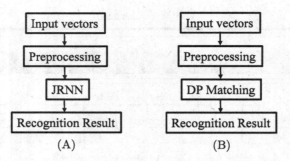

Fig. 6. (A) JRNN and (B) DP matching.

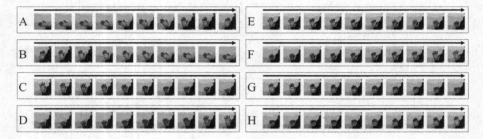

Fig. 7. Gestures used in the simulation.

sets α and β were prepared by different individuals. Each data set consisted of images taken in the morning and evening. Images taken in the morning was used for training and recognition tests were conducted by using images taken in the evening. The size of the image was 128×128. The feature vector dimension was 32. For the moving average N_r in Eq. (3) was set to be there. The dimension of the posture sequence vector was 20 ($F = 10$). The number of neurons in SOM used in the posture recognition and gesture recognition were both 16×16. As shown in Fig. 7, the number of classes to be recognizes was $H = 8$.

Figure 8 shows the vector distance transitions for eight different gestures. Since all gestures were defined by 10 frames, distances take minimum values, i.e., bottoms, at every 10 frames. However, they also exhibit local minima between the bottoms. By employing the moving average, the transitions become smoother as shown in Fig. 9, which helps the system to find the bottoms accurately.

Using the data sets α and β, the recognition accuracy of the proposed method was examined. Table 1 summarizes the results. The average recognition rate of the proposed method for data sets α and β is 95.8 %. The table also includes recognition accuracies of JRNN and DP matching. In terms of the recognition accuracy, the table reveals that the performance of the proposed method is the best among the three algorithms.

It should be noted the recognition accuracy of JRNN significantly varies for gestures. For example, the recognition rate for gesture 'E' in Table 1 (B) is 0 %

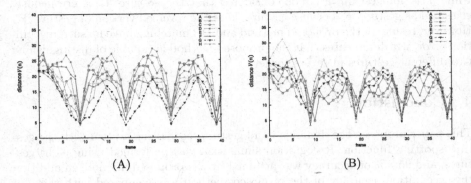

Fig. 8. Transition of vector distance, (A) data set α, (B) data set β.

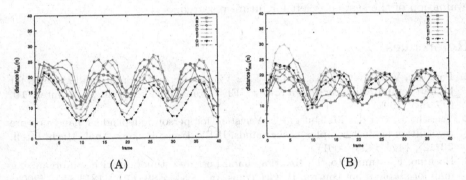

Fig. 9. Transition of moving average of vector distance, (A) data set α, (B) data set β.

Table 1. Recognition accuracies, (A) data set α, (B) data set β.

(A)				(B)			
Gesture	proposed	JRNN	DP	Gesture	proposed	JRNN	DP
A	**100.00%**	54.00%	98.00%	A	**100.00%**	96.00%	97.00%
B	92.00%	**100.00%**	86.00%	B	**100.00%**	98.00%	99.00%
C	**98.00%**	92.00%	30.00%	C	**100.00%**	0.00%	93.00%
D	82.00%	**100.00%**	92.00%	D	90.00%	82.00%	**91.67%**
E	**100.00%**	**100.00%**	80.00%	E	92.00%	0.00%	**93.00%**
F	98.00%	**100.00%**	94.00%	F	**96.00%**	**96.00%**	**96.00%**
G	90.00%	34.00%	92.00%	G	**100.00%**	76.00%	97.00%
H	98.00%	**100.00%**	89.00%	H	98.00%	74.00%	60.67%
Average	**94.75%**	85.00%	82.63%	Average	**97.00%**	65.25%	90.92%

while 96 % for gesture 'F'. It is considered that the gesture 'E' is erroneously classified as gesture 'F' because gesture 'E' is the reversed version of gesture 'F'. However, results of the proposed method and DP matching have no such recognition error, which indicates that the proposed method is capable of distinguishing two different gestures in reverse order.

4   Conclustion

This paper proposed the dynamic hand gesture recognition system with the gesture spotting function. Recognition simulation was performed using eight gestures, and 95.8 % of accuracy was attained by the proposed method. In addition the recognition accuracy of the proposed method was compared with those of JRNN and DP matching. Simulation results revealed that the proposed method outperforms the both algorithm in terms of recognition accuracy.

Development of real-time gesture recognition system is underway, and the evaluation of the system is left for future research.

References

1. Pavlovic, V.I., Sharma, R., Huang, T.S.: Visual interpretation of hand gestures for human-computer interaction: a review. IEEE Trans. Pattern Anal. Mach. Intell. **19**, 677–695 (1997)
2. Triesch, J., von der Malsburg, C.: A system for person-independent hand posture recognition against complex backgrounds. IEEE Trans. Pattern Anal. Mach. Intell. **23**(12), 1449–1453 (2001)
3. Hoshino, K., Tanimoto, T.: Realtime hand posture estimation with self-organizing map for stable robot control. IEICE Trans. Inf. Syst. **E89–D**(6), 1813–1819 (2006)
4. Bobick, A.F., Wilson, A.D.: A state-based approach to the representtion and recognition of gesture. IEEE Trans. Pattern Anal. Mach. Intell. **19**(12), 1325–1337 (1997)
5. Yang, H.-H., Ahuja, N., Tabb, M.: Extraction of 2D motion trajectories and its application to hand gesture recogntion. IEEE Trans. Pattern Anal. Mach. Intell. **24**(8), 1061–1074 (2002)
6. Kohonen, T.: Self-Organizing Maps. Information Sciences, vol. 30, 3rd edn. Springer, New York (2001)
7. Hikawa, H., Kaida, K.: Novel FPGA implementation of hand sign recognition system with SOM-hebb classifier. IEEE Trans. Circ. Syst. Video Technol. **25**(1), 153–166 (2015)

Cross-Database Facial Expression Recognition via Unsupervised Domain Adaptive Dictionary Learning

Keyu Yan[1,2], Wenming Zheng[1(✉)], Zhen Cui[1(✉)], and Yuan Zong[1]

[1] Key Laboratory of Child Development and Learning Science
of Ministry of Education, Research Center for Learning Science,
Southeast University, Nanjing 210096, Jiangsu, China
{wenming_zheng,zhen.cui}@seu.edu.cn
[2] School of Information Science and Engineering, Southeast University,
Nanjing 210096, Jiangsu, China

Abstract. Dictionary learning based methods have achieved state-of-the-art performance in the task of conventional facial expression recognition (FER), where the distributions between training and testing data are implicitly assumed to be matched. But in the practical scenes this assumption is usually broken, especially when testing samples and training samples come from different databases, a.k.a. the cross-database FER problem. To address this problem, we propose a novel method called unsupervised domain adaptive dictionary learning (UDADL) to deal with the unsupervised case that all samples in target database are completely unlabeled. In UDADL, to obtain more robust representations of facial expressions and to reduce the time complexity in training and testing phases, we introduce a dual dictionary pair consisting of a synthesis one and an analysis one to mutually bridge the samples and their codes. Meanwhile, to relieve the distribution disparity of source and target samples, we further integrate the learning of unlabeled testing data into UDADL to adaptively adjust the misaligned distribution in an embedded space, where geometric structures of both domains are also encourage to be preserved. The UDADL model can be solved by an iterate optimization strategy with each sub-optimization in a closed analytic form. The extensive experiments on Multi-PIE and BU-3DFE databases demonstrate that the proposed UDADL is superior over most widely-used domain adaptation methods in dealing with cross-database FER, and achieves the state-of-the-art performance.

Keywords: Cross-database facial expression recognition · Domain adapatation · Transfer learning · Dictionary learning

1 Introduction

Facial expression recognition (FER) is one of the most popular research topics of affective computing, pattern recognition, computer vision. The goal of a typical

© Springer International Publishing AG 2016
A. Hirose et al. (Eds.): ICONIP 2016, Part II, LNCS 9948, pp. 427–434, 2016.
DOI: 10.1007/978-3-319-46672-9_48

FER task is to learn a classifier based on labeled facial images such that the classifier is able to predict the expression categories of the unlabeled ones as well as most vision classification tasks [9]. During past decades, a variety of effective FER methods have been proposed and achieved promising performance. It is notable that these methods are always developed and tested based on a single facial expression database and there is actually a prior assumption that the training and testing samples obey the same distribution. In practical scenario, however, the above assumption is often not satisfied and the performance of these methods may degrade. A simple example is that the training and testing samples are from two databases recorded by different equipments under different environments. This thus creates a challenging and meaningful topic called cross-database FER problem. Obviously, cross-database FER can be referred to as the domain adaptation (DA) [8] problem which has recently drawn much attention in the computer vision community. DA aims to utilize plenty of labeled data from a source domain to learn a classifier which is also applicable to the data from a target domain where the target data usually has a different distribution from the source one. Generally, DA methods can be classified into two categories according to the availability of labeled data in the target domain: (1) semi-supervised DA and (2) unsupervised DA. Semi-supervised DA leverages a few of label information of the target data to reduce the divergence between the source and target domain, while unsupervised DA is inherently a more challenging problem without any labeled target data to build relation between two domains.

In this paper, we will investigate the unsupervised cross-database FER problem, i.e., the expression label information of target database is completely unknown. Consequently, this problem is quilt difficult and challenging. Nevertheless, there are still numerous methods have been developed to deal with cross-database FER or its related topics. Recently, in the work of [1], Chu et al. proposed a novel method called selective transfer machine (STM) for personalized facial action units detection. STM utilizes the testing samples to learn weights for the training sample such that the classifier could also be suitable for testing samples. Sangineto et al. [10] proposed another method for cross-domain facial expression recognition problem by using classifier parameter transfer approach and achieved satisfactory results. More recently, Zheng et al. [15] proposed a transductive transfer linear discriminant analysis (TTLDA) method to deal with cross-pose FER problem. TTLDA selects some of target samples to serve as the auxiliary set to jointly train the LDA model with source samples. As a result, the learnt LDA model is able to accurately predict the expression labels of target samples in a discriminative subspace.

In this paper, we propose a novel cross-database FER method called unsupervised domain adaptive dictionary learning (UDADL) to handle the unsupervised case that all samples in target database are completely unlabeled. We first employ a dictionary pair consisting of a synthesis one and an analysis one to mutually bridge the samples and their codes. Then, by using analysis dictionary to re-represent original samples, we propose a novel criterion, which is derived from common space criterion and is able to reduce the distribution mismatch between source and target databases, for DL to build the UDADL model. What is more, the UDADL model can efficiently be solved by an iterate optimization strategy and each sub-optimization of UDADL has a closed analytic form.

The rest of the paper is organized as follows: Sect. 2 presents our UDADL method and describes how it works for cross-database FER problem. Section 3 gives the detailed optimization method for the proposed UDADL model. Extensive experiments and discussions are conducted in Sect. 4. Finally, we conclude the paper in the last section.

2 Proposed Method

2.1 Notations

In this section, we will address UDADL model detailedly. For convenience, we first give some notations which are used throughout the paper in advance. Let $\mathbf{X}_s \in \mathbb{R}^{n \times N_s}$, $\mathbf{X}_t \in \mathbb{R}^{n \times N_t}$ be the feature matrices belonging to the source and target facial expression databases, respectively, where n is the dimension of the expression features, and N_s and N_t denote the sample number of the source and target databases, respectively. Let $\mathbf{D}$ be the dictionary learnt from $\mathbf{X}$, where $\mathbf{D} \in \mathbb{R}^{n \times n}$, and $\mathbf{X} = [\mathbf{X}_s, \mathbf{X}_t] \in \mathbb{R}^{n \times (N_t + N_s)}$.

2.2 Unsupervised Domain Adaptive Dictionary Learning (UDADL)

Similar with conventional dictionary learning, the UDADL model aims to learn a dictionary $\mathbf{D}$ based on the source and target samples $\mathbf{X}$ and the optimal $\mathbf{D}$ can be obtained by solving the following optimization problem:

$$\{\mathbf{D}^*, \mathbf{A}^*\} = \arg \min_{\mathbf{D}, \mathbf{A}} \|\mathbf{X} - \mathbf{D}\mathbf{A}\|_F^2 = \arg \min_{\mathbf{D}, \mathbf{A}_s, \mathbf{A}_t} \|[\mathbf{X}_s, \mathbf{X}_t] - \mathbf{D}[\mathbf{A}_s, \mathbf{A}_t]\|_F^2 \quad (1)$$

where $\mathbf{A} = [\mathbf{A}_s, \mathbf{A}_t]$ is the coding coefficient matrix of $\mathbf{X}$ with respect to dictionary $\mathbf{D}$. In addition to this, UDADL also aims to force the coding coefficient matrices $\mathbf{A}_s$ and $\mathbf{A}_t$ of source and target samples to have similar distributions. For this purpose, we will borrow the basic idea of the common subspace criterion proposed by Kan et al. [6] and present a reduced criterion version to determine the optimal coding coefficient matrices, where coding coefficient matrices $\mathbf{A}_s$ and $\mathbf{A}_t$ are the solution of the following optimization problem:

$$\{\mathbf{W}_{ts}^*, \mathbf{W}_{st}^*\} = \arg \min_{\mathbf{A}_s, \mathbf{A}_t} \|\mathbf{A}_s - \mathbf{A}_t \mathbf{W}_{ts}\|_F^2 + \|\mathbf{A}_t - \mathbf{A}_s \mathbf{W}_{st}\|_F^2 \quad (2)$$

where $\mathbf{W}_{ts}$ and $\mathbf{W}_{st}$ are representation coefficient matrices. By imposing the objective function of Eq. (2) on the objective function of Eq. (1), we will arrive at our desired UDADL model:

$$\{\mathbf{D}^*, \mathbf{A}_s^*, \mathbf{A}_t^*\} = \arg \min_{\mathbf{D}^*, \mathbf{A}_s, \mathbf{A}_t} \|[\mathbf{X}_s, \mathbf{X}_t] - \mathbf{D}[\mathbf{A}_s, \mathbf{A}_t]\|_F^2$$
$$+ \lambda(\|\mathbf{A}_s - \mathbf{A}_t \mathbf{W}_{ts}\|_F^2 + \|\mathbf{A}_t - \mathbf{A}_s \mathbf{W}_{st}\|_F^2) \quad (3)$$

where λ is the trade-off parameter.

However, it is unfortunate that the UDADL with the above definition has no close analytic solutions for both $\mathbf{A}_s$ and $\mathbf{A}_t$ as well as ideal approximate solutions. To solve this problem, we will introduce recent projective dictionary pair learning [3] for building a solvable UDADL model. According to the work of [3], a dictionary pair consisting of an analysis dictionary and a synthesis one is able to better bridge original data and its corresponding code. Followed by dictionary pair framework, instead of synthesis dictionary $\mathbf{D}$, we predefine a analysis dictionary $\mathbf{P}$ as well such that $\mathbf{A} = \mathbf{PX}$ or $\mathbf{A} \approx \mathbf{PX}$ and hence the optimal $\mathbf{P}$ must satisfy the following optimization problem:

$$\mathbf{P}^* = \arg\min_{\mathbf{P}} \|\mathbf{PX} - \mathbf{A}\|_F^2 = \|\mathbf{P}[\mathbf{X}_s, \mathbf{X}_t] - [\mathbf{A}_s, \mathbf{A}_t]\|_F^2 \tag{4}$$

By use of analysis dictionary, we are able to improve the criterion of Eq. (2) such that UDADL could be efficiently solved. Specifically, we propose a new criterion as below:

$$\{\mathbf{P}^*, \mathbf{W}_{ts}^*, \mathbf{W}_{st}^*\} = \arg\min_{\mathbf{P}, \mathbf{A}_s, \mathbf{A}_t} \lambda\|\mathbf{A}_s - \mathbf{PX}_t\mathbf{W}_{ts}\|_F^2 + \|\mathbf{A}_t - \mathbf{PX}_s\mathbf{W}_{st}\|_F^2)$$
$$+ \tau(\|\mathbf{P}[\mathbf{X}_s, \mathbf{X}_t] - [\mathbf{A}_s, \mathbf{A}_t]\|_F^2) \tag{5}$$

where λ and τ are the trade-off parameters to balance the two terms. Thus, we are able to get the final UDADL formulations follows:

$$\min_{\mathbf{D}^*, \mathbf{P}^*, \mathbf{A}_s^*, \mathbf{A}_t^*, \mathbf{W}_{st}^*, \mathbf{W}_{ts}^*} \|[\mathbf{X}_s, \mathbf{X}_t] - \mathbf{D}[\mathbf{A}_s, \mathbf{A}_t]\|_F^2 + \mu(\|\mathbf{A}_s\|_F^2 + \|\mathbf{A}_t\|_F^2)$$
$$+ \lambda(\|\mathbf{A}_s - \mathbf{PX}_t\mathbf{W}_{ts}\|_F^2 + \|\mathbf{A}_t - \mathbf{PX}_s\mathbf{W}_{st}\|_F^2)$$
$$+ \tau(\|\mathbf{PX}_s - \mathbf{A}_s\|_F^2 + \|\mathbf{PX}_t - \mathbf{A}_t\|_F^2) \tag{6}$$

where λ, μ and τ are trade-off parameters. Note that we also select the collaborative representation term [13] for our UDADL such that more efficient $\mathbf{A}_s$ and $\mathbf{A}_t$ can be obtained. The UDADL can be solved by an simple iterate optimization strategy and the detailed solution method will be given in next section.

2.3   Cross-Database FER Based on UDADL

Once the optimal solution of Eq. (5) are obtained, we are able to use the new representation $\mathbf{A}_s^*$ of original source samples with respect to $\mathbf{D}^*$ to learn a classifier, e.g., support vector machine (SVM). Then, the target samples' expression categories can be determined by the learnt classifier and their representations $\mathbf{A}_t^*$.

3   Optimization

Although the objective function of Eq. (6) is generally non-convex, it is convex with respect to each model parameter when the others are fixed. Consequently, we are able to solve the optimization problem by updating one of model parameters $\{\mathbf{P}^*, \mathbf{D}^*, \mathbf{A}_s^*, \mathbf{A}_t^*, \mathbf{W}_{st}^*, \mathbf{W}_{ts}^*\}$ and fixing others iteratively until convergence. More specifically, the detailed updating rule consist of the following four steps:

(1) Fix $\mathbf{P}^*, \mathbf{W}_{st}^*, \mathbf{W}_{ts}^*, \mathbf{A}_s^*, \mathbf{A}_t^*$ and update $\mathbf{D}$. We can calculate the optimal synthesis dictionary $\mathbf{D}$ as:

$$\mathbf{D} = \mathbf{X}\mathbf{A}^T(\mathbf{A}\mathbf{A}^T)^{-1}$$

(2) Fix $\mathbf{D}^*, \mathbf{P}^*, \mathbf{A}_s^*, \mathbf{A}_t^*$ and update $\mathbf{W}_{ts}, \mathbf{W}_{st}$. Similar to step (2), the closed-form solution of $\mathbf{W}_{ts}$ and $\mathbf{W}_{st}$ can be expressed as:

$$\mathbf{W}_{ts} = (\mathbf{X}_t^T\mathbf{P}^T\mathbf{P}\mathbf{X}_t)^{-1}\mathbf{X}_t^T\mathbf{P}^T\mathbf{A}_s$$
$$\mathbf{W}_{st} = (\mathbf{X}_s^T\mathbf{P}^T\mathbf{P}\mathbf{X}_s)^{-1}\mathbf{X}_s^T\mathbf{P}^T\mathbf{A}_t$$

(3) Fix $\mathbf{P}^*, \mathbf{D}^*, \mathbf{W}_{st}^*, \mathbf{W}_{ts}^*$ and update $\mathbf{A}_s, \mathbf{A}_t$. Then, the optimization problem would reduce to be a standard least squares problem and we have the optimal solutions of $\mathbf{A}_s$ and $\mathbf{A}_t$ as follows:

$$\mathbf{A}_t = [\mathbf{D}^T\mathbf{D} + (\tau + \mu + \lambda)\mathbf{I}]^{-1}(\mathbf{D}^T\mathbf{X}_t + \tau\mathbf{P}\mathbf{X}_t + \lambda\mathbf{P}\mathbf{X}_s\mathbf{W}_{st})$$
$$\mathbf{A}_s = [\mathbf{D}^T\mathbf{D} + (\tau + \mu + \lambda)\mathbf{I}]^{-1}(\mathbf{D}^T\mathbf{X}_s + \tau\mathbf{P}\mathbf{X}_s + \lambda\mathbf{P}\mathbf{X}_t\mathbf{W}_{ts})$$

(4) Fix $\mathbf{D}^*, \mathbf{W}_{st}^*, \mathbf{W}_{ts}^*, \mathbf{A}_s^*, \mathbf{A}_t^*$ and update $\mathbf{P}$. In this case, we can get the closed-form solution of the analysis dictionary by setting the derivative of objective with respect to $\mathbf{P}$ to be 0 as below:

$$\mathbf{P} = (\tau\mathbf{A}\mathbf{X}^T + \lambda\mathbf{A}_s\mathbf{W}_{ts}^T\mathbf{X}_t^T + \lambda\mathbf{A}_t\mathbf{W}_{st}^T\mathbf{X}_s^T)[\tau\mathbf{X}\mathbf{X}^T$$
$$+ \lambda\mathbf{X}_t\mathbf{W}_{ts}\mathbf{W}_{ts}^T\mathbf{X}_t^T + \lambda\mathbf{X}_s\mathbf{W}_{st}\mathbf{W}_{st}^T\mathbf{X}_s^T]^{-1}$$

4   Experiments

In this section, we will conduct extensive cross-database FER experiments to evaluate the proposed UDADL method. Two widely-used expression databases, i.e., BU-3DFE [12] and Mulit-PIE [2], are adopted in this paper. BU-3DFE database is built by Yin et al. [12] and consists of 2500 three-dimensional facial expression models from 100 subjects corresponding to seven basic facial expressions, i.e., anger(AN), disgust(DI), fear(FE), happiness(HA), sadness(SA), surprise(SU), and neutral(NE). Multi-PIE database developed by Gross et al. [2] covers six facial expressions including disgust(DI), scream(SC), smile(SM), squint(SQ), surprise(SU), and neutral(NE). We select the samples belonging to four common expressions (NE, HA/SM, SU, and DI) from both two databases for our experiments, respectively. Note that for Multi-PIE database, the facial images with the facial views of $0°$ and $30°$ are chosen. Meanwhile, for BU-3DFE database, we use OpenGL [14] software to convert the 3D facial expression models to numerous 2D gray facial images with facial views of $0°$ and $30°$ as well.

Fig. 1. Examples of the 49 landmark points of the 0° & 30° views.

In the experiments, we choose one of BU-3DFE and Multi-PIE with facial view of $0°$ or $30°$ to serve as the source database and target database, respectively. Thus, our experiments can be divided into two categories, i.e., cross-database (same-view) and cross-database (cross-view). In the former case, the facial images from source and target databases have the same facial pose ($0°$ or $30°$), while for the latter one, the facial view of the source and target samples are different, namely, the source samples's view is $0°$ and the target samples's view is $30°$ or otherwise. Furthermore, regarding expression features, we employ SIFT features to describe facial images. Specifically, given a facial image, we first locate 49 facial landmarks which are located around the positions of brows, eyes, and mouths as shown in Fig. 1. Then, we be able to extract the SIFT features of these 49 facial landmarks and concatenate them into a feature vector.

Finally, for comparison purpose, we choose three state-of-the-art domain adaptation methods which have been widely-used in dealing with various cross-domain problems, i.e., the kernel mean matching (KMM) [5], the Kullback-Leibler importance estimation procedure (KLIEP) [11], and the unconstrained least-squares importance fitting (uLSIF) [7], to conduct the same experiments with the proposed UDADL. The parameters setup of these methods are the same as the suggestion of the work of [4]. Furthermore, it is noted that we use linear SVM as the classifier for all the domain adaptation methods including UDADL and the results of SVM with no domain adaptation are also included in the comparison to serve as the baseline. In the process of optimization, these important parameters τ, μ, λ was empirically fixed at $\{3, 1, 4, 7\}$, $\{3, 3, 5, 12\}$, and 0.005, respectively.

Tables 1 and 2 depict the experimental results of all the methods. From the results, it is clear to see that the proposed UDADL achieves better overall recognition accuracies than other methods in most cases. However, we observe that when Multi-PIE with $30°$ is used as source database, the results of KMM are better than that of the proposed UDADL. This may be because KMM is more suitable to bridge source and target domain in this case. In addition, it is interesting to see that compared with SVM, the recognition accuracies of three domain adaptation methods except UDADL using BU-3DFE as source database are seriously less than the results of using Multi-PIE as source database. In other

Table 1. Recognition rate of cross-database under the view of 0° vs 0° and 30° vs 30°

Source domain	Target domain	Accuracy (%)				
		SVM	KMM	KLIEP	uLSIF	UDADL
BU-3DFE (0°)	Multi-PIE (0°)	25.50	30.25	25.42	25.75	**63.5**
Multi-PIE (0°)	BU-3DFE (0°)	46.96	62.87	61.17	48.75	**67.01**
BU-3DFE (30°)	Multi-PIE (30°)	26.04	30.67	26.80	26.04	**59.09**
Multi-PIE (30°)	BU-3DFE (30°)	58.45	67.84	55.34	56.93	**64.67**

Table 2. Recognition rate of cross-database and cross-view under the view of 0° vs 30° and 30° vs 0°

Source domain	Target domain	Accuracy (%)				
		SVM	KMM	KLIEP	uLSIF	UDADL
BU-3DFE (30°)	Multi-PIE (0°)	25.52	27.50	26.75	26.00	**62.00**
Multi-PIE (0°)	BU-3DFE (30°)	45.17	54.61	53.15	45.12	**64.40**
BU-3DFE (0°)	Multi-PIE (30°)	26.00	30.67	25.00	26.03	**62.37**
Multi-PIE (30°)	BU-3DFE (0°)	58.15	**69.24**	46.53	55.59	63.96

word, this case that BU-3DFE is served as source database is more challenging and difficult and these three methods are seemingly ineffective. Nevertheless, the proposed UDADL always achieves a significant improvement of recognition accuracy over the SVM regardless of source database.

5   Conclusion

In this paper, we have presented a novel domain adaptation method called unsupervised domain adaptive dictionary learning (UDADL) for cross-database facial expression recognition (FER) problem. In UDADL, we first introduce a analysis dictionary together with conventional one named synthesis dictionary to build a mutual relation between the samples and their codes. By use of analysis dictionary, UDADL could learn a reliable synthesis dictionary for both source and target samples such that the distribution mismatch between two different domains could be alleviated in the code space with respect to the learnt dictionary. Extensive cross-database FER experiments on BU-3DFE and Multi-PIE databases demonstrate that the proposed UDADL has a promising performance and outperforms recent popular domain adaptation methods.

Acknowledgement. This work was supported in part by the National Basic Research Program of China under Grant 2015CB351704, in part by the National Natural Science Foundation of China (NSFC) under Grants 61231002 and 61572009, and in part by the Natural Science Foundation of Jiangsu Province under Grant BK20130020.

References

1. Chu, W.S., Torre, F., Cohn, J.: Selective transfer machine for personalized facial action unit detection. In: Proceedings of the IEEE Conference on Computer Vision and Pattern Recognition, pp. 3515–3522 (2013)
2. Gross, R., Matthews, I., Cohn, J., Kanade, T., Baker, S.: Multi-pie. Image Vis. Comput. **28**(5), 807–813 (2010)
3. Gu, S., Zhang, L., Zuo, W., Feng, X.: Projective dictionary pair learning for pattern classification. In: Advances in Neural Information Processing Systems, pp. 793–801 (2014)
4. Hassan, A., Damper, R., Niranjan, M.: On acoustic emotion recognition: compensating for covariate shift. IEEE Trans. Audio Speech Lang. Process. **21**(7), 1458–1468 (2013)
5. Huang, J., Gretton, A., Borgwardt, K.M., Schölkopf, B., Smola, A.J.: Correcting sample selection bias by unlabeled data. In: Advances in Neural Information Processing Systems, pp. 601–608 (2006)
6. Kan, M., Wu, J., Shan, S., Chen, X.: Domain adaptation for face recognition: targetize source domain bridged by common subspace. Int. J. Comput. Vis. **109**(1–2), 94–109 (2014)
7. Kanamori, T., Hido, S., Sugiyama, M.: A least-squares approach to direct importance estimation. J. Mach. Learn. Res. **10**, 1391–1445 (2009)
8. Pan, S.J., Yang, Q.: A survey on transfer learning. IEEE Trans. Knowl. Data Eng. **22**(10), 1345–1359 (2010)
9. Rubinstein, R., Bruckstein, A.M., Elad, M.: Dictionaries for sparse representation modeling. Proc. IEEE **98**(6), 1045–1057 (2010)
10. Sanagineto, E., Zen, G., Ricci, E., Sebe, N.: We are not all equal: personalizing models for facial expression analysis with transductive parameter transfer. In: Proceedings of the ACM International Conference on Multimedia, pp. 357–366. ACM (2014)
11. Sugiyama, M., Nakajima, S., Kashima, H., Buenau, P.V., Kawanabe, M.: Direct importance estimation with model selection and its application to covariate shift adaptation. In: Advances in Neural Information Processing Systems, pp. 1433–1440 (2008)
12. Yin, L., Wei, X., Sun, Y., Wang, J., Rosato, M.J.: A 3D facial expression database for facial behavior research. In: 7th International Conference on Automatic Face and Gesture Recognition, FGR 2006, pp. 211–216. IEEE (2006)
13. Zhang, C., Liu, J., Tian, Q., Xu, C., Lu, H., Ma, S.: Image classification by nonnegative sparse coding, low-rank and sparse decomposition. In: 2011 IEEE Conference on Computer Vision and Pattern Recognition (CVPR), pp. 1673–1680. IEEE (2011)
14. Zheng, W., Tang, H., Lin, Z., Huang, T.S.: A novel approach to expression recognition from non-frontal face images. In: 2009 IEEE 12th International Conference on Computer Vision, pp. 1901–1908. IEEE (2009)
15. Zheng, W., Zhou, X.: Cross-pose color facial expression recognition using transductive transfer linear discriminat analysis. In: IEEE International Conference on Image Processing, pp. 1935–1939. IEEE (2015)

Adaptive Multi-view Semi-supervised Nonnegative Matrix Factorization

Jing Wang[1], Xiao Wang[2], Feng Tian[1(✉)], Chang Hong Liu[3],
Hongchuan Yu[4], and Yanbei Liu[5]

[1] Faculty of Science and Technology, Bournemouth University, Poole, UK
{jwang,ftian}@bournemouth.ac.uk
[2] Department of Computer Science and Technology,
Tsinghua University, Beijing, China
wangxiao_cv@tju.edu.cn
[3] Department of Psychology, Bournemouth University, Poole, UK
liuc@bournemouth.ac.uk
[4] National Centre for Computer Animation,
Bournemouth University, Poole, UK
hyu@bournemouth.ac.uk
[5] School of Electronic Information Engineering,
Tianjin University, Tianjin, China
liuyanbei@tju.edu.cn

Abstract. Multi-view clustering, which explores complementary information between multiple distinct feature sets, has received considerable attention. For accurate clustering, all data with the same label should be clustered together regardless of their multiple views. However, this is not guaranteed in existing approaches. To address this issue, we propose Adaptive Multi-View Semi-Supervised Nonnegative Matrix Factorization (AMVNMF), which uses label information as hard constraints to ensure data with same label are clustered together, so that the discriminating power of new representations are enhanced. Besides, AMVNMF provides a viable solution to learn the weight of each view adaptively with only a single parameter. Using $L_{2,1}$-norm, AMVNMF is also robust to noises and outliers. We further develop an efficient iterative algorithm for solving the optimization problem. Experiments carried out on five well-known datasets have demonstrated the effectiveness of AMVNMF in comparison to other existing state-of-the-art approaches in terms of accuracy and normalized mutual information.

Keywords: Nonnegative matrix factorization · Multi-view learning · Semi-supervised learning

1 Introduction

Real data are often comprised of multiple views (features) [3]. For example, color and texture information can be utilized as different views of images and videos; a document may be translated into multiple languages, and a web page may be

© Springer International Publishing AG 2016
A. Hirose et al. (Eds.): ICONIP 2016, Part II, LNCS 9948, pp. 435–444, 2016.
DOI: 10.1007/978-3-319-46672-9_49

represented by multiple contents and hyperlinks. In these examples, each view describes a specific perspective of the data. Therefore, it becomes natural to integrate multiple views and discover the underlying data structures.

Multi-view clustering has attracted increasing interests and been explored in several studies. For example, Kumar et al. [7] proposed two objectives to regularize the Laplacian embeddings between different views to be similar and spectral analysis is employed for parameter learning. Tzortzis et al. [14] learned a unified kernel through a weighted combination of kernels of all the views. MultiNMF [12] was proposed to obtain a common consensus matrix, which was designed to reflect the latent clustering structure shared by different views. Subsequently, Wang et al. [15] proposed a regression-like objective, which conducts multi-view clustering and feature selection at the same time. Zhang et al. [16] also developed MMNMF which attempted to preserve intrinsic geometrical structure of data across multiple views. Cao et al. [5] utilized a diversity constraints on subspaces to enhance the complementarity among multiple views.

Various existing methods indeed improve the clustering performance for multi-view data, nevertheless, some challenges remain. Firstly, in reality, supervised information, e.g., the labels of data or the pairwise information (must-link and cannot-link constraints) between data, are often available. They have been integrated into the single-view learning and demonstrated the effectiveness. While for multi-view learning, we notice that the supervised information usually has consistency across multiple views. If we can guarantee data with same label but come from various views are still grouped into the same cluster, this will improve the clustering accuracy [1,17]. Therefore, how to utilize this discriminative information for guiding the multi-view learning is of great value. Secondly, when we cluster data across multiple views, each view may have different contributions. We can consider each view has the same weight in a straightforward way, but this oversimplified assumption may be not always satisfied in the real-world application. Taking the face clustering as an example, a frontal or a three-quarter view is a better representation for faces than a profile view [2,9]. So, the weight of each view should be determined automatically rather than being treated equally. Finally, outliers or noisy data are ubiquitous, and thus, a robust multi-view learning approach is required for practical applications.

To address these challenges altogether, we propose a new multi-view clustering approach based on non-negative matrix factorization (NMF) [8], called Adaptive Multi-View Semi-Supervised Nonnegative Matrix Factorization (AMVNMF). The overall advantages of this approach are as follows:

1. By taking the label information as hard constraints, AMVNMF guarantees that data sharing the same label will have the same new representation and be mapped into the same class in the low-dimensional space regardless whether they come from the same view.
2. To our best knowledge, this is the first attempt to introduce a single parameter to control the distribution of weighting factors for NMF-based multi-view clustering. Consequently, the weight factor of each view can be assigned automatically depending on the dissimilarity between each new representation matrix and the consensus matrix.

3. Using the structured sparsity-inducing, $L_{2,1}$-norm, AMVNMF is robust against noises and hence can achieve more stable clustering results.

2   A Brief Review of NMF and CNMF

Given N data $\mathbf{X} = [x_1, x_2, ..., x_N] \in \mathbb{R}^{P \times N}$, each data x_i is represented by P-dimensional feature vector. NMF [8] aims to find two nonnegative matrix factors $\mathbf{W} \in \mathbb{R}^{P \times K}$ and $\mathbf{H} \in \mathbb{R}^{N \times K}$ where the product of the two factors can well approximate the original matrix, represented as $\mathbf{X} \approx \mathbf{W}\mathbf{H}^T$. In particular, $\mathbf{H}$ can be considered as the new representations of data in terms of the basis $\mathbf{W}$. NMF measures the dissimilarity between $\mathbf{X}$ and $\mathbf{W}\mathbf{H}^T$ by F-norm, which is defined as

$$\|\mathbf{X} - \mathbf{W}\mathbf{H}^T\|_F^2, \tag{1}$$

To extend the traditional unsupervised NMF to a semi-supervised learning approach, CNMF [10] builds a label constraint matrix which incorporates the label information as hard constraints so that the data sharing the same label have the same new representation. In particular, assuming the first l data points are labeled with c classes, then an indicator matrix $\mathbf{C}$ can be constructed, where $c_{i,j} = 1$ if v_i is labeled with jth class; or $c_{i,j} = 0$ otherwise. Then, the label constraint matrix $\mathbf{A}$ can be defined as follows,

$$\mathbf{A} = \begin{pmatrix} \mathbf{C}_{l \times c} & 0 \\ 0 & \mathbf{I}_{N-l} \end{pmatrix}, \tag{2}$$

where $\mathbf{I}_{N-l}$ is a $(N - l) \times (N - l)$ identity matrix. Recall that NMF maps each data point x_i to its new representation h^i from P-dimensional space to K-dimensional space, where h^i represents the ith row of $\mathbf{H}$. To incorporate label information, we introduce an auxiliary matrix $\mathbf{Z}$ with $\mathbf{H} = \mathbf{A}\mathbf{Z}$. As we can see from $\mathbf{A}$, if x_i and x_j have the same label, then the ith row and jth row of $\mathbf{A}$ must be the same, and so $h^i = h^j$, which guarantees that data sharing the same label have the same new representation. Therefore, the objective function can be written as follows,

$$\min_{\mathbf{W} \geq 0, \mathbf{Z} \geq 0} \|\mathbf{X} - \mathbf{W}\mathbf{Z}^T\mathbf{A}^T\|_F^2. \tag{3}$$

3   Adaptive Multi-view Semi-supervised Nonnegative Matrix Factorization(AMVNMF)

3.1   AMVNMF Model

Let $\mathbf{X}^{(v)} \in \mathbb{R}^{P_v \times N}$ denote the features in vth view, $\mathbf{W}^{(v)} \in \mathbb{R}^{P_v \times K}$ and $\mathbf{Z}^{(v)} \in \mathbb{R}^{K \times N}$ be the basis and auxiliary matrix in vth view, respectively. Since the matrix $\mathbf{A}$ above is constructed only based on the label information and consistent for different features, which means different features share the same constraint matrix $\mathbf{A}$. Thus, given M types of heterogeneous features, $v = 1, 2, ...M$, we naturally integrate all these view together and propose the objective function as follows,

$$\min_{\mathbf{W} \geq 0, \mathbf{Z} \geq 0} \sum_{v=1}^{M} \|\mathbf{X}^{(v)} - \mathbf{W}^{(v)}(\mathbf{Z}^{(v)})^T \mathbf{A}^T\|_F^2. \tag{4}$$

Assuming that, new representation matrices of M views are regularized towards a common consensus matrix $\mathbf{H}^*$, we aim to obtain $\mathbf{H}^*$, which uncovers the common latent structure shared by multiple views. With the constraint matrix $\mathbf{A}$ and a consensus auxiliary matrix $\mathbf{Z}^*$, we have $\mathbf{H}^* = \mathbf{A}\mathbf{Z}^*$. Since $\mathbf{A}$ is known, we turn the problem of finding $\mathbf{H}^*$ into the problem of finding $\mathbf{Z}^*$. The objective function can be rewritten as follows,

$$\min_{\mathbf{W}^{(v)}, \mathbf{Z}^{(v)}, \mathbf{Z}^* \geq 0} \sum_{v=1}^{M} \|\mathbf{X}^{(v)} - \mathbf{W}^{(v)}(\mathbf{Z}^{(v)})^T \mathbf{A}^T\|_F^2 + \sum_{v=1}^{M} \lambda_v \|\mathbf{Z}^{(v)} - \mathbf{Z}^*\|_F^2, \tag{5}$$

where λ_v is the weight factor for vth view.

Note that different views may not be comparable at the same scale, hence, without loss of generality, we assume $\|\mathbf{X}^{(v)}\|_1 = 1$. Also, in order to make different $\mathbf{Z}^{(v)}$ comparable and meaningful, we need to constrain $\|\mathbf{W}\|_1 = 1$. To do so, we introduce

$$\mathbf{Q}^{(v)} = Diag(\sum_{i=1}^{M} \mathbf{W}_{i,1}^{(v)}, \sum_{i=2}^{M} \mathbf{W}_{i,2}^{(v)}, ..., \sum_{i=1}^{M} \mathbf{W}_{i,K}^{(v)}) \tag{6}$$

to normalize $\mathbf{W}$ by using $\mathbf{W} = \mathbf{W}\mathbf{Q}^{-1}$. In this way, we can approximately constrain $\|(\mathbf{Z}^{(v)})^T \mathbf{A}^T\|_1 = 1$ so that $\mathbf{Z}^{(v)}$ is within the same range [12]. Due to $\mathbf{W}^{(v)}\mathbf{Z}^{(v)T}\mathbf{A}^T = \mathbf{W}^{(v)}\mathbf{Q}^{-1}(\mathbf{Z}^{(v)}\mathbf{Q})^T \mathbf{A}^T$, (5) could then be written as

$$\min_{\mathbf{W}^{(v)}, \mathbf{Z}^{(v)}, \mathbf{Z}^* \geq 0} \sum_{v=1}^{M} \|\mathbf{X}^{(v)} - \mathbf{W}^{(v)}(\mathbf{Z}^{(v)})^T \mathbf{A}^T\|_F^2 + \sum_{v=1}^{M} \lambda_v \|\mathbf{Z}^{(v)}\mathbf{Q}^{(v)} - \mathbf{Z}^*\|_F^2. \tag{7}$$

Normally, for all M views, each parameter λ_v need to be specified manually which reflects each view's importance. Apparently, it is hard to decide which view contributes more or less with no prior knowledge. To address this limitation, we use a single parameter γ to control the distribution of weight factors $\alpha^{(v)}$ in all M views, such that the important views are assigned bigger weights. Then we have

$$J = \min_{\mathbf{W}^{(v)}, \mathbf{Z}^{(v)}, \mathbf{Z}^*, \alpha^{(v)} \geq 0} \sum_{v=1}^{M} \|\mathbf{X}^{(v)} - \mathbf{W}^{(v)}(\mathbf{Z}^{(v)})^T \mathbf{A}^T\|_F^2$$

$$+ \sum_{v=1}^{M} (\alpha^{(v)})^\gamma \|\mathbf{Z}^{(v)}\mathbf{Q}^{(v)} - \mathbf{Z}^*\|_F^2$$

$$s.t. \sum_{v=1}^{M} \alpha^{(v)} = 1. \tag{8}$$

Note that the first term in (8) is the least square loss function, which is very sensitive to outliers in real word data, as the error for each data is squared and can easily dominate the objective function. Instead, we propose a more robust formulation as the following:

$$J = \min_{\mathbf{W}^{(v)}, \mathbf{Z}^{(v)}, \mathbf{Z}^*, \alpha^{(v)} \geq 0} \sum_{v=1}^{M} \|\mathbf{X}^{(v)} - \mathbf{W}^{(v)}(\mathbf{Z}^{(v)})^T \mathbf{A}^T\|_{2,1}$$

$$+ \sum_{v=1}^{M} (\alpha^{(v)})^\gamma \|\mathbf{Z}^{(v)} \mathbf{Q}^{(v)} - \mathbf{Z}^*\|_F^2$$

$$s.t. \sum_{v=1}^{M} \alpha^{(v)} = 1. \tag{9}$$

In this objective function, the error for each data is not squared and thus the impact of large errors is weaken greatly. Correspondingly, the effects of data outliers are reduced and the robustness of NMF is improved.

3.2 Algorithm of AMVNMF model

To solve the optimization problem (9), we propose an iterative update procedure. When $\mathbf{Z}^*$ is fixed, for each given v, the computation of $\mathbf{W}^{(v)}$ or $\mathbf{Z}^{(v)}$ does not depend on $\mathbf{W}^{(v')}$ or $\mathbf{Z}^{(v')}$, where $v \neq v'$. Therefore, we use $\mathbf{X}$, $\mathbf{W}$, $\mathbf{Z}$, and $\mathbf{Q}$ to represent $\mathbf{X}^{(v)}$, $\mathbf{W}^{(v)}$, $\mathbf{Z}^{(v)}$ and $\mathbf{Q}^{(v)}$ for brevity. The objective function can be simplified as

$$J = \min_{\mathbf{W}, \mathbf{Z}, \mathbf{Z}^*, \alpha^{(v)} \geq 0} \|\mathbf{X} - \mathbf{W}\mathbf{Z}^T \mathbf{A}^T\|_{2,1} + (\alpha^{(v)})^\gamma \|\mathbf{Z}\mathbf{Q} - \mathbf{Z}^*\|_F^2. \tag{10}$$

The following multiplicative updating rules for $\mathbf{W}$, $\mathbf{Z}$ and $\mathbf{D}$ are applied to update their values sequentially and iteratively.

(1) Fixing $\mathbf{Z}^*$, $\mathbf{Z}$, $\mathbf{D}$ and $\alpha^{(v)}$, update $\mathbf{W}$

Let $\Phi_{i,k}$ be the Lagrange multiplier matrix for the constraint $\mathbf{W}_{i,k} \geq 0$, and $\mathbf{\Phi} = [\Phi_{i,k}]$. The Lagrange function is $L_1 = J + Tr(\mathbf{\Phi}\mathbf{W})$, we only care the terms that are relevant to $\mathbf{W}^{(v)}$.

$$L_1 = Tr(-2\mathbf{X}\mathbf{D}\mathbf{A}\mathbf{Z}\mathbf{W}^T + \mathbf{W}\mathbf{Z}^T\mathbf{A}^T\mathbf{D}\mathbf{A}\mathbf{Z}\mathbf{W}^T)$$
$$+ (\alpha^{(v)})^\gamma Tr(\mathbf{Z}\mathbf{Q}\mathbf{Q}^T\mathbf{Z}^T - 2\mathbf{Z}\mathbf{Q}(\mathbf{Z}^*)^T) + Tr(\mathbf{\Phi}\mathbf{W}). \tag{11}$$

Taking the derivatives of L_1 with respect to $\mathbf{W}$ and using the Karush-Kuhn-Tucker condition $\Phi_{i,k}\mathbf{W}_{i,k} = 0$, we get the update rule as follows,

$$\mathbf{W}_{i,k} = \mathbf{W}_{i,k} \cdot \frac{(\mathbf{X}\mathbf{D}\mathbf{A}\mathbf{Z})_{i,k} + (\alpha^{(v)})^\gamma \sum_{j=1}^{N-l+c} \mathbf{Z}_{j,k}\mathbf{Z}_{j,k}^*}{(\mathbf{W}\mathbf{Z}^T\mathbf{A}^T\mathbf{D}\mathbf{A}\mathbf{Z})_{i,k} + (\alpha^{(v)})^\gamma \sum_{f=1}^{d_v}\mathbf{W}_{f,k}\sum_{j=1}^{N-l+c}\mathbf{Z}_{j,k}^2}. \tag{12}$$

(2) Fixing $\mathbf{Z}^*$, $\mathbf{W}$, $\mathbf{D}$ and $\alpha^{(v)}$, update $\mathbf{Z}$

For each view, we first normalize the column vectors of $\mathbf{W}$ using $\mathbf{Q}$ as in (6), then

$$\mathbf{W} \leftarrow \mathbf{W}\mathbf{Q}^{-1}, \mathbf{Z} \leftarrow \mathbf{Z}\mathbf{Q}. \tag{13}$$

Thus, the object function equals

$$\min_{\mathbf{W},\mathbf{Z},\mathbf{Z}^*,\alpha^{(v)} \geq 0} \|\mathbf{X} - \mathbf{W}\mathbf{Z}^T\mathbf{A}^T\|_{2,1} + (\alpha^{(v)})^\gamma \|\mathbf{Z} - \mathbf{Z}^*\|_F^2. \tag{14}$$

Let Ψ be the Lagrange multiplier matrix for the constraint $\mathbf{Z} \geq 0$, and $\Psi = [\Psi_{j,k}]$. Similarly,

$$L_2 = Tr(-2\mathbf{X}\mathbf{D}\mathbf{A}\mathbf{Z}\mathbf{W}^T + 2\mathbf{W}\mathbf{Z}^T\mathbf{A}^T\mathbf{D}\mathbf{A}\mathbf{Z}\mathbf{W}^T)$$
$$+(\alpha^{(v)})^\gamma Tr(2\mathbf{Z}\mathbf{Z}^T - 2\mathbf{Z}(\mathbf{Z}^*)^T) + Tr(\Psi\mathbf{Z}). \tag{15}$$

Taking derivative of L_2 with respect to $\mathbf{Z}$ and using the Kuhn-Tucker condition $\Psi_{j,k}\mathbf{Z}_{j,k} = 0$, we have

$$\mathbf{Z}_{j,k} = \mathbf{Z}_{j,k} \cdot \frac{(\mathbf{A}^T\mathbf{D}\mathbf{X}^T\mathbf{W})_{j,k} + (\alpha^{(v)})^\gamma \mathbf{Z}_{j,k}^*}{(\mathbf{A}^T\mathbf{D}\mathbf{A}\mathbf{Z}\mathbf{W}^T\mathbf{W})_{j,k} + (\alpha^{(v)})^\gamma \mathbf{Z}_{j,k}}. \tag{16}$$

(3) Fixing $\mathbf{Z}^*$, $\mathbf{W}$, $\mathbf{Z}$ and $\alpha^{(v)}$, update $\mathbf{D}$
$\mathbf{D} \in \mathbb{R}^{N \times N}$ is the diagonal matrix with the diagonal elements given by

$$D_{ii} = \frac{1}{\|\mathbf{X}_i - \mathbf{W}(\mathbf{Z}^T\mathbf{A}^T)_i\|}. \tag{17}$$

(4) Fixing $\mathbf{W}$, $\mathbf{Z}$, $\mathbf{D}$ and $\alpha^{(v)}$, update $\mathbf{Z}^*$
Taking the derivative of the objective function J in (8), we get

$$\mathbf{Z}^* = \frac{\sum_{v=1}^M (\alpha^{(v)})^\gamma \mathbf{Z}^{(v)}\mathbf{Q}^{(v)}}{\sum_{v=1}^M (\alpha^{(v)})^\gamma}. \tag{18}$$

(5) Fixing $\mathbf{Z}^*$, $\mathbf{W}$, $\mathbf{Z}$ and $\mathbf{D}$ update $\alpha^{(v)}$
We only consider the term that relevant to α, thus, it is equal to minimize

$$J = \sum_{v=1}^M (\alpha^{(v)})^\gamma \|\mathbf{Z}^{(v)}\mathbf{Q}^{(v)} - \mathbf{Z}^*\|_F^2. \tag{19}$$

By setting $\mathbf{G}^{(v)} = \|\mathbf{Z}^{(v)}\mathbf{Q}^{(v)} - \mathbf{Z}^*\|_F^2$, and due to $\sum_{v=1}^M \alpha^{(v)} = 1$, Then, the Lagrange function of (19) is

$$J = \sum_{v=1}^M (\alpha^{(v)})^\gamma \mathbf{G}^{(v)} - \lambda \sum_{v=1}^M (\alpha^{(v)} - 1). \tag{20}$$

The solution can be obtained as

$$\alpha^{(v)} = \frac{(\gamma\mathbf{G}^{(v)})^{\frac{1}{1-\gamma}}}{\sum_{v=1}^M (\gamma\mathbf{G}^{(v)})^{\frac{1}{1-\gamma}}}. \tag{21}$$

4   Experiments

We use five benchmark datasets, SensIT[1], ORL [6], Reuters[2], CiteSeer[3] and Cora[4] to assess the performance of AMVNMF. SensIT uses two sensors to classify three types of vehicle. We randomly sample 100 data for each class, and then conduct experiments on 2 views and 3 classes. ORL contains 10 different images of each of 40 people. The images are represented by two views, raw pixel values and GIST [13]. Reuters contains 1200 documents over the 6 labels. Each sample is translated into five languages. We experiment on the English, French and German views same as [12]. Citeseer and Cora are composed of publications. These publications are linked via citations. Both of them take contents and citations as two views.

We compare our AMVNMF with several representative multi-view clustering methods and their modifications.

1. Best Single View (BSV): Using the most informative view which achieves the best performance with our AMVNMF.
2. ConCNMF: The method firstly concatenates the features of all views and applies CNMF [10] to extract the low dimensional subspace representation.
3. MultiNMF: The NMF-based multi-view clustering method proposed in [12].
4. RMKMC: The multi-view k-means proposed in [4].
5. CoRegSPC: The co-regularized pairwise multi-view spectral clustering method proposed in [7].

Since clustering performances depend on the initializations, we repeat each method 10 times with random initializations and report the average performance.

The performances are measured with two widely used evaluation metrics, accuracy (AC) and normalized mutual information (NMI) [10]. For all the metrics, the higher value indicates better clustering quality. To compare the performance of semi-supervised approaches, same as [11], 30 % of labeled data are randomly picked up as priors.

Table 1 summarizes the clustering performances of different algorithms on the five datasets. It is clear to see that AMVNMF outperforms the second best algorithm in all cases. Furthermore, BSV always gets the second best performance. It outperforms other multi-view methods greatly, i.e., 7.99 %/10.27 % on SensIT and 3.10 %/5.38 % on Reuters in terms of AC and NMI, respectively. This is mainly due to AMVNMF guarantees that all the data sharing the same labels are grouped together, regardless they are come from the same or different views. Therefore, both AMVNMF and BSV (running AMVNMF with single view) produce superior results.

We show the parameter tuning and algorithm convergence on SensIT, ORL and Reuters as examples in Figs. 1 and 2, respectively. The parameter γ controls

---

[1] https://www.csie.ntu.edu.tw/~cjlin/libsvmtools/datasets/multiclass.html.
[2] http://multilingreuters.iit.nrc.ca.
[3] http://linqs.umiacs.umd.edu/projects//projects/lbc/index.html.
[4] http://linqs.umiacs.umd.edu/projects//projects/lbc/index.html.

Table 1. Clustering results on five real-world datasets (%)

Metrics	Datasets	BSV	ConCNMF	MultiNMF	RMKMC	CoRegSPC	AMVNMF
AC	SensIT	69.66	52.30	55.04	60.07	61.67	**71.33**
	ORL	74.3	49.59	54.6	45.5	78.20	**80.5**
	Reuters	57.50	49.59	51.87	39.80	54.40	**59.88**
	Citeseer	50.08	40.70	34.36	43.21	47.42	**53.14**
	Cora	33.42	32.42	44.83	43.90	37.20	**48.71**
NMI	SensIT	30.14	15.67	19.87	14.84	17.75	**31.73**
	ORL	89.29	51.32	75.23	65.34	90.84	**91.73**
	Reuters	41.95	30.37	36.14	21.82	36.57	**42.75**
	Citeseer	21.38	13.34	20.97	20.61	21.10	**26.13**
	Cora	26.73	9.87	27.95	21.27	15.44	**34.59**

the distribution of weight factors $\alpha^{(v)}$ for different views. More preciously, when $\gamma \to \infty$, the weight for all views is equal. When $\gamma \to 1$, the weight factor of 1 is assigned to the most important view whose $\mathbf{G}^{(v)}$ value is the smallest and 0 is assigned to the weights of the other views. Hence, this strategy allows well adjusting the ratio of each view and saves the cost of tuning multiple parameters. As shown in Fig. 1, AMVNMF performs stably with varying γ (from 2 to 902). Please note that even the worst results of AMVNMF are always better than other approaches in most cases. The corresponding convergence curves together with performances are shown in Fig. 2. The blue solid line shows the value of the objective function and the red dashed line indicates the accuracy. It can be seen that the value of the objective function decreases steadily with more iterations and converges after around 20 times.

Since AMVNMF is a semi-supervised method, we also randomly pick up 10 % and 20 % labeled data to further demonstrate the benefits of priors. Notice that ORL has only 10 images for each category, thus 10 % gives one image only.

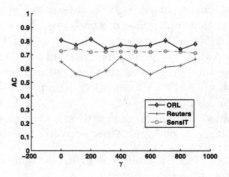

Fig. 1. Performance of AMVNMF w.r.t. parameter γ.

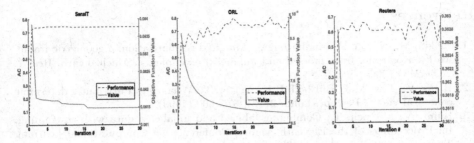

Fig. 2. Convergence and corresponding performance curve. (Color figure online)

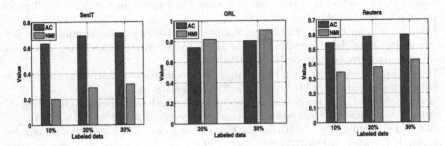

Fig. 3. Performance of AMVNMF w.r.t. labeled data.

However, one label is meaningless for AMVNMF since this algorithm maps the images with the same label onto the same coordinate in the new representation space. Thus, we omit the result with 10 % labeled data. From Fig. 3, it can be seen that both AC and NMI are improved with more labeled data. Also, it is worth pointing out that even with only 10 % labeled data, AMVNMF performs better than other approaches when 30 % labeled data are applied. For example, for the SensIT dataset, AMVNMF achieves 62 % AC and 20 % NMI with 10 % labeled data, which is better than the best performance of other approaches, i.e., 61.67 % AC and 19.87 % NMI (as shown in Table 1).

5   Conclusion

A novel NMF-based multi-view method, AMVNMF, is proposed in this paper. It efficiently learn the underlying clustering structure embedded in multiple views, by regularizing the new representation matrices learnt from different views towards a common consensus. The advantages of AMVNMF are shown in three aspects. First, it guarantees that labeled data come with multiple views can be clustered into the same low-dimension space. Second, it learns each view's corresponding weight adaptively with a single parameter γ. Third, it handles the noises more effectively. For future work, a sparse regulation may be introduced into AMVNMF to obtain more accurate new representation matrix, with which the clustering performance is expected to be further improved.

References

1. Belkin, M., Niyogi, P., Sindhwani, V.: Manifold regularization: a geometric framework for learning from labeled and unlabeled examples. J. Mach. Learn. Res. **7**, 2399–2434 (2006)
2. Blanz, V., Tarr, M.J., Bülthoff, H.H., Vetter, T.: What object attributes determine canonical views? Percept.-Lond. **28**(5), 575–600 (1999)
3. Blum, A., Mitchell, T.: Combining labeled and unlabeled data with co-training. In: Proceedings of the Eleventh Annual Conference on Computational Learning Theory, pp. 92–100. ACM (1998)
4. Cai, X., Nie, F., Huang, H.: Multi-view k-means clustering on big data. In: Proceedings of the Twenty-Third International Joint Conference on Artificial Intelligence, pp. 2598–2604. AAAI Press (2013)
5. Cao, X., Zhang, C., Fu, H., Liu, S., Zhang, H.: Diversity-induced multi-view subspace clustering. In: Proceedings of the IEEE Conference on Computer Vision and Pattern Recognition, pp. 586–594 (2015)
6. Hidru, D., Goldenberg, A.: Equinmf: graph regularized multiview nonnegative matrix factorization. arXiv preprint arXiv:1409.4018 (2014)
7. Kumar, A., Rai, P., Daume, H.: Co-regularized multi-view spectral clustering. In: Advances in Neural Information Processing Systems, pp. 1413–1421 (2011)
8. Lee, D.D., Seung, H.S.: Learning the parts of objects by non-negative matrix factorization. Nature **401**(6755), 788–791 (1999)
9. Liu, C.H., Chaudhuri, A.: Reassessing the 3/4 view effect in face recognition. Cognition **83**(1), 31–48 (2002)
10. Liu, H., Wu, Z., Li, X., Cai, D., Huang, T.S.: Constrained nonnegative matrix factorization for image representation. IEEE Trans. Pattern Anal. Mach. Intell. **34**(7), 1299–1311 (2012)
11. Liu, H., Yang, G., Wu, Z., Cai, D.: Constrained concept factorization for image representation. IEEE Trans. Cybern. **44**(7), 1214–1224 (2014)
12. Liu, J., Wang, C., Gao, J., Han, J.: Multi-view clustering via joint nonnegative matrix factorization. In: Proceedings of SDM, vol. 13, pp. 252–260. SIAM (2013)
13. Oliva, A., Torralba, A.: Modeling the shape of the scene: a holistic representation of the spatial envelope. Int. J. Comput. Vis. **42**(3), 145–175 (2001)
14. Tzortzis, G., Likas, A.: Kernel-based weighted multi-view clustering. In: 2012 IEEE 12th International Conference on Data Mining (ICDM), pp. 675–684. IEEE (2012)
15. Wang, H., Nie, F., Huang, H.: Multi-view clustering and feature learning via structured sparsity. In: Proceedings of the 30th International Conference on Machine Learning (ICML-13), pp. 352–360 (2013)
16. Zhang, X., Zhao, L., Zong, L., Liu, X., Yu, H.: Multi-view clustering via multi-manifold regularized nonnegative matrix factorization. In: 2014 IEEE International Conference on Data Mining (ICDM), pp. 1103–1108. IEEE (2014)
17. Zhu, X., Ghahramani, Z., Lafferty, J., et al.: Semi-supervised learning using Gaussian fields and harmonic functions. In: ICML, vol. 3, pp. 912–919 (2003)

Robust Soft Semi-supervised Discriminant Projection for Feature Learning

Xiaoyu Wang[1,2], Zhao Zhang[1,2(✉)], and Yan Zhang[1,2]

[1] School of Computer Science and Technology and Joint International Research Laboratory of Machine Learning and Neuromorphic Computing, Soochow University, Suzhou 215006, China
cszzhang@gmail.com
[2] Collaborative Innovation Center of Novel Software Technology and Industrialization, Nanjing 210023, China

Abstract. Image feature extraction and noise/outlier processing has received more and more attention. In this paper, we first take the full use of labeled and unlabeled samples, which leads to a semi-supervised model. Based on the soft label, we combine unlabeled samples with their predicted labels so that all the samples have their own soft labels. Our ratio based model maximizes the soft between-class scatter, as well as minimizes the soft within-class scatter plus a neighborhood preserving item, so that our approach can explicitly extract discriminant and locality preserving features. Further, to make the result be more robust to outliers, all the distance metrics are configured as L1-norm instead of L2-norm. An effective iterative method is taken to solve the optimal function. Finally, we conduct simulation experiments on CASIA-HWDB1.1 and MNIST handwriting digits datasets. The results verified the effectiveness of our approach compared with other related methods.

Keywords: Semi-supervised learning · Soft label · Locality discriminant projection · L1-norm · Robust image representation

1 Introduction

High-dimensional data in our daily life always have redundant or unfavorable features [9,10]. Traditional feature learning methods which solve this problem can be described as supervised or unsupervised learning methods. Typical supervised learning algorithms such as Linear Discriminant Analysis (LDA) [1], Local Fisher Discriminant Analysis [7] can produce discriminant features. These models make full use of the labeled data, but in fact labeled data usually cannot be easily got, and these methods may suffer from overfitting. Several popular unsupervised learning methods include Principal Component Analysis (PCA) [1], Neighborhood Preserving Embedding (NPE) [2] and so on. This kind of algorithms study the relation and structure between the data but cannot use the label information. Therefore, semi-supervised learning methods have been

© Springer International Publishing AG 2016
A. Hirose et al. (Eds.): ICONIP 2016, Part II, LNCS 9948, pp. 445–453, 2016.
DOI: 10.1007/978-3-319-46672-9_50

proposed recent years, which overcomes the shortcomings of the above traditional learning methods. Typical algorithm such as [12] can usually get better results than the previous two types of learning methods.

On the other hand, all these traditional methods use L2-norm based criteria, which has been proven to be sensitive to outliers. To reduce the impact of outliers, L1-norm based algorithms have been proposed, such as L1-Norm maximization based Principal component Analysis (PCA-L1) [3] and L1-Norm Maximization based Discriminant Locality Preserving Projections (DLPP-L1) [13]. As can be seen, these algorithms usually get better solutions than the corresponding L2-norm based ones. However, the methods mentioned above are either supervised or unsupervised learning methods, which still suffer from similar disadvantages.

In this paper, we propose an L1-norm based semi-supervised model for feature learning named Robust Semi-Supervised Discriminant Projection (RS^{3}DP-L1) which combines the advantages of discriminating and locality preserving. Our model is formulated based on Soft Label based Linear Discriminant Analysis (SL-LDA) [12] and NPE [2]. SL-LDA integrates the unlabeled samples into the model by soft labels. Moreover, NPE preserves the local manifold structure which enhances the performance of our approach. It is also important that L1-norm based metric is regularized on the discriminating and local scatter matrices in our model, so that our RS^{3}DP-L1 is robust to outliers. Further to enhance the performance of soft label, we replace the soft label matrix in SL-LDA with that in Nonnegative Sparse Neighborhood Propagation (SparseNP) [11], which ensures the soft labels to meet the terms of being discriminative and roust to noise.

In Sect. 2, the related work is briefly reviewed. Our method is proposed in Sect. 3. Section 4 shows the experimental results and analysis. Finally Sect. 5 draws the conclusion.

2 Related Work

2.1 Soft Label Based Linear Discriminant Analysis (SL-LDA)

SL-LDA is a semi-supervised dimensionality reduction method [12]. "Soft label" means the probability of belonging to some class, compared to "hard label" as the absolute label like the label of a labeled sample. In other words, if we use hard label, the probability of a sample belonging to some class is either 0 or 1. Suppose there are l labeled samples, u unlabeled samples and c classes, the "soft labels" are described in a matrix $\mathbf{F} \in \mathbb{R}^{(c+1)\times(l+u)}$, in which the element $f_{i,j}(i = 1, 2, \ldots, c+1, j = 1, 2, \ldots, l+u)$ shows the probability of the j-th sample belonging to i-th class, while the $(c+1)$-th class represents the outliers. The soft label based within-class and between-class scatter matrices are as follows.

$$\tilde{\mathbf{S}}_w = \sum_{i=1}^{c}\sum_{j=1}^{l+u} f_{i,j}(\mathbf{x}_j - \tilde{\boldsymbol{\mu}}_i)(\mathbf{x}_j - \tilde{\boldsymbol{\mu}}_i)^T = \mathbf{X}(\mathbf{E} - \mathbf{F}_c^T\mathbf{G}^{-1}\mathbf{F}_c)\mathbf{X}^T, \qquad (1)$$

$$\tilde{\mathbf{S}}_b = \sum_{i=1}^{c}\sum_{j=1}^{l+u} f_{i,j}(\tilde{\boldsymbol{\mu}}_i - \tilde{\boldsymbol{\mu}})(\tilde{\boldsymbol{\mu}}_i - \tilde{\boldsymbol{\mu}})^T, \qquad (2)$$

where $\mathbf{E} \in \mathbb{R}^{(l+u) \times (l+u)}$ and $\mathbf{G} \in \mathbb{R}^{c \times c}$ are two diagonal matrices, and each element satisfies $E_{j,j} = \sum_{i=1}^{c} f_{i,j}$ and $G_{i,i} = \sum_{j=1}^{l+u} f_{i,j}$. $\mathbf{F}_c \in \mathbb{R}^{c \times (l+u)}$ is formed by the first c rows of soft label matrix $\mathbf{F}$. Moreover, The soft means of i-th class and all classes are defined as:

$$\tilde{\mu}_i = \sum_{k=1}^{l+u} f_{i,k} \mathbf{x}_k \bigg/ \sum_{k=1}^{l+u} f_{i,k} , \tag{3}$$

$$\tilde{\mu} = \sum_{i=1}^{c} \sum_{k=1}^{l+u} f_{i,k} \mathbf{x}_k \bigg/ \sum_{i=1}^{c} \sum_{k=1}^{l+u} f_{i,k} . \tag{4}$$

The model based on the idea of maximizing the soft between-class scatter matrix while minimizing the soft within-class scatter matrix is shown as follows.

$$J(\mathbf{V}) = \max_{\mathbf{V}} Tr \left((\mathbf{V}^T \tilde{\mathbf{S}}_w \mathbf{V})^{-1} \mathbf{V}^T \tilde{\mathbf{S}}_b \mathbf{V} \right). \tag{5}$$

2.2 Neighborhood Preserving Embedding (NPE)

NPE is an unsupervised algorithm which aims at preserving the local neighborhood structure on the data manifold [2]. Given a set of data points in the ambient space, the weight matrix $\mathbf{M}$ describes the linear combination of the neighboring data points and the combination coefficients between the data points. The objective function is demonstrated in Eq. (6).

$$\min \ \phi(\mathbf{P}) = \sum_i \| \mathbf{P}^T \mathbf{x}_i - \sum_j M_{i,j} \mathbf{P}^T \mathbf{x}_j \|^2, \quad s.t. \sum_j M_{i,j} = 1, \tag{6}$$

where $\| \cdot \|$ denotes L2-norm, i.e. for matrix $\mathbf{S} = [s_{i,j}]$, $\|\mathbf{S}\| = \sqrt{\sum_{i,j} s_{i,j}^2}$.

2.3 Nonnegative Sparse Neighborhood Propagation (SparseNP)

SparseNP is an improvement to the existing neighborhood propagation which ensures soft labels to be discriminative and robust to noise [11]. The model can be defined as:

$$\mathbf{F}^S = \arg \min_{\mathbf{F}} \mathfrak{R}(\mathbf{F}) = \sum_{i=1}^{l+u} \left\| \mathbf{f}_i - \sum_{j : \mathbf{x}_j \in \mathbb{N}(\mathbf{x}_i)} W_{i,j} \mathbf{f}_j \right\|^2 + \sum_{i=1}^{l+u} \mu_i D_{ii} \| \mathbf{f}_i - \mathbf{y}_i \|^2 + \psi \sum_{i=1}^{l+u} \| \hat{\mathbf{f}}^i \|, \tag{7}$$

$$s.t. \ f_{j,i} \geq 0, \ \sum_{i=1}^{c} f_{j,i} = 1, \ (i = 1, 2, \dots, l+u, \ j = 1, 2, \dots, c)$$

where $\mathbf{f}_i \in \mathbb{R}^{c \times 1}$ is the i-th column of $\mathbf{F}$. $\mathbb{N}(\mathbf{x}_i)$ is the K-neighbor set of $\mathbf{x}_i$. Other parameters definitions can be seen in [11].

3   Problem Formulation

3.1   RS³DP-L1 Framework

Suppose we have N samples denoted as $\mathbf{X} = [\mathbf{X}_L, \mathbf{X}_U] \in \mathbb{R}^{d \times N}$ in which we have l labeled samples denoted as $\mathbf{X}_L = [\mathbf{x}_1, \mathbf{x}_2, \ldots, \mathbf{x}_l] \in \mathbb{R}^{d \times l}$ and u unlabeled samples denoted as $\mathbf{X}_U = [\mathbf{x}_{l+1}, \mathbf{x}_{l+2}, \ldots, \mathbf{x}_{l+u}] \in \mathbb{R}^{d \times u}$. It is to say that $l + u = N$ and every sample is a d-dimensional vector. The label vector corresponding to those l samples is denoted as $\mathbf{y} = [y_1, y_2, \ldots, y_l]$, where $y_i \in \{1, 2, \ldots, c\}, (i = 1, 2, \ldots, l)$ in total c classes. Feature extraction problem is to find an effective model to obtain a projection matrix $\mathbf{P} \in \mathbb{R}^{d \times n}$, where $n \ll d$.

First of all, according to the idea of combining discriminating and locality manifold preserving with soft label, we describe our L2-norm based learning model as follows:

$$\mathbf{P}_{opt} = \arg \max_{\mathbf{P}} H(\mathbf{P}), \ s.t. \mathbf{P}^T \mathbf{P} = \mathbf{I}_n, \tag{8}$$

where

$$H(\mathbf{P}) = \frac{\sum\limits_{i=1}^{c} \sum\limits_{j=1}^{N} f_{i,j}^S \|\mathbf{P}^T(\tilde{\boldsymbol{\mu}}_i - \tilde{\boldsymbol{\mu}})\|^2}{\sum\limits_{i=1}^{c} \sum\limits_{j=1}^{N} f_{i,j}^S \|\mathbf{P}^T(\mathbf{x}_j - \tilde{\boldsymbol{\mu}}_i)\|^2 + \gamma \sum\limits_{i} \|\mathbf{P}^T \mathbf{x}_i - \sum_j M_{i,j} \mathbf{P}^T \mathbf{x}_j\|^2}, \tag{9}$$

where $\mathbf{F}^S$ is the soft label matrix which is mentioned in Eq. (7) computed by SparseNP, $\tilde{\boldsymbol{\mu}}_i$ is the soft class average as shown in Eq. (3) and $\tilde{\boldsymbol{\mu}}$ is the soft average of all samples as shown in Eq. (4). According to Eq. (1), $\mathbf{E} - \mathbf{F}_c^T \mathbf{G}^{-1} \mathbf{F}_c$ is similar to the Laplacian matrix [8], so in this paper we consider $\mathbf{F}_c^T \mathbf{G}^{-1} \mathbf{F}_c$ as the weight matrix for measuring the similarities. Therefore, the soft label within class item can be rewritten as Eq. (10). Similarly, according to [8], neighborhood preserving term in NPE can also be rewritten as Eq. (11).

$$\sum_{i=1}^{c} \sum_{j=1}^{N} f_{i,j}^S \|\mathbf{P}^T(\mathbf{x}_j - \boldsymbol{\mu}_i)\|^2 = \frac{1}{2} \sum_{i,j=1}^{N} \tilde{W}_{i,j}^{(w_1)} \|\mathbf{P}^T(\mathbf{x}_i - \mathbf{x}_j)\|^2, \tag{10}$$

$$\sum_{i} \|\mathbf{P}^T \mathbf{x}_i - \sum_{j} M_{i,j} \mathbf{P}^T \mathbf{x}_j\|^2 = \frac{1}{2} \sum_{i,j=1}^{N} \tilde{W}_{i,j}^{(w_2)} \|\mathbf{P}^T(\mathbf{x}_i - \mathbf{x}_j)\|^2, \tag{11}$$

where $\tilde{W}^{(w_1)} = (\mathbf{F}^S)^T \mathbf{G}^S \mathbf{F}^S$, $G_{i,i}^S = \sum_{j=1}^{N} f_{i,j}^S$ and $\tilde{W}^{(w_2)} = \mathbf{M} + \mathbf{M}^T - \mathbf{M}^T \mathbf{M}$. Thus, the objective function can be simplified as:

$$H(\mathbf{P}) = \sum_{i=1}^{c} \sum_{j=1}^{N} f_{i,j}^S \|\mathbf{P}^T(\tilde{\boldsymbol{\mu}}_i - \tilde{\boldsymbol{\mu}})\|^2 \Bigg/ \sum_{i,j=1}^{N} \tilde{W}_{i,j}^{(w)} \|\mathbf{P}^T(\mathbf{x}_i - \mathbf{x}_j)\|^2, \tag{12}$$

where $\tilde{\mathbf{W}}^{(w)} = \frac{1}{2}(\tilde{\mathbf{W}}^{(w_1)} + \gamma \tilde{\mathbf{W}}^{(w_2)})$.

Since L2-norm is sensitive to outliers, we consider the L1-norm based form:

$$H(\mathbf{P}) = \sum_{i=1}^{c} \sum_{j=1}^{N} f_{i,j}^{S} \|\mathbf{P}^{T}(\tilde{\boldsymbol{\mu}}_i - \tilde{\boldsymbol{\mu}})\|_1 \Big/ \sum_{i,j=1}^{N} \tilde{W}_{i,j}^{(w)} \|\mathbf{P}^{T}(\mathbf{x}_i - \mathbf{x}_j)\|_1 , \qquad (13)$$

where $\| \cdot \|_1$ denotes L1-norm. Then we consider the condition of $n = 1$ in order to solve the problem. The converted model is as follows:

$$\mathbf{p}^* = \arg\max_{\mathbf{p}} H(\mathbf{p}), \ s.t. \mathbf{p}^T \mathbf{p} = 1, \qquad (14)$$

where

$$H(\mathbf{p}) = \sum_{i=1}^{c} \sum_{j=1}^{N} f_{i,j}^{S} |\mathbf{p}^{T}(\tilde{\boldsymbol{\mu}}_i - \tilde{\boldsymbol{\mu}})| \Big/ \sum_{i,j=1}^{N} \tilde{W}_{i,j}^{(w)} |\mathbf{p}^{T}(\mathbf{x}_i - \mathbf{x}_j)| . \qquad (15)$$

3.2 Optimization

The solving process of L1-norm based problem is much difficult than L2-norm based one. An effective optimization approach is based on an iterative process [13]. Therefore, all the functions in this subsection contain the parameter denoted as t to represent the iterative times.

We first establish two polar functions shown in Eqs. (16) and (17), which can help us eliminate the absolute value function in Eq. (15).

$$q_{i,j}^{x}(t) = \begin{cases} 1, & \mathbf{p}^{T}(t)(\mathbf{x}_i - \mathbf{x}_j) \geqslant 0, \\ -1, & \mathbf{p}^{T}(t)(\mathbf{x}_i - \mathbf{x}_j) < 0, \end{cases} (i, j = 1, 2, \ldots, N). \qquad (16)$$

$$q_{i}^{\mu}(t) = \begin{cases} 1, & \mathbf{p}^{T}(t)(\tilde{\boldsymbol{\mu}}_i - \tilde{\boldsymbol{\mu}}) \geqslant 0, \\ -1, & \mathbf{p}^{T}(t)(\tilde{\boldsymbol{\mu}}_i - \tilde{\boldsymbol{\mu}}) < 0, \end{cases} (i = 1, 2, \ldots, c). \qquad (17)$$

By substituting (16) and (17) into (15), we have the optimal function as shown in Eq. (18). And the iterative update value is calculated by in Eq. (19).

$$H(\mathbf{p}(t)) = \frac{\sum_{i=1}^{c} \sum_{j=1}^{N} f_{i,j}^{S} q_{i}^{\mu}(t) \mathbf{p}^{T}(t)(\tilde{\boldsymbol{\mu}}_i - \tilde{\boldsymbol{\mu}})}{\sum_{i,j=1}^{N} \tilde{W}_{i,j}^{(w)} q_{i,j}^{x}(t) \mathbf{p}^{T}(t)(\mathbf{x}_i - \mathbf{x}_j)}. \qquad (18)$$

$$\boldsymbol{\delta}(t) = \frac{\sum_{i=1}^{c} \sum_{j=1}^{N} f_{i,j}^{S} q_{i}^{\mu}(t)(\tilde{\boldsymbol{\mu}}_i - \tilde{\boldsymbol{\mu}})}{\sum_{i=1}^{c} \sum_{j=1}^{N} f_{i,j}^{S} |\mathbf{p}^{T}(t)(\tilde{\boldsymbol{\mu}}_i - \tilde{\boldsymbol{\mu}})|} - \frac{\sum_{i,j=1}^{N} \tilde{W}_{i,j}^{(w)} q_{i,j}^{x}(t)(\mathbf{x}_i - \mathbf{x}_j)}{\sum_{i,j=1}^{N} \tilde{W}_{i,j}^{(w)} |\mathbf{p}^{T}(t)(\mathbf{x}_i - \mathbf{x}_j)|}. \qquad (19)$$

It is easy to update $\mathbf{p}(t+1)$ by $\mathbf{p}(t+1) = \mathbf{p}(t) + \beta\boldsymbol{\delta}(t)$, where β is the update rate taking a small positive value close to 0. If $H(\mathbf{p}(t+1))$ cannot increase significantly, the iteration is terminated and the local optimal solution $\mathbf{p}^* = \mathbf{p}(t)$ can be output. Otherwise, update until $H(\mathbf{p}(t+1))$ is almost stable. The algorithm for $n = 1$ is demonstrated in Algorithm 1.

Algorithm 1. RS^{3}DP-L1 optimization for $n = 1$

Input: Sample matrix $\mathbf{X}$, label vector $\mathbf{y}$.
Output: Local optimal projection vector $\mathbf{p}$.

1) **_Preparation:_** Use SparseNP method to compute the soft label matrix $\mathbf{F}^S$. Then calculate the LLE weight matrix $\mathbf{M}$ and derive the weight matrix $\tilde{\mathbf{W}}^{(w)}$.

2) **_Initialization:_** Set $t = 0$, initialize $\mathbf{p}(0) = \arg\max_{\mathbf{m}_i} \|\mathbf{m}_i\|$, where $\mathbf{m}_i$ denotes the mean vector of the sample $\mathbf{x}$ in i-th class, and rescale it to unit length as $\mathbf{p}(0) = \mathbf{p}(0)/\|\mathbf{p}(0)\|$.

3) **_Calculation:_** Compute $q_{i,j}^x(t), q_i^\mu(t)$ by Eq. (16)(17) and $H(\mathbf{p}(t))$ by Eq. (18).

4) **_Updating:_** Calculate $\boldsymbol{\delta}(t)$ by equation (19) and update $\mathbf{p}(t+1)$. Then rescale $\mathbf{p}(t+1)$ to unit length and set $t = t+1$.

5) **_Iteration:_** If $H(\mathbf{p}(t))$ cannot increase significantly, the iteration terminates and the local optimal $\mathbf{p}^* = \mathbf{p}(t)$ can be output. Otherwise, go to Step 3.

Algorithm 2. RS^{3}DP-L1 optimization for $n > 1$

Input: Sample matrix $\mathbf{X}$, label vector $\mathbf{y}$.
Output: Local optimal projection matrix $\mathbf{P} \in \mathbb{R}^{d \times n}$.

1) Set $\mathbf{p}_0 = \mathbf{0}$, $(\mathbf{x}_i)_0 = \mathbf{x}_i (i = 1, 2, \ldots, N)$, $k = 1$.

2) $(\tilde{\boldsymbol{\mu}})_k = (\tilde{\boldsymbol{\mu}})_{k-1} - \mathbf{p}_{k-1}\mathbf{p}_{k-1}^T(\tilde{\boldsymbol{\mu}})_{k-1}$.
   For $i = 1$ to N **do** $(\mathbf{x}_i)_k = (\mathbf{x}_i)_{k-1} - \mathbf{p}_{k-1}\mathbf{p}_{k-1}^T(\mathbf{x}_i)_{k-1}$.
   For $j = 1$ to c **do** $(\tilde{\boldsymbol{\mu}}_j)_k = (\tilde{\boldsymbol{\mu}}_j)_{k-1} - \mathbf{p}_{k-1}\mathbf{p}_{k-1}^T(\tilde{\boldsymbol{\mu}}_j)_{k-1}$.

3) Apply Algorithm 1 to extract $\mathbf{p}_k$ using the samples $(\mathbf{x}_i)_k$.

4) If $k < n$, set $k = k+1$ and go to Step 2); otherwise, output $\mathbf{P} = (\mathbf{p}_1, \mathbf{p}_2, \ldots, \mathbf{p}_n)$.

3.3   Extension to $n > 1$

Here, we want to obtain the orthogonal projection matrix $\mathbf{P}$. It is obvious that the projection vector $\mathbf{p}$ is the linear combination of $\mathbf{x}_i(i = 1, 2, \ldots, N)$, $\tilde{\boldsymbol{\mu}}_j(j = 1, 2, \ldots, c)$ and $\tilde{\boldsymbol{\mu}}$ according to Eq. (19). So effective update to those three variables leads to our result. The detailed algorithm is shown in Algorithm 2. Since the proof is similar to that in [13], we omit it due to page limitation.

4   Simulation Results

In this section, we conduct simulations on two databases to evaluate the effectiveness of our RS^{3}DP-L1 method. The first one is handwriting digits in CASIA-HWDB1.1 database [5] and the second one is MNIST handwriting digits database [4]. The performance is compared with the following algorithms: Semi-Supervised Maximum Margin Criterion (SSMMC) and Semi-Supervised Linear Discriminant Analysis (SSLDA) [6], DLPP-L1 [13] and PCA-L1 [3]. In each experiment, we randomly select 50 images in each class for training, i.e. $N = 500$, and in that the number of labeled samples is $Lab = 18, 24, 30, 36$, while 1000 samples for testing. To test the robustness, we added Gaussian noise on half of the images. The parameter $\gamma = 0.2$. Finally, we use oneNN method to classify the unlabeled samples and evaluate the DR methods (Figs. 1 and 2).

(a) Original images of digits in HWDB1.1   (a) Original images in MNIST

(b) Images of digits added noise in HWDB1.1   (b) Images added noise in MNIST

Fig. 1. Some samples in HWDB1.1   **Fig. 2.** Some samples in MNIST

Table 1. Performance comparisons in HWDB1.1 $(n = 10)$

	$N = 500, Lab = 18$		$N = 500, Lab = 24$		$N = 500, Lab = 30$		$N = 500, Lab = 36$	
	Mean	Best	Mean	Best	Mean	Best	Mean	Best
SSMMC	77.26%	79.20%	79.76%	81.40%	80.94%	83.60%	81.38%	84.10%
SSLDA	76.84%	80.90%	78.37%	81.70%	79.56%	82.10%	79.78%	83.60%
DLPP-L1	85.12%	86.80%	86.30%	87.30%	87.22%	88.80%	87.80%	89.50%
PCA-L1	82.78%	84.70%	84.32%	86.30%	85.41%	86.50%	85.96%	88.80%
RS^{3}DP-L1	**86.14%**	**88.20%**	**86.87%**	**88.60%**	**87.74%**	**89.30%**	**87.96%**	**90.20%**
SSMMC (noise)	70.88%	74.40%	71.65%	74.80%	74.78%	79.50%	75.73%	79.80%
SSLDA (noise)	77.76%	79.50%	79.09%	81.20%	80.49%	82.60%	82.62%	84.80%
DLPP-L1 (noise)	82.01%	83.80%	83.77%	85.40%	84.81%	86.60%	86.29%	87.80%
PCA-L1 (noise)	77.74%	82.80%	79.71%	81.70%	81.44%	82.60%	81.99%	83.70%
RS^{3}DP-L1 (noise)	**82.40%**	**84.20%**	**83.94%**	**85.50%**	**85.13%**	**87.10%**	**86.67%**	**88.60%**

Table 2. Performance comparisons in MNIST $(n = 10)$

	$N = 500, Lab = 18$		$N = 500, Lab = 24$		$N = 500, Lab = 30$		$N = 500, Lab = 36$	
	Mean	Best	Mean	Best	Mean	Best	Mean	Best
SSMMC	69.60%	72.10%	71.28%	72.80%	73.16%	76.30%	75.26%	77.30%
SSLDA	69.16%	71.20%	71.68%	74.00%	72.02%	74.80%	73.37%	76.30%
DLPP-L1	73.28%	75.50%	75.73%	77.40%	75.94%	77.50%	77.73%	79.30%
PCA-L1	71.66%	75.50%	73.24%	76.40%	74.86%	77.00%	75.50%	78.00%
RS^{3}DP-L1	**73.89%**	**75.90%**	**76.36%**	**78.00%**	**77.01%**	**79.00%**	**77.80%**	**79.80%**
SSMMC (noise)	66.73%	68.60%	68.84%	71.20%	70.56%	73.90%	72.93%	76.50%
SSLDA (noise)	68.38%	72.50%	70.94%	74.00%	72.52%	75.00%	75.03%	78.70%
DLPP-L1 (noise)	69.71%	73.30%	72.59%	75.30%	74.16%	78.10%	76.64%	78.90%
PCA-L1 (noise)	68.94%	71.50%	70.20%	72.30%	71.88%	76.40%	74.41%	76.60%
RS^{3}DP-L1 (noise)	**70.00%**	**73.80%**	**72.89%**	**76.10%**	**74.36%**	**78.40%**	**76.96%**	**79.60%**

In the simulations, we can see that the performance of each method improves when the number of labeled images increases. Secondly, promising results are exhibited by our proposed RS^{3}DP-L1, since it performs better in most cases (Tables 1 and 2).

5 Conclusion

A Robust Soft Semi-Supervised Discriminant Projection for feature learning is proposed. For enhancing the data representations, our formulation seamlessly integrates the merits of soft discriminating &locality preserving plus the robustness of L1-norm to noise. For feature learning, our method aims at minimizing the L1-norm based local soft within-class divergence and maximizing the L1-norm based local soft between-class divergence at the same time in addition to possessing the L1-norm based locality preserving power during the training stage. The experimental results show the effectiveness of our work. In our future work, we will investigate more efficient algorithm to speed up the L1-norm based models including existing and our method.

Acknowledgment. This work is partially supported by the National Natural Science Foundation of China (61402310, 61373093), Major Program of Natural Science Foundation of Jiangsu Higher Education Institutions of China(15KJA520002), Special Funding of China Postdoctoral Science Foundation (2016T90494), Postdoctoral Science Foundation of China (2015M580462), Postdoctoral Science Foundation of Jiangsu Province of China (1501091B), Natural Science Foundation of Jiangsu Province of China (BK20140008 and BK20141195), and the Undergraduate Student Innovation Program of Soochow University (2014xj034).

References

1. Duda, R.O., Hart, P.E., Stork, D.G.: Pattern Classification. Wiley, Hoboken (2012)
2. He, X., Cai, D., Yan, S., Zhang, H.J.: Neighborhood preserving embedding. In: Tenth IEEE International Conference on Computer Vision, 2005. ICCV 2005, vol. 2, pp. 1208–1213. IEEE (2005)
3. Kwak, N.: Principal component analysis based on l1-norm maximization. IEEE Trans. Pattern Anal. Mach. Intell. **30**(9), 1672–1680 (2008)
4. LeCun, Y., Bottou, L., Bengio, Y., Haffner, P.: Gradient-based learning applied to document recognition. Proc. IEEE **86**(11), 2278–2324 (1998)
5. Liu, C.L., Yin, F., Wang, D.H., Wang, Q.F.: Online and offline handwritten chinese character recognition: benchmarking on new databases. Pattern Recogn. **46**(1), 155–162 (2013)
6. Song, Y., Nie, F., Zhang, C., Xiang, S.: A unified framework for semi-supervised dimensionality reduction. Pattern Recogn. **41**(9), 2789–2799 (2008)
7. Sugiyama, M.: Local fisher discriminant analysis for supervised dimensionality reduction. In: Proceedings of the 23rd International Conference on Machine Learning, pp. 905–912. ACM (2006)
8. Yan, S., Xu, D., Zhang, B., Zhang, H.J., Yang, Q., Lin, S.: Graph embedding and extensions: a general framework for dimensionality reduction. IEEE Trans. Pattern Anal. Mach. Intell. **29**(1), 40–51 (2007)
9. Zhang, Z., Chow, T.W., Zhao, M.: M-Isomap: orthogonal constrained marginal isomap for nonlinear dimensionality reduction. IEEE Trans. Cybern. **43**(1), 180–191 (2013)
10. Zhang, Z., Yan, S., Zhao, M.: Pairwise sparsity preserving embedding for unsupervised subspace learning and classification. IEEE Trans. Image Process. **22**(12), 4640–4651 (2013)

11. Zhang, Z., Zhang, L., Zhao, M., Jiang, W., Liang, Y., Li, F.: Semi-supervised image classification by nonnegative sparse neighborhood propagation. In: Proceedings of the 5th ACM on International Conference on Multimedia Retrieval, pp. 139–146. ACM (2015)
12. Zhao, M., Zhang, Z., Chow, T.W., Li, B.: A general soft label based linear discriminant analysis for semi-supervised dimensionality reduction. Neural Netw. **55**, 83–97 (2014)
13. Zhong, F., Zhang, J., Li, D.: Discriminant locality preserving projections based on l1-norm maximization. IEEE Trans. Neural Netw. Learn. Syst. **25**(11), 2065–2074 (2014)

A Hybrid Pooling Method for Convolutional Neural Networks

Zhiqiang Tong[1]([✉]), Kazuyuki Aihara[1,2], and Gouhei Tanaka[1,2]

[1] Graduate School of Engineering, The University of Tokyo, Tokyo 113-8656, Japan
{tong,aihara,gouhei}@sat.t.u-tokyo.ac.jp
[2] Institute of Industrial Science, The University of Tokyo, Tokyo 153-8505, Japan

Abstract. The convolutional neural network (CNN) is an effective machine learning model which has been successfully used in the computer vision tasks such as image recognition and object detection. The pooling step is an important process in the CNN to decrease the dimensionality of the input image data and keep the transformation invariance for preventing the overfitting problem. There are two major pooling methods, i.e. the max pooling and the average pooling. Their performances depend on the data and the features to be extracted. In this study, we propose a hybrid system of the two pooling methods to improve the feature extraction performance. We randomly choose one of them for each pooling zone with a fixed probability. We show that the hybrid pooling method (HPM) enhances the generalization ability of the CNNs in numerical experiments with the handwritten digit images.

Keywords: Convolutional neural networks · Deep learning · Pooling method · Image recognition

1 Introduction

1.1 Convolutional Neural Networks

Convolutional neural networks (CNNs) have drawn much attention due to their excellent performance in the computer vision applications, such as image recognition and classification [1]. The CNN is a feedforward neural network consisting of alternate arrangements of convolution layers and pooling layers, each of which is formed by a two-dimensional array of artificial neurons as illustrated in Fig. 1. Compared with the general fully-connected feedforward neural networks, the inter-layer connections in the CNN are sparse because of the characteristics such as the local receptive field, the shared weights, and the sub-sampling. Namely, the number of free parameters in the CNN is much smaller than that in the fully connected layered neural networks. Therefore, the CNN can produce high generalization ability to extract features of image data, which are invariant under transformations such as rotation and shift. However, it is still a problem that the learning from a large number of dataset requires the CNNs with a deep structure, causing overtraining. Therefore, it is significant to improve the performance of CNNs.

A. Hirose et al. (Eds.): ICONIP 2016, Part II, LNCS 9948, pp. 454–461, 2016.
DOI: 10.1007/978-3-319-46672-9_51

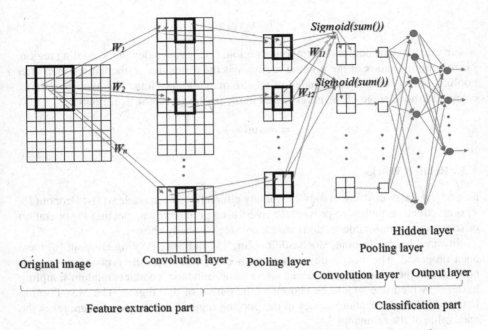

Fig. 1. An illustration showing the structure of a convolutional neural network.

The CNN includes two parts, the feature extraction part and the classification part, as illustrated in Fig. 1. The feature extraction part consists of alternate arrangements of convolution layers and pooling layers. At the end of the feature extraction part, the feature map will unfold to form the input to the classification part. The classification part is a fully-connected feedforward neural network.

In the convolution layer, the convolution kernel with a quadrangle window is used to scan the grey scale image and calculate the convolution of the pixel values in the window. In the pooling layer, the pooling method is used to downsize the feature map. At the end of the CNN, the difference between the output of the CNN and the required output is computed, and then, the backpropagation algorithm [2] is employed to adjust the parameters in the convolution kernel and the bias parameters of the nodes to minimize the output difference.

1.2 Pooling Method

The pooling process is a very important step in the CNN, because it can decrease the dimension of the feature map which indicates the filtered image in the convolution layer. It is helpful for the transformation invariance such as rotation, translation, and expansion, and therefore, it can enhance the generalization ability of the CNN.

There are two major pooling methods which are widely used in the pooling step of the CNN, i.e., the max pooling and the average pooling.

The max pooling calculates the maximum of the pixel values in the pooling window and takes it as the output of the pooling window as follows:

$$y_{vw} = max(x_{ij})$$ (1)

where y denotes the output of the pooling region, i is the row index of the pooling region, j is the column index of the pooling region, v is the row index of the feature map after pooling, and w is the column index of feature map after pooling. The average pooling calculates the average of the pixel values in the pooling window as follows:

$$y_{vw} = mean(x_{ij})$$ (2)

1.3   Related Works

To prevent the overfitting in the CNN, many efforts have been made so far. Dropout [3, 4] is an effective method to prevent the overfitting problem by neglecting the operation of some nodes and/or connections which are stochastically chosen.

Inspired by the dropout, stochastic pooling [5] and Max-Pooling Dropout [6] have been proposed. The stochastic pooling method [5] calculates the probability of each node within the pooling region based on its value, and then, conducts random sampling to decide which one should be chosen as the output of that region. The Max-Pooling Dropout [6] neglects some values in the pooling region randomly, and then, takes the max value of the remnant.

Boureau [7] performed a theoretical analysis of the max pooling and the average pooling and compared their performances with numerical experiments. This paper proposed a generalized pooling including the max and average methods as extreme cases.

1.4   Motivation

Some methods based on the dropout, such as stochastic pooling [5], Max-Pooling Dropout [6], maxout networks [8], and DropConnect [9], use the stochastic process to make some nodes or connection weights inactive temporarily in each training step. The inactive nodes or connection weights are different for different training steps. Therefore, the stochasticity produces a variety of responses in the information processing in the same CNN and the weighted average of them can be used as the output. This concept is effective to reduce the learning cost as well as improve the generalization ability.

In the pooling stage of the CNN, the max pooling merely takes the largest pixel value in the pooling region and neglects all the others. The average pooling takes all the pixel values within the field of the pooling window into account and calculates the average of them.

We propose to switch the above two pooling methods stochastically. We call it a hybrid pooling method. We compare the performance of the hybrid method with those of the two existing methods in the numerical experiments with monochrome images.

2   The Proposed Method

In this section, we introduce a hybrid pooling method (HPM) combining the maximum and average methods in the pooling stage of the CNN.

We do not use the same pooling method for all the feature map. In the training stage, we separate the convolution feature map into two parts of pooling regions, one of which uses the max pooling and the other of which uses the average pooling. For each pooling region, the maximum method is applied with probability p and the average method is applied with probability $1-p$. Therefore, in the training stage, the network is trained by the mix of the two pooling methods.

In the testing stage, the output of the pooling region is given by the weighted average of the two methods as follows:

$$y_{vw} = p^* max(x_{ij}) + (1 - p)^* mean(x_{ij}) \tag{3}$$

An example is shown in Fig. 2. The left figure corresponds to the existing pooling methods, max and average pooling, for the pooling region at the left top corner, whereas the right figure shows how to calculate of the pooling value in the testing stage.

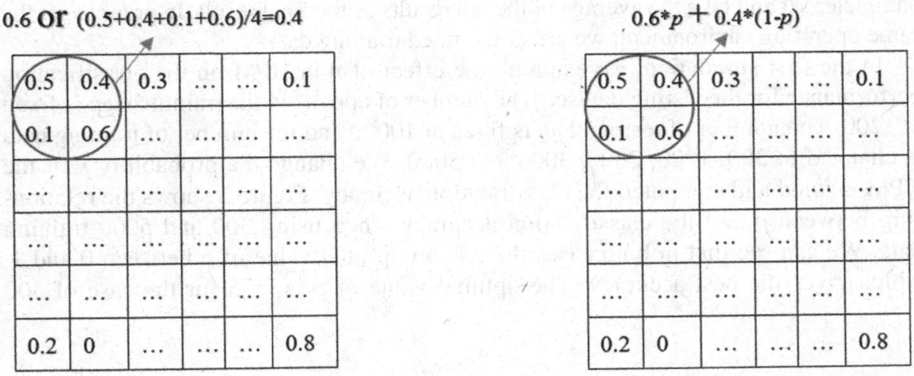

Fig. 2. (Left) The normal pooling processes based on the max or average function. (Right) The proposed pooling process based on the hybrid method.

3   Results

3.1   Dataset for Classification Problems

We will use the MNIST image dataset [1] in our simulations. The MNIST dataset includes 60000 training data and 10000 testing data. Each data in MNIST is a 28×28 pixel grayscale image, which takes some different handwritten digits from number 0 to 9. We normalize all the pixels in the range [0, 1] before using them as inputs.

3.2 Simulation Setting

The convolutional neural network used here has five feature extraction layers, including one input layer, two convolution layers, and two pooling layers. The output image of the feature extraction part is rasterized as the input of the fully-connected feedforward neural network. The output layer consists of ten output units. We use the MATLAB software to do the program.

The structure is set as follows: the input layer contains 28×28 inputs; the first convolution layer has $6 \times 24 \times 24$ feature maps; the second convolution layer has $12 \times 8 \times 8$ feature maps; the size of the filter for each convolution layer is 5×5; the pooling region for the two pooling layers is 2×2; the stride is 2.

We use the sigmoid function for all the nodes in the feature extraction and the fully-connected layers. The model is trained by the stochastic mini-batch gradient descent method with batch size 50. The momentum is set at 0.99. The learning rate is set at 1.

3.3 Simulation Results

Due to the stochasticity of our algorithm, we run every simulation ten times for each parameter set and take the average of the ten results as the final result. In order to set the same operating environment, we select the fixed training data.

In the first simulation, we examine the effect of p in HPM on the classification performance for the testing dataset. The number of epochs in the training step is fixed at 1200. The number of testing data is fixed at 10000 and the number of training data is changed to 300, 1000, 2000, 3000 and 5000. We change the probability p of the HPM method and computed the classification accuracy. Figure 3 shows the relationship between p and the classification accuracy when using 300 and 5000 training data. We can see that in both cases there is an optimal value of p between 0 and 1, which gives the best accuracy. The optimal value of p is 0.75 for the case of 300

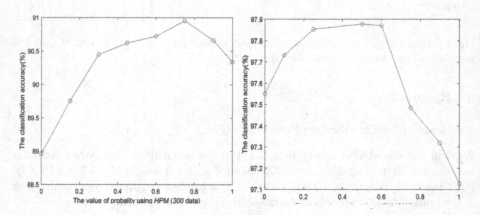

Fig. 3. The relationship between p and the classification accuracy when using 300 training data (left) and 5000 training data (right).

training data. In other words, the HPM with an appropriate value of p can be better than the max pooling ($p = 1$) and the average pooling ($p = 0$).

In the second simulation, we investigate the effect of the length of training steps. The number of parameter update is called an epoch. In this simulation, we fix the value of p at $p = 0.75$, the number of training data at 300, and the number of testing data at 10000. Figure 4 shows how the classification accuracy depends on the number of epochs in the training phase. As the number of epochs is increased, the classification accuracy tends to be improved in all the three methods. The result indicates that the HPM and the max pooling are better than the average method for all the epochs tested. The HPM is comparable to the max pooling.

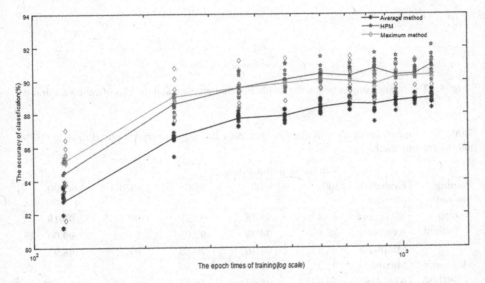

Fig. 4. The relationship between the number of epochs in the training phase and the classification accuracy

In the third simulation, we tested how the number of training data influences the performance. We set $p = 0.75$ for the cases with 1 to 150, 300, 450, 600, and 750 training data, $p = 0.4$ for the cases with 1 to 1000 and 2000 data, $p = 0.25$ for the case with 1 to 3000 data, $p = 0.5$ for the case with 1 to 5000 data, $p = 0.25$ for the case with 1 to 60000 data from the training dataset. The number of epochs is fixed at 1200 and the number of testing data is fixed at 10000. Figure 5 shows the comparison of the performance among the three pooling methods. The HPM is better than the other two existing methods on average, suggesting that the combination of the two methods can enhance the generalization ability of the CNN. In Table 1, the performance is measured by the maximum, average, and minimum of the classification accuracy values in 10 trials. We find that the HPM can yield the best performance in terms of all the three measures when using the 300, 1000, 3000, 5000 training data, and can get the best performance in terms of average and minimum measurement when using the 60000 data.

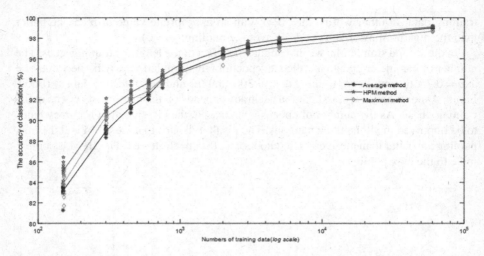

Fig. 5. The relationship between the number of training data and the classification accuracy.

Table 1. Comparison of the classification accuracy (%) between the proposed method (HPM) and the existing methods.

Pooling method	Evaluation measure	Number of training data				
		300	1000	3000	5000	60000
Average method	Maximum	89.48	95.18	97.28	97.77	**99.18**
	Average	88.96	94.79	97.05	97.55	99.00
	Minimum	88.35	94.46	96.88	97.35	98.93
Maximum method	Maximum	91.31	95.12	96.89	97.29	98.74
	Average	90.34	94.62	96.67	97.13	98.675
	Minimum	88.94	94.10	96.49	96.78	98.64
HPM	Maximum	**92.13**	**95.99**	**97.72**	**98.06**	99.15
	Average	**90.95**	**95.41**	**97.43**	**97.88**	**99.10**
	Minimum	**90.00**	**95.04**	**97.17**	**97.77**	**98.99**

4   Conclusions

In this study, we have proposed a hybrid pooling method using the stochastic mixture of the average pooling method and the maximum pooling method for convolutional neural networks. Numerical results for the MNIST image database have shown that the proposed method can effectively enhance the generalization ability of the convolutional neural network if the mixing probability is appropriately adjusted. The effectiveness of our method needs to be further verified with other datasets such as CIFAR-10, CIFAR-100 and ImageNet, and compared with other pooling methods.

Applications of our method to deep neural networks and other types of information processing tasks are the issue to be addressed in the future. For example, the medical image processing [10, 11] is a good target which the proposed pooling method should be well applied to.

Acknowledgments. This work was partially supported by the Otsuka Toshimi Scholarship Foundation (ZT) and JSPS KAKENHI Grant Number 16K00326 (GT).

References

1. LeCun, Y., Bottou, L., Bengio, Y., Haffner, P.: Gradient-based learning applied to document recognition. Proc. IEEE **86**(11), 2278–2324 (1998)
2. Rumelhart, D.E., Hinton, G.E., Williams, R.J.: Learning representations by back-propagating errors. Cogn. Model. **5**(3), 1 (1988)
3. Hinton, G.E., Srivastava, N., Krizhevsky, A., et al.: Improving neural networks by preventing co-adaptation of feature detectors (2012)
4. Srivastava, N., Hinton, G., Krizhevsky, A., et al.: Dropout: a simple way to prevent neural networks from overfitting. J. Mach. Learn. Res. **15**(1), 1929–1958 (2014)
5. Zeiler, M.D., Fergus, R.: Stochastic pooling for regularization of deep convolutional neural networks. In: ICLR (2013)
6. Wu, H., Gu, X.: Max-pooling dropout for regularization of convolutional neural networks. In: Arik, S., Huang, T., Kin Lai, T., Liu, Q. (eds.) Neural Information Processing, vol. 9486, pp. 46–54. Springer, Cham (2015)
7. Boureau, Y.L., Ponce, J., LeCun, Y.: A theoretical analysis of feature pooling in visual recognition. In: Proceedings of the 27th International Conference on Machine Learning (ICML-10), pp. 111–118 (2010)
8. Goodfellow, I.J., Warde-Farley, D., Mirza, M., et al.: Maxout networks. In: Proceedings of International Conference on Machine Learning (ICML) (2013)
9. Wan, L., Zeiler, M., Zhang, S., et al.: Regularization of neural networks using dropconnect. In: Proceedings of the 30th International Conference on Machine Learning, ICML-13, pp. 1058–1066 (2013)
10. Cha, K.H., Hadjiiski, L., Samala, R.K., et al.: Urinary bladder segmentation in CT urography using deep-learning convolutional neural network and level sets. Med. Phys. **43**(4), 1882–1896 (2016)
11. Wolterink, J.M., Leiner, T., de Vos, B.D., et al.: Automatic coronary artery calcium scoring in cardiac CT angiography using paired convolutional neural networks. Med. Image Anal. (2016)

Multi-nation and Multi-norm License Plates Detection in Real Traffic Surveillance Environment Using Deep Learning

Amira Naimi[1,2(✉)], Yousri Kessentini[1,3], and Mohamed Hammami[1]

[1] MIRACL Laboratory, University of Sfax,
Tunis Road Km 10 BP. 242, 3021 Sfax, Tunisia
naimiamira@gmail.com, mohamed.hammami@fss.rnu.tn
[2] Digital Research Center of Sfax, Technopole of Sfax,
PO Box 275, Sakiet Ezzit, 3021 Sfax, Tunisia
[3] LITIS Laboratory EA 4108, St Etienne du Rouvray, France
yousri.kessentini@litislab.eu

Abstract. This paper aims to highlight the problems of license plate detection in real traffic surveillance environment. We notice that existing systems require strong assumptions on license plate norm and environment. We propose a novel solution based on deep learning using self-taught features to localize multi-nation and multi-norm license plates under real road conditions such poor illumination, complex background and several positions. Our method is insensitive to illumination (day, night, sunrise, sunset,...), translation and poses. Despite the low resolution of images collected from real road surveillance environment, a series of experiments shows interesting results and the fastest time processing comparing with traditional algorithms.

Keywords: License plate detection · Deep learning · Multi-nation · Multi-norm · Real road surveillance · Low resolution · Poor illumination

1 Introduction

In the modern society, number of cars in city shows a massive increase due to the rapid economic development and the social progress. This increases urban traffic management difficulties. Since our world suffers from the increased rate of violence, criminality and terrorism in public road, the traffic management and surveillance become one of the most active field in both research and industry. To solve this problem of road traffic security monitoring, the idea is to identify cars through their license plates (**LP**). Many works have been proposed to detect license plate from traffic images. We classify those existing works into two categories of methods depending on their definition of the license plate; either as a predefined object, or as an homogeneous text zone. Object detection methods use different features to localize the object "License plate" such; color using color-based method [1], or shapes detected by edge-based method [5].

© Springer International Publishing AG 2016
A. Hirose et al. (Eds.): ICONIP 2016, Part II, LNCS 9948, pp. 462–469, 2016.
DOI: 10.1007/978-3-319-46672-9_52

Other method in this category, define the region containing license plate, even as region rich in contrast information and with high edge density like morphology-based method [9], or as region containing high density of key points detected with scale invariant feature transform (SIFT) descriptor [6] where the LP is localized through a template matching process. Also, the detection of LP as an object, could be defined as a boosting problem such [10] where vertical edge are detected with a set of weak classifiers and the AdaBoost classifier select the best one. Text detection methods consider the detection task as a problem of text-region localization. Indeed, the text present a regular signature based on similarity in texture, geometrical features (alignment and orientation) and textual information such regularity and directionality. Citing the use of color texture based method [3], the Improved Connective Hough Transform(ICHT) [7] and the use of the Conditional Random Field (CRF) [4]. However, one main drawback of published methods is the fact of making strong assumptions about norm and context variability of the license plate. Thus, they present a good result only for limited car-License plate models (generally for each country there is a special solution) and under controlled conditions. Nevertheless, traffic surveillance environment present different difficulty such complex backgrounds, poor image quality caused by various weather conditions, variant illumination and many license plate model, size and poses, which make the license plate detection a not completely solved problem. In fact, the license plate detection is a subfield of the object detection. In recent years, the use of deep convolutional neural networks for object detection improves the performance of such detection systems. Furthermore, all license plate detection methods are based on hand-crafted features. As a consequence, just a few pertinent license-plate-features are remain. In this work, we propose to break this standard and use a new LP localization approach stand on self-taught features.

Motivated by this new approach for object detection on one hand, and by the needs of the Tunisian Ministry of the Interior on other hand, we propose a novel license plate detection system based on deep learning for the complex traffic surveillance environment.

Therefore, our main contribution is to apply deep neural networks to detect muli-nation and multi-norm license plates in real scene and under challenging conditions such bad illumination and low resolution images.

2   Proposed Approach

Unlike state-of-the art approach [4, 5, 7, 10] which are based on hand-crafted features, we propose to use self-taught features which ensure the selection of most discriminative LP features automatically in order to take in consideration the different appearance characteristics and properties of different LP norms and nations. Indeed, our approach uses a deep learning model that automatically extract the most discriminating LP features. Such model uses stronger criteria for voting true presence of LP region among regions of interest (ROI), which decrease the false positive. Since we aims to runtime application, and we need a high detection accuracy, we based our approach on the state-of-art of the object detection: Faster-RCNN [8], which assure both high accuracy and low time cost. The recent success of this model was driven by the success of the combination between the region proposals network (RPN) and the convolutional neural networks(CNN).

Our system is a region proposals neural network that localizes the LP in an any size image through three steps; first it extracts region proposals using RPN, then extracts features from each region through CNN, and as a last step, it classifies those regions. In order to improve the cost and the accuracy by sharing convolution feature maps, we used the technique proposed by [8] to establish a unified network composed of RPN and CNN. This unified one combine both networks. Thus, our approach is divided into two steps as it is presented in the Fig. 1; Region proposals and features extraction step and the classification step.

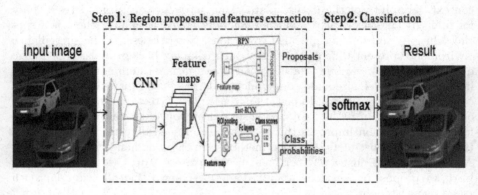

Fig. 1. Illustration of our license plate detection system based on Faster-RCNN [8]. This shows the two steps of our system: The first step present the unified network; after initialization of both RPN and Fast-RCNN [2] networks, the training began by extracting features with CNN and provide the feature maps. Those feature maps are used by the RPN and the Fast-RCNN. In the second step, the classifier (*softmax*) provide the bonding boxes of the License Plates.

2.1   Region Proposals and Features Extraction

The region proposals extraction is the step that allows the generation of a set of rectangular object proposals with theirs objectness score (measures membership to a set of object classes vs background) from the input image. This prediction is realized through a region-proposal neural networks RPN, which is a fully convolutional neural networks trained end-to-end that predicts high-quality region proposal. The RPN is the network used to enable almost cost-free region proposals by sharing the full-image convolutional feature maps with the detection network. The detector network used is the Fast-RCNN [2]. To establish this sharing process, we initialize both networks separately, then start the training of the unified network. So the first step of our system is the initialization of the RPN by our trained model. This model is constructed from our private base offered by the Tunisian Ministry of the Interior collected from different radar images (presented in the Sect. 4). The output is a set of regions defined as regions contains LP, or part of it. This same set are used to train the detection network which is also initialized by our trained model. Clearly, the RPN and the detection network doesn't share convolutional layers yet. During the next task, which is training, the feature maps constructed by the detector network is used to initialize RPN training, while fixing the shared convolutional layers. Therewith, fine-tuning only the layers unique to RPN, then fine-tuning the fully connected layers of the detection network. At this point, both networks become a unified one as shown in the first step of the Fig. 1. This network use a multi-task loss to train networks in a single training stage. This learning process allows to, simultaneously, learns to classify object proposals and refine spatial locations.

2.2 Classification:

To identify the LP localization, we need to classify the detected regions. For this final task, SVM and softmax are generally used, but it have demonstrated in [2] that the softmax classifier slightly outperforming linear SVMs. As a consequence, we used the softmax classifier in our solution to localize the LP as the second step shown in the Fig. 1.

3 Experimental Results

Experiments were conducted on a Core i7 PC with 16 GB of RAM and a GTX 980 GPU. The involved solution where implemented in C++ and Matlab.

Before presenting the results, we describe the used dataset, then the evaluation of performance and finally the comparison of our system with the-state-of-art of license plate detection systems.

3.1 The Dataset:

Our dataset is composed of 5000 JPEG color images taken from radars in different Tunisian roads. The images have low resolution (96 dpi) and were taken from real scene (complex background) under various light condition (day, night, sunshiny, raining,...), from different angel and positions and for several license plate shapes (square and rectangle), nations (Tunisian, Algerian, Libyan and European)and norms (the Tunisian LP has 26 norms and 4 colors possible). Moreover, each image contains from 1 to 3 different license plates and many vehicle types such car, truck, minivan and ambulance. The license plate ground-truth of this dataset was semi-automatically made using the Viper-GT annotation tool. This dataset was divided into three sets; 2000 test images, 1500 train images and 1500 evaluation images. Since our system need images as many as possible, and to enhance the training process, we used a flip operation to double the number of images used in train and evaluation from 3000 to 6000 images. We divided our test set into three datasets to best evaluate our solution. The set 1 is composed of 688 night images (early night light and dark night), set 2 of 1312 day images with different lightness condition (sunrise, sunset, different contrast) and the set 3 is the whole test set to get the average results. All the images have complex background and different vehicle types (car, truck, minivan, ambulance). In our experiments, no condition are imposed on the images or the license plate norm or location, which is the case in real traffic situations. Since this dataset is a private one offered by the Tunisian Ministry of the Interior, it is confidential which explain the blurred license plate characters in the images presented in this paper.

3.2   Performance Evaluation and Comparison:

To evaluate our solution, we used eight criteria for license plate detection. We used the true positive rate **TPR**: all bonding box containing the LP or most of it is considered as a true positive. The second criteria is the exact detection rate **EDR** which compute only the bonding boxes containing exactly the whole LP. To make the evaluation more significant, we used the classical criteria such the false negative rate **FNR**, the false positive rate **FPR**, the precision **P** and the recall **R**. Moreover, we presented the test time consumed for each image. The Table 1 present the results of our solution in the three sets.

Table 1. Detection performance

Dataset	TPR	EDR	FPR	FNR	P	R
Set 1	82.84 % (570/688)	76.5 % (527/688)	0.87 % (6/688)	16.13 % (111/688)	0.99	0.83
Set 2	79.6 % (1044/1312)	70.58 % (972/1312)	0.22 % (3/1312)	20.04 % (263/1312)	0.99	0.79
Set 3	80.7 % (1614/2000)	72 % (1454/2000)	0.45 % (9/2000)	18.7 % (374/2000)	0.99	0.81
Processing Time = 0.06 s/Image						

We notice that our system has a good TPR where most of them are exactly detected (only 6 % to 9 % aren't exactly detected). Since we get a low FPR we have a very high precision (0.99). However, our system present an average recall of 0.81 which is explained by an FNR up to 18.7 %. For the processing time, our system present a very interesting result of 0.06 second per image.

To make our results significant, we try to compare it to the existing systems results. Yet this task is a difficult one; there is no public base of LP since this filed has a confidential aspect. In addition, as we present in the introduction, the existing system are totally linked to an LP norm, which make it same time impossible to test it on other datasets. Indeed, no existing system could detect

Table 2. Comparaison of performance

Method	Our	Edge	Boosting	ICHT	CRF
Test data base (Image)	**2000**	636	636	636	484
LP	**Multi-nation**	European	European	European	Chinese
Time (s/Image)	**0.06**	0.2	0.12	0.14	0.767
R	0.81	0.87	0.93	0.96	**0.99**
FPR	**0.45**	7.9	3.1	1.7	0.63

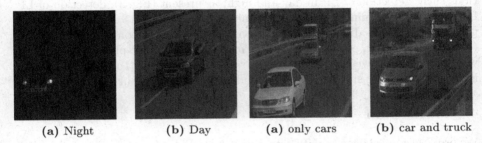

(a) Night	(b) Day	(a) only cars	(b) car and truck

Fig. 2. Detection in images from set1:Night and set2:Day

Fig. 3. Detection from different vehicles

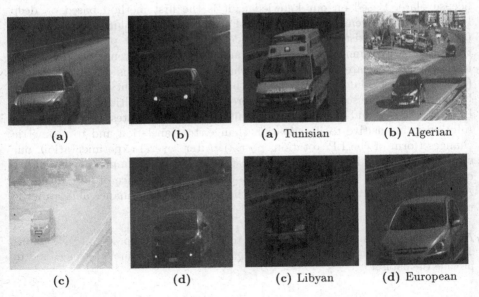

(a)	(b)	(a) Tunisian	(b) Algerian
(c)	(d)	(c) Libyan	(d) European

Fig. 4. Detection under different illumination conditions

Fig. 5. Detection of multi-nation plates

the Tunisian LP. As a consequence, we compared the obtained results with others state-of-art systems run on their own datasets as presente the Table 2. We compared our method to the Boosting-based method [10], Edge-based method [5], Improved Connective Hough Transform (ICHT) [7] and CRF method [4]. As result, this comparison isn't very significant for recall rate, since our base is multi-nation and multi-norm, but still have meaning for processing time.

Through this table, we notice that we have the fastest LP detection system comparing to those methods. Despite the hard data set that we use, our system present the lowest FPR. Our solution overcome real traffic challenging conditions such; detection of LP in day and night as present (Fig. 2), detection of LP from cars, trucks, minivans and ambulance as present (Figs. 3 and 5a). Also, we get good results in different illuminations (Fig. 4) and in complex background (Fig. 5b). Moreover, our system detects multi-nation license plate (Fig. 5). But, we still have the lowest recall rate. This rate is explained by the FNR. This rate is acceptable under such challenging condition pretense in our private dataset.

4   Conclusion

We propose a novel license plate detection solution based on deep learning for the complex traffic surveillance environment. Such method is based on self-taught features which make our system capable of learning all pertinent features of license plate. Based on our knowledge, it is the first method based on deep Learning for the license plate detection. Furthermore, our solution overcome the problems of the existing license plate detection system; it present no limitation on the environment of the image and the norm of license plates. Hence, it detect license plate not only under challenging conditions of the real traffic (day, night, bad illumination, complex background, different vehicles type,...), but also with low resolution images. Moreover, it detect multi-nation (Tunisian, Libyan, Algerian and European) and multi-norm license plates. In addition, our solution is insensitive to illumination change, to translation and to geometric changes (form of the LP, rotation, post...). After several experimentation, our system present good detection rate, lowest false alarms and impressively reduced cost. However, our solution has limitations such the presence of false negative. As a future objective, we will improve the detection performance and start the recognition step.

Acknowledgments. Authors would like to express their deepest gratitude to the Tunisian Ministry of the Interior for the opportunity to cooperate with them and to provide the dataset.

References

1. Ashtari, A.H., Nordin, M.J., Fathy, M.: An Iranian license plate recognition system based on color features. IEEE Trans. Intell. Transp. Syst. **15**(4), 1690–1705 (2014)
2. Girshick, R.: Fast R-CNN. In: Proceedings of the IEEE International Conference on Computer Vision, pp. 1440–1448 (2015)

3. Kim, S., Kim, D., Ryu, Y., Kim, G.: A robust license-plate extraction method under complex image conditions. In: Proceedings of the 16th International Conference on Pattern Recognition, vol. 3, pp. 216–219. IEEE (2002)

4. Li, B., Tian, B., Li, Y., Wen, D.: Component-based license plate detection using conditional random field model. IEEE Trans. Intell. Transp. Syst. **14**(4), 1690–1699 (2013)

5. Mahini, H., Kasaei, S., Dorri, F.: An efficient features-based license plate localization method. In: 18th International Conference on Pattern Recognition, ICPR 2006, vol. 2, pp. 841–844. IEEE (2006)

6. Ng, H.S., Tay, Y.H., Liang, K.M., Mokayed, H., Hon, H.W.: Detection and recognition of malaysian special license plate based on sift features. arXiv preprint arXiv:1504.06921 (2015)

7. Nguyen, C.D., Ardabilian, M., Chen, L.: Robust car license plate localization using a novel texture descriptor. In: Sixth IEEE International Conference on Advanced Video and Signal Based Surveillance, AVSS 2009, pp. 523–528. IEEE (2009)

8. Ren, S., He, K., Girshick, R., Sun, J.: Faster R-CNN: towards real-time object detection with region proposal networks. In: Advances in Neural Information Processing Systems, pp. 91–99 (2015)

9. Wu, H.H.P., Chen, H.H., Wu, R.J., Shen, D.F.: License plate extraction in low resolution video. In: 18th International Conference on Pattern Recognition, ICPR 2006, vol. 1, pp. 824–827. IEEE (2006)

10. Zhang, H., Jia, W., He, X., Wu, Q.: Real-time license plate detection under various conditions. In: Ma, J., Jin, H., Yang, L.T., Tsai, J.J.-P. (eds.) UIC 2006. LNCS, vol. 4159, pp. 192–199. Springer, Heidelberg (2006)

A Study on Cluster Size Sensitivity of Fuzzy c-Means Algorithm Variants

László Szilágyi[1,2,3(✉)], Sándor M. Szilágyi[2,4], and Călin Enăchescu[4]

[1] Faculty of Technical and Human Science of Tîrgu Mureş,
Sapientia - Hungarian Science University of Transylvania, Tîrgu Mureş, Romania
lalo@ms.sapientia.ro
[2] Department of Control Engineering and Information Technology,
Budapest University of Technology and Economics, Budapest, Hungary
[3] University of Canterbury, Christchurch, New Zealand
[4] Department of Informatics, Petru Maior University of Tîrgu Mureş,
Tîrgu Mureş, Romania

Abstract. Detecting clusters of different sizes represents a serious difficulty for all c-means clustering models. This study investigates the set of various modified fuzzy c-means clustering algorithms within the bounds of the probabilistic constraint, from the point of view of their sensitivity to cluster sizes. Two numerical frameworks are constructed, one of them addressing clusters of different cardinalities but relatively similar diameter, while the other manipulating with both cluster cardinality and diameter. The numerical evaluations have shown the existence of algorithms that can effectively handle both cases. However, these are difficult to automatically adjust to the input data through their parameters.

Keywords: Fuzzy clustering · Cluster size sensitivity · Improved partition · Suppressed partition

1 Introduction

Fuzzy clustering algorithms are unsupervised learning classification algorithms that employ fuzzy membership functions to describe the partition. The fuzzy c-means (FCM) algorithm introduced by Dunn [1] and generalized by Bezdek [2] is probably the most popular fuzzy clustering technique, due to its simplicity and the fine partition it usually produces. However, that fine property is relative and conditioned by several aspects. The probabilistic constraint causes several undesired phenomena, including the sensitivity to outlier data, and the multi-modality of the fuzzy membership functions produced by FCM [3]. The latter also causes difficulties when the cardinality of clusters differs strongly, or the physical size (diameter) of the clusters is different. This is usually referred to as cluster size sensitivity.

The work of S.M. Szilágyi was supported by the János Bolyai Fellowship Program of the Hungarian Academy of Sciences.

© Springer International Publishing AG 2016
A. Hirose et al. (Eds.): ICONIP 2016, Part II, LNCS 9948, pp. 470–478, 2016.
DOI: 10.1007/978-3-319-46672-9_53

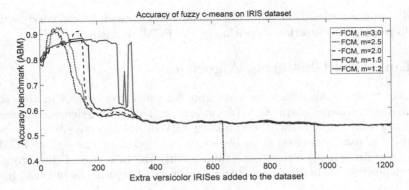

Fig. 1. Limited accuracy of FCM when the input data consist of the IRIS data set extended with synthetic versicolor items.

Problem Formulation. Figure 1 presents some accuracy benchmarks of the FCM algorithm, using a modification of the IRIS data set [4]. The original IRIS data set consists of 150 data vectors, each describing four different physical measures of individual iris flowers. As ground truth, these 150 vectors are divided into three clusters of 50 items each, named after the species of iris flowers: setosa, versicolor, and virginica. The modification consisted in generating further versicolor data vectors by averaging all possible couples of versicolor irises of the original data set. This way we have obtained $50 \times 49/2 = 1225$ further data vectors that are also supposed to belong to the versicolor class. These synthetic data vectors were gradually added to the original set of 150 vectors and fed to FCM at various settings of the fuzzy exponent m. Figure 1 shows us how the accuracy of FCM evolved, plotted against the number of synthetic vectors included into the input data. The formula of the employed benchmark indicator is given in Eq. (13). Obviously, as the number if synthetic data grows, the size of the clusters gets less and less balanced. For example at $m = 3$, 150 extra versicolor items are enough to cause the crash of the algorithm, as it is not able to establish the true boundary between the versicolor and virginica clusters. Further on, around 650 extra versicolor items, the boundary between setosa and versicolor is also mistaken. At lower values of m the accuracy is somewhat better, but there is no possible setting of the FCM algorithm which could accurately handle 400 extra versicolor vectors.

This study intends to provide a comparison of several existing extensions of the FCM algorithm that address the multimodality of the fuzzy membership functions, from the point of view cluster size sensitivity. The literature contains a wide spectrum of such algorithms including the suppressed FCM (s-FCM) [5], and its generalization gs-FCM [6], the FCM algorithm with improved partition (IFP-FCM) [7] and its generalized version GIFP-FCM [8], the FCMA algorithm [9], and the penalized FCM algorithm [10]. Literature also contains two variants of so-called cluster size insensitive FCM algorithms [11,12], which were not included in the comparison, as in our consideration they have deviated far from the alternative optimization algorithm of the FCM algorithm.

2 Employed Clustering Algorithms

All fuzzy c-means algorithm variants and derivations employed in this study cluster a set of object data $\mathbf{X} = \{\boldsymbol{x}_1, \boldsymbol{x}_2, \ldots, \boldsymbol{x}_n\}$ into a predefined number of clusters denoted by c, through minimizing a quadratic objective function. Most of these algorithms derived their objective function from the one of FCM by adding certain penalty terms that would modify the behavior of the algorithm. In the following, we briefly present the repository of algorithms involved in this study.

Fuzzy c-means. The FCM algorithm minimizes

$$J_{\mathrm{FCM}} = \sum_{i=1}^{c} \sum_{k=1}^{n} u_{ik}^m \|\boldsymbol{x}_k - \boldsymbol{v}_i\|^2 = \sum_{i=1}^{c} \sum_{k=1}^{n} u_{ik}^m d_{ik}^2, \tag{1}$$

constrained by the probabilistic condition $\sum_{i=1}^{c} u_{ik} = 1$, where v_i $(i = 1 \ldots c)$ represent the cluster prototypes, u_{ik} $(i = 1 \ldots c,\ k = 1 \ldots n)$ are the fuzzy membership functions that describe the degree to which vector $\boldsymbol{x}_k$ belongs to the cluster Ω_i represented by $\boldsymbol{v}_i$, and $m > 1$ is the fuzzy exponent [2]. The minimization of J_{FCM} is achieved via alternately applying the following partition and cluster prototype update formulas:

$$u_{ik} = \frac{d_{ik}^{-2/(m-1)}}{\sum_{j=1}^{c} d_{jk}^{-2/(m-1)}} \qquad \begin{array}{l} \forall i = 1 \ldots c \\ \forall k = 1 \ldots n \end{array}, \tag{2}$$

$$\boldsymbol{v}_i = \frac{\sum_{k=1}^{n} u_{ik}^m \boldsymbol{x}_k}{\sum_{k=1}^{n} u_{ik}^m} \qquad \forall i = 1 \ldots c. \tag{3}$$

Fuzzy c-means with Improved Partition. The fuzzy c-means algorithm with improved partition were introduced by Höppner and Klawonn [7], and generalized by Zhu et al. [8]. In its generalized form (GIFP-FCM), the algorithm minimizes

$$J_{\mathrm{GIFP-FCM}} = \sum_{i=1}^{c} \sum_{k=1}^{n} u_{ik}^m d_{ik}^2 + \sum_{k=1}^{n} a_k \sum_{i=1}^{c} u_{ik}(1 - u_{ik}^{m-1}), \tag{4}$$

subject to the probabilistic constraint, where a_k $(k = 1 \ldots n)$ are penalty terms. The optimization is achieved via alternately applying the partition update formula

$$u_{ik} = \frac{(d_{ik}^2 - a_k)^{-1/(m-1)}}{\sum_{j=1}^{c} (d_{jk}^2 - a_k)^{-1/(m-1)}} \qquad \begin{array}{l} \forall i = 1 \ldots c \\ \forall k = 1 \ldots n \end{array}, \tag{5}$$

and the prototype update formula, which is identical with Eq. (3). The values of a_k terms are chosen at the beginning of each loop using the formula $a_k = \omega \times \min\{d_{ik}^2, i = 1 \ldots c\}$, with $\omega \in [0.9, 0.99]$.

Suppressed Fuzzy c-means Algorithms. Suppressed FCM (s-FCM) [5] was not introduced through the objective function it minimizes. Instead of that, s-FCM by definition performs an extra step between the application of Eqs. (2) and (3) that modifies the partition according to the formula:

$$u_{ik} \leftarrow \begin{cases} \alpha u_{ik} & \text{if } i \neq w_k \qquad \forall i = 1 \ldots c \\ 1 - \alpha + \alpha u_{ik} & \text{if } i = w_k \qquad \forall k = 1 \ldots n \end{cases} \tag{6}$$

where $w_k = \arg\min_j\{d_{jk}^2, j = 1 \ldots c\}$, and $\alpha \in [0, 1]$ is the so-called suppression rate. Suppressed FCM also received several generalization schemes denoted by gs-FCM [6], which made the suppression rate context sensitive, that is, dependent on k. Such algorithms also have a single extra parameter varying in the interval $[0, 1]$ which governs the choice of suppression rates α_k ($k = 1 \ldots n$). Based on previous performance analysis, in this study we have chosen to include the gs-FCM algorithm of type ξ, which defines the suppression rate as: $\alpha_k = [1 - (sin\frac{\pi u_{wk}}{2})^\xi][1 - u_{wk}]^{-1}$, with parameter $\xi \in [0, 1]$, where u_{wk} stands for the largest fuzzy membership function value of vector $\boldsymbol{x}_k$ provided by the FCM algorithm. Further details of the algorithm can be found in [6]. The objective function optimized by all suppressed FCM algorithms is given in [13].

FCMA by Miyamoto and Kurosawa. The FCMA algorithm optimizes

$$J_{\text{FCMA}} = \sum_{i=1}^{c} \sum_{k=1}^{n} \alpha_i^{1-m} u_{ik}^m d_{ik}^2, \tag{7}$$

subject to the probabilistic constraint of the fuzzy memberships, and of the extra parameters α_i ($i \ldots c$): $\sum_{i=1}^{c} \alpha_i = 1$ [9]. The minimization of J_{FCMA} is achieved via alternately applying the partition updating formula:

$$u_{ik} = \frac{\alpha_i d_{ik}^{-2/(m-1)}}{\sum_{j=1}^{c} \alpha_j d_{jk}^{-2/(m-1)}} \qquad \begin{array}{l} \forall i = 1 \ldots c \\ \forall k = 1 \ldots n \end{array}, \tag{8}$$

the prototype update formula given in Eq. (3), and the extra formula:

$$\alpha_i = \frac{\sqrt[m]{\sum_{k=1}^{n} u_{ik}^m d_{ik}^2}}{\sum_{j=1}^{c} \sqrt[m]{\sum_{k=1}^{n} u_{jk}^m d_{jk}^2}} \qquad \forall i = 1 \ldots c. \tag{9}$$

Penalized FCM by Yang. The Penalized FCM (PFCM) algorithm optimizes

$$J_{\text{PFCMA}} = \sum_{i=1}^{c} \sum_{k=1}^{n} u_{ik}^m [d_{ik}^2 - \lambda \log \alpha_i], \tag{10}$$

subject to the probabilistic constraint of the fuzzy memberships, and of the extra parameters α_i ($i \ldots c$): $\sum_{i=1}^{c} \alpha_i = 1$ [10]. The minimization of J_{PFCM} is achieved via alternately applying the partition updating formula:

$$u_{ik} = \frac{[d_{ik}^2 - \lambda \log \alpha_i]^{-1/(m-1)}}{\sum_{j=1}^{c} [d_{jk}^2 - \lambda \log \alpha_j]^{-1/(m-1)}} \qquad \begin{array}{l} \forall i = 1 \ldots c \\ \forall k = 1 \ldots n \end{array}, \tag{11}$$

the prototype update formula given in Eq. (3), and the extra formula:

$$\alpha_i = \frac{\sum_{k=1}^n u_{ik}^m}{\sum_{j=1}^c \sum_{k=1}^n u_{jk}^m} \qquad \forall i = 1 \dots c. \tag{12}$$

3   Results and Discussion

The algorithms enumerated in Sect. 2 underwent thorough numerical evaluation using two different scenarios. The first test employed the IRIS data set with synthetic extension of a centrally located versicolor cluster, as described in Sect. 1. Data vectors were initially normalized, namely the values in each dimension of the feature vectors were linearly mapped upon the $[0, 1]$ interval. In case of all algorithms, we investigated the conditions of relatively good accuracy, meaning that the majority of setosa, versicolor, and virginica irises are assigned to three different clusters. In order to characterise the accuracy with a single numerical value, we propose the accuracy benchmark index defined as:

$$\mathrm{ABM} = \max_{p \in P_c} \left\{ \sum_{i=1}^c \frac{|\Lambda_i \cap \Omega_{p(i)}|^2}{|\Lambda_i| \times |\Omega_{p(i)}|} \right\}, \tag{13}$$

where Λ_i stands for the class i according to the grand truth, Ω_i represents the cluster with index i, P_c is the set of all possible permutations of numbers $1, 2, \dots, c$, and $|\Psi|$ stands for the cardinality of set Ψ. ABM can range from 0 to 1: the maximum value indicates perfect separation of the ground true classes, while any deviance from the perfect separation is penalized.

Figure 2 exhibits the obtained ABM characteristics for various algorithms and settings, plotted against the number of synthetic versicolor irises added to the input data. All FCM algorithms with improved or suppressed partition have their limit around 350-400 synthetic vectors, above which they fail to distinguish the three clusters. On the other hand, for FCMA and PFCM there exists such a setting which can provide acceptable accuracy even in case of 1225 extra items. However, finding automatically these settings is not a trivial job. Table 1 shows some examples of confusion matrices obtained during the tests. The algorithms are ranked according to their performance. The fact that FCMA can stand at the top ($m = 1.5$) or the bottom ($m = 2$) of the ranking, clearly justifies the importance of well chosen parameter values. PFCM has an optimal λ value for both $m = 1.5$ and $m = 2$, but that is impossible to guess. There is no wide interval for λ, where PFCM has the ideal behavior, see Table 2.

While the first test employed classes of different cardinality without significantly changing the diameter of expected clusters, in the second example we have a different case. Let us define two collections of two-dimensional vectors, centered in $(-2, 0)$ and $(2, 0)$. The first circular group has a fixed radius of $r_1 = 1$ unit, and contains 100 randomly generated vectors distributed with uniform density. On the other hand, for the second circular group we will gradually change the radius r_2 from 1 to 3, while keeping the density of vectors constant.

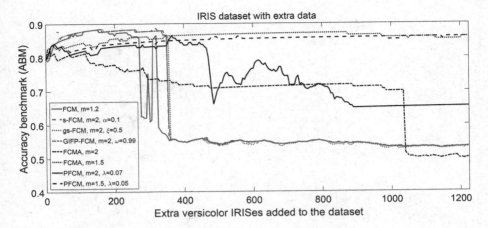

Fig. 2. Accuracy benchmark values obtained by various tested algorithm plotted against the number of extra versicolor irises.

Table 1. Confusion matrices obtained by various algorithms on the IRIS data set with 50, 150, 350, and 850 extra versicolor irises. Zeros are omitted

Algorithm and parameters	Ground truth	IRIS + 50			IRIS + 150			IRIS + 350			IRIS + 850		
		Ω_1	Ω_2	Ω_3	Ω_1	Ω_2	Ω_3	Ω_1	Ω_2	Ω_3	Ω_1	Ω_2	Ω_3
FCMA $m = 1.5$	setosa	50			50			50			50		
	versicolor	100			200			400			900		
	virginica		17	33		20	30		18	32		18	32
PFCM $m = 1.5$ $\lambda = 0.05$	setosa	50			50			50			50		
	versicolor	100			200			400			900		
	virginica		18	32		20	30		20	30		20	30
PFCM $m = 2.0$ $\lambda = 0.07$	setosa	50			50			50			50		
	versicolor	100			200			400			899	1	
	virginica		18	32		23	27		22	28		45	5
s-FCM $m = 2.0$ $\alpha = 0.1$ | gs-FCM $m = 2.0$ $\xi = 0.5$	setosa	50			50			50			50		
	versicolor	100			200			396	4		488	412	
	virginica		14	36		14	36		14	36		2	48
GIFP-FCM $m = 2.0$ $\omega = 0.99$	setosa	50			50			50			50		
	versicolor	99	1		200			395	5		474	426	
	virginica		14	36		14	36		14	36		2	48
FCM $m = 1.2$	setosa	50			50			50			50		
	versicolor	99	1		200			208	192		484	416	
	virginica		14	36		14	36		3	47		2	48
FCMA $m = 2.0$	setosa	50			50			49	1		48	2	
	versicolor	100			200			400			900	416	
	virginica		21	29		27	23		35	15		38	12

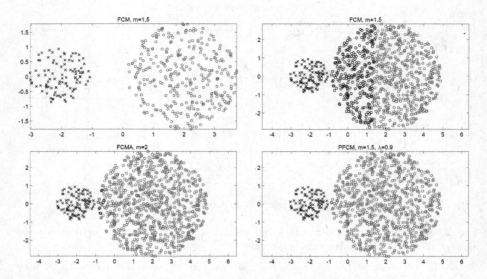

Fig. 3. Case of two clusters of different sizes: at $r_2 = 1.5$ even FCM can be accurate, but a more significant difference in cluster sizes require more sophisticated solutions, which are possible with FCMA and PFCM.

Table 2. Behavior of various tested algorithms in case of two clusters of different sizes

| Algorithm | Parameters | | Maximum r_2 with perfect accuracy | Misclassifications at | | |
	m	other		$r_2 = 2$ out of 500	$r_2 = 2.5$ out of 725	$r_2 = 2.9$ out of 941
FCM	2.0		1.7	33	178	316
FCM	1.2		1.9	14	153	302
s-FCM	2.0	$\alpha = 0.1$	1.9	6	141	304
gs-FCM	2.0	$\xi = 0.8$	1.9	10	137	289
GIFP-FCM	2.0	$\omega = 0.9$	1.9	10	141	296
GIFP-FCM	1.5	$\omega = 0.9$	1.9	7	146	305
FCMA	2.0		2.7	0	0	16
FCMA	1.5		2.6	0	0	150
PFCM	2.0	$\lambda = 0.8$	2.4	0	4	127
PFCM	2.0	$\lambda = 0.9$	2.6	0	0	36
PFCM	2.0	$\lambda = 1.0$	2.1	0	5	11
PFCM	1.5	$\lambda = 0.8$	2.7	0	0	23
PFCM	1.5	$\lambda = 0.9$	2.8	0	0	1
PFCM	1.5	$\lambda = 1.0$	2.4	0	2	12

We will investigate, how the tested algorithms react to different sized clusters, and what circumstances or settings are required to achieve best accuracy. Figure 3 shows some examples of the clustering outcome, while Table 2 exhibits numerical information of the best performance achieved by each algorithm. FCM can provide two accurately separated clusters up to $r_2 = 1.7$, while improved and suppressed partitions can extend this behavior up to $r_2 = 1.9$. To achieve perfect separation of the two groups at $r_2 > 2$, one needs to turn to PFCM or FCMA. In the extreme case of $r_2 = 2.9$, the best result is achieved by PFCM with a single misclassification. Although the best outcome is provided by PFCM, FCMA can also be useful because it has no extra parameter compared to FCM, so it is much easier to tune than PFCM.

4 Conclusions

This study performed a comparative analysis of several modified and enhanced fuzzy c-means clustering in terms of sensitivity to cluster sizes. Two numerical tests were proposed, to assess the behavior of the algorithms both in case of clusters with different cardinality but relatively similar diameter, and in case of clusters that differ in both cardinality and diameter. The best performance was achieved by FCMA and PFCM, but even these are difficult to automatically tune to the input data. All other tested algorithms are significantly less effective.

References

1. Dunn, J.C.: A fuzzy relative of the isodata process and its use in detecting compact well-separated clusters. Cybern. Syst. **3**(3), 32–57 (1973)
2. Bezdek, J.C.: Pattern recognition with fuzzy objective function algorithms. Plenum, New York (1981)
3. Komazaki, Y., Miyamoto, S.: Variables for Controlling Cluster Sizes on Fuzzy c-Means. In: Torra, V., Narukawa, Y., Navarro-Arribas, G., Megías, D. (eds.) MDAI 2013. LNCS, vol. 8234, pp. 192–203. Springer, Heidelberg (2013)
4. Anderson, E.: The irises of the Gaspe peninsula. Bull. Am. Iris Soc. **59**, 2–5 (1935)
5. Fan, J.L., Zhen, W.Z., Xie, W.X.: Suppressed fuzzy c-means clustering algorithm. Patt. Recogn. Lett. **24**, 1607–1612 (2003)
6. Szilágyi, L., Szilágyi, S.M.: Generalization rules for the suppressed fuzzy c-means clustering algorithm. Neurocomput. **139**, 298–309 (2014)
7. Höppner, F., Klawonn, F.: Improved fuzzy partition for fuzzy regression models. Int. J. Approx. Reason. **5**, 599–613 (2003)
8. Zhu, L., Chung, F.L., Wang, S.: Generalized fuzzy c-means clustering algorithm with improved fuzzy partition. IEEE Trans. Syst. Man Cybern. B. **39**, 578–591 (2009)
9. Miyamoto, S., Kurosawa, N.: Controlling cluster volume sizes in fuzzy c-means clustering. In: SCIS and ISIS, Yokohama, Japan, pp. 1–4 (2004)
10. Yang, M.S.: On a class of fuzzy classification maximum likelihood procedures. Fuzzy Sets Syst. **57**(3), 365–375 (1993)

11. Noordam, J., Van Den Broek, W., Buydens, L.: Multivariate image segmentation with cluster size insensitive fuzzy c-means. Chemom. Intell. Lab. Syst. **64**(1), 65–78 (2002)

12. Lin, P.L., Huang, P.W., Kuo, C.H., Lai, Y.H.: A size-insensitive integrity-based fuzzy c-means method for data clustering. Patt. Recogn. **47**(5), 2024–2056 (2014)

13. Szilágyi, L.: A Unified Theory of Fuzzy c-Means Clustering Models with Improved Partition. In: Torra, V., Narukawa, T. (eds.) MDAI 2015. LNCS, vol. 9321, pp. 129–140. Springer, Heidelberg (2015)

Social Networks

Influence Spread Evaluation and Propagation Rebuilding

Qianwen Zhang[1], Cheng-Chao Huang[1,2], and Jinkui Xie[1(✉)]

[1] Department of Computer Science and Technology, East China Normal University,
Shanghai 200241, China
{zqw1005,ecnucchuang}@126.com, jkxie@cs.ecnu.edu.cn
[2] Institute of Systems Science, East China Normal University,
Shanghai 200241, China

Abstract. In social networks, studies about influence maximization mainly focus on the algorithm of finding seed nodes, but ignore the intrinsic properties of influence propagation. In this paper, we consider the relationship between seed sets & influence spread. For static propagation, we reasonably abstract the relationship between the size of the seed set and the influence spread in influence maximization problem as a logarithmic function. We also provide experiments on large collaboration networks, showing the rationality and the accuracy of the proposed function. For dynamic influence propagation, we rebuild it as a continuous linear dynamical system called 3DS, which is based on Newton's law of cooling. Furthermore, we give an efficient method to compute the influence spread function of time without much loss of accuracy. Its efficiency is demonstrated by complexity analysis.

Keywords: Social networks · Influence maximization · Influence spread evaluation · Propagation rebuilding

1 Introduction

In nearly a decade, social networks gain unprecedented popularity. Many social networks, such as Facebook, Google+, Flickr, Weibo, and Youtube, help strengthen individuals' relationships online. Meanwhile, they are also becoming huge dissemination and marketing platforms, allowing information and ideas to influence a large population, also called *influence propagation*. This phenomenon has been found useful for marketing purpose. For example, a company may give some free products to a few influential users, with the hope to trigger a large scale of adoptions. The problem of identifying the set of individuals we should target at was defined as *Influence Maximization Problem* (IM problem) by Kempe et al. [9]. In IM problem, a social network is modeled as a graph $G = (V, E)$ with vertices representing users and edges representing their relationships. For every $e_{ij} \in E$, p_{ij} denotes the influence probability that ith vertex activates jth vertex. Let S be the subset of vertices selected to start the influence propagation, which we call seed set. Let $\sigma(S)$ denote the expected number of vertices

© Springer International Publishing AG 2016
A. Hirose et al. (Eds.): ICONIP 2016, Part II, LNCS 9948, pp. 481–490, 2016.
DOI: 10.1007/978-3-319-46672-9_54

influenced by S, which we call *influence spread*. Then we describe the problem formally as below:

Problem 1 (IM Problem). For a graph $G = (V, E)$ and a positive integer k, compute the seed set

$$S^* = \arg\max_{S \subset V \wedge |S| = k} \sigma(S).$$

In 2003, Kempe et al. proved IM problem is NP-hard and provided a greedy approximation algorithm [9]. Later, many studies tried to tackle the efficiency issue. In Ref. [11], Leskovec et al. presented a "Lazy-forward" optimization in selecting new seeds. In Ref. [4], Chen et al. further improved the greedy algorithm, greatly decreasing the time while keeping closed influence spread. They continued to propose PMIA model in Ref. [3], maintaining good balance between efficiency and effectiveness, which is a popular algorithm to select seed sets.

While studying influence propagation, there are two basic models: *Linear Threshold Model* proposed by Granovetter and Schelling [8,17], and *Independent Cascade Model* (IC model) proposed by Goldenberg et al. [5,6]. These models are considered to be static, which regard influence propagation as a instantaneous procedure and influenced probabilities of nodes as constants (such as in Ref. [1,3,4,9]).

In this paper, we study the relationship between seed sets and influence spread. For static propagation, Chen et al. have proved computing the influence spread by a seed set is in #P-hard (a superset of NP-hard) [3]. In IM problem, for a given seed set, the influence spread can be computed by Monte Carlo simulation. Here, we are interested in evaluating the influence spread only by the size of the seed set (#seeds). We abstract the maximal influence spread as a univariate logarithmic function of the size of the seed set found by mainstream methods for solving IM problem. We further provide experiments on large collaboration networks, showing the rationality and the accuracy of the proposed function.

Furthermore, for a given seed set, we also consider the influence spread in dynamic influence propagation. In Ref. [7] Goyal et al. learnt the influence probabilities from a log of past propagations, and in Ref. [10] Lei et al. learnt the probability information by a multiple-trial approach. Besides, thermal theories of heat transfer are widely used in dynamic modeling in many fields, such as economics, control theory, bioinformatics, and information science [14,19]. In 2013 Zhou and Liu also introduced heat diffusion into influence graphs [20]. Inspired by these works, we rebuild dynamic influence propagation as a continuous linear dynamical system and compute the influence spread function, whose rationale is based on the theories of both thermal physics and IC model. We also give an approximated method to compute the influence spread function fast, whose efficiency is guaranteed by the complexity analysis.

Organization. In Sect. 2, for static IM problem, we propose the relational function between #seeds & influence spread and show our experiments. In Sect. 3, we rebuild dynamic influence propagation, further give a speedup method to compute the influence spread function and analyze the complexity. We conclude the paper in Sect. 4.

2   Relation Between #Seeds and Influence Spread

In this section, we aim to give the relationship between the size of the seed set and the influence spread of a given graph G for static propagation. To this end, we first recall the basic conceptions and properties of IC model and the greedy algorithm for IM problem. Then we abstract the relationship as a logarithmic function and show the corresponding experiments on large collaboration networks.

2.1   Relational Function

The Independent Cascade model is a popular diffusion model used to model the influence propagation. Given a seed set S, the IC model works as follows. Let S_n be the set of vertices that are activated in the nth round, with $S_0 = S$. In the $n + 1$th round, each newly activated vertex v_i may activate its neighbor v_j which is not yet activated with an independent probability of p_{ij}. This process is repeated until S_n is empty. Note that each activated vertex only has one chance to activate its neighbors.

Let X represents a set of vertices, $f(X)$ denotes the vertices activated by X. During the random process of propagation in the IC model, given a seed set S, the influence spread can be written as follows:

$$\begin{cases} I^0(S) = \emptyset, \quad I^1(S) = S, \\ I^{n+2}(S) = I^{n+1}(S) \cup f(I^{n+1}(S) \backslash I^n(S)) \quad \text{for } n \geq 1. \end{cases}$$

where $I^n(S)$ denotes the set of activated vertices at nth round by the seed set S, and let level(i) denotes the index of the round in which the ith vertex activated. The expected influence spread $\sigma(S)$ is $|I^\infty(S)|$.

Given an input k, the IM problem of finding the seed set is NP-hard, but a constant-ratio approximation algorithm is available. Algorithm 1 describes the general greedy algorithm solving the IM problem.

Algorithm 1. Greedy(k, f) [3]

1: initialize $S^* = \emptyset$
2: **for** $i = 1$ to k **do**
3:     select $u = \arg\max_{w \in V \backslash S^*} (\sigma(S^* \cup \{w\}) - \sigma(S^*))$
4:     $S^* = S^* \cup \{u\}$
5: **end for**
6: **return** S^*;

In Ref. [9], it is shown that Algorithm 1 solves the IM problem with an approximation ratio of $1 - 1/e$, using the way of propagation mentioned in IC model. Kempe et al. guaranteed the approximation ratio of the algorithm by proving the following lemma:

Lemma 1. *(Theorem 2.2 of [9]) for an arbitrary instance of the IC model, the resulting influence function $\sigma(\cdot)$ is submodular.*

A function f is submodular if $f(S \cup \{v\}) - f(S) \geq f(T \cup \{v\}) - f(T)$, for all $v \in V$ and all pairs of subsets $S \subseteq T \subseteq V$. That means the incremental value of the function decreases as the size of the input set increases.

Then we define $\phi(\kappa)$ as the function of the maximal influence spread when the seed set contains κ vertices (namely $\kappa = |S^*|$). According to Lemma 1 and the fact that $\Delta\phi(\kappa)$ is decreasing exponentially by the κ increasing ($\Delta\phi(\kappa) > 0$), we say that there exists a $w > 0$ such that $\Delta\phi(\kappa) - w$ tends to zero exponentially fast. Therefore, we abstract the relationship between κ and $\phi(\kappa)$ as

$$exp(\alpha(\Delta\phi(\kappa) - w)) - 1 = \frac{1}{\kappa + \beta}$$

for some parameters $\alpha, \beta, w > 0$. Clearly, it is a one-order difference equation. We solve it and obtain the closed form solution of $\phi(\kappa)$ as below:

$$\phi(\kappa) = w\kappa + a\ln(b\kappa + 1),$$

where a, b are the reciprocals of α, β respectively.

2.2 Experiments

We conduct experiments on several real-life networks. Our experiments aim at illustrating the effective of proposed function $\phi(\kappa)$.

Datasets. We use three real social network datasets. Two collaboration networks NetHEPT and NetPHY are obtained from arXiv.org in the High Energy Physics Theory and Physics domains respectively. Another dataset is DBLP, which is an academic collaboration network. In these datasets, each vertex in the network represents an author, and an edge between a pair of vertices represents their co-authorship. These datasets are commonly used in the literature of influence maximization [3,4,9,10]. Table 1 shows the details of these datasets.

Table 1. Datasets.

Dataset	#Vertices	#Edges	Avg. degree
NetHEPT	15233	58891	7.73
NetPHY	37154	231584	12.24
DBLP	100000	373645	7.47

In static IM problem, weighted cascade model are usually adopted to obtain the probabilities [9]. We set $p_{ij} = 1/d(j)$ for an edge e_{ij}, where $d(j)$ is the in-degree of jth vertex. We choose to use algorithm PMIA to select the seed set, whose results is very close to the greedy algorithm. In experiments, for each #seeds $\kappa = 100 \times N$ ($N = 0, 1, \ldots, 49$) we compute the seed sets and further obtain the corresponding influence spread (denoted by $\Phi(\kappa)$) by Monte Carlo simulation. Then we pick up three samples with small-scale to determine the undetermined coefficients w, a, b and finally obtain the form of $\phi(\kappa)$. The curves of $\phi(\kappa)$, $\Phi(\kappa)$, and the sample points are showed in Fig. 1. Intuitively, the curve of $\phi(\kappa)$ accurately fit the curve of $\Phi(\kappa)$ with the mean absolute percentage errors 1.29%, 1.58%, and 0.97% respectively for three datasets. It is noteworthy that comparing with the whole range of κ from 0 to 4900, the samples picked up are very small. In other word, we can efficiently obtain an accurate relationship between the #seeds and the maximal influence spread only by computing several small-scale samples. Therefore, one of the potential applications of this technique is to predict the maximal influence spread of large seed sets.

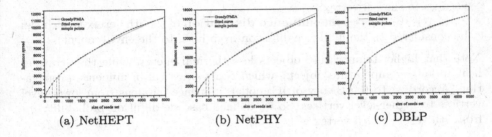

(a) NetHEPT   (b) NetPHY   (c) DBLP

Fig. 1. Experimental Results

3   Influence Propagation Rebuilding

Most traditional models of influence propagation in social networks are seemed to be static, ignoring time cost of influence propagation and ideally regarding influence propagation as instantaneous procedure. In reality, however, the influence propagation not only cost time, but also dissipate over time. In this section, we aim to rebuild dynamic influence propagation. For a given graph G with a certain seed set S, we construct a much more practical influence propagation model—Dynamic Diffusion & Dissipation System (3DS for short), which takes time into consideration. In essence, the model is a continuous linear dynamical system (continuous LDS), whose rationale is based on the theories of both Thermal physics and IC model. Furthermore, we also give an efficient method to compute the influence spread function of time.

3.1   Dynamic Diffusion and Dissipation System

In what follows, we denote by $x_i(t)$ the probability of that the ith vertex is influenced at time t, and let $X(t)$ denote the column vector consisting of each

$x_i(t)$, viz. $X(t) = [x_1(t), x_2(t), \ldots, x_n(t)]^{\mathrm{T}}$. According to the *Newton's Law of Cooling* (heat-transfer version of Newton's law) [2], the derivative of the thermal energy of an object is proportional to the difference of temperature between other objects and itself. That is

$$\frac{dQ(t)}{dt} = \sum_k h_k \cdot (T_k(t) - T(t)),$$

where h_k is the heat transfer coefficient between the kth object and itself. By analogy between heat-transfer and influence propagation, for the ith vertex, we have the following equation

$$\frac{dx_i(t)}{dt} = \sum_j \left[\nu p_{ji}(x_j(t) - x_i(t)) - \mu(x_i(t) - 0) \right], \tag{1}$$

with positive constants ν and μ as heat transfer coefficients. The right side of this equation consists of two parts:

- the sum of $\nu p_{ji}(x_j(t) - x_i(t))$ denotes the **influence diffusion** from other vertices to the ith vertex.
- $-\mu(x_i(t) - 0)$ denotes the **influence dissipation** of the ith vertex (which can be considered as the influence diffusion from itself to the environment).

Note that higher-temperature objects lose thermal energy while they transfer heat to lower-temperature objects, which is not practical in influence propagation. According to the theory of IC model, influence happens from lower-level vertices to higher-level vertices. So we defined cascade diffusion probability q_{ij} from jth vertex to ith vertex as

$$q_{ij} = \begin{cases} \nu p_{ji} & \text{if } \mathrm{level}(j) < \mathrm{level}(i) \\ 0 & \text{otherwise} \end{cases},$$

Then the equation (1) can be revised as

$$\frac{dx_i(t)}{dt} = \sum_j \left[q_{ij}(x_j(t) - x_i(t)) - \mu x_i(t) \right].$$

Namely, we have

$$\dot{X}(t) = (Q - R - \mu I) \cdot X(t), \tag{2}$$

in which

- $Q = \{q_{ij}\}_{n \times n}$ is a matrix consisting of cascade diffusion probabilities,
- R is a diagonal matrix where the (k, k)-entry is the sum of all entries of the kth row in Q.

It is clear that $Q - R - \mu I$ is a constant matrix, thus the equation (2) is exactly a continuous LDS. In other words, we use a continuous LDS to describe dynamic influence propagation. Note that the system is solvable and the closed form solutions of each coordinates of $X(t)$ can be obtained. Furthermore, all these computations aforeméntioned (both constructing the system and solving it) can be finished in polynomial time.

Closed Form Solution. We here rewrite the 3DS (2) as

$$\dot{X}(t) = \mathcal{B} \cdot X(t) \tag{3}$$

where $\mathcal{B} = Q - R - \mu I$, with the initial value

$$x_i(t) = \begin{cases} 1 & \text{if } i \in S, \\ 0 & \text{otherwise.} \end{cases}$$

For an arbitrary coordinate of $X(t)$, the closed form solution can be expressed as follow [13]:

$$x_i(t) = \sum_{k=1}^{s} f_{ik}(t) \exp(\lambda_k t) \qquad (i = 1, \ldots, n), \tag{4}$$

where λ_k $(k = 1, \ldots, s)$ is the kth eigenvalue of matrix $\mathcal{B}$ with multiplicity m_k, $f_{ik}(t)$ is a polynomial with degree m_k and coefficients $C_{k0}^{(i)}, \ldots, C_{km_k}^{(i)}$. Then we define an observing function of the system

$$\Psi(t) = \sum_{i=1}^{n} x_i(t) = \sum_{k=1}^{s} F_k(t) \exp(\lambda_k t)$$

as the influence spread function of time $t \in \mathbb{R}^+$, where $F_k(t) = \sum_{i=1}^{n} f_{ik}(t)$.

Two Important Properties.

Property 1 (Dissipativity). The influence spread $\Psi(t)$ is exponentially tending to zero with the time goes by.

Proof. Denote by $\mathcal{B}_{ij}$ the (i, j)-entry of matrix $\mathcal{B}$. Recall that $\mathcal{B} = Q - R - \mu I$. For each ith row, $\sum_{j \neq i} |\mathcal{B}_{ij}| = R_{ii}$ and $\mathcal{B}_{ii} = -R_{ii} - \mu$. According to *Gershgorin's circle theorem*, the real part of an arbitrary eigenvalue of $\mathcal{B}$, denoted by $\mathfrak{R}(\lambda_k)$, is strictly negative. Then the property directly follows. □

This property shows that the influence in the graph is generally decreasing, which essentially due to the influence dissipation (i. e. the term $-\mu I$ in matrix $\mathcal{B}$).

Property 2 (Peak-Value). The influence spread $\Psi(t)$ will appear peak-values at some time points.

Proof. Clearly, $\Psi(t)$ is analytical over its domain $t \in \mathbb{R}_{\geq 0}$, and $\Psi(0) = |S|$, $\lim_{t \to +\infty} \Psi(t) = 0$ (by the dissipativity). Therefore, there exist finitely many $t^* \in \mathbb{R}_{\geq 0}$ such that $\Psi(t^*) = \inf \Psi(t)$. Then the property follows. □

Furthermore, these t^* can be obtained by computing the real zeros of $d\Psi(t)/dt$. For practical applications, it is meaningful to show when the influence is most conspicuous in the whole procedure of propagation.

3.2 Approximation for Speedup

We do have constructed a continuous linear dynamical system—3DS to rebuild dynamic influence propagation in a graph. Specifically, it is in the form (3) with the closed form solution as (4). It is clear that in order to obtain the closed form solution of a 3DS, we need to:

(I) compute all eigenvalues of $\mathcal{B}$ with their multiplicities;
(II) compute all constant coefficients $C_{kj}^{(i)}$.

By conventional ways, for (I), we need to solve the polynomial equation $\mathrm{Det}(\lambda I - \mathcal{B}) = 0$. For (II), we need to solve a n^2-variables linear equations. Both (I) and (II) can be done in time complexity $\mathcal{O}(n^6)$ by standard algebraic methods. However, for large scale graphs, the efficiency of the aforementioned procedure is unacceptable.

In what follows, we improve the efficiency of this procedure by approximation method from two perspectives. We give the approximated closed form solutions of the system, and resort to numerical methods to reduce the complexity. Moreover, we also reduce the size of graphs without much loss of information by a proposed method.

Approximated Closed Form Solution. Recall the form of (4), and we define

$$\Lambda = \{\lambda_i : |\Re(\lambda_i)| \geq \Re(|\lambda_j|) \quad \forall i, j = 0, 1, \ldots, n\}.$$

Namely, Λ is the set of the eigenvalues with real part of maximum modulus (called *dominant eigenvalues*). Then we have the terms containing dominant eigenvalues dominate the other terms, since for any $\lambda_k \in \Lambda$ and $\lambda_k' \notin \Lambda$, $f_{ik'}(t) \exp(\lambda_k' t)/f_{ik}(t) \exp(\lambda_k t)$ is tending to zero exponentially fast [18]. Therefore, we approximate (4) by

$$x_i(t) \approx \sum_{\lambda_k \in \Lambda} f_{ik}(t) \exp(\lambda_k t) \qquad (i = 1, \ldots, n), \tag{5}$$

Compute the Eigenvalues. For computing the eigenvalues efficiently, we use *harmonic restarted Arnoldi algorithm* which can not only compute the complex eigenvalues but also determine the multiplicity of them [12]. (*Power Iteration* method only computes the largest real eigenvalue without its multiplicity.)

Fast Compute $\Psi(t)$. After giving the approximated closed form solution, only the dominant eigenvalues are reserved. At the same time, the number of coefficients decrease from n^2 to $|\Lambda|n$, where $|\Lambda| = \sum_{\lambda_k \in \Lambda}(m_k)$. It implies that computing these coefficients has time complexity $\mathcal{O}(|\Lambda|^3 n^3)$ by Gaussian elimination, which is much smaller than $\mathcal{O}(n^6)$ since $|\Lambda| \ll n$ in general.

Actually, we do care rather the whole influence effect (viz. $\Psi(t)$) than the influence on the individual vertices in a graph. It means that we don't need to compute all closed form solution of $X(t)$, if we have another method to obtain $\Psi(t)$. In what follows, we give an efficient method to compute $\Psi(t)$ base on numerical computation.

According to (5), we obtain the approximation of $\Psi(t)$ as

$$\Psi(t) \approx \sum_{\lambda_k \in \Lambda} F_k(t) \exp(\lambda_k t) , \qquad (6)$$

which contains only $|\Lambda|$ undetermined coefficients. These coefficients can be computed if we have $|\Lambda|$ pairs of $(t, \Psi(t))$. For the continuous LDS (3), by numerical integration, we can easily compute $|\Lambda| - 1$ pairs of $(t, X(t))$ from the initial point $(0, X(0))$, and further obtain the $|\Lambda|$ pairs of initial points $(t_i, \Psi(t_i))$ corresponding to these $(t_i, X(t_i))$. Then, we can compute all coefficients in $\Psi(t)$ by solving a $|\Lambda|$-variables linear equations. Note that the time of numerical integration here is mainly spent on the inner product of a matrix and a vector in dimension n, which has the time complexity $\mathcal{O}(n^2)$. Finally, the complexity of computing the coefficients in $\Psi(t)$ has been significantly reduced from $\mathcal{O}(n^6)$ to $\mathcal{O}(n^2 + |\Lambda|^3)$. Moreover, initial points obtained by numerical integration would not bring about an uncontrollable error of $\Psi(t)$, since it is *non-chaotic*. Namely, the system is not highly sensitive to its initial conditions.

Reduce the Size of Graphs. Note that the closed forms of both the solution of 3DS and $\Psi(t)$ depend on the eigenvalues (specifically on dominate eigenvalues) of the graph. (Recent work [15] also showed that for almost any propagation model, important characteristics are captured by the spectrum of the graph.) For our aim, we expect a method to reduce the size of the graph without changing the dominate eigenvalues. Thanks to the work of Purohit et al. [16], we are provided an algorithm to find the set of node pairs which when merged lead to small changes of eigenvalues. It enables us to reduce the graph by 90 % without much loss of information, which will significantly reduce the size of the graph and make the procedure of rebuilding efficiently.

4   Conclusion

In this paper, we proposed the relational function to evaluate the influence spread in static IM problem and showed its accuracy by experiments. Besides, we rebuilt dynamic influence propagation as a continuous LDS and gave an efficient method to compute the influence spread function.

For further work, we are interested in how to determine the heat transfer coefficients ν, μ of situations in reality. Moreover, the problem of choosing seed sets for 3DS is also attractive. Specifically, we intend to study the strategy of choosing the initial value of the 3DS such that the influence spread (viz. $\Psi(t)$) may satisfy one of the following properties (called *Conditional Initial-Value Problem*):

(1) achieve its maximal peak-value;
(2) achieve its peak-value in minimal time;
(3) has the maximal mean value over a given closed time interval.

References

1. Aslay, C., Barbieri, N., Bonchi, F., Baeza-Yates, R.: Online topic-aware influence maximization queries. In: International Conference on Extending Database Technology (2014)
2. Burmeister, L.C.: Convective Heat Transfer, 2nd edn. Wiley, Hoboken (1993)
3. Chen, W., Wang, C., Wang, Y.: Scalable influence maximization for prevalent viral marketing in large-scale social networks. In: KDD, pp. 1029–1038 (2010)
4. Chen, W., Wang, Y., Yang, S.: Efficient influence maximization in social networks. In: KDD, pp. 199–208 (2009)
5. Goldenberg, J., Libai, B.: Using complex systems analysis to advance marketing theory development: modeling heterogeneity effects on new product growth through stochastic cellular automata. Acad. Mark. Sci. Rev. (2001)
6. Goldenberg, J., Libai, B., Muller, E.: Talk of the network: a complex systems look at the underlying process of word-of-mouth. Mark. Lett. **12**(3), 211–223 (2001)
7. Goyal, A., Bonchi, F., Lakshmanan, L.V.S.: Learning influence probabilities in social networks. In: Proceeding of WSDM, pp. 241–250 (2010)
8. Granovetter, M.: Threshold models of collective behavior. Am. J. Sociol. **83**(6), 1420–1443 (1978)
9. Kempe, D., Kleinberg, J., Tardos, E.: Maximizing the spread of influence through a social network. In: Proceedings of the 9th ACM SIGKDD, pp. 137–146 (2003)
10. Lei, S., Maniu, S., Mo, L., Cheng, R., Senellart, P.: Online influence maximization. In: Proceedings of the 20th ACM SIGKDD, pp. 645–654 (2014)
11. Leskovec, J., Krause, A., Guestrin, C., Faloutsos, C., Vanbriesen, J.M., Glance, N.S.: Cost-effective outbreak detection networks. In: KDD Proceedings of the 13rd ACM SIGKDD, pp. 780–782 (2007)
12. Morgan, R.B., Zeng, M.: A harmonic restarted arnoldi algorithm for calculating eigenvalues and determining multiplicity. Linear Algebr. Appl. **415**(1), 96–113 (2006)
13. Morris, R.D., Hirsch, W., Smale, S.: Differential Equations, Dynamical Systems, An Introduction to Chaos, 2nd edn. Academic, Boston (2004)
14. Patel, V.K., Savsani, V.J.: Heat transfer search (hts): a novel optimization algorithm. Inf. Sci. **324**, 217–246 (2015)
15. Prakash, B.A., Chakrabarti, D., Valler, N.C., Faloutsos, M., Faloutsos, C.: Threshold conditions for arbitrary cascade models on arbitrary networks. Knowl. Inf. Syst. **33**(3), 537–546 (2011)
16. Purohit, M., Prakash, B.A., Kang, C., Zhang, Y., Subrahmanian, V.S.: Fast influence-based coarsening for large networks. In: Proceedings of the 20th ACM SIGKDD, pp. 1296–1305 (2014)
17. Schelling, T.: Micromotives and Macrobehavior. Norton, New York (1978)
18. Xu, M., Huang, C.-C., Li, Z.-B., Zeng, Z.: Analyzing ultimate positivity for solvable systems. Theoret. Comput. Sci. **609**, 395–412 (2016)
19. Yun, T.Q.: Applications of Heat Diffusion Equation and G-Contractive Mapping. LAP LAMBERT Academic, Saarbrücken (2013)
20. Zhou, Y., Liu, L.: Social influence based clustering of heterogeneous information networks. In: Proceedings of the 19th ACM SIGKDD, pp. 338–346 (2013)

A Tag Probability Correlation Based Microblog Recommendation Method

Di Zhang, Huifang Ma[✉], Junjie Jia, and Li Yu

College of Computer Science and Engineering, Northwest Normal University,
Lanzhou 730070, China
nwnuzhangd@yeah.net, mahuifang@nwnu.edu.cn

Abstract. In order to improve users' experience it is necessary to recommend valuable and interesting content for users. A tag probability correlation based microblog recommendation method (TPCMR) is presented via analyzing microblog features and the deficiencies of existing microblog recommendation algorithm. Firstly, our method takes advantage of the probability correlation between tags to construct the tag similarity matrix. Then the weight of the tag for each user is enhanced based on the relevance weighting scheme and the user tag matrix can be constructed. The matrix is updated using the tag similarity matrix, which contains both the user interest information and the relationship between tags and tags. Experimental results show that the algorithm is effective for microblog recommendation.

Keywords: Probability correlation · Microblog recommendation · User-tag matrix · Tag weight

1 Introduction

As a symbol of Web 2.0, microblog has developed rapidly and has been extensively used in recent years. As a social network platform for information share, spread and acquisition based on user relations, it not only can expand the interpersonal circle and facilitate social contact, but also plays an important role for people to obtain the latest information and comments from others.

Currently, numerous researches on microblog recommendation are focusing on improving the representation of its content. Microblogs are in general much nosier, and sparser, therefore many traditional text representation techniques cannot be directly applied to microblogs. Some researchers consider the characteristics of microblog and try to extend text feature [1], others take advantage of external knowledge (such as Wikipedia) to expand semantic information [2,3]. Microblogs are characterized by shortness and complex in the text length, which not only contain the information of users' interests, but also include chats and emotional expression that unrelated to users' interests. This will result in the difficulty in analyzing and investigating users' interests. Tag can be taken as the keyword that depicts the personal occupation, characteristic and interests,

© Springer International Publishing AG 2016
A. Hirose et al. (Eds.): ICONIP 2016, Part II, LNCS 9948, pp. 491–499, 2016.
DOI: 10.1007/978-3-319-46672-9_55

which can reveal users' underlying interests. However, the existing researches concentrate on the content and relationship with users' information of tags. As for microblog recommendation, some researchers try to use tags to recommend microblog information for users [4–6], Nevertheless, they do not investigate the probability correlation between multi-tags, which is the main point in our paper.

In this paper, we propose a microblog recommendation method based on tag probability correlation. Firstly, we calculate the probability correlation between tags and construct the tag similarity matrix. And then the weight of the tag for each user can be enhanced based on the relevance weighting scheme and the user tag matrix can therefore be constructed. In addition, we update the user tag matrix of multi-user with the tag similarity matrix, which contains both the user interest information and the relationship between multi-tags.

The rest of this paper is organized as follows. We illustrate the probabilistic correlation of tags and tag similarity matrix construction in Sect. 2. Section 3 presents our probabilistic correlation-based weighting scheme of tag, as well as user-tag matrix and the microblog recommendation function. In Sect. 4, we discuss the effectiveness of our approach from two perspectives. Finally, we conclude this paper in Sect. 5.

2 Tag Probability Correlation and Similarity Matrix

Traditionally, tags added by microblog users are equally important and mutually independent. Besides, there are no subjective logical relationships between these tags. However, in practice, this is not the case. There is a potential relationship between these tags from the objective point of view, which makes it indicating distinct significance to users. The polysemy of tag will lead to ambiguity thus cannot describe the users' interests. However, we can get the tendency of users by calculating the probability of correlation between tags.

2.1 Tag Probability Correlation

Let the microblog user $U = \{u_1, u_2, \ldots, u_i, \ldots, u_N\}$, N is the number of user. The set of microblog posted by user u_i is $D_i = \{d_{i1}, d_{i2}, \ldots, d_{iM_i}\}$, where M_i denotes that the number of the microblog that posted by user u_i, $i = (1, 2, \ldots, N)$, and the set of microblog that posted by all users is $D = \bigcup_{i=1}^{N} D_i$. The tag set of microblog user u_i is $T_i = \{t_{i1}, t_{i2}, \ldots, t_{in_i}\}$, n_i denotes that the number of the tag added by user u_i, the tag set $T = \bigcup_{i=1}^{N} T_i$, and the total number of tag in tag set T is K.

From the entire microblog corpus' point of view, if the two tags are always added by users, then there is a correlation between these two tags. From another point of view, if a tag appears, the other tag has a large probability to appear, then these two tags are considered to have a strong co-occur relationship. This relationship is defined as the Eq. (1):

$$p(t_i|t_j) = \frac{p(t_it_j)}{p(t_j)} \tag{1}$$

where $p(t_j)$ denotes the probability that tag t_i appear, and $p(t_it_j)$ denotes the probability that tag t_i and tag t_j appear together, which can be estimated as $p(t_it_j) \approx uf(t_it_j)/N$, $uf(t_it_j)$ denotes the number of tag t_i and tag t_j added by the same user. In other words, $p(t_it_j)$ is a ratio that the number of tag t_i and tag t_j are added by the same user divided by the total number of user in microblog corpus. Therefore, the conditional probability of tag t_i and tag t_j can also be described as Eq. (2):

$$p(t_i|t_j) = \frac{uf(t_it_j)}{uf(t_j)} \tag{2}$$

where $uf(t_j)$ denotes the number of user with tag t_j in microblog corpus.

The conditional probability between tags t_i and t_j is asymmetric, i.e., $p(t_i|t_j) \neq p(t_j|t_i)$. While the similarity between tags is often regarded as a symmetric relationship, we define the correlation of tags in a symmetric way as:

$$cor(t_i, t_j) = p(t_i|t_j) \times p(t_j|t_i) \tag{3}$$

2.2 Tag Similarity Matrix

Conventionally, microblog users are represented via the well-known Vector Space Model, which models a collection of users by a tag-user matrix, with each entry w_{ij} indicating the importance of the user u_i in j-th tag. In other words, tag can be treated as a user vector $[u_{i,1}, u_{i,2}, \ldots, u_{i,N}]$. In order to measure the correlation between two tags, a common approach is to compute the similarity of the user vector. However, user vector representation also has the extremely sparse problem. Therefore, we use tag correlation vector to characterize the tag rather than the traditional user vector.

According to the probability correlation between tags, tag t_i can be represented as tag correlation vector in the microblog corpus. Specifically, tag t_i is considered as a tag correlation vector $[t_{i,1}, t_{i,2}, \ldots, t_{i,n}, \ldots, t_{i,K}]$, $t_{i,n}$ is the probability correlation between tag t_i and tag t_n, defined as shown in Eq. (4):

$$[cor(t_i, t_1), cor(t_i, t_2), \ldots, cor(t_i, t_n), \ldots, cor(t_i, t_K)] \tag{4}$$

In order to construct the tag similarity matrix, we used the classical method of cosine similarity to measure the similarity between tags. It is defined as:

$$sim(t_1, t_2) = \frac{t_1 \cdot t_2}{\|t_1\| \cdot \|t_2\|} = \frac{\sum\limits_{i=1}^{K} cor(t_1, t_i)cor(t_2, t_i)}{\sqrt{\sum\limits_{i=1}^{K} cor(t_1, t_i)^2}\sqrt{\sum\limits_{i=1}^{K} cor(t_2, t_i)^2}} \tag{5}$$

The tag similarity matrix S is constructed based on the tag correlation vector and S_{ij} denotes that the cosine similarity of tag t_i and tag t_j, the formula is as:

$$S_{ij} = \begin{cases} 1, \ i = j \\ sim(t_i, t_j), \ i \neq j \end{cases} \tag{6}$$

3 Microblog User Interest Representation and Recommendation Algorithm

3.1 Microblog User Interest Representation

The tags marked by user can often be viewed as important description of its own characteristics and interests. The more similar the users' tag, the more similar the microblog content, and vice versa. This suggests that tag can reflect the content of microblog to a certain extent. We construct a user tag matrix with the tag correlation weights, and then get the tag weight distribution of each user.

User Tag Weight. The tags marked by users show the morphology of power-law distribution in distribution [7], i.e. only a few tags will appear frequently while a large number of tags appear very few times. Here, we propose a new weighting scheme of tags with probability correlation between tags, namely correlation weight inspired from literature [4]. Since we use the conditional probability as correlations between tags, tags with more and higher correlations to the others are more likely to represent the user's interests and can be treated as an important local feature.

The tag correlation weighted scheme is based on degree that correlation between tags of the same user. If a tag with higher co-occurrence relations to other tag for the same user, it implies that this tag is more important for the user than other tags. The tag correlation weight can be defined as Eq. (7):

$$cow(u_k, t_i) = \frac{\sum_{t_j \in u_k} cor(t_i, t_j)}{|u_k|} \tag{7}$$

where $|u_k|$ denotes that the number of the tag of user u_k. Tag t_i and tag t_j are marked by user u_k. $cor(t_i, t_j)$ denotes the probability correlation between tag t_i and tag t_j.

In order to represent the tag weight more accurately, we need to be further consider the representative of the tag for user in the entire microblog corpus. Besides, we take the logarithmic ratio of the total number of users and the number of user marked by u_i, namely the inverse user frequency IUF, defined as shown in Eq. (8):

$$iuf(t_i) = \log_2(\frac{N}{uf(t_i)} + 1) \tag{8}$$

where $uf(t_i)$ denotes that the number of users with tag t_i. Altogether if a tag is marked by the user and this tag has been marked rarely by other users, this tag is more important for user. So the weight of tag t_i in user u_k is defined as:

$$w_{ki} = cow(u_k, t_i) \times iuf(t_i) \tag{9}$$

User Tag Matrix. A tag weight vector $\overrightarrow{V_i} = (w_{i1}, w_{i2}, \ldots, w_{iK})$ is constructed to store the weight of the tag. Based on the above users' weight vector, we create a $N \times K$ matrix M, which is defined as:

$$M = \begin{bmatrix} \overrightarrow{V_1} \\ \overrightarrow{V_2} \\ \cdots \\ \overrightarrow{V_N} \end{bmatrix} = \begin{bmatrix} w_{11} & w_{12} & \cdots & w_{1K} \\ w_{21} & w_{22} & \cdots & w_{2K} \\ \vdots & \vdots & \vdots & \vdots \\ w_{N1} & w_{N2} & \cdots & w_{NK} \end{bmatrix} \tag{10}$$

where N denotes the number of microblog users, K is the number of tag in microblog corpus. Consequently, the matrix is a $N \times K$ user-tag matrix and w_{ij} is the weight of the j-th tag for the i-th user in matrix.

We then take advantage of the probability correlation between tags to construct the tag similarity matrix. The matrix is updated with the tag similarity matrix, which not only can solve the problem of high dimension and sparse of the original matrix, but also own more abundant semantic information and can better express the user's interest:

$$M' = M \times S \tag{11}$$

where M denotes the initial user-tag matrix of the microblog user, S denotes tag similarity matrix, M' is $N \times K$ the updated user-tag matrix that. In order to better understand that matrix is no longer sparse, we can decompose matrix M' into two matrices.

Because each tag is certainly similar to itself, so the diagonal entry of M' must be greater than zero. In addition, we assume that there are at least a pair of tags are similar in the corpus, which can guarantee that the right result in the above formula is a non-zero matrix. Since all the tags appear in the microblog corpus do exist probability correlation, so all elements in user vector are non-zero. Therefore, the space vector of each user will become a less sparse vector that has non-zero entries after the mapping.

3.2   The Microblog Recommendation Algorithm

Given a microblog d_p, the ranking function $f(u_i, d_p)$ for user u_i, is defined as [8]:

$$f(u_i, d_p) = \overrightarrow{E_p} \times \overrightarrow{V_i'} \tag{12}$$

which denotes the similarity between microblog d_p and user u_i. Each microblog d_p is represented as vector $\overrightarrow{E_p} = (w_{p1}, w_{p2}, \ldots, w_{pn})$, if d_p contains tag t_j then $w_{pj} = 1, w_{pj} = 0$ otherwise. The vector $\overrightarrow{V_i'} = (w_{i1}, w_{i2}, \ldots, w_{in})$ is the updated tag weight vector for user u_i. The ranking function f is then used to measure the relevance of microblog. We predefine a threshold r_i, if $f(u_i, d_p) > r_i$, then the microblog d_p will be recommended to the user u_i.

4   Experimental Evaluation

In this section, we present experimental results on two tasks. First, we test the proposed tag weighing strategy and compare the performances with other methods. Second, we evaluate the influence of the construction method of the user similarity matrix for the recommendation algorithm.

4.1   Experiment Settings

Dataset. Experiments are conducted on the dataset collected from Sina Microblog platform. The database contains 621 users with a large number of microblog from October 16th to October 20th, 2015. We preprocessed these microblogs dataset via removing the noise information at first, which contains the forwarding signs, emoticons, links, segmentation and so on. The experimental data set is divided into 16 categories, including 36878 training samples and 10700 test samples, as shown in Table 1:

Table 1. Experimental data set statistics

Category	#Train	#Test	Category	#Train	#Test
Military	2285	800	Sports	3967	1100
Parenting	1976	500	Technology	2983	900
Environment	4315	1200	Education	2369	800
Health	2951	900	Stocks	1863	500
Travel	4562	1200	Emotion	2451	800
Medicine	3577	1000	Entertainment	3579	1000

Effectiveness Metrics. In order to evaluate the quality of the microblog recommendation algorithm, we adopt the *F-measure* as the criteria to evaluate the effectiveness of our method, which is defined as:

$$precision = \frac{N_{rs}}{N_s}, recall = \frac{N_{rs}}{N_r}, F\text{-}measure = \frac{2 precision \cdot recall}{precision + recall}$$

where N_{rs} denotes the number of relevant recommended microblogs, N_s denotes that the number of recommended microblogs and N_r denotes that the number of relevant microblog in the entire microblog corpus.

4.2   Results and Analysis

Comparison with that of 2 Other Algorithms on Tag Weighing. This experiment is designed to examine the performance of our tag weighing method

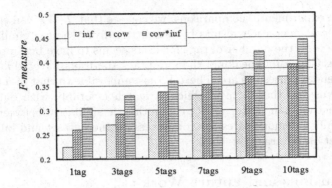

Fig. 1. F-meaure of different tag weighting strategies for recommendation algorithm

$cow * iuf$ and compare it with two baseline methods iuf and cow. Our method considers both the local and global feature for weighting while the baseline methods only capture one feature. For each scheme, we select users with $\{1, 3, 5, 7, 9, 10\}$ tags for comparison. The recommendation accuracies are reported in Fig. 1. For comparison, the horizontal line plotted in the figure indicates the *F-measure* of 0.3785 achieved by Maximum Entropy classifier as baseline [9].

From the results, we make the following observations. First, user with more tags (from 1 tag to 7 tags) generally lead to better recommendation result. The improvement becomes very minor when users with more than 7 tags. Besides, among all of the above strategies, $cow * iuf$ exhibits it best performance when users are with 3 tag. Overall, $cow * iuf$ achieves the highest performance on the dataset. This observation verifies the effectiveness of our method.

Comparison of Tag Similarity Matrix Construction. As previously mentioned, there are two strategies for tag similarity matrix construction. The first one is User Vector Similarity (UVS for short), which take the tags as a User Vector $[u_{i,1}, u_{i,2}, \ldots, u_{i,N}]$, while our method represents the tag as a Tag Correlation Vector ($TCVS$ for short) $t_i = [cor(t_i, t_1), cor(t_i, t_2), \ldots, cor(t_i, t_n), \ldots, cor(t_i, t_K)]$ for the construction of tag similarity matrix. We Choose 6 group of users with $\{1, 3, 5, 7, 9, 10\}$ tags and update M' according to formula (12). All the experimental are obtained by averaging 20 runs. The influence of the tag similarity matrix construction on the recommendation algorithm is shown in Table 2.

Table 2. The F-measure of the construction method of similarity matrix on the recommendation algorithm

Method	Tag-1	Tag-3	Tag-5	Tag-7	Tag-9	Tag-10
UVS	0.282	0.302	0.322	0.331	0.342	0.358
TCVS	0.308	0.331	0.361	0.384	0.419	0.445

From the experimental comparisons, we can see that $TCVS$ can enhance the microblog recommendation results by benefitting from the tag similarity computation. Besides, the number of tags for users seems to have big impact on the performance. Generally speaking, the more tags the users have add, the better the recommendation results are. The UVS is comparable to that of $TCVS$ when the users owns less than 5 tags, which is match favorably with our hypotheses as our method can better reflect the authentic similarity between users. In summary, $TCVS$ demonstrates the robustness on the dataset and improves the recommendation accuracy.

5   Conclusions and Future Work

Given the short nature of the microblog posts and the tags marked by user can be viewed as important description of its own characteristics and interests, in this paper, we explore the performance of a probabilistic tag correlation based approach for microblog recommendation. Concretely, the tag similarity matrix is adopted to update the user-tag matrix. This matrix contains both the user interest information and the relationship between tags and tags, which is considered as the basis for ranking function. Nevertheless, microblogs offer several other information that we have not yet discussed or explored. Future work aims at using content of the posts to find some "implicit" tags.

Acknowledgments. This work is supported by the National Natural Science Foundation of China (No. 61363058, 61165002), Youth Science and technology support program of Gansu Province(145RJZA232, 145RJYA259).

References

1. Huang, X., Chen, H., Liu, Y., Xiong, L.: A novel algorithm for feature selectionon micro-blog short text. Comput. Eng. Sci. **37**(9), 1761–1767 (2015)
2. Sun, A.: Short text classification using very few words. In: The 35th International ACM SIGIR Conference on Research and Development in Information Retrieval (SIGIR 2012), Portland, pp. 1145–1146 (2012)
3. Liu, W., Quan, X., Feng, M., Qiu, B.: A short text modeling method combining semantic and statistical information. Inf. Sci. **180**(20), 4031–4041 (2010)
4. Xing, Q., Liu, L., Liu, Y., Zhang, M., Ma, S.: Study on user tags in Weibo. J. Softw. **26**(7), 1626–1637 (2015)
5. Yamaguchi, Y., Amagasa, T., Kitagawa, H.: Tag-based user topic discovery using Twitter lists. In: International Conference on Advances in Social Networks Analysis and Mining (ASONAM 2011), Kaohsiung, pp. 13–20 (2011)
6. Ma, H., Jia, M., Xie, M., Lin, X.: A microblog recommendation algorithm based on multi-tag correlation. In: Zhang, S., Wirsing, M., Zhang, Z. (eds.) KSEM 2015. LNCS, vol. 9403, pp. 483–488. Springer, Heidelberg (2015). doi:10.1007/978-3-319-25159-2_43

7. Song, S., Zhu, H., Chen, L.: Probabilistic correlation-based similarity measure on text records. Inf. Sci. **289**(1), 8–24 (2014)
8. Zhou, X., Wu, S., Chen, C., Chen, G., Ying, S.: Real-time recommendation for microblogs. Inf. Sci. **279**(279), 301–325 (2014)
9. Phan, X., Nguyen, L., Horiguchi, S.: Learning to classify short and sparse text and web with hidden topics from large-scale data collections. In: The 17th International Conference on World Wide Web (WWW2008), Beijing, pp. 91–100 (2008)

A New Model and Heuristic for Infection Minimization by Cutting Relationships

Rafael de Santiago[1,2]([✉]), Wellington Zunino[1], Fernando Concatto[1], and Luís C. Lamb[2]

[1] Universidade do Vale do Itajaí, Itajaí, Brazil
rsantiago@univali.br
[2] Federal University of Rio Grande do Sul, Porto Alegre, Brazil

Abstract. Models of infection spreading have been used and applied to economic, health, and social contexts. Seeing them as an optimization problem, the spreading can be maximized or minimized. This paper presents a novel optimization problem for infection spreading control applied to networks. It uses as a parameter the number of relations (edges) that must be cut, and the optimal solution is the set of edges that must be cut to ensure the minimal infection over time. The problem uses the states of SEIS nodes, which is based on the SEIR and SIS models. We refer to the problem as Min-SEIS-Cluster. The model also considers that the infections occurred over different probabilities in different clusters of individuals (nodes). We also report a heuristic to solve Min-SEIS-Cluster. The analysis of the obtained results allows one to observe that there exists a positive correlation between the proportion of removed edges and relative increase of mitigation effectiveness.

Keywords: Heuristic · Contagion mitigation · Contagion spreading

1 Introduction

The first mathematical model of contagious of diseases dates from 1760 [8]. The use of a time window in the epidemic analysis was probably first used in 1927 [8]. These models have led to the main known mathematical representation of contagious spreading and the understanding of the dynamics of an epidemic. Models of spread of contagious infection are not only used to understand disease behavior, but also to identify other kinds of contagion such as culture, information, and behavior [8,9]. For example, a model could be applied to maximize the spread of an idea, like a preference, or to understand the mitigation of a social behavior [9]. These mathematical models of contagious spreading can be transformed in search problems to maximize or minimize the contagion.

Seeing the contagious mitigation as an optimization problem, it can be seen as the reduction of a total number of infected individuals. The meaning of such effect could be read as a reduction of affected (ill) people and the money spent to control the epidemic. During political events, a model could be used to minimize problems in an unstable region.

© Springer International Publishing AG 2016
A. Hirose et al. (Eds.): ICONIP 2016, Part II, LNCS 9948, pp. 500–508, 2016.
DOI: 10.1007/978-3-319-46672-9_56

This paper presents a novel optimization model to minimize the contagious spread in a static network. The suggested model considers that an individual could be in one of the following states: (*i*) Susceptible, when the individual is not infected; (*ii*) Exposed, when the individual is contaminated, but is not infectious; (*iii*) Infected, when an infected individual is contaminated and infectious. After the duration of the Infected state, the individual becomes Susceptible again. This model is based on the Susceptible-Infected-Susceptible (SIS) and Susceptible-Exposed-Infected-Recovered (SEIR) definition, and it uses a multilayer feature in which clusters are assigned to individuals to represent the dynamics of contagious inter and intra-groups. The motivation for this multilayer concept is that some clusters are more susceptible to the infection if one of its internal individuals is infected. We call this optimization model "Min-SEIS-Cluster".

This novel model aims to minimize the number of infected individuals for a period. The required inputs for the problem are the static network, the clusters and their internal contagious dynamics, the outside cluster infection dynamics, and the number k of relations that the problem is limited to cut to assure the minimal spreading.

This paper is divided into five more sections. The second section describes the background and related works. The third section describes the optimization problem proposed in this paper. After that, a heuristic and its design properties are presented. The fifth section reports experiments and results of experimental analysis of the heuristic. The last section summarizes contributions and directions for further investigations.

2   Related Work

There are several models of contagious spreading. Some of them are classic in literature such as SIR, SIS, SIRS and SEIR [3,8]. Susceptible-Infected-Recovered (SIR) is a model where each individual is susceptible to the infection, and when the individuals are infected, they try to infect adjacent individuals. After a period, the infected individual is removed. The Susceptible-Infected-Susceptible (SIS) model is a very similar to the SIR model, but the infected individual becomes Susceptible after the period of infection. Susceptible-Infected-Recovered-Susceptible (SIRS) is like SIR, but after being removed, the individual is Susceptible again. This model is useful to simulate birth and low-high immunity. In the Susceptible-Exposed-Infected-Recovered (SEIR) model, the individual infected has a period of exposition to the disease or idea, defined as the Exposed state. While Exposed, the individual does not affect the others. After the exposition period, the individual becomes infected and tries to infect its adjacencies. After the infection, the individual is removed or immunized.

Static models do not have differences in their network structures in any time window. They can present changes on [8]: (*i*) degree, in which the individuals have the probability of being linked with a fixed number other individuals; (*ii*) clusters, where individuals are divided into groups or communities; (*iii*) weights, where the strength of each relation is measured by a specific value;

and (iv) direction, in which the infection happens only in defined orientations, from one specific source individual to the target individual.

Goyal and Kearns [6] presented a contagious tournament model between two players using game theory concepts. Each player tries to infect the maximal number of possible individuals. In [7], the tournament happens with a designer and an adversary, in which the game consists of a competition between access to the resources provided by the designer and attacks assigned by the attacker.

A minimization of contagious spreading in a network tries to reduce the number of infected individuals. Tsai et al. [10] report on heuristics for the game known as Influence Blocking. Given a graph $G = (V, E)$, V is a set of leaders and E is the set of edges between pairs of leaders. Each leader has a probability for influencing the other leaders to which it is connected. The players play this game as an attacker or a defender. The attacker tries to maximize the number of leaders supporting a "cause" that is the infectious idea. The defender tries to mitigate the attacker's influence. The game starts with some leaders supporting the "cause". The tested heuristics are Double Oracle, LSMI Oracle, and PageRank Oracle. As results, the authors reported that the defender's effectiveness is not scalable for large networks.

Gil et al. [5] developed a computational model of minimization of a contagious spreading. This model was constructed to detect clearer metrics for security in computer networks. The model is inspired on the probability of an individual developing a disease or genetic mutation, assigning the computers and services to potential threats.

3   Min-SEIS-Cluster Problem

In this section, we present the new Min-SEIS-Cluster problem. This problem uses the infection model SEIS (Susceptible-Exposed-Infected-Susceptible) that is based on the states of SEIR, but instead of removing the Infected individuals, the individuals become Susceptible again. As the problem was created to mitigate influences, an individual is susceptible to an idea. If he is in contact with another infected individual, he can become exposed to the idea. On the Exposed state, the individual does not yet spread the idea. After that, the individual is considered Infected and starts spreading the idea for the duration of the infection. Then, the individual no longer spreads the idea and is considered Susceptible again.

Min-SEIS-Cluster considers that each individual belongs to one cluster. These clusters are used to represent the dynamics of infection and exposition for each social group within the network. In groups of friends, the infection could be spread more often than in groups composed of co-workers. There are several other models for cluster detection in the literature, e.g. [4,11].

The model also utilizes a parameter k to define the limit of relationships that must be cut off to lead to the minimal infection over the execution time. So, search methods can find the set of k edges that minimize the infection when they are cut off.

The problem Min-SEIS-Cluster is defined as a tuple $(G, C, T, \delta, \chi, \phi, k, \epsilon, \lambda)$, where $G = (V, E)$ is the network composed of a set of individuals V and the set of edges between them E; $C = \{\{v\}|v \in V\}$ is the cluster set, composed of subsets of V; the total time of spreading T; $\chi = \{P_i|i \in |C|\}$ is the set of probability functions of an individual infecting an adjacent when both are inside the same cluster i, where $P_i : t \to [0,1]$ for each $P_i \in \chi$, where $t \in [T]$; $\phi : t \to [0,1]$ is the probability function of an individual infecting an adjacent when they are not within the same cluster, where $t \in [T]$; k is the maximal number of edges that could be cut from G; the exposition time function $\epsilon : t \to \mathbb{Z}^+$ calculates the Exposition state duration for each period, where $t \in [T]$; the infection time function $\lambda : t \to \mathbb{Z}^+$ calculates the duration of the infection, where $t \in [T]$.

The objective function of a method that must find the best k edges to be cut off can be found in function 1a below, where a_{vt} is 1 if the node v is Infected at the period of time t, otherwise, a_{vt} is 0. Ignoring clusters and the spreading dynamics, the constraint 1b restricts the value of the amount of edges that must be cut to find the best solution. The variable $e_{\{u,v\}}$ is assigned to each edge of the input graph G. If $e_{\{u,v\}} = 0$, the edges $\{u,v\}$ is cut, otherwise $\{u,v\}$ is not cut. A search method for Min-SEIS-Cluster must find the set of k edges that must be cut to assure the minimal infection.

$$\min \sum_{t\in[T]} \sum_{v\in V} a_{vt} \tag{1a}$$

$$\text{subject to:} \sum_{\{u,v\}\in E} e_{\{u,v\}} = |E| - k \tag{1b}$$

$$a_{vt} \in \{0,1\}, \forall t \in [T], v \in V$$

$$e_{\{u,v\}} \in \{0,1\}, \forall u, v \in E.$$

4   A Novel Heuristic Model

The method used to develop heuristic is based on the Monte Carlo concept, massive sampling based results at edges are removed randomly. The Algorithm 1 shows the Monte Carlo heuristic used to solve the Min-SEIS-Cluster problem. The variables *bestValue* and *bestSolution* store the best solution found during the search. At the start, the *bestValue* is zero and the *bestSolution* is defined as an empty set. In line 3, a loop starts repeating *attempts* times. Each attempt is a new random selected set of edges that are removed from the original graph G. The set *edgesRemoved* is composed of the edges removed from G, so at each iteration of this loop, a new solution is tested. Inside this loop, simulations are performed *replications* times over the solution. Before the start of each simulation, the sets *susceptibles*, *exposeds*, *infecteds* are defined. Then for each time $t \in [T]$, the method *simulateInfection* is executed to simulate the spreading. At the end of the simulation of an entire solution, the value *worstValue* is updated if it the largest value of a summation of the number of Infected individuals for all $t \in [T]$ is found. At the end of all replications assigned to a solution, the *worstValue* is compared to the best-found *worstValue* of a solution tested in the past (*bestValue*). So, this is considered as the best solution found (the one which has presented the worst case better than the others) and presentes less Infected people for all $t \in [T]$.

Algorithm 1. Min-SEIS-Cluster heuristic method

Input   : $G(V, E)$, *replications, attempts, T, k, initialInfected, clusters,* χ, ϕ, ϵ, λ

1 *bestValue* $\leftarrow \infty$
2 *bestSolution* $\leftarrow \{\}$
3 **foreach** *attempt* $\in$ *[attempts]* **do**
4   *edgesRemoved* $\leftarrow$ *removeRandomEdges*(G, k)
5       $E' \leftarrow E \backslash edgesRemoved$
6   *worstValue* $\leftarrow -\infty$
7   *worstSolution* $\leftarrow \{\}$
8   **foreach** *rep* $\in$ *[replications]* **do**
9   *infecteds* $\leftarrow$ *initialInfected*
10   *susceptibles* $\leftarrow V \backslash initialInfected$
11   *exposeds* $\leftarrow \{\}$
12   *value* $\leftarrow 0$
13   **foreach** $t \in [T]$ **do**
14   *simulateInfection*$(G(V, E'), \chi, \phi, \epsilon, \lambda, clusters, infecteds,$
                susceptibles, exposeds)
15   *value* $\leftarrow$ *value* $+ |infecteds|$
16   **if** *worstValue* $<$ *value* **then**
17   *worstValue* $\leftarrow$ *value*

18   **if** *bestValue* $>$ *worstValue* **then**
19   *bestValue* $\leftarrow$ *worstValue*
20   *bestSolution* $\leftarrow$ *edgesRemoved*

21 **return** $\{bestSolution, bestValue\}$

The *simulateInfection* method is explained in Algorithm 2. It is divided into two phases. The first phase iterates over all Infected individuals. At each iteration of the loop that starts in line 1, the infected individual is checked if it remains infected (at line 2). If it does not, the infected individual becomes Susceptible. If the individual remains Infected, it tries to infect every adjacent Susceptible node. If the targeted neighbor is in the same cluster of the Infected individual, the probability that it becomes Exposed is defined by the function χ, otherwise by ϕ. If the target did become Exposed, the exposition time is calculated by the function ϵ and stored. In the second phase, the Exposed individuals are checked if they become Infected. If they do, the duration of this new status is calculated by the function λ.

5   Experiments and Results

This section describes the experiments run, the results obtained with the use of the Min-SEIS-Cluster model (Sect. 3) and the presented heuristic (Sect. 4).

Table 1 shows the instances used in experiments. The column "Graph" shows the instance name [1], the column "#Clusters" shows the number of clusters found in the instance, and "#Nodes" and "#Edges" show the amount of nodes

---

Algorithm 2. Simulate infection

Input   : G, χ, ϕ, ϵ, λ, *clusters*, *infecteds*, *susceptibles*, *exposeds*, t

1 **foreach** *infected* $\in$ *infecteds* **do**
2   **if** *infected.time* $\leq t$ **then**
3   *infecteds* $\leftarrow$ *infecteds*\\{*infected*}
4   *susceptibles* $\leftarrow$ *susceptibles* $\cup$ {*infected*}
5   **else**
6   **foreach** *target* $\in$ *susceptibles* $\cap$ N(*infected*) **do**
7   *chance* $\leftarrow 0$
8   **if** $C_{target} = C_{infected}$ **then**
9   *chance* $\leftarrow \chi_{C_{target}}(t)$
10   **else**
11   *chance* $\leftarrow \phi(t)$
12   **if** *random*() $<$ *chance* **then**
13   *target.time* $= t + \epsilon(t)$
14   *susceptibles* $\leftarrow$ *susceptibles*\\{*target*}
15   *exposeds* $\leftarrow$ *exposeds* $\cup$ {*target*}

16 **foreach** *exposed* $\in$ *exposeds* **do**
17   **if** *exposed.time* $\leq t$ **then**
18   *exposed.time* $\leftarrow t + \lambda(t)$
19   *exposeds* $\leftarrow$ *exposeds*\\{*exposed*}
20   *infecteds* $\leftarrow$ *infecteds* $\cup$ {*exposed*}

---

and edges of the instance, respectively. The clusters used in the tests were obtained by the Louvain method heuristic for Modularity Maximization community detection problem [2].

Table 1. Instances used in the experiments

Graph	#Clusters	#Nodes	#Edges
Karate	2	34	78
Dolphins	7	62	159
Celegans	6	297	3592
Email	43	1113	5451

To depict our results, we developed a visual representation of simulations by using the best-found solutions. This visual representation has all the initial information of the infection, and also the changes that have occurred in the graph each time unit when specific k edges are removed. In that representation, red nodes are individuals Infected, yellow nodes are Exposed and green nodes

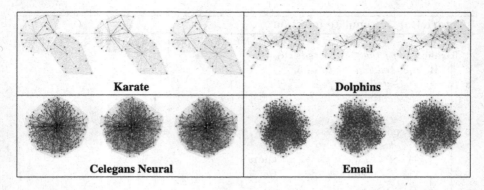

Fig. 1. First test results, where it was used the parameters $T = 100$, *attempts* = 20, *replications* = 20, 5 nodes infected initially, and $k = 0.1|E|$

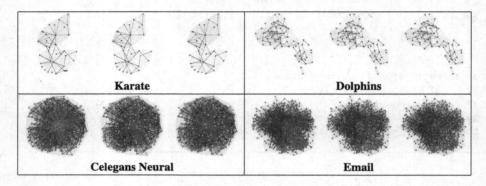

Fig. 2. Second test results, where it was used the parameters $T = 100$, *attempts* = 20, *replications* = 20, 5 nodes infected initially, and $k = 0.5|E|$

represent Susceptible individuals. The shaded areas in the images represent the clusters of the graph, each with its randomly defined color.

The tests can be divided into two parts, and the results can be seen in Figs. 1 and 2. For each test, it was specified an infection model. They have a success rate of infection from 1 % to 20 % for nodes that belong to the same cluster, and 5 % for nodes that belong to different clusters. The first test (Fig. 1) was performed with 100 time units (parameter T), 20 attempts, 20 replications, 5 nodes infected initially, and 10 % of edges removed (parameter k). The second test (Fig. 2) used almost the same parameters of the first test, except by k that allows removing 50 % of the edges.

Based in the experiments presented above, one can notice a default behavior of infections initially due to higher rates of intra-cluster infections. The infection can settle better in the area where are initially infected nodes, and then it expands rapidly to other regions of the graph. It may also be noted that the success of infection is not decreased even with a small number of nodes infected initially.

Considering the first test, we can note a large spread of infection, and it happens because of a large amount of visited adjacent nodes, causing all nodes try to infect all adjacent nodes. Removing only 10 % of edges resulted in a not very effective mitigation. The second test generated different results than the first. The spreading was less accelerated, because the k parameter was increased.

6 Conclusions

The work reported in this paper aimed to develop an optimization problem to reduce the spreading of an infection by removing nodes. Thus, the main results of this paper are the Min-SEIS-Cluster problem (Sect. 3), and a heuristic that solves it effectively (Sect. 4).

The tests were divided into two, and they are graphically shown in Figs. 1 and 2. The results show that there is a positive correlation between the proportion of the number of removed edges and the effectivity of spreading control. The model has flexibility in its application and can be used in public health, economics, social behavior, and problems that can be adapted to the required inputs. As future work, we could consider the application of heuristics in real models and extensions of the model to applications where spreading is highly relevant or encouraged.

Acknowledgments. This work is partly supported by the Brazilian Research Council CNPq, the *Universidade do Vale do Itajaí*, and government of the State of Santa Catarina (Brazil).

References

1. Batagelj, V., Mrvar, A.: Pajek datasets (2006). http://vlado.fmf.uni-lj.si/pub/networks/data/
2. Blondel, V.D., Guillaume, J.L., Lambiotte, R., Lefebvre, E.: Fast unfolding of communities in large networks. J. Stat. Mech.: Theory Exp. **2008**(10), 10008 (2008)
3. Easley, D., Kleinberg, J.: Networks, Crowds, and Markets. Cambridge University Press, Cambridge (2010)
4. Fortunato, S., Castellano, C.: Community structure in graphs. In: Meyers, R.A. (ed.) Computational Complexity, pp. 490–512. Springer, New York (2012)
5. Gil, S., Kott, A., Barabási, A.L.: A genetic epidemiology approach to cybersecurity. Sci. Rep. **4**, 5659 (2014)
6. Goyal, S., Kearns, M.: Competitive contagion in networks. In: Proceedings of the 44th Symposium on Theory of Computing - STOC 2012. p. 759. ACM, New York (2012)
7. Goyal, S., Vigier, A., Jong, M.D., Elliot, M., Galeotti, A., Gallo, E., Gagnan, J., Goenka, A., Hoyer, B., Jackson, M., Kovenock, D., Levy, G., Meyer, M., Nava, F., Pancs, R., Prummer, A., Razin, R., Reich, B., Rutsaert, P.: Attack, defense and contagion in networks. Rev. Econ. Stud. (2014)

8. Pastor-Satorras, R., Castellano, C., Van Mieghem, P., Vespignani, A.: Epidemic processes in complex networks. Rev. Mod. Phys. **87**(3), 925–979 (2015)
9. Pionitti, P.Y.A., Gomes, M.F.D.C., Samay, N., Perra, N., Vespignani, A.: The infection tree of global epidemics. Netw. Sci. **2**(01), 132–137 (2014)
10. Tsai, J., Weller, N., Tambe, M.: Analysis of heuristic techniques for controlling contagion. In: AAAI Fall Symposium: Social Networks and Social Contagion, pp. 69–75 (2012)
11. Xie, J., Kelley, S., Szymanski, B.K.: Overlapping community detection in networks. ACM Comput. Surv. **45**(4), 1–35 (2013)

Sentiment and Behavior Analysis of One Controversial American Individual on Twitter

J. Eliakin M. de Oliveira[1], Moshe Cotacallapa[1], Wilson Seron[2],
Rafael D.C. dos Santos[1], and Marcos G. Quiles[2(✉)]

[1] Instituto Nacional de Pesquisas Espaciais (INPE), São José dos Campos, SP, Brazil
{joao.eliakin,frank.moshe,rafael.santos}@inpe.br
[2] Federal University of São Paulo (UNIFESP), São José dos Campos, Brazil
{wilsonseron,quiles}@unifesp.br

Abstract. Social media is a convenient tool for expressing ideas and a powerful means for opinion formation. In this paper, we apply sentiment analysis and machine learning techniques to study a controversial American individual on Twitter., aiming to grasp temporal patterns of opinion changes and the geographical distribution of sentiments (positive, neutral or negative), in the American territory. Specifically, we choose the American TV presenter and candidate for the Republican party nomination, Donald J. Trump. The results acquired aim to elucidate some interesting points about the data, such as: what is the distribution of users considering a match between their sentiment and their relevance? Which clusters can we get from the temporal data of each state? How is the distribution of sentiments, before and after, the first two Republican party debates?

1 Introduction

Social media have hardened a new communication paradigm, in such a way, that comments happen to be valued even for influencing offline events. One example is the case of the Greek athlete Voula Papachristou, expelled from the 2012 Olympic Games (BBC News[1]) after a racist *joke* published on Twitter.

Sentiment analysis is a powerful tool to analyse and interpret online textual data, i.e., an analysis of the attitudes, emotions or opinions, that could be extracted from comments about a particular subject and their classification on, at least, three categories: positive, neutral or negative [1–3]). Moreover, since sentiments are sensitive to different events, it certainly could change over time, so it is important to track these changes and formulate their causes.

In this paper, we concentrate attentions on Twitter[2], applying, for the American territory, a state by state study on the behavioral patterns of sentiments about Donald J. Trump, an American businessman and TV presenter, currently running for the Republican party nomination, in view of the next presidential election, in November 2016.

---

[1] http://www.bbc.com/news/world-europe-18987678.
[2] https://twitter.com.

© Springer International Publishing AG 2016
A. Hirose et al. (Eds.): ICONIP 2016, Part II, LNCS 9948, pp. 509–518, 2016.
DOI: 10.1007/978-3-319-46672-9_57

Donald Trump has been acknowledged, since his formal announcement as a presidential candidate in June 16, 2015, for his controversial remarks on sensitive subjects, such as immigration policies. At first, these observations were considered by the media as sort of self-promotion. However, since his intention of being a presidential candidate remounts from the 1988's electoral process (USA Today), we kept track of comments related to him on Twitter (tweets), expecting for new explosive assertions or something unusual to happen. After almost two months of data acquisition, much has happened, not only his remarks became gradually sharper, but this way of campaigning has provided good results, putting him first on each and every poll since August 2015.

Several studies related to sentiment analysis on politics and specific events have been published recently [4–7]; among others). Three of them are more related to this paper, Soelistio and Surendra [8] implemented a method based on digital articles and newspapers to infer sentiment over-determined Indonesian political figures. Wang et al. [9] developed a system for real-time sentiment analysis on Twitter for the 2012 American presidential election and Mejova et al. [10] studied sentiment analysis on Twitter concerning the Republican party primaries for the 2012 nomination. Although in terms of context these papers could be taken as related works, our approach has its novelty on the temporal analysis of each and every action of a determined person who happens to be a presidential candidate, in other words, this paper has as objective to forge a mindset for analysing a specific candidate depicting different means of data representation and by them reach non-trivial conclusions.

This paper is structured as follows: in Sect. 2 our dataset and the sentiment analysis are discussed, in Sect. 3 the experiments and major the outcomes are described. Finally, in Sect. 4, some concluding remarks are drawn.

2 Dataset and Sentiment Analysis

The classical Naïves Bayes classifier is adopted and, as can be seen in Sect. 3, it has provided a very consistent and meaningful classification results. It is one of the simplest classifiers in the literature according to Taheri et al. [11], albeit very effective in particular situations, such as the one considered in this work.

2.1 The Dataset

The dataset used for the experiments was composed of tweets acquired from the Twitter streaming API (Application Program Interface). The API provides access to the global flow of tweets on circulation in real-time, without any constraints on the number of requests. However, the API provides up to 1 % of the entire public stream of tweets. In Morstatter et al. [12] it is shown that this limitation is not a crucial constraint if the targeted subject is very specific. In total, we have collected more than 4 millions of tweets related to Donald Trump from August 4 to September 23 of 2015. Among them, only a set of 65044 were, for certain, tweeted in the American territory (georeferenced).

Next, we describe how the tweets were preprocessed. Having a tweet at hand, it may present numerous spurious characters and strings that may difficult the work of an algorithm. Thus, what follows, aims to prepare the raw tweet into a friendly format to be used as a classifier input.

The first step is the removal of *URLs*; also the elimination of citations is performed, they are always prefixed by an @ character. After, all the remaining characters are put in lowercase. Next, we apply a sentiment-aware tokenizer. In comments the appearance of emoticons is common (e.g. :), :(, etc.). Here, emoticons are identified as a unity (token).

Finally, the tokens are separated by dashes (−). Moreover, to improve the tokenization, one might be careful with words assuming different meanings. Thus, we apply the same methodology employed in [13,14], which appends the "_NEG" suffix to every word appearing between a negation and a clause-level punctuation mark.

As an example, we have the following tweet acquired in August 20, 2015:

"@XXXX I advocate for free speech and must say, if Trump wins were no longer a free country. https://XXXX"

After the preprocessing we obtain the following tokens:

i - advocate - for - free - speech - and - must - say -   if - trump - wins - were - no - longer_NEG - a_NEG - free_NEG - country_NEG

Observe that the word free is in two different contexts, one positive, in "free speech" and the other negative in "free country". This is the most important characteristic of this approach since the classifier will learn the proper meaning of two identical words when placed in very different scenarios.

Christopher Potts [15] showed that the sentiment-aware tokenizer, combined with the negation mark approach provides substantial gains independent to the amount of data on the training set, and it is well fitted when applied to short texts like tweets, overcoming several tokenization approaches for a Naïve Bayes classifier. Having this into account, we then did not apply the removal of stop words.

2.2 Training Set

The Naïve Bayes classifier will have a high accuracy, only if the training set accurately represents the whole dataset, i.e., the training set of tweets must represent precisely the data space.

Here, instead of manually labeling a subset of samples, we perform the classification of the training set by using the special Twitter strings, named hashtags (prefixed by #). Thus, the tweets are categorized as positive or negative accordingly to their hashtags towards Donald Trump. The hashtags were captured from Twitter users openly in favor and against specific Donald Trump controversial remarks.

To be considered a good instance for the training set, a tweet is labeled as positive if it presents positive hashtags but no negative ones; the contrary is also applied.

For neutral tweets, since there was no set of hashtags that could fit for a neutral sentiment, we have chosen a group of Twitter users, which solely tweets unbiased breaking news with no clear sentiment related to Donald Trump.

Once having accomplished the creation of the training set for each class, the Naïve Bayes training is performed, the output of the algorithm will be a learned classifier γ which will be employed to classify the remaining tweets of the whole dataset.

According to Jiang et al. [16] for consistent classification using a Naïve Bayes classifier, the amount of information in the training set is important. In general, more is better, however, having meaningful tweets (criteriously chosen and properly preprocessed) at hand usually provides good results. In the experiments we deal with 65044 tweets, certainly originated in the American territory. Out of this set, 2000 samples of each class have been selected and labeled, which corresponds to nearly 9.2 % of the whole dataset. The accuracy of the classifier, by applying a 10-fold cross validation was about 91 %, which represents a quite satisfactorily accuracy for this domain.

3   Experiments and Results

3.1   Social Relevance

In Twitter, we have two attributes concerning a connection for each user i: the number of followers (number of users following i) and the number of friends (number of users followed by i). Since the occurrence of a *to follow* connection is independent of the consent of the user being followed, it is natural that a relevant user will have much more followers than friends. Thus, its tweets will likely have a more relevant impact.

Aiming to avoid users which tweet a lot but have no impact, we formulate the relevance attribute R using the following equation:

$$R^i = \frac{F_o^i - F_r^i}{F_o^i + F_r^i} \tag{1}$$

for a user i we have the number of followers F_o^i and the number of friends F_r^i. If a user has no followers neither friends, the R^i value is set to 0 (zero).

In the limit, when $F_o^i \to \infty$, $R^i \to 1$. And when $F_r^i \to \infty$, $R_i \to -1$. So the value of R indeed provides us with a consistent normalization rule where a very relevant user have $R \approx 1$ and a classical source node has very low relevance, expressed by $R \approx -1$. In the middle case, when $R \approx 0$ we have users with quite balanced rate of followers and friends.

3.2   Average Temporal Sentiment

Human personality is complex, hardly someone with free will would agree completely with someone else in every subject. When a user tweets throughout a period, there is the possibility that not all the tweets carry the same sentiment.

To avoid an inflexible analysis over a determined user, we employ in our experiments, the concept of average sentiment.

Consider, for a period Δ_t, the temporally ordered set S^i of sentiments related to each tweet made by the user i. The average temporal sentiment of i, supposing an amount of n tweets in Δ_t, can be formulated as follows:

$$\overline{S^i} = \frac{\sum_{j=1}^{n} S_j^i}{n} \tag{2}$$

in Eq. 2, the average sentiment ($\overline{S^i}$) will be the arithmetic sum of each sentiment (S_j^i, $j = 1, \ldots, n$) related to every tweet made by user i throughout Δ_t, divided by the total amount of tweets.

To represent numerically the sentiments we set them in the interval $[-1, 1]$. Particularly, -1 express deep *anti-Trump*) sentiment, 0 an undoubtedly neutral sentiment, and $\overline{S^i} = 1$, a strong *pro-Trump* sentiment.

By using both measures (Eqs. (1) and (2)), we present, in Table 1, the top-5 neutral, positive and negative users, with the highest relevance. For positive and negative users, their Twitter usernames were omitted to preserve identities, only a brief characterization, extracted from their respective Twitter profiles has been given.

An important point to highlight is that every neutral user is related to the press[3,4] and none of them were in the group created for the training set, which is a good indication that, in practice, the classifier knew how to identify this type of users.

Table 1. Top 5 positive, neutral and negative users, by relevance

Positive users	Neutral users	Negative users
Fox News reporter	chrislhayes	Environment
Dad-husband-patriot	WashingtonCP	Latino reporter
Trump Golf Club	CGasparino	Comedian
Patriot-war veteran	CBSRadioNews	S-Up Comic
Mother-Christian	mckaycoppins	I. Journalist

For the positives, apart from his feud with Fox News reporter Megyn Kelly (see Sect. 3.3), Donald Trump has been very welcomed at Fox News since his announcement as a presidential candidate, hence having a Fox News reporter as the most relevant user with a solid positive sentiment could not be considered a surprise. In the sequence, we have very conservative user profiles, in which the words family, patriotism, and war veteran are very prevalent. An interesting

---

[3] CP in WashingtonCP means City Press.
[4] chrislhayes and CGasparino are journalists; mckaycoppins is a political writer.

point to note is that the user standing for "Trump Golf Club" is, in fact, one of Trump's campaign advisers.

For the negatives, the user profiles has followed the trends of Trump's controversial remarks, ranging from an environmentalist, a Latino reporter to comedians and an independent journalist. In the next section, we discuss some of these controversial statements what will certainly elucidate why these user profiles are very pertinent.

3.3   Number of Users and Their Sentiments

Figure 1 shows the number of users tweeting, in the American territory, related to Donald Trump. It is obvious the existence of peaks and besides the two first Republican party debates, the others are deeply related to the mentioned controversial remarks. Below we give a brief explanation of the causes of some of these peaks (follow the geometric marks in Fig. 1 from left to right).

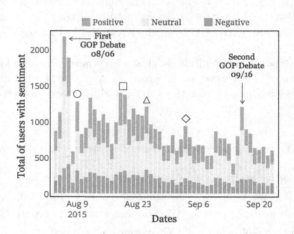

Fig. 1. The total number of users with sentiment revealed over time. The polygons over a peak denotes a relevant event which is described in this section. The colors identify the proportions of positive (green), neutral (yellow) and negative (red) tweets, daily. (Color figure online)

– ◯: The controversial remark towards the reporter Megyn Kelly, during the first debate organized by Fox News. Due to its misogynistic allusion, added to a follow up on Twitter [5], has lead to the appearing of the negative hashtag #waronwomen.
– □: In August 18, Donald Trump has assured, on Twitter[6], that, whether elected, he would immediately approve the construction of the Keystone XL

---

[5] https://twitter.com/realdonaldtrump/status/629997060830425088.
[6] https://twitter.com/realdonaldtrump/status/633739970985897984.

oil pipeline. According to environmentalists[7], this pipeline would lead to a significant adverse impact on the environment.

- △: During his first speech as a presidential candidate in June 16, Donald Trump apparently put himself against illegal immigrants and stated that Mexico usually does not send their best people to live in the United States, generalizing them as criminals[8]. The reverberation of this very controversial remark had one of its climaxes on August 26, when the Mexican reporter Jorge Ramos (Univision TV channel), was expelled from a press conference in Iowa, after trying to ask Trump about his immigration proposals.
- ◇: In September 3, Donald Trump did something not necessarily controversial but relevant to the electoral process, he signed his loyalty pledge to the Republican party[9].

3.4 Behavior Analysis

Concentrating efforts only on tweets generated in the American territory permitted us to carry out a behavioral analysis by states. In this particular study, the sentiments of the tweets are yet fundamental, but they are interpreted under a temporal series. Thus, their variation will lead us to conclusions.

In Fig. 2(a), we can see a graph showing the number of users tweeting per state. Knowing the population distribution in the US, it is not a surprise that California, Texas, New York and Florida are the states with the highest number of tweets. Thus, a straight comparison between states would lead to unreliable results. In this context, we have normalized the data per day according to Eq. (2).

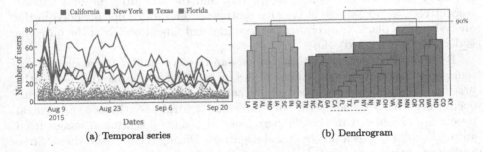

(a) Temporal series   (b) Dendrogram

Fig. 2. Behavioral clustering study. (a) The number of users tweeting per day separated by states. (b) A dendrogram showing two clusters of states with similar behaviors to Trump's remarks. The threshold is set on 10 % of likelihood, which is a very loose constraint. Thus Kentucky (KY) has certainly a very unusual behavior. It is worth noting that the most populous states, highlighted by dashes under their initials, have very similar behavior even for a tighter threshold. (Color figure online)

---

[7] http://www.nrdc.org/energy/keystone-pipeline/.
[8] via NBC News: http://nbcnews.to/1GXruPb.
[9] https://twitter.com/realdonaldtrump/status/639517204434800640.

Here, we have considered the Euclidean distance to compare the time series formed by the daily sentiment of each state. The dendrogram in Fig. 2(b) illustrates the clustering among the states according to their similarities.

Due to the lack of data, the following states were not included in this analysis: Alaska, Hawaii, Idaho, Montana, Wyoming, Utah, North Dakota, South Dakota, Kansas, New Mexico, Wisconsin, Arkansas, Mississippi, West Virginia, Delaware, Vermont, New Hampshire, Connecticut and Rhode Island.

One can see in Fig. 2 that the most populous states, i.e., California (CA), Florida (FL), Texas (TX), Illinois (IL) and New York (NY), have quite similar sentiment patterns. This is very insightful considering how the presidential electoral process works in the United States. The winner is not who earns the biggest amount of votes in the overall, but who have the majority of votes in the electoral college.

Having this into account, it is relevant to understand what is happening in the most populous states. Once they had similar reactions throughout the time, it is very probable that if Trump gives another controversial remark, the reactions to it, in these states, will be very similar.

3.5   An Analysis of the Debates Through Maps

In this subsection, we explore the distribution of sentiments, per state, around the two first Republican party debates.

The first debate, held by Fox News on August 6, has provoked the largest peak in a number of users tweeting about Donald Trump (see Fig. 1). In Figs. 3(a)–(c) it is shown maps of the United States depicting the average sentiment distribution, by state, the day before, in the day, and the day after the first debate, respectively, as percentage of positive sentiment. The states not included in the analysis are in black. Concerning the second debate, held by CNN, the outcome is depicted in Figs. 3(d)–(f).

By analyzing both sets of maps, one can observe similar variations on the sentiment distribution. On August 5 and on September 15, one day before the first and second debate, respectively, it can be seen the presence of states with overall negative sentiments (in red) and some neutral ones. Indeed, no neutral tweets are considered in these experiments; the neutral sentiment comes from an equilibrium between positives and negatives. During the debates, the positive sentiment shows prevalence. However, on the following day, the sentiments have the tendency to go negative again.

The most populous states had a very similar disposition of sentiments each day, agreeing with the dendrogram. The interesting case is a particular sentiment transition from the day of the second debate to the day after (Figs. 3(e) – (f)). The state of Kentucky (KY) had the most notorious shift on behavior, from very positive to very negative. It was probably due to a clash between Donald Trump and the Kentucky Senator Rand Paul during the debate.

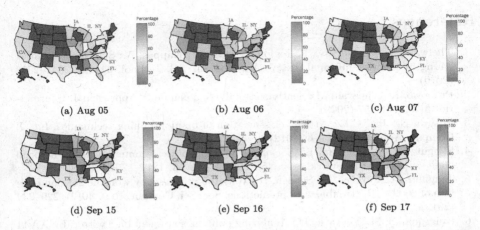

(a) Aug 05   (b) Aug 06   (c) Aug 07

(d) Sep 15   (e) Sep 16   (f) Sep 17

Fig. 3. The sentiment distribution over US states. (a)–(c) a day before, on the day, and a day after the first Republican party debate, respectively; (d)–(f) a day before, on the day, and a day after the second Republican party debate, respectively

4 Concluding Remarks

In this paper, we have discussed sentiment and behavior analysis of one controversial individual on Twitter. Applying the orthodox Naïve Bayes classifier we were able to draw fascinating and practical insights about the analyzed data. Apart from the high accuracy of the classifier, reaching about 91 %, the qualitative analysis of its performance through the *relevance versus sentiment* graph, revealed that the classifier has been very stable on whom is considered to take part of each sentiment class.

We have also compared the temporal signature for each state and perform a hierarchical clustering. There, clusters represent states with similar reactional behavior to Donald Trump's remarks. The most insightful result from this analysis was the fact that the most populous states had very similar behavior, which is certainly a non-trivial information that any candidate would desire to have as feedback.

A map analysis around the period of the debates was also insightful. The case of Kentucky, which did not take part of any cluster, has had the biggest overnight sentiment shift.

This paper certainly drew much more insights than conclusions, but undoubtedly the near future will demand a reliable framework for political trend analysis. Social media will be increasingly more popular and decisive at influencing the mood of society towards a candidate. Those who could be able to grasp the dynamics around the sentiments will certainly have a serious advantage.

Acknowledgments. The authors would like to thank the CNPq, CAPES, and FAPESP (Proc. 2011/18496-7), for financial support.

References

1. Thelwall, M., Buckley, K., Paltoglou, G., Cai, D., Kappas, A.: Sentiment in short strength detection informal text. J. Am. Soc. Inf. Sci. Technol. **61**(12), 2544–2558 (2010)
2. Prabowo, R., Thelwall, M.: Sentiment analysis: a combined approach. J. Informetrics **3**(2), 143–157 (2009)
3. Mishra, N., Jha, C.K.: Article: classification of opinion mining techniques. Int. J. Comput. Appl. **56**(13), 1–6 (2012)
4. Stieglitz, S., Dang-Xuan, L.: Social media and political communication: a social media analytics framework. Soc. Netw. Anal. Min. **3**, 1277–1291 (2012)
5. Jungherr, A., Jürgens, P., Schoen, H.: Why the pirate party won the german election of 2009 or the trouble with predictions. Soc. Sci. Comput. Rev. **30**(2), 229–234 (2012)
6. Ringsquandl, M., Petkovic, D.: Analyzing political sentiment on Twitter. In: AAAI Spring Symposium: Analyzing Microtext (2013)
7. Seron, W., Zorzal, E., Quiles, M.G., Basgalupp, M.P., Breve, F.A.: #Worldcup2014 on Twitter. In: Gervasi, O., Murgante, B., Misra, S., Gavrilova, M.L., Rocha, A.M.A.C., Torre, C., Taniar, D., Apduhan, B.O. (eds.) ICCSA 2015. LNCS, vol. 9155, pp. 447–458. Springer, Heidelberg (2015)
8. Soelistio, Y.E., Surendra, M.R.S.: Simple text mining for sentiment analysis of political figure using naive bayes classifier method. CoRR abs/1508.05163 (2015)
9. Wang, H., Can, D., Kazemzadeh, A., Bar, F., Narayanan, S.: A system for real-time Twitter sentiment analysis of 2012 U.S. presidential election cycle. In: ACL 2012 System Demonstrations, ACL 2012, pp. 115–120 (2012)
10. Mejova, Y., Srinivasan, P., Boynton, B.: Gop primary season on Twitter: "popular" political sentiment in social media. In: Sixth ACM International Conference on Web Search and Data Mining, WSDM 2013, pp. 517–526 (2013)
11. Taheri, S., Mammadov, M., Bagirov, A.M.: Improving naive bayes classifier using conditional probabilities. In: Ninth Australasian Data Mining Conference, AusDM 2011, vol. 121, pp. 63–68 (2011)
12. Morstatter, F., Pfeffer, J., Liu, H., Carley, K.: Is the sample good enough? Comparing data from Twitter's streaming API with Twitter's firehose. In: International AAAI Conference on Weblogs and Social Media (2013)
13. Das, S., Chen, M.: Yahoo! for Amazon: extracting market sentiment from stock message boards. In: Asia Pacific Finance Association Annual Conference (APFA) (2001)
14. Pang, B., Lee, L., Vaithyanathan, S.: Thumbs up?: Sentiment classification using machine learning techniques. In: ACL-02 Conference on Empirical Methods in Natural Language Processing EMNLP, pp. 79–86 (2002)
15. Potts, C.: http://sentiment.christopherpotts.net/lingstruc.html. Accessed 09 Dec 2015
16. Jiang, L., Wang, D., Cai, Z., Yan, X.: Survey of improving naive bayes for classification. In: Alhajj, R., Gao, H., Li, X., Li, J., Zaïane, O.R. (eds.) ADMA 2007. LNCS (LNAI), vol. 4632, pp. 134–145. Springer, Heidelberg (2007)

Brain-Machine Interface

Emotion Recognition Using Multimodal Deep Learning

Wei Liu[1], Wei-Long Zheng[1], and Bao-Liang Lu[1,2,3(✉)]

[1] Department of Computer Science and Engineering,
Center for Brain-like Computing and Machine Intelligence, Shanghai, China
{liuwei-albert,weilong}@sjtu.edu.cn
[2] Key Laboratory of Shanghai Education Commission for Intelligent
Interaction and Cognition Engineering, Shanghai, China
[3] Brain Science and Technology Research Center,
Shanghai Jiao Tong University, Shanghai, China
bllu@sjtu.edu.cn

Abstract. To enhance the performance of affective models and reduce the cost of acquiring physiological signals for real-world applications, we adopt multimodal deep learning approach to construct affective models with SEED and DEAP datasets to recognize different kinds of emotions. We demonstrate that high level representation features extracted by the Bimodal Deep AutoEncoder (BDAE) are effective for emotion recognition. With the BDAE network, we achieve mean accuracies of 91.01% and 83.25% on SEED and DEAP datasets, respectively, which are much superior to those of the state-of-the-art approaches. By analysing the confusing matrices, we found that EEG and eye features contain complementary information and the BDAE network could fully take advantage of this complement property to enhance emotion recognition.

Keywords: EEG · Emotion recognition · Multimodal deep learning · Auto-encoder

1 Introduction

Nowadays, many human machine interface (HMI) products are used by ordinary people and more HMI equipments will be needed in the future. Since emotional functions of HMI products play an important role in our daily life, it is necessary for HMI equipments to be able to recognize humans emotions automatically.

Many researchers studied emotion recognition from EEG. Liu *et al.* used fractal dimension based algorithm to recognize and visualize emotions in real time [1]. Li and Lu used EEG signals of gamma band to classify two kinds of emotions, and their results showed that gamma band was suitable for emotion recognition [2].

Duan *et al.* found that differential entropy features are more suited for emotion recognition tasks [3]. Wang *et al.* compared three different kinds of EEG features and proposed a simple approach to track the trajectory of emotion changes

© Springer International Publishing AG 2016
A. Hirose et al. (Eds.): ICONIP 2016, Part II, LNCS 9948, pp. 521–529, 2016.
DOI: 10.1007/978-3-319-46672-9_58

with time [4]. Zheng and Lu employed deep neural network to classify EEG signals and examined critical bands and channels of EEG for emotion recognition [5].

To fully use information from different modalities, Yang *et al.* proposed an auxiliary information regularized machine, which treats different modalities with different strategies [6].

In [7], the authors built a single modal deep autoencoder and a bimodal deep autoencoder to generate shared representations of images and audios. Srivastava and Salakhutdinov extended the methods developed in [7] to bimodal deep Boltzmann machines to handle multimodal deep learning problems [8].

As for multimodal emotion recognition, Verma and Tiwary carried out emotion classification experiments with EEG singals and peripheral physiological signals [9]. Lu *et al.* used two different fusion strategies for combining EEG and eye movement data: feature level fusion and decision level fusion [10]. Liu *et al.* employed bimodal deep autoencoders to fuse different modalities and the authors tested the framework on multimodal facilitation, unimodal enhancement, and crossmodal learning tasks [11].

To our best knowledge, there is no research work reported in the literature to deal with emotion recognition from multiple physiological signals using multimodal deep learning algorithms. In this paper, we propose a novel multimodal emotion recognition method using multimodal deep learning techniques. In Sect. 2, we will introduce the bimodal deep autoencoder. Section 3 presents data pre-proessing, feature extraction and experiment settings. The experiment results are described in Sect. 4. Following discusses in Sect. 5, conclusions and future work are in Sect. 6.

2   Multimodal Deep Learning

2.1   Restricted Boltzmann Machine

A restricted Boltzmann machine (RBM) is an undirected graph model, which has a visible layer and a hidden layer. Connections exist only between visible layer and hidden layer and there is no connection either in visible layer or in hidden layer. Assuming visible variables $\mathbf{v} \in \{0,1\}^M$ and hidden variables $\mathbf{h} \in \{0,1\}^N$, we have the following energy function E:

$$E(\mathbf{v}, \mathbf{h}; \theta) = -\sum_{i=1}^{M}\sum_{j=1}^{N} W_{ij} v_i h_j - \sum_{i=1}^{M} b_i v_i - \sum_{j=1}^{N} a_j h_j \tag{1}$$

where $\theta = \{\mathbf{a}, \mathbf{b}, \mathbf{W}\}$ are parameters, W_{ij} is the symmetric weight between visible unit i and hidden unit j, and b_i and a_j are bias terms of visible unit and hidden unit, respectively. With energy function, we can get the joint distribution over the visible and hidden units:

$$p(\mathbf{v}, \mathbf{h}; \theta) = \frac{1}{\mathcal{Z}(\theta)} \exp(E(\mathbf{v}, \mathbf{h}; \theta)), \text{and}$$

$$\mathcal{Z}(\theta) = \sum_{\mathbf{v}}\sum_{\mathbf{h}} \exp(E(\mathbf{v}, \mathbf{h}; \theta)) \tag{2}$$

where $\mathcal{Z}(\theta)$ is the normalization constant.

Given a set of visible variables $\{\mathbf{v}_n\}_{n=1}^{N}$, the derivative of log-likelihood with respect to weight $\mathbf{W}$ can be calculated from Eq. (2):

$$\frac{1}{N}\sum_{i=1}^{N}\frac{\partial \log p(\mathbf{v}_n;\theta)}{\partial W_{ij}} = \mathbb{E}_{P_{data}}[v_i h_j] - \mathbb{E}_{P_{model}}[v_i h_j]$$

Various algorithms can be used to train a RBM, such as Contrastive Divergence (CD) algorithm [12]. In this paper, Bernoulli RBM is used. We treat the visual layer as the probabilities and we use CD algorithm to train RBMs.

2.2 Model Construction

The proposed multimodal emotion recognition framework using deep learning is depicted in Fig. 1. There are three steps in total. The first step is to train the BDAE network.

We call this step feature selection. The second step is supervised training, and we use the extracted high level features to train a linear SVM classifier. And the last step is a testing process, from which the recognition results are produced.

The BDAE training procedures, including encoding part and decoding part, are shown in Fig. 2. In encoding part, we first train two RBMs for EEG features

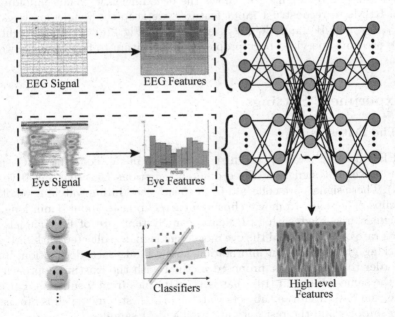

Fig. 1. The proposed multimodal emotion recognition framework. Here the BDAE network is used to generate high level features from low level features or original data and a linear SVM is trained with extracted high level features.

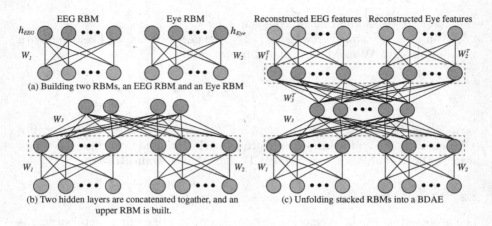

Fig. 2. The structure of Bimodal Deep AutoEncoder.

and eye movement features, respectively. As shown in Fig. 2(a), EEG RBM is for EEG features and eye RBM is for eye movement features. Hidden layers are indicated by h_{EEG} and h_{Eye}, and W_1, W_2 are the corresponding weight matrices. After training these two RBMs, hidden layers, h_{EEG} and h_{Eye}, are concatenated together. The concatenated layer is used as the visual layer of an upper RBM, as depicted in Fig. 2(b). Figure 2(c) shows the decoding part. When unfolding the stacked RBMs to reconstruct input features, we keep the weight matrices tied, and W_1, W_2, and W_3 and W_1^T, W_2^T, and W_3^T in Fig. 2(c) are tied weights. At last, we used unsupervised back-propagation algorithm to fine-tune the weights and bias.

3   Experiment Settings

3.1   The Datasets

The SEED dataset[1], which was first introduced in [5], contains EEG signals and eye movement signals of three different emotions (positive, negative, and neutral). These signals are collected from 15 subjects during watching emotional movie clips. There are 15 movie clips and each clip lasts about 4 min long. The EEG signals, recorded with ESI NeuroScan System, are of 62 channels at a sampling rate of 1000 Hz and the eye movement signals, collected with SMI ETG eye tracking glasses, contain information about blink, saccade fixation, and so on. In order to compare our proposed method with the existing approach [10], we use the same data as in [10], that is, 27 data files from 9 subjects. For every data file, the data from the subjects watching the first 9 movie clips are used as training samples and the rest ones are used as test samples.

---

[1] http://bcmi.sjtu.edu.cn/~seed/index.html.

The DEAP dataset was first introduced in [13]. The EEG signals and peripheral physiological signals of 32 participants were recorded when they were watching music videos. The dataset contains 32 channel EEG signals and 8 peripheral physiological signals. The emotional music videos include 40 one-minute long small clips and subjects were asked to do self-assessment by assigning values from 1 to 9 to five different status, namely, valence, arousal, dominance, liking, and familiarity. In order to compare the performance of our proposed method with previous results in [14,15], we did not take familiarity into consideration. We divided the trials into two different classes according to the assigned values. The threshold we chose is 5, and the tasks can be treated as four binary classification problems, namely, high or low valence, arousal, dominance and liking. Among all of the data, 90% samples were used as training data and the rest 10% samples were used as test data.

3.2 Feature Extraction

For SEED dataset, both Power Spectral Density (PSD) and Differential Entropy (DE) features were extracted from EEG data. These two kinds of features contain five frequency bands: δ (1–4 Hz), θ (4–8 Hz), α (8–14 Hz), β (14–31 Hz), and γ (31–50 Hz). For every frequency band, the extracted features are of 62 dimensions and there are 310 dimensions for EEG features in total. As for eye movement data, we used the same features as in [10], and there are 41 dimensions in total including both PSD and DE features. The extracted EEG features and eye movement features were then rescaled to [0,1] and the rescaled features were used as the inputs of BDAE network.

For DEAP dataset, we used the downloaded preprocessed data directly as the inputs of BDAE network to generate shared representations of EEG signals and peripheral physiological signals. First, the EEG signals and peripheral physiological signals were separated and then the signals were segmented into 63 s. After segmentation, different channel data of the same time period (one second) are combined to form the input signals of BDAE network. And then, shared representation features were generated by the BDAE network.

3.3 Classification

The shared representation features generated by BDAE network are used to train a linear SVM classifier. Because of the variance between EEG signals collected from different people at different time, the BDAE model is data-specified, which means that we will build a BDAE model for each data and there are 27 BDAE models built for SEED dataset and 32 BDAE models for DEAP dataset. Network parameters, including hidden neuron numbers, epoch numbers, and learning rate, are chosen with grid searching.

4   Results

We compare our model with two other experimental settings. When only single modality is available, we classify different emotions with PSD and DE features by

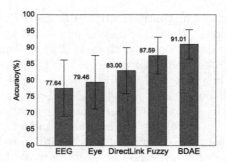

Fig. 3. Multimodal facilitation results on SEED dataset. Here the first two bars denote single modality, the rest bars denote multimodal with different fusion strategies and the fourth Fuzzy bar denotes the best result in [10].

Table 1. Accuracies of BDAE model on SEED dataset (%).

Feature		δ + eye	θ + eye	α + eye	β + eye	γ + eye	All
PSD	Ave.	**85.12**	83.89	83.18	83.23	82.92	85.10
	Std.	**11.09**	13.13	12.68	13.65	13.59	11.82
DE	Ave.	85.41	84.64	84.58	86.55	88.01	**91.01**
	Std.	14.03	11.03	12.78	10.48	10.25	**8.91**

SVM. When multimodal information is available, features of different modalities are linked directly and different emotions are recognized with the concatenated features by SVM.

SEED Results. Figure 3 shows the summary of multimodal facilitation experiment results. As can be seen from Fig. 3, the BDAE model has the best accuracy (91.01 %) and the smallest standard deviation (8.91 %).

Table 1 is the detailed experimental results of the BDAE model. The last column means that we linked all five frequency bands of EEG features and eye movement features directly. We examined the BDAE model three times and the recognition accuracies shown in Table 1 were averaged.

DEAP Results. In the literature, Rozgic et al. treated the EEG signals as a sequence of overlapping segments and a novel non-parametric nearest neighbor model was employed to extract response-level feature from these segments [14]. Li et al. used Deep Belief Network (DBN) to automatically extract high-level features from raw EEG signals [15].

The experimental results on the DEAP dataset are shown in Table 2. Besides baselines mentioned above, we also compared the BDAE results with results in [15] and [14]. As can be seen from Table 2, the BDAE model improved recognition accuracies in all classification tasks.

Table 2. Comparison of six different approaches on DEAP dataset (Accuracy, %).

Method	Valence	Arousal	Dominance	Liking
EEG only	52.6	53.01	55.0	55.0
Others only	63.9	59.6	62.5	60.7
Linking	61.5	58.6	59.7	60.0
Rozgic et al. [14]	76.9	69.1	73.9	75.3
Li et al. [15]	58.4	64.3	65.8	66.9
Our Method	**85.2**	**80.5**	**84.9**	**82.4**

5   Discussion

From the experimental results, we have demonstrated that the BDAE network can be used to extract shared representations from different modalities and the extracted features have better performance than other features.

From Table 3(a), we can see that EEG features are good for positive emotion recognition but are not good for negative emotions. As a complement, eye features have advantage in negative emotion recognition which can be seen from Table 3(b). When linking EEG and eye features directly, positive emotion accuracy is improved compare with situation where only eye features exist and negative emotion accuracy is also enhanced compared with when only EEG features are used. The BDAE framework achieves an even better result. The BDAE model has the highest accuracies in all three kinds of emotions, indicating that the BDAE model can fully use both EEG features and eye features.

Table 3. Confusing matrices of single modality and different modality merging methods

(a) EEG

	Positive	Neutral	Negative
Positive	**93.72%**	0.94%	5.34%
Neutral	5.56%	**81.35%**	13.09%
Negative	14.24%	29.49%	**56.27%**

(b) Eye

	Positive	Neutral	Negative
Positive	**81.92%**	7.41%	10.67%
Neutral	14.81%	**74.08%**	11.11%
Negative	9.38%	11.59%	**79.03%**

(c) Linking

	Positive	Neutral	Negative
Positive	**93.69%**	3.42%	2.89%
Neutral	7.06%	**77.62%**	15.32%
Negative	6.11%	16.72%	**77.17%**

(d) BDAE

	Positive	Neutral	Negative
Positive	**99.03%**	0.00%	0.97%
Neutral	3.70%	**90.26%**	6.04%
Negative	11.25%	3.57%	**85.18%**

6   Conclusions and Future Work

This paper has shown that the shared representations extracted from the BDAE model are good features to discriminate different emotions. Compared with other existing feature extraction strategies, the BDAE model is the best with accuracy of 91.01 % on SEED dataset. For DEAP dataset, the BDAE network largely improves recognition accuracies on all four binary classification tasks. We analysed the confusing matrices of different methods and found that EEG features and eye features contain complementary information. The BDAE framework could fully take advantage of the complementary property between EEG and eye features to improve emotion recognition accuracies.

Our future work will focus on invesgating the complementarity between EEG features and eye movement features and explaining the mechanism of multimodal deep learning for emotion recognition from EEG and other physiological signals.

Acknowledgments. This work was supported in part by the grants from the National Natural Science Foundation of China (Grant No. 61272248), the National Basic Research Program of China (Grant No. 2013CB329401) and the Major Basic Research Program of Shanghai Science and Technology Committee (15JC1400103).

References

1. Liu, Y., Sourina, O., Nguyen, M.K.: Real-time EEG-based human emotion recognition and visualization. In: 2010 International Conference on Cyberworlds, pp. 262–269. IEEE (2010)
2. Li, M., Lu, B.-L.: Emotion classification based on gamma-band EEG. In: Annual International Conference of the IEEE Engineering in Medicine and Biology Society, EMBC 2009, pp. 1223–1226. IEEE (2009)
3. Duan, R.-N., Zhu, J.-Y., Lu, B.-L.: Differential entropy feature for eeg-based emotion classification. In: 2013 6th International IEEE/EMBS Conference on Neural Engineering, pp. 81–84. IEEE (2013)
4. Wang, X.-W., Nie, D., Bao-Liang, L.: Emotional state classification from EEG data using machine learning approach. Neurocomputing **129**, 94–106 (2014)
5. Zheng, W.-L., Bao-Liang, L.: Investigating critical frequency bands and channels for EEG-based emotion recognition with deep neural networks. IEEE Trans. Auton. Mental Dev. **7**(3), 162–175 (2015)
6. Yang, Y., Ye, H.-J., Zhan, D.-C., Jiang, Y.: Auxiliary information regularized machine for multiple modality feature learning. In: IJCAI 2015, pp. 1033–1039. AAAI Press (2015)
7. Ngiam, J., Khosla, A., Kim, M., Nam, J., Lee, H., Ng, A.Y.: Multimodal deep learning. In: ICML 2011, pp. 689–696 (2011)
8. Srivastava, N., Salakhutdinov, R.: Multimodal learning with deep boltzmann machines. J. Mach. Learn. Res. **15**(1), 2949–2980 (2014)
9. Verma, G.K., Tiwary, U.S.: Multimodal fusion framework: a multiresolution approach for emotion classification and recognition from physiological signals. NeuroImage **102**, 162–172 (2014)
10. Lu, Y., Zheng, W.-L., Li, B., Lu, B.-L.: Combining eye movements and EEG to enhance emotion recognition. In: IJCAI 2015, pp. 1170–1176 (2015)

11. Liu, W., Zheng, W.-L., Lu, B.-L.: Multimodal emotion recognition using multi-modal deep learning. arXiv preprint arXiv:1602.08225 (2016)
12. Hinton, G.E.: Training products of experts by minimizing contrastive divergence. Neural Comput. 14(8), 1771–1800 (2002)
13. Koelstra, S., Mühl, C., Soleymani, M., Lee, J.-S., Yazdani, A., Ebrahimi, T., Pun, T., Nijholt, A., Patras, I.: Deap: a database for emotion analysis using physiological signals. IEEE Trans. Affect. Comput. 3(1), 18–31 (2012)
14. Rozgic, V., Vitaladevuni, S.N., Prasad, R.: Robust EEG emotion classification using segment level decision fusion. In: 2013 IEEE International Conference on Acoustics, Speech and Signal Processing, pp. 1286–1290. IEEE (2013)
15. Li, X., Zhang, P., Song, D., Yu, G., Hou, Y., Hu, B.: EEG based emotion identification using unsupervised deep feature learning. In: SIGIR2015 Workshop on Neuro-Physiological Methods in IR Research, August 2015

Continuous Vigilance Estimation Using LSTM Neural Networks

Nan Zhang[1], Wei-Long Zheng[1], Wei Liu[1], and Bao-Liang Lu[1,2,3(✉)]

[1] Center for Brain-like Computing and Machine Intelligence,
Department of Computer Science and Engineering,
Shanghai Jiao Tong University,Shanghai, China
vincentzn4@outlook.com, {weilong,liuwei-albert,bllu}@sjtu.edu.cn
[2] Key Laboratory of Shanghai Education Commission for Intelligent Interaction
and Cognition Engineering, Shanghai Jiao Tong University, Shanghai, China
[3] Brain Science and Technology Research Center,
Shanghai Jiao Tong University, Shanghai, China

Abstract. In this paper, we propose a novel continuous vigilance estimation approach using LSTM Neural Networks and combining Electroencephalogram (EEG) and forehead Electrooculogram (EOG) signals. We combine these two modalities to leverage their complementary information using a multimodal deep learning method. Moreover, since the change of vigilance level is a time dependent process, temporal dependency information is explored in this paper, which significantly improves the performance of vigilance estimation. We introduce two LSTM Neural Network architectures, the F-LSTM and the S-LSTM, to encode the time sequences of EEG and EOG into a high level combined representation, from which we can predict the vigilance levels. The experimental results demonstrate that both of the two LSTM multimodal structures can improve the performance of vigilance estimation models in comparison with the single modality models and non-temporal dependent models.

Keywords: EEG · Vigilance estimation · Multimodal · Deep learning · Recurrent neural network

1 Introduction

Brain-computer interaction (BCI) aims to translate brain activity or state into control signals for computer devices [1]. A lot of studies have been done on the assessment of human's brain states such as vigilance and emotion in order to develop affective brain-computer interaction systems [2]. Vigilance or alertness, which means the ability to maintain sustained concentration, is an important kind of mental state for user aware BCI systems. High vigilance is usually required for some occupations such as truck drivers or pilots. In these cases, low vigilance may bring tragedy to both themselves and other people. For example driving fatigue is believed to be a significant reason for most of the fatal traffic accidents [3]. Therefore a robust vigilance estimation system is desired to improve the transportation safety.

© Springer International Publishing AG 2016
A. Hirose et al. (Eds.): ICONIP 2016, Part II, LNCS 9948, pp. 530–537, 2016.
DOI: 10.1007/978-3-319-46672-9_59

Various approaches have been proposed to estimate the vigilance level over the past years. Different kinds of signals are utilized such as video [4], speech [5] and physiological signals [3]. In these signals, EEG is considered as a good indicator of the transition from wakefulness to sleepiness. Eoh et al. showed that the proportion of different spectral components in EEG is related to the alertness level [3]. Davidson et al. introduced a warning system capable of detecting lapse with high temporal resolution [4]. In addition to EEG, EOG signal also contains information that has close relationship with vigilance status. Eye movement features such as slow eye movements (SEM) and blinks [8] have been shown to be good indicators of vigilance level. The traditional EOG are collected from electrodes placed around the eyes, which may distract subjects and cause discomfort. Zhang et al. proposed to place the electrodes on the forehead and extract features from forehead EOG to detect driving fatigue [9].

However, most of these methods ignore the time dependency property of the vigilance changing process. The subject's mental states are treated as independent points and the temporal dependency information are discarded in these models. Recurrent Neural Network (RNN) is a kind of artificial neural network where connections between units form a cycle which makes it suitable to process sequence data. RNN has been successfully applied to research domains such as machine translation [10] and speech recognition [11]. In this paper, we introduce the RNN as a multimodal encoder which can incorporate the temporal changes of EEG and EOG features to help with the estimation of vigilance. The mental state sequence is encoded into a fixed-dimensional vector representation which contains meaningful information to decode the vigilance level.

This paper is organized as follows. In Sect. 2, we describe the method used to build vigilance estimation system. Section 3 gives a detailed description about the setup of our simulated driving experiment and how we collect our data. In Sect. 4 we discuss the experiment results using different models. Finally in Sect. 5, conclusions are presented.

2 LSTM Neural Networks

Vigilance changing is a dynamic process. To incorporate the time dependency information, we introduce the Recurrent Neural Network (RNN) model as a temporal encoder. RNN contains cyclical connections in its hidden layers and can remember the history of its input. For a length T input sequence x, at time t in forward pass, the hidden units states are updated as:

$$h_t = f(Wx_t + Uh_{t-1} + b) \tag{1}$$

where h_t and x_t are respectively the output vector and input vector of a hidden layer at time t, f is the activation function, W and U are weight matrices, and b is the bias vector.

The problem of this simple RNN architecture is that only small range of contextual information can be used from the input sequence which will cause the vanishing gradient problem when applying to longtime sequence. Since we need to learn information from longtime EEG/EOG sequences, the Long Short Time Memory (LSTM) neural network is applied. LSTM neural network is a RNN with LSTM blocks as units in hidden layers. Each LSTM block contains memory cells and input gate, output gate and forget gate, which provide write, read and reset operations for the cells. In this way, the LSTM cells can store states over long periods of time. The state of memory cells is updated at every time step t as follows:

Input Gate:

$$i_t = f\left(W_i x_t + U_i h_{t-1} + b_i\right) \tag{2}$$

Forget Gate:

$$f_t = f\left(W_f x_t + U_f h_{t-1} + b_f\right) \tag{3}$$

Cells update:

$$\overline{C_t} = g\left(W_c x_t + U_c h_{t-1} + b_c\right) \tag{4}$$

$$C_t = i_t * \overline{C_t} + f_t * C_{t-1} \tag{5}$$

Output Gate:

$$o_t = f\left(W_o x_t + U_o h_{t-1} + b_o\right) \tag{6}$$

$$h_t = o_t * k\left(C_t\right) \tag{7}$$

where f, g and k are all activation functions, i_t, f_t and o_t are outputs of gates, and $\overline{C_t}$ is the candidate of cells' state.

The EEG and EOG feature sequences need to be adapted to the input of RNN architecture. First, the data is normalized to zero mean and unit variance, then the whole data sequence is divided into many short data sequences. Each data sequence is nearly one minute which, as we show in the experiment result, is long enough to estimate vigilance levels. In order to learn from multi modalities, we propose two LSTM network architectures that can fuse information from EEG and EOG sequences. One is to use two independent LSTM encoders to encode EEG and EOG sequences respectively and then combine their representations into one feature vector (F-LSTM) shown in Fig. 1(a). The other is to concatenate the feature vectors of EEG and EOG at each time step and then use stacked LSTM layers to encode the feature sequence into a compact feature vector (S-LSTM) shown in Fig. 1(b).

We implement our model using python theano and decide all the hyper parameters by cross validation. In S-LSTM, we use 3 stacked hidden LSTM layers as encoder and one sigmoid neuron as output layer. Each LSTM layer has half number neurons comparing to the input layer. The internal weights in LSTM units are initialized from a standard Gaussian distribution followed by a SVD orthogonalization. The other weights are initialized from a uniform distribution with scale parameter determined by Xavier algorithm. The bias value of forget gates are initialized with ones. The activation functions of all the gates are sigmoid while tanh is used otherwhere in LSTM units. In F-LSTM, we append a

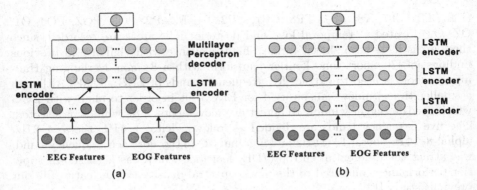

Fig. 1. Two LSTM structures adopted in this paper. Figure (a) depicts the F-LSTM model which combines two LSTM encoders designed respectively for EEG and EOG. Figure (b) depicts the S-LSTM model which merges the EEG and EOG at feature level.

Multilayer Perceptron (MLP) model after the last LSTM layer as a decoder. The activation function used in MLP is ReLU. In order to generalize our model, dropout with a probability 0.5 is added before the output layer. In training, RMSProp method is used instead of basic stochastic gradient descent method to optimize the loss function. Early stopping strategy is adopted when no improvement appears on the performance on validation set after 10 epochs.

3 Experiment Setup and Data Processing

3.1 Experiment Setup

The experiments were performed in a simulated driving environment. A four-lane high way scene was shown in front of a car. The subjects drove the car just like driving a real car outdoors. There were in total 21 subjects (12 men 9 women) at the age between 20 and 25, who participated in the experiments. Before the experiments started, a warm up session was performed to ensure every participant was familiar with the operation. All the experiments were conducted after lunch from 13:00 pm to 15:00 pm or after dinner from 21:00 pm to 23:00 pm. The participants were asked to drive the car for 2 h in the simulated driving environment. Both of the EOG and EEG signals were recorded simultaneously using the Neuroscan system with a 1000 Hz sampling rate. At the same time, eye movement data was recorded using SMI ETG eye tracking glasses.

3.2 Feature Extraction

EEG Signals: The EEG signals are down-sampled to 200 Hz to reduce computational complexity and preprocessed with a band-pass filter between 1 Hz and 75 Hz to reduce noise and artifacts. EEG signals from 17 channels

(FT7, FT8, T7, T8, TP7, TP8, CP1, CP2, P1, PZ, P2, PO3, POZ, PO4, O1, OZ, O2) located at temporal lobe and posterior lobe areas are recorded, since these areas have been shown to have high relevance with vigilance in previous findings [6] [7]. Short-time Fourier transform with a 8 s non-overlapping Hanning window is used to extract five frequency bands of EEG signals. Although a smaller time window can be used for EEG, but in order to align with EOG which needs a bigger window to detect eye movements, a 8 s window is selected. The five frequency bands are divided as follows, delta: 1–4 Hz, theta: 4–8 Hz, alpha: 8–14 Hz, beta: 14–31 Hz and gamma: 31–75 Hz. For each frequency band, we extract the differential entropy (DE) features, which has been shown superior performance compared to the power spectral density (PSD) features in our previous study [12].

EOG signals: EOG features are also extracted with a 8 s non-overlapping window on EOG signals. For traditional EOG, the electrodes are placed around eyes as shown in Fig. 2(a). This will distract subject from the driving process and bring discomfort to the subject. In this work, all electrodes are placed on the forehead as Fig. 2(b) and we extract features from forehead EOG. For traditional EOG, the vertical EOG (VEO) and horizontal EOG (HEO) are extracted by subtracting electrodes four from three and electrodes one from two, respectively.

For forehead EOG, forehead VEO is extracted from electrodes four and seven by using independent component analysis (ICA). Forehead HEO is extracted by simply subtraction from electrode five and six. After preprocessing forehead EOG signals, we detect eye movements such as blinks and saccades using wavelet transform method. Continuous wavelet coefficients at scale 8 with Mexican mother wavelet are calculated. The blinks and saccades are then detected from VEO and HEO, respectively. The statistical parameters such as blink/saccade duration, mean, maximum, variance and derivative are extracted as EOG features. A total of 36 EOG features are used in this paper. The detailed descriptions of EOG features are described in [9].

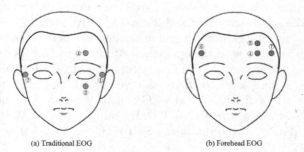

(a) Traditional EOG   (b) Forehead EOG

Fig. 2. Electrode placements for the traditional and forehead EOG setups. The yellow and blue colors indicate the electrode placements of the traditional EOG and forehead EOG, respectively. The electrode four is the shared electrode of both setups. (Color figure online)

3.3 Vigilance Labeling

In this study, the ground truth vigilance values are obtained using eye tracking glasses proposed in [13]. PERCLOS, which refers to the percentage of eyelid closure over time, is used as the index of alertness level. PERCLOS is defined as [13]:

$$PERCLOS = \frac{blink + CLOS}{interval} \tag{8}$$

$$interval = blink + fixation + saccade + CLOS \tag{9}$$

where 'CLOS' denotes the duration of eye closure. We calculate the PERCLOS values using eye tracking glasses as the labels of vigilance levels. It should be noted here that although the eye tracking glasses can estimate the vigilance level precisely, it's not a good choice to use in real world applications due to its expensive cost and longtime delay. So we only use it as a vigilance labeling method and obtain the labels to train our models.

4 Experiment Results

We use the support vector regression (SVR) with radial basis function (RBF) kernel as a baseline in this paper. To evaluate the experiment results, we divided our whole data sequence from one experiment into five segments and evaluated the performance with 5-fold cross validation. The Root Mean Square Error (RMSE) and Correlation Coefficient (COR) are used as metrics for the experiment results.

First we investigate whether multiple modalities are helpful for the result of vigilance estimation. We used the S-LSTM model for the two single modalities, which means instead of the concatenation of EEG and EOG features either the EEG or EOG feature was used as input to S-LSTM model. For multiple modalities, the S-LSTM and F-LSTM network architectures were used to fuse the two modalities. The mean and standard deviation for RMSE and COR are shown in Table 1. We can see that both of the two multimodalities models achieved better results than single modality methods.

Next we will examine the importance of time dependency information in estimating vigilance. The SVR model used doesn't take time dependency into

Table 1. Experiment results for different models. Each single modality uses S-LSTM model in first two columns. Last three columns are models fusing mulimodalities

Model		EEG	EOG	S-LSTM	F-LSTM	SVR
COR	Mean	0.8237	0.8203	0.8329	**0.8363**	0.7958
	Std	**0.0831**	0.1191	0.0961	0.1009	0.1131
RMSE	Mean	0.0927	0.0935	0.0816	**0.0807**	0.1186
	Std	0.0259	0.0215	0.0189	**0.0135**	0.0515

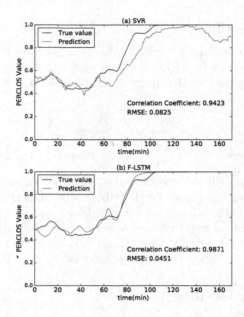

Fig. 3. The vigilance level prediction curves obtained by SVR and F-LSTM models.

consideration. The input of SVR model is the concatenation of EEG and EOG features. The mean and standard deviation for RMSE and COR are shown in Table 1. We can see from the results that LSTM models can significantly improve the estimation results compared to SVR. Figure 3 shows the vigilance prediction curves of SVR and M-LSTM models. The curves of M-LSTM model is more smooth comparing to SVR model. This means incorporating the time dependency information to vigilance estimation can make the system more robust to noise and predict the trend of vigilance levels more accurately.

5   Conclusion

In this paper, we have proposed a vigilance estimation approach combining two modalities and incorporating time dependency information. Two LSTM neural network structures were proposed to encode longtime signal sequences. The experimental results show that our proposed multimodal LSTM based methods can achieve significant improvement on vigilance estimation comparing to the traditional models.

Acknowledgment. This work was supported in part by the grants from the National Natural Science Foundation of China (Grant No. 61272248), the National Basic Research Program of China (Grant No. 2013CB329401) and the Major Basic Research Program of Shanghai Science and Technology Committee (15JC1400103).

References

1. Brunner, C., et al.: BNCI horizon 2020 – towards a roadmap for brain/neural computer interaction. In: Stephanidis, C., Antona, M. (eds.) UAHCI 2014, Part I. LNCS, vol. 8513, pp. 475–486. Springer, Heidelberg (2014)
2. Lu, Y., Zheng, W.-L., Li, B., Lu, B.-L.: Combining eye movements and EEG to enhance emotion recognition. In: IJCAI 2015, pp. 1170–1176 (2015)
3. Eoh, H.J., Chung, M.K., Kim, S.-H.: Electroencephalographic study of drowsiness in simulated driving with sleep deprivation. Int. J. Ind. Ergon. **35**(4), 307–320 (2005)
4. Davidson, P.R., Jones, R.D., Peiris, M.T.R.: EEG-based lapse detection with high temporal resolution. IEEE Trans. Biomed. Eng. **54**(5), 832–839 (2007)
5. Krajewski, J., Batliner, A., Golz, M.: Acoustic sleepiness detection: framework and validation of a speech-adapted pattern recognition approach. Behav. Res. Methods **41**(3), 795–804 (2009)
6. Khushaba, R.N., Kodagoda, S., Lal, S., Dissanayake, G.: Driver drowsiness classification using fuzzy wavelet-packet-based feature extraction algorithm. IEEE Trans. Biomed. Eng. **58**(1), 121–131 (2011)
7. Shi, L.-C., Bao-Liang, L.: EEG-based vigilance estimation using extreme learning machines. Neurocomputing **102**, 135–143 (2013)
8. Ma, J.-X., Shi, L.-C., Lu, B.-L.: Vigilance estimation by using electrooculographic features. In: 32nd Annual International Conference of the IEEE Engineering in Medicine and Biology Society, pp. 6591–6594 (2010)
9. Zhang, Y.-F., Gao, X.-Y., Zhu, J.-Y., Zheng, W.-L., Lu, B.-L.: A novel approach to driving fatigue detection using forehead EOG. In: 2015 7th International IEEE/EMBS Conference on Neural Engineering, pp. 707–710 (2015)
10. Sutskever, I., Vinyals, O., Le, Q.V.: Sequence to sequence learning with neural networks. In: Advances in Neural Information Processing Systems, pp. 3104–3112 (2014)
11. Deng, L., Li, J., Huang, J.-T., et al.: Recent advances in deep learning for speech research at Microsoft. In: 2013 IEEE International Conference on Acoustics, Speech and Signal Processing, pp. 8604–8608 (2013)
12. Shi, L.-C., Jiao, Y.-Y., Lu, B.-L.: Differential entropy feature for EEG-based vigilance estimation. In: 2013 35th Annual International Conference of the IEEE Engineering in Medicine and Biology Society, pp. 6627–6630 (2013)
13. Gao, X.-Y., Zhang, Y.-F., Zheng, W.-L., Lu, B.-L.: Evaluating driving fatigue detection algorithms using eye tracking glasses. In: 2015 7th International IEEE/EMBS Conference on Neural Engineering, pp. 767–770 (2015)

Motor Priming as a Brain-Computer Interface

Tom Stewart[1,2], Kiyoshi Hoshino[2], Andrzej Cichocki[1],
and Tomasz M. Rutkowski[1,3,4(✉)]

[1] RIKEN Brain Science Institute, Wako-shi, Saitama, Japan
{tom,tomek}@bci-lab.info
[2] University of Tsukuba, Tsukuba, Ibaraki, Japan
[3] The University of Tokyo, Tokyo, Japan
[4] Saitama Institute of Technology, Saitama, Japan
http://bci-lab.info/

Abstract. This paper reports on a project to overcome a difficulty associated with motor imagery (MI) in a brain–computer interface (BCI), in which user training relies on discovering how to best carry out the MI given only open-ended instructions. To address this challenge we investigate the use of a motor priming (MP), a similar mental task but one linked to a tangible behavioural goal. To investigate the efficacy of this approach in creating the changes in brain activity necessary to drive a BCI, an experiment is carried out in which the user is required to prepare and execute predefined movements. Significant lateralisations of *alpha* activity are discussed and significant classification accuracies of movement preparation versus no preparation are also reported; indicating that this method is promising alternative to motor imagery in driving a BCI.

Keywords: Brain-computer interface (BCI) · Motor priming · EEG · Neurophysiological information processing and classification

1 Introduction

A brain–computer interface (BCI) offers a unique mode of communication between a user and an environment by interpreting and acting upon signals arising directly from the brain [10]. In doing so, BCI bypasses the body's usual output pathway which requires motor signals to be sent from the brain, through the spinal cord and peripheral nerves to reach, and contract skeletal muscles [10]. Through a number of conditions, most notably spinal cord injury and ALS, this pathway can be interrupted and in extreme cases, the body can lose control of all of its skeletal musculature. In those cases, BCI is the only means available for a patient to interact with his/her environment. As such, one critical application for BCI is to control assistive devices such as wheelchairs and neuroprosthesis.

1.1 Motor Imagery

Current BCI's aimed at neuroprosthesis control are usually based around a mental task known as motor imagery [10]. To perform motor imagery, a subject is

© Springer International Publishing AG 2016
A. Hirose et al. (Eds.): ICONIP 2016, Part II, LNCS 9948, pp. 538–545, 2016.
DOI: 10.1007/978-3-319-46672-9_60

instructed to imagine moving a particular part of their body such as opening and closing one of their hands. The act of imagining a movement is thought to engage the same neural circuits as those that would be involved in planning an actual movement [5] and has been shown to cause patterns in the amplitude of $8 - 12\,Hz$ brainwave EEG activity. The nature of these patterns is consistent with a topographical map of the body found in the brain's sensorimotor cortex known as the homunculus. In general, motor imagery relating to a particular body part will cause a suppression of μ−rhythms in the corresponding region of the homunculus. To harness this phenomenon and create a BCI, machine learning algorithms are used to classify these brainwave patterns and to output an estimate of which body part's movement, if any, was being imagined based on a user's on−going EEG readings. The result of this classification can then be used to create a command which may operate a computer or some form of assistive device [5]. As a mental task, motor imagery (MI) is well suited to brain computer interface for a number of reasons. Most importantly, it can be used to communicate different commands depending on the part of the body which the subject chooses to imagine moving. Secondly, MI represents a mental state that the brain can enter into entirely of it's own volition and without the need for external stimulus. This means that subjects can potentially drive the BCI at their own pace [10]. Finally, MI has been shown to be possible even when the subject has no control over the body part with which the movement is being imagined meaning that it can be utilised even by subjects with severe paralysis [10]. To date, MI−based BCI have been successfully applied to a number of applications including navigating through virtual worlds [6] flying quadrocopters [3] and even controlling the exoskeleton that was used to deliver the opening kick at the 2014 FIFA World Cup.

However, despite it's potential, this type of BCI has yet to reach the level of robustness and reliability necessary to be applied practically outside of laboratory conditions [8]. One widely recognized challenge associated with this type of BCI is training users in effectively utilising MI to create strong and reliable patterns in their brainwave activity. Motor Imagery is an abstract mental visualisation task without a tangible behavioural goal and it's therefore difficult to give explicit instructions on how it should be properly carried out. The instructions that are given are quite open to interpretation and usually in the form of "imagine opening and closing your left hand". This allows for the possibility that different subjects may interpret these instructions differently and arrive at varying strategies on how to carry out the MI task. If these differing strategies bring with them differing levels of efficacy then this might account for the reported variability in subject performance of MI [2].

1.2   Motor Priming

This research aims to address this challenge by investigating the use of a motor priming (MP) as an alternative to MI in controlling a brain computer interface. MP, also referred to as a motor attention [7], is the mental process of covertly preparing to execute a movement. This concept has been studied extensively in

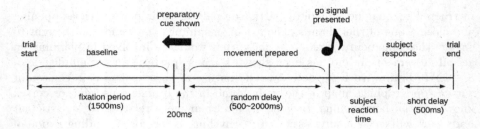

Fig. 1. Experiment trial timing.

the field of cognitive neuroscience to elucidate the neural mechanisms involved in the preparation and execution of voluntary movement [1]. It has been found that preparation to make a voluntary hand movement gives rise to an effect known as a motor related amplitude asymmetry in the *alpha* band ($8 - 12\,\text{Hz}$) of the EEG spectrum [1]. The orientation of this asymmetry is dependent on which hand is being prepared for the movement. Left hand movement preparation is accompanied by a left hemespheric increase in *alpha* activity and a right hemispheric decrease. During right hand movement preparation the reverse can be observed. This effect is most prominent over the $CP3/CP4$ electrode pair, which are situated over the motor cortex. This lateralisation of *alpha* activity closely resembles effects of MI. Specifically, hand motor imagery has also been found to cause similar contralateral reductions in *alpha* band activity [2]. These too can be best measured over motor cortex electrodes, typically $C3$ and $C4$.

Despite these parallels with MI, MP has received little, if any attention from the BCI community. Never the less, motor priming has several key characteristics which might make it an appealing alternative to motor imagery in voluntarily generating the brain activity patterns necessary to control a BCI. Most importantly, motor priming is tied to a clear behavioural goal, that is, to execute a prepared movement. Therefore in order to explain to a subject how to carry out MP, it's only necessary to instruct them to make a predefined movement as quickly as possible. It can then be reasoned that the subject will instinctively place themselves in the required mental state in order to minimise their reaction time and in so doing create the necessary changes in brain activity to drive the BCI. This bypasses the need to give subjects instructions on how to carry out a mental task which may be difficult to convey and prone to misinterpretation. Furthermore, since the execution of a covertly prepared movement is common in daily activities, it's reasoned that subjects might be able to draw on instinct or past experience to engage in motor priming, and might therefore be capable of carrying it out with less training than motor imagery. Finally, since motor priming is linked to a subject's reaction time in carrying out a movement, that reaction time can be used as an unambiguous metric for subject performance. This allows for clear feedback during user training as well as a means of identifying how engaged the user is in carrying out the mental task.

Table 1. Preparatory cues and associated movements

Preparatory cue	Movement
→	Press right button with right hand
←	Press left button with left hand
↓	Press foot pedal with right foot
○	No movement

2 Materials and Methods

To investigate whether different motor priming tasks could be classified on a single trial basis and thus serve as an EEG control strategy, an experiment was carried out in which seven subjects were asked to covertly prepare, and then subsequently execute a movement as quickly as possible upon the presentation of a "go" signal.

The experiments with human subjects reported in this paper were conducted with an approval of RIKEN Brain Science Institute Ethical Committee permission. Subjects were seated comfortably in front of a computer monitor. Subjects were asked to place their hands on two arcade–style buttons and fixate their gaze on a small dot shown in the middle of the computer monitor. The experiment consisted of 300 trials with a rest period after each trial and an extended rest period every 50 trials. Each trial began with a 1500 ms interval to be used as a baseline in the EEG data analysis. Immediately after the baseline period, the fixation dot on the computer monitor was replaced with one of three cues for a duration of 200 ms. The cue informed the subject of which movement to prepare to execute. A left arrow indicated a left button press, a right arrow indicated a right button press and a circle indicated that the subject should not prepare any movement at all. Trials in which no movement was prepared are hereafter referred to as "no–go" trials whilst left and right hand movement trials were collectively referred to as "go" trials. These cues are summarised in Table 1. After random delay of between 500 and 2000 ms a go signal was presented to the subject in the form of an auditory tone. Subjects were asked to respond to the go signal by making the cued movement as quickly as possible. To encourage the subject to minimise their reaction time, feedback was given upon the user's response in accordance with their reaction time. This feedback was in the form of increasingly congratulatory sounds depending on the speed of the subject reaction. The timing of the experiment is shown in Fig. 1.

2.1 Data Collection

Behavioural data in the form of reaction times and EEG activity were collected throughout each experimental session. EEG activity was measured using a g.HIamp amplifier by gtec Medical Instruments GmbH, Austria, which recorded activity from 62 scalp locations conforming to a 10/10 electrode montage. EEG measurement was carried out at 512 Hz using a right ear reference and active electrodes.

2.2 Data Analysis

Reaction time was taken as the time between the onset of the go signal and the subject's response. Accuracy was defined as the percentage of correct responses where incorrect movements or failures to respond were taken as errors. EEG data analysis was carried out using Matlab in conjunction with the EEG/MEG data analysis toolbox Fieldtrip [4]. The continuous EEG data was first preprocessed into epochs time–locked to the onset of the preparatory cue. In the present analysis, only trials with a delay period of 2000 ms were considered which constituted 50 % of the total. The remaining trials were included to ensure that the subject could not predict the "go" signal and therefore had to focus their attention for the duration of the random delay. Artifact rejection was carried out manually by omitting trials and channels on the basis of EEG signal variance. All analysis was focused on the *alpha* band of the EEG spectrum $(8 - 12\,\text{Hz})$ which was extracted using a Morlet wavelet transform. Two analysis approaches were used to investigate how effectively a classification might be made between "go" vs. "no–go" trials as well as left hand movement vs. right hand movement trials. To investigate left hand vs. right hand trials an area under the ROC curve (AUC) analysis was carried out using a threshold placed on the log lateralisation index as a binary classifier. To calculate the lateralisation index, the *alpha* band power from each channel in the left hemisphere (α_L) was matched to counterpart in the right hemisphere (α_R) and the index was calculated for the pair using $LI = \log(\alpha_l/\alpha_R)$. AUC scores were calculated for each channel pair and time within the period between the cue and the go signal. The statistical significance of the scores was assessed using a permutation test to generate a population of 300 AUC scores against which the original score was compared using a two-tailed t-test. The "go" vs. "no–go" trials weren't expected to exhibit any lateralisation so instead logistic regression was used to discriminate between left and right hand movement and no movement trials. The classifier was trained for each time in the cue / go signal delay period using the instantaneous *alpha* band power measured from all channels as a feature vector. To reduce the effects of overfitting and perform dimensionality reduction a regularisation component was also included and set to $\alpha = 0.2$. The accuracy of the resulting classifiers was evaluated using five–fold cross–validation.

3 Results

The average reaction time across all subjects was $416 \pm 112\,\text{ms}$ while the accuracy was $97.9 \pm 2.2\,\%$. Classification accuracy between "go" and "no–go" trials exhibited a high degree of variability depending on the latency at which the classifier was trained. All subjects temporarily reached periods of statistically significant classification however only the subjects #1, #2 and #5 showed sustained periods high classification $(> 100\,\text{ms})$. By inspecting the topography of weights learned by classifier during these periods it was found that these classifications were made primarily on the basis of an *alpha* desynchronisations in the occipital region. Figure 2a shows the classification accuracy left hand movement

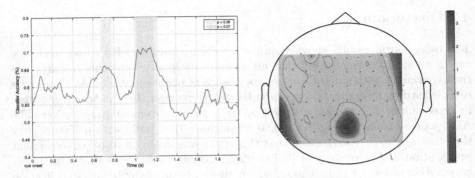

Fig. 2. Subject #1 accuracies (a) and electrode importance map (b) results. (a) Left hand / no movement classification accuracy vs. time. (b) Left hand / no movement classifier importance map.

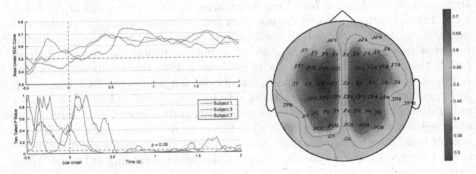

Fig. 3. Subjects #1, 3 and 7 AUC scores and statistical significances (a) subject #3 AUC score topography. (b). (a) Statistical significances of AUC scores vs. time. (b) AUC score topography in left hand / right hand discrimination.

vs. no movement by the subject #1 with respect to time. Figure 2b shows the importance map of the weights learned by the classifier during the periods of statistically significant classification.

In the AUC analysis, subjects #1, 3 &7 showed statistically significant lateralisations of *alpha* activity in response to left vs. right hand movement priming. These manifested as an increase in the AUC score almost immediately after the onset of the cue and reaching statistical significance ($p < 0.05$) at roughly t = 500 ms. This effect was observed for the duration of the trial and peaked in significance over the $C3 - C4$ and $CP1 - CP2$ electrode pairs.

Figure 3a shows the changes in AUC score with respect to time for the most highly discriminating electrode pairs in subjects #1, 3 &7. Figure 3a shows the statistical significance of these AUC scores in a two tailed t-test. Figure 3b shows the topography of the AUC scores for each electrode pair, note that the right hemisphere scores were set to $1 - AUC$ for the corresponding electrode pair.

4   Discussion

The behavioural results showed similar reaction times and high task accuracy across all subjects suggesting that all seven participants were able to carry out the experiment with similar, satisfactory levels of proficiency. The AUC analysis revealed strong lateralization in *alpha* band activity over the sensorimotor cortex in three subjects in response to left or right hand movement preparation. The AUC scores reached strong statistical significance ($p < 0.01$) within 500 ms after the onset of the cue; suggesting that this effect was indeed a result of movement preparation. In the context of BCI control, this results suggests that a user may issue either a left or right command on the basis of *alpha* band lateralisation over $C3 - C4/CP1 - CP2$ electrode pairs through MP in the same way as is possible through the use of MI.

Additionally, the "go" vs "no–go" trials, showed spurious but statistically significant discrimination based on desynchronisations in occipital *alpha* activity. These desynchronisations did not show any consistent spatial characteristics between the different movement conditions therefore it's likely that this distinction is a result of a heightened state of attention during the movement trials rather than any motor specific activity. Nevertheless, this result suggests that it might be possible to distinguish intentional left and right BCI commands from ambient lateralisation in rolandic activity. The question remains as to why only three of the seven subjects exhibited significantly elevated AUC scores. The observed reaction times suggested all subjects were actively engaged with the mental task, however a more fine grained analysis may reveal underlying differences to suggest reduced concentration in the non–lateralising subjects. Alternatively, similar research on both motor priming [1] as well as visual spatial attention [9] has found that different subjects show varying degrees of lateralisation which might account for this inter-subject variability. If this is the case then a more general feature extraction method such as a common spatial pattern might yield better separation of left vs. right hand trials.

5   Conclusions

This research demonstrates that by carrying out a simple movement preparation task given only basic instructions, it's possible for subjects to achieve significant changes in *alpha* band EEG activity that may be used to drive a brain computer interface. These changes in brain activity have similar characteristics to with those associated with motor imagery in terms of a contralateral decrease and an ipsilateral increase in *alpha* power relative to the hand being prepared for movement. Unlike MI, MP is linked to a tangible behavioural goal and should therefore be easy and intuitive for new users to acquire. The is hypothesis is supported by the fact that the three subjects who achieved significant lateralisation did so in one experimental session despite having no prior experience with MP tasks.

Further investigation into the reaction time characteristics of the non-laterali-sing subjects along with more general EEG feature selection is expected to account for the observed variability in subject performance.

References

1. Deiber, M.P., Sallard, E., Ludwig, C., Ghezzi, C., Barral, J., Ibañez, V.: EEG alpha activity reflects motor preparation rather than the mode of action selection. Front. Integrat. Neurosci. **6**(59) (2012)
2. LaFleur, K., Cassady, K., Doud, A., Shades, K., Rogin, E., He, B.: Quadcopter control in three-dimensional space using a noninvasive motor imagery-based brain-computer interface. J. Neural Eng. **10**(4), 046003 (2013)
3. Lotte, F., Larrue, F., Mühl, C.: Flaws in current human training protocols for spontaneous brain-computer interfaces: lessons learned from instructional design. Front. Hum. Neurosci. **7**, 568 (2013)
4. Oostenveld, R., Fries, P., Maris, E., Schoffelen, J.M.: Fieldtrip: open source software for advanced analysis of MEG, EEG, and invasive electrophysiological data. Comput. Intell. Neurosci. **2011**, Article no. 1 (2010)
5. Pfurtscheller, G., Brunner, C., Schlögl, A., da Silva, L.: Mu rhythm (de) synchronization and EEG single-trial classification of different motor imagery tasks. Neuroimage **31**(1), 153–159 (2006)
6. Pfurtscheller, G., Neuper, C.: Motor imagery and direct brain-computer communication. Proc. IEEE **89**(7), 1123–1134 (2001)
7. Rushworth, M.F., Krams, M., Passingham, R.E.: The attentional role of the left parietal cortex: the distinct lateralization and localization of motor attention in the human brain. J. Cogn. Neurosci. **13**(5), 698–710 (2001)
8. Scherer, R., Lee, F., Schlogl, A., Leeb, R., Bischof, H., Pfurtscheller, G.: Toward self-paced brain-computer communication: navigation through virtual worlds. IEEE Trans. Biomed. Eng. **55**(2), 675–682 (2008)
9. Van Gerven, M., Bahramisharif, A., Heskes, T., Jensen, O.: Selecting features for BCI control based on a covert spatial attention paradigm. Neural Netw. **22**(9), 1271–1277 (2009)
10. Wolpaw, J., Wolpaw, E.W.: Brain-Computer Interfaces: Principles and Practice. Oxford University Press, Oxford (2012)

Discriminating Object from Non-object Perception in a Visual Search Task by Joint Analysis of Neural and Eyetracking Data

Andrea Finke[✉] and Helge Ritter

Neuroinformatics Group and CITEC, Bielefeld University,
Inspiration 1, 33619 Bielefeld, Germany
{afinke,helge}@techfak.uni-bielefeld.de

Abstract. The single-trial classification of neural responses to stimuli is an essential element of non-invasive brain-machine interfaces (BMI) based on the electroencephalogram (EEG). However, typically, these stimuli are artificial and the classified neural responses only indirectly related to the content of the stimulus. Fixation-related potentials (FRP) promise to overcome these limitations by directly reflecting the content of visual information that is perceived. We present a novel approach for discriminating between single-trial FRP related to fixations on objects versus on a plain background. The approach is based on a source power decomposition that exploits fixation parameters as target variables to guide the optimization. Our results show that this method is able to classify object versus non-object epochs with a much better accuracy than reported previously. Hence, we provide a further step to exploiting FRP for more versatile and natural BMI.

1 Introduction

Fixation-related potentials (FRP) are not only a valuable means to study cognitive processes, but promise to open a path for more capable and versatile brain-machine interface (BMI) based on electroencephalography (EEG). FRP belong to the class of event-related potentials (ERP). Unlike the standard ERP that are time-locked to stimuli (e.g., the popular P300) in usually simple and restricted experimental paradigms, FRP are time-locked to fixation onsets, and can thus be assessed by the simultaneous acquisition of EEG and eyetracking data. Moreover, they allow for a direct insight into the cognitive processes that correspond to the perception of particular visual information, for example, whether an object is perceived or irrelevant background. To be useful as a means for interaction technology, e.g. in a BMI or a multi-modal interface, FRP must be classified on a single-trial basis. Furthermore, it is essential to not only discriminate single-trial epochs in a binary fashion whether they contain an FRP or not, but to additionally discriminate epochs based on the visual content of the fixated information. Recently, we have presented a study where we applied a gaze-contingent visual search task to acquire FRP data [2]. There, we classified single-trial EEG epochs

© Springer International Publishing AG 2016
A. Hirose et al. (Eds.): ICONIP 2016, Part II, LNCS 9948, pp. 546–554, 2016.
DOI: 10.1007/978-3-319-46672-9_61

based on whether a target object, a non-target object or the plain background was fixated. Our results showed that while target epochs could be discriminated from both remaining classes (non-target and background) quite well, the discrimination of non-target versus background epochs gave significantly inferior results. This indicates a general difficulty to discriminate between object and non-object content of the fixation, which is an issue for the exploitation of FRP in more complex and natural scenarios.

We present a novel approach for the single-trial classification of FRP based on whether the respective epochs belongs to an object or non- object fixation. We exploit a method for source separation, Source Power Comodulation (SPoC), which was recently presented by Daehne and colleagues [1] and is capable of exploiting a target variable to guide the source separation optimization. We apply the fixation durations as target variable and thus SPoC provides a truly bi-modal approach to feature extraction and classification of FRP.

2   Methods

2.1   Experiment

Gaze-Contingent Search Task. We designed a gaze-contingent search task to record single-trial FRP data. Gaze-contingency refers to a technique, where not a full stimulus image is visible, but only a small part which corresponds to a circle around the current gaze position, a "keyhole". This technique allows for unrestricted search scenarios while at the same time full experimental control is retained over the exact piece of visual information that is processed by the participant at a time. We used 112 stimuli (equals the number of trials), each of which contained 12 everyday objects on plain white background. In each trial, one specific object had to be found and a key pressed when the task was accomplished. Please refer to [2] for more in-depth information on gaze-contingency and our experimental design. We conducted the experiment with ten participants aged 21 to 34 (mean 26.8 ± 3.7, 7 female).

Data Acquisition. EEG data was recorded with a gUSBamp 16-channel amplifier (Guger Technologies, Graz, Austria) at the locations Fz, Cz, Pz, Oz, F4, C4, P4, PO8, F3, C3, P3, PO7 together with four additional EOG channels (electrooculogram). EEG data was sampled at 256 Hz and highpass filtered at 0.1 Hz to remove the DC and the slow drift over time. Eyetracking data was acquired at 120 Hz (interlaced) with a binocular remote eyetracker that allows for head movements (EyeFollower™, LC Technology, Clearwater, USA). Synchronization of EEG and eyetracking data was achieved by logging mutual time stamps for both streams.

2.2　Eye Movements and Ocular Artifacts

Fixation Detection. The onsets of fixations serve as time stamps for the segmentation of the continuous EEG data stream and need to be computed from the raw eyetracking data. Further relevant measures are fixation duration and the location of the fixation on the stimulus image. The latter were divided into three groups: on a target object, on a non-target object and on the plain background. Here, we only consider the latter two groups for the classification task. The algorithm used a spatial and a temporal threshold criterion: A fixation was considered valid if at least three consecutive samples (50 ms) fell into a radius of 2deg visual angle (this corresponds to the foveal vision).

Semi-automatic Removal of Ocular Artifacts. Eye movements are the major sources of artifacts in EEG data and can easily lead to spurious effects that could be misinterpreted as neural activity. Hence, we need to remove those signal portions that originate from eye movements. An effective and common method to remove ocular artifacts is Independent Component Analysis (ICA), a blind source separation method that computes a demixing matrix $\mathbf{W}$ representing independent sources of activity. Artifact removal is typically done by manually inspecting the components (i.e., the columns of $\mathbf{W}$) based on their scalp topology, spectral properties and the inter-trial variability. This process is time-consuming and not suited for a fully automated data analysis.

We evaluated a recently proposed method for automatic removal of artifacts [7] based on supervised classification. First, we segmented the continuous EEG data based on the previously computed fixation onsets with 1 s epoch length. Then, we grouped the epochs based on the fixation location classes: non-target and background. ICA is known to capture sources of higher frequency noise (typically from muscular artifacts) very well [4]. However, this kind of noise can be easily eliminated by lowpass filtering the signal. ERP/FRP components of interest reside below 10 Hz. Ocular artifacts, however, occupy a similar frequency band, which is the reason why they are so delicate for FRP analysis and have to be dealt with very carefully. Thus, we applied a lowpass filter with a cut-off frequency of 30 Hz prior to the ICA decomposition. MARA was used in a first, automated step to classify components likely to represent sources of artifacts. In a second, manual step the ICA components were visually inspected as explained before and the results of the automated classification evaluated. We noticed that the algorithm was not able to detect all artifactual sources. In fact, for 6 out of the 10 data sets the algorithm did not identify "bad" components at all, although there were clearly some that largely originated from ocular rather than neural sources. This is probably due to the mixed nature of the components, which contain largely, however not exclusively, eye movements. We marked those components for rejection as well, before projecting all epochs into source space based on the reduced demixing matrix. The number of removed components varied between one and four, depending on the individual dataset.

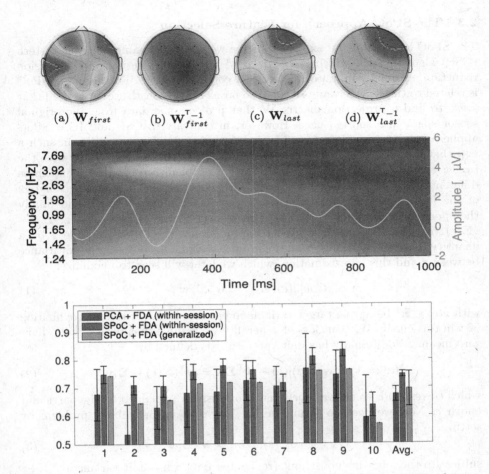

Fig. 1. *Top:* SPoC eigenvectors plotted topographically for a representative dataset (no. 8). Left: First component (a) spatial filter, (b) spatial pattern. Right: Last component (c) spatial filter, (d) spatial pattern. *Middle:* Fixation-related potentials can be seen as slow waves: The yellow line shows the averaged activation for all target epochs at electrode Cz in a representative dataset (no. 9). The image plot shows the spectral representation of this signal obtained by a continuous wavelet transformation (CWT) with a complex morlet wavelet. One can nicely see, how the FRP is reflected as a slow frequency component in the range between 3 to 5 Hz. This representation supports the application of a narrow bandpass filter (1–7 Hz) prior to computing the SPoC decomposition. *Bottom:* Comparison of the previous and the new results for discriminating non-target from background related epochs. The two blue bars show the average for a ten-fold cross-validation in the within-session condition. The error bars indicate the standard deviation. The green bars show the more challenging between-session generalized approach, where training and testing data stem from different sessions and participants. In this condition no cross-validation is needed and thus there are no error bars available for individual data sets. Please note that the y-axis starts at chance level. (Color figure online)

2.3   The SPoC Approach for Feature Selection

The SPoC approach is an extension of the well known common spatial pattern (CSP) algorithm [5], which is widely used for spatial filtering and dimension reduction, especially in motor imagery-based BMI systems. Furthermore, SPoC is related to the ICA decomposition in so far as it also constitutes a method that seeks to find a projection matrix $\mathbf{W}$ that projects the data from the original sensor space into source space. However, in contrast to CSP and ICA, SPoC applies a target variable z to guide and optimize the decomposition in such a way that the power modulations in the source space co-vary maximally with the target variable. To express this optimization problem, we need to define three different types of covariance matrices. The global or mean covariance matrix of the full dataset: $\boldsymbol{\Sigma}$. The epoch-wise covariance matrices $\boldsymbol{\Sigma}(e)$, and finally the epoch-wise covariance matrices with respect to the target variable: $\boldsymbol{\Sigma}_z := \langle \boldsymbol{\Sigma}(e)z(e) \rangle$. Furthermore, we need to approximate the target variable z by a quantity that can be derived from the projected signal to compute the covariance between z and this approximation, which we later will seek to maximize:

$$Cov[\hat{z}(e), z(e)] = \mathbf{w}^\mathsf{T} \boldsymbol{\Sigma}_z \mathbf{w} , \tag{1}$$

with $z(e) \in \mathbb{R}^1$ being the target variable in epoch e and $\hat{z}(e)$ the approximation of z in this epoch. Without loss of generality we assume that both z and $\hat{z}$ have zero mean. z additionally has unit variance. $\hat{z}$ is defined by

$$\hat{z}(e) := Var[\mathbf{w}^\mathsf{T} \mathbf{x}(t)](e) - \mathbf{w}^\mathsf{T} \boldsymbol{\Sigma} \mathbf{w} = \mathbf{w}^\mathsf{T}(\boldsymbol{\Sigma}(e) - \boldsymbol{\Sigma})\mathbf{w} , \tag{2}$$

which corresponds to the variance that is exclusively explained by the particular epoch e. Now we need to formulate the above as an optimizing problem by setting

$$f_\lambda = Cov[\hat{z}(e), z(e)] = \mathbf{w}^\mathsf{T} \boldsymbol{\Sigma}_z \mathbf{w} , \tag{3}$$

subject to the following constraint (remember only z has unit variance)

$$Var[\mathbf{w}^\mathsf{T} \mathbf{x}(t)] = \mathbf{w}^\mathsf{T} \boldsymbol{\Sigma} \mathbf{w} \overset{!}{=} 1 . \tag{4}$$

This optimization problem can be solved efficiently by using Lagrange multipliers and setting the derivation of the Lagrangian to zero. One then obtains a generalized eigenvalue problem of the following form:

$$\boldsymbol{\Sigma}_z \mathbf{w} = \lambda \boldsymbol{\Sigma} \mathbf{w} . \tag{5}$$

Solving this generalized eigenvalue problem gives the desired projection matrix $\mathbf{W}$, with the eigenvectors in the columns. If the columns are sorted in descending order with respect to the corresponding eigenvalues λ, one obtains the eigenvector with the largest positive correlation in the first and with the largest negative correlation in the last column of $\mathbf{W}$. Figure 1 (top) shows these two eigenvectors (the filters, used for projecting the original data) and the corresponding spatial patterns (the neurologically plausible activations), which can be obtained by $\mathbf{W}^{\mathsf{T}-1}$. This relationship is also known from the CSP approach [5].

The SPoC approach was originally developed to cope with EEG components that are defined by oscillation in a particular frequency band. One prominent example is data from motor imagery-based brain-machine interfaces, where the component of interest, event-related (de)synchronization (ERD) constitutes a decrease in mu- and beta-band power over the sensorimotor cortex areas during the imagination of moving a particular limb, e.g., one hand [5]. For FRP an interpretation in terms of oscillatory properties is not as straight forward, because ERP in general are usually defined by the polarity and latency of the amplitude deflection. However, each single trial ERP can be considered as a slow wave at a particular frequency. For FRP, this frequency is around 3 to 4 Hz. Therefore, we applied a narrow bandpass filter (1–7 Hz) prior to running the SPoC algorithm (but after artifact removal with ICA). Additionally, the data was downsampled by factor ten, to reduce the amount of redundant information. Figure 1 (middle) depicts the analogy of the FRP potential and a slow wave. This correspondence is further reflected in the first SPoC-computed pattern in Fig. 1 (top, b), where a central positivity is the dominant activation. Hence, the SPoC approach is able to capture and amplify the major property of FRP in an appropriate fashion.

2.4   Classification

Classification was achieved using Fisher's Discriminative Analysis (FDA) in a binary fashion. FDA seeks to find a one-dimensional projection of the feature vectors onto a weight vector $\mathbf{w}$ that maximizes the distance between the projected means of the two classes. In feature space, the optimization criterion is the so called Fisher criterion $J(\mathbf{w}) = \mathbf{S}_b/\mathbf{S}_w$, where $\mathbf{S}_b$ is the between-class scatter matrix and $\mathbf{S}_w$ the within-class scatter matrix. $J(\mathbf{w})$ has thus to be maximized to obtain the optimal class separating hyperplane. Estimating well-conditioned covariance (scatter) matrices on training data requires to use a regularized approach as postulated by the Ledoit-Wolf theorem [3]: $\Sigma^* = \lambda\widetilde{\Sigma} + (1 - \lambda)\,\Sigma$, with Σ being the sample covariance matrix, $\widetilde{\Sigma}$ the sample covariance of a sub-model, and $\lambda \in [0, 1]$ denoting the shrinkage intensity. As proposed in [6], we use a diagonal matrix where all diagonal elements are equal, that is, all variances are equal and all covariances are zero.

Performance Assessment. The number of non-target and background epochs varies per participant, resulting in an imbalanced distribution of classes. Thus, the performance of the approach was assessed using the area under the receiver operator characteristics (ROC), the AUC. The ROC is obtained by mapping all real-valued classifier outputs $y_i = \mathbf{x} \cdot \mathbf{w}$ onto the interval $[0, 1]$ and testing $\hat{y}_i = y_i + b_k$ for $b_1 = 0$ and $b_K = 1$ with $k = 100$ and the b_k monotonically increasing and equally spaced. Then, the number of true positives is expressed as a function of the number of false positives. Consequently, the chance level for the AUC is always 0.5.

3 Results

Our aim in this work was to find a suitable method that is capable of improving the previously low discrimination between non-target and background-fixation related epochs. Although previously clearly above chance level, the results are not satisfactory with respect to potential online applications in BMI systems. We distinguish between two cases (or conditions) of different complexity: A *within-session*, and *between-session generalized* approach. The former refers to the case, where training and testing data originate from the same dataset, recorded from one participant in one continuous session. The latter denotes the case where all training data is from different participants and/or sessions.

3.1 Within-Session Cross-Validation

Classifier performance and variance was tested using a ten-fold cross-validation. Each dataset was divided randomly into ten disjoint subsets that contained the same proportion of samples from both classes as present in the overall data set. Neither the training nor the test sets were balanced, that is, the number of samples (or the apriori probability) in each class reflected the true distribution of classes in the data set. This distribution varied per set due to the individual search pattern of each subject and consequently directly reflects the number of fixations made by a participant. Nine out of the ten sets were merged and used for computing the SPoC projection matrix and to train the FDA classifier. The remaining tenth set was used for testing. This procedure was iterated until each set had served once as test set. We evaluated our previous results using PCA (please refer to [2] for details on the method) against the proposed novel approach based on the SPoC decomposition. While with the former method, the binary classification of non-targets versus background yielded only an average AUC of 0.67, the latter, novel approach resulted in an average AUC of 0.75. This improvement is clearly statistically significant. A paired, two-tailed t-test gives a p-value of 0.02 (t=2.50). Figure 1 (bottom, blue bars) shows the detailed results for each data set.

3.2 Generalization Between Sessions and Participants

Procedure. In this condition nine out of the ten available datasets (recorded from *different* participants) were merged to one large training set. Here, no cross-validation was needed. As before, the class distribution was imbalanced and all available epochs used for training. Testing and performance assessment was then done on the tenth dataset that stemmed from a different participant. This procedure was iterated until each individual dataset had been classified once. For this case no previous AUCs were available. The reason is that with our previous approach the results in this more demanding condition hardly exceeded the chance level. Thus, we need to compare the newly obtained results with the previous ones from the less complex case (the within-session condition). The resulting AUCs are also given in Fig. 1 (bottom, green bars). As one can see, the AUCs are now very clearly above chance with an average AUC of 0.69. For six out of the ten datasets, the AUC of the SPoC-based between-session generalized condition is even higher than previously in the simpler condition. There is obviously a very clear advantage of our novel approach compared to the previous one.

4 Discussion and Conclusion

The discrimination of different categories based on FRP during natural vision, for example, during a search or exploration task, promises to allow for BMI that can be applied in more natural and less constrained settings than it is the case today. While we could already discriminate between fixations on target objects and fixations on other locations with a high accuracy in our previous work, the discrimination of non-target objects and plain background remained, although above chance, on a low level. The more challenging and very important approach for practical applications, the between-session generalized training scheme, could not be accomplished with accuracies above chance. In this paper, we showed that by applying a different method for extracting features from single-trial FRP data that make additional use of fixation parameters (in our case the durations), we could significantly increase the classification accuracy in the within-session cross-validation condition and even achieve reasonable accuracies, clearly above chance, in the between-session generalized condition. For more than half of the datasets these were even higher than in the simpler condition. Hence, we made another step forward towards exploiting FRP in BMI systems that do not depend on artificial stimuli any more, but can operate in natural, complex real-world environments. The next steps on this path need to relax the constraints on the experimental setting even further and investigate how the proposed methods can cope with the data acquired in such scenarios. Ultimately, the overall goal is to develop human-machine interfaces that exploit FRP to understand and interpret the neural correlates of human visual perception.

Acknowledgments. This research/work was partially supported by the Cluster of Excellence Cognitive Interaction Technology 'CITEC' (EXC 277) at Bielefeld University, which is funded by the German Research Foundation (DFG).

References

1. Daehne, S., Meinecke, F.C., Haufe, S., Hoehne, J., Tangermann, M., Mueller, K.R., Nikulin, V.V.: SPOC: A novel framework for relating the amplitude of neuronal oscillations to behaviorally relevant parameters. NeuroImage **86**, 111–122 (2014)
2. Finke, A., Essig, K., Marchioro, G., Ritter, H.: Toward FRP-based brain-machine interfaces: single-trial classification of fixation-related potentials. PLoS ONE **11**(1), 1–19 (2016)
3. Ledoit, O., Wolf, M.: A well-conditioned estimator for large-dimensional covariance matrices. J. Multivar. Anal. **88**(2), 365–411 (2004)

4. Pignat, J.M., Koval, O., Ville, D.V.D., Voloshynovskiy, S., Michel, C., Pun, T.: The impact of denoising on independent component analysis of functional magnetic resonance imaging data. J. Neurosci. Methods **213**(1), 105–122 (2013)
5. Ramoser, H., Mueller-Gerking, J., Pfurtscheller, G.: Optimal spatial fitering of single trial EEG during imagined hand movement. IEEE Trans. Rehab. Eng. **8**(4), 441–446 (2000)
6. Schaefer, J., Strimmer, K.: A shrinkage approach to large-scale covariance matrix estimation and implications for functional genomics. Stat. Appl. Genet. Mol. Biol. **4**(1), Article no. 32 (2005)
7. Winkler, I., Haufe, S., Tangermann, M.: Automatic classification of artifactual ICA-components for artifact removal in EEG signals. Behav. Brain Funct. **7**(1), 1–15 (2011)

Assessing the Properties of Single-Trial Fixation-Related Potentials in a Complex Choice Task

Dennis Wobrock[✉], Andrea Finke, Thomas Schack, and Helge Ritter

Neuroinformatics Group and Neurocognition and Action Group,
CITEC, Inspiration 1, 33619 Bielefeld, Germany
{dwobrock,afinke,helge}@techfack.uni-bielefeld.de,
thomas.schack@uni-bielefeld.de

Abstract. Event-related potentials (ERP) are usually studied by means of their grand averages, or, like in brain-machine interfaces (BMI), classified on a single-trial level. Both approaches do not offer a detailed insight into the individual, qualitative variations of the ERP occurring between single trials. These variations, however, convey valuable information on subtle but relevant differences in the neural processes that generate these potentials. Understanding these differences is even more important when ERP are studied in more complex, natural and real-life scenarios, which is essential to improve and extend current BMI. We propose an approach for assessing these variations, namely amplitude, latency and morphology, in a recently introduced ERP, fixation-related potentials (FRP). To this end, we conducted a study with a complex, real-world like choice task to acquire FRP data. Then, we present our method based on multiple-linear regression and outline, how this method may be used for a detailed, qualitative analysis of single-trial FRP data.

1 Introduction

In non-invasive brain-machine interface (BMI) research based on electroencephalography (EEG), event-related potentials (ERP) count among the most common signal components. The best studied and widely applied component in this context is the P300 potential (e.g., [3]). In BMI, ERP have to be classified on a single-trial level to implement fast and reliable interfaces. On a more general and fundamental level, ERP are typically studied by their *grand averages*. The latter refers to averaging epochs locked to one stimulus category over all trials and all participants from one experiment. The properties of interest, that is, the amplitude, latency and morphology (the duration) of the studied potential are then assessed based on this grand average. While the general idea - the detection of effects that are stable over a whole population - is reasonable, this method cannot assess the between-session and between-participant variations that may contain valuable information on subtle effects that occur on the single-trial level. In a classification task in contrast, using single-trial epochs is the standard case. There, however, one is mainly interested in finding the best features with respect

© Springer International Publishing AG 2016
A. Hirose et al. (Eds.): ICONIP 2016, Part II, LNCS 9948, pp. 555–563, 2016.
DOI: 10.1007/978-3-319-46672-9_62

to the discrimination task. Hence, qualitative aspects of the variations mentioned above have rarely been explored. Recently, fixation-related potentials (FRP) have been investigated as a novel means for more intuitive BMI and classified on a single-trial level, e.g., in simple [1] and more complex gaze-contingent [2] search tasks. Also they have been shown to be modulated by temporal uncertainty in appearance of moving stimuli [8]. FRP are ERP that occur at the onset of an eye fixations during task-driven scene exploration. Future BMI based on FRP will, however, require FRP properties to persist during complex, natural and information rich scenarios. Investigating the single-trial variation of FRP properties can potentially give insight into the mechanisms underlying complex scene exploration and help us to improve future BMI systems.

In this paper, we present a novel approach to assess and study the single-trial properties of FRP in a choice task that is much more complex than the scenarios used in the studies quoted above. To this end, we seek to determine the amplitude, latency and morphology of single-trial FRP. Our approach is an adaptation and extension of a method introduced by Hu et al. [5,6], MLRd - multiple linear regression with a discriminant term. This approach serves as an intermediate step between the high level grand averages and single-trial classification. The method is able to extract the above mentioned parameters from single-trial epochs by means of a model generated from the grand average. While Hu and colleagues used their original method on data from a simple and very controlled experiment resulting in prominent ERP with a high signal-to-noise ratio (laser-evoked potentials), our modifications are essential for our much more complex setting resulting in more noisy data.

2   Study

2.1   Task

The experiment was designed to study how FRP properties vary in real-life like situations. Departing from a previous study [3], we incorporated complexity on both the task and the stimulus level. Conceptually, the participants were required to group objects into subsets based on a given task. To use a common and frequent setting, we chose a kitchen scenario. The experiment consisted of 40 trials, in each of which a stimulus image was presented that contained 6 to 10 typical and everyday objects that are commonly found in a kitchen (e.g., orange, knife, towel). The images were photographs that we took in our institutes's lab kitchen (see Fig. 3). Each trial, i.e. each image, was associated with a specific task (e.g., peel an orange, make a coffee). Each image contained objects that were relevant or useful for the task (one to five) and some that were not. The participants were instructed to select in each trial those objects that they deemed to be useful and to memorize these items. The 40 trials were randomized for each participant. In total, we tested 11 participants (5 female). Each trial started with the task shown on the screen. The participants had to press a key once they had understood the task and were then presented with the stimulus image. There was no time limit for accomplishing the choice task.

Subsequently, the memorized and chosen items had to be indicated on a blurred version of the stimulus image by clicking on them. This strategy was used to firstly, remove the need for memorizing too many details of the chosen objects, and secondly, to discourage participants from changing their selection during this second phase. Each trial resulted in a set of *chosen* and *non-chosen* objects. We will use these terms throughout this paper to label epochs according to whether they belong to a fixation on a chosen, or a non-chosen object. Additionally, we will use *background*, to refer to epochs associated with fixations on non-object background of the image (Fig. 1).

Fig. 1. The presented kitchen scene associated with the task "peel an orange". The circles correspond to fixations, circle size represents the relative fixation duration. Green and red circles mark the first and last fixation during scene exploration. (Color figure online)

2.2 Apparatus

We used two synchronized g.USBamp 16-channel EEG amplifiers for the study. Sixteen EEG channels were recorded at the locations Fz, F3, F4, F7, F8, Cz, C3, C4, T7, T8, Pz, P3, P4, PO7, PO8 and Oz, referenced to the mastoids. Impedances were kept below $5\,k\Omega$. The devices sampled data with a frequency of $256\,Hz$ and applied a built-in $0.1\,Hz$ high-pass Butterworth filter. The eye movements were recorded with the EyeFollower^{TM}, a binocular remote eye tracker sampling at $120\,Hz$ (interlaced) with an accuracy of $< 0.4°$.

3   Methods

3.1   Fixation Detection and Artifact Removal

Fixation detection was done using a threshold-based algorithm proposed by the manufacturer of the eye tracking device. The algorithm considered that a fixation was valid when 6 gaze points (corresponding to 100 milliseconds) were present within a 2 degree foveal radius from each other. Resulting fixations were labeled as chosen, non-chosen or background depending on their distance to selected and non-selected objects in the scene. This labeling is used to be able to categorize the EEG epoch resulting from the segmentation. This was done to explore potential FRP property differences between these conditions.

The recorded EEG data was segmented into epochs of 1 second duration (256 samples) according to previously calculated fixation onsets. Eye movement artifacts were removed using an independent component analysis (ICA) based rejection. The artifact-free EEG data was bandpass filtered with cutoff frequencies of 0.5 and 10 Hz. As to consider only early components, epochs were shortened to 400 ms (102 samples) duration and grouped according to their associated fixation labels. Lastly, to ensure using only epochs which clearly present FRP components, we retained only those associated with the first fixation on an inspected object. All other epochs were discarded.

3.2   The MLRd Method

The MLRd method, as proposed by Hu et al. [5], is an approach consisting of recovering the location of relevant peaks, which were identified in the grand average, in every epoch of a dataset. By locating these peaks, their properties, such as amplitude or latency, can be extracted. Consequently, acquiring these properties across the entire dataset enables visualization of their distribution.

The MLRd method is composed of a variety of steps, starting by identifying those peaks in the grand average that provide meaningful task-related information and thus serve as templates in the subsequent regression. These peaks will be called *model peaks*. We will present all these steps in detail and outline our modifications.

Peak Selection. To identify all potentially relevant peaks in FRPs, we visually inspect the grand average of each channel (see Fig. 2a). The only notable peaks are potentials commonly attributed to visual scene exploration, such as an occipital P100 or a fronto-central N100/P200 cluster [7]. However, since we do not know if these peaks correspond to the relevant peaks in this experiment, we take a more in depth look at the squared differences (r^2) between the grand averages of epochs from the chosen and non-chosen conditions (see Fig. 2b).

Maxima of r^2 curves identify local differences between conditions. These relate to peaks presenting high variability across epochs. To associate r^2 differences with grand average extrema, we compare temporal distances between the two.

Here Mr_i is the ith maximum of the r^2 curve, Ep_j is the jth extremum in the grand Average and Pm_i is the peak corresponding to Mr_i. The set of Pm peaks will serve as templates and be referred to as model peaks.

$$Pm_i = Ep_j| \min_j |latency(Ep_j) - latency(Mr_i)|$$

Extensions to MLRd. To facilitate localizing model peaks in every epoch, the MLRd method simplifies each epoch through regression. This regression uses the first three principal components, from variability matrices generated

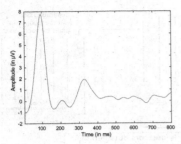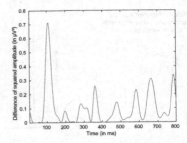

Fig. 2. Analysis of the Grand Averages corresponding to the Oz Channel. (a) Grand Average over all epochs. (b) r^2 curve comparing averages from chosen and non chosen conditions.

from model peak regions, as regressors [5]. To test if peak localization can be improved, we explore the effects different sets of regressors have on the quality of peak location. Indeed, while Hu et al. relied on regressors generated from the prominent extrema regions only [5], those do not guaranteed a proper fit, and thus, an appropriate localization of extrema, for FRPs.

We created three different sets of regressors (see Fig. 3), which we separately tested on the data. These sets contained different amounts of regressors. They were designed to give us insight into the effects of epoch over- or underfitting on peak localization quality. The first set (R1) contained regressors generated from all peaks regions present in the grand average, regardless if they are model peaks or not. The second set (R2) contained regressors generated from all model peaks. The third set (R3) contained only regressors generated from one of the model peaks. While the R1 and R2 sets can be used to localize all model peaks in an epoch, the R3 can distinctly localize only one. Consequently, multiple R3 sets were created, one for each model peak, and were separately used on the data.

Once an epoch is regressed, the model peaks can be located. All extrema present in the fitted epoch are given a set of weights that represent their distances to the model peak. These coefficients are calculated by multiplying the distances in amplitude and latency between the considered extremum and a model peak in the grand average. The extremum with the highest weight value is marked as corresponding to the related model peak. Here, Ef_j is the jth extremum present in the fitted epoch and El_i is extremum corresponding to the ith model peak. The set of El is called the located extrema.

$$El_i = Ef_j \mid \min_j |(amplitude(Ef_j) - amplitude(Pm_i))$$
$$(latency(Ef_j) - latency(Pm_i))|$$

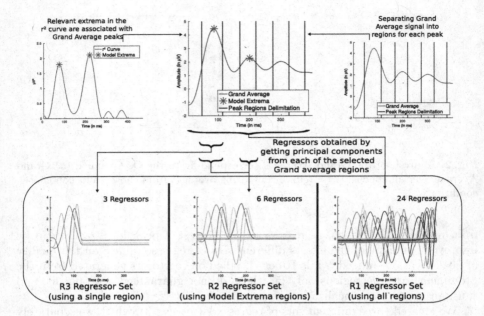

Fig. 3. Three different sets of regressors were created from extremum regions in the grand averages. The curves were generated from the Oz channel of one participant. These specific three sets were used to localize the first model peak in all epochs of Oz.

If the direction (minimum or maximum) or chronological order of the located extrema is not the same as for the model peaks, the located extremum, which disrupts this order, is considered to be a false positive. Consequently, the associated model peak is labeled absent in the epoch. This also occurs if the highest weight value does not exceed a specified threshold, which sets a limit on the distance to the model peaks.

3.3 Qualitative Evaluation

Having located the model extrema for each epoch in the dataset, we now extract the desired properties. Each located extremum is identified by its temporal location, i.e. the sample at which it occurs. This corresponds to the latency. Amplitude is then extracted by getting the magnitude at this sample in the corresponding fitted epoch. Finally, to obtain the morphology, we locate the closest samples to the left and right of the located extremum, where the second derivative is zero. The morphology is the distance between these two samples. After extracting these properties for all single trials, we can plot the respective distributions as histograms. Also, we can compare the distributions between two distinct groups of epochs. Figure 4 shows an overlay of distributions for both conditions. This representation allows comparison as differences can be spotted quickly. For the histograms, we chose a number of bins corresponding the

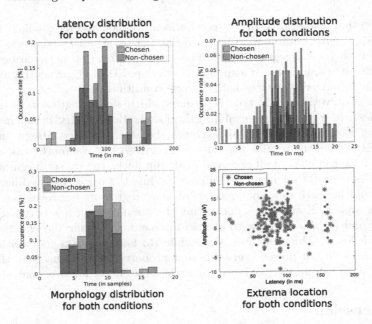

Fig. 4. Histograms showing the probabilistic distribution of properties for the first model peak (P100) in a single subject Oz Channel, extracted from epochs generated with R2 regressors. The bottom-right plot presents peak location in all epochs.

number of unique values a property could take. As our goal is to assess the inter-trial variability, this choice allows to compare the smallest measurable changes.

Latency and amplitude histograms present one or multiple large maxima that differ from the rest of the data. The amplitude histogram, unconstrained by sampling rate, presents the largest amount of bins and the largest range of variation (from -13 to $20\,\mathrm{V}$ for the peak presented in Fig. 4). Latency and morphology present a lower amount of variations (from 12 to 160 ms and 12 to 66 ms in Fig. 4 respectively). To compare the distributions of two experimental conditions, we propose to calculate the areal overlap of the histograms. This bin-by-bin comparison method accentuates differences in peaks.

4   Results

Analyzing the histograms resulting from our experiment, we notice the presence of multiple maxima in the distribution. This indicates that, while variation is present and can occur with a large spread, certain specific values of amplitude, latency and morphology are vastly more common than the rest. Amplitude presents the highest spread on single trial level. Added to the high number of bins, it makes this property the most likely to present notably different distributions, compared to the other properties, between distinct subsets of epochs. This is confirmed by the calculation of the overlap between chosen fixations and non chosen fixation histograms. In this comparison, latency and morphology showed high overlap, with minimum of 70 % overlap across all observed peaks (values obtained with the R2 regressor set). The amplitude overlap between both conditions was, however, lower (mean: 55.6 % overlap, SD: 5.4), presenting multiple values that occurred more frequently in one condition than the other.

These observations hold true regardless which set of regressors is applied. Nonetheless, it is possible to evaluate the localization quality between sets by comparing the number of model extrema that were labeled absent in fitted epochs. Analyzing the results, the R1 set of regressors shows more peak absence than the R1 set in 63 % of all observed peak. Similarly, the R2 set has more peak absence than the R3 set in 64,5 % of cases. While the performance difference is low, it could indicate that the presence of more regressors improves the chance of finding a peak. However, by comparing the previous overlaps between conditions, R2 and R3 showed close to identical percentages, while the results for R1 were much lower. This indicates, that while R3 locates peaks more easily, the located peaks do not present very different property distribution than the ones found by R2. Overall, the R2 regressor is the best choice for this analysis.

5   Discussion

The MLRd method offers a powerful tool to get insight into the different types of variability present within FRPs. Our modifications allow for identification of a common set of early properties across all epochs and observe their variation throughout an experimental session. To improve the identification of ERP of interest, we studied the effects of three different sets of regressors. Our results indicate that the set generated from model peak regions provides the best compromise between quantity and quality of peak localization. The distributions (i.e., the histograms) showed that certain distinct values (i.e., property measures) occur much more frequently than others. This hints at potential subtle categories on how objects are perceived during the experiment. Furthermore, while variations of amplitude are to be expected [3], their presence in the latency and morphology distributions could be due to the increased complexity of the presented experimental task. In this study, we have extended the MLRd method and its application to the, so far, little studied FRP. The MLRd is a complex method with many aspects that can be improved to further broaden its range of application. For example, modifying the procedure to create the variability matrices in such a way that the analysis is no longer limited to regions containing prominent peaks, could further improve the assessment of relevant FRP properties in noisy data from complex tasks.

In general, the qualitative analysis of single-trial epochs as an intermediate step between grand average ERP analysis and single trial classification has, to date, been only little explored. Such approaches, like the one presented in this work, may have two-fold benefits. On the one hand, they can help to advance our understanding of neural processes underlying complex, natural tasks, and on the other hand, to inform classification methods for brain-machine interfaces in order to become more adaptive with respect to individual variations on the single-trial level.

Acknowledgments. This research/work was supported by the Cluster of Excellence Cognitive Interaction Technology 'CITEC' (EXC 277) at Bielefeld University, which is funded by the German Research Foundation (DFG).

References

1. Brouwer, A.M., Reuderink, B., Vincent, J., van Gerven, M.A.J., van Erp, J.B.F.: Distinguishing between target and nontarget fixations in a visual search task using fixation-related potentials. J. Vis. **13**(3), 1–10 (2013)
2. Finke, A., Essig, K., Marchioro, G., Ritter, H.: Toward FRP-based brain-machine interfaces: single-trial classification of fixation-related potentials. PLoS ONE **11**(1), 1–19 (2016)
3. Finke, A., Lenhardt, A., Ritter, H.: The mindgame: a P300-based brain-computer interface game. Neural Netw. **22**, 1329–1333 (2009)
4. Hayhoe, M., Ballard, D.: Eye movements in natural behavior. TRENDS Cogn. Sci. **9**(4), 188–194 (2005)
5. Hu, L., Liang, M., Mouraux, A., Wise, R., Hu, Y., Iannetti, G.: Taking into account latency, amplitude and morphology: improved estimation of single-trial ERPS by wavelet filtering and multiple linear regression. J. Neurophysiol. **106**, 3216–3229 (2011)
6. Hu, L., Mouraux, A., Hu, Y., Iannetti, G.: A novel approach for enhancing the signal-to-noise ratio and detecting automatically event-related potentials (ERPS) in single trials. NeuroImage **50**(1), 99–111 (2010)
7. Luck, S.J.: An Introduction to the Event Related Potential Technique. MIT Press, Cambridge (2005)
8. Uscumlic, M., Blankertz, B.: Active visual search in non-stationary scenes: coping with temporal variability and uncertainty. J. Neural Eng. **13**(1), 016015 (2016)

Computer Vision

Unconstrained Face Detection from a Mobile Source Using Convolutional Neural Networks

Shonal Chaudhry[1] and Rohitash Chandra[2(✉)]

[1] School of Computing, Information and Mathematical Sciences,
University of the South Pacific, Suva, Fiji
[2] Artificial Intelligence and Cybernetics Research Group Software Foundation,
Nausori, Fiji
c.rohitash@gmail.com
http://aicrg.softwarefoundationfiji.org

Abstract. We present unconstrained mobile face detection using convolutional neural networks which have potential application for guidance systems for visually impaired persons. We develop a dataset of videos captured from a mobile source that features motion blur and noise from camera shakes. This makes the application a very challenging aspect of unconstrained face detection. The performance of the convolutional neural network is compared with a cascade classifier. The results show promising performance in daylight and artificial lighting conditions while the challenges lie for moonlight conditions with the need for reduction of false positives in order to develop a robust system.

Keywords: Assistive system · Face detection · Convolutional neural networks · Mobile computing

1 Introduction

It is well known that computer vision applications perform some of the most computationally intensive tasks [20]. Even simple applications such as those that perform object detection require a significant amount of computation power [5,12]. Modern computers are able to execute these programs in real-time without any major issues, however, they cannot be used in situations where portability is a high priority and hence computer vision remains a major challenge for mobile computing applications.

Some examples of mobile object detection are assistive systems for disabled persons [14] and iris recognition systems [10]. Mobile object detection systems such as these can have a wide range of applications due to portability [10,14]. While detection of static objects in general is a relatively easier task, the detection of moving objects is more challenging [20]. The inclusion of motion in computer vision applications incorporates major difficulties which can include blur, constant scale and position changes, obstructions, and illumination changes [5]. Advanced detection methods are required to account for these challenges with the hope to achieve satisfactory performance.

© Springer International Publishing AG 2016
A. Hirose et al. (Eds.): ICONIP 2016, Part II, LNCS 9948, pp. 567–576, 2016.
DOI: 10.1007/978-3-319-46672-9_63

Convolutional neural networks (CNNs) are suitable for detection tasks since they are robust and have promising performance for difficult computer vision applications [11,16]. Although, CNNs are commonly used for image recognition and other pattern recognition problems, they can also be applied to real-time object or face detection problems [9]. Although unconstrained face recognition has been getting popular [6], we gather that there has not been much work done in the area of mobile face detection [8,19]. Mobile face detection consists of detection from a mobile source on stationary and moving subjects which leads to input that contains motion blur and noise.

This paper presents unconstrained mobile face detection using CNN with potential application for guidance systems for visually impaired persons. We develop a dataset of videos captured from a mobile source that features motion blur and noise from camera shakes which makes the application very challenging aspect of unconstrained face detection. The video data set consists of several videos for mobile face detection applications. The performance of the CNN is compared with a related detection method that feature a cascade classifier. The proposed approach contributes to a larger system designed to aid visually impaired persons through mobile face detection [1].

The rest of the paper is organised as follows. Section 2 presents the methodology and Sect. 3 gives the experimental design and results. Section 4 presents a discussion of the results while Sect. 5 concludes the paper with a discussion of future work.

2   Unconstrained Mobile Face Detection Using Convolutional Neural Networks

As mentioned earlier, the proposed unconstrained mobile face recognition using CNNs contributes to a framework which provides a face detection service through a mobile application that is designed to assist visually impaired persons [1].

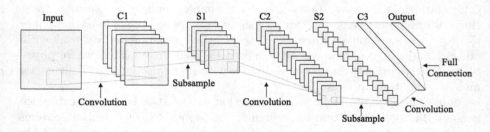

Fig. 1. An example of a CNN architecture with 3 convolution stages and 2 sub-sampling stages

Note that CNNs are essentially feedforward neural networks in which synaptic organisation between the neurons are inspired by the organization of the animal visual cortex [7]. CNNs are trainable multi-stage architectures (Fig. 1) which consist of multiple stages of image processing [11]. The input and output of each stage are sets of arrays known as feature maps. At the output, each feature map represents a particular feature extracted at all locations on the input. Each stage in the CNN uses an $n \times n$ kernel[1] of a specific size for processing the input image. In every stage, there are three layers which are mathematical functions; a filter bank layer, a non-linearity layer and a feature pooling layer. A typical CNN can be made up of one, two or three such 3-layer stages, followed by a classification module.

Each of the three layers in a stage of a CNN uses different methods to perform computations. In the filter bank layer, a number of filters can be used to convolve the input image [11]. The definition of these filters depends on the problem. Low-level filters in the first convolution stage are normally edges and may represented as diagonal, horizontal and vertical lines. The non-linearity layer uses mathematical functions such as the hyperbolic tangent, sigmoid, and rectifier functions to make the learned features more robust. The pooling layers typically use approaches such as average or maximum pooling. They compute the average or maximum of a particular feature over a specific region of the image in order to increase invariance.

In our CNN implementation for mobile face detection, we employed an architecture that incorporates three convolution stages and two sub-sampling stages (Fig. 1). Stages $C1$, $C2$ and $C3$ consist of a 5×5 kernel ($n = 5$) while stages $S1$ and $S2$ have a 2×2 kernel ($n = 2$). Each stage has an additive bias and a hyperbolic tangent (tanh) sigmoid function for the non-linearity layer. The convolution stages $C1$ and $C2$ also feature a subtractive normalisation layer with a 5×5 kernel ($n = 5$).

2.1 Data Collection

The custom video data set[2] was created and used for the experiments as there are no existing data sets that fulfilled the requirements of our experiments. The data set consists of 11 videos recorded at a resolution of 1920×1088 at 29 frames-per-second (FPS) using a mobile phone camera placed inside a shirt pocket of the moving person. The videos are not recorded in controlled environments (which is the case for many other data sets), and hence provides real-world conditions where various challenges are present. Each video features a moving source (camera) and tests for face detection in difficult conditions which include different lighting conditions that includes motion blur, obstructions, rotations and scale changes as shown in Fig. 2.

---

[1] A kernel is a small matrix in image processing used for performing various operations on an image such as blurring or edge detection.

[2] http://aicrg.softwarefoundationfiji.org/open-source-software/ aicrg-moving-object-video-dataset.

The different lighting conditions include artificial light, daylight and moonlight (Fig. 2a, 2b, 2c). A total of six videos were recorded for artificial lighting and daylight conditions with each condition having three video recordings. An additional three videos were taken in both artificial lighting and daylight while two videos were recorded in both artificial lighting and moonlight. Videos with two lighting conditions (Fig. 2d) include transitions from one condition to another. For example, a video with artificial lighting and daylight has the video begin in artificial lighting and then transition to daylight as the camera moves from one location to another.

Artificial   Daylight   Moonlight   Artificial
Lighting   and   Artificial   Lighting   and
                                      Lighting   Daylight

Fig. 2. Different lighting conditions of the dataset.

3   Simulation and Results

This section gives details of the experiment design for CNNs and cascade classifiers. The methodology involves the collection of video data using video camera from a mobile phone for various conditions by a moving person. The subjects were both mobile and stationary with conditions described in Sect. 2. The dataset is then further processed by extraction of frames from the video as images. These are later used by CNNs for face detection. The frames from each video were extracted using the *scene filter* feature of *VLC Media Player*[3]. A total of 6111 frames were extracted from all of the videos. Each frame was then converted to a resolution of 283×500 to decrease computation time while keeping detection rates high. The Lanczos re-sampling algorithm was used for converting the images since it balances conversion speed and image quality [4].

---

[3] http://www.videolan.org/vlc/index.html.

3.1 Experiment Design

The CNN was implemented using the *EBLearn library* [3] which is an object-oriented *C++ library* that also features various machine learning methods.

The CNN training was executed using face images from the cropped greyscale version of the *Labeled Faces in the Wild* [6] database (LFWcrop [15]) and background images from the *Caltech background image dataset* [18]. A Linux computer with a *2.2 Giga-hertz dual-core processor* was used to run the experiments. The true and false positives from the output image files were manually verified after all experiments were finished.

The cascade classifier was implemented using functions from the *open source computer vision* (OpenCV) library[4]. This implementation uses an *Extensible Markup Language* (XML) classifier that utilises Haar features [17] for detection. The XML classifier is trained on face and background images from the LFWcrop database and Caltech background image data set. In the data compilation step, the images are split into positive and negative images (which refer to faces and backgrounds).

Correct - Video 3   Correct - Video 6   Correct - Video 8

Fig. 3. Examples of correct face detection for Videos 3, 6 and 8 indicated by the blue lines. (Color figure online)

As each frame from a video dataset was processed, the output frame (with or without a detection using CNN) was captured. This process was repeated until a total of 30 independent experimental runs were made on all the frames. The same approach was used for the cascade classifier.

3.2 Results

The results highlight *detection* rate in a video with the emphasis on *correct* and *incorrect* detection cases. The results are shown in Table 1 where each frame was processed on average in 1.61 s.

---

[4] http://opencv.org/.

Incorrect - Video   Incorrect - Video   Incorrect - Video
3   6   8

Fig. 4. Examples of false detections for Videos 3, 6 and 8 indicated by the blue lines. (Color figure online)

We observe that the performance is poor in artificial lighting (Video 1) especially when there are only a few faces in all of the frames. When lighting conditions are more suitable, average performance is achieved in some videos (Videos 2–3) (Fig. 3a). In normal daylight, however, the performance is good and depends on motion blur, the angle of the face and obstructions (Videos 4–5). These factors represent the difficulty of recognition of facial features available in the input image. A high performance is attainable in videos with relatively easy detection scenarios while performance is lower in videos with the highlighted difficulties in recognition of features. Note that Video 6 provides the best detection performance in daylight (Fig. 3b) with some false detections reducing the detection rate (Fig. 4b). This video consisted of faces from 7 different individuals with a high detection rate, however, there was difficulty in detection in some frames due to motion blur.

The videos which feature transitions between daylight and artificial lighting (Videos 8–9) also show good performance (Fig. 3c). The video with most difficult detection scenario (Video 7) gives poor performance. This is due to a combination of constant rotation changes and motion blur in each frame caused by camera shakes. The detection rate is also reduced by false detections such as the one shown in Fig. 4c). Finally, videos that include moonlight and artificial lighting (Videos 10–11) gives failure in detection.

Therefore, in general the results shows that the proposed system only works well in situations with adequate lighting (artificial lighting and daylight videos).

The results from the cascade classifier is shown in Table 2. The average computation time for each frame was 265 ms. The performance of the cascade classifier was also poor in artificial lighting conditions (Videos 1–2). Among the artificial lighting videos, the best performance was achieved in bright light (Video 3).

We observe that poor performance is achieved in daylight conditions (Videos 4–5). The only video that had little noise (Video 6) show capable performance.

Table 1. CNN detection results

Video	Condition	Frames	Detection	Correct	Incorrect	Detection Rate (%)	Error (%)
1	Artificial Light	296	15	2	13	13.33	1.49
2		518	14	9	5	64.29	1.49
3		359	9	6	3	66.67	1.49
4	Daylight	229	46	27	19	58.70	1.49
5		122	10	6	4	60.00	1.49
6		224	56	53	3	94.64	1.49
7	Artificial	862	27	12	15	44.44	0.78
8	Light &	774	191	172	19	90.05	0.78
9	Daylight	1206	124	105	19	84.68	1.49
10	Moonlight &	1047	16	0	16	0.00	0
11	Artificial Light	474	16	2	14	12.50	1.49

The cascade classifier performance is also poor in transitional videos (Videos 7 and 9). We observe that good performance is only achieved when the faces are nearer to the camera (Video 8). There is failure in detection for the videos with moonlight and artificial lighting conditions (Videos 10–11). It seems that minimal lighting on faces and motion blur leads to a high number of false-positives.

The detection performance of both methods together with the error are shown in Fig. 5.

Table 2. Cascade classifier detection results

Video	Condition	Frames	Detection	Correct	Incorrect	Detection Rate (%)	Error (%)
1	Artificial Light	296	69	23	46	33.33	1.65
2		518	78	18	60	23.08	1.34
3		359	84	45	39	53.57	1.48
4	Daylight	229	146	59	87	40.41	1.28
5		122	83	40	43	48.19	1.46
6		224	82	48	34	58.54	1.34
7	Artificial	862	62	17	45	27.42	0.80
8	Light and	774	214	158	56	73.83	0.76
9	Daylight	1206	428	141	287	32.94	1.49
10	Moonlight and	1047	191	9	182	4.71	2.06
11	Artificial Light	474	119	11	108	9.24	2.1

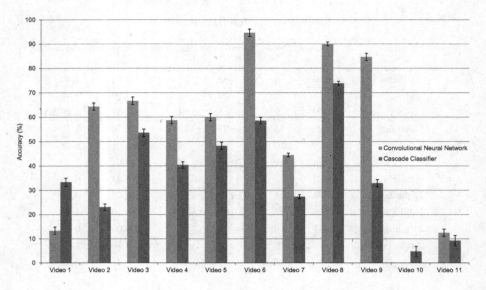

Fig. 5. Performance of the detection programs

4   Discussion

The implementation of the face detection program is presently limited to only well-lit conditions. This is shown in the experiment results where a maximum detection rate of 94.64% is achieved in daylight conditions while the detection rate in general is lower for other conditions. The low light on people's faces, motion blur and rotation changes makes the application very challenging. This can be improved by using night vision and infra-red camera for moonlight conditions. However, motion blur will continue to remain a major challenge even if night vision techniques are used.

We note that object detection using CNNs are accurate, however, they also require a significant amount of computation power [2] which prevents implementation in mobile devices. This could be addressed through cloud-based computation that enable mobile devices to act as input devices for mobile computer vision applications [13].

In addition to the computational requirements, energy limitations of mobile devices makes it impractical for the system to be implemented on high-end mobile devices with multi-core processors at this stage. This could be possible in future as mobile computing is progressing with better technologies for battery and renewable energy options.

5   Conclusion and Future Work

We presented an approach to unconstrained mobile face recognition using convolutional neural networks. The results show promising performance in daylight and artificial lighting by convolutional neural networks while the challenges lie in detection of faces in moonlight conditions. The challenge is in reducing the number of false positives. We compared the performance of convolutional neural networks with cascade classifiers and find that in general, they were mostly reliable for only the simple detection tasks which include proper lighting conditions.

We developed a specific video dataset which featured a mobile camera that provided video recordings that featured mobile subjects. The lack of benchmark datasets has motivated us to develop and make the dataset openly available for further research.

In the future, improved performance can be accomplished by performing more advanced data pre-processing that includes different feature extraction methods. Moreover, training using a larger dataset and more sophisticated detection methods that include transfer and multi-task learning can also be explored. The proposed method be expanded to include face recognition that takes place after the detection. Furthermore, an implementation for wearable devices that include video camera input in eye glasses can be used to make the system convenient for users.

References

1. Chaudhry, S., Chandra, R.: Design of a mobile face recognition system for visually impaired persons (2015). arXiv:1502.00756
2. Cireşan, D.C., Meier, U., Masci, J., Gambardella, L.M., Schmidhuber, J.: Flexible, high performance convolutional neural networks for image classification. In: Proceedings of the Twenty-Second International Joint Conference on Artificial Intelligence, IJCAI 2011, vol. 2, pp. 1237–1242. AAAI Press (2011)
3. Computational and Biological Learning Laboratory, New York University: Eblearn home (2015). http://eblearn.sourceforge.net/doku.html
4. Duchon, C.E.: Lanczos filtering in one and two dimensions. J. Appl. Meteorol. **18**(8), 1016–1022 (1979)
5. Ess, A., Leibe, B., Schindler, K., Van Gool, L.: Moving obstacle detection in highly dynamic scenes. In: 2009 IEEE International Conference on Robotics and Automation, ICRA 2009, pp. 56–63, May 2009
6. Huang, G.B., Ramesh, M., Berg, T., Learned-Miller, E.: Labeled faces in the wild: a database for studying face recognition in unconstrained environments. Technical report, 07–49, University of Massachusetts, Amherst (2007)
7. Hubel, D.H., Wiesel, T.N.: Receptive fields, binocular interaction and functional architecture in the cat's visual cortex. J. Physiol. **160**(1), 106 (1962)
8. Jafri, R., Arabnia, H.R.: A survey of face recognition techniques. J. Inf. Process. Syst. **5**(2), 41–68 (2009)
9. Karpathy, A., Toderici, G., Shetty, S., Leung, T., Sukthankar, R., Fei-Fei, L.: Large-scale video classification with convolutional neural networks. In: 2014 IEEE Conference on Computer Vision and Pattern Recognition (CVPR), pp. 1725–1732. IEEE (2014)
10. Latman, N.S., Herb, E.: A field study of the accuracy and reliability of a biometric iris recognition system. Sci. Justice **53**(2), 98–102 (2013)

11. LeCun, Y., Kavukvuoglu, K., Farabet, C.: Convolutional networks and applications in vision. In: Proceedings of International Symposium on Circuits and Systems (ISCAS 2010). IEEE (2010)
12. Lu, C., Adluru, N., Ling, H., Zhu, G., Latecki, L.J.: Contour based object detection using part bundles. Comput. Vis. Image Underst. **114**(7), 827–834 (2010)
13. Miettinen, A.P., Nurminen, J.K.: Energy efficiency of mobile clients in cloud computing. In: Proceedings of the 2nd USENIX Conference on Hot Topics in Cloud Computing, HotCloud 2010, p. 4. USENIX Association, Berkeley (2010)
14. Moreno, M., Shahrabadi, S., José, J., du Buf, J., Rodrigues, J.: Realtime local navigation for the blind: detection of lateral doors and sound interface. Proc. Comput. Sci. **14**, 74–82 (2012)
15. Sanderson, C.: LFWcrop Face Dataset (2014). http://conradsanderson.id.au/lfwcrop/
16. Sun, Y., Wang, X., Tang, X.: Deep convolutional network cascade for facial point detection. In: Proceedings of the 2013 IEEE Conference on Computer Vision and Pattern Recognition, CVPR 2013, pp. 3476–3483. IEEE Computer Society, Washington, DC (2013)
17. Viola, P., Jones, M.: Rapid object detection using a boosted cascade of simple features. In: Proceedings of the 2001 IEEE Computer Society Conference on Computer Vision and Pattern Recognition, CVPR 2001, vol. 1, pp. 511–518. IEEE (2001)
18. Weber, M.: Background image dataset (2014). http://www.vision.caltech.edu/archive.html
19. Zhang, C., Zhang, Z.: A survey of recent advances in face detection. Technical report, MSR-TR-2010-66. http://research.microsoft.com/apps/pubs/default.aspx?id=132077
20. Zhou, J., Gao, D., Zhang, D.: Moving vehicle detection for automatic traffic monitoring. IEEE Trans. Veh. Technol. **56**(1), 51–59 (2007)

Driver Face Detection Based on Aggregate Channel Features and Deformable Part-Based Model in Traffic Camera

Yang Wang[1(✉)], Xiaoma Xu[2], and Mingtao Pei[1]

[1] Beijing Laboratory of Intelligent Information Technology,
School of Computer Science, Beijing Institute of Technology,
Beijing 100081, People's Republic of China
{wangyangbit,peimt}@bit.edu.cn
[2] Research Institute of Petroleum Exploration
and Development-Northwest, PetroChina,
No. 535 Yanerwan Street, Lanzhou 730020, Gansu, China
xuxiaoma@bit.edu.cn

Abstract. We explore the problem of detecting driver faces in cabs from images taken by traffic cameras. Dim light in cabs, occlusion and low resolution make it a challenging problem. We employ aggregate channel features instead of a single feature to reduce the miss rate, which will introduce more false positives. Based on the observation that most running vehicles have a license plate and the relative position between the plate and driver face has an approximately fixed pattern, we refer to the concept of deformable part-based model and regard a candidate face and a plate as two deformable parts of a face-plate couple. A candidate face will be rejected if it has a low confidence score. Experiment results demonstrate the effectiveness of our method.

Keywords: Driver face detection · Aggregate channel features · Deformable part-based model · Face-plate couple

1 Introduction

Face detection has drawn much attention these years [1–4]. In-depth review of this domain can be consulted to researches by Viola and Jones, et al. [5–8]. In traditional face detection framework, most application are restricted to indoor scenarios. When it comes to driver face detection, the related work mainly concentrate on vehicle front seat occupancy detection and driver fatigue detection. The former is used in the examination at traffic checkpoints along high-occupancy vehicle lanes to improve the transportation efficiency [9,10]. The latter detects and analyzes a drivers facial behaviors via an in-car camera and sending out fatigue warning if necessary [11–14].

Driver face detection has its own difficulties and characteristics. Generally speaking, as shown in Fig. 1, conditions in vehicle cabs are different from those

© Springer International Publishing AG 2016
A. Hirose et al. (Eds.): ICONIP 2016, Part II, LNCS 9948, pp. 577–584, 2016.
DOI: 10.1007/978-3-319-46672-9_64

Fig. 1. A driver face and a license plate can be defined as two deformable parts in a face-plate ouple which has a special pattern

in indoor: because of dim light and occlusion in cabs, reflection caused by windscreens, long distances between vehicles and traffic cameras et al., driver faces in images usually have low resolution and low image quality, particularly in daytime.

In this paper, we employ aggregate channel features (histogram of gradient & LBP) to detect driver faces [15], which gives a high Recall as well as a low Precision. To filter out false positive driver faces, we refer to the concept of deformable part-based model (DPM) [16–18], where we regard a license plate and a candidate driver face as two deformable parts of a face-plate couple. The reason for doing so is that when captured by a traffic camera in Chinese Mainland, the face part is ought to be at the top right position neither too near nor too far from the plate part (as illustrated in Fig. 1).

2   Aggregate Channel Features

A channel is a registered map of an original input image [19], Aggregate channel features (ACF) can offer more powerful capacity in representing faces than a single channel [15]. However, detecting driver faces in cabs has its own characteristics that make it different from detecting faces in door. As a result, some channels like RGB, etc. may be inefficient in our work.

2.1   Channel Types

Considering the efficiency of Local Binary Pattern (LBP) in face detection domain, we introduce LBP as a channel into aggregate channel features. That is we compute LBP from corresponding patches of each pixel and convert it to a decimal number as the pixel value. Therefore, our ACF consists of four types of channels, which are gradient magnitude channel, histogram of gradient channels (6 orientation bins are used to generate 6 channels) [15,19], Gabor channels (2 scales with 5 orientations are used to generate 10 channels) and LBP channel. Since some channels may be useless and some channels may be redundant, our goal is to experiment and select the most effective combination of channel types for driver faces.

Fig. 2. ACF detection stage performance

2.2   Detector Training

Referring to the training pipeline in [15], Our training data is normalized to 60×60 images. In this way, each channel dimension is $\left(\frac{60}{4}\right) \times \left(\frac{60}{4}\right) = 225$, where 4 is a subsampling factor. So if we aggregate **n** channels, the driver face detector will have a resulting feature dimension of $\mathbf{n} \times 225$. The classifier learning method followed [19], in which 1024 depth-2 decision trees act as weak classifiers in soft-cascade structured Adaboost.

2.3   ACF Driver Face Detection

In the ACF detection stage, we adopt the sliding window strategy over a set of image pyramids. Every window is put into the boosted classifier. All positive windows are predicted to be candidate driver faces. For each positive window, we use the score $H(\mathbf{x})$ from the last cascade layer to measure the confidence of this predicted candidate driver face:

$$H(\mathbf{x}) = \sum_{i=1}^{N} \alpha_i h_i(x), \tag{1}$$

where $\mathbf{x}$ is the input aggregate channel features, $h_i(x)$ is an indicator function (1 for a driver face, 0 for a nonface), N is the number of weak classifiers, α_i is a coefficient. We applied non-maximal suppression (NMS) to suppress and merge multiple overlapped detections.

After that, we preliminarily detect out driver faces from the whole image. In the ACF detection stage, our goal is to obtain a low miss rate, which will bring some false positives including passengers, wing mirrors, taxi LOGOs, headlights and so on, as illustrated in Fig. 2.

3   Filter out False Positives

3.1   Employing License Plates and DPM

Generally speaking, when a running vehicle is captured by a traffic camera in Chinese Mainland, it should have exactly one license plate and one driver face.

Fig. 3. License plate detection can help the driver face detection. The numbers are the deformation scores S_f of candidate face parts

The face is ought to be at the top right location not far away from the corresponding license plate. So far, license plate detection is a relatively mature technology and some commercial applications come out, which give robust detection results. In our work, we use our previous work on license plate detection [20], as illustrated in Fig. 3.

After detecting out candidate driver faces and license plates, we consult the concept of deformable part-based model [16–18], and alternately regard each license plate combined with each candidate driver face as two deformable parts of a face-plate couple.

3.2   Extracting the Feature for Face-Plate Couples

We propose the following scenario of designing a robust relative location feature f for face-plate couples. A driver face center coordinate (X_{face}, Y_{face}) and the corresponding license plate center coordinate (X_{plate}, Y_{plate}) construct a "feature rectangular" whose width and height are:

$$Width_{rect} = X_{face} - X_{plate},$$
$$Height_{rect} = Y_{plate} - Y_{face}. \tag{2}$$

Then, our face-plate couple feature f consists of two dimensions. The first dimension is defined as:

$$f_1 = \frac{Width_{rect}}{Width_{plate}}, \tag{3}$$

where $Width_{plate}$ is the width of a license plate. f_1 deals with different deformation among different vehicle types and restrict relative position between a plate part and a face part in a face-plate couple. The second dimension is defined as:

$$f_2 = \frac{Height_{rect}}{Width_{plate}}, \tag{4}$$

which is used to ensure the feature rectangular is neither too large nor too small to filter out wrong patterns.

3.3 Part Confidence Score in the Face-Plate Couple

The feature f can be modeled by an elliptical Gaussian joint probability density function (pdf), which is defined as:

$$pdf = \frac{1}{2\pi|\Sigma_f|^{\frac{1}{2}}} e^{-\frac{1}{2}(f-\mu_f)^{\tau}\Sigma_f^{-1}(f-\mu_f)}, \tag{5}$$

where mean value μ_f and covariance matrix Σ_f are the distribution parameters, which can be estimated by doing maximum likelihood estimation (MLE) from training data.

For every detected license plate, we alternately couple it with every candidate driver face to construct the face-plate couple. Then we extract f and compute the corresponding Gaussian probability P_f. The deformation score of a face part is defined as:

$$S_f = 1 - P_f. \tag{6}$$

A final driver face confidence score will be the ACF score in Eq. (1) add up the part deformation score in Eq. (6):

$$S_{face} = \alpha \times H(\mathbf{x}) + \beta \times S_f, \tag{7}$$

where α and β are weight factors. A candidate face is regarded as a true positive driver face only if its S_{face} is above a threshold thd_{face} (illustrated in Fig. 3).

4 Experiments and Analyses

4.1 Driver Face Dataset

To demonstrate the effectiveness of our method, we collected 1402 1600×1200 vehicle front view images (683 daytime and 719 nighttime) taken by traffic surveillance cameras, which formed a "driver face in cabs" dataset named as "DFIC". Since some images contain two or more drivers, the faces number in "DFIC" is 1680. Ground truth license plates and driver faces are annotated manually. All experiments were conducted on an Intel(R) Core(TM)2 CPU 2.8 GHz PC with 4.0 GB RAM.

4.2 Training Data

Driver Face Data for ACF. In ACF stage, the LFW database (13232 60×60 face images) combined with $\frac{1}{5}$ the number of DFIC driver faces (we call them "AugDF", the left $\frac{4}{5}$ DFIC images are used for testing) is used as positive samples, whose total number is $13232 + 1680 \times \frac{1}{5} = 13568$. Our negative samples also consist of two parts: (1) randomly cropped 30000 60×60 "nonfaces" from the PASCAL VOC 2008 database; (2) manually cropped 1836 60×60 "road images" form AugDF, which include green belts, traffic signs, wing mirrors, vehicle LOGOs, headlights, et al.

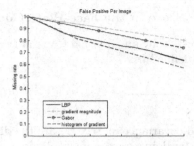

Fig. 4. FPPI shows that histogram of gradient performs best

Driver Face Data for the DPM Face-Plate Couple. We used ground truth license plates and driver faces in the AugDF to construct DPM training data. So there are $1680 \times \frac{1}{5} = 336$ samples for training the face-plate couple. We extracted relative location feature f and did MLE to obtain μ_f and Σ_f of the Gaussian *pdf*.

4.3   Comparison of Different Channel Features

As mentioned, we experimented on DFIC and selected the most effective aggregation of channel types for driver faces. False Positive Per Image (FPPI) shows that histogram of gradient performs best, which followed by LBP, Gabor and gradient magnitude (illustrated as Fig. 4). We select the histogram of gradient channels (6 orientation bins are used to generate 6 channels) aggregated with LBP. That is our ACF driver face detector contains 7 channels which have a resulting feature dimension of $7 \times 225 = 1575$.

4.4   Quantitative Comparison

We compare our ACF feature (histogram of gradient plus LPB) with the widely used HOG and LBP for driver face detection. ACF outperforms the other two and achieves the highest Recall as we want. However, in the ACF stage, a low Precision results in a low $F_1 = \frac{2 \times Precision \times Recall}{Precision + Recall}$, which indicates the comprehensive performance. As shown in Tables 1 and 2, merely a ACF for driver face detection is not acceptable.

Table 1. Quantitative evaluation of our method and other approaches in daytime

	Precision	Recall	F_1
HOG	25.12 %	86.76 %	38.96 %
LBP	22.29 %	82.24 %	35.07 %
ACF	27.23 %	89.15 %	41.71 %
Ours (ACF+DPM)	94.27 %	86.17 %	90.04 %

Table 2. Quantitative evaluation of our method and other approaches in nighttime

	Precision	Recall	F_1
HOG	27.28 %	94.64 %	42.35 %
LBP	25.84 %	89.68 %	40.12 %
ACF	30.45 %	97.80 %	46.44 %
Ours (ACF+DPM)	98.70 %	96.21 %	97.44 %

In our DPM Filter Stage, the Precision is significantly raised to higher than 90 % with a negligible decrease in Recall. The weight factors in Eq. (7) and the threshold thd_{face} affect the detection results. In our experiment, we set $\alpha = 0.4$, $\beta = 0.6$ and $thd_{face} = 0.5$. Experiment results indicate that a license plate part can be employed to effectively filter out false positive driver face parts.

5   Conclusion

We have shown that when faced with images taken by traffic cameras, the aggregate channel features (histogram of gradient & LBP) can be applied to driver face detection in cabs. The concept of deformable part-based model can be employed to filer out false positives in the ACF detection stage and significantly improve Precision with little influence on Recall.

Acknowledgments. This work was supported in part by the Natural Science Foundation of China (NSFC) under Grant No. 61472038 and No. 61375044.

References

1. Zhu, X., Ramanan, D.: Face detection, pose estimation, and landmark localization in the wild. In: 2012 IEEE Conference on Computer Vision and Pattern Recognition (CVPR), pp. 2879–2886. IEEE (2012)
2. Lei, Z., Yi, D., Li, S.Z.: Local gradient order pattern for face representation and recognition. In: 2014 22nd International Conference on Pattern Recognition (ICPR), pp. 387–392. IEEE (2014)
3. Bilaniuk, O., Fazl-Ersi, E., Laganiere, R., Xu, C., Laroche, D., Moulder, C.: Fast LBP face detection on low-power SIMD architectures. In: 2014 IEEE Conference on Computer Vision and Pattern Recognition Workshops (CVPRW), pp. 630–636. IEEE (2014)
4. Shen, X., Lin, Z., Brandt, J., Wu, Y.: Detecting and aligning faces by image retrieval. In: 2013 IEEE Conference on Computer Vision and Pattern Recognition (CVPR), pp. 3460–3467. IEEE (2013)
5. Viola, P., Jones, M.J.: Robust real-time face detection. Int. J. Comput. Vis. **57**(2), 137–154 (2004)
6. Viola, P., Jones, M.: Rapid object detection using a boosted cascade of simple features. In: Proceedings of the 2001 IEEE Computer Society Conference on Computer Vision and Pattern Recognition, 2001. CVPR 2001, vol. 1, pp. I–511. IEEE (2001)

7. Hsu, R.L., Abdel-Mottaleb, M., Jain, A.K.: Face detection in color images. IEEE Trans. Pattern Anal. Mach. Intell. **24**(5), 696–706 (2002)

8. Yang, M.H., Kriegman, D.J., Ahuja, N.: Detecting faces in images: a survey. IEEE Trans. Pattern Anal. Mach. Intell. **24**(1), 34–58 (2002)

9. Hao, X., Chen, H., Li, J.: An automatic vehicle occupant counting algorithm based on face detection. In: 2006 8th International Conference on Signal Processing, vol. 3. IEEE (2006)

10. Xu, B., Paul, P., Artan, Y., Perronnin, F.: A machine learning approach to vehicle occupancy detection. In: 2014 IEEE 17th International Conference on Intelligent Transportation Systems (ITSC), pp. 1232–1237. IEEE (2014)

11. Smith, P., Shah, M., Lobo, N.D.V.: Determining driver visual attention with one camera. IEEE Trans. Intell. Transp. Syst. **4**(4), 205–218 (2003)

12. Wang, Q., Yang, J., Ren, M., Zheng, Y.: Driver fatigue detection: a survey. In: The Sixth World Congress on Intelligent Control and Automation, 2006. WCICA 2006, vol. 2, pp. 8587–8591. IEEE (2006)

13. Eriksson, M., Papanikolopoulos, N.P.: Driver fatigue: a vision-based approach to automatic diagnosis. Transp. Res. Part C: Emerg. Technol. **9**(6), 399–413 (2001)

14. Ji, Q., Zhu, Z., Lan, P.: Real-time nonintrusive monitoring and prediction of driver fatigue. IEEE Trans. Veh. Technol. **53**(4), 1052–1068 (2004)

15. Yang, B., Yan, J., Lei, Z., Li, S.Z.: Aggregate channel features for multi-view face detection. In: 2014 IEEE International Joint Conference on Biometrics (IJCB), pp. 1–8. IEEE (2014)

16. Felzenszwalb, P., McAllester, D., Ramanan, D.: A discriminatively trained, multiscale, deformable part model. In: IEEE Conference on Computer Vision and Pattern Recognition, 2008. CVPR 2008, pp. 1–8. IEEE (2008)

17. Felzenszwalb, P.F., Girshick, R.B., McAllester, D., Ramanan, D.: Object detection with discriminatively trained part-based models. IEEE Trans. Pattern Anal. Mach. Intell. **32**(9), 1627–1645 (2010)

18. Felzenszwalb, P.F., Girshick, R.B., McAllester, D.: Cascade object detection with deformable part models. In: 2010 IEEE conference on Computer vision and pattern recognition (CVPR), pp. 2241–2248. IEEE (2010)

19. Dollár, P., Tu, Z., Perona, P., Belongie, S.: Integral channel features. In: BMVC, vol. 2, p. 5 (2009)

20. Wang, Y., Pei, M., Jia, Y.: License plate detection based on multiple features. J. Image Graph. **19**(003), 471–475 (2014)

Segmentation with Selectively Propagated Constraints

Peng Han[1], Guangzhen Liu[1], Songfang Huang[2], Wenwu Yuan[1],
and Zhiwu Lu[1]([✉])

[1] Beijing Key Laboratory of Big Data Management and Analysis Methods,
School of Information, Renmin University of China,
Beijing 100872, China
luzhiwu@ruc.edu.cn
[2] IBM China Research Lab, Beijing, China

Abstract. This paper presents a novel selective constraint propagation method for constrained image segmentation. In the literature, many pairwise constraint propagation methods have been developed to exploit pairwise constraints for cluster analysis. However, since these methods mostly have a polynomial time complexity, they are not much suitable for segmentation of images even with a moderate size, which is equal to cluster analysis with a large data size. In this paper, we thus choose to perform pairwise constraint propagation only over a selected subset of pixels, but not over the whole image. Such a selective constraint propagation problem is then solved by an efficient graph-based learning algorithm. Finally, the selectively propagated constraints are exploited based on L_1-minimization for normalized cuts over the whole image. The experimental results show the promising performance of the proposed method.

Keywords: Constrained image segmentation · Pairwise constraint propagation · Graph-based learning

1 Introduction

Image segmentation is very important for image content analysis. Despite many years of research [10,11], general-purpose image segmentation is still challenging because segmentation is inherently ill-posed. To improve the results of image segmentation, much attention has been paid to constrained image segmentation [2,3,12,13], where certain constraints are initially provided for image segmentation. In this paper, we focus on constrained image segmentation using pairwise constraints, which can be derived from the initial labels of selected pixels.

The main challenge in constrained image segmentation is how to effectively exploit a limited number of pairwise constraints for image segmentation. A sound solution is to perform pairwise constraint propagation to generate more pairwise constraints. Although many pairwise constraint propagation methods [7–9] have been developed for constrained clustering [5,6], they mostly have a polynomial

A. Hirose et al. (Eds.): ICONIP 2016, Part II, LNCS 9948, pp. 585–592, 2016.
DOI: 10.1007/978-3-319-46672-9_65

time complexity and thus are not much suitable for segmentation of images even with a moderate size, which is actually equivalent to clustering with a large data size. For constrained image segmentation, we need to develop more efficient pairwise constraint propagation methods, instead of directly utilizing the existing methods like [7–9]. Here, it is worth noting that even the simple assignment operation incurs a large time cost of $O(N^2)$ if we perform pairwise constraint propagation over all the pixels, since the number of all possible pairwise constraints is $N(N-1)/2$. The unique choice is to propagate the initial pairwise constraints *only to a selected subset of pixels*, but not across the whole image.

Finally, the selectively propagated constraints obtained by our selective constraint propagation are exploited to adjust the original weight matrix based on optimization techniques, in order to ensure that the new weight matrix is as consistent as possible with the selectively propagated constraints. In this paper, we formulate such weight adjustment as an L_1-*minimization* problem, which can be solved efficiently due to the limited number of selectively propagated constraints. The obtained new weight matrix is then applied to normalized cuts [11] for image segmentation.

It should be noted that the present work is distinctly different from previous work on constrained image segmentation [2,3,12,13]. In [13], only linear equality constraints (analogous to must-link constraints) are exploited for image segmentation based on normalized cuts. In [2,3,12], although more types of constraints are exploited for image segmentation, the linear inequality constraints analogous to cannot-link constraints are completely ignored just as [13]. In contrast, our selective constraint propagation method exploits both must-link and cannot-link constraints for normalized cuts. More notably, as shown in our later experiments, our method *obviously outperforms* [13] due to extra consideration of cannot-link constraints for image segmentation.

2   Selective Constraint Propagation

2.1   Problem Formulation

In this paper, our goal is to propagate the initial pairwise constraints to a selected subset of pixels for constrained image segmentation. To this end, we need to first select a subset of pixels for our selective constraint propagation. Although there exist different sampling methods in statistics, we only consider random sampling for efficiency purposes, i.e., the subset of pixels are selected randomly from the whole image. In the following, the problem formulation for selective constraint propagation over the selected subset of pixels is elaborated from a graph-based learning viewpoint.

Let $\mathcal{M} = \{(i,j) : l_i = l_j, 1 \leq i,j \leq N\}$ denote the set of initial must-link constraints and $\mathcal{C} = \{(i,j) : l_i \neq l_j, 1 \leq i,j \leq N\}$ denote the set of initial cannot-link constraints, where l_i is the region label assigned to pixel i and N is the total number of pixels within an image. The set of constrained pixels is thus denoted as $P_c = \{i : (i,j) \in \mathcal{M} \cup \mathcal{C}, 1 \leq j \leq N\} \cup \{i : (j,i) \in \mathcal{M} \cup \mathcal{C}, 1 \leq j \leq N\}$ with $N_c = |P_c|$. Moreover, we randomly select a subset of pixels $P_s \subset \{1, 2, ..., N\}$

with $N_s = |P_s|$, and then form the final selected subset of pixels used for our selective constraint propagation as $P_u = P_s \cup P_c$ with $N_u = |P_u|$.

In this paper, we construct a k-nearest neighbors (k-NN) graph over all the pixels so that the normalized cuts for image segmentation can be performed efficiently over this k-NN graph. Let its weight matrix be $W = \{w(i,j)\}_{N \times N}$. We define the weight matrix $W_u = \{w_u(i,j)\}_{N_u \times N_u}$ over the selected subset of pixels P_u as:

$$w_u(i,j) = w(P_u(i), P_u(j)),\tag{1}$$

where $P_u(i)$ denotes the i-th member of P_u. The normalized Laplacian matrix is then given by

$$L_u = I - D^{-1/2} W_u D^{-1/2},\tag{2}$$

where I is an identity matrix and D is a diagonal matrix with its i-th diagonal entry being the sum of the i-th row of W_u. Moreover, for convenience, we represent the two sets of initial pairwise constraints $\mathcal{M}$ and $\mathcal{C}$ using a single matrix $Z_u = \{z_u(i,j)\}_{N_u \times N_u}$ as follows:

$$z_u(i,j) = \begin{cases} +1, & (P_u(i), P_u(j)) \in \mathcal{M}; \\ -1, & (P_u(i), P_u(j)) \in \mathcal{C}; \\ 0, & \text{otherwise.} \end{cases}\tag{3}$$

Based on the above notations, the problem of selective constraint propagation over the selected subset of pixels P_u can be formulated from a graph-based learning viewpoint:

$$\min_{F_v, F_h} \|F_v - Z_u\|_F^2 + \mu \text{tr}(F_v^T L_u F_v) + \|F_h - Z_u\|_F^2$$
$$+ \mu \text{tr}(F_h L_u F_h^T) + \gamma \|F_v - F_h\|_F^2,\tag{4}$$

where μ and γ denote the positive regularization parameters, $\|\cdot\|_F$ denotes the Frobenius norm of a matrix, and $\text{tr}(\cdot)$ denotes the trace of a matrix. Here, it is worth noting that the above problem formulation actually imposes both *vertical and horizontal* constraint propagation upon the initial matrix Z_u. That is, each column (or row) of $\acute{Z}_u$ can be regraded as the initial configuration of a *two-class label propagation* problem, which is formulated just the same as [14]. Moreover, in this paper, we assume that the vertical and horizontal constraint propagation have the same importance for constrained image segmentation.

The objective function given by Eq. (4) is further discussed as follows. The first and second terms are related to the *vertical* constraint propagation, while the third and fourth terms are related to the *horizontal* constraint propagation. The fifth term then ensures that the solutions of these types of constraint propagation are as approximate as possible. Let F_v^* and F_h^* be the best solutions of vertical and horizontal constraint propagation, respectively. The best solution of our selective constraint propagation is defined as:

$$F_u^* = (F_v^* + F_h^*)/2.\tag{5}$$

As for the second and fourth terms, they are known as Laplacian regularization [14,15], which means that F_v and F_h should not change too much between similar pixels. Such Laplacian regularization has been widely used for different graph-based learning problems in the literature.

2.2 Efficient SCP Algorithm

Let $\mathcal{Q}(F_v, F_h)$ denote the objective function in Eq. (4). The alternate optimization technique can be adopted to solve $\min_{F_v, F_h} \mathcal{Q}(F_v, F_h)$ as follows: (1) Fix $F_h = F_h^*$, and perform the vertical propagation by $F_v^* = \arg\min_{F_v} \mathcal{Q}(F_v, F_h^*)$; (2) Fix $F_v = F_v^*$, and perform the horizontal propagation by $F_h^* = \arg\min_{F_h} \mathcal{Q}(F_v^*, F_h)$.

Vertical Propagation: For F_h fixed at F_h^*, the first subproblem can be solved by setting $\frac{\partial \mathcal{Q}(F_v, F_h^*)}{\partial F_v}$ to zero:

$$(I + \hat{\mu}L_u)F_v = (1 - \beta)Z_u + \beta F_h^*, \tag{6}$$

where $\hat{\mu} = \mu/(1 + \gamma)$ and $\beta = \gamma/(1 + \gamma)$. Since $I + \hat{\mu}L_u$ is positive definite, the above linear equation has a solution:

$$F_v^* = (I + \hat{\mu}L_u)^{-1}((1 - \beta)Z_u + \beta F_h^*). \tag{7}$$

However, this analytical solution is not efficient at all for constrained image segmentation, since the matrix inverse incurs a large time cost of $O(N_u^3)$. In fact, this solution can be *efficiently found using the label propagation technique* [14] based on k-NN graph over P_u.

Horizontal Propagation: For F_v fixed at F_v^*, the second subproblem can be solved by setting $\frac{\partial \mathcal{Q}(F_v^*, F_h)}{\partial F_h}$ to zero:

$$F_h(I + \hat{\mu}L_u) = (1 - \beta)Z_u + \beta F_v^*. \tag{8}$$

This linear equation can also be efficiently solved using the label propagation technique [14] based on k-NN graph, similarly to what we do for Eq. (6).

Since the wight matrix W_u over P_u is derived from the original weight matrix W of the k-NN graph over all the pixels according to Eq. (1), W_u can be regarded as the weight matrix of a k-NN graph constructed over P_u. Hence, we can adopt the label propagation technique [14] to efficiently solve both Eqs. (6) and (8). Moreover, to speed up our selective constraint propagation, we also discard those less important propagated constraints during both vertical and horizontal propagation. That is, the two matrices F_v and F_h are forced to become sparser and thus less computation load is needed during iteration. In summary, our SCP algorithm has a time cost of $O(kN_u^2)$ ($N_u \ll N$).

3   Constrained Image Segmentation

In this section, we discuss how to exploit the selectively propagated constraints stored in the output F_u^* of our SCP algorithm for image segmentation based on normalized cuts. The basic idea is to adjust the original weight matrix W_u over the selected subset of pixels P_u using these selectively propagated constraints. The problem of such weight adjustment can be formulated as:

$$\min_{\tilde{W}_u \geq 0} \frac{1}{2}||\tilde{W}_u - F_u^*||_F^2 + \lambda||\tilde{W}_u - W_u||_1, \tag{9}$$

where $\tilde{W}_u \in R^{N_u \times N_u}$ is the new weight matrix over P_u, and λ is the regularization parameter. It is worth noting that the new weight matrix $\tilde{W}_u$ is actually derived as a *nonnegative fusion* of F_u^* and W_u by solving the above L_1-minimization problem. More notably, the L_1-*norm regularization* term $||\tilde{W}_u - W_u||_1$ can force the new weight matrix $\tilde{W}_u$ not only to approach W_u but also to become as sparse as W_u, given that W_u can be regarded as the weight matrix (thus sparse) of a k-NN graph constructed over P_u.

The problem in Eq. (9) can be solved based on the basic L_1-minimization technique. It has an explicit solution:

$$\tilde{W}_u^* = \text{soft\_thr}(F_u^*, W_u, \lambda), \tag{10}$$

where $\text{soft\_thr}(\cdot, \cdot, \lambda)$ is a soft-thresholding function. Here, we directly define $z = \text{soft\_thr}(x, y, \lambda)$ as:

$$z = \begin{cases} z_1 = \max(x - \lambda, y), & f_1 \leq f_2 \\ z_2 = \max(0, \min(x + \lambda, y)), & f_1 > f_2 \end{cases}, \tag{11}$$

where $f_1 = (z_1 - x)^2 + 2\lambda|z_1 - y|$ and $f_2 = (z_2 - x)^2 + 2\lambda|z_2 - y|$. Since the L_1-minimization problem given by Eq. (9) is limited to P_u, finding the best new weight matrix $\tilde{W}_u^*$ incurs a time cost of $O(N_u^2)$ ($N_u \ll N$).

Once we have found the best new weight matrix $\tilde{W}_u^* = \{\tilde{w}_u^*(i', j')\}_{N_u \times N_u}$ over the selected subset of pixels P_u, we can derive the new weight matrix $\tilde{W} = \{\tilde{w}(i, j)\}_{N \times N}$ over all the pixels from the original weight matrix $W = \{w(i, j)\}_{N \times N}$:

$$\tilde{w}(i, j) = \begin{cases} \tilde{w}_u^*(i', j'), & i, j \in P_u, P_u(i') = i, P_u(j') = j \\ w(i, j), & \text{otherwise} \end{cases},$$

where $P_u(i')$ denotes the i'-th member of P_u. This new weight matrix $\tilde{W}$ over all the pixels is then applied to normalized cuts for image segmentation.

4   Experimental Results

4.1   Experimental Setup

For segmentation evaluation, we select 25 images from the Berkeley segmentation database [10], and some example images are shown in Fig. 1. In general, the selected images have *confusing backgrounds* (e.g. the penguin image), and the segmentation over these images is very challenging. Furthermore, we consider a vector of color and texture features for each pixel of an image just as [1].

The segmentation results are measured by the adjusted Rand (AR) index [4], and a higher AR score is better. We compare our normalized cuts with selective constraint propagation (NCuts_SCP) with three closely related methods: normalized cuts with linear equality constraints (NCuts_LEC) [13], normalized cuts with spectral learning (NCuts_SL) [5], and standard normalized cuts (NCuts) [11]. Here, we do not compare with other constraint propagation methods [7–9], since they have a polynomial time complexity.

We randomly select a small set of labeled pixels to infer the initial linear equality constraints (for NCuts_LEC) and pairwise constraints (for NCuts_SCP and NCuts_SL). Moreover, we set the parameters of our NCuts_SCP as: $k = 60$, $\alpha = 0.9$, $\beta = 0.1$, $\epsilon = 10^{-7}$, and $\lambda = 0.001$. The parameters of other methods are also set their respective optimal values.

4.2   Segmentation Results

We first show the effect of N_s (i.e. the number of randomly selected pixels) on the performance of our NCuts_SCP in Table 1. Here, we measure the performance of our NCuts_SCP for constrained image segmentation by both AR index and running time averaged over all the images. In particular, we collect the running time by running our NCuts_SCP (Matlab code) on a computer with 3 GHz CPU and 32 GB RAM. From Table 1, it can be clearly observed that our NCuts_SCP requires more running time but leads to higher AR index when N_s takes a larger value. Considering the tradeoff between the effectiveness and efficiency of constrained image segmentation, we select $N_s = 2,500$ for our NCuts_SCP. This setting is used throughout the following experiments.

The comparison between different image segmentation methods is listed Table 2, and some example results are shown in Fig. 1. Here, the segmentation results are evaluated by both AR index and running time averaged over all the images, i.e., we take both effectiveness and efficiency into account. The immediate observation is that our NCuts_SCP significantly outperforms the other three methods in terms of AR index. Since our NCuts_SCP incurs a time cost comparable to closely related methods, it is preferred for constrained image segmentation by an *overall consideration*. In addition, it can be observed that NCuts_SCP, NCuts_LEC, and NCuts_SL lead to better results than the standard NCuts due to the use of constraints for image segmentation.

Table 1. The average segmentation results achieved by our NCuts_SCP with a varied number (i.e. N_s) of pixels being selected randomly

N_s	1,000	1,500	2,000	2,500	3,000	3,500
AR index	0.47	0.48	0.48	0.50	0.50	0.50
Time (sec.)	31	32	33	35	36	37

Table 2. The average segmentation results obtained by different image segmentation methods in terms of both effectiveness and efficiency

Methods	NCuts	NCuts_SL	NCuts_LEC	NCuts_SCP
AR index	0.36	0.39	0.40	0.50
Time (sec.)	28	31	25	35

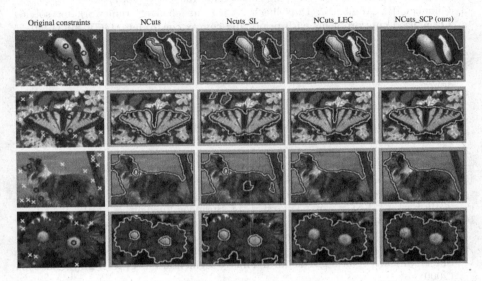

Fig. 1. Example results of constrained image segmentation. A small set of labeled pixels (marked by blue 'o' or yellow 'x') are provided to infer the initial linear equality constraints and pairwise constraints (Color figure online)

5   Conclusions

In this paper, we have investigated the challenging problem of pairwise constraint propagation in constrained image segmentation. We first choose to perform pairwise constraint propagation only over a selected subset of pixels, and then solve such selective constraint propagation problem by developing an efficient graph-based learning algorithm. The selectively propagated constraints are finally used to adjust the weight matrix based on L_1-minimization for image segmentation. The experimental results have shown the promising performance of the proposed algorithm in constrained image segmentation. For future work, we will extend the proposed algorithm to other challenging tasks such as semantic image segmentation and multi-face tracking.

Acknowledgments. This work was partially supported by National Natural Science Foundation of China (61573363 and 61573026), 973 Program of China (2014CB340403 and 2015CB352502), the Fundamental Research Funds for the Central Universities and the Research Funds of Renmin University of China (15XNLQ01), and IBM Global SUR Award Program.

References

1. Carson, C., Belongie, S., Greenspan, H., Malik, J.: Blobworld: image segmentation using expectation-maximization and its application to image querying. TPAMI **24**(8), 1026–1038 (2002)
2. Eriksson, A., Olsson, C., Kahl, F.: Normalized cuts revisited: a reformulation for segmentation with linear grouping constraints. In: ICCV (2007)
3. Ghanem, B., Ahuja, N.: Dinkelbach NCUT: an efficient framework for solving normalized cuts problems with priors and convex constraints. IJCV **89**(1), 40–55 (2010)
4. Hubert, L., Arabie, P.: Comparing partitions. J. Classif. **2**(1), 193–218 (1985)
5. Kamvar, S., Klein, D., Manning, C.: Spectral learning. In: IJCAI, pp. 561–566 (2003)
6. Kulis, B., Basu, S., Dhillon, I., Mooney, R.: Semi-supervised graph clustering: a kernel approach. In: ICML, pp. 457–464 (2005)
7. Li, Z., Liu, J., Tang, X.: Pairwise constraint propagation by semidefinite programming for semi-supervised classification. In: ICML, pp. 576–583 (2008)
8. Lu, Z., Carreira-Perpinan, M.: Constrained spectral clustering through affinity propagation. In: CVPR (2008)
9. Lu, Z., Ip, H.H.S.: Constrained spectral clustering via exhaustive and efficient constraint propagation. In: Daniilidis, K., Maragos, P., Paragios, N. (eds.) ECCV 2010, Part VI. LNCS, vol. 6316, pp. 1–14. Springer, Heidelberg (2010)
10. Martin, D., Fowlkes, C., Tal, D., Malik, J.: A database of human segmented natural images and its application to evaluating segmentation algorithms and measuring ecological statistics. In: ICCV, vol. 2, pp. 416–423 (2001)
11. Shi, J., Malik, J.: Normalized cuts and image segmentation. TPAMI **22**(8), 888–905 (2000)
12. Xu, L., Li, W., Schuurmans, D.: Fast normalized cut with linear constraints. In: CVPR. pp. 2866–2873 (2009)
13. Yu, S., Shi, J.: Segmentation given partial grouping constraints. TPAMI **26**(2), 173–183 (2004)
14. Zhou, D., Bousquet, O., Lal, T., Weston, J., Schölkopf, B.: Learning with local and global consistency. In: NIPS, vol. 16, pp. 321–328 (2004)
15. Zhu, X., Ghahramani, Z., Lafferty, J.: Semi-supervised learning using Gaussian fields and harmonic functions. In: ICML, pp. 912–919 (2003)

Gaussian-Bernoulli Based Convolutional Restricted Boltzmann Machine for Images Feature Extraction

Ziqiang Li, Xun Cai[✉], and Ti Liang

School of Computer Science and Technology, Shandong University,
Jinan 250101, Shandong, China
ziqiang_li@hotmail.com, caixunzh@sdu.edu.cn

Abstract. Image feature extraction is an essential step in image recognition. In this paper, taking the benefits of the effectiveness of Gaussian-Bernoulli Restricted Boltzmann Machine (GRBM) for learning discriminative image features and the capability of Convolutional Neural Network (CNN) for learning spatial features, we propose a hybrid model called Convolutional Gaussian-Bernoulli Restricted Boltzmann Machine (CGRBM) for image feature extraction by combining GRBM with CNN. Experimental results implemented on some benchmark datasets showed that our model is more effective for natural images recognition tasks than some popular methods, which is suggested that our proposed method is a potential applicable method for real-valued image feature extraction and recognition.

Keywords: Deep learning · Feature extraction · Convolutional Gaussian-Bernoulli Restricted Boltzmann Machines · Image classification

1 Introduction

Image recognition has been wildly used in many applications. Image feature extraction is one of the most important steps in the procedure of image recognition system, which directly affects the accuracy of final results. As we all know, the traditional image features are colour, shape, texture and so on. However, these raw features are not applicable for image recognition. More abstractive features, such as Histogram of Oriented Gradient (HOG) [1] and Scale-Invariant Feature Transform (SIFT) [2] have been successfully used for some specific image recognition domains. However, as these hand-crafted features are fixed, they are not adaptive to complex situations. With the development of deep learning in recent year, more general and efficient features are learned by multiple level unsupervised methodology of deep learning, which show more applicable in visual recognition tasks.

Among deep learning methods, Deep Belief Network (DBN) [4] is one of the most classic deep learning models, which consists of multi-layers of

© Springer International Publishing AG 2016
A. Hirose et al. (Eds.): ICONIP 2016, Part II, LNCS 9948, pp. 593–602, 2016.
DOI: 10.1007/978-3-319-46672-9_66

Restricted Boltzmann Machine (RBM) [3]. Many successful variants of RBM which is the element of DBN have been proposed in order to further improve model representation ability of different types of data. For example, Conditional Restricted Boltzmann Machine (CRBM) [5] was proposed for collaborative filtering, Gaussian-Bernoulli ·Restricted Boltzmann Machine (GRBM) [6] was asserted to be suitable for real-valued continuous data, and Recurrent Temporal Boltzmann Machine (RTBM) [7] was developed for representing high-dimensional sequence data, such as speech data and video data.

As mentioned before, GRBM is more suitable for real-valued data. Therefore, it is more applicable to model non-binary image data than tradition RBM. However, the full-connected structure of GRBM is inherited from traditional RBM, which is easily involved over-fitting in feature learning and led to large computation costs.

In recent years, Convolutional Neural Network (CNN) [8] have shown the extraordinary ability of image feature representation in image recognition field [9]. The architecture of CNN consists of several multi-level hierarchies of convolution and pooling. By this architecture, CNN can efficiently extract the spatial information from images and greatly decrease the amount of computations.

In this paper, we propose a novel generative model called Convolutional GRBM(CGRBM) by combining CNN with GRBM. CGRBM extracts meaningful features though a full-size image by generative convolution filters, which reduce quite a number of connecting weights and learn the spatial informations from neighbouring image patches effectively.

The rest of this paper is organized as follows: In Sect. 2, we briefly review the GRBM. Our proposed CGRBM will be shown in Sect. 3. The pseudo-code of training procedure of CGRBM is given in Sect. 4. Experimental results are shown in Sect. 5 and some conclusion will be given in Sect. 6.

2   Related Work

GRBM is a common case of RBM in which the energy function includes a group of real-valued visible neurons $\mathbf{v}$ and a group of binary hidden neurons $\mathbf{h}$. A weight matrix W between two layers represents symmetric connections. The energy function of GRBM [6] is defined as follows:

$$E\left(\mathbf{v},\mathbf{h}\right) = \sum_i \frac{\left(v_i - b_i\right)^2}{2\sigma_i^2} - \sum_{i,j} W_{ij} h_j \frac{v_i}{\sigma_i} - \sum_j c_j h_j \tag{1}$$

where real-valued states of i-th visible unit and binary states of j-th hidden unit are defined as v_i and h_j, respectively. b_i is visible unit bias and c_j is hidden unit bias, σ notes the standard deviation for Gaussian distribution. Then, the joint probability distribution of a state $(\mathbf{v},\mathbf{h})$ is

$$p\left(\mathbf{v},\mathbf{h}\right) = \frac{1}{Z} e^{-E(\mathbf{v},\mathbf{h})} \tag{2}$$

where Z is the partition function. Since GRBM is derived from RBM, the conditional probabilities of visible and hidden units are performed by Gibbs sampling and given as follows:

$$p\left(h_j = 1|\mathbf{v}\right) = sigmoid\left(\sum_i W_{ij}\frac{v_i}{\sigma_i} + c_j\right) \tag{3}$$

$$p\left(v_i = v|\mathbf{h}\right) = \mathcal{N}\left(v; \sum_j h_j W_{ij} + b_i, \sigma_i^2\right) \tag{4}$$

where $\mathcal{N}\left(\cdot; \mu, \sigma^2\right)$ denotes the Gaussian distribution with mean μ and variance σ^2. v is reconstructed value with the Gaussian distribution.

3 Proposed Convolutional GRBM

3.1 Notation

As the Fig. 1 shown, the architecture of proposed model consists of two main parts: the first part is feature extraction, the second part is image reconstruction. In first part, the feature maps which are denoted by H are initially generated from visible layer which are denoted by V. In second part, feature maps are zero-padded and then are convoluted to reconstruct the visible layer. In visible layer, $N_V \times N_V$ pixels of each image are accepted by visible units and share a bias b. In hidden layer, there are K "group" $N_H \times N_H$ feature maps so as to total $K \times N_H{}^2$ hidden units. Each feature map corresponds to a $N_W \times N_W$ convolution kernel and a bias c_k. Unlike connected mode of GRBM, pixels of each feature map are connected with blocks of image in visible layer though a certain generative kernel. Here, for calculating conveniently, we constant the size of input and feature map to be square.

We denote convolution as $*$, dot production as $\bullet$, and the operation of flipping the array horizontally and vertically as $rot180°\left(\cdot\right)$.

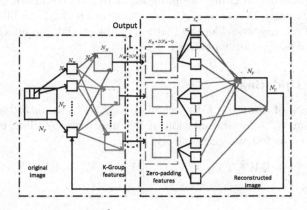

Fig. 1. The architecture of proposed CGRBM model

3.2 Energy Function of CGRBM

Taking the convolutional filters between visible units and hidden units into account, we modify the energy function of GRBM in Eq. (1) as follows:

$$E\left(\mathbf{v},\mathbf{h}\right) = \sum_{i,j=1}^{N_V} \frac{\left(v_{ij}-b\right)^2}{2\sigma_{ij}^2} - \sum_{k=1}^{K}\sum_{i,j=1}^{N_H}\sum_{r,s=1}^{N_W} h_{ij}^k W_{rs}^k \frac{v_{i+r-1,j+s-1}}{2\sigma_{i+r-1,j+s-1}^2}$$
$$-\sum_{k=1}^{K}\sum_{i,j=1}^{N_H} c_k h_{ij}^k. \tag{5}$$

Based on the definitions of operations described in Sect. 3.1, Eq. (5) can be simplified as follows:

$$E\left(\mathbf{v},\mathbf{h}\right) = \sum_{i,j} \frac{\left(v_{ij}-b\right)^2}{2\sigma_{ij}^2} - \sum_{k} h^k \bullet \left(rot180^\circ\left(W^k\right) * \frac{v}{2\sigma^2}\right) - \sum_{k}\sum_{i,j} c_k h_{i,j}^k. \tag{6}$$

During feature extract part, based on Eq. (2), the conditional probability of ij-th unit of k-th feature map is derived from Eq. (6) and formulated as follows:

$$P(h_{ij}^k = 1|\mathbf{v}) = sigmoid\left(c_k + \left(rot180^\circ\left(W^k\right) * \frac{v}{2\sigma^2}\right)_{ij}\right). \tag{7}$$

Then, for image reconstruction stage, unlike C-RBM [11], we add zero-padding to ensure that reconstructed images have same size as original images. Thus, the Gaussian conditional probability of probability distribution of ij-th visible unit is given as follows:

$$P\left(v_{ij} = v|\mathbf{h}\right) = \mathcal{N}\left(v; b + \sum_{k} W^k * h^k, \sigma_i^2\right). \tag{8}$$

Since the frequency of Gibbs sampling for CGRBM should not be infinite, we adopt the Contrastive Divergence (CD) learning method [12] to biased estimate the gradient of parameters. The CD method is quite valid and effectively reduce the deviation of estimation when $k=1$ [12].

3.3 Weight Updating

To speed up rate of updating k-th weight matrix, W^k, we introduce a new the momentum item W_i^k. The formulation of W_i^k and corresponding updating rule for W^k are given as follows, respectively:

$$W_i^k = \lambda W_{i-1}^k - \nabla W^k/(N_H + N_V) + \left\|W^k\right\|_1/N_W^2 - W^k/(N_H + N_V), \tag{9}$$

$$W^k = W^k + W_i^k. \tag{10}$$

In Eq. (9), i denotes the iteration index, λ is defined as momentum factor for controlling. ∇W^k represents the gradient of W^k and is normalized by $(N_H + N_V)$. $\|\cdot\|_1$ denotes the L_1-norm in the third term and used as sparse penalty. Unlike the method proposed in [11] just reconstructed a part of visible units and used a large sparse item, we normalized sparse penalty by N_W^2. By referring to [9], we introduce the weight decay to reduce the training error in the fourth term.

For calculating the ∇W^k, since the traditional C-RBM [11] is regardless of variance of training samples which greatly affect on the conditional mean of visible units. Hereby, we modify the gradient computation rules [4] by dividing the variance of training samples, which is given as follows:

$$\nabla W^k = \eta \left\langle \frac{rot180° \, (h^k) * v}{2\sigma_i^2} \right\rangle_{data} - \eta \left\langle \frac{rot180° \, (h^k) * v}{2\sigma_i^2} \right\rangle_{recon} \tag{11}$$

$$\nabla b = \eta \sum_i \left(\left\langle \frac{v_i}{2\sigma_i^2} \right\rangle_{data} - \left\langle \frac{v_i}{2\sigma_i^2} \right\rangle_{recon} \right) / N_v^2 \tag{12}$$

$$\nabla c_k = \eta \sum_j \left(\langle h_j^k \rangle_{data} - \langle h_j^k \rangle_{recon} \right) / N_h^2 \tag{13}$$

$$\nabla \sigma_{ij} = \eta \left\langle \frac{(v_{ij} - b)^2}{\sigma_{ij}^3} - \sum_{k=1}^{K} h^k \bullet rot180°(W^k) * \frac{v}{2\sigma^2} \right\rangle_{data}$$
$$-\eta \left\langle \frac{(v_{ij} - b)^2}{\sigma_{ij}^3} - \sum_{k=1}^{K} h^k \bullet rot180°(W^k) * \frac{v}{2\sigma^2} \right\rangle_{recon} \tag{14}$$

where η represents learning rate, $\langle \cdot \rangle_{data}$ and $\langle \cdot \rangle_{recon}$ denote the desired values of data distribution and reconstruction distribution, respectively. ∇ denotes the gradient of following parameter.

4   Procedure

The pseudo-code of training procedure of CGRBM is given in Algorithm 1.

5   Experimental Results

To evaluate the performance of the proposed model, We implement several experiments on two image benchmark datasets and compare the results with RBM [3], GRBM [6], Auto-encoder (AE) [14], CNN [8]. In these experiments, features are extracted by all the models and fed into a linear SVM for recognition.

5.1   Datasets and Experiment Setting

These two benchmark datasets we selected are Fifteen Scene Category dataset (Scene-15) contains 4485 different scene images which are categorized into

---

Algorithm 1. training procedure of CGRBM

Input: an original image V_{data}, variance σ^2, learning rate η, momentum item λ.
Output: filter W^k, filter bias c_k, image bias b, variance σ^2.
  Initialization: W^k, c_k, b, σ^2.
  for $i = 1 \rightarrow k$ **do**
    Calculate $P(H_{data}^k)$ using Eq. (7)
    $H_{data} = Bernoulli\,(P(H_{data}^k))$
  end for
  Calculate $P(V_{recon})$ using Eq. (8)
  $V_{recon} = Gaussian\,(P(V_{recon}))$
  for $i = 1 \rightarrow k$ **do**
    Calculate $P(H_{recon}^k)$ using Eq. (7)
    $H_{recon} = Bernoulli\,(P(H_{recon}^k))$
  end for
  Calculate gradient of W^k, b, c_k, σ^2 using Eq. (11), (12), (13), (14)
  Update b, c_k, σ^2 with learning rate η
  Update W^k with learning rate η and momentum λ using Eq. (9), (10).

---

15 classes. CIFAR-10 dataset is one of the most challenging object recognition datasets. This dataset contains 60000 colourful natural images which are categorized into 10 classes equally. Both datasets involve complex background, different levels of illumination and occlusion.

For Fifteen Scene Category dataset, we manually resize the total samples to be equal size, as 128×128. We keep the original size for CIFAR-10 dataset. All the images in each dataset are whitened by Zero-phase Component Analysis (ZCA) [12] in order to eliminate the correlation of neighbouring pixels. Therefore, the initial σ of CGRBM and GRBM is set to 1. After feature extraction of CGRBM, we add a 2×2 probabilistic max pooling [13] in order to obtain the shift-invariance pooled features. For RBM, GRBM and AE, we set the number of hidden units to be 1000 and keep the original setting for CNN (LeNet-2, LeNet-5). According to [14], initial parameters of weights are sampled from $U \sim \left(-\frac{\sqrt{6}}{\sqrt{N_H}+\sqrt{N_V}}, \frac{\sqrt{6}}{\sqrt{N_H}+\sqrt{N_V}}\right)$. The biases of total models are initialized by zero. The learning rate η and momentum factor λ are fixed to 0.05 and 0.9 respectively.

5.2   The Results of Scene-15 Datasets

On Scene-15 dataset, in order to evaluate the performance of CGRBM with different sizes of filters, we tested the reconstructed errors with the sizes of filters changed from 5 to 12 and set the number of filters to be 9. For learning method, CD-1 (for one step) is adopted due to its fast speed for training. The reconstructed errors are listed in Table 1. It is obvious that the too large or small sizes of filters will lead to large reconstruction errors due to feature loss.

We evaluated five popular learning methods of CGRBM, Parallel Tempering (PT-10) [16], Persistent Contrastive Divergence (PCD-1, PCD-10) [15], Contrastive Divergence (CD-1, CD-10). The experimental results listed in Fig. 2

Table 1. The reconstruction error for different size of filters after 30 epochs

Filter size	$N_W=5$	$N_W=6$	$N_W=7$	$N_W=8$	$N_W=9$	$N_W=10$	$N_W=11$	$N_W=12$
R.e	1.283	1.257	1.249	1.216	1.179	1.201	1.212	1.243

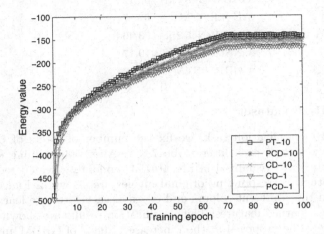

Fig. 2. Negative energy values of CGRBM with different training methods

shows that PT-10 is performed best and CD-1 performed worst, while its computational cost is almost 6 times than that of CD-1. However, computational cost of PT-10 is almost 6 times than that of CD-1. To balance between cost and speed, the PCD-1 is best choice to train the CGRBM for image classification.

Based on PCD-1 training methods, the random selected image and corresponding pooled features are shown in Fig. 3. The recognition accuracy of our proposed model compared with that of RBM, GRBM, AE and CNN is shown in Table 2. It is obvious that our proposed method is better than RBM and its derived models, among all compared models, performance of CGRBM is mostly close to deep CNN(LeNet-5) model.

Fig. 3. The *left* image shows the whitened original random image. The *right* image shows the 9 pooled features, each feature contains 60×60 pixels

Table 2. 5-fold cross validation for each method on Fifteen Scene category dataset

Methods	Accuracy(%)
RBM [3]-SVM	58.79
GRBM [6]-SVM	65.12
AE [14]-SVM	67.87
CNN(LeNet-2)[8]	73.40
CNN(LeNet-5)[8]	79.17
CGRBM-SVM	75.53

5.3   CIFAR-10 Dataset

For CIFAR-10 classification task, we fix the number of filters of CGRBM to be 40 with the size of each filter to be 7×7. After 500 training epochs, the learned 40 filters are visualized in Fig. 4(a). 4 randomly selected reconstructed images with their corresponding original images are shown in Fig. 4(b). It can be easily seen that different colours and edges of each filter are learned clearly. Based on these learned features, The classification results are shown in Table 3. The accuracy of the proposed method increase 11.08 % of GRBM and 2.54 % of LeNet-2, respectively.

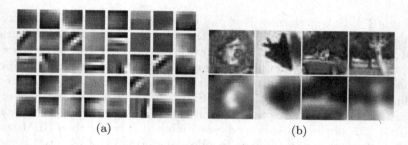

(a)   (b)

Fig. 4. (a): 40 learned filters in CGRBM after 500 training epochs. (b): *top*: 4 random original images, *bottom*: 4 reconstructed images corresponding to original images (Color figure online)

Table 3. 5-fold cross validation for each method on CIFAR-10 dataset

Methods	Accuracy(%)
RBM [3]-SVM	48.04
GRBM [6]-SVM	56.73
AE [14]-SVM	57.90
CNN(LeNet-2) [8]	65.27
CNN(LeNet-5) [8]	71.68
CGRBM-SVM	67.81

6   Conclusion

In this paper, we present a novel probability generation model, CGRBM, based on Gaussian-Bernoulli Restricted Boltzmann Machine. The proposed model combines CNN model with GRBM by taking the advantages of the effectiveness of GRBM for learning discriminative image features and the capability of CNN for learning spatial features. After features extraction by CGRBM and probabilistic max pooling, a group of shift-invariance features are fed into a discriminative classifier. From the experimental results on two benchmarking datasets, we can conclude that CGRBM can effectively improve the performance of feature extraction by the introduced variance and convolution. However, as the bias estimation is performed during CGRBM training, it inevitably generates bias in parameter estimation and affect final classification. In addition, the binary intermediate hidden units also lead some information loss to reconstructed images.

In the future, we will further improve the feature extraction ability of CGRBM by training deeper CGRBMs. We believe that the classification accuracy will be improved.

Acknowledgments. This work was supported by National Natural Science Foundation of China under grant no. 51473088 and National Key Research and Development Plan of China under grant no. 2016YFC0301400.

References

1. Dalal, N., Triggs, B.: Histograms of oriented gradients for human detection. In: 2005 IEEE International Computer Vision and Pattern Recognition (CVPR), pp. 886–893. IEEE (2005)
2. Lowe, D.G.: Distinctive image features from scale-invariant keypoints. Int. J. Comput. Vis. **60**, 91–110 (2004)
3. Rumelhart, D.E., McClelland, J.L., PDP Research Group: Parallel Distributed Processing: Explorations in the Microstructure of Cognition, vol. 1–2 (1986)
4. Hinton, G., Salakhutdinov, R.: Reducing the dimensionality of data with neural networks. Science **313**(5786), 504–507 (2006)
5. Salakhutdinov, R., Mnih, A., Hinton, G.: Restricted Boltzmann machines for collaborative filtering. In: Proceedings of the 24th International Conference on Machine Learning, pp. 791–798. ACM (2007)
6. Wang, N., Melchior, J., Wiskott, L.: Gaussian-binary restricted Boltzmann machines on modeling natural image statistics. arXiv preprint (2014) arXiv:1401.5900
7. Sutskever, I., Hinton, G.E., Taylor, G.W.: The recurrent temporal restricted Boltzmann machine. In: Advances in Neural Information Processing Systems, pp. 1601–1608 (2008)
8. LeCun, Y., Bottou, L., Bengio, Y., Haffner, P.: Gradient-based learning applied to document recognition. Proc. IEEE **86**(11), 2278–2324 (1998)
9. Krizhevsky, A., Sutskever, I., Hinton, G.E.: Imagenet classification with deep convolutional neural networks. In: Advances in Neural Information Processing Systems, pp. 1097–1105 (2012)

10. Karpathy, A., Toderici, G., Shetty, S., Leung, T., Sukthankar, R., Fei-Fei, L.: Large-scale video classification with convolutional neural networks. In: CVPR (2014)
11. Norouzi, M., Ranjbar, M., Mori, G.: Stacks of convolutional restricted Boltzmann machines for shift-invariant feature learning. In: CVPR (2009)
12. Hinton, G.: A practical guide to training restricted Boltzmann machines. Momentum **9**(1), 926 (2010)
13. Lee, H., Grosse, R., Ranganath, R., Ng, A.Y.: Convolutional deep belief networks for scalable unsupervised learning of hierarchical representations. In: ICML (2009)
14. Bengio, Y., Lamblin, P., Popovici, D., Larochelle, H., Montral, U.D., Qubec, M.: Greedy layer-wise training of deep networks. In: NIPS. MIT Press (2007)
15. Tieleman, T.: Training restricted Boltzmann machines using approximations to the likelihood gradient. In: Proceedings of the 25th International Conference on Machine Learning. ICML 2008, pp. 1064–1071. ACM Press, New York (2008)
16. Cho, K.H., Raiko, T., Ilin, A.: Enhanced gradient for training restricted Boltzmann machines. Neural Comput. **25**(3), 805–831 (2013)

Gaze Movement Control Neural Network Based on Multidimensional Topographic Class Grouping

Wenqi Zhong[1], Jun Miao[1(✉)], and Laiyun Qing[2]

[1] Key Lab of Intelligent Information Processing of Chinese Academy of Sciences (CAS), Institute of Computing Technology, CAS, Beijing 100190, China
jmiao@ict.ac.cn
[2] School of Computer and Control Engineering,
University of Chinese Academy of Sciences, Beijing 100049, China

Abstract. Target search is an important ability of the human visual system. One major problem is that the real human visual cognitive process, which requires only few samples for learning, has abilities of inference with obtained knowledge for searching when he meets the new target. Based on the Topographic Class Grouping (TCG) [1] and a series of models of Visual Perceiving and Eyeball-Motion Controlling Neural Networks [2–5], we make effective improvements to the models, by incorporating the cerebral self-organizing feature mapping function in terms of multidimensional TCG. In this paper, we propose the gaze movement control neural network based on multidimensional TCG. Experiments show that gaze movement control neural network by adding a block of multidimensional TCG and by self-organizing visual field image features-spatial relationship clustering achieves the visual inference and stable results on the target search tasks.

Keywords: Target searching · Self-organizing maps · Gaze movement control

1 Introduction

Target search is an important ability of the human visual system. The main issues of the search target are "where" the target location prediction mechanism, and "what"-target identification mechanism. According to above biological mechanisms, Weng et al. proposed the "Where-What Network" model which simulate human ventral and dorsal visual pathways for visual target searching [6]. In their newest system [7], any new pair of objects can be detected. Walther's model [8] simulates the layer-wise processing of target information and expression of the "what" pathway. In addition, under the influence of bio-inspired visual searching, target searching also involves the inference of target location. In 1982, according to the human brain biology, physiology and human psychology, Kohonen proposed Self-Organizing Maps (SOM) [9]. Based on two-dimensional thin film

© Springer International Publishing AG 2016
A. Hirose et al. (Eds.): ICONIP 2016, Part II, LNCS 9948, pp. 603–610, 2016.
DOI: 10.1007/978-3-319-46672-9_67

structure of the cerebral cortex, SOM is constructed for simulating brain system's self-organizing function [10,11]. Furthermore, Luciw et al. proposed Topographic Class Grouping (TCG) [1], which simulates autonomous mental development by using the top-down information and bottom-up information. The top-down connections can reduce uncertainty at the lower layer with respect to the features in the higher layer, which enables relevant information to be uncovered at the lower layer so that irrelevant information can preferentially be discarded.

Some work has been done to simulate oculomotor mechanism. Nevertheless, only a few researches are based on the cognitive physiological principles and corresponding neuron inter-connecting structure. In [2], Miao et al. proposed a model named Visual Perceiving and Eyeball-Motion Controlling Neural Network, which searches target by reasoning with visual context encoding mechanism. In [3,4], they improved the network by using the population-cell-coding mechanism. They substantially reduce the amount of encoding quantity of the visual context. However, the disadvantage of this network is that the neurons of larger responding in the third layer were involved in gaze motion synthesis in the fourth layer. In [5], a new movement coding neuron layer, which groups connection weights form the population coding neurons to the movement control neurons in the fourth layer, is inserted between the third layer and the fourth layer in the previous structure. Comparing experiments on target searching showed that the improved system made the significant improvement by Self-Organizing gaze movement information. The architecture of [5] is shown in Fig 1.

In this paper, we use multidimensional TCG to simulate brain system's SOM functions with top-down connection, and make effective improvements to the system in [5]. According to this improvement, we proposed gaze movement control

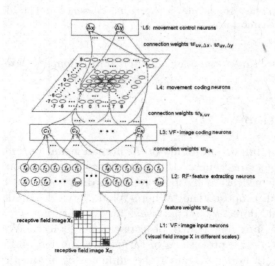

Fig. 1. Grouped population cell coding structure with five layers of neurons for visual field image representation and gaze movement controlling (VF: Visual Field, RF: Receptive Field)

neural network. In the following, this paper is organized as follows: Sect. 2 describes the whole architecture of gaze movement control neural network Based on Multi-dimensional TCG. The detailed cognizing and learning mechanism are discussed in Sect. 3. In Sect. 4, comparative experiments for gaze movement control neural network and improved model on real image database are discussed. Finally, conclusion and future study are given in the last section.

2   Architecture of Gaze Movement Control Neural Network

In target search, the improved system in [5] uses visual context and grouped population-cell-coding mechanism, and achieves high locating precision. However, the system ignores something. In the process of human visual perception and cognition, human will not directly draw a conclusion through his memory. Nevertheless, human will infer the result based on his previous knowledge. Therefore, we add Topographic Class Grouping, of which role is simulating the human abstraction process after visual information obtaining. To this system, the whole architecture is shown in Fig. 2.

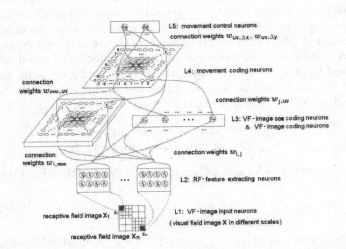

Fig. 2. Architecture of Gaze Movement Control Neural Network Based on Multidimensional Topographic Class Grouping (VF: Visual Field, RF: Receptive Field, SOM: Self-Organizing Maps)

2.1   Receptive Field Image Encoding

In the second layer of the improved system [5], Miao et al. extracted extended LBP (local binary pattern) features which combine original LBP coding with coding response. However, extended LBP coding is not very flexible to learn robust feature. In order to settle this problem and adapt to weight adjusting of VF-image

SOM coding neurons, an approximate method is put forward: using a 8dimension vector, which represents 8 pairs of sign of the differences between 8 surrounding pixels and the central pixel in its receptive field image of 3×3 input neurons, to replace the extended LBP coding feature. The approximate LBP feature is show in Eqs. (1) and (2):

$$F_i = (x_{i0} - x_{i8}, x_{i1} - x_{i8}, \ldots, x_{i7} - x_{i8}) \tag{1}$$

$$similarity(F_i, F_j) = \begin{cases} F_i^T F_j & , sign(F_i) = sign(F_j) \\ 0 & , sign(F_i) \neq sign(F_j) \end{cases} \tag{2}$$

where the vector $X_i = (x_{i0}, x_{i1}, \ldots, x_{i8})^T$ represents the ith image block of 3×3 pixels or the ith receptive field image of 3×3 input neurons; the term F_i represents the approximate LBP feature vector of ith receptive field image of 3×3 input neurons; the $sign(F_i)$ represents the sign of feature vector in each dimension. The $F_i^T F_j$ represents the inner-product of two vectors.

2.2   Visual Field Image Memory

The role of the VF-image coding neurons is encoding the visual field image features and the spatial relationship from the center of the visual field image to the center of target. The neurons in the VF-image coding neurons are using dynamical generation mechanism like [5]. The connection weight between VF-image coding neurons and receptive field feature extracting neurons are learned by Hebbian rule. For the approximate LBP feature of the ith receptive field image F_i, we define the connection weight w_{ij}. Based on the Hebbian rule, the learning rule for the connection weight w_{ij} of VF-image coding neurons is formularized in Eq. (3):

$$\begin{cases} w_{i,j}(0) = 0, \Delta w_{i,j}(0) = \alpha R_a R_b = \alpha F_i R_j \\ w_{i,j}(1) = w_{i,j}(0) + \Delta w_{i,j}(0) = \alpha F_i R_j \end{cases} \tag{3}$$

Both α and R_j are set to 1 for simplifying computation, and then Eq. (3) is changed to Eq. (3a):

$$w_{i,j}(1) = F_i \tag{3a}$$

Similarly, the connection weight $w_{j,uv}$ between the jth neuron in VF-image coding neurons and the uv^{th} movement coding neuron can be computed with Hebbian rule, in which the learning rate and the response of the movement coding neuron are set to 1 for simplifying computation:

$$\begin{cases} w_{j,uv}(0) = 0, \Delta w_{j,uv}(0) = \beta R_a R_b = 1 \\ w_{j,uv}(1) = w_{j,uv}(0) + \Delta w_{j,uv}(0) = 1 \end{cases} \tag{4}$$

When the new visual field image stimulates a VF-image coding neuron, the neuron's response is computed by Eq. (5):

$$R_j = \sum_{i=1}^{I} similarity(F_i, w_{i,j}) \tag{5}$$

where F_i represents the approximate LBP feature of ith receptive field image, I represents the total number of receptive field image in the new visual field images; R_j represent the response of the jth VF-image coding neuron.

2.3 Visual Field Inference

The basic structure of VF-image SOM coding neurons is built by Topographic Class Grouping (TCG) [1], which simulates the SOM functions of brain systems with top-down connections. By analyzing the sampling methods, the topological structure of TCG must satisfies the expression requirement, in which at least three essential variables are needed to express: both horizontal and vertical retina cell arrays and different objects. Therefore, VF-image SOM coding neurons is a multi-dimensional topological structure that could be built by a multidimensional TCG.

In the learning procedure, the detail learning method is similar to TCG: The neuronal precompetitive response function is weighted sum of top-down connections from the fourth layer and the button-up connections from the second layer, as described in Eq. (6). However, because VF-image SOM coding neurons is the multidimensional TCG, the Lateral Excitation is replaced by hypercube neighborhoods.

$$
\begin{aligned}
R_{mno} &= (1 - \gamma)\hat{R}_{mno}^{Second} + \gamma\hat{R}_{mno}^{Fourth} \\
\hat{R}_{mno}^{Second} &= \sum_{i=1}^{I} \frac{similarity(F_i, w_{i,mno})}{||F_i|| ||w_{i,mno}||} \\
\hat{R}_{mno}^{Fourth} &= \frac{\sum_u \sum_u R_{uv} w_{mno,uv})}{\sqrt{\sum_u \sum_u R_{uv}^2}\sqrt{\sum_u \sum_u w_{mno,uv}^2}}
\end{aligned}
\tag{6}
$$

where R_{mno} is the mno^{th} neurons response in TCG; $0 \leq \gamma \leq 1$ controls the relative influence of top-down to bottom-up; $\hat{R}_{mno}^{Second}$ is the bottom-up response; $\hat{R}_{mno}^{Fourth}$ is the top-down response; R_{uv} is the uv^{th} neurons response in the fourth layer.

2.4 Decoding of Spatial Relationship for Gaze Movement Control

Gaze movement control, which includes the fourth layer-movement coding neurons and the fifth layer-movement control neurons, is directly responsible for visual object research. In the fourth layer-movement coding neurons, neurons can gather the memory information from coding neurons and inference information from VF-image SOM coding neurons. In [5], Miao et al. use populatio-cell-coding mechanism and similarity factor $P = \frac{P_m}{P_l}$ to control M, where P_m and P_l are the Mth largest and first largest responses of VF-image coding neurons respectively. In our research, we separately use similarity factor to control the responding neurons number of VF-image coding neurons and VF-image SOM coding neurons by $P^{code} = \frac{P_m^{code}}{P_l^{code}}$ and $P^{som} = \frac{P_m^{som}}{P_l^{som}}$, which led to produce M^{code} and M^{som}. In order to balance VF-image coding neurons and VF-image SOM coding neurons, the final parameter $M = \min\{M^{code}, M^{som}\}$.

While the neurons in the fourth layer separately receive the first M image coding neurons' responses and the first M image SOM coding neurons' responses, it may produce L different movement estimations and grouped to L sets ($1 \leq L \leq$

$16 \times 16 = 256$, movement coding neurons). Each image coding neuron. Let M_l represent the population number of image coding neurons in the lth set, then we have:

$$M = \sum_{l=1}^{L} M_l \tag{7}$$

And the response of the $u_l v_l^{th}$ movement coding neuron is:

$$R_{u_l v_l} = \sum_{j_{l'}=1}^{M_l} w_{j_{l'},u_l v_l} R_{j_{l'}} + \sum_{m_{l'}=1}^{M_l} \sum_{n_{l'}=1}^{M_l} \sum_{o_{l'}=1}^{M_l} w_{m_{l'} n_{l'} o_{l'}, u_l v_l} R_{m_{l'} n_{l'} o_{l'}} \tag{8}$$

where $R_{j_{l'}}$ is the response of the $j_{l'}^{th}$ image coding neuron; $w_{j_{l'},u_l v_l}$ is the connection weight from the $j_{l'}^{th}$ image coding neuron to the $u_l v_l^{th}$ movement coding neuron in the fourth layer; $R_{m_{l'} n_{l'} o_{l'}}$ is the response of the $m_{l'} n_{l'} o_{l'}^{th}$ image SOM coding neuron; $w_{m_{l'} n_{l'} o_{l'}, u_l v_l}$ is the connection weight from the $m_{l'} n_{l'} o_{l'}^{th}$ image coding neuron to the $u_l v_l^{th}$ movement coding neuron in the fourth layer. Substituting Eq. (3a) into Eq. (8), we get:

$$R_{u_l v_l} = \sum_{j_{l'}=1}^{M_l} R_{j_{l'}} + \sum_{m_{l'}=1}^{M_l} \sum_{n_{l'}=1}^{M_l} \sum_{o_{l'}=1}^{M_l} w_{m_{l'} n_{l'} o_{l'}, u_l v_l} R_{m_{l'} n_{l'} o_{l'}} \tag{9a}$$

In the fourth layer, each $u_l v_l^{th}$ movement coding neuron is influenced by local lateral excitation and the global lateral inhibition. Local lateral excitation involves spreading of responses to the adjacent neighbors (in 5×5 neighborhoods) of the $u_l v_l^{th}$ movement coding neuron. Global lateral inhibition is winner-take-all mechanism. The following structure just like [5].

3   Cognizing and Learning Mechanism

Based on human visual system cognizing theory, a system can be designed to use a gradual search strategy and finally predicts the movement through the movement of inference in each resolution of visual field images. What's more, when the human sees with the new object which is similar to the familiar things in memory, the existing memory is strengthened. Based on this biological inspiration, the learning process of VF-image coding neurons has been designed with the similar mechanism: If the error between system prediction and ground truth is less than the threshold Th, the neurons in VF-image coding neurons will response and transfer the information to the next layer when the system is learning a new image. After that, these neurons connection weights between the second layer and the VF-image coding neurons will be strengthened in 1.2 times.

4   Experiment for Target Searching

In this section, we investigate the effectiveness of the proposed gaze movement control neural network through comparative experiments with respect to target searching. In order to be close to the real human cognitive processes and compare with [5], which require only few samples for learning, we set up two experiments: "30 vs. 270" and "90 vs. 210". We conduct experiments on a public databases: University of Bern. University of Bern has 300 images with 30 people in ten different poses (ten images each person). The image size is 320×214 pixels. The average radius of the eyeballs of these 30 persons is 4.02 pixels.

In "30 vs. 270" and "90 vs. 210" experiments, we compare the improved system in [5] and the gaze movement control neural network proposed in this paper. This experiment explores the ability of gaze movement control neural network's learning in few samples. Table 1 list the comparative results in the mean and the standard deviation of locating errors and the comprehensive test error.

Table 1. Performance comparison on learning in few samples

Experiment	System (P, Th)	Locating errors (unit: pixel)		
		Mean (mn)	Standard deviation (std)	Comprehensive test error $(\sqrt{mn^2 + std^2})$
30 vs. 270	The system [5] $(P = 0.7, Th = 0.06)$	**2.1**	4.93	5.36
	This paper $(P = 0.5, Th = 0.06)$	2.59	**1.89**	**3.2**
90 vs. 210	The system [5] $(P = 0.7, Th = 0.08)$	**1.5**	2.83	3.2
	This paper $(P = 0.5, Th = 0.08)$	2.09	**0.94**	**2.3**

From table 1, it can be learned that the improved system [5] achieved better accuracy of target search and the gaze movement control neural network achieved better standard deviation of locating errors. Though the improved system achieved better accuracy, the proposed model can get much more stable results in an acceptable level (The average radius of the eyeballs of these 30 persons is 4.02 pixels).

5   Conclusion and Discussion

In this paper, we proposed a novel neural network, called gaze movement control neural network that improves the stability of the improved system of [5] in an acceptable level. Compared with the fifth layers in the improved system proposed in [5], the multidimensional SOM topological structure using TCG is inserted between the second layer and the fourth layer for providing the ability of inference. Experiment results indicated that though learning in few samples, the gaze movement control neural network can get much more stable result in an acceptable level. In the future, the research will be pay close attention to how to simulate the process of visual cognizing for better searching results.

Acknowledgements. This research is partially sponsored by Natural Science Foundation of China (Nos. 61272320, 61472387 and 61572004) and Beijing Natural Science Foundation (Nos. 4152005 and 4162058).

References

1. Luciw, M., Weng, J.: Top-down connections in self-organizing hebbian networks: topographic class grouping. IEEE Trans. Auton. Ment. Dev. **2**, 248–261 (2010)
2. Miao, J., Chen, X., Gao, W., Chen, Y.: A visual perceiving and eyeball-motion controlling neural network for object searching and locating. In: International Joint Conference on Neural Networks, pp. 4395–4400 (2006)
3. Miao, J., Zou, B., Qing, L., Duan, L., Fu, Y.: Learning internal representation of visual context in a neural coding network. In: Diamantaras, K., Duch, W., Iliadis, L.S. (eds.) ICANN 2010, Part I. LNCS, vol. 6352, pp. 174–183. Springer, Heidelberg (2010)
4. Miao, J., Qing, L., Zou, B., Duan, L., Gao, W.: Top-down gaze movement control in target search using population cell coding of visual context. IEEE Trans. Auton. Ment. Dev. **2**, 196–215 (2010)
5. Miao, J., Duan, L., Qing, L., Qiao, Y.: An improved neural architecture for gaze movement control in target searching. In: IEEE International Joint Conference on Neural Networks, pp. 2341–2348 (2011)
6. Ji, Z., Weng, J., Prokhorov, D.: Where-what network 1: where and what assist each other through top-down connections. In: IEEE International Conference on Development and Learning, pp. 61–66 (2008)
7. Guo, Q., Wu, X., Weng, J.: WWN-9: cross-domain synaptic maintenance and its application to object groups recognition. In: 2014 International Joint Conference on Neural Networks, pp. 716–723 (2014)
8. Walther, D.: Interactions of visual attention and object recognition: computational modeling, algorithms, and psychophysics. Walther Dirk (2006)
9. Kohonen, T.: Self-organized formation of topologically correct feature maps. Biol. Cybern. **43**(1), 59–69 (1982)
10. Kohonen, T.: Self-organizing map. Proc. IEEE **78**(9), 1464–1480 (1990)
11. Kohonen, T.: Self-Organizing Maps. Springer, Heidelberg (2001)

Incremental Robust Nonnegative Matrix Factorization for Object Tracking

Fanghui Liu[1], Mingna Liu[2,3], Tao Zhou[3], Yu Qiao[1], and Jie Yang[1(✉)]

[1] Institute of Image Processing and Pattern Recognition,
Shanghai Jiao Tong University, Shanghai, China
`lfhsgre@gmail.com`, {`zhou.tao,qiaoyu,jieyang`}`@sjtu.edu.cn`
[2] Shanghai Institute of Spaceflight Control Technology, Shanghai, China
`mingnal@126.com`
[3] Infrared Detection Technology Research and Development Center,
China Aerospace Science and Technology Corporation, Shanghai, China

Abstract. Nonnegative Matrix Factorization (NMF) has received considerable attention in visual tracking. However noises and outliers are not tackled well due to Frobenius norm in NMF's objective function. To address this issue, in this paper, NMF with $L_{2,1}$ norm loss function (robust NMF) is introduced into appearance modelling in visual tracking. Compared to standard NMF, robust NMF not only handles noises and outliers but also provides sparsity property. In our visual tracking framework, basis matrix from robust NMF is used for appearance modelling with additional ℓ_1 constraint on reconstruction error. The corresponding iterative algorithm is proposed to solve this problem. To strengthen its practicality in visual tracking, multiplicative update rules in incremental learning for robust NMF are proposed for model update. Experiments on the benchmark show that the proposed method achieves favorable performance compared with other state-of-the-art methods.

Keywords: Incremental robust NMF · Appearance model · Visual tracking

1 Introduction

Appearance modelling is an overriding concern in visual tracking and has been studied for several years [12]. One widespread adoption of appearance modelling is with generative method [11,17], which aims to search the most similar candidate to the target with minimizing reconstruction error. The representative generative methods include but are not limited to subspace learning [8], sparse representation [5,16].

Recently, Nonnegative Matrix Factorization (NMF) based on subspace learning has been successfully used in visual tracking with variety works [13,15]. Wang et al. [9] utilize projective NMF for appearance modelling. Liu *et al.* [7] propose an inverse coding view for visual tracking, in which NMF is served as a feature coder. NMF seeks for two nonnegative matrices $\mathbf{U}$ (the basis matrix) and $\mathbf{V}$

© Springer International Publishing AG 2016
A. Hirose et al. (Eds.): ICONIP 2016, Part II, LNCS 9948, pp. 611–619, 2016.
DOI: 10.1007/978-3-319-46672-9_68

(the coefficient matrix) to represent the original data matrix $\mathbf{X}$ ($\mathbf{X} \approx \mathbf{UV}$). The basis matrix $\mathbf{U}$ is a representation to the target in a low-dimensional space, like PCA. The objective function in standard NMF uses the least square error function which is well known but not stable. Because real data may contain many undesirable noises and outliers due to partial occlusions in visual tracking. Though these above methods can deal with outliers by imposing additive various sparse constraints on $\mathbf{U}$ or $\mathbf{V}$, they could not tackle this problem in essence, and suffer from significant performance degradation. To address this issue, data reconstruction function is formulated in the form of robust matrix norms such as ℓ_1 norm or $L_{2,1}$ norm [6,14]. Robust NMF (RNMF) not only handles outliers and noises but also incurs almost the same computation cost as standard NMF.

Motivated by the above issues, this paper introduces RNMF into visual tracking framework. Specially, incremental learning for RNMF should be taken into consideration for online tracking. The main novelties of the proposed method include: (a) We present an iterative algorithm including the standard nonnegative least square method for target coefficients and the soft-threshold method to obtain sparse error coefficients. (b) Multiplicative update rules for incremental RNMF are provided, as the indispensable parts of online visual tracking. (c) Model update scheme (First-in and First-out) for visual tracking is proposed with incremental RNMF learning. Experiments on different videos illustrate our tracker can handle partial occlusion and other challenging factors.

2   RNMF and Its Incremental Learning

2.1   Review: Robust NMF with $L_{2,1}$ norm

Given a data matrix $\mathbf{X} = [\mathbf{x}_1, \mathbf{x}_2, ..., \mathbf{x}_N] \in \mathbb{R}^{M \times N}$, robust NMF is defined as,

$$\min_{\mathbf{U},\mathbf{V}} \|\mathbf{X} - \mathbf{UV}\|_{2,1}$$
$$s.t. \quad \mathbf{U} \geq 0, \quad \mathbf{V} \geq 0 \tag{1}$$

where the basis matrix $\mathbf{U} \in \mathbb{R}^{M \times K}$, and $L_{2,1}$ norm is defined as $\|\mathbf{X}\|_{2,1} = \sum_{i=1}^{N} \sqrt{\sum_{j=1}^{M} x_{ij}^2} = \sum_{i=1}^{N} \|\mathbf{x}_i\|$ and $\mathbf{x}_i$ is the i-th column of $\mathbf{X}$. Because the objective function is not convex on both $\mathbf{U}$ and $\mathbf{V}$, and $L_{2,1}$ norm is harder to solve than standard NMF with Frobenius norm. The corresponding objective function can be rewritten as in [6],

$$\mathcal{O}(\mathbf{U}, \mathbf{V}) = Tr[(\mathbf{X} - \mathbf{UV})\mathbf{D}(\mathbf{X} - \mathbf{UV})^\top] \tag{2}$$

where $\mathbf{D}$ is a diagonal matrix and its diagonal element is given by $D_{ii} = 1/\|\mathbf{x}_i - \mathbf{Uv}_i\|$. The iteratively updating algorithm, proposed by Kong $etal$. [6], obtains a local minimum value of robust NMF. The iteration formula is following:

$$u_{jk}^{t+1} \leftarrow u_{jk}^t \frac{(\mathbf{XDV}^\top)_{jk}}{(\mathbf{UVDV}^\top)_{jk}}, v_{ki}^{t+1} \leftarrow v_{ki}^t \frac{(\mathbf{U}^\top\mathbf{XD})_{ki}}{(\mathbf{U}^\top\mathbf{UVD})_{ki}} \tag{3}$$

2.2   Incremental Learning for Robust NMF

The implicit assumption behind the standard incremental NMF [1] is that the previous coefficient matrix $\mathbf{V}$ has no effect on the incremental process when a new sample $\mathbf{x}$ is added: $[\mathbf{X}, \mathbf{x}] \approx \mathbf{U} \times [\mathbf{V}, \mathbf{v}]$. Similarly, in our incremental updating for RNMF, define that $\mathbf{X}_{t+1} = [\mathbf{X}_t, \mathbf{x}]$, $\mathbf{V}_{t+1} = [\mathbf{V}_t, \mathbf{v}]$, $\mathbf{U}_{t+1}$, and $\mathbf{D}_{t+1} = diag(\mathbf{D}_t, d_{t+1})$ are the corresponding matrices when the $(t+1)$-th sample arrives. The corresponding objective function is formulating as,

$$
\begin{aligned}
\mathcal{O}(\mathbf{U}_{t+1}, \mathbf{V}_{t+1}) &= \mathcal{O}(\mathbf{U}_{t+1}, \mathbf{v}) \\
&= Tr[(\mathbf{X}_{t+1} - \mathbf{U}_{t+1}\mathbf{V}_{t+1})\mathbf{D}_{t+1}(\mathbf{X}_{t+1} - \mathbf{U}_{t+1}\mathbf{V}_{t+1})^{\top}] \\
&= Tr(\mathbf{X}_{t+1}\mathbf{D}_{t+1}\mathbf{X}_{t+1}^{\top}) - 2Tr(\mathbf{X}_{t+1}\mathbf{D}_{t+1}\mathbf{V}_{t+1}^{\top}\mathbf{U}_{t+1}^{\top}) \\
&\quad + Tr(\mathbf{U}_{t+1}\mathbf{V}_{t+1}\mathbf{D}_{t+1}\mathbf{V}_{t+1}^{\top}\mathbf{U}_{t+1}^{\top})
\end{aligned}
\tag{4}
$$

We expend $\mathbf{D}_{t+1}$ and $\mathbf{V}_{t+1}$ to separate $d_{t+1} = 1/\|\mathbf{x} - \mathbf{U}_{t+1}\mathbf{v}\|$ and $\mathbf{v}$ for convenience,

$$
\begin{aligned}
\mathcal{O}(\mathbf{U}_{t+1}, \mathbf{v}) &= Tr(\mathbf{X}_t\mathbf{D}_t\mathbf{X}_t^{\top}) + d_{t+1}Tr(\mathbf{x}\mathbf{x}^{\top}) \\
&\quad - 2Tr(\mathbf{X}_t\mathbf{D}_t\mathbf{V}_t^{\top}\mathbf{U}_{t+1}^{\top}) - 2d_{t+1}Tr(\mathbf{x}\mathbf{v}^{\top}\mathbf{U}_{t+1}^{\top}) \\
&\quad + Tr(\mathbf{U}_{t+1}\mathbf{V}_t\mathbf{D}_t\mathbf{V}_t^{\top}\mathbf{U}_{t+1}^{\top}) + d_{t+1}Tr(\mathbf{U}_{t+1}\mathbf{v}\mathbf{v}^{\top}\mathbf{U}_{t+1}^{\top})
\end{aligned}
\tag{5}
$$

Incremental Update Rules on $\mathbf{U}_{t+1}$: Given $\mathbf{V}_{t+1}$, let ψ_{ij} be a Langrange Multiplier for the nonnegative constraint on $\mathbf{U}_{t+1}$. The relevant Langrange function corresponding to Eq. (5): $\mathcal{L}_{\mathbf{U}_{t+1}} = \mathcal{O}(\mathbf{U}_{t+1}, \mathbf{v}) + Tr(\psi\mathbf{U}_{t+1})$. The partial derivatives of $\mathcal{L}_{\mathbf{U}}$ with respect to $\mathbf{U}_{t+1}$ is:

$$
\begin{aligned}
\frac{\partial \mathcal{L}_{\mathbf{U}}}{\partial \mathbf{U}_{t+1}} &= \mathbf{x}^{\top}\mathbf{x}\frac{\partial d_{t+1}}{\partial \mathbf{U}_{t+1}} - 2\mathbf{X}_t\mathbf{D}_t\mathbf{V}_t^{\top} - 2Tr(\mathbf{x}\mathbf{v}^{\top}\mathbf{U}_{t+1}^{\top})\frac{\partial d_{t+1}}{\partial \mathbf{U}_{t+1}} \\
&\quad - 2d_{t+1}\mathbf{x}\mathbf{v}^{\top} + 2\mathbf{U}_{t+1}\mathbf{V}_t\mathbf{D}_t\mathbf{V}_t^{\top} + 2d_{t+1}\mathbf{U}_{t+1}\mathbf{v}\mathbf{v}^{\top} \\
&\quad + \mathbf{v}^{\top}\mathbf{U}_{t+1}^{\top}\mathbf{U}_{t+1}\mathbf{v}\frac{\partial d_{t+1}}{\partial \mathbf{U}_{t+1}} + \psi
\end{aligned}
\tag{6}
$$

where $\frac{\partial d_{t+1}}{\partial \mathbf{U}_{t+1}} = d_{t+1}^3(\mathbf{x} - \mathbf{U}_{t+1}\mathbf{v})\mathbf{v}^{\top}$. Using the KKT condition for each elements in ψ and $\mathbf{U}_{t+1}$ to satisfy $\psi_{ij}u_{ij} = 0$, we get the equations for u_{ij}. And then these equations lead to the following update rule shown on $\mathbf{U}_{t+1}$ in Eq. (8), where $S_1 = 2d_{t+1}^3 Tr(\mathbf{x}\mathbf{v}^{\top}\mathbf{U}_{t+1}^{\top})$, $S_2 = d_{t+1}^3\mathbf{v}^{\top}\mathbf{U}_{t+1}^{\top}\mathbf{U}_{t+1}\mathbf{v}$ and $S_3 = \mathbf{U}_{t+1}\mathbf{v}\mathbf{v}^{\top}$.

Incremental Update Rules on $\mathbf{v}$: Let ϕ_j be a Langrange Multiplier for the nonnegative constraint on $\mathbf{v}$. The relevant Langrange function with respect to Eq. (5): $\mathcal{L}_{\mathbf{v}} = \mathcal{O}(\mathbf{U}_{t+1}, \mathbf{v}) + Tr(\phi\mathbf{v})$. The partial derivatives of $\mathcal{L}_{\mathbf{v}}$ with respect to $\mathbf{v}$ is:

$$
\begin{aligned}
\frac{\partial \mathcal{L}_{\mathbf{v}}}{\partial \mathbf{v}} &= \mathbf{x}^{\top}\mathbf{x}\frac{\partial d_{t+1}}{\partial \mathbf{v}} - 2d_{t+1}\mathbf{U}_{t+1}^{\top}\mathbf{x} - 2Tr(\mathbf{x}\mathbf{v}^{\top}\mathbf{U}_{t+1}^{\top})\frac{\partial d_{t+1}}{\partial \mathbf{v}} \\
&\quad + 2d_{t+1}\mathbf{U}_{t+1}^{\top}\mathbf{U}_{t+1}\mathbf{v} + \mathbf{v}^{\top}\mathbf{U}_{t+1}^{\top}\mathbf{U}_{t+1}\mathbf{v}\frac{\partial d_{t+1}}{\partial \mathbf{v}} + \phi
\end{aligned}
\tag{7}
$$

where $\frac{\partial d_{t+1}}{\partial \mathbf{v}} = d_{t+1}^3 \mathbf{U}_{t+1}^\top (\mathbf{x} - \mathbf{U}_{t+1}\mathbf{v})$. Using the KKT condition for each element in $\mathbf{v}$ with $\phi_j v_j = 0$, we get the following update rule for $\mathbf{v}$ shown in Eq. (9), where $S_4 = \mathbf{U}_{t+1}^\top \mathbf{U}_{t+1}\mathbf{v}$. Therein, we give relatively detailed description about incremental learning for robust NMF, which can be used in the proposed visual tracking method effectively.

3   Proposed Robust NMF Tracker

3.1   Particle Filter in Visual Tracking Framework

Particle filter has been widely applied to visual tracking for state estimation. The implicit rationale behind particle filter is to estimate the posterior distribution

$$(\mathbf{U}_{t+1})_{ij} \leftarrow (\mathbf{U}_{t+1})_{ij} \cdot \frac{\{2\mathbf{X}_t \mathbf{D}_t \mathbf{V}_t^\top + d_{t+1}^3 \|\mathbf{x}\|^2 \mathbf{U}_{t+1} \mathbf{v}\mathbf{v}^\top + S_1 \mathbf{x}\mathbf{v}^\top + 2d_{t+1}\mathbf{x}\mathbf{v}^\top + S_2 S_3\}_{ij}}{\{d_{t+1}^3 \|\mathbf{x}\|^2 \mathbf{x}\mathbf{v}^\top + S_1 S_3 + 2\mathbf{U}_{t+1} \mathbf{V}_t \mathbf{D}_t \mathbf{V}_t^\top + 2d_{t+1}S_3 + S_2 \mathbf{x}\mathbf{v}^\top\}_{ij}} \tag{8}$$

$$v_i \leftarrow v_i \cdot \frac{\{d_{t+1}^3 \|\mathbf{x}\|^2 S_4 + S_1 \mathbf{U}_{t+1}^\top \mathbf{x} + 2d_{t+1}\mathbf{U}_{t+1}^\top \mathbf{x} + S_2 S_4\}_i}{\{d_{t+1}^3 \|\mathbf{x}\|^2 \mathbf{U}_{t+1}^\top \mathbf{x} + S_1 S_4 + S_2 \mathbf{U}_{t+1}^\top \mathbf{x} + 2d_{t+1}S_4\}_i} \tag{9}$$

$p(\mathbf{x}_t|\mathbf{Y}_{1:t})$ approximately by a finite set of random sampling particles. Given some observed image patches at t-th frame $\mathbf{Y}_{1:t} = \{\mathbf{y}_1, \mathbf{y}_2, ..., \mathbf{y}_{t-1}\}$, the state of the target $\mathbf{x}_t$ can be estimated recursively.

$$p(\mathbf{x}_t|\mathbf{Y}_{1:t}) \propto p(\mathbf{y}_t|\mathbf{x}_t) \int p(\mathbf{x}_t|\mathbf{x}_{t-1}) p(\mathbf{x}_{t-1}|\mathbf{Y}_{1:t-1}) \mathrm{d}\mathbf{x} \tag{10}$$

where $p(\mathbf{x}_t|\mathbf{x}_{t-1})$ denotes state transition between two consecutive frames, and $p(\mathbf{y}_t|\mathbf{x}_t)$ denotes observation model that estimates the likelihood of observing $\mathbf{y}_t$ at state $\mathbf{x}_t$. The optimal state at t-th frame is obtained by maximizing the approximate posterior probability:

$$\mathbf{x}_t^* = \underset{\mathbf{x}_t}{\mathrm{argmax}}\, p(\mathbf{y}_t|\mathbf{x}_t) p(\mathbf{x}_t|\mathbf{x}_{t-1}) \tag{11}$$

3.2   Observation Model

The observation model in our proposed tracking method is shown in Fig. 1. At the first m frames, target templates are collected to be as the initial data matrix $\mathbf{X}$. RNMF decomposes it into the basis matrix $\mathbf{U}$ composed of several basis vectors. Compared to standard NMF, the learned basis matrix $\mathbf{U}$ not only provides sparse representation property but also retains the Euclidean property of data vectors attribute to $L_{2,1}$ norm. Motivated by [10], the estimated target can be modeled by a linear combination of basis vectors and trivial templates.

$$\mathbf{y} = \mathbf{U}\mathbf{z} + \mathbf{e} = [\,\mathbf{U}\,\mathbf{I}\,] \begin{bmatrix} \mathbf{z} \\ \mathbf{e} \end{bmatrix} \tag{12}$$

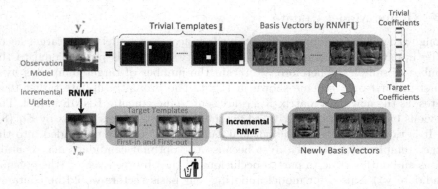

Fig. 1. Illustration of the proposed method by RNMF

where $\mathbf{y}$ denotes an observation vector, $\mathbf{z}$ indicates the coefficients of basis vectors, and $\mathbf{e}$ can be assumed to sparse error. The objective function in our proposed method is formed as,

$$\min_{\mathbf{z},\mathbf{e}} \frac{1}{2}\|\mathbf{y} - \mathbf{Uz} - \mathbf{e}\|_2^2 + \lambda\|\mathbf{e}\|_1 \quad s.t. \quad \mathbf{z} \geq 0 \tag{13}$$

Obviously, there is no closed-form solution for the optimization problem with Eq. (13). The iterative algorithm to compute $\mathbf{z}$ and $\mathbf{e}$ is proposed.

When the optimal $\hat{\mathbf{e}}$ is obtained, the optimization problem inverts to $\mathcal{O}(\mathbf{z}) = \frac{1}{2}\|\hat{\mathbf{y}} - \mathbf{Uz}\|_2^2$, $s.t.$ $\mathbf{z} \geq 0$, where $\hat{\mathbf{y}} = \mathbf{y} - \hat{\mathbf{e}}$. It is a typical nonnegative least square problem and easily solved with several iterations. When the optimal $\hat{\mathbf{z}}$ is obtained, Eq. (13) is equal to $\mathcal{O}(\mathbf{e}) = \frac{1}{2}\|\bar{\mathbf{y}} - \mathbf{e}\|_2^2 + \|\mathbf{e}\|_1$, where $\bar{\mathbf{y}} = \mathbf{y} - \mathbf{U}\hat{\mathbf{z}}$. It has a closed-form solution by soft-threshold operation which is defined as $S_\lambda(x) = \mathrm{sgn}(x)(|x| - \lambda)$. The algorithm for solving $\mathbf{z}$ and $\mathbf{e}$ is shown in **Algorithm** 1. When target coefficients $\mathbf{z}_t$ obtained at t-frame, the observation likelihood can be measured by the reconstruction error of each observed image patch.

$$p(\mathbf{y}_t^i|\mathbf{x}_t^i) \propto \exp(-\|\mathbf{y}_t^i - \mathbf{U}\mathbf{z}_t^i\|_2^2) \tag{14}$$

Algorithm 1. Algorithm for $\hat{\mathbf{z}}$ and $\hat{\mathbf{e}}$.

Input: An observation vector $\mathbf{y}$, RNMF basis matrix $\mathbf{U}$, the regularization parameter is $\lambda = 0.05$

Output: The optimal $\hat{\mathbf{z}}$ and $\hat{\mathbf{e}}$

1 Set: stop error ε.
2 Initialize $i = 0$ and $\mathbf{e}$ with random positive values.
3 **Repeat**
4   Update $\mathbf{z}^{i+1} = \min_{\mathbf{z}} \frac{1}{2}\|\hat{\mathbf{y}} - \mathbf{Uz}\|_2^2$, $s.t.$ $\mathbf{z} \geq 0$;
5   Update $\mathbf{e}^{i+1} = S_\lambda(\bar{\mathbf{y}} - \mathbf{U}\mathbf{z}^{i+1})$;
6       $i := i + 1$;
7 **Until** $\frac{\|\mathbf{z}^{i+1} - \mathbf{z}^i\|_2}{\|\mathbf{z}^i\|_2} \leq \varepsilon$;

3.3   Online Model Update

Along with a new tracking result obtained, appearance model of the target needs to be updated. The computational complexity of NMF is proportional to the number of samples. Therefore, we retain the number of target templates by a "First-in First-out" scheme shown in Fig. 1. The promptly update target templates, as the new data matrix $\mathbf{X}_{t+1}$, are sent to incremental RNMF model. The previous basis matrix is substituted for the recalculated basis matrix by Eq. (8).

It is worth noting that the optimal candidate $\mathbf{y}^*$ can be not added into the target template set directly. Just because the optimal candidate may contain noises and outliers due to partial occlusions. If the observed vector (the optimal candidate $\mathbf{y}^*$) is used for model updating, our basis vectors would be contaminated by occlusions (the book) shown in Fig. 1.

Considering robust NMF can effectively handle noises and outliers, RNMF is used to address this issue. At t-th frame, the corresponding encoding coefficient vector $\mathbf{v}_t^*$ is utilized to represent the reconstruction sample $\mathbf{y}_{res} = \mathbf{U}_t \mathbf{v}_t^*$ with the positive templates $\mathbf{U}_t$ of RNMF. It is added into the target template set to obtain the newly basis vectors by incremental learning for RNMF. For the tradeoff between the computation and the model fitness, the basis matrix $\mathbf{U}$ is recalculated every five frames in our test.

4   Experiments

In this section, we test the proposed model on Object Tracking Benchmark (OTB) [12] with 29 trackers and 51 sequences. Besides, CN [2], SST [16] and KCF [4] are taken into comparison.

Setup: The proposed tracker was implemented in MATLAB on a PC with Intel Xeon E5506 CPU (2.13 GHz) and 24 GB memory. The following parameters were used for our tests: each patch of image was normalized to 32×32 pixels; the number of initial target templates was $N = 50$ respectively from the first 5 frames; the number of basis vectors was set to 16; kNN was fixed to 5.

We show the overall performance of OPE for our tracker and compare it with some other state-of-the-arts (ranked within top 10) as shown in Fig. 2(a), (b).

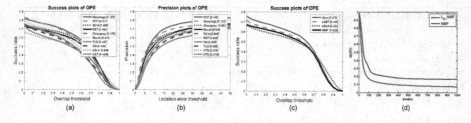

Fig. 2. (a,b): Plots of OPE. The performance score for each tracker is shown in the legend. For each figure, the top 10 trackers are presented for clarity; (c): Success plot of OPE for self-validation; (d) Convergence analysis for RNMF and NMF

Our method with HOG feature ranks the first on the success plot and the second on precision plot followed by KCF tracker. The proposed tracker with raw feature shows a comparable performance.

Specially, in Fig. 2(c), ASLA, LSST and NMF method are also provided just for self-validation. In NMF tracker, basis vectors are obtained by NMF and updated by incremental NMF. Results on OPE demonstrate that our tracker with robust NMF is with improvements than the conventional NMF tracker. Compared to LSST with PCA vectors, basis vectors obtained by robust NMF have a higher representation ability to capture the object's appearance.

Results: Figure 3 shows screen shots of tracking results from different trackers including Struck [3], ASLA [5], SCM [17], SST [16] and KCF [4].

Fig. 3. Representative frames of some sampled tracking results. And subfigures from left to right in the first row are *David3*, *Sylvester*. In the second row, these subfigures are *Basketball* and *Freeman1*. In the second row, these subfigures are *Freeman1*, *Carscale* and *Pedestrian*.

Occlusion: Occlusion is the most important attributes which lead to tracking drifts. In *David3* sequence, when David passes through the tree, SST and Struck suffer from severe drifts. ASLA and SST lose tracking accuracy to some extent. In *Basketball* sequence, SST, LSST and the proposed method accurately track the basketball player when the player is occluded by others. Results on *Occlusion* attribute show that our method can effectively handle outliers due to occlusions.

Deformation: Trackers based on NMF have natural advantage on this attribute because of part-based representation. Many methods including SST, Struck and ASLA are not adapt to appearance changes in *David3*, *Sylvester* and *Basketball* sequences. They show difficult in capturing appearance changes, where the appearance becomes much dissimilar to their initial one. Our method shows more robust and performs well without drifts.

Other attributes: There are others remaining attributes we do not refer to. The *Freeman1* sequence contains *out-of-plane rotation* and the challenging issues in

Carscale are *occlusion* and *scale variation*. The *Deer* sequence is accompanied with *fast motion* and *motion blur*. The proposed method not only locates the target to obtain accurate appearance model but also covers the target accurately.

Besides, our tracker also performs well in an infrared video *Pedestrian* than other trackers, such as KCF, SCM and SST. In summary, our test results on these videos have shown that the proposed tracker are effective and robust to heavy occlusions and deformation.

Convergence analysis: We compare the convergence properties (i.e. speed, objective function value) between RNMF and NMF on these sequences shown in Fig. 2(d). Compared with the standard NMF with Frobenius norm, RNMF with $L_{2,1}$ norm has better precision and is relatively faster to reach steady state. However, the computation time cost on RNMF is higher than NMF due to the updating rule on the diagonal matrix $\mathbf{D}$ in each iteration. It is worth mentioning that computational complexity of these two methods is both $\mathcal{O}(tMNK)$ when the multiplicative update is supposed to stop after t iterations.

5   Conclusion

This paper proposes an incremental learning for robust NMF method in visual tracking. Multiplicative update rules are derived with convergence verification. RNMF and its incremental update rules are incorporated into visual tracking framework. The optimal problem can be solved by an iterative algorithm with nonnegative least square method and soft-threshold operator. Quantitative and qualitative comparisons with other state-of-the-art methods on OTB have demonstrated the effectiveness and robustness of the proposed tracker.

Acknowledgment. This research is partly supported by NSFC, China (No: 6127 3258), 863 Plan, China (No. 2015AA042308), USCAST2015-10 and USCAST2013-07.

References

1. Bucak, S.S., Gunsel, B.: Incremental subspace learning via non-negative matrix factorization. Pattern Recognit. **42**(5), 788–797 (2009)
2. Danelljan, M., Khan, F.S., Felsberg, M., Weijer, J.V.D.: Adaptive color attributes for real-time visual tracking. In: IEEE Conference on Computer Vision and Pattern Recognition (2014)
3. Hare, S., Saffari, A., Torr, P.H.S.: Struck: structured output tracking with kernels. In: IEEE International Conference on Computer Vision, pp. 263–270 (2011)
4. Henriques, J.F., Caseiro, R., Martins, P., Batista, J.: High-speed tracking with kernelized correlation filters. IEEE Trans. Pattern Anal. Mach. Intell. **37**(3), 583–596 (2015)
5. Jia, X., Lu, H., Yang, M.H.: Visual tracking via adaptive structural local sparse appearance model. In: Proceedings of the IEEE Computer Society Conference on Computer Vision and Pattern Recognition, pp. 1822–1829 (2012)

6. Kong, D., Ding, C., Huang, H.: Robust nonnegative matrix factorization using l21-norm. In: Proceedings of the 20th ACM International Conference on Information and Knowledge Management, pp. 673–682. ACM (2011)

7. Liu, F., Zhou, T., Yang, J., Gu, I.Y.: Robust visual tracking via inverse nonnegative matrix factorization. In: 2016 IEEE International Conference on Acoustics, Speech and Signal Processing (ICASSP). IEEE (2016)

8. Ma, L., Zhang, X., Hu, W., Xing, J., Lu, J., Zhou, J.: Local subspace collaborative tracking. In: Proceedings of the IEEE International Conference on Computer Vision, pp. 4301–4309 (2015)

9. Wang, D., Lu, H.: On-line learning parts-based representation via incremental orthogonal projective non-negative matrix factorization. Signal Process. **93**(6), 1608–1623 (2013)

10. Wang, D., Lu, H., Yang, M.H.: Online object tracking with sparse prototypes. IEEE Trans. Image Process. **22**(1), 314–325 (2013)

11. Wang, N., Wang, J., Yeung, D.Y.: Online robust non-negative dictionary learning for visual tracking. In: Proceedings of the IEEE International Conference on Computer Vision, pp. 314–325 (2013)

12. Wu, Y., Lim, J., Yang, M.H.: Object tracking benchmark. IEEE Trans. Pattern Anal. Mach. Intell. **38**(4), 1–1 (2015)

13. Wu, Y., Shen, B., Ling, H.: Visual tracking via online nonnegative matrix factorization. IEEE Trans. Circ. Syst. Video Technol. **24**(3), 374–383 (2014)

14. Zhang, H., Zha, Z.J., Yang, Y., Yan, S., Chua, T.S.: Robust (semi) nonnegative graph embedding. IEEE Trans. Image Process. **23**(7), 2996–3012 (2014)

15. Zhang, H., Hu, S., Zhang, X., Luo, L.: Visual tracking via constrained incremental non-negative matrix factorization. IEEE Signal Process. Lett. **22**(9), 1350–1353 (2015)

16. Zhang, T., Liu, S., Xu, C., Yan, S., Ghanem, B., Ahuja, N., Yang, M.H.: Structural sparse tracking. In: Proceedings of the IEEE Conference on Computer Vision and Pattern Recognition, pp. 150–158 (2015)

17. Zhong, W., Lu, H., Yang, M.H.: Robust object tracking via sparse collaborative appearance model. IEEE Trans. Image Process. **23**, 2356–2368 (2014)

High Precision Direction-of-Arrival Estimation for Wideband Signals in Environment with Interference Based on Complex-Valued Neural Networks

Kazutaka Kikuta[✉] and Akira Hirose

Department of Electrical Engineering and Information Systems,
The University of Tokyo, 7-3-1 Hongo, Bunkyo-ku, Tokyo 113-8656, Japan
kikuta@eis.t.u-tokyo.ac.jp, ahirose@ee.t.u-tokyo.ac.jp
http://www.eis.t.u-tokyo.ac.jp/

Abstract. We propose a null steering scheme for wideband acoustic pulses and narrow band interference (NBI) using direction of arrival (DoA) estimation based on complex-valued spatio-temporal neural networks (CVSTNNs) and power inversion adaptive array (PIAA). For acoustic imaging, pulse spectrum should be wide to make the pulses less invasive. When a pulse has a wide frequency band, narrow band interference causes DoA errors in conventional CVSTNN. We use the weights of PIAA as the initial weights of CVSTNN to achieve higher precision in DoA estimation. Simulations demonstrate that the proposed method realizes accurate DoA estimation than the conventional CVSTNN method.

Keywords: Null steering · Power inversion · Complex-valued spatio-temporal neural network (CVSTNN)

1 Introduction

Acoustic imaging is a low-invasive imaging technique widely used in the fields such as echo diagnosis and non-destructive inspection. This technique is often used under water and inside solid object where optical methods have difficulty. In usual, ultrasound imaging uses a technique named beamforming, which utilizes adaptive transducer array. There are a number of beamforming techniques. The delay and sum (DAS) method, which is the most basic one, as well as the so-called Capon's method [1] obtain images by scanning the beam. In recent years, other imaging methods have emerged, such as MUSIC [2] and ESPRIT [3] algorithms which enable higher angle resolution by employing super-resolution techniques. In these schemes, imaging is performed by steering nulls, the minimal gain points among lobes. Since a null is sharper than a beam, higher resolution can be achieved than normal beamforming.

In all these methods, the steering angle depends on the wavelength of acoustic waves. Therefore, the acoustic signals are limited within narrow bandwidth. This

© Springer International Publishing AG 2016
A. Hirose et al. (Eds.): ICONIP 2016, Part II, LNCS 9948, pp. 620–626, 2016.
DOI: 10.1007/978-3-319-46672-9_69

problem is solved by introducing a neural network. We proposed acoustic imaging based on a complex-valued spatio-temporal neural network (CVSTNN) [4,5]. Wideband singals makes lower energy spectrum, hence, it is less invasive. The short pulse length leads to a higher resolution and less speckle noise.

The distortion from narrow band interference (NBI) is often seriously harmful in wideband imaging. Wide working band makes the disturbance from interference much larger. The learning results will not converge to optimum directions in null steering because of NBI as shown in the experiment section in this paper.

In this paper, we propose a novel null steering scheme where both pulses and NBI direction are steered. With a set of initial weights that makes nulls directing to NBI realized by power inversion adaptive array (PIAA), this method can adjust the weight fine and precisely. Since we use PIAA algorithm to identify NBI direction, we can achieve high speed null steering with small amount of calculation.

2   Null Steering Based on PIAA

We propose a high-precision null steering scheme to direct nulls to NBI and wideband pulses based on CVSTNN with initial weights determined by PIAA. We assume discrete objects for imaging targets. We steer nulls to the targets to estimate the direction of arrival (DoA) for imaging. We introduce the idea of power inversion adaptive array (PIAA) [6] for wideband imaging to reduce the disturbance from the NBI. PIAA points nulls of the array to the direction of large-power signals. Since desired and interference waves are distinguished only by input power, PIAA does not require a priori knowledge about signal structures or arrival angles [7].

2.1   NBI Null Steering Using Power Inversion

We describe the null steering method based on power inversion below. We consider an N-element linear array as shown in Fig. 1. When we define a complex input vector $X(t)$ as $X(t) = [x_1, x_2, ..., x_N]^T$, the correlation matrix is defined as $R_{xx} = E[X(t)X^H(t)]$. The optimal weight and the steering vector is calculated as $W_{opt} = R_{xx}^{-1}S$ when the steering vector is $S = [1, 0, ..., 0]^T$. This weight $W_{opt} = [w_{opt1}, w_{opt2}, ..., w_{optN}]$ is regulated to minimize the output power. When a desired signal is a wideband pulse and interference is narrowband, the spectrum density of the pulse is much smaller than that of the interference. By taking advantage of this characteristic, turning nulls to the NBI is possible and null steering is realized effectively.

2.2   Learning Method of Complex-Valued Spatio-Temporal Neural Network

Construction. We previously presented a wideband null steering method using CVSTNN [4,5]. In this section, we describe CVSTNN null steering method.

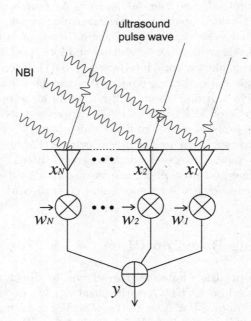

Fig. 1. The structure of N-element linear array.

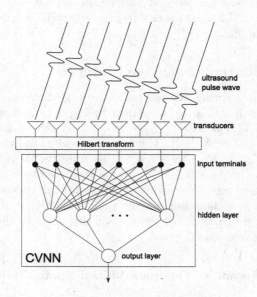

Fig. 2. Basic construction of CVNN

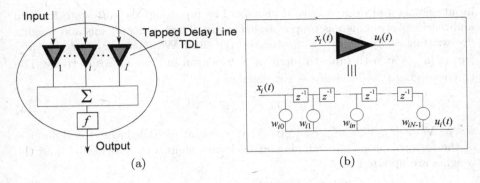

Fig. 3. (a) Neuron structure and (b) tapped delay line (TDL).

We propose the combination of the NBI mitigation scheme mentioned above with the steering method.

Figures 2 and 3 show the construction of the CVSTNN. The neural network has an input terminal layer, a hidden neuron layer and an output neuron layer. In Fig. 2, each neurons are shown in circles. Signals detected at sensors are converted into analytical signals by Hilbert transform. Hilbert transform makes the phase of positive- and negative-frequency components to advance and retardant by 90°, respectively. The analytical signal is expressed as $(1 + j\mathcal{H})\,x$. This signal is inputted to the neural network and processed.

Figure 3(a) shows the process in each neuron. Every signal goes through tapped delay lines (TDLs) which work as neural weights. The results are summed up to a neural internal state u and finally to an output restricted to the range of -1 to 1 by the activation function $f(u) = \tanh(|u|)\exp(i\arg(u))$. Figure 3(b) shows the structure of the TDL working as a synaptic weight. The z^{-1} represents a unit delay, and $w_{in} = a_{in}\exp(j\theta_{in})$ is a weight at the n-th tap for i-th input. The output of the TDL is the sum of the time-sequential signals existing in the TDL. The internal state of the neuron u is expressed as

$$u_i(t) = \sum_{n=0}^{N-1} x_i(t-n)w_{in} \tag{1}$$

$$u(t) = \sum_i u_i(t) \tag{2}$$

The initial value of the weight $w(n)$ is set to $w_{in} = w_{\mathrm{opti}}$ to adopt the weight gained by the power inversion. This weight makes nulls turn to the direction of NBI and the output signal becomes almost wideband pulse only. By adjusting the delay, amplitude and phase values by updating $w(n)$ through learning, wideband null steering is realized.

Learning Dynamics. The learning process of CVSTNN [8] is explained as follows. First we prepare a set of teacher signals, which is the combination of input signals and desired output signals. The input signals propagate forward and teacher output signals propagate backward. The teacher signals and weights are written as $\hat{\boldsymbol{y}}^l = [\hat{y}_j^l]^T \equiv [|\hat{y}_j^l| \exp\{j\hat{\theta}_j^l\}]^T$ and $\mathbf{W}^l = [\boldsymbol{w}_1^l \ \ldots \ \boldsymbol{w}_j^l \ \ldots \ \boldsymbol{w}_J^l]$ ($\boldsymbol{w}_j^l \equiv [w_{jin}^l]$ at n-th tap i-th input of j-th neuron in layer l), respectively. The teacher signal $\hat{\boldsymbol{y}}^{l-1}$ in layer $l-1$ is given as

$$(\hat{\boldsymbol{y}}^{l-1})^* = f(\ (\hat{\boldsymbol{y}}^l)^*\ \mathbf{V}^l\ ) \tag{3}$$

where $\mathbf{V}^l = [v_{jin}]^l \equiv w_{jin}^l / |w_{jin}^l|^2$. At the output of each layer, the difference of the forward propagate signals and backward signals are calculated, and the weights are updated as

$$
\begin{aligned}
|w_{jin}|^{\text{new}} \\
= |w_{jin}|^{\text{old}} - K\Big\{ & (1 - |y_j|^2) \\
& \Big(|y_j| - |\hat{y_j}| \cos\Big(\theta_j - \hat{\theta}_j\Big)\Big) |x_{in}| \cos\theta_{jin}^{\text{rot}} \\
& - |y_j||\hat{y_j}| \sin\Big(\theta_j - \hat{\theta}_j\Big) \frac{|x_{in}|}{|u_j|} \sin\theta_{jin}^{\text{rot}} \Big\}
\end{aligned}
\tag{4}
$$

$$
\begin{aligned}
\theta_{jin}^{\text{new}} \\
= \theta_{jin}^{\text{old}} - K\Big\{ & (1 - |y_j|^2) \\
& \Big(|y_j| - |\hat{y_j}| \cos\Big(\theta_j - \hat{\theta}_j\Big)\Big) |x_{in}| \sin\theta_{jin}^{\text{rot}} \\
& + |y_j||\hat{y_j}| \sin\Big(\theta_j - \hat{\theta}_j\Big) \frac{|x_{in}|}{|u_j|} \cos\theta_{jin}^{\text{rot}} \Big\}
\end{aligned}
\tag{5}
$$

where $w_{jin} \equiv |w_{jin}| \exp\{\theta_{jin}\}$ (layer index l is omitted), $(\cdot)^{\text{old}}$ and $(\cdot)^{\text{new}}$ stand for before and after the update, $\theta_{jin} \equiv \arg(w_{jin})$, $\theta_j \equiv \arg(y_j)$ and $\theta_{jin}^{\text{rot}} \equiv \theta_j - \theta_i - \theta_{jin}$.

In the iteration of the above weight updates, the learning CVSTNN progresses and the output gets closer to the teacher output. The nulls turn for input wideband signals when the output teacher signals set to zero.

In the null learning, we have to avoid the trivial solution $\boldsymbol{w}_{ji} = 0$ (for $\forall(j,i)$). Therefore, we set one connection in each layer fixed. The fixed weights are the initial weights obtained from the NBI power inversion.

3   Simulation Evaluation

3.1   Simulation Setup

We evaluate the performance of our proposed imaging method in a simulation. We estimate DoA of a pulse signal in one-dimensional array antenna. The CVSTNN has 3 input terminals and 8 hidden neurons. Each TDL has 29 taps of 10^{-7} s unit delay. The array has three elements and the spacing is 2.0 mm. We suppose one scatterer is positioned at $30°$ direction.

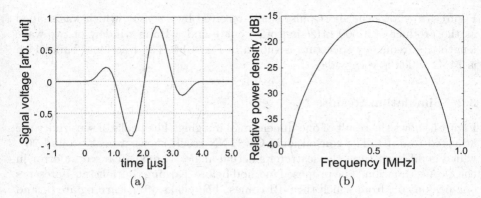

Fig. 4. Generated pulse wave represented in (a) time and (b) frequency domains, respectively.

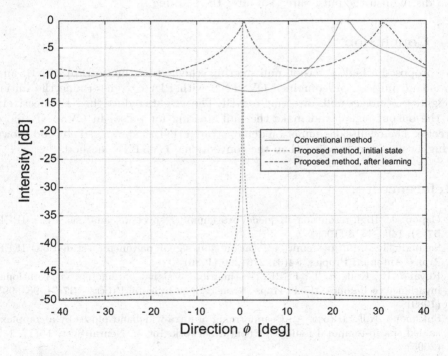

Fig. 5. One-dimensional imaging results of CVSTNN beamformers.

When an ultrasonic short pulse is radiated in the space, it is scattered and received at the array to generate a pulse at respective elements. The incident wave contains wideband signal and one NBI, the frequency of which is 500 kHz. The incident wave is fed to the neural network as a set of input teacher signals, as well as a zero output teacher signal, to generate a null to the scatterer direction.

Figure 4(a) shows the short pulse shape received by a sensor, which was shaped as the product of a 500 kHz sinusoidal curve and a Hann window of two-wave length. Its frequency spectrum is shown in Fig. 4(b). The fractional bandwidth is 74 %, which is very wide.

3.2 Simulation Results

Figure 5 shows the result of one-dimensional imaging. The peaks of sensitivity are adjusted to 0 dB. The conventional CVSTNN result shows a peak around 20°, which is different from the scatterer's actual angle. The NBI caused an error in the DoA estimation. The proposed method before learning (initial state) presents one peak at 0°, from which the NBI comes. The peaks after learning are 0° and 30° which are the directions of the NBI and the pulse. The pulse direction is estimated correctly. Since we already identified the NBI before the learning, we can distinguish the pulse direction after the learning.

4   Conclusion

We proposed a high-precision null steering scheme to estimate DoA of NBI and wideband pulses. We combined CVSTNN with PIAA to determine the initial weights that makes nulls directing to NBI. The weights reduce the NBI effectively in the output signal and make the null steering for pulses in CVSTNN more precise. The results of simulation demonstrated that the proposed method shows a higher accuracy imaging than the conventional CVSTNN method.

References

1. Capon, J.: High-resolution frequency-wavenumber spectrum analysis. Proc. IEEE **57**(8), 1408–1418 (1969)
2. Schmidt, R.: Multiple emitter location and signal parameter estimation. IEEE Trans. Antennas Propag. **34**(3), 276–280 (1986)
3. Royand, R., Kailath, T.: ESPRIT-estimation of signal parameters via rotational invariance techniques. IEEE Trans. Acoust. Speech Signal Process. **37**(7), 984–995 (1989)
4. Suksmono, A.B., Hirose, A.: Beamforming of ultra-wideband pulses by a complex-valued spatio-temporal multilayer neural network. Int. J. Neural Syst. **15**(1), 1–7 (2005)
5. Terabayashi, K., Natsuaki, R., Hirose, A.: Ultrawideband direction-of-arrival estimation using complex-valued spatiotemporal neural networks. IEEE Trans. Neural Netw. Learn. Syst. **25**(9), 1727–1732 (2014)
6. Compton, R.: The power-inversion adaptive array: concept and performance. IEEE Trans. Aerosp. Electron. Syst. AES-15 **6**, 803–814 (1979)
7. Kikuta, K., Hirose, A.: Interference mitigation in UWB receivers utilizing the differences in direction of arrival and power spectrum density. In: 2015 Asia-Pacific Microwave Conference (APMC), Nanjing (2015)
8. Hirose, A.: Complex-Valued Neural Networks, 2nd edn. Springer, Heidelberg (2012)

Content-Based Image Retrieval
Using Deep Search

Zhengzhong Zhou and Liqing Zhang[✉]

Key Laboratory of Shanghai Education Commission for Intelligent Interaction
and Cognitive Engineering, Department of Computer Science and Engineering,
Shanghai Jiao Tong University, Shanghai, China
tczhouzz@sjtu.edu.cn, zhang-lq@cs.sjtu.edu.cn

Abstract. The aim of Content-based Image Retrieval (CBIR) is to find
a set of images that best match the query based on visual features.
Most existing CBIR systems find similar images in low level features,
while Text-based Image Retrieval (TBIR) systems find images with relevant tags regardless of contents in the images. Generally, people are
more interested in images with similarity both in contours and high-
level concepts. Therefore, we propose a new strategy called *Deep Search*
to meet this requirement. It mines knowledge from the similar images of
original queries, in order to compensate for the missing information in
feature extraction process. To evaluate the performance of *Deep Search*
approach, we apply this method to three different CBIR systems (HOF
[5], HOG and GIST) in our experiments. The results show that *Deep
Search* greatly improves the performance of original algorithms, and is
not restricted to any particular methods.

Keywords: CBIR · Deep search · Image semantics

1 Introduction

Content-based Image Retrieval (CBIR) explores image similarity in low level
features such as color, texture or shape [1,2]. It finds a list of images that best
matches the query based on visual features. However, people are interested in
the high-level concepts as well. Take clothing retrieval as an example. Existing
methods including *Hierarchical Orientation Feature* (HOF) [5], HOG and GIST
still focus on primitive features, which do not reflect characteristics like the
clothing materials and styles. We can provide images in data with labels to
indicate their semantic concepts. While such information is not available for
query images. Thus, it limits practical applications of CBIR system.

In this paper, we develop an strategy called *Deep Search* (DS) to refine the
order of retrieved images with high semantic similarity to the query. The main
idea is to find similar images of the original query, then use machine learning to
mine their high-level concepts, which are not available in image feature extraction
process.

© Springer International Publishing AG 2016
A. Hirose et al. (Eds.): ICONIP 2016, Part II, LNCS 9948, pp. 627–634, 2016.
DOI: 10.1007/978-3-319-46672-9_70

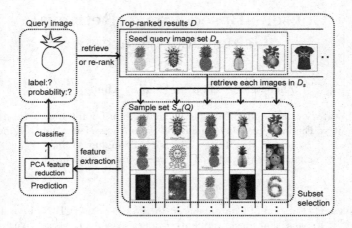

Fig. 1. Framework of DS.

We apply DS to three different CBIR systems (HOF, HOG and GIST) and evaluate their performance in our experiments. The results show that our strategy achieves a significant improvement compared with the original algorithms. Furthermore, it is not restricted to any particular methods, and can be accelerated by parallel processing.

2   Framework

In this section, we introduce the framework of our proposed strategy and explain why it is effective for optimizing the CBIR systems. Table 1 summarizes the notations used in this paper.

Table 1. Symbols and semantic.

Symbols	Semantic.
I_k	An image in the dataset.
$c(.)$	The class label of an image.
C	The class label set of dataset.
c_i	A class label, $c_i \in C$.
Q	The query image.
D	The retrieved image set, $D = \{X_1, X_2, \cdots, X_M\}$.
D_s	The seed query image set, $D_s = \{D_1, D_2, \cdots, D_m\}$.
$N_k(D_i)$	The k-nearest neighbors of D_i.
$S_m(Q)$	The sample set generated by D_s.
K	The size of $S_m(Q)$, $K = m * k$.

A large-scale CBIR system usually indexes millions of images. Each image labels at least one category. We denote I_k as an image in the dataset, $c(I_k)$ as the class label of I_k. The class label set of dataset is denoted by $C = \{c_1, c_2, \cdots\}$ ($\forall I_k \in dataset, c(I_k) \in C$). The common retrieval procedure is described as follows: First, the system translates the query image Q into a vector of visual features. Then it evaluates the similarity between Q and database images in the feature space. Finally, top-ranked similar images and their class labels will be presented to users. We see that the performance of CBIR system is heavily determined by the effectiveness of image feature representation.

As we discussed, existing CBIR algorithms are not applicable to retrieving same type of images with large variance in object shape, size, location and background. The semantic information that users express may be not available in the process of feature extraction. Generally, it is difficult to identify these information from single image during search. While it becomes easier to do it from a group of retrieved images, because they may contain consistent labels. Therefore, we aim to find a set of retrieved images which are similar to query Q in the sense of its high-level concept.

Figure 1 shows the flowchart of DS. It contains three main parts: *subset selection, prediction* and *re-ranking*. First, we perform image search based on certain visual features using Q as query image. The first M retrieved images are denoted by $D = \{X_1, X_2, \ldots, X_M\}$. The subset selection is to select a number of images in D as seed query images, denoted by $D_s = \{D_1, D_2, \ldots, D_m\}$. Next, we perform image search again using D_i as the query image and its top-k retrieved images are denoted by $N_k(D_i)$, $i = 1, \ldots, m$. We set $S_m(Q) = \bigcup_i N_k(D_i), D_i \in D_s$ and K as the size of $S_m(Q)$, $K = m * k$. This step can be implemented by parallel processing, for retrieving each D_i is independent. In the prediction stage, we classify Q using the set of retrieved images $S_m(Q)$. Finally, we develop a new measure to re-rank the retrieved images by combining the visual features similarity and the predicted categories. All images in $S_m(Q)$ are retrieved from the second round of searches. Also, we can perform next round search by using images in $S_m(Q)$ as seed query image set. Thus such a strategy is called as *Deep Search*.

The basic assumption for DS is that a query image usually contains partial visual features of its category due to object variance in shape, size, location and background as well. As is shown in Fig. 2, DS provides a new dimension to find similar images with same type of objects in it but difference in object size or location. For example, if two images contain same object while their locations are quite far. The visual features of them are rather different due to the big difference in location. We cannot find one of these images by using the other as the query through only one search. If the object locations in two images are quite close, we can retrieve one image from another as query image, because it is reasonable to assume that small variant images are proximal in visual feature space. DS provides a new approach to bridging images with big difference in object size or location through a link of images, of which each pair of images has only a small variance.

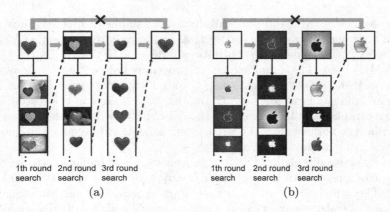

1th round   2nd round   3rd round   1th round   2nd round   3rd round
search   search   search   search   search   search

(a)   (b)

Fig. 2. DS bridges objects with big variance in (a) location or (b) size through multiple round searches. Each search finds similar images with a small variance.

2.1   Subset Selection

The selection of D_s is critical for the image classification and retrieval performance. Here, we define three approaches to select the seed query images for DS:

Visual-Based Deep Search (VDS). we select m-nearest samples of Q as our seed query set ($m << K$). In another words, $D_s = N_m(Q)$. This method selects the seed query images in the order of the visual feature similarity.

Class-Based Deep Search (CDS). Unlike VDS, we focus on the semantic similarity of images in D. We count the number of samples by class in the retrieved image set, and ignore those minor classes. We ensure the proportion of each category in D_s is consistent with their sample sizes in D. For example, $M = 1000$ and $m = 10$. 300 images are labeled as "Cat" in the retrieved image set D. Then we add 3 images labeled with "Cat" which are most similar to Q into D_s.

K-Nearest Neighborhood (KNN). This approach omits the step of retrieving the images in D_s and directly sets $S_m(Q) = D_s = D$. It is a special kind of VDS, where $K = m = M$.

We compare KNN with VDS and CDS. KNN does not need to conduct search again using D_is, thus reducing its computational cost. However, the quality of the sample set generated by VDS or CDS is much higher. According to the manifold hypothesis, data of different semantic concepts are expected to concentrate in the vicinity of different manifolds. Unlike KNN, the sample space formulated by VDS or CDS is not spherical, it can describe more complex space. Figure 3 shows the distributions of KNN, VDS and CDS in the feature space of real image data.

For a new query image Q, we evaluate the performance of each approach by calculating the entropy:

$$Entropy(S_m(Q)) = \sum_{c_i \in C} - \lg P(c_i|S_m(Q))P(c_i|S_m(Q)), \tag{1}$$

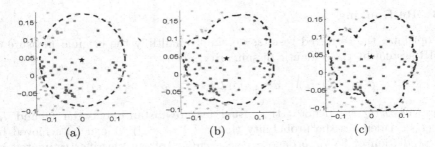

Fig. 3. Comparison of (a) KNN, (b) VDS and (c) CDS. The black stars are the features of original query. The squares represent D_s. Red and Blue mark sets are the two manifolds of candidate categories. Green marks are image features with other categories in the dataset. The dashed line enclosed areas are the sample spaces generated by different approaches. (Color figure online)

where $P(c_i|S_m(Q))$ is the proportion of images labeled c_i in $S_m(Q)$. In the view of information theory, $S_m(Q)$ with lower information entropy contains less ambiguous samples and could better express a particular semantic concept. Therefore, we determine the best approach and the appropriate parameter m by minimizing $Entropy(S_m(Q))$. This procedure can be self-adaptive, and we call it min-$Entropy$(minE). Additionally, DS performs the dataset $m+1$ times search for each query image. We adopt parallel computing to improve the processing speed. However, finding minimum $Entropy(S_m(Q))$ for large m is unnecessary. We set $m \leq 20$ to meet our hardware environment.

2.2   Prediction

In this stage, we estimate the category and corresponding probability of the query image through machine learning. We use the images in $S_m(Q)$ and their class labels to train a prediction model. Then, we utilize it to classify Q. Our method naturally supports incremental learning and is applicable to the growing dataset.

We adopt *Support Vector Machine* (SVM) to train prediction models. Its complexity does not depend on the feature dimensions, which ensures the performance of our estimate in different features. In this paper, we use gaussian radial basis function (RBF) as the kernel of SVM. Besides the possible categories of Q, we also predict the probabilities that Q belongs to them. These probabilities help us to express our estimation more clearly and will be used in the re-ranking procedure.

Actually, most classes contain few samples in $S_m(Q)$. The possibility that Q belongs to these classes is very low. These small samples are considered as interference and should be removed from the retrieved image set. We keep samples with top-n majority classes for similarity re-ranking. Generally, we set $n = 3$. The rationality of such a choice will be verified in the experimental section.

2.3   Re-Ranking

To optimize the retrieved results, we simply multiply the original measure of CBIR system by the semantic dissimilarity:

$$dis_{new}(Q, I_k) = dis_{old}(Q, I_k) * (1 - \rho * P(c(I_k)|Q)), \tag{2}$$

where $dis_{old}(Q, I_k)$ denotes the visual feature distance between Q and I_k. $P(c(I_k)|Q)$ denotes the probability that $c(I_k) = c(Q)$. It can be achieved by the prediction of SVM. ρ denotes the accuracy of our classifier estimated by 5-fold cross validation.

3   Experiments

In this section, we evaluate the performance of DS by performing online image retrieval on a large scale image dataset. We compare KNN, VDS, CDS and minE with the baseline systems and illustrate the results by the ground truth experiment.

DS is not restricted to any particular visual features. We conduct experiments on three different CBIR systems (HOF, HOG and GIST), and treat them as baselines. All of them could be accelerated by parallel processing.

3.1   Dataset and Implementation Details

We build a dataset containing 1.3 million images crawled from Google and Flickr. These images are divided into 560 categories including objects, scenes, people, buildings, etc. Our experiments focus on three different features. Unlike HOF, HOG and GIST do not have their own index structures. Hence we create the *Winner Take All* (WTA) [4] hash tables for them to improve the runtime.

Our system runs on the sever with 2 Intel Xeon 2.66 GHz six-core processors and 64 GB memory. By parallel processing, we reduce the average responsible time to within 3 s for a single query.

3.2   Evaluation of DS

In this part, we analyze how the effectiveness of DS changes with the corresponding parameter. We evaluate the performance by measuring *Mean Average Precision* (MAP) [3]. 4 subjects are invited to design 100 query tasks for each CBIR system.

We design these query tasks according to the characteristic of visual features. GIST feature is mainly used in scene categorization, thus we use images of attractions and landscapes as the query images. HOF is proposed for sketch-based image retrieval, so the query images are hand-draw sketches. HOG feature is suitable for retrieving images with objects and people. We select such kind of images as the queries.

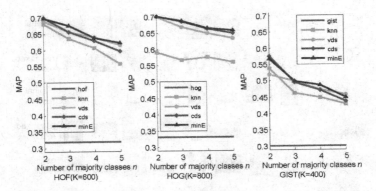

Fig. 4. MAPs of KNN, VDS, CDS and minE on three features with respect to the sample size K.

Figure 4 shows the effect of DS on HOF, HOG and GIST, when changing the sample set size K. From experimental results, we observe that DS strategy greatly improves the performance of original systems. The MAPs of baseline is about 0.3 on different features, while DS achieves MAP values between $[0.6, 0.7]$ on HOF, HOG, and $[0.4, 0.5]$ on GIST. Besides, we can see that minE performs better than others in most situations. This is in consistent with the previous analysis.

3.3 Experiment for Ground Truth

To evaluate the performance of DS on the ground truth situation, we merge all the test queries together. For a given query, we perform searches on 3 CBIR systems (HOF, HOG and GIST) and obtain their retrieved image set Ds, respectively. We generate $S_m(Q)$s from each D and use DS strategy to re-rank the retrieved images D having minimal $entropy(S_m(Q))$ as our final result (Table 2).

Table 2. MAPs of HOF, HOG, GIST and DS on the ground truth situation.

Method	HOF	HOG	GIST	DS
MAP	0.24	0.17	0.13	0.64

For general image search, the three types of features do not achieve comparable performance as in retrieving special categories. This indicates that the existing visual features are designed for special applications, and could not perform well on the ground truth situation. Nevertheless, the performance of DS is still better than the three baseline methods. We demonstrate that our strategy is not restricted to any particular features or search tasks. Figure 5 shows some examples of our ground truth experiment.

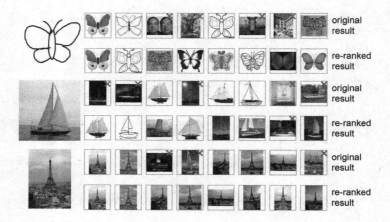

Fig. 5. Examples of the re-ranked results on the ground truth situation. Irrelevant retrieved images are marked by crosses.

4   Conclusion

In this paper, we propose a retrieval strategy called Deep Search to enhance the performance of CBIR system. It is applicable to CBIR algorithms which extract one-dimensional feature vectors from images. The experiment results show that it greatly improves the semantic similarity of original system. Besides, using parallel processing could accelerate our strategy.

Acknowledgements. The work was supported by the National Natural Science Foundation of China (61272251), the Key Basic Research Program of Shanghai Municipality, China (15JC1400103) and the National Basic Research Program of China (2015CB856004).

References

1. Cao, Y., Wang, C., Zhang, L., Zhang, L.: Edgel index for large-scale sketch-based image search, pp. 761–768 (2011)
2. Sun, Z., Wang, C., Zhang, L., Zhang, L.: Query-adaptive shape topic mining for hand-drawn sketch recognition. In: ACM International Conference on Multimedia, pp. 519–528 (2012)
3. Turpin, A., Scholer, F.: User performance versus precision measures for simple search tasks. In: SIGIR, pp. 11–18 (2006)
4. Yagnik, J., Strelow, D., Ross, D.A., Lin, R.: The power of comparative reasoning. In: ICCV, pp. 2431–2438 (2011)
5. Zhou, R., Chen, L., Zhang, L.: Sketch-based image retrieval on a large scale database. In: ACM MM, pp. 973–976 (2012)

Robust Part-Based Correlation Tracking

Xiaodong Liu and Yue Zhou[✉]

Institute of Image Processing and Pattern Recognition,
Shanghai Jiao Tong University, Shanghai, China
{lxd7714059,zhouyue}@sjtu.edu.cn

Abstract. Visual tracking is a challenging task where the target may undergo background clutters, deformation, severe occlusion and out-of-view in video sequences. In this paper, we propose a novel tracking method, which utilizes representative parts of the target to handle occlusion situations. For the sake of efficiency, we train a classifier for each part using correlation filter which has been used in visual tracking recently due to its computational efficiency. In addition, we exploit the motion vectors of reliable parts between two consecutive frames to estimate the position of the object target and we utilize the spatial relationship between representative part and target center to estimate the scale of the target. Furthermore, part models are adaptively updated to avoid introducing errors which can cause model drift. Extensive experiments show that our algorithm is comparable to state-of-the-art methods on visual tracking benchmark in terms of accuracy and robustness.

Keywords: Visual tracking · Part-based model · Correlation filter · Target estimation · Model update

1 Introduction

Visual tracking is an important research topic in computer vision with numerous applications, such as surveillance, driverless car, human computer interaction, etc. Given the position of the target in a frame, trackers are expected to locate the accurate position of target in the subsequent frames. However, there are many factors that affect the performance of a tracking algorithm, such as abrupt object motion, non-rigid object deformation, out-of-plane rotation, occlusion, etc.

Up to now, there are many tracking algorithms, most of which can be divided into generative methods and discriminative methods. Generative algorithms represent target as a series of templates and track object by searching the most similar region to the templates [1,2,17]. However, discriminative methods treat tracking as a binary classification problem [6,7,10]. It trains a classifier using samples obtained from target object and background with labels 1 or 0. Discriminative methods are also called tracking-by-detection methods. Recently, correlation filters are widely used in the visual tracking problem owing to its computational efficiency [3,8,15]. For example, the KCF tracker [8], which labels the samples from 1 to 0 gradually, achieves a very good performance with very high

© Springer International Publishing AG 2016
A. Hirose et al. (Eds.): ICONIP 2016, Part II, LNCS 9948, pp. 635–642, 2016.
DOI: 10.1007/978-3-319-46672-9_71

speed. However, the KCF tracker may fail when target undergo heavy occlusion, deformation and scale change.

To handle partial occlusion and deformation in visual tracking, many part-based tracking algorithms have been proposed [9,18]. They model the object with multiple parts of different structures, such as regular grid [9], star model [18], etc. It is obvious that some visible parts are still representative and useful for locating the object even when the target is partial occluded and deformed.

In this paper, we utilize the idea of part-based tracker for its robustness to partial occlusion and select parts based on keypoints extraction and clustering. In addition, we train a classifier for each part using correlation filter. Furthermore, the motion vectors and spatial relationships of parts are used for estimating the state of the target. Our main contributions include: (i) representative parts selection based on keypoints; (ii)part classifiers using correlation filters; and (iii) a novel target state estimation method.

2   Proposed Method

In this section, we present the proposed part-based tracking algorithm which is robust to occlusion, scale variation and deformation. Firstly, we describe a representative parts selection method based on keypoints extraction. Secondly, we simply introduce the KCF tracker which is the basis of our part trackers. Thirdly, we estimate the state of the target by studying the motion information and spatial relationships of reliable parts. Finally, we adaptively update our part trackers corresponding to their confidence scores.

2.1   Parts Initialization

Part initialization is very important to part-based trackers since proper parts segmentation can represent the target well and result in good tracking results. To obtain representative parts of the object, we extract keypoints using SIFT [14] and cluster these keypoints into n categories which are regarded as parts and are represented with rectangles. In addition, the central point of all parts is the same as target center which are defined as $\frac{1}{n}\sum_i P_i = P_t$, where P_i is the center of the ith part and P_t is the center of the target.

These parts are trackable since they contain discriminative information and they are robust to partial occlusion as the parts will not be occluded in the same time.

2.2   Correlation Tracking

Part-based trackers are usually slow since the computational complexity. Therefore, we train part trackers using correlation filters which are used for tracking problem because of the computational efficiency.

In this section, we simply introduce the correlation tracking, readers can get more details in [8]. Giving an initial target $\mathbf{x}$ of $M \times N$ pixels, a correlation tracker considers all circular shifts $\mathbf{x}_{m,n}$, $(m, n) \in \{0, 1, \ldots, M-1\} \times \{0, 1, \ldots, N-1\}$, as training samples with labels $y(m, n)$. The label of sample is obtained by Gaussian function corresponding to the distance or similarity of the target, which takes value of 1 for the original target and smoothly decrease to 0. The samples are then used for training a classifier using ridge regression to find a function $f(\mathbf{z}) = \mathbf{w}^T \mathbf{z}$ that minimizes the squared error over samples and their labels $y(m, n)$,

$$\min_{\mathbf{w}} \sum_{m,n} \|\varphi(\mathbf{x}_{m,n}) \cdot \mathbf{w} - y(m, n))\|^2 + \lambda \| \mathbf{w} \|^2 , \tag{1}$$

where the φ is the mapping to a kernel space and λ is the regularization parameter.

After transformed into Fourier domain, the $\mathbf{w}$ can be minimized as $\mathbf{w} = \sum_{m,n} \alpha(m, n)\varphi(\mathbf{x}_{m,n})$, and the coeficient α ia obtained by

$$\alpha = \mathcal{F}^{-1}\left(\frac{\mathcal{F}(y)}{\mathcal{F}(\varphi(\mathbf{x}) \cdot \varphi(\mathbf{x})) + \lambda}\right) , \tag{2}$$

where $\mathcal{F}$ and $\mathcal{F}^{-1}$ donate the discrete Fourier transformation and its inverse.

In the tracking process, tracker searches for the patches around the target in the previous frame with same size of $\mathbf{x}$, and computes the confidence score as

$$\hat{\mathbf{y}} = \mathcal{F}^{-1}(\mathcal{F}(\varphi(\mathbf{z}) \cdot \varphi(\hat{\mathbf{x}}) \odot \mathcal{F}(\alpha)) , \tag{3}$$

where the $\hat{\mathbf{x}}$ is the learned target appearance and $\odot$ is the Hadamard product. Then the new position of the target is located with the maximal value of $\hat{\mathbf{y}}$.

2.3   State Estimation

After parts segmentation and part classifiers training, we track parts in the following frames individually. According to the tracking results, we can then estimate the target state including position and scale.

Position Estimation. In the tracking process, target may undergo partial occlusion and deformation. Different parts have different response value, so they are not equal in representing the target, some parts with low confidence scores may even cause inaccurate results. We exploit those parts whose confidence scores are higher than a threshold T_v to vote for the central point of the target.

$$P_t^t = P_t^{t-1} + \frac{\sum_i w_i^t V_i^t}{\sum_i w_i^t} , \tag{4}$$

where the P_t^t is the center of the target in t-th frame, the $V_i^t = P_i^t - P_i^{t-1}$ is the movement vector of ith part between tth and $(t-1)$th frame, and the w_i^t is the weight of ith part corresponding to its confidence score.

Algorithm 1. Proposed tracking algorithm.

Require: Initial target bounding box x_0,
Ensure: Estimated object state x_t in the following frames.
 1: **for** $t = 1 : m$ (m is the number of frames of a video) **do**
 2:     **if** $t = 1$ **then**
 3:         Divide the target into n parts and train a tracker for each part;
 4:     **end if**
 5:     Compute coffidence map and locate new positons of parts;
 6:     Estimate the positon of the target using Eq. (5);
 7:     Estimate the scale of the target using Eq. (6);
 8:     **for** $i = 1 : n$ **do**
 9:         **if** $w_i > T_u$ **then**
10:             Update the model of ith part using Eq. (7,8);
11:         **end if**
12:     **end for**
13: **end for**

Scale Estimation. Parts are on behalf of local appearance of target, they have special spatial relations which can be used for estimating the scale of the object. We record the position of each part and calculate the distance between the part and the central point of the target as $D_i^t = |P_t^t - P_i^t|$. In addition, we only take parts with confidence scores higher than T_v into consideration, and the size of the target is obtained by

$$Size(P_t^t) = \frac{\sum_i w_i^t \frac{D_i^t}{D_i^{t-1}}}{\sum_i w_i^t} \cdot Size(P_t^{t-1}) . \tag{5}$$

2.4 Model Update

Different from the KCF tracker whose model is update frame by frame, we update each part tracker adaptively to avoid introducing errors which can lead to model drift. Parts with higher detection scores should be update more while those with lower response values ought to be update less or even not be updated. The model update mechanism is defined as:

$$\hat{\mathcal{F}}(\alpha)_i^t = \begin{cases} (1 - kw_i^t)\mathcal{F}(\alpha)_i^{t-1} + kw_i^t\mathcal{F}(\alpha)_i^t \ if \ w_i^t > T_u \\ \mathcal{F}(\alpha)_i^{t-1} \qquad\qquad\qquad\qquad else \end{cases} , \tag{6}$$

$$\hat{x}_i^t = \begin{cases} (1 - kw_i^t)\hat{x}_i^{t-1} + kw_i^t x_i^t \ if \ w_i^t > T_u \\ \hat{x}_i^{t-1} \qquad\qquad\qquad\quad else \end{cases} , \tag{7}$$

where the k is the learning rate and T_u ia the update threshold.

The outline of our method is in Algorithm 1 and the flowchart of our method is in Fig. 1.

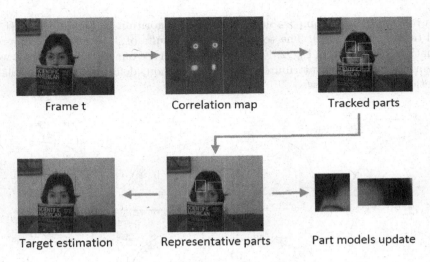

Fig. 1. The flowchart of the proposed algorithm.

Table 1. Comparisons with state-of-art trackers in location error at a threshold of 20 pixels and overlap score at an overlap threshold 0.5. The best two results are shown in red and green. (Color figure online)

	Ours	KCF	SCM	TGPR	CT	TLD	Struck	VTD	VTS	CXT
LE(%)	82.2	73.4	64.3	75.9	39.4	59.8	64.6	56.6	56.5	56.9
OS(%)	75.1	65.6	63.9	68.7	37.5	55.8	58.8	52.2	52.3	51.8

3 Experimental Results

To evaluate our algorithm, we test it on the tracking benchmark dataset [16], which contains 50 video sequences with 11 attributes, including illumination variation, scale variation, occlusion, deformation, motion blur, etc. We also compare our tracker with 31 state-of-art tracking algorithms, KCF, TGPR [5] and other 29 trackers obtained from the anthors' websites.

3.1 Quantitative Evaluation

All the trackers are evaluated with two metrics, (i) precision plot, which shows the percentage of frames whose estimated location is within the given threshold distance of the ground truth; (ii) success plot, which indicates the percentage of frames whose bounding box overlap is large than a threshold t_o.

Giving the bounding box of the target B_t and the bounding box of the ground truth B_g, the overlap score is defined as $S = \frac{Aera(B_t \cap B_g)}{Aera(B_t \cup B_g)}$, and we set t_o to 0.5. Due to limitation of pages, we only list the results of KCF [8], SCM [19], TGPR [5], CT [17], TLD [11], Struck [6], VTD [12], VTS [13], CXT [4]. Table 1 shows that our algorithm is comparable to state-of-art algorithms in location error (LE) and overlap score (OS). The top 10 tracking results of 31 trackers in different situations are shown in Fig. 2. From Fig. 2, we can see our method performs well in several challenging attributes, such as occlusion, deformation, out-of-plane rotation and out-of-view.

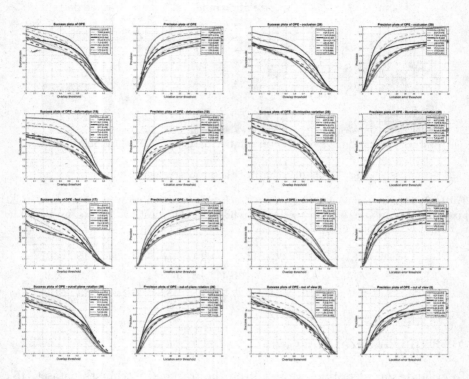

Fig. 2. Success and precision plots of the benchmark. From left to right and top to bottom are overall performance, occlusion, deformation, illumination variation, fast motion, scale variation, out-of-plane rotation and out-of-view.

3.2   Qualitative Evaluation

We also plot the results of top five trackers (Ours, KCF [8], SCM [19], TGPR [5] and Struck [6]) on eight challenging sequences for qualitative comparison. From Fig. 3, we can see that our algorithm can handle heavy occlusion (Coke, Jogging-1 and Lemming) well while the other four trackers fails in some videos. Furthermore, the results indicate that our method can track the object well when the target undergo scale change, deformation, illumination variation, background clutters, fast motion, etc.

Fig. 3. Tracking results of top 5 trackers on 8 challenging sequences. From left to right and top to bottom are Bolt, Coke, David3, Jogging-1, Lemming, Shaking, Trellis and Singer2.

4   Conclusion

In this paper, we proposed a part-based tracker based on correlation filter for robust visual tracking. We divide the target into several parts based on SIFT and train a classifier for each part using correlation filter. All parts are tracked in the following frames and reliable parts are selected according to their response values. Target is estimated by exploiting the movement information and spatial relationships of reliable parts. Furthermore, we update part models adaptively according to their weights. Extensive experiments have been done to verify the good performance of the proposed algorithm.

Acknowledgments. The work is supported by National High-Tech R&D Program (863 Program) under Grant 2015AA016402 and Shanghai Natural Science Foundation under Grant 14Z111050022.

References

1. Avidan, S.: Ensemble tracking. IEEE Trans. Pattern Anal. Mach. Intell. **29**(2), 261–271 (2007)

2. Babenko, B., Yang, M.H., Belongie, S.: Visual tracking with online multiple instance learning. In: IEEE Conference on Computer Vision and Pattern Recognition, 2009. CVPR 2009, pp. 983–990. IEEE (2009)
3. Bolme, D.S., Beveridge, J.R., Draper, B.A., Lui, Y.M.: Visual object tracking using adaptive correlation filters. In: 2010 IEEE Conference on Computer Vision and Pattern Recognition (CVPR), pp. 2544–2550. IEEE (2010)
4. Dinh, T.B., Vo, N., Medioni, G.: Context tracker: exploring supporters and distracters in unconstrained environments. In: 2011 IEEE Conference on Computer Vision and Pattern Recognition (CVPR), pp. 1177–1184. IEEE (2011)
5. Gao, J., Ling, H., Hu, W., Xing, J.: Transfer learning based visual tracking with Gaussian processes regression. In: Fleet, D., Pajdla, T., Schiele, B., Tuytelaars, T. (eds.) ECCV 2014, Part III. LNCS, vol. 8691, pp. 188–203. Springer, Heidelberg (2014)
6. Hare, S., Saffari, A., Torr, P.H.: Struck: structured output tracking with kernels. In: 2011 IEEE International Conference on Computer Vision (ICCV), pp. 263–270. IEEE (2011)
7. Henriques, J.F., Caseiro, R., Martins, P., Batista, J.: Exploiting the circulant structure of tracking-by-detection with kernels. In: Fitzgibbon, A., Lazebnik, S., Perona, P., Sato, Y., Schmid, C. (eds.) ECCV 2012, Part IV. LNCS, vol. 7575, pp. 702–715. Springer, Heidelberg (2012)
8. Henriques, J.F., Caseiro, R., Martins, P., Batista, J.: High-speed tracking with kernelized correlation filters. IEEE Trans. Pattern Anal. Mach. Intell. 37(3), 583–596 (2015)
9. Jia, X., Lu, H., Yang, M.H.: Visual tracking via adaptive structural local sparse appearance model. In: 2012 IEEE Conference on Computer Vision and Pattern Recognition (CVPR), pp. 1822–1829. IEEE (2012)
10. Kalal, Z., Matas, J., Mikolajczyk, K.: PN learning: bootstrapping binary classifiers by structural constraints. In: 2010 IEEE Conference on Computer Vision and Pattern Recognition (CVPR), pp. 49–56. IEEE (2010)
11. Kalal, Z., Mikolajczyk, K., Matas, J.: Tracking-learning-detection. IEEE Trans. Pattern Anal. Mach. Intell. 34(7), 1409–1422 (2012)
12. Kwon, J., Lee, K.M.: Visual tracking decomposition. In: 2010 IEEE Conference on Computer Vision and Pattern Recognition (CVPR), pp. 1269–1276. IEEE (2010)
13. Kwon, J., Lee, K.: Tracking by sampling trackers. In: 2011 IEEE International Conference on Computer Vision (ICCV), pp. 1195–1202. IEEE (2011)
14. Lowe, D.G.: Distinctive image features from scale-invariant keypoints. Int. J. Comput. Vis. 60(2), 91–110 (2004)
15. Ma, C., Yang, X., Zhang, C., Yang, M.H.: Long-term correlation tracking. In: Proceedings of the IEEE Conference on Computer Vision and Pattern Recognition, pp. 5388–5396 (2015)
16. Wu, Y., Lim, J., Yang, M.H.: Online object tracking: a benchmark. In: Proceedings of the IEEE Conference on Computer Vision and Pattern Recognition, pp. 2411–2418 (2013)
17. Zhang, K., Zhang, L., Yang, M.-H.: Real-time compressive tracking. In: Fitzgibbon, A., Lazebnik, S., Perona, P., Sato, Y., Schmid, C. (eds.) ECCV 2012, Part III. LNCS, vol. 7574, pp. 864–877. Springer, Heidelberg (2012)
18. Zhang, L., van der Maaten, L.J.: Preserving structure in model-free tracking. IEEE Trans. Pattern Anal. Mach. Intell. 36(4), 756–769 (2014)
19. Zhong, W., Lu, H., Yang, M.H.: Robust object tracking via sparsity-based collaborative model. In: 2012 IEEE Conference on Computer Vision and Pattern Recognition (CVPR), pp. 1838–1845. IEEE (2012)

A New Weight Adjusted Particle Swarm Optimization for Real-Time Multiple Object Tracking

Guang Liu[1(⊠)], Zhenghao Chen[1], Henry Wing Fung Yeung[1],
Yuk Ying Chung[1], and Wei-Chang Yeh[2]

[1] School of Information Technologies,
University of Sydney, Sydney, NSW 2006, Australia
guang.liu@sydney.edu.au,
zche2021@uni.sydney.edu.au
[2] Department of Industrial Engineering and Engineering Management,
National Tsing Hua University, P.O. Box 24-60, Hsinchu 300, Taiwan, R.O.C.

Abstract. This paper proposes a novel Weight Adjusted Particle Swarm Optimization (WAPSO) to overcome the occlusion problem and computational cost in multiple object tracking. To this end, a new update strategy of inertia weight of the particles in WAPSO is designed to maintain particle diversity and prevent pre-mature convergence. Meanwhile, the implementation of a mechanism that enlarges the search space upon the detection of occlusion enhances WAPSO's robustness to non-linear target motion. In addition, the choice of Root Sum Squared Errors as the fitness function further increases the speed of the proposed approach. The experimental results has shown that in combination with the model feature that enables initialization of multiple independent swarms, the high-speed WAPSO algorithm can be applied to multiple non-linear object tracking for real-time applications.

Keywords: Object tracking · Particle swarm optimization · Root sum squared errors · Multiple object tracking

1 Introduction

Recently, object tracking has become a hot topic of research in the field of computer vision. It has diverse application areas including surveillance system, computer-human interaction and traffic monitoring.

Object tracking is defined as the estimation of the trajectory of a target in a sequence of video frames. In the recent decade Particle Swarm Optimization (PSO) has become a popular choice in object tracking due to its accuracy and fast convergence nature [4–8]. PSO, proposed by Eberhart and Kennedy, is an evolutionary computation method mimicking the behavior of natural swarms such as bird flocking and fish schooling [1–3]. During the optimization process, a swarm of particles is at first randomly generated in a search space located inside the frame. The particles are then updated based on their own experience as well as the experience of the whole swarm. Iteration of the process allows the particles to continuously evolve and thus quickly

© Springer International Publishing AG 2016
A. Hirose et al. (Eds.): ICONIP 2016, Part II, LNCS 9948, pp. 643–651, 2016.
DOI: 10.1007/978-3-319-46672-9_72

converge to the optimum. However, owing to its fast converging property, the traditional PSO has some drawbacks, most notably pre-mature convergence and loss in particle diversity. PSO is also found to be ineffective in situations involving occlusion and unexpected disappearance of the target objects. Furthermore, the performance of PSO is severely constrained by the efficiency of its fitness function, yet most of the existing modifications made to the traditional model overlooked the need for a better alternative to the widely adopted histogram based approach.

In this paper, we propose a novel Weight Adjusted Particle Swarm Optimization algorithm (WAPSO) that facilitates convergence to global optimal while maintaining particle diversity. In addition, Root Sum Squared Errors (RSSE) fitness function is chosen as an alternative for the common histogram based approach. Section 2 presents a brief description on the traditional PSO algorithm. Section 3 introduces the WAPSO and the RSSE fitness function in detail. Section 4 presents the experimental results while Sect. 5 concludes the paper.

2 The Traditional PSO for Object Tracking

In PSO, a set of particles, defined by their x and y coordinates, are generated randomly across a search space with pre-defined width and height located within the image. Each particle is the center of a search window which is of the same size as the target object. All particles with their corresponding search windows are considered as potential candidates of the solution to the tracking problem. Their fitness values are evaluated by a fitness function and subsequent updates will apply to renew their positions in the search space. After sufficient iterations, the particles will converge to the optimum and particle with the best fitness value will be chosen as the solution.

The traditional PSO comprises of two equations and one fitness function. The first Eq. (1) governs the change in velocity of each particle whereas the second Eq. (2) updates the position of each particle for a particular iteration.

$$v_n^{t+1} = \omega \cdot v_n^t + C_1 \cdot R_1 \left(Pbest_n^t - x_n^t \right) + C_2 \cdot R_2 \cdot \left(Gbest^t - x_n^t \right) \tag{1}$$

$$x_n^{t+1} = x_n^t + v_n^{t+1} \tag{2}$$

In Eqs. (1) and (2), x_n^t and v_n^t denote respectively the position and velocity of the nth particles in the tth iteration. The position that gives the best fitness value among all past and present positions of the nth particle is denoted as $Pbest_n^t$ while the best fitting position among all past and present positions of all the particles is given by $Gbest^t$. R_1 and R_2 are random variables with range 0 to 1. C_1 and C_2, which usually sum to 4, are defined respectively as the individual and social factor. ω is the inertia weights that describes the degree of path dependency of the particle.

3 The Proposed Approach

3.1 Weight Adjusted Particle Swarm Optimization (WAPSO)

WAPSO proposed in this paper aims to solve the problem of diversity loss and pre-mature convergence which are the most notable shortcomings of the traditional PSO. During tracking, it is necessary to maintain a diversified cohort of particles in order to fully utilize the whole search space. Diversity loss refers to a situation which particles process very similar properties. This situation leaves us with a limited choice of potential solutions, thus hinders the performance of the algorithm in locating the global optimum. In the case of diversity loss, we are expected to observe similar distance to global best in most of the particles, or in mathematical terms, a low deviation of distance (D_n). The deviation of the distance to global best is captured by

$$\sigma_D = \frac{Min(D_n, \bar{D})}{Max(D_n, \bar{D})} \qquad (3)$$

which is a strictly decreasing function of the difference between D_n and $\bar{D}$ for all n particles. Capturing the deviation in distance alone may not be sufficient since equidistant particles to the global best can possibly process very different fitness value. Therefore, it is necessary to have another mechanism to measure the deviation of the fitness values (F_n) from mean $(\bar{F})$ for all n particles, given by

$$\sigma_F = \frac{Min(F_n, \bar{F})}{Max(F_n, \bar{F})} \qquad (4)$$

which is a strictly decreasing function of the difference between F_n and $\bar{F}$.

In the traditional PSO, pre-mature convergence refers to the situation which the particles converge to a solution without thorough search in the given space. For example, if the particles located at a local optimum instead of the global optimum in the first iteration, all particles will converge pre-maturely to the local optimum. In the ideal case, particles should remain spread out to search for the global optimum during early iterations and coverage when the loop approaches the end. Therefore, we proposed a mechanism that manages the tendency to converge (τ), defined as

$$\tau = 1 - \frac{I_c}{I_t} \qquad (5)$$

where I_c and I_t are the current and total number of iteration respectively.

The WAPSO algorithm introduces an update equation for the inertia weight ω_n which incorporates Eqs. (3) to (5). It is given by

$$\omega_n = \tau \cdot (\omega_i + \alpha \cdot \sigma_D + \beta \cdot \sigma_F) \qquad (6)$$

where ω_i is initial inertia weight. α and β are parameters with range 0 to 1. The proposed Eq. (6) shows that ω_n is a decreasing function of iterations and particle diversity.

When particle diversity decreases, the impact of inertia weight for particle n is increased, encouraging the particle to explore more area. Furthermore, the inertia weight ω_n is kept high at early iterations to avoid pre-mature convergence. As iteration number increases, the inertia weight will decrease, allowing for convergence of the particles.

In addition to Eq. (6), a mechanism is implemented in the WAPSO aiming to detect occlusion and disappearance of the target object and relocate it once it reappears. This is accomplished by setting a fitness value threshold (F_T). If the fitness value of the global best particle falls below F_T, the search space will enlarge to cover the whole image, allowing for global search to recapture the target object.

Given the above-mentioned advantages of WAPSO, it can achieve more accurate results than the traditional PSO and is more robust to handle occlusion and disappearance problems. Figure 1 below provides an overall description of the procedure for the proposed WAPSO.

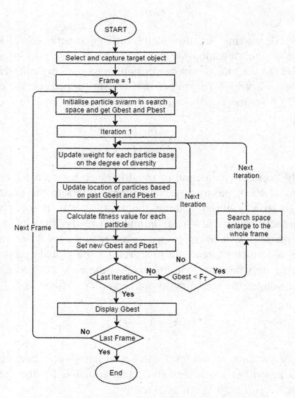

Fig. 1. Flowchart of the WAPSO algorithm

3.2 Fitness Function

In WAPSO, the fitness value (F) is measured by a fitness function to evaluate the similarity of each candidate particle with the target object. Most previous research adopts a histogram comparison approach as a measure of similarity between the particle and the target object [4–7]. The chosen alternative in this paper given is by equations:

$$RSSE(P_s, P_t) = \sqrt{\sum_i \sum_j \left(P_{s(ij)} - P_{t(ij)}\right)^2} \qquad (7)$$

$$F = f\left(\frac{RSSE(P_s, P_t)}{M \cdot N}\right) \qquad (8)$$

where $RSSE(P_s, P_t)$ is the Root Sum Squared Errors, $P_{s(ij)}$ denotes the pixel in the ith row and j^{th} column of the image captured by the search window of a particle and $P_{t(ij)}$ denotes the pixel in the ith row and j^{th} column of the target image. Each pixel is presented in HSV form. The RSSE value is divided by M and N, which represent the row and column of the input image respectively, and is then normalized to the range of 0 to 1. RSSE is chosen for its simplicity which substantially reduces computation and allows for real-time application.

3.3 Multiple Object Tracking

Tracking multiple objects is a challenging task because it usually requires higher computational cost comparing to tracking a single object. In addition, the observations of different target objects may overlap during occlusion which may result in low tracking performance. To overcome these difficulties, we introduce multiple swarms for multiple object tracking. By selecting multiple target objects to be tracked from the screen, multiple object models are established. Each of the models is associated with a particle swarm. The swarms then parallel process the target searching. The parallel feature of this method renders the searching speed while the occlusion problem can be well-handled as each swarm only focuses on its own target object independently.

4 Experiment Results

The proposed framework was tested and compared with the traditional PSO and the CPSO proposed by Sha (2015) [6]. The CPSO is a PSO based tracking algorithm that shares similar trait to our proposed WAPSO. It serves as a benchmark to assess our proposed framework. The first testing video assessed the performance of WAPSO using a real world traffic environment. Tracking was performed on a car with neither occlusion nor disappearance from screen. The second video tested the ability of WAPSO in multiple tracking of non-linear targets using 3 balls with different colors. We set $\omega_i = 1$, $\alpha = \beta$ $\beta = 0.5$ and $C_1 = C_2 = 2$ in WAPSO. Same values of C_1 and C_2 were also applied in both PSO and CPSO. The value of F_T in WAPSO is very sensitive as it

Table 1. Video – car (pixels: 1920 * 1080)

Frame	Test 1: WAPSO	Test2 : CPSO	Test 3: PSO
2			
19			
140			

Table 2. Video - color balls (pixels: 1920 * 1080)

Frame	Test 1: WAPSO	Test2 : CPSO	Test 3: PSO
2			
35			
60			
116			
166			

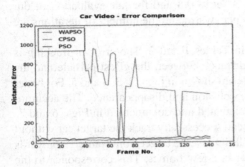

Fig. 2. Accuracy comparison - car

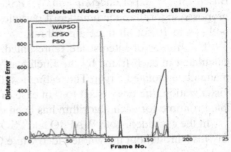

Fig. 3. Accuracy comparison – blue ball

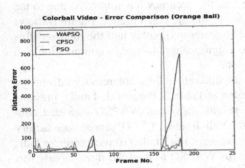

Fig. 4. Accuracy comparison – orange ball

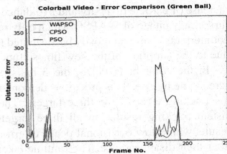

Fig. 5. Accuracy comparison – green ball

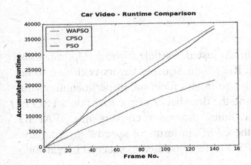

Fig. 6. Runtime comparison - car

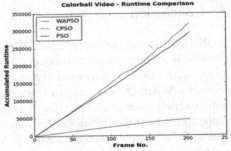

Fig. 7. Runtime comparison – color balls

determines whether the particles should conduct a global search. Lower value of F_T means more global search and usually requires more number of iterations and particles to cover the larger search space and maintain the tracking accuracy. Higher value of F_T would result in less global search and therefore less number of iterations and particles, which can increase the tracking speed but sometimes fail to handle occlusion problems.

Based on our experiences, F_T in WAPSO is set to 0.8 and for fair evaluation of our proposed algorithm, the number of particles per swarm and the number of iterations are both set to 10 for all three algorithms.

The frames of interest are represented in Tables 1 and 2. The error of tracking is calculated in each frame by the Euclidean distance between the gBest particle and the manually annotated target. The testing results are shown in Figs. 2, 3, 4 and 5. For better visualization, the error is set to 0 in case of occlusion and disappearance. The accumulated runtime for each algorithm has been calculated and documented in Figs. 6 and 7.

In the car video, both WAPSO and PSO can successfully track the target car without any loss throughout the whole video while CPSO lost the target from frame 19 onwards and only managed to recapture the target car in a few frames. This corresponds to the relatively mild fluctuation of errors for WAPSO and PSO and the sharp variation of errors for CPSO as shown in Fig. 2. Moreover, the error of the WAPSO fell below that of the PSO in almost all frames, suggesting that the accuracy has improved due to the increase of particle diversity and avoidance of pre-mature convergence. Figure 6 shows further advantage of WAPSO which only required approximately half the runtime of its counterparts. This can mostly be attributed to a significant reduction in computation cost due to the adoption of the new fitness function.

In the color balls video, the 3 balls with different colors are removed from the screen one by one. This process is demonstrated in Table 2. Figures 3, 4 and 5 presents the tracking error of the three targets. The results show that WAPSO provided consistent tracking results on all three targets in all frames. CPSO scored satisfactory results with a few occasional peaks of errors whereas PSO failed to recapture the balls after their disappearance. The runtime taken for WAPSO to track all three targets was approximately a sixth of the other frameworks, allowing WAPSO to perform real-time multiple object tracking.

5   Conclusion

In this paper, we present a novel Weight Adjusted Particle Swarm Optimization (WAPSO) for multiple object tracking. The Root Sum Squared Errors is chosen to be an alternative to the widely used histogram based fitness function. Experimental results show that WAPSO has successfully overcome the drawbacks of the traditional PSO in rediscovering objects after occlusion and avoiding premature convergence. WAPSO outperforms both the traditional PSO and the CPSO in terms of speed and accuracy. Due to its substantial speed advantage, WAPSO is demonstrated to succeed in real-time tracking of multiple objects.

For future work, we intend to further improve the tracking performance by evaluating the influence of each WAPSO parameter and adaptively changing them according to the motion state of the target object.

References

1. Eberhart, R.C., Kennedy, J.: A new optimizer using particle swarm theory. In: Proceedings of the Sixth International Symposium on Micro Machine and Human Science, vol. 1, pp. 39–43 (1995)
2. Kennedy, J., Eberhart, R.C.: A discrete binary version of the particle swarm algorithm, systems, man, and cybernetics. IEEE Int. Conf. Comput. Cybern. Simul. **5**(12–15), 4104–4108 (1997)
3. Eberhart, R.C., Shi, Y.: Particle swarm optimization: developments, application and resources. In: Proceedings of the 2001 Congress on Evolutionary Computation, Seoul, South Korea, vol. 1, pp. 81–86 (2001)
4. Zheng, Y., Meng, Y.: The PSO-based adaptive window for people tracking. In: IEEE Symposium on Computational Intelligence in Security and Defense Applications, 2007. CISDA 2007, pp. 23–29. IEEE (2007)
5. Hsu, C., Dai, G.T.: Multiple object tracking using particle swarm optimization. World Acad. Sci. Eng. Technol. **68**, 41–44 (2012)
6. Sha, F., Bae, C., Liu, G., et al.: A categorized particle swarm optimization for object tracking. In: 2015 IEEE Congress on Evolutionary Computation (CEC), pp. 2737–2744. IEEE (2015)
7. Sha, F., Bae, C., Liu, G., et al.: A probability-dynamic particle swarm optimization for object tracking. In: 2015 International Joint Conference on Neural Networks (IJCNN), pp. 1–7. IEEE (2015)
8. Zhang, L., Tang, Y., Hua, C., et al.: A new particle swarm optimization algorithm with adaptive inertia weight based on Bayesian techniques. Appl. Soft Comput. **28**, 138–149 (2015)

Fast Visual Object Tracking
Using Convolutional Filters

Mingxuan Di, Guang Yang, Qinchuan Zhang, Kang Fu, and Hongtao Lu[⊠]

Key Laboratory of Shanghai Education Commission
for Intelligent Interaction and Cognitive Engineering,
Department of Computer Science and Engineering,
Shanghai Jiao Tong University, Shanghai 200240, China
dimxcs@gmail.com, yangguangpkpk@gmail.com,
{qinchuan.zhang,fukang1993,htlu}@sjtu.edu.cn

Abstract. Recently, a class of tracking techniques called synthetic exact filters has been shown to give promising results at impressive speeds. Synthetic exact filters are trained using a large number of training images and associated continuous labels, however, there is not much theory behind it. In this paper, we theoretically explain the reason why synthetic exact filters based methods work well and propose a novel visual object tracking algorithm based on convolutional filters, which are trained only by training images without labels. Compared with the prior methods such as synthetic exact filters which are trained by training images and labels, advantages of the convolutional filters training include: faster and more robust than synthetic exact filters, insensitive to parameters and simpler in pre-processing of training images. Convolutional filters are theoretically optimal in terms of the signal-to-noise ratio. Furthermore, we utilize spatial context information to improve robustness of our tracking system. Experiments on many challenging video sequences demonstrate that our convolutional filters based tracker is competitive with the state-of-the-art trackers in accuracy and outperforms most trackers in efficiency.

1 Introduction

Visual tracking is a challenging problem and has many significant applications in computer vision. The basic framework of visual tracking is that an object is identified manually in the first frame of a video sequence (using a rectangular bounding box), then a tracker is constructed to predict the location of the object in the subsequent frames by estimating its trajectory as it moves around. Many factors, such as heavy occlusions, abrupt motion and cluttered background, would interfere tracker and lead to unexpected results. Although

Electronic supplementary material The online version of this chapter (doi:10. 1007/978-3-319-46672-9_73) contains supplementary material, which is available to authorized users.

A. Hirose et al. (Eds.): ICONIP 2016, Part II, LNCS 9948, pp. 652–662, 2016.
DOI: 10.1007/978-3-319-46672-9_73

there have been some successes in building trackers to suppress these interferences and achieve perfect performance in some dataset, tracking generic objects remains hard.

Current tracking algorithms could be categorized into generative and discriminative models. Generative models learn an appearance model to represent the target and formulate the tracking problem as searching for the regions with the highest likelihood. Recent examples include: visual tracking decomposition (VTD) [11], adaptive correlation filters tracking (MOSSE) [2] and spatio-temporal context tracking (STC) [15]. Unlike the generative approach, discriminative methods train classifier to discriminate the target from the background and formulate tracking as a classification problem. Recent examples include: structured output tracking (Struck) [8], multiple instance learning tracking (MIL) [1] and tracking-learning-detection (TLD) [10]. A typical tracking system consists of three components: (1) an appearance model, which can predict the likelihood that the object of interest is at some particular location based on the local image appearance; (2) a location model, which evaluates the prior probability that the object is present at a particular location; and (3) a search strategy for finding the maximum a posteriori location of the object in the current frame.

In this paper, we first extend the 1-D matched filter to the 2-D optimal spacial filter. Then we introduce the 2-D convolutional filter and analyse synthetic exact filter in theory. Finally, based on this 2-D optimal convolutional filter, we propose a novel tracking system. Our goal is to develop a simple yet robust way of constructing an appearance model. The appearance model of our tracking system is based on the convolutional filters. When a new frame inputs, the updated appearance model convolves with the image of the searching window and outputs confidence map of the searching region. The location of maximum value in confidence map indicates the position of the target. Based on the new target location, we train the convolutional filters and update the appearance model iteratively.

Convolutional filter is a theoretical optimal filter and can output the maximum value at the location of the target when convolving with the image of searching window. We introduce a novel method to train convolutional filters, which are trained using a training image in tracking window without labels, and thus robustness is improved and the computational complexity is reduced. The tracking window of frame t is determined by the target location of frame t, the searching window of frame t is determined by the target location of frame $t-1$.

2   Related Work

In recent years, synthetic exact filter and its variants are successfully employed for visual tracking, such as Average of Synthetic Exact Filters (ASEF) [3,15] and Minimum Output Sum of Squared Error (MOSSE) filters [2,9]. Numerous experiments have proved that synthetic exact filter is a useful method for visual tracking.

ASEF filter [3] is constructed by training pairs f_i, g_i consisting of a training image f_i and an associated desired discrete label g_i. The label g_i is generated with a bright peak at the center of target and smaller values everywhere else. Generally, the label is defined to be a two dimensional Gaussian centered at the target location. In [15], filter is also trained by training image and associated label, but the label is defined to be a two dimensional Laplacian.

MOSSE is an algorithm for producing ASEF-like filters from fewer training images and labels. MOSSE aims to find a filter that minimizes the sum of squared error between the actual output of the convolution and the desired output of the convolution. In [2], a tracker consisting of MOSSE filters and running average is used for updating tracker. In [9], the filters proposed in [2,3] are kernelized and used to achieve more stable results.

However, so far there is still lack of reliable theory to support this method. Moreover, all these synthetic exact filters are learned by training images and their corresponding labels. Labeling is time-consuming and is sensitive to parameters. We believe that the method we propose solves the problems above. We analyse synthetic exact filter in theory and based on this theory, we introduce a robust and efficient tracker based on convolutional filters.

The contributions of this paper are as follows:

1. We introduce convolutional filter and theoretically explain the reason why the synthetic exact filter could work well.
2. Based on this theory, we introduce the convolutional filters based tracker, which is theoretically optimal and is different from the above methods [2,3, 9,15].
   (a) Convolutional filter is constructed only by training image without label, which significantly improves robustness and reduces computational complexity.
   (b) Due to the robustness of convolutional filters, pre-processing of training images is simpler than the methods in [2,3,15].
   (c) Moreover, many filter based trackers improve their performance by using affine perturbations to initialize the first filter, which leads to inefficiency. Different from them, our tracker performs well without initializing the first filter.

3   Convolutional Filter Based Tracking

In this section, we extend the 1-D matched filter to an 2-D convolutional filter, which reveals the principle of the traditional synthetic exact filters. Based on this principle, we introduce our tracking algorithm, which includes appearance model based on convolutional filter and other techniques.

3.1   Pre-processing

One issue with the Fast Fourier Transform (FFT) is that it does not respect the image boundaries. In Fourier domain, the image is periodic. In other words, FFT connects the left edge of the image to the right edge, and the top to the bottom. In order to reduce the effects of FFT, [2, 15] have outlined different preprocessing steps, which usually include a log function and a window function.

Due to the robustness of convolutional filters, we only need to use a simpler method without log function to preprocess training images. The original $M \times N$ image is multiplied by a Blackman window, which is defined as:

$$w(x,y) = [0.42 - \cos(\frac{2\pi x}{M-1}) + 0.08\cos(\frac{4\pi x}{M-1})]R_M(x)$$
$$\cdot [0.42 - \cos(\frac{2\pi y}{N-1}) + 0.08\cos(\frac{4\pi y}{N-1})]R_N(y)$$

The pixel values near the edge are reduced to zero. This puts more emphasis near the center of the target.

3.2 Convolutional Filters

Traditional synthetic exact filters are produced by a set of training images f_i and training labels g_i. Generally, g_i can take any shape. In [2], g_i has a compact 2D Gaussian shaped peak centered on the target in training image. Experiments demonstrate that synthetic exact filter based tracker is robust and efficient. However, label is hard to be defined and is sensitive to parameter, which easily leads to over-fitting. Now we need to ask a deeper and practically more useful question: is there a better filter to replace the synthetic exact filter? We will answer this question in theory and introduce a new convolutional filter step by step.

Our idea is motivated by 1-D matched filter, which is a widely used method to detect patterns in signals and allows us train filters without labels. It has close relations to convolutional neutral networks and is a kind of filter that maximizes the signal-to-noise ratio (SNR). In [7], a special 1-D matched filter is derived from calculus of variations.

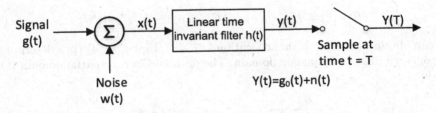

Fig. 1. Overview of 1-D matched filter. Signal $g(t)$ is processed by filter $h(t)$ and the output of filter is $g_0(t)$. $n(t)$ is output associated with noise $w(t)$.

Now we attempt to extend the 1-D matched filter to an 2-D optimal spatial filter. When a new frame comes, target in this frame can be modeled by

$g(x, y)$, which is pixel values of the tracking window. Interference factors of target including illumination variation, occlusion, deformation, fast motion and etc., can be regarded as noise. We can model the noise by additive white Gaussian noise (AWGN). $W(u, v) = \frac{N_0}{2}$ describes AWGN in frequency domain. The new constructed 2-D optimal spatial filter could generate confidence map, where the value corresponding to location of target is maximum. In order to suppress interference factors, the proposed filter should maximize the signal-to-noise ratio at location of target. In the following, we will show how to construct an 2-D optimal spatial filter.

First, the Fourier transform of the input image: $G(u, v) = \mathcal{F}[g(x, y)]$ is computed. The output of filter at location (x, y) in spatial domain is:

$$g_0(x, y) = \sum_{u=0}^{M-1} \sum_{v=0}^{N-1} H(u, v)G(u, v)e^{j2\pi(\frac{ux}{M} + \frac{vy}{N})}, \tag{1}$$

where $H(u, v)$ is the Fourier transform of spatial filter $h(x, y)$, which is to be determined. Given the location of target (x_T, y_T), we can calculate the signal-to-noise ratio η at this location:

$$\eta = \frac{|g_0(x_T, y_T)|^2}{E[n^2(x, y)]} = \frac{|\sum_{u=0}^{M-1} \sum_{v=0}^{N-1} H(u, v)G(u, v)e^{j2\pi(\frac{ux_T}{M} + \frac{vy_T}{N})}|^2}{\sum_{u=0}^{M-1} \sum_{v=0}^{N-1} |\frac{N_0}{2} H(u, v)|^2} \tag{2}$$

Applying the Schwarz's inequality to Eq. (2), we obtain the inequality:

$$\begin{aligned}
&|\sum_{u=0}^{M-1} \sum_{v=0}^{N-1} H(u, v)G(u, v)e^{j2\pi(\frac{ux_T}{M} + \frac{vy_T}{N})}|^2 \\
&\leq \sum_{u=0}^{M-1} \sum_{v=0}^{N-1} |H(u, v)|^2 \sum_{u=0}^{M-1} \sum_{v=0}^{N-1} |G(u, v)|^2
\end{aligned} \tag{3}$$

When Eq. (3) satisfies the following condition:

$$H(u, v) = H_{opt}(u, v) = kG^*(u, v)e^{-j2\pi(\frac{ux_T}{M} + \frac{vy_T}{N})}, \tag{4}$$

we can obtain maximum ratio at location (x_T, y_T). Equation (4) thus determines the optimal filter in frequency domain. The optimal filter in spatial domain thus reads,

$$\begin{aligned}
h_{opt}(x, y) &= k \sum_{u=0}^{M-1} \sum_{v=0}^{N-1} G^*(u, v)e^{-j2\pi(\frac{ux_T}{M} + \frac{vy_T}{N})} \cdot e^{j2\pi(\frac{ux}{M} + \frac{vy}{N})} \\
&= kg(x_T - x, y_T - y)
\end{aligned} \tag{5}$$

A complete derivation is in the supplemental materials.

From Eq. (5), we can explain the reason why synthetic exact filter is a useful method for visual tracking. We can get a filter just using target image based on

Eq. (5). Elements of this filter are the pixel values of the inverse target image and have a offset which is determined by the location of the target. When a new frame comes, the filter convolves with the image of searching window and outputs the convolution results. Assuming there is a target in the searching window, we can see that output results have a maximum value at the center position of the target and small values elsewhere. The shape of the output results is completely identical with the manually designed output of synthetic exact filter, which is usually a 2D Gaussian. This similarity reveal the reason why synthetic exact filter could give promising results at impressive speeds. The synthetic exact filter which is constructed using manually designed labels is a variant of the optimal spatial filter. The optimal spatial filter in Eq. (5) is named convolutional filter because it is similar to convolutional layer in CNN.

In our tracking system, convolutional filter is used to construct and update appearance model, which generates confidence map by convolving with image of searching window. Location corresponding to the maximum value of confidence map indicates the new position of the target. In order to cope with distortions in the object appearance and construct a more robust tracking system, spatial context is also adopted.

3.3 Spatial Context

One commonly recognized agreement is that context plays a significant role in image understanding task. The contribution of contextual information in object detection is evaluated by [6]. Some trackers also utilize contextual information to improve performance of visual tracking [5,14]. In this paper, spatial context is used to handle occlusion and rotation. Experiments on many challenging sequences indicate that spatial context is beneficial to construct robust tracker.

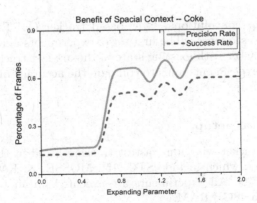

Fig. 2. This figure shows change of precision and success rate when expanding parameter varies in video sequence Coke [13]. Precision and success rate are introduced in Sect. 4.1

Spatial context, also called local pixel context, captures the basic notion that image pixels around the region of interest carry useful information. Increasing the size of a tracking window to include surrounding pixels is a classic trick to extract spatial contextual information. Our tracker employs this trick. The expanding parameter in our tracker is set to 1.2. Figure 2 shows that spatial context adopted in our tracker plays an important role in accuracy and robustness. There is a trade-off between accuracy and efficiency. Bigger expanding parameter is beneficial to accuracy of tracker, but it also leads to loss of efficiency.

3.4 Online Update

In order to adapt to the varying appearance of target, we need to update appearance model of tracker. Update strategy commonly used in filter based trackers has been outlined in [2,15]. In this paper, we employ the running average scheme to update our appearance model.

$$H_t = \alpha H_{opt}^t + (1 - \alpha)H_{t-1} \tag{6}$$

where H_{opt}^t is the convolutional filter learned from frame t, which is the Fourier transform of $h_{opt}(x, y)$ in Eq. (5), α is learning rate and H_t is the appearance model that would be used in frame $t+1$. Learning rate α lets the effect of previous frames decay over time. Larger α is beneficial to tracker to adapt to varying appearance, but leads to bad performance of handling occlusion. Experiments of grid search on many challenging video sequences indicate that α equals to 0.05 allows the filter to quickly adapt to varying appearance while still being able to handle heavy occlusions.

4 Experiments

In this section, our convolutional filter based tracker (MFT) is evaluated on many famous video sequences. The first set of experiments aims to compare the efficiency of our tracker with existing state-of-the-art fast trackers listed in [13]. The second set of experiments aims to compare the accuracy of our tracker with these existing trackers.

4.1 Experimental Setup

We compare our tracker with four fastest trackers listed in the [13] and other filter based trackers, which include STC [15], MOSSE [2], KMS [4], CPF [12], CSK [9] and CT [16]. All experiments are executed on an Intel Xeon X5650 2.66GHz CPU with 48GB RAM.

Datasets: We evaluate our method on a recent public benchmark dataset [13] consisting of 50 video sequences which have a wide range of challenging aspects including illumination variations, scale variations, occlusions, deformations, fast motions and (out-of-plane) rotations. However, a filters based tracker failed on five sequences using its open source, so we show quantitative comparison results on remaining 45 sequences[1].

Evaluation Methodology: In order to quantitatively evaluate our tracker performance, we used two widely accepted evaluation metrics, precision and success rate [13]. The precision rate is defined as the percentage of frames whose estimated location was within the given threshold distance of the ground truth. The success rate shows the ratios of successful frames where the bounding box overlap exceeds a threshold $t \in [0, 1]$. We run the One-Pass Evaluation (OPE) on the benchmark and use the area under curve (AUC) to rank the tracking algorithms. Both the precision and success plots show the mean scores over all the sequences.

4.2 Real Time Performance

In this section we evaluate the efficiency of our tracker MFT. Due to perfect performance of Fast Fourier Transform, our convolutional filter based trackers easily outperform most fast tracking systems. We ran source codes of state-of-the-art fast trackers on our machine and recorded the FPS (frames per second) of these trackers on every benchmark sequence in the supplemental materials. Then we calculated average tracking speed and showed the statistics of the tracking speed in Table 1.

Table 1. Average tracking speed on benchmark sequences. FPS: average frame per second

Method	MFT	STC	MOSSE	CSK	CT	CPF	KMS
A-FPS	204.2	84.6	42.4	190.1	57.3	70.0	2092.3

The computational complexity of the convolutional filter based tracker is $O(PlogP)$, where P is the number of pixels in the tracking window. Compared with other filter based trackers, no tracking initialization and no need to process labels further accelerate tracking speed. Table 1 indicates that our tracker is faster than other filter based trackers and trackers outlined in [13] except for KMS [4]. High-speed trackers are beneficial to identify the influence of small changes to the tracker parameters and make it easy to perform qualitative analysis of the tracker performance for different tracking conditions.

4.3 Accuracy Performance

To evaluate the ability of our algorithm to maintain tracking, we compared our tracker with above state-of-the-art fast trackers. The comparison reveals the

---

[1] In the supplemental material we show quantitative comparison results on all 50 sequences without failing tracker.

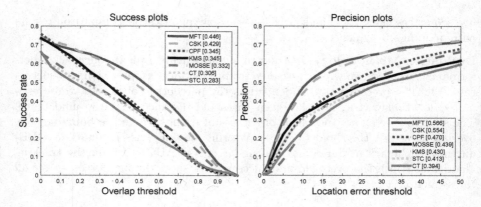

Fig. 3. This figure shows precision plot and success plot of our tracker MFT, CSK [9], CPF [12], KMS [4], MOSSE [2], CT [16] and STC [15].

potential benefit of using convolutional filters when tracking a target. Figure 3 indicates that our tracker outperforms other trackers on average.[2]

Illumination Variation: Tests on $Car4, CarDark, Fish, Shaking$ sequences indicate that our method has perfect performance when it undergoes illumination variation. For example, in the $Fish$ sequence, the CPF, CSK, KMS and STC tracker drift away from the target when illumination dramatically changes, but MOSSE tracker and our method perform well. This can be attributed to the use of filters which are insensitive to illumination. Filters extract differences between pixels, which changes slightly when illumination dramatically changes.

Occlusion and Rotation: When the target rotates or is occluded, appearance of target changes dramatically. The proposed tracker exploits spatial context and extracts auxiliary objects, which is helpful to recognize and track the target. In the $Tiger1$ sequence, numerous methods fail to track the target because it is often occluded by leafs. Our tracker does not drift away when the target reappears. In the $MountainBike$ sequence, the target rotates all the time. While a few trackers are able to maintain tracks to the end, the proposed algorithm achieves lowest center location error.

Background Clutters: The $CarDark$ sequence is challenging as there are numerous cars of different shapes and colors, target and backgrounds have similar color. The CPF, CT and KMS trackers drift to background or other cars whereas our tracker keeps track of the target to the end. For $David3$ sequence, the background near the target has the similar texture as the target. While most trackers drift away from the target, the STC tracker and our method keep track of the target throughout the sequence, but our method has higher precision rate.

---

[2] We record precision and success rate of each tracker on every benchmark sequence in the supplemental materials.

Table 2. We list precision rate of seven trackers over 12 sequences, which support the qualitative analysis in illumination variation, occlusion and rotation, background clutters.(*mBike** denotes mountainBike.)

Sequence	CPF	CSK	CT	KMS	MOSSE	MFT	STC
Basketball	0.643	0.860	0.317	0.515	0.134	0.864	0.500
Car4	0.329	0.615	0.241	0.253	0.919	0.735	0.222
CarDark	0.202	0.926	0.015	0.559	0.949	0.957	0.935
David3	0.634	0.631	0.376	0.801	0.490	0.893	0.865
Deer	0.091	0.892	0.040	0.484	0.184	0.772	0.042
Dudek	0.498	0.735	0.470	0.432	0.123	0.759	0.087
fish	0.233	0.183	0.781	0.421	0.883	0.926	0.356
Freeman3	0.165	0.539	0.274	0.462	0.365	0.769	0.173
Girl	0.689	0.616	0.626	0.155	0.793	0.883	0.615
*mBike**	0.123	0.863	0.158	0.575	0.862	0.863	0.852
Shaking	0.158	0.653	0.088	0.245	0.090	0.737	0.801
Tiger1	0.409	0.272	0.404	0.380	0.198	0.533	0.271

5 Conclusion

In this paper, we analyse synthetic exact filter in theory and reveal the reason why synthetic exact filters based methods can work well. We introduce convolutional filter and present a novel tracking system based on this filter. Unlike existing synthetic exact filter, convolutional filter is robust and theoretical optimal. It allows us to learn filters without labeling training images, which makes it easy to train filters and pre-process training images. Similar to many tracking systems, our tracker employs spatial context information to improve robustness. We measured quantitative performance of our tracker and compared it to a number of state-of-the-art algorithms on many challenging sequences; the experimental results show that our tracker is competitive with the state-of-the-art trackers in accuracy and outperforms most trackers in efficiency.

References

1. Babenko, B., Yang, M.H., Belongie, S.: Robust object tracking with online multiple instance learning. IEEE Trans. Pattern Anal. Mach. Intell. **33**(8), 1619–1632 (2011)
2. Bolme, D.S., Beveridge, J.R., Draper, B.A., Lui, Y.M.: Visual object tracking using adaptive correlation filters. In: IEEE Conference on Computer Vision and Pattern Recognition, pp. 2544–2550 (2010)
3. Bolme, D.S., Draper, B.A., Beveridge, J.R.: Average of synthetic exact filters. In: IEEE Conference on Computer Vision and Pattern Recognition, pp. 2105–2112 (2009)
4. Comaniciu, D., Ramesh, V., Meer, P.: Kernel-based object tracking. IEEE Trans. Pattern Anal. Mach. Intell. **25**(5), 564–577 (2003)

5. Dinh, T.B., Vo, N., Medioni, G.: Context tracker: exploring supporters and distracters in unconstrained environments. In: IEEE Conference on Computer Vision and Pattern Recognition, pp. 1177–1184 (2011)
6. Divvala, S.K., Hoiem, D., Hays, J.H., Efros, A.A., Hebert, M.: An empirical study of context in object detection. In: IEEE Conference on Computer Vision and Pattern Recognition, pp. 1271–1278 (2009)
7. Duda, R.O., Hart, P.E., Stork, D.G.: Pattern Classification (2012)
8. Hare, S., Saffari, A., Torr, P.H.: Struck: structured output tracking with kernels. In: IEEE International Conference on Computer Vision, pp. 263–270 (2011)
9. Henriques, J.F., Caseiro, R., Martins, P., Batista, J.: Exploiting the circulant structure of tracking-by-detection with kernels. In: Fitzgibbon, A., Lazebnik, S., Perona, P., Sato, Y., Schmid, C. (eds.) ECCV 2012, Part IV. LNCS, vol. 7575, pp. 702–715. Springer, Heidelberg (2012)
10. Kalal, Z., Mikolajczyk, K., Matas, J.: Tracking-learning-detection. IEEE Trans. Pattern Anal. Mach. Intell. 34(7), 1409–1422 (2012)
11. Kwon, J., Lee, K.M.: Visual tracking decomposition. In: IEEE Conference on Computer Vision and Pattern Recognition, pp. 1269–1276. IEEE (2010)
12. Pérez, P., Hue, C., Vermaak, J., Gangnet, M.: Color-based probabilistic tracking. In: Heyden, A., Sparr, G., Nielsen, M., Johansen, P. (eds.) ECCV 2002, Part I. LNCS, vol. 2350, pp. 661–675. Springer, Heidelberg (2002)
13. Wu, Y., Lim, J., Yang, M.H.: Online object tracking: a benchmark. In: IEEE Conference on Computer Vision and Pattern Recognition, pp. 2411–2418 (2013)
14. Yang, M., Wu, Y., Hua, G.: Context-aware visual tracking. IEEE Trans. Pattern Anal. Mach. Intell. 31(7), 1195–1209 (2009)
15. Zhang, K., Zhang, L., Liu, Q., Zhang, D., Yang, M.-H.: Fast visual tracking via dense spatio-temporal context learning. In: Fleet, D., Pajdla, T., Schiele, B., Tuytelaars, T. (eds.) ECCV 2014, Part V. LNCS, vol. 8693, pp. 127–141. Springer, Heidelberg (2014)
16. Zhang, K., Zhang, L., Yang, M.-H.: Real-time compressive tracking. In: Fitzgibbon, A., Lazebnik, S., Perona, P., Sato, Y., Schmid, C. (eds.) ECCV 2012, Part III. LNCS, vol. 7574, pp. 864–877. Springer, Heidelberg (2012)

An Effective Approach for Automatic LV Segmentation Based on GMM and ASM

Yurun Ma, Deyuan Wang, Yide Ma[✉], Ruoming Lei, Min Dong,
Kemin Wang, and Li Wang

School of Information Science and Engineering,
Lanzhou University, Lanzhou 730000, China
yidema14@126.com

Abstract. In this paper, we propose a novel approach for automatic left ventricle (LV) segmentation in cardiac magnetic resonance images (CMRI). This algorithm incorporates three key techniques: (1) the mid-ventricular coarse segmentation based on Gaussian mixture model (GMM); (2) the mid-slice endo-/epi-cardial initialization based on geometric transformation; (3) the myocardium tracking based on active shape models (ASM). Experiment results tested on a standard database demonstrate the effectiveness and competitiveness of the proposed method.

Keywords: Cardiac cine MR images · Left ventricle · Automatic segmentation · Gaussian-mixture model · Active shape model

1 Introduction

Left ventricle (LV) segmentation is used to measure some quantitative indicators for LV function evaluation. Manual segmentation is tedious, labor-intensive and time-consuming, so the computer-aided LV segmentation has been a research focus. It includes the endo- and epi-cardial contour delineation. Endocardium segmentation is limited by grey inhomogeneities and the presence of the papillary muscles (PM) having the same intensity as myocardium. Epicardium segmentation is difficult because of the low contrast between backgrounds and myocardium. Besides, there is great variability of the images among slices. For example, myocardial structures are unclosed in some basal slices due to LV outflow tract (LVOT), and the resolution of LV cavity is too low in some apical slices.

In the past two decades, lots of semi-automatic and automatic LV segmentation methods were proposed, which can be classified into two main categories: (1) methods with no priori or weak priori; (2) methods with strong priori. Weak priori information includes some anatomical assumptions: (1) LV is almost located in image center; (2) The geometry shape of LV is similar to a circle. These assumptions are integrated into the image-driven frameworks [1–5] and deformable models [6–10]. A simple approach based on the multilevel thresholding and region growing was proposed in [2]. Even though it was fast, its accuracy and robustness was not ideal. Active contour models had been widely used in medical image segmentation since Kass et al. firstly developed it [11]. Wang et al. integrated a circle constraint into the gradient vector flow

© Springer International Publishing AG 2016
A. Hirose et al. (Eds.): ICONIP 2016, Part II, LNCS 9948, pp. 663–672, 2016.
DOI: 10.1007/978-3-319-46672-9_74

(GVF) snake to extract endo-/epi-cardium [6]. This work was completed by replacing GVF with the gradient vector convolution (GVC) snake [9]. However, it was difficult to set parameters uniformly for a variety of images in snake models. Compared with aforementioned methods, active shape model (ASM) and active appearance model (AAM) obtained better performance because of the application of some strong priori such as statistical shape and local texture features [12–15]. Qin et al. proposed an improved ASM approach to increase segmentation accuracy [15]. However, the initial contour of most ASM/AAM models was set manually. To date, LV segmentation is still an open problem [16].

In this paper, we propose a coarse-to-fine LV segmentation method via combining Gaussian mixture model (GMM) with ASM. GMM and geometric transform are employed to initialize ASM, and endo-/epi-cardium is segmented via ASM. Rest of this paper is organized as follows. Theory background is introduced in Sect. 2. The proposed method is described in detail in Sect. 3. Database, validation and experiment environment are introduced in Sect. 4. Result analysis and comparison are given in Sect. 5. Finally, our work is concluded in Sect. 6.

2 Theory Background

2.1 Gaussian Mixture Model

Regarding an image as a matrix consisting of random variables, the pixel values will have a probability distribution. If we assume pixel $\mathbf{x} = (x_1, \cdots, x_n)^{\mathrm{T}} \in R^2$ can be classified into m classes, the probability density of x (i.e. $f(\mathbf{x})$) will be approximated by a parametric family of finite mixture densities [17] as follows:

$$f(\mathbf{x}) = \sum_{i=1}^{m} P_i N(\mathbf{x}|\phi_i), \mathbf{x} = (x_1, \cdots, x_n)^{\mathrm{T}} \in R^2 \tag{1}$$

where P_i is the probability that x belongs to the ith class C_i. It means $P_i = P(\mathbf{x} \in C_i), 0 < P_i < 1, \sum_{i=1}^{m} P_i = 1$. $N(\mathbf{x}|\phi_i)$ represents the probability distribution function parameterized by ϕ_i. When we assume that x is composed of a number of Gaussian distributions, the model in Eq. (1) is called GMM, that is:

$$N(\mathbf{x}|\phi_i) = N(\mathbf{x}|\mu_i, \sigma_i)$$
$$= \frac{1}{\sigma_i\sqrt{2\pi}}\exp[-\frac{1}{2\sigma_i^2}(\mathbf{x} - \mu_i)^2] \tag{2}$$

where μ_i and σ_i stand the mean and the standard deviation of C_i, respectively. The solution of GMM is computing the optimal parameters of $\phi_i = (\mu_i, \sigma_i)$ [18].

2.2 Active Shape Model

ASM is a deformable method including two steps: shape model training and object point searching.

(1) Shape model training

This step is to produce the profile statistical characteristics. Firstly, manual delineated contours are collected and aligned [19]. Record the aligned shapes as $S = [S_1, S_2, \ldots, S_N]$. N is the number of shapes, each shape is $S_i = [x_1^i, x_2^i, \ldots, x_k^i, y_1^i, y_2^i, \ldots, y_k^i]^T$. k is the number of landmarks. The mean and covariance of S are computed as follows:

$$\overline{S} = \frac{1}{N} \sum_{i=1}^{N} S_i \tag{3}$$

$$C = \frac{1}{N-1} \sum_{i=1}^{N} (S_i - \overline{S})(S_i - \overline{S})^T \tag{4}$$

And then, principal component analysis (PCA) [20] is applied to process S. We calculate the eigenvectors $\mathbf{P}(p_1, p_2, \ldots p_j)$ and eigenvalues $\lambda(\lambda_1, \lambda_2, \ldots \lambda_j)$ of C, and only preserve the eigenvectors $\mathbf{P_t}(p_1, p_2, \ldots p_t)$ which match t largest eigenvalues $\lambda_t(\lambda_1, \lambda_2, \ldots \lambda_t)$. Thus, any shape S_i can be represented by $\mathbf{S_i} \approx \overline{\mathbf{S}} + \mathbf{P_t b_t}$, and the vector b_t can be computed by $b_t = P_t^T \cdot (S_i - \overline{S})$. Here, the selection of t is limited by $\sum_{i=1}^{t} \lambda_i \Big/ \sum_{i=1}^{j} \lambda_i > f_v \sum_{i=1}^{j} \lambda_i$ ($f_v = 90$ %), and b_t is limited in the range of $\pm 3\sqrt{\lambda_t}$.

Besides of above shape statistical features, the grey gradient features are also computed. For each landmark P_{in}, we extend l ($l = 3$) points on its both sides along the line perpendicular to the manual contour. And, the normalized grey gradient of these points is calculated and recorded as $g_i = [g_1, g_2, \ldots, g_l]^T$. Using all training shapes, we obtain a matrix $g = [g_1, g_2, \ldots g_N]^T$ whose mean and covariance (named $\bar{g}$ and C_g) are computed to represent the final gradient features.

(2) Object point searching

In this step, initial contour deforms iteratively according the afore-obtained statistical features (both the shape and the gradient). The point of object contour is finally determined by minimizing the Mahanobolis distance defined in Eq. (5).

$$f(\mathbf{g}) = (\mathbf{g} - \bar{g})^T C_{\mathbf{g}}^{-1} (\mathbf{g} - \bar{g}) \tag{5}$$

3 LV Myocardium Segmentation

3.1 Rough Segmentation Using GMM

Before segmentation, a region of interest (ROI) is defined. It is a square frame whose size and location are respectively 120 × 120 and (128, 128) on the original image (shown in Fig. 1(a, b)). This definition was tested on 18 cases of patients including 3327 images in all, and the detection rate was 100 %.

We segment LV from middle slice to two ends (base and apex). Given an ROI in middle slice at end diastole (ED) phase, we firstly conduct coarse segmentation as follows. Firstly, ROI is classified into three classes based on GMM (shown in Fig. 1c). Subsequently, a circle-binary mask located in ROI center is used to determine LV cavity: the area of the overlap between circle mask and classified ROI is measured, and only the region with maximum overlap area is retained to be LV cavity (shown in Fig. 1e). And then, the convex hull of LV cavity is calculated and smoothed to be the coarse segmentation result (shown in Fig. 1f). We record the coarse segmentation result as S_1 and use it to initialize ASM. For the end systole (ES) phase of middle slice, we use the detected LV cavity of ED to be the binary mask.

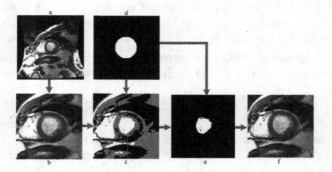

Fig. 1. Coarse segmentation: (a) Original image; (b) ROI; (c) GMM classification Result; (d) circle mask; (e) Determined LV cavity; (f) Coarse segmentation Result S_1 (green) (Color figure online)

Table 1. Information for 2D-shape modelling

2D ASM	Definition	No. images (N)	No. landmarks (k)
$S_{b\_endo\_ed}$	Basal slice endocardial model at ED phase	13	45
$S_{b\_endo\_es}$	Basal slice endocardial model at ES phase	13	45
$S_{m\_endo\_ed}$	Mid-slice endocardial model at ED phase	45	45

(*Continued*)

Table 1. (*Continued*)

2D ASM	Definition	No. images (N)	No. landmarks (k)
$S_{m\_endo\_es}$	Mid-slice endocardial model at ES phase	45	45
$S_{a\_endo\_ed}$	Apical slice endocardial model at ED phase	18	25
$S_{a\_endo\_es}$	Apical slice endocardial model at ES phase	18	25
$S_{b\_epi\_ed}$	Basal slice epicardial model at ED phase	13	45
$S_{m\_epi\_ed}$	Mid-slice epicardial model at ED phase	45	45
$S_{a\_epi\_ed}$	Apical slice epicardial model at ED phase	18	25

3.2 Myocardium Extraction Using ASM

(1) 2D Shape modeling

In this step, we calculate the shape model based on ASM. All training and testing data are from the database published by MICCAI 2009 [21]. The HF-I and HF-NI cases from training set are used to generate nine 2D-shape models whose detail information are listed in Table 1.

(2) Contour initialization and myocardium segmentation

Curve is limited by the trained mean shape, so S_1 cannot be directly used as initial contour. Thus, we firstly transform mean shape $\overline{S_{m\_endo\_ed}}$ according to S_1.

a. Distortion

The distortion makes $\overline{S_{m\_endo\_ed}}$ has a similar size with S_1. We calculate the height and the width of S_1 and $\overline{S_{m\_endo\_ed}}$ respectively, and record them as (H_1, W_1) and (H_m, W_m). Thus, we can obtain a new contour S_{new} by Eqs. (6)–(8).

$$\begin{cases} xnew = (W1/Wm) \cdot xm \\ ynew = (H1/Hm) \cdot ym \end{cases} \tag{6}$$

$$\begin{cases} H1 = \max(ys) - \min(ys) \\ W1 = \max(xs) - \min(xs) \end{cases} \tag{7}$$

$$\begin{cases} Hm = \max(ym) - \min(ym) \\ Wm = \max(xm) - \min(xm) \end{cases} \tag{8}$$

where (x_s, y_s), (x_m, y_m) and (x_{new}, y_{new}) respectively represent the point coordinates of S_1, $\overline{S_{m\_endo\_ed}}$ and S_{new}.

b. Shift

This step is to make S_{new} have the same position with S_I. The center positions of S_I and S_{new} are calculated and recorded as C_I $(C_{x1},\ C_{y1})$ and C_{new} $(C_{xnew},\ C_{ynew})$, respectively. And, the final initial contour S_{ini} can be obtained by Eqs. (9)–(11).

$$\begin{cases} xini = xnew - (Cxnew - Cx1) \\ yini = ynew - (Cynew - Cy1) \end{cases} \tag{9}$$

$$\begin{cases} Cx1 = 0.5 * [\max(xs) - \min(xs)] + \min(xs) \\ Cy1 = 0.5 * [\max(ys) - \min(ys)] + \min(ys) \end{cases} \tag{10}$$

$$\begin{cases} Cxnew = 0.5 * [\max(xnew) - \min(xnew)] + \min(xnew) \\ Cynew = 0.5 * [\max(ynew) - \min(ynew)] + \min(ynew) \end{cases} \tag{11}$$

where $(x_{ini},\ y_{ini})$ is the point coordinates of S_{ini}. Figure 2a shows the endocardium initialization.

After obtaining S_{ini}, endocardium is searched near it. The searching range is set to be fourteen points (including the landmark point), in which four points are opposite to the normal direction $(L_1 = 4)$, nine points are following with the normal direction $(L_2 = 9)$ (Fig. 2b). All landmark points (P_{in}) are adjusted iteratively based on ASM. And, convergence condition is majority of points (about 95 %) do not change their locations significantly or the algorithm attains maximum iteration (it is ten in our study). Finally, we smooth the final contour to obtain endocardium (Fig. 2c).

We segment epicardium based on the same principle mentioned above. Notice that, there are three different points: (1) for initialization, the epicardial coarse segmentation S_2 (obtained by expanding the segmented endocardium along its normal direction) is used; (2) for searching, $L_1 = 9$ and $L_2 = 4$; (3) ROI is modified by setting the pixel intensity in the segmented LV cavity to be one.

For the segmentation of other slices/phases, the segmentation of previous slice is used for initialization, and the searching range is $L_1 = L_2 = 7$.

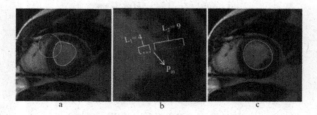

Fig. 2. Endocardium segmentation: (a) Searching range; (b) Initialization, $\overline{S_{m\_endo\_ed}}$ (blue), S_I (yellow), S_{new} (green), and S_{ini} (red); (c) Segmented endocardium (green) and ground truth (red) (Color figure online)

4   Validation

Ten cases of data from MICCAI 2009 are used to evaluate the proposed algorithm. The used quantitative metrics are APD (the average perpendicular distance), ODM (the overlapping dice metric) and PGC (the percentage of good contours) which are respectively defined as follows. APD is the distance from the automatic contour to the corresponding ground truth, averaged over all contour points [21]. Only the good contours (APD is less than 5 mm) are considered for metric calculation. PGC is calculated by:

$$PGC = \frac{N_{\text{good}}}{Nall} * 100\%$$ (12)

where N_{good} and N_{all} respectively stand the number of good contours and that of all tested contours. ODM is the region overlapping proportion between the automatic delineation and the ground truth, and it is calculated by:

$$ODM = \frac{2|A \cap B|}{|A| + |B|}$$ (13)

where $|A|$ (or $|B|$) is the pixel number in area A (or B), and $|A \cap B|$ represents the intersection of both areas. For the cardiac function evaluation, we calculate EF and LVM by:

$$EF = \frac{\left(V_{\text{endo}}^{ED} - V_{\text{endo}}^{ES}\right)}{V_{\text{endo}}^{ED}} * 100\%$$ (14)

$$LVM = 1.05 * \left(V_{epi}^{ED} - V_{\text{endo}}^{ED}\right)$$ (15)

where V_{epi}^{ED} and V_{endo}^{ED} respectively stand for the epi-/endo-cardial volume at ED phase, and V_{endo}^{ES} is the endocardial volume at ES phase.

All experiments are completed on matlab7.10.0 and the hardware system platform is Pentium(R) Dual-Core CPU E5300 @ 2.60 GHz,2G DDRII with Windows XP.

5   Results and Discussion

5.1   Qualitative Assessments

Figure 3 shows the segmentation results of SC-HF-I-05 (drawn with red) on various slices at ED/ES phase, which demonstrates the new algorithm can not only segment middle slices but also capture the myocardium of apical and basal slices. The mid-slice segmentation (from the slices 3-6) shows the great ability of our method for overcoming interference from PM. Because of LVOT, part of myocardium is missing in apical slices (slice 1 and slice 2). Even so, the proposed algorithm also can successfully delineate LV. Although the contrast between myocardium and backgrounds is very low on some slices such as slices 1–3, the proposed algorithm still segment epicardium successfully.

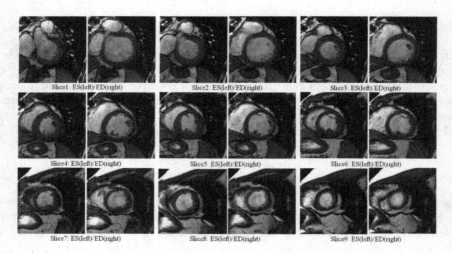

Fig. 3. Segmentation results of SC-HF-I-05 (Color figure online)

5.2   Quantitative Assessments and Comparison

Ten cases results compared with manual segmentation are listed in Table 2. Our method achieves good results with a PGC of 93.92 %, an APD of 2.20 mm and an ODM of 0.91. Epicardium segmentation performs better than endocardium. Due to some poor segmentation in apical slices, the endocardium result of SC-HF-I-07 and SC-HF-NI-31 are slightly worse than others. Meanwhile, Table 2 shows the EF and LVM comparison between our method and expert segmentation: the average absolute errors between these two methods are 2.93 % and 16.08 g respectively. It demonstrates that this automatic segmentation should be workable in clinical practice.

We compare our method with three state-of-art methods [1, 3, 22] based on PGC in Table 3. For the overall performance, the proposed method is better than Ref. [1] and

Table 2. Evaluation results on ten cases from MICCAI database

Cases	PGC (%)		APD (mm)		ODM		EF (%)		LVM (g)	
	Endo	Epi	Endo	Epi	Endo	Epi	Auto	Expert	Auto	Expert
SC-HF-I-05	100	100	1.89	2.14	0.92	0.94	35.63	33.03	130.23	115.45
SC-HF-I-06	90.91	90.91	2.13	2.14	0.91	0.94	26.48	25.78	152.32	147.34
SC-HF-I-07	81.25	100	2.97	1.91	0.85	0.94	25.21	28.18	132.58	114.12
SC-HF-NI-07	95.83	100	2.07	1.86	0.89	0.94	14.52	12.91	139.25	130.54
SC-HF-NI-11	100	100	2.17	2.15	0.91	0.94	16.08	14.84	175.65	158.25
SC-HF-NI-31	78.95	100	2.86	2.22	0.88	0.94	42.68	35.59	147.03	127.38
SC-HF-NI-33	100	90	2.05	2.31	0.86	0.92	60.67	58.35	151.13	130.79
SC-HYP-37	84.62	85.71	1.93	2.35	0.86	0.90	73.48	71.68	110.77	125.38
SC-N-06	92.31	100	2.27	2.12	0.88	0.93	48.53	54.59	84.89	64.02
All	91.54	96.29	2.26	2.13	0.88	0.93				
Overall	93.92		2.20		0.91					

Table 3. Comparison between the proposed method and the state-of-art methods

Cases	Ref. [1]		Ref. [22]		Ref. [3]		Proposed	
	Endo	Epi	Endo	Epi	Endo	Epi	Endo	Epi
SC-HF-I-05	83.33	88.89	100	88.89	100	100	**100**	**100**
SC-HF-I-06	95.45	100	95.45	90.91	95.45	90.91	**90.91**	**90.91**
SC-HF-I-07	87.50	100	87.50	100	87.50	100	**81.25**	**100**
SC-HF-NI-07	87.50	91.67	66.67	83.33	95.83	100	**95.83**	**100**
SC-HF-NI-11	100	80	100	80	100	90	**100**	**100**
SC-HF-NI-31	63.16	70	63.16	80	100	100	**78.95**	**100**
SC-HF-NI-33	83.33	90	94.44	80	100	90	**100**	**90**
SC-HYP-37	46.15	57.14	84.62	85.71	92.31	71.43	**84.62**	**85.71**
SC-N-06	84.62	85.71	92.31	71.43	92.31	85.71	**92.31**	**100**
All	81.23	84.82	87.13	84.47	95.93	92.01	**91.54**	**96.29**
Overall	83.03		85.80		93.97		**93.92**	

Ref. [22] and comparable to Ref. [4]. Besides, our method achieves the highest PGC for epicardium segmentation (about 96.29 %). This comparison demonstrates the proposed method is very competitive.

6   Conclusion

In this paper, a fully automatic method based on GMM and ASM was developed for the LV segmentation in cardiac cine MR images. In the proposed algorithm, we respectively built the training model for basal, middle and apical slices. Both the rough segmentation based on GMM and the geometric transformations are employed to initialize ASM. Experimental results show that the new method achieves good performance for both the endocardium and the epicardium segmentation. In the future, we consider adding more effective local features in ASM to improve the precision of our method.

Acknowledgment. The authors would like to thank the reviewers for their comments that have helped improve this paper. We also thank Sunnybrook Health Sciences Centre for providing the clinical image data, ground truths and evaluation software for us. This paper is jointly supported by National Natural Science Foundation of China (No. 61175012) and Natural Science Foundation of Gansu Province (No. 1208RJZA265).

References

1. Lu, Y., Radau, P., Connelly, K., Dick, A., Wright, G.A.: Segmentation of left ventricle in cardiac cine MRI: an automatic image-driven method. In: MIDAS J-Card MR Left Vent Segmentation Chall, pp. 339–347 (2009)
2. Huang, S., Liu, J., Lee, L.C., Venkatesh, S.K., Teo, L.L.S., Au, C., et al.: An image-based comprehensive approach for automatic segmentation of left ventricle from cardiac short axis cine MR images. J. Digit. Imaging **24**, 598–608 (2011)

3. Hu, H., Liu, H., Gao, Z., Huang, L.: Hybrid segmentation of left ventricle in cardiac MRI using gaussian-mixture model and region restricted dynamic programming. Magn. Reson. Imaging **31**, 575–584 (2013)

4. Dakua, S.P.: AnnularCut: a graph-cut design for left ventricle segmentation from magnetic resonance images. IET Image Process. **8**, 1–11 (2014)

5. Wang, L., Pei, M., Codella, N.C.F., Kochar, M., Weinsaft, J.W., Li, J., et al.: Left ventricle: fully automated segmentation based on spatiotemporal continuity and myocardium information in cine cardiac magnetic resonance imaging (LV-FAST). Biomed. Res. Int. **2015**, 9 (2015). http://dx.doi.org/10.1155/2015/367583. Article ID 367583

6. Wang, Y., Jia, Y.: Segmentation of the left venctricle from MR images via snake models incorporating shape similarities. In: Proceedings - International Conference Image Process. ICIP, pp. 213–226. IEEE Press, Atlanta (2006)

7. Grosgeorge, D., Petitjean, C., Caudron, J., Fares, J., Dacher, J.N.: Automatic cardiac ventricle segmentation in MR images: a validation study. Int. J. Comput. Assist. Radiol. Surg. **6**, 573–581 (2011)

8. Kadir, K., Gao, H., Payne, A., Soraghan, J., Berry, C.: LV wall segmentation using the variational level set method (LSM) with additional shape constraint for oedema quantification. Phys. Med. Biol. **57**, 6007–6023 (2012)

9. Wu, Y., Wang, Y., Jia, Y.: Segmentation of the left ventricle in cardiac cine MRI using a shape-constrained snake model. Comput. Vis. Image Underst. **117**, 990–1003 (2013)

10. Tufvesson, J., Hedström, E., Steding-Ehrenborg, K., Carlsson, M., Arheden, H., Heiberg, E.: Validation and development of a new automatic algorithm for time-resolved segmentation of the left ventricle in magnetic resonance imaging. Biomed. Res. Int. **2015**, 1–12 (2015)

11. Kass, M., Witkin, A., Terzopoulos, D.: Snakes: active contour models. Int. J. Comput. Vis. **1**, 321–331 (1988)

12. O'Brien, S.P., Ghita, O., Whelan, P.F.: A novel model-based 3D+time left ventricular segmentation technique. IEEE Trans. Med. Imag. **30**, 461–474 (2011)

13. Inamdar, R.S., Ramdasi, D.S.: Active appearance models for segmentation of cardiac MRI data. In: International Conference on Communication Signal Processing, pp. 96–100. IEEE Press, Melmaruvathur (2013)

14. Faghih, Roohi S., Aghaeizadeh, Zoroofi R.: 4D statistical shape modeling of the left ventricle in cardiac MR images. Int. J. Comput. Assist. Radiol. Surg. **8**, 335–351 (2013)

15. Qin, X., Tian, Y., Yan, P.: Feature competition and partial sparse shape modeling for cardiac image sequences segmentation. Neurocomputing **149**, 904–913 (2015)

16. Petitjean, C., Dacher, J.N.: A review of segmentation methods in short axis cardiac MR images. Med. Image Anal. **15**, 169–184 (2011)

17. Redner, R.A., Walker, H.F.: Mixture densities, maximum likelihood and the EM algorithm. SIAM Rev. **26**(2), 195–239 (1984). http://www.jstor.org/stable/2030064

18. Farnoosh, R., Zarpak, B.: Image restoration with Gaussian mixture models. Wseas Trans. Math. **4** (2004)

19. Goodall, Colin: Procrustes methods in the statistical analysis of shape. J. R. Stat. Soc. B **53**, 285–339 (1991)

20. Turk, M., Pentland, A.: Eigenfaces for recognition. J. Cogn. Neurosci. **3**, 71–86 (1991)

21. Radau, P., Lu, Y., Connelly, K., Paul, G., Dick, A.J., Wright, G.A.: Evaluation framework for algorithms segmenting short axis cardiac MRI. MIDAS J.-Card MR Left Vent Segmentation Chall (2009). http://hdl.handle.net/10380/3070

22. Cousty, J., Najman, L., Couprie, M., Clément-Guinaudeau, S., Goissen, T., Garot, J.: Segmentation of 4D cardiac MRI: automated method based on spatio-temporal watershed cuts. Image Vis. Comput. **28**(8), 1229–1243 (2010)

Position Gradient and Plane Consistency Based Feature Extraction

Sujan Chowdhury[✉], Brijesh Verma, and Ligang Zhang

Centre for Intelligent Systems, Central Queensland University,
Brisbane, Australia
{s.chowdhury2, b.verma, l.zhang}@cqu.edu.au

Abstract. Labeling scene objects is an essential task for many computer vision applications. However, differentiating scene objects with visual similarity is a very challenging task. To overcome this challenge, this paper proposes a position gradient and plane consistency based feature which is designed to distinguish visually similar objects and improve the overall labeling accuracy. Using the proposed feature we can differentiate objects with the same histogram of the gradient as well as we can differentiate horizontal and vertical objects. Integrating the proposed feature with low-level texture features and a neural network classifier, we achieve a superior performance (82 %) compared to state-of-the-art scene labeling methods on the Stanford background dataset.

Keywords: Scene labeling · Feature extraction · Neural networks · Position gradient · Plane consistency

1 Introduction

Scene labeling is a complex problem. The primary objective of scene labeling is to detect objects from the scene and recognize the corresponding class for each object, i.e. labeling each pixel to one object [1, 2]. There are wide varieties of applications for scene labeling. However, the task is not very easy as there are a number of problems. Based on previous research, we found two key problems affecting the overall performance in object identification. First one is differentiating visually similar objects and the second one is differentiating vertical and horizontal objects. Because of those key problems, it is very difficult to design appropriate features.

To overcome these problems, many researchers proposed different approaches including a number of feature extraction techniques. But no approach can fulfill the requirements as there are still two key factors affecting the overall performance [3]. Appropriate feature representation [4, 5] of objects is one of the first and most important challenges that needs to be solved. This can help to effectively differentiate visually similar objects. Some common objects that we found in real-world cause the problem e.g. distinguishing grass from tree, road from water and building from other objects. In order to find out a proper representation of features we see the emergence of many feature extraction techniques such as SIFT [6], HOG [7], GIST [8] proposed by different researchers. While creating the new features, researchers focused on domain specific problems. That's why although those techniques show promising performance

© Springer International Publishing AG 2016
A. Hirose et al. (Eds.): ICONIP 2016, Part II, LNCS 9948, pp. 673–681, 2016.
DOI: 10.1007/978-3-319-46672-9_75

in some applications e.g. pedestrian detection, but show bad performances in some other applications e.g. scene labeling.

With existing super pixel features, it is difficult to differentiate some portion of grass with trees, building from roads and horizontal objects with vertical objects. Hence, we propose new features to solve this issue. Our contributions are two folds. We propose a position based histogram of gradients which solves the intra-class variations and a plane consistency estimation which solves the issue of differentiating horizontal and vertical objects.

We explore the existing works for scene labeling [4] in two aspects. One aspect is proposing new representation based features while another aspect of exploring is an improvement of learning features. Graph model [9] and multiscale feature learning [10] are two most popular approaches dealing with these two aspects. The problem occurs during classification and investigates the relationship between confusing objects. The low-level visual feature is one of the important reasons for misclassification. Using such features it is difficult to distinguish "water" from "road", "grass" from "tree or forest" and "sky" from "ocean". Markov Random Field (MRF) [11, 12] and Conditional Random Field (CRF) are two approaches proposed to overcome such problem which are constructed by the combination of low-level features and global dependencies using a graph-based model. Segmentation is a prerequisite process to further proceed with those models. Hence, we use super pixel [13] based approach to avoid segmentation problem.

Recently, Deep learning and Convolution Neural Network (CNN) [14] approaches gain popularity due to their success on a wide variety of applications. Recurrent Convolution Neural Networks (RCNNs) [15] and Long Short-Term Memory (LSTM) [16] based recurrent neural network have also been proposed for further improvement on pixel label accuracy. However, the specific problems as mentioned above for objects with similar visual features have not been addressed. Therefore, in this paper, we focus on investigating new features which can address the problem of distinguishing visually similar objects. The existing features include color, texture, shape and location features [17]. While existing GIST, SIFT and HOG gain popularity and most of the recent works incorporate those high-level features. Some of the improved features like DSIFT, Histogram of visual SIFT and Dense HOG were also investigated to improve the global label accuracy. In this paper, we propose new features and incorporate them with some of the existing features such as LBP, histogram, location, entropy and SIFT to improve the scene labeling accuracy.

The remainder of the paper is organized as follows. Section 2 presents new feature extraction technique. Section 3 presents the experimental results and a comparative analysis. Finally, the concluding remarks are presented in Sect. 4.

2   Proposed Model

Figure 1 shows an overview of the proposed model. The main focus of this paper is to propose a position gradient feature and a plane consistency estimation feature as highlighted using red box in Fig. 1. We use a linear iterative clustering [18] based super pixel extraction technique rather than using pixel-based technique. This technique is computationally efficient. Here for a given image I with a width and height $M \times N$ is assigned a discrete label D. So the output image is denoted by $S \in D^{M \times N}$. Initially, raw image I is decomposed into multi-scale super pixels $S : S = f(I, \theta)$ with appropriate parameter Θ, where Θ represents the ration and number of super pixel that is used for decomposition. Choosing appropriate super pixel size is an iterative procedure and hard to define. After some analysis, proposed model choose seventy (70) super pixels for scene decomposition. From those super pixels we extract super pixel features and further build proposed features. The proposed features are used to train a neural network and after training it is used for testing on test image sets. Before training, each feature is individually normalized for a better learning. The parameters for training network were chosen after several trails.

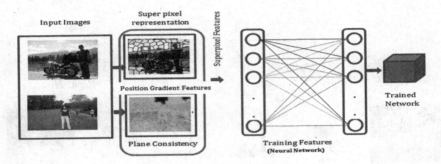

Fig. 1. Block diagram of proposed framework

In this section, we addressed the limitations which we found within existing feature extraction techniques for scene labeling. We start with Histogram of Gradient (HOG) feature which is one of the most popular methods for feature extraction and has been successfully used in a wide variety of applications. One limitation we figured out is loss of positions within pixel gradients during feature extraction. As a result, two different superpixels with different positions may have the very similar histograms. Figure 2 shows a simple example within two different superpixels with same histograms. From Fig. 2, it is clear that in both cells the position of orientation is different but the numbers of components are same. Nevertheless, they have the same number of components in the objects belonging to different class. This is the reason why the position plays a vital role in distinguishing between same types of gradient histograms. Another vital feature is similarity between two objects. Although they have the same position of HOG, they are differentiable in respect of a plane which is done by plane consistency estimation. By combining these two novel ideas we are expected to increase the overall scene labeling performance. The whole idea is described in the following two subsections.

2.1   Position Gradient Histogram (PosGH)

Suppose we have a super pixel $s1$ with $(m1 \times n1)$ pixels. Within each super pixel, we create N = 9 small sub-blocks. From N super pixels we choose B as one of the block. As we are adding additional features, the first step is choosing $\theta(x, y, i)$ on position (x, y) from the N pixels. Orientation in each direction calculated by Eqs. 1 and 2. Equation 3 is used to calculate the interval on eight orientations and update the range each time with the interval.

$$Lower\ Range = -\pi + 2 * \pi/9; \tag{1}$$

$$Upper\ Range = -\pi + 2 * pi/9 + 2 * \pi/9; \tag{2}$$

$$Interval = 2 * \pi/9; \tag{3}$$

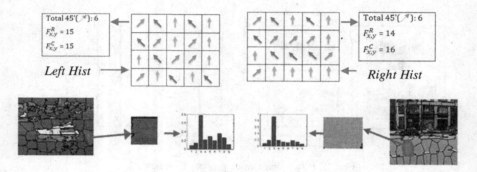

Fig. 2. Illustration of using Gradient Histogram to distinguish superpixels with similar visual appearance.

Next step is to accumulate the position information which can be calculated using the following equations.

$$T_{x,y}^r = R||\theta(x, y, i) == H(x, y, i) \tag{4}$$

$$T_{x,y}^c = C||\theta(x, y, i) == H(x, y, i) \tag{5}$$

Here $H(x, y, i)$ is the range for finding position and can be expressed as $H(x, y, i) \in \theta(x, y, i)$. Initially the value of $H(x, y, i)$ is determined from lower range and upper range. For the corresponding orientation $H(x, y, i)$, we got row and column positions information $T_{x,y}^r$ and $T_{x,y}^c$. Feature value is calculated by multiplying the row positions $P_{x,y}^r$ with number of items on each row position R_x. Finally we sum the values for the whole block and we get $G_{x,y}^R$. To calculate the column position value we do the

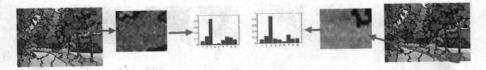

Fig. 3. Histogram equality between two different objects

same thing and get $F_{x,y}^C$. Later a position gradient feature is calculated by merging $G_{x,y}^R$ and $G_{x,y}^C$ within histogram of gradient value. Figure 2 shows a reason for adding the new value. In Fig. 2, two charts present two different representations for two patches taken from two images. If we consider in terms of gradient of histogram, it is same for both charts e.g. for 45 degree rotation the value is 6 in both cases. But if we focus on their position there is a significant change we noticed and if we calculate the value position wise using Eqs. 6 and 7, we find out the difference. From Fig. 2, for left hist the row and column values are 15 but on right hist they are 14 and 16 respectively. Hence, introducing position creates the feature vector more powerful for differentiating similar objects. Figure 3 shows another example of visual similarity and importance of adding a new feature value for better representation of feature value.

$$G_{x,y}^R = \sum_{x=1}^{R_x} \sum_{y=1}^{R_y} R_x + x * P_{x,y}^r \tag{6}$$

$$G_{x,y}^C = \sum_{x=1}^{C_x} \sum_{y=1}^{C_y} C_x + y * P_{x,y}^c \tag{7}$$

2.2 Plane Consistency Estimation (PCE)

Finding objects with respect to plane always create confusion during object classification. In the real world, some objects are found which are difficult to distinguish using existing features. After analyzing those objects we found that they are only differentiable in respect to the plane. Some objects consistent horizontally with plane whereas others are consistent with respect to vertical plane. Hence to differentiate vertical and horizontal objects we introduced a new technique named as PCE which helps to distinguish between numerous confusing objects like tree and grass, road and building. These objects have the similar visual appearance but different directions in respect to the plane.

Figure 3 shows an example of texture feature extraction and importance of introducing the new PCE technique. From the Fig. 3, it is depicted that for tree and grass texture properties are identical. They are only differentiable if we consider their consistency with respect to the plane. Grasses are horizontally consistent with respect to

(a)   (b)   (c)   (d)

Fig. 4. (a) Original Image (b) Entropy Image (c) Binary Image after Opening (d) Binary Image after Closing

the plane while trees are vertically consistent with respect to the plane. Hence, the idea for adding the orientation of object is a new finding of our research. In the real world, we also face the same problem for some other objects like road and building, road with foreground horizontal objects. Figure 4 shows the steps for PCE technique implementation.

Here an original image I showed in Fig. 4(a). After calculating entropy function on the input image I corresponding output is shown in Fig. 4(b). Equation 8 illustrates the calculation process for entropy.

$$E = \sum_{i=1}^{n} P(x_i)I(x_i) = -\sum_{i=1}^{n} P(x_i).log_2 P(x_i) \tag{8}$$

Here E denotes the entropy value for an image I and P represents the histogram counts for the image. In the next step, entropy image is converted into an intensity image which takes the range between the min entropy value and max entropy value using Eqs. 9, 10 and 11. Using linear combination of pixel width and height along with delta (Eqs. 12 and 13), we calculate the final intensity mage G.

$$x = \frac{x(j) - x(i)}{size(C, 2) - 1}; j = 2, \ i = 1 \tag{9}$$

$$y = \frac{y(j) - y(i)}{size(C, 1) - 1}; j = 2, \ i = 1 \tag{10}$$

$$limits = [min(E(:)) \ max(E(:))] \tag{11}$$

$$delta = \frac{1}{limits(2) - limits(1)} \tag{12}$$

$$G = x * delta + y * E - limit(1) * delta \tag{13}$$

$$\begin{cases} BW = 1; & if \ p > T \\ BW = 0; & if \ p < T \end{cases} \tag{14}$$

Next step is converting intensity image into binary image and find out the proper vertical and horizontal image information. Selecting the threshold value is one of the challenging issues for this technique. After going through a long iterative process, we found the threshold value. If the input value is greater than 0.9 we replace it with the value 1 (white) and replace all other pixels with the value 0 (black). Finally, we apply opening and closing operations to find out the actual vertical and horizontal regions from the target image as shown in Figs. 4(c) and (d) respectively.

3   Experimental Results and Comparisons

This section presents the training setup for our proposed model and conducts experiments on Stanford background dataset. To evaluate the performance of the proposed approach the whole dataset [4] is divided into five folds [4]. There are 715 images of outdoor scenes composed of 8 classes; sky, tree, grass, road, water, building, mountain and foreground with image size 320 × 240 pixels. During training 572 images were used while for testing 143 images were used. Table 1 shows the individual pixel-wise accuracy for each class. We construct our neural network with 70 neurons in hidden layer and an output layer with 8 neurons to predict the class labels. Based on some trails, we set 300 iterations and goal RMS error 0.001.

Table 1. Class wise pixel accuracy

Sky	Tree	Road	Grass	Water	Building	Mountain	Foreground
94.5	83.2	88.1	83.3	72.9	79.1	37.1	76.4

Fig. 5. (a) Original Image (b) Ground Truth (c) Output Image

Table 2. Comparison of proposed method with other existing methods

Method	Accuracy
Gould et al. [4]	76.4 %
Munoz et al. [8]	76.9 %
Kumar and Koller [11]	79.4 %
Socher et al. [10]	78.1 %
Lempitsky et al. [19]	81.9 %
Pinheiro and Collobert [20]	80.2 %
Byeon et al. [16]	78.56 %
Proposed method [This Paper]	82.64 %

We also compare the overall accuracy of existing methods which is shown in Table 2 and it shows our proposed approach outperforms other existing methods. Figure 5 demonstrates some results with respect to the original image and ground truth image.

4   Conclusion

In this paper, we proposed an effective feature extraction method for scene labeling. The proposed feature extraction method solved a number of problems with distinguishing visually similar objects in scene labeling. The position gradient feature solved the problem of differentiating visually similar object issues while plane consistency feature solved the problems with vertical and horizontal differentiation. The experimental results show that the proposed feature extraction effectively classifies challenging objects and outperformed the existing state-of-the art methods. Despite the successful results, some misclassification occurred due to illumination and shadow. We are currently investigating how to further improve the method to overcome this misclassification problem.

References

1. Shotton, J., Winn, J., Rother, C., Criminisi, A.: Textonboost for image understanding: multi-class object recognition and segmentation by jointly modeling texture, layout, and context. Int. J. Comput. Vis. **81**, 2–23 (2009)
2. Farabet, C., Couprie, C., Najman, L., LeCun, Y.: Learning hierarchical features for scene labeling. IEEE Trans. Pattern Anal. Mach. Intell. **35**, 1915–1929 (2013)
3. Bu, S., Han, P., Liu, Z., Han, J.: Scene parsing using inference Embedded Deep Networks. Pattern Recogn. **59**, 188–496 (2016)
4. Gould, S., Fulton, R., Koller, D.: Decomposing a scene into geometric and semantically consistent regions. In: 12th International Conference on Computer Vision, pp. 1–8 (2009)
5. Grangier, D., Bottou, L., Collobert, R.: Deep convolutional networks for scene parsing. In: ICML 2009 Deep Learning Workshop (2009)

6. Lowe, D.G.: Distinctive image features from scale-invariant keypoints. Int. J. Comput. Vis. **60**, 91–110 (2004)
7. Dalal, N., Triggs, B.: Histograms of oriented gradients for human detection. In: IEEE Computer Society Conference on Computer Vision and Pattern Recognition, pp. 886–893 (2005)
8. Munoz, D., Bagnell, J., Hebert, M.: Stacked hierarchical labeling. In: Daniilidis, K., Maragos, P., Paragios, N. (eds.) ECCV 2010, Part VI. LNCS, vol. 6316, pp. 57–70. Springer, Heidelberg (2010)
9. He, X., Zemel, R.S., Carreira-Perpiñán, M.Á.: Multiscale conditional random fields for image labeling. In: Proceedings of the 2004 IEEE Computer Society Conference on Computer Vision and Pattern Recognition, vol. 2, pp. II-695–II-702 (2004)
10. Socher, R., Huval, B., Bath, B., Manning, C.D., Ng, A.Y.: Convolutional-recursive deep learning for 3D object classification. In: Advances in Neural Information Processing Systems, pp. 665–673 (2012)
11. Kumar, M.P., Koller, D.: Efficiently selecting regions for scene understanding. In: Computer Vision and Pattern Recognition (CVPR), pp. 3217–3224 (2010)
12. Tighe, J., Lazebnik, S.: SuperParsing: scalable nonparametric image parsing with superpixels. In: Daniilidis, K., Maragos, P., Paragios, N. (eds.) ECCV 2010, Part V. LNCS, vol. 6315, pp. 352–365. Springer, Heidelberg (2010)
13. Yan, J., Yu, Y., Zhu, X., Lei, Z., Li, S.Z.: Object detection by labeling superpixels. In: IEEE Conference on Computer Vision and Pattern Recognition, pp. 5107–5116 (2015)
14. Jun, W., Chaolliang, Z., Shirong, L., Jian, W.: Outdoor scene labeling using deep convolutional neural networks. In: Control Conference (CCC), pp. 3953–3958 (2015)
15. Liang, M., Hu, X., Zhang, B.: Convolutional neural networks with intra-layer recurrent connections for scene labeling. In: Advances in Neural Information Processing Systems, pp. 937–945 (2015)
16. Byeon, W., Breuel, T.M., Raue, F., Liwicki, M.: Scene labeling with LSTM recurrent neural networks. In: Computer Vision and Pattern Recognition, pp. 3547–3555 (2015)
17. Gould, S.: Multiclass pixel labeling with non-local matching constraints. In: 2012 IEEE Conference on Computer Vision and Pattern Recognition (CVPR), pp. 2783–2790 (2012)
18. Achanta, R., Shaji, A., Smith, K., Lucchi, A., Fua, P., Susstrunk, S.: SLIC superpixels compared to state-of-the-art superpixel methods. IEEE Trans. Pattern Anal. Mach. Intell. **34**, 2274–2282 (2012)
19. Lempitsky, V., Vedaldi, A., Zisserman, A.: Pylon model for semantic segmentation. In: Advances in Neural Information Processing Systems, pp. 1485–1493 (2011)
20. Pinheiro, P.H., Collobert, R.: Recurrent convolutional neural networks for scene parsing, arXiv preprint arXiv:1306.2795 (2013)

Fusion of Multi-view Multi-exposure Images with Delaunay Triangulation

Hanyi Yu and Yue Zhou[✉]

Institute of Image Processing and Pattern Recognition, Shanghai Jiao Tong University,
Shanghai, China
yhydtc1@gmail.com, zhouyue@sjtu.edu.cn

Abstract. In this paper, we present a completely automatic method for multi-view multi-exposure image fusion. The technique adopts the normalized cross-correlation (NCC) as the measurement of the similarity of interest points. With the matched feature points, we divide images into a set of triangles by Delaunay triangulation. Then we apply affine transformation to each matched triangle pairs respectively to get the registration of multi-view images. After images aligned, we partition the image domain into uniformed regions and select the images that provides the most information with certain blocks. The selected images are fused together under monotonically blending functions.

Keywords: Multi-view · Multi-exposure · Delaunay triangulation · Image registration · Image fusion

1   Introduction

It's usual condition for many people to capture images with their camera or smart phone set to auto-exposure and without a tripod. Therefore the images we get will be multi-exposure and multi-view. Even with the help of tripod, slight shifts still happen sometimes. How to cope with such images and get a well-exposed synthetic result is a meaningful and practical problem. Recovering response functions, exposures, and depth maps with multi-view stereo algorithm is a valid means. But in practice, we very often have to cope with images whose source is unknown, which means hardly possible to get depth maps. In these cases, using image registration techniques to transform multi-view images to single view is also an effective choice.

If light irradiance across scene varies vastly, no matter how long the exposure time is used, the image captured will always have under- or over-exposed regions. It's obvious that the under- or over-exposed regions carry less information and provide less details than the well-exposed regions. One way to get well-exposed image is to use high dynamic-range reconstruction (HDR-R) techniques to reconstruct a set of low dynamic-range (LDR) image to a HDR image [1]. As for displaying on a monitor, the HDR image is compressed by tone-mapping (TM) [2] algorithm. Another way is image fusion (IF), which combine multi-exposed images into a well-exposed image directly.

In this paper we propose a method to address the problem about the registration of images taken at different viewpoints and with different exposures, and the fusion of

A. Hirose et al. (Eds.): ICONIP 2016, Part II, LNCS 9948, pp. 682–689, 2016.
DOI: 10.1007/978-3-319-46672-9_76

aligned images. The rest of this paper is organized as follows. Section 2 reviews previous methods related to multi-view multi-expose problem. Section 3 presents our methods in detail. Section 4 discusses our experimental performances. And finally Sect. 5 gives our conclusions.

2 Related Work

Multi-exposure image fusion methods aim to combine information from different images of a scene into a single image which will be well-exposed. Goshtasby [3] partitions the image domain into uniform blocks and for each block select the image containing the most information with the block. Then the selected blocks are blended together by using monotonically decreasing blending functions which centered at the blocks. The method determines the optimal block size and blending function width by using a gradient-ascent algorithm to maximize the fused image information. Shen et al. [4] propose a novel probabilistic model-based fusion technique. They aims to achieve an optimal balance between local contrast and color consistency, while combining the well-exposed details from different images. A generalized random walks framework is proposed to calculate a globally optimal solution constrained by the two quality measures.

Troccoli et al. [5] adapts multi-stereo from different exposure images to simultaneously recover dense depth and high dynamic details. They use an exposure-invariant similarity statistic to establish correspondences, through which they are able to extract the camera response function and the image exposures. Then they can convert all images into radiance space and select optimal radiance data to recover high dynamic range detail. Greg Ward [6] accelerate image operations and avoid problems with the multi-exposure images by employing percentile threshold bitmaps. They construct an image pyramid from gray exposure images and these are converted to bitmaps aligned horizontally and vertically using inexpensive shift and difference operations. Anna and Radoslaw [7] search for key points in consecutive images and use these points to find matrices to transform images to a single coordinate system and eliminate global misalignments.

3 Approach

The input of our system is a set of images captured by two or more cameras whose response function and exposure is unknown. Our goal is to align these multi-view images and fuse aligned images into a well-exposed image. To achieve the goal, we combine image registration technique with multi-exposure image fusion algorithm.

We first extract local feature points from images and match them with SIFT algorithm. Then we refine the matching result by using normalized cross-correlation to measure the similarity of matched pairs. With the refined feature points, we divide the image into a set of triangular region by Delaunay Triangulation algorithm. Instead of transform one image to the viewpoint of another directly, we transform triangular regions in one image to coordinates of their corresponding triangular regions in another image respectively. Finally we run fusion algorithm to turn the aligned images into a well-exposed image.

3.1   Feature Points Matching of Multi-exposed Images

As mentioned above, many techniques like SIFT have been used in the field of multi-exposure image registration. However in the experiments we find that in some case the same objects in image has different local gradient variance due to different exposure and difference viewpoint. As Fig. 1 shows, the SIFT feature points in two images are all matched incorrectly. Though it is an extreme example, it demonstrates the necessity to check the similarity between two matched feature points furthermore.

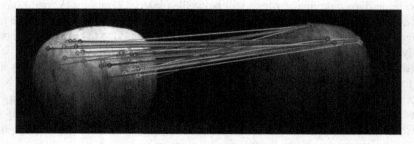

Fig. 1.   Incorrect matches of SIFT feature points in multi-exposure multi-view images

In our method, to eliminate incorrect matches, we choose normalized cross-correlation (NCC) as a means to measure the similarity between feature point pairs. There are many image matching methods based on NCC [8, 9]. And Troccoli et al. [5] note its value for multi-exposure image and verify the invariance of normalized cross-correlation under different exposure. Then we can make use of this property to calculate similarity and refine matching result.

Let $m_1 = I_1(x_1, y_1)$ be an interest point in an image and $m_2 = I_2(x_2, y_2)$ be an interest point in another image. W_1 and W_2 are two correlation windows of size $(2w + 1) \times (2w + 1)$ centered on each interest point. Then W_1 and W_2 can be described as two sets of pixel intensities A and B:

$$
\begin{aligned}
A_{uv} &= I_1(x_1 + u, y_1 + v), \\
B_{uv} &= I_2(x_2 + u, y_2 + v),
\end{aligned}
\tag{1}
$$

where $u, v \in [-w, w]$. And the similarity between the two interest points can be measured by normalized cross-correlation between A and B:

$$
S_{1,2} = \frac{\sum\limits_{u=-w}^{w} \sum\limits_{v=-w}^{w} \left(A_{uv} - \overline{A}\right)\left(B_{uv} - \overline{B}\right)}{\sigma(A)\sigma(B)},
\tag{2}
$$

where $\overline{A}$ is the average and $\sigma(A)$ is the standard deviation of all points in A, and so is B. If the two interest points are exactly the same, it's obvious that A_{uv} equals to B_{uv} and then the similarity $S_{1,2}$ has a value of 1.

3.2   Image Registration

With refined matching result, we can apply registration to multi-view images. It is a common method to calculate the projective transformation matrix with matched feature point pairs, and transform two images into same viewpoint. However the method is only effective when objects in the scene of images are approximately in the same plane. If objects have different depth, the performance of projective transformation will not be ideal, which we are going to exhibit in Sect. 4. A solution [5] to this problem is to use images and their corresponding camera matrices to calculate their depth maps. With the depth maps, a well-exposed image can be generated by view interpolation. But in many cases, images are given with their sources unknown, then it will be impossible to get corresponding camera matrices. So it is necessary to seek a new algorithm to cope with such situations.

Then we propose the idea to partition each image into a set of triangular regions according to feature points, and transform these regions respectively. In this way, image transformation is less possible to have shift because feature points are fixed. It is obvious that with a set of points there will be a lot of ways to composite triangles. Then there remains a question that how to partition image into triangular regions can make the transformation result best.

As shown in Fig. 2, for the point set with four points *abcd*, there are two different ways of triangulation. We can see in the left one the points in triangles are less dispersed, which means the scene in the triangular region is less likely to be distorted after transformation. And this kind of triangulation is called Delaunay which satisfies the empty circumcircle property: the circumcircle of a triangle in the triangulation does not contain any input points in its interior. There are many methods to construct Delaunay triangulation such as incremental insertion algorithm [10], sweep-line algorithm [11], divide and conquer algorithm [12] and so forth. In our paper, we adopt incremental algorithm to construct Delaunay triangulation. The pseudo code can be describe as follow:

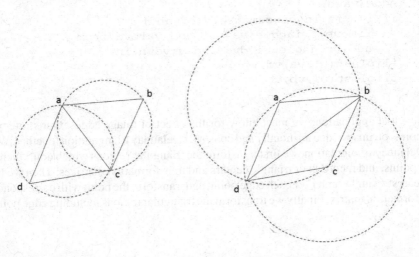

Fig. 2. Two ways to triangulate points set *abcd*. Only the left one is Delaunay triangulation.

```
Algorithm Incremental Delaunay
Input   two-dimension point set V with n points
Output  Delaunay Triangulation DT
Var     i:  1..n
        vi: the ith point in V
        va, vb, vc: certain point in V

begin.
    add an appropriate triangle bouding box whose vertices
    are a, b, c to contain V;
    initialize DT(a,b,c) as triangle abc;
    for i=1:n.
        find the triangle vavbvc which contains vi;
        add triangle vavbvi, vbvcvi, vcvavi into DT;
        flipTest(DT,va,vb,vi);
        flipTest(DT,va,vb,vi);
        flipTest(DT,va,vb,vi);
    end.
    Remove the bounding box and relative triangle which
    contains a, b or c;
end.


Algorithm flipTest(DT,va,vb,vi)
Input   DT: Delaunay Triangulation
        va, vb, vc: point in V

begin.
    find the third vertex (vd) of triangle which contains
    edge vavb;
    if (vi is in circumcircle of abd)
        remove triangle vavbvi and vavbvd from DT;
        add triangle vivdva and vivdvb into DT;
    flipTest(DT,va,vd,vi);
    flipTest(DT,vb,vd,vi);
end.
```

In actual application, we must allow for the effect of image edges. Our strategy is add points on image edge artificially and construct Delaunay triangulation together. With the Delaunay triangulation, we first transform the triangular regions which don't include edge points, and record their triangle centers and transformation matrices. Then we find the nearest triangle center for each edge point, and transform the point with corresponded transformation matrix. Finally we transform the triangular regions including edge points.

3.3 Multi-exposed Image Fusion

In our method, we use [3] as a reference and fuse well-exposed image from best exposed blocks. First the images are divided into block of size $d \times d$, where d is a parameter having influence on the fusion result. Larger d may cause loss of detail information, while smaller d increases runtime and makes the result easy to be impacted by noise.

Fig. 3. (. Top) two images from Moebius sequence. (Middle left) fusion result by method in [7]. (Middle right) fusion result by our method. (Bottom) details of above results

Then we calculate entropy for each block of each image and select the select the image which has the highest entropy in certain block to represent corresponding block in the output. If the output image is just created from the above procedures, it will be likely to have sharp discontinuity on the boundary between blocks. To eliminate the discontinuity, rather than cutting and pasting the local region to fuse a new image, we blend image with Gaussian smooth functions.

4   Experiments

We implement our method on sets of real images and evaluate its performance. We apply the method to samples from the data set download from the website: http://vision.middlebury.edu/stereo/data/scenes2005. We can see that in the top row of Fig. 3, two photographs (1390 × 1110 pixels, stored in PNG format) of the same scene are depicted. The two images have different exposure and viewpoint. The images are fused using algorithm in [7] and our algorithm respectively. The fusion results are given in the middle row of Fig. 3. And in the bottom row of Fig. 3, we depict the detail close-ups of the fusion results, where our algorithm aligns two images perfectly. While the algorithm to transform image directly generates obvious ghosting and blur.

5   Conclusion

In this paper, we give a solution to the problem that in some conditions traditional registration algorithms cannot match multi-exposed images correctly. We use normalized cross correlation as evaluation to refine the matches among images. And we adopt Delaunay triangulation algorithm to divide and align images. Experimental results demonstrate that our method are effective on images captured with different exposures and from different viewpoints.

References

1. Debevec, P.E., Malik, J.: Recovering high dynamic range radiance maps from photographs. In: ACM SIGGRAPH 2008 Classes, p. 31. ACM (2008)
2. Reinhard, E., Heidrich, W., Debevec, P., Pattanaik, S., Ward, G., Myszkowski, K.: High dynamic range imaging: acquisition, display, and image-based lighting. Morgan Kaufmann, Burlington (2010)
3. Goshtasby, A.A.: Fusion of multi-exposure images. Image Vis. Comput. 23(6), 611–618 (2005)
4. Shen, R., Cheng, I., Shi, J., Basu, A.: Generalized random walks for fusion of multi-exposure images. IEEE Trans. Image Process. 20(12), 3634–3646 (2011)
5. Troccoli, A., Kang, S.B., Seitz, S.: Multi-view multi-exposure stereo. In: Third International Symposium on 3D Data Processing, Visualization, and Transmission, pp. 861–868. IEEE (2006)
6. Ward, G.: Fast, robust image registration for compositing high dynamic range photographs from hand-held exposures. J. Graph. Tools 8(2), 17–30 (2003)

7. Tomaszewska, A., Mantiuk, R.: Image registration for multi-exposure high dynamic range image acquisition (2007)

8. Pilu, M.: A direct method for stereo correspondence based on singular value decomposition. In: 1997 IEEE Computer Society Conference on Computer Vision and Pattern Recognition, 1997. Proceedings, pp. 261–266. IEEE (1997)

9. Zhao, F., Huang, Q., Gao, W.: Image matching by normalized cross-correlation. In: 2006 IEEE International Conference on Acoustics Speech and Signal Processing Proceedings, vol. 2, pp. II–II. IEEE (2006)

10. Guibas, L., Stolfi, J.: Primitives for the manipulation of general subdivisions and the computation of voronoi. ACM Trans. Graph. (TOG) 4(2), 74–123 (1985)

11. Fortune, S.: A sweepline algorithm for voronoi diagrams. Algorithmica 2(1–4), 153–174 (1987)

12. Su, P., Drysdale, R.L.S.: A comparison of sequential delaunay triangulation algorithms. In: Proceedings of the Eleventh Annual Symposium on Computational Geometry, pp. 61–70. ACM (1995)

Detection of Human Faces
Using Neural Networks

Mozammel Chowdhury[1(✉)], Junbin Gao[2], and Rafiqul Islam[1]

[1] School of Computing and Mathematics, Charles Sturt University,
Bathurst, Australia
{mochowdhury, mislam}@csu.edu.au
[2] Discipline of Business Analytics,
The University of Sydney Business School, Sydney, Australia
junbin.gao@sydney.edu.au

Abstract. Human face detection is a key technology in machine vision applications including human recognition, access control, security surveillance and so on. This research proposes a precise scheme for human face detection using a hybrid neural network. The system is based on visual information of the face image sequences and is commenced with estimation of the skin area depending on color components. In this paper we have considered HSV and YCbCr color space to extract the visual features. These features are used to train the hybrid network consisting of a bidirectional associative memory (BAM) and a back propagation neural network (BPNN). The BAM is used for dimensional reduction and the multi-layer BPNN is used for training the facial color features. Our system provides superior performance comparable to the existing methods in terms of both accuracy and computational efficiency. The low computation time required for face detection makes it suitable to be employed in real time applications.

Keywords: Face detection · Neural networks · Access control · Security surveillance · Biometric authentication

1 Introduction

Human face and facial features are the most distinctive and widely used key components towards a person's identity. In computer vision, many researchers have been doing research on automatic detection of human faces and facial features from image sequences. The problem of estimating human faces and facial parts in images has become a popular area of research due to the emerging applications in human-computer interaction, person identification in surveillance and security systems, robot navigation, financial transactions, forensic applications, pedestrian detection, secure access control and so on [1].

Face detection and facial features extraction is concerned with identification of the human faces and facial parts in the image sequences. The problems associated with face detection and facial features estimation are attributed to variations in pose, illuminations, locations, orientation, occlusions, and backgrounds [2, 3].

© Springer International Publishing AG 2016
A. Hirose et al. (Eds.): ICONIP 2016, Part II, LNCS 9948, pp. 690–698, 2016.
DOI: 10.1007/978-3-319-46672-9_77

Numerous approaches for face detection and facial feature extraction have been reported in literature over the last few decades. Some of the techniques include: geometric modeling [4], auto-correlation [5], neural networks [6], principal component analysis [7], genetic algorithm [8], adaptive boosting (AdaBoost) [9], support vector machines (SVM) [10], and so on. Model based approaches assume that the initial location of the face is known. Color based approaches reduce the search space in face detection algorithm. The neural network-based approaches require a large number of face and non-face training examples, and are designed primarily to locate frontal faces in gray scale images. Sung and Poggio [11] have developed an example-based approach for locating frontal views of human face in complex scenes. Since this method has been developed for vertical frontal view faces, faces with other orientations cannot be detected. Rowley et al. [6] have developed a neural network-based frontal face detection system where a retinal connected neural network has been employed to justify the small windows of size 20 × 20 of an image whether it contains a face or not. Lam and Yan [12] have used snake model for detecting face boundary. Although the snake provides good results in boundary detection, but the main problem is to find the initial position. Moghaddam and Pentland [13] have employed principal component analysis for describing the face pattern with lower-dimensional feature space. Shen et al. [14] propose a method to detect faces by image retrieval. Li et al. [15] use a convolutional neural network (CNN) for face detection. However, these systems suffer from high computation cost and cannot deal better with images of complex backgrounds. Hence, there is still a lot of space for improvement.

This research aims to explore a robust and fast approach for face detection based on neural networks which integrates the detection of human faces by skin color segmentation. The system works with visual information of the human face and detects the face area using the similarity measure of the color components in the images based on HSV and YCbCr color histograms. The proposed technique can treat images with different lighting conditions and complex backgrounds.

The rest of the paper is organized as follows. Section 2 presents our proposed architecture of the face detection approach. Experimental results are reported in Sect. 3. Finally, Sect. 4 concludes the paper.

2   Proposed System Architecture

Face detection is concerned with determining which part of an image contains face. In our face detection approach, the image sequences are first pre-processed for noise elimination and smoothing. The scene is then enhanced using histogram equalization because, the captured scenes are usually of low resolution and poor contrast due to the limitation of lighting conditions. The skin like color components are extracted using HSV and YCbCr color histograms. The neural network is used to teach the extracted color components. The face skeleton is detected from the largest connected area of the skin colour segmented region. The approach considers the frontal view of the faces in colour image sequences. The overview of the proposed face detection method is shown in Fig. 1.

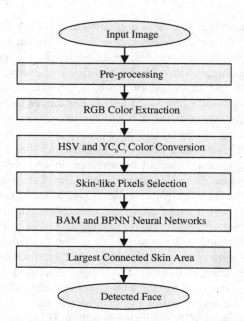

Fig. 1. Architecture of the proposed face detection technique.

2.1 Preprocessing of the Images

In real time vision systems, the captured image sequences may be contaminated by significant amount of noise during their acquisition due to the differences in camera orientation and lighting conditions. For this reason, we employ a fuzzy filtering [16] for smoothing the captured scenes. The filter is very effective for removal of impulsive noise which is usually injected during image capturing.

2.2 Image Enhancement

The captured images may be of poor contrast and low resolution because of the limitations of the lighting conditions. Therefore, histogram equalization is used to compensate for the lighting conditions and improve the contrast of the image, as shown in Fig. 2. We employ the histogram equalization technique proposed by Cheng and Shi [17] for enhancement purpose.

2.3 Color Segmentation

Face detection is achieved by means of skin color segmentation. This section presents a color segmentation approach for determining the face region in a scene. The input scene is typically a RGB color image. However, The RGB color space does not consider the luminance effect on skin color which may provide erroneous results. It has weakness in representing shading effects or rapid illumination changing and hence face detection

(a) Original image   (b) Histogram equalized image

Fig. 2. Histogram equalization of a face image.

may fail if the lighting condition changes from image to image [18]. Therefore, we use a combination of HSV and YCbCr model for color segmentation in face detection.

(a) HSV Color Model: In the HSV color model a color is described by three attributes Hue, Saturation and Value. Hue is the attribute of visual sensation that corresponds to color perception associated with the dominant colors, saturation implies the relative purity of the color content and value measures the brightness of a color. The HSV space classifies similar colors under similar hue orientations. The image content is converted from RGB to HSV color space using the following equations:

$$H = \cos^{-1}\left\{ \frac{\frac{1}{2}[(R-G)+(R-B)]}{\sqrt{(R-G)^2+(R-B)(G-B)}} \right\} \tag{1}$$

Ranging $[0, 2\pi]$, where $H = H_1$ if $B \leq G$; otherwise $H = 360^\circ - H_1$;

$$S = \frac{\max(R,G,B) - \min(R,G,B)}{\max(R,G,B)} \tag{2}$$

$$V = \frac{\max(R,G,B)}{255} \tag{3}$$

Where R, G, B are the red, green and blue component values which exist in the range [0,255]. From iterative experimental results, it is found that we can segment skin color from a color scene considering the thresholds with hue value: $0^0 \leq H_{\text{face}} \leq 50^0$, and saturation value: $0.20 \leq S_{\text{face}} \leq 0.68$ (Fig. 3).

 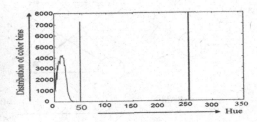

Fig. 3. Hue histogram of a typical color image.

(b) YC$_b$C$_r$ Color Model: YC$_b$C$_r$ color model segments the image into luminance and chrominance components. Luminance is the amount of light. It is encoded in the Y and the blueness and the redness are encoded in C$_b$C$_r$. The skin portion of an image is extracted using chrominance values. The conversion from RGB to YC$_b$C$_r$ color space can be done by,

$$\left.\begin{array}{l} Y = 0.257 \times R + 0.504 \times G + 0.098 \times B + 16 \\ C_b = 0.148 \times R - 0.291 \times G + 0.439 \times B + 128 \\ C_r = 0.439 \times R - 0.368 \times G - 0.071 \times B + 128 \end{array}\right\} \tag{4}$$

The threshold value for the skin color is empirically found as,

$$(140 < C_r < 160) \text{ and } (105 < C_b < 140) \tag{5}$$

2.4   Neural Networks for Detecting Face Area

After colour segmentation, the extracted H, S, Y, C_r and C_b components are fed into the neural networks for training purposes. Face recognition is achieved by employing a hybrid neural network, consisting of two networks, (i) Bidirectional Associative Memory (BAM) for dimensional reduction of the feature matrix to make the recognition faster and more efficient, and (ii) Multilayer Perceptron with backpropagation algorithm for training the network. The architecture of the hybrid neural network is illustrated in Fig. 4. The first layer receives the skin-like color features. The number of nodes in this layer is, equal to the dimension of the feature vector incorporating the color features. The number of nodes in the output layer equals to the number of individual faces the network is required to recognize. The number of epochs for this

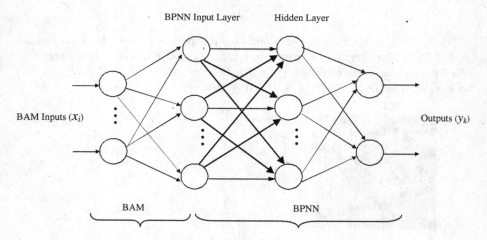

Fig. 4. Hybrid Neural network for detecting face area.

experiment was 35,000 and the minimum error margin was 0.001. The detected face skeletons are cropped and registered with a dimension of 160 × 140 pixels.

3 Experimental Evaluation

The effectiveness and robustness of this approach is justified using real images captured by a camera with different positions, expressions and lighting conditions. In order to evaluate our proposed method, we have also used two standard image datasets [19, 20] with different poses and different illumination conditions. Experiments are carried out on a computer with 2.2 GHz Intel Core i5 processor and 4 GB RAM. The algorithm has been implemented using Visual C++.

The face images are analyzed to demonstrate the feasibility of the proposed detection method. When a complex image is subjected in the input, the face detection result highlights the facial part of the image. Figure 5 shows the face detection results with our proposed method for real image sequences. Images of different persons are taken at different environments both in shiny and gloomy weather. To evaluate our proposed method we consider images with different expressions, pose, orientation, structural components and illumination. The system can also cope with the problem of partial occlusion.

Our system demonstrates better performance in case of the frontal face images in simple background while provides worst results for the images in complex background. We perform experiments to compare our proposed algorithm with other face detection methods which are reported in Fig. 6 and Table 1. From the experimental evaluation,

(a) Original image   (b)Detected face image   (a) Original image   (b)Detected face image

Fig. 5. Face detection process: (a) the original image, and (b) detected face image.

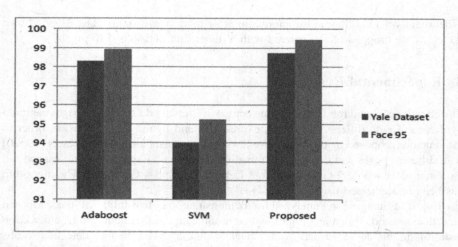

Fig. 6. Detection accuracy (%) for different face detection methods with standard datasets.

Table 1. Performance of face detection methods for real image sequences in terms of accuracy

Type of face image	No. of images	Detection accuracy (%)			
		RGB color based	YIQ color based	HSV color based	Proposed method
Frontal	50	82.6	92.25	93.31	98.74
Tilted	50	81.2	90	91.45	97.21
Partial occluded	20	80.5	88.4	89.78	93.61
Complex background	30	76.75	82.5	83.13	87.86

Table 2. Computation efficiency for different methods

Detection method	Computation time (second)
RGB color based	1.23
YIQ color based	0.87
HSV color based	0.89
Convolutional NN	0.77
Proposed method	0.56

we find that our method provides almost similar or better result (for Face 95 dataset) comparable to a state-of-the-art method [15] that uses convolutional neural network. However, our algorithm achieves more computation efficiency compared to that method which is reported in Table 2. Experimental outcomes confirm the robustness of our proposed face detection algorithm comparable to others methods.

4   Conclusion

This paper presents a neural network based face detection scheme with skin-color features extracted from HSV and YCbCr color histograms. Our system can cope with pose variations and illumination changes. The effectiveness of the proposed algorithm has been justified using real image sequences and standard datasets with complex and simple backgrounds. Experimental evaluation confirms that our proposed method achieves superior performance in terms of both computation accuracy and efficiency comparable to other existing methods. Our next approach is to extend the algorithm for multi-face detection.

References

1. Darrell, T., Tollmar, K., Bentley, F., Checka, N., Morency, L.-P., Rahimi, A., Oh, A.: Face-responsive interfaces: from direct manipulation to perceptive presence. In: International Conference of Ubiquitous Computing (2002)
2. Li, H., Lin, Z., Shen, X., Brandt, J., Hua, G.: A convolutional neural network cascade for face detection. In: CVPR 2015, pp. 5325–5334
3. Yang, M.-H., Kriegman, D.J., Ahuja, N.: Detecting faces in images: a survey. IEEE Trans. PAMI **24**(1), 34–58 (2002)
4. Shinn-Ying, H., Hui-Ling, H.: Facial modeling from an uncalibrated face image using a coarse-to-fine genetic algorithm. Pattern Recogn. **34**(8), 1015–1031 (2001)
5. Goudail, F., Francois, E., Lange, T., Iwamoto, K.Kazuo, Otsu, N.: Face recognition system using local autocorrelations and multiscale integration. IEEE Trans. Pattern Anal. Mach. Intell. **18**(10), 1024–1028 (1996)
6. Rowley, H., Baluja, S., Kanade, T.: Neural network-based face detection. IEEE Trans. Pattern Anal. Mach. Intell. **20**(1), 23–37 (1998)
7. Heisele, B., Serre, T., Poggio, T.: A component-based framework for face detection and identification. Int. J. Comput. Vis. **74**(2), 167–181 (2007)
8. Uddin, J., Mondal, A.M., Hoque Chowdhury, M.M., Bhuiyan, M.A.-A.: Face detection using genetic algorithm. In: Proceedings of 6th International Conference on Computer and Information Technology, Dhaka, Bangladesh, pp. 41–46, December 2003
9. Talele, K.T., Kadam, S., Tikare, A.: Efficient face detection using adaboost. In: IJCA Proceedings of International Conference in Computational Intelligence (2012)
10. Kukenys, I., McCane, B.: Support vector machines for human face detection. In: Proceedings of New Zealand Computer Science Research Student Conference (2008)
11. Sung, K., Pogio, T.: Example-based learning for view-based human face detection. IEEE Trans. Pattern Anal. Mach. Intell. **20**(1), 39–50 (1998)
12. Lam, K., Hong, Y.: Locating and extracting the eye in human face images. Pattern Recogn. **29**(5), 771–779 (1996)
13. Moghaddam, B., Pentland, A.: Face recognition using view-based modular eigenspaces. In: Proceedings of Automatic Systems for the Identification and Inspection of Humans, SPIE, vol. 2277, July 1994
14. Shen, X., Lin, Z., Brandt, J., Wu, Y.: Detecting and aligning faces by image retrieval. In: Proceedings of IEEE Conference on Computer Vision and Pattern Recognition (2013)
15. Li, H., Lin, Z., Shen, X., Brandt, J., Hua, G.: A convolutional neural network cascade for face detection. In: IEEE Conference on Computer Vision and Pattern Recognition (2015)

16. Chowdhury, M., Gao, M., Islam, R.: Fuzzy logic based filtering for image de-noising. In: IEEE International Conference on Fuzzy Systems (FUZZ-IEEE 2016), Vancouver, Canada, pp. 24–29, July 2016
17. Cheng, H.D., Shi, X.J.: A simple and effective histogram equalization approach to image enhancement. Digit. Signal Process. **14**, 158–170 (2004)
18. Hjelmas, E., Low, B.K.: Face detection: a survey. Comput. Vis. Image Underst. **83**, 236–274 (2001)
19. Extended Yale Face Database B. http://vision.ucsd.edu/~iskwak/ExtYaleDatabase/ExtYaleB.html
20. http://cswww.essex.ac.uk/mv/allfaces/faces95.html

Compound PDE-Based Image Restoration Algorithm Using Second-Order and Fourth-Order Diffusions

Tudor Barbu[⊠]

Institute of Computer Science of the Romanian Academy, Iaşi, Romania
tudor.barbu@iit.academiaromana-is.ro

Abstract. A hybrid nonlinear diffusion-based image restoration technique is proposed in this article. The novel compound PDE denoising model combines nonlinear second-order and fourth-order diffusions to achieve a more effective image enhancement. The weak solution of the combined PDE, representing the restored digital image, is determined by developing a robust explicit numerical approximation scheme using the finite-difference method. The performed denoising tests and method comparison are also described in this paper.

Keywords: Hybrid image restoration · Nonlinear PDE model · Second-order diffusion · Fourth-order diffusion · Finite-difference method · Numerical approximation scheme

1 Introduction

Constructing an image restoration technique that removes successfully the Gaussian noise while preserving the essential features and avoiding the unintended effects, represents still a very challenging image processing task. The conventional smoothing methods, such as Average, Gaussian or Median filters, reduce the amount of noise but also generate the undesired blurring effect that destroy the edges and other image details [1]. For this reason, more performant denoising approaches, like those based on Wavelets [2] and those based on nonlinear partial differential equations (PDE) [3], have been developed in the last decades.

Numerous nonlinear second-order diffusion-based techniques, in the PDE or variational form, have been developed since Perona and Malik introduced their anisotropic diffusion scheme [4]. Another influential filtering model is the TV Denoising, elaborated by Rudin et al. [5], many PDE variational approaches being derived from it.

Although these second-order PDE denoising models overcome successfully the image blurring and preserve the features, by performing the filtering along but not across the boundaries, they may also generate the staircase effect. For this reason, some improved second-order nonlinear PDE and variational techniques that reduce the image staircasing, like Adaptive TV denoising [6], TV-L1 model [7] and TV minimization with Split Bregman [8], have been proposed. We have also developed some robust second-order PDE-based restoration methods that remove the Gausian noise and alleviate considerably the staircase effect [9, 10].

© Springer International Publishing AG 2016
A. Hirose et al. (Eds.): ICONIP 2016, Part II, LNCS 9948, pp. 699–706, 2016.
DOI: 10.1007/978-3-319-46672-9_78

However, the nonlinear fourth-order PDE-based denoising models represent a much better staircasing removing solution than second-order diffusion schemes. Some influential fourth-order PDE restoration schemes are the isotropic diffusion algorithm proposed by You and Kaveh in 2000 [11] and the fourth order PDE-based denoising model introduced by Lasaker et al. in 2003 [12]. We also proposed some effective fourth-order PDE-based approaches for image restoration in our past works [13, 14].

While these nonlinear fourth-order diffusion models overcome successfully the staircase effect and produce more natural restored images, they perform a weaker deblurring and may generate speckle noise. Therefore, the hybrid PDE-based restoration schemes have been introduced as a solution to this problem. Some compound denoising models combine nonlinear second-order and fourth-order diffusions [15], in order to enjoy the restoration benefits of both categories. Other hybrid denoising solutions combine PDEs and Wavelets [16].

The nonlinear PDE-based hybrid restoration scheme proposed in this paper combine second and fourth order diffusions to achieve effective smoothing results and overcome all the undesired effects. This novel compound diffusion-based model is described in the next section.

A consistent numerical approximation algorithm, developed for the proposed PDE-based model by applying the finite-difference method, is described in the third section. The explicit iterative discretization scheme converges fast to the solution of the combined PDE, representing the enhanced image.

Our image restoration experiments and method comparison are discussed in the fourth section. This article finalizes with a conclusions section, acknowledgements and a reference list.

2   Nonlinear Hybrid Diffusion-Based Restoration Model

A novel compound differential model for image restoration is proposed in this section. It is composed of a nonlinear PDE combining second-order and fourth-order diffusions, and a set of boundary conditions. Thus, the developed denoising model is expressed as following:

$$\begin{cases} \frac{\partial u}{\partial t} = div(\varphi_u(|\nabla u|)\nabla u) - \Delta(\psi_u(\|\nabla u\|)\nabla^2 u) - \delta(u - u_0) \\ \quad u(0, x, y) = u_0 \\ \quad u(t, x, y) = 0, \text{on } \partial\Omega, \forall t \geq 0 \end{cases} \tag{1}$$

where the domain $\Omega \subseteq (0, \infty) \times R^2$, $\partial\Omega$ represents its frontier, $\Delta = \nabla^2$, $\delta \in (0, 0.5]$, and u_0 is the initial image, affected by the Gaussian noise.

We consider the following diffusivity functions for this fourth-order PDE model: $\phi_u, \psi_u : [0, \infty) \rightarrow (0, \infty)$, given as

$$\begin{cases} \varphi_u(s) = \lambda \left(\dfrac{\xi(u)}{\gamma |\log_{10}(s^k + \xi(u)^r| + \beta} \right)^{1/3} \\ \psi_u(s) = \dfrac{\alpha}{\left(\frac{s}{\eta(u)} \right)^k + \eta(u) \left| \ln \left(\frac{s}{\eta(u)} \right)^r \right|} \end{cases} \tag{2}$$

where $\lambda, \gamma \in (0,1]$, $\beta, k, r \in (0,5)$, $\alpha \in (2,7)$ and the conductance parameters are constructed as functions of the current state of the image, using some statistics of it:

$$\begin{cases} \xi(u) = |v \cdot \mu(\|\nabla u\|)| + \rho \\ \eta(u) = |v \cdot median(\|\Delta u\|) - \varepsilon| \end{cases} \tag{3}$$

where $v, \rho, \varepsilon \in (0,1)$, μ returns the average value and *median* returns the median value.

The diffusivity functions provided by (2) are properly modeled for an effective image restoration. They satisfy the conditions required for a good edge-stopping function [3]. Obviously, they are always positive: $\phi_u(s), \psi_u(s) > 0, \forall s \in [0, \infty)$. Also, both of them are monotonically decreasing, since we have: $\forall s_1 \leq s_2$, $\varphi_u(s_1) \geq \varphi_u(s_2)$ and $\psi_u(s_1) \geq \psi_u(s_2)$. These functions also converge to zero: $\lim_{s \to \infty} \varphi_u(s) = \lim_{s \to \infty} \psi_u(s) = 0$.

The proposed denoising model enjoy the restoration advantages of both second-order and fourth-order PDEs. Thus, the first smoothing term in (1), representing a second-order anisotropic diffusion, overcomes the image blurring and preserves the details. The second smoothing term performs the fourth-order diffusion, overcoming the staircase effect and providing a natural denoised image. So, our hybrid PDE-based scheme achieves an effective restoration and removes the unintended effects, because of the diffusion combination.

The combined PDE model given by (1)–(3) admits a weak solution that would represent the restored image, u. This solution of the PDE is determined by constructing a robust numerical approximation scheme. That discretization of the nonlinear diffusion model is described in the next section.

3   Consistent Finite-Difference Based Numerical Approximation

We develop a consistent numerical approximation scheme for the proposed PDE-based model. The proposed discretization algorithm is based on the finite-difference method [17].

Thus, let us consider a space grid size of h and a time step Δt. The space and time coordinates will be quantized as:

$$x = ih, y = jh, t = n \Delta t, \forall i \in \{0, 1, \dots, I\}, j \in \{0, 1, \dots, J\}, n \in \{0, 1, \dots, N\} \tag{4}$$

The partial differential equation from (1) can be re-written in the following form:

$$\frac{\partial u}{\partial t} - \varphi_u(\|\nabla u\|)\Delta u - \nabla(\varphi_u(\|\nabla u\|)) \cdot \nabla u + \Delta(\psi_u(\|\nabla u\|)\nabla^2 u) + \delta(u - u_0) = 0 \quad (5)$$

Each of these components of the above equation is then discretized by using finite differences [17]. The first approximation, corresponding to $\frac{\partial u}{\partial t} + \delta(u - u_0)$, is computed as:

$$A_1 = \frac{u_{i,j}^{n+\Delta t} - u_{i,j}^n}{\Delta t} + \delta\left(u_{i,j}^n - u_{i,j}^0\right) = \frac{u_{i,j}^{n+\Delta t} + (\delta\Delta t - 1)u_{i,j}^n - \delta\Delta t u_{i,j}^0}{\Delta t} \quad (6)$$

The second approximation, corresponding to $\varphi_u(\|\nabla u\|)\Delta u$, is performed by using the discrete Laplacian:

$$A_2 = \varphi_u\left(\sqrt{\frac{\left(u_{i+h,j}^n - u_{i-h,j}^n\right)^2}{4h^2} + \frac{\left(u_{i,j+h}^n - u_{i,j-h}^n\right)^2}{4h^2}}\right)\frac{u_{i+h,j}^n + u_{i-h,j}^n + u_{i,j+h}^n + u_{i,j-h}^n - 4u_{i,j}^n}{h^2}$$

$$= \varphi_u\left(\frac{\sqrt{\left(u_{i+h,j}^n - u_{i-h,j}^n\right)^2 + \left(u_{i,j+h}^n - u_{i,j-h}^n\right)^2}}{2h}\right)\frac{u_{i+h,j}^n + u_{i-h,j}^n + u_{i,j+h}^n + u_{i,j-h}^n - 4u_{i,j}^n}{h^2} \quad (7)$$

The component $\nabla(\varphi_u(\|\nabla u\|)) \cdot \nabla u$ from (5) could be re-written as following:

$$\nabla(\varphi_u(\|\nabla u\|)) \cdot \nabla u = \left(\frac{\partial}{\partial x}\varphi_u\left(\sqrt{\left(\frac{\partial u}{\partial x}\right)^2 + \left(\frac{\partial u}{\partial y}\right)^2}\right), \frac{\partial}{\partial y}\varphi_u\left(\sqrt{\left(\frac{\partial u}{\partial x}\right)^2 + \left(\frac{\partial u}{\partial y}\right)^2}\right)\right) \cdot \left(\frac{\partial u}{\partial x}, \frac{\partial u}{\partial y}\right)$$

$$= \frac{\partial\varphi_u}{\partial s}(\|\nabla u\|)\frac{\left(\frac{\partial u}{\partial x}\right)^2\frac{\partial^2 u}{\partial x^2} + \frac{\partial u}{\partial x}\frac{\partial u}{\partial y}\frac{\partial^2 u}{\partial x\partial y} + \left(\frac{\partial u}{\partial y}\right)^2\frac{\partial^2 u}{\partial y^2} + \frac{\partial u}{\partial x}\frac{\partial u}{\partial y}\frac{\partial^2 u}{\partial x\partial y}}{\sqrt{\left(\frac{\partial u}{\partial x}\right)^2 + \left(\frac{\partial u}{\partial y}\right)^2}} \quad (8)$$

One may perform more approximations on (8), and obtain the following relation:

$$\nabla(\varphi_u(\|\nabla u\|)) \cdot \nabla u \approx \frac{\partial\varphi_u}{\partial s}(\|\nabla u\|)\frac{\frac{\partial^2 u}{\partial x\partial y}\left(\frac{\partial u}{\partial x} + \frac{\partial u}{\partial y}\right)^2}{\sqrt{\left(\frac{\partial u}{\partial x}\right)^2 + \left(\frac{\partial u}{\partial y}\right)^2}} \approx \varphi_u'\left(\sqrt{u_x^2 + u_y^2}\right)u_{xy}(u_x + u_y) \quad (9)$$

where $u_x = \partial u / \partial x$. By applying the finite-difference method [17], we get the following approximation for (9):

$$A_3 = \phi'_u \left(\frac{\sqrt{\left(u^n_{i+h,j} - u^n_{i-h,j}\right)^2 + \left(u^n_{i,j+h} - u^n_{i,j-h}\right)^2}}{2h} \right) \frac{\left(u^n_{i+h,j+h} - u^n_{i+h,j-h} - u^n_{i-h,j+h} + u^n_{i-h,j-h}\right)\left(u^n_{i+h,j} - u^n_{i-h,j} + u^n_{i,j+h} - u^n_{i,j-h}\right)}{8h^3} \tag{10}$$

The last component, $\Delta(\psi_u(\|\nabla u\|)\nabla^2 u)$, is approximated by applying the discrete Laplacian to the approximation of $\psi_u(\|\nabla u\|)\Delta u$ that is obtained as in (7). So, we get:

$$A_4 = \frac{\psi^n_{i+h,j} + \psi^n_{i-h,j} + \psi^n_{i,j+h} + \psi^n_{i,j-h} - 4\psi^n_{i,j}}{h^2} \tag{11}$$

where $\psi_{i,j} = \psi_u \left(\dfrac{\sqrt{\left(u^n_{i+h,j} - u^n_{i-h,j}\right)^2 + \left(u^n_{i,j+h} - u^n_{i,j-h}\right)^2}}{2h} \right) \dfrac{u^n_{i+h,j} + u^n_{i-h,j} + u^n_{i,j+h} + u^n_{i,j-h} - 4u^n_{i,j}}{h^2}$.

Obviously, the final discretization of (5) is $A_1 - A_2 - A_3 + A_4 = 0$. If one considers $h = 1$ and $\Delta t = 1$, then we get the next implicit numerical approximation scheme:

$$\begin{aligned}
&u^{n+1}_{i,j} + (\delta - 1)u^n_{i,j} - \varphi_u \left(\frac{\sqrt{\left(u^n_{i+1,j} - u^n_{i-1,j}\right)^2 + \left(u^n_{i,j+1} - u^n_{i,j-1}\right)^2}}{2} \right) \left(u^n_{i+1,j} + u^n_{i-1,j} + u^n_{i,j+1} + u^n_{i,j-1} - 4u^n_{i,j} \right) \\
&- \varphi'_u \left(\frac{\sqrt{\left(u^n_{i+1,j} - u^n_{i-1,j}\right)^2 + \left(u^n_{i,j+1} - u^n_{i,j-1}\right)^2}}{2} \right) \frac{\left(u^n_{i+1,j+1} - u^n_{i+1,j-1} - u^n_{i-1,j+1} + u^n_{i-1,j-1}\right)\left(u^n_{i+1,j} - u^n_{i-1,j} + u^n_{i,j+1} - u^n_{i,j-1}\right)}{8} \\
&- \delta u^0_{i,j} + \psi^n_{i+1,j} + \psi^n_{i-1,j} + \psi^n_{i,j+1} + \psi^n_{i,j-1} - 4\psi^n_{i,j} = 0
\end{aligned} \tag{12}$$

which leads to the following explicit numerical approximation of (1):

$$\begin{aligned}
&u^{n+1}_{i,j} = (1-\delta)u^n_{i,j} + \varphi_u \left(\frac{\sqrt{\left(u^n_{i+1,j} - u^n_{i-1,j}\right)^2 + \left(u^n_{i,j+1} - u^n_{i,j-1}\right)^2}}{2} \right) \left(u^n_{i+1,j} + u^n_{i-1,j} + u^n_{i,j+1} + u^n_{i,j-1} - 4u^n_{i,j} \right) \\
&+ \varphi'_u \left(\frac{\sqrt{\left(u^n_{i+1,j} - u^n_{i-1,j}\right)^2 + \left(u^n_{i,j+1} - u^n_{i,j-1}\right)^2}}{2} \right) \frac{\left(u^n_{i+1,j+1} - u^n_{i+1,j-1} - u^n_{i-1,j+1} + u^n_{i-1,j-1}\right)\left(u^n_{i+1,j} - u^n_{i-1,j} + u^n_{i,j+1} - u^n_{i,j-1}\right)}{8} \\
&+ \delta u^0_{i,j} - \psi^n_{i+1,j} - \psi^n_{i-1,j} - \psi^n_{i,j+1} - \psi^n_{i,j-1} + 4\psi^n_{i,j} = 0
\end{aligned} \tag{13}$$

where $\psi_{i,j} = \psi_u \left(\dfrac{\sqrt{\left(u_{i+1,j}^n - u_{i-1,j}^n\right)^2 + \left(u_{i,j+1}^n - u_{i,j-1}^n\right)^2}}{2} \right) \left(u_{i+1,j}^n + u_{i-1,j}^n + u_{i,j+1}^n + u_{i,j-1}^n - 4u_{i,j}^n\right)$

and the derivative ϕ_u' is determined from (2). This obtained iterative restoration algorithm receives the degraded $[I \times J]$ image and applies the operation (13), for each n from 0 to N. This consistent iterative discretization scheme converges fast to the solution of the PDE model, representing the optimal image denoising, therefore the number of steps, N, is quite low.

4   Experiments

We have performed numerous restoration experiments using the proposed PDE-based filtering method. It has been successfully tested on high image databases, such as the USC - SIPI database, composed of 4 volumes. The original images from those volumes were corrupted with various amount of Gaussian noise, then the restoration scheme being applied. The proposed compound differential model achieves satisfactory noise removal results while preserving the boundaries and other important features very well. The undesired effects are also avoided. The next set of parameters of the PDE model, determined by trial and error, lead to the best restoration results:

$$\delta = 0.2, \lambda = 0.8, \gamma = 0.7, \beta = 4, \alpha = 2.5, k = 2, r = 3, v = 0.7, \rho = 0.8, \varepsilon = 0.9, N = 11 \quad (14)$$

We used various performance measures to asses this approach, like Peak Signal-to-Noise Ratio (PSNR), Norm of the Error (NE) or Structural Similarity Image Metric (SSIM). It outperforms many classic and PDE-based methods, getting better average values of these metrics. See the highest PSNR value in Table 1, got by our scheme. It outperforms conventional filters, by overcoming the blurring, second-order nonlinear PDE schemes, like Perona-Malik and TV Denoising, by removing the staircasing, and fourth-order diffusion models, such as the You-Kaveh method, by overcoming the speckle noise and providing a better deblurring. Also, the proposed hybrid algorithm executes faster than these nonlinear PDE-based methods, achieving the optimal smoothing in few iterations and having a running time of less than 1 s. One can see in Fig. 1 the restoration results obtained by our technique and the other schemes on the $[512 \times 512]$ *Peppers* image, corrupted by Gaussian noise with $\mu = 0.21$ and *var* = 0.03.

Table 1. Method comparison: the average PSNR values obtained by several models

Our model	Averaging	Gaussian	Median	P-M	TV	Y-K
27.15(dB)	24.32(dB)	23.91(dB)	24.85(dB)	25.72(dB)	26.23(dB)	26.84(dB)

Fig. 1. Restoration results achieved by various denoising methods

5   Conclusions

A compound PDE-based image denoising scheme mixing nonlinear second and fourth order diffusions has been proposed in this article. This hybrid diffusion model represents the major contribution of this paper. Because of its two diffusion components, it produces an effective detail-preserving Gaussian noise reduction, while overcoming all the unintended effects (image blurring, staircase effect and speckle noise).

A consistent explicit numerical approximation scheme has been constructed for this nonlinear diffusion model. The iterative discretization algorithm proposed here is based on the finite-difference method and converges fast to the solution of the PDE.

Given its compound character, this denoising technique outperforms numerous state of the art second-order and fourth-order PDE models and avoids their drawbacks, as resulting also from the described experiments and method comparison.

Acknowledgments. This research is supported by project PN-II-RU-TE-2014-4-0083 (UEFSCDI Romania) and Institute of Computer Science of the Romanian Academy.

References

1. Jain, A.K.: Fundamentals of Digital Image Processing. Prentice Hall, Upper Saddle River (1989)
2. Jansen, M.: Noise Reduction by Wavelet Thresholding. Springer, New York (2001)
3. Weickert, J.: Anisotropic Diffusion in Image Processing, European Consortium for Mathematics in Industry. B.G. Teubner Stuttgart, Germany (1998)
4. Perona, P., Malik, J.: Scale-space and edge detection using anisotropic diffusion. In: Proceedings of the IEEE Computer Society Workshop on Computer Vision, pp. 16–22, November 1987
5. Rudin, L., Osher, S., Fatemi, E.: Nonlinear total variation based noise removal algorithms. Phys. D: Nonlinear Phenom. **60**(1), 259–268 (1992)
6. Chen, Q., Montesinos, P., Sun, Q., Heng, P., Xia, D.: Adaptive total variation denoising based on difference curvature. Image Vis. Comput. **28**(3), 298–306 (2010)
7. Micchelli, C.A., Shen, L., Xu, Y., Zeng, X.: Proximity algorithms for image models II: L1/TV denosing. Adv. Comput. Math. **38**(2), 401–426 (2013)
8. Cai, J.F., Osher, S., Shen, Z.: Split Bregman methods and frame based image restoration. Multiscale Model. Sim. **8**(2), 337–369 (2009)
9. Barbu, T., Barbu, V., Biga, V., Coca, D.: A PDE variational approach to image denoising and restorations. Nonlinear Anal. RWA **10**, 1351–1361 (2009)
10. Barbu, T., Favini, A.: Rigorous mathematical investigation of a nonlinear anisotropic diffusion-based image restoration model. Electr. J. Diff. Eqn. **129**, 1–9 (2014)
11. You, Y.L., Kaveh, M.: Fourth-order partial differential equations for noise removal. IEEE Trans. Image Process. **9**, 1723–1730 (2000)
12. Lysaker, M., Lundervold, A., Tai, X.C.: Noise removal using fourth-order partial differential equation with applications to medical magnetic resonance images in space and time. IEEE Trans. Image Process. **12**(12), 1579–1590 (2003)
13. Barbu, T.: A PDE based Model for Sonar Image and Video Denoising. Analele Ştiinţifice ale Universităţii Ovidius, Constanţa, Seria Matematică **19**(3), 51–58 (2011)
14. Barbu, T.: Nonlinear fourth-order hyperbolic PDE-based image restoration scheme. In: Proceedings of EHB 2015, Iaşi, Romania, pp. 19–21. IEEE, November 2015
15. Wang, H., Wang, Y., Ren, W.: Image denoising using anisotropic second and fourth order diffusions based on gradient vector convolution. Comput. Sci. Inf. Syst. **9**, 1493–1512 (2012)
16. Wu, J.Y., Ruan, Q.: Combining adaptive PDE and wavelet shrinkage in image denoising with edge enhancing property. In: Proceedings of the 18th International Conference on Pattern Recognition, vol. 3, pp. 718–721 (2006)
17. Johnson, P.: Finite Difference for PDEs. School of Mathematics, University of Manchester (2008)

Multi-swarm Particle Grid Optimization for Object Tracking

Feng Sha[1]([⊠]), Henry Wing Fung Yeung[1], Yuk Ying Chung[1],
Guang Liu[1], and Wei-Chang Yeh[2]

[1] School of Information Technologies, University of Sydney,
Sydney, NSW 2006, Australia
feng.sha@sydney.edu.au
[2] Department of Industrial Engineering and Engineering Management,
National Tsing Hua University, P.O. Box 24-60, Hsinchu 300, Taiwan, R.O.C.

Abstract. In recent years, one of the popular swarm intelligence algorithm Particle Swarm Optimization has demonstrated to have efficient and accurate outcomes for tracking different object movement. But there are still problems of multiple interferences in object tracking need to overcome. In this paper, we propose a new multiple swarm approach to improve the efficiency of the particle swarm optimization in object tracking. This proposed algorithm will allocate multiple swarms in separate frame grids to provide higher accuracy and wider search domain to overcome some interferences problem which can produce a stable and precise tracking orbit. It can also achieve better quality in target focusing and retrieval. The results in real environment experiments have been proved to have better performance when compare to other traditional methods like Particle Filter, Genetic Algorithm and traditional PSO.

Keywords: Object tracking · Multi-swarm · PSO · Color histogram

1 Introduction

Swarm Intelligence (SI) applications have become popular in recent years because of its faster operation in speed and accurate solutions in video processing. One of the powerful SI is Particle Swarm Optimization (PSO). It demonstrated to have high potential in various research and practical problems such as mathematical optimization, neural network, artificial intelligence and image processing. It uses less resource and can provide significant improvements in object tracking. Also, it shows to be a powerful utility for tracking objects in complex environment.

For object tracking in video, its process includes three components [1]: Observation Model, Motion Model and Search Strategy. The observation model defines the properties of the target. The motion model provides definition for the target movement trend. And the search strategy consists of proper algorithm to predict the solution of tracking. In this paper, we choose color histogram [2, 3] to represent target's feature description as observation model. This definition could be used to distinguish the likelihood between actual target and possible solution to quantify pixel numbers for intensity values.

© Springer International Publishing AG 2016
A. Hirose et al. (Eds.): ICONIP 2016, Part II, LNCS 9948, pp. 707–714, 2016.
DOI: 10.1007/978-3-319-46672-9_79

Furthermore, we have defined (x, y) point as the central point for the pre-defined target with a scale of rectangle in two-dimensional frames for motion model [4].

There are many different approaches proposed for different searching strategies including PSO [5] and its enhancements. In this paper, a new multi-swarm gridding approach with weighted PSO algorithm has been proposed to handle the complicated target movement pattern such as fast movement speed, target disappearance and retrieval. It can separate helper swarms in frame grids to cover most of the frame search region, and assign smaller weight value according to particle numbers for each swarm to generate better particle convergence. Therefore, this new approach could overcome the drawbacks of traditional PSO with more accurate tracking result in real environment.

This paper is organized in five sections. First, it gives a brief introduction about the object tracking and provides general information about new methods. In Sect. 2, the introduction to PSO algorithm and its improvements have been reviewed. It includes linearly decreasing inertia weight [6] and constriction factor [7] approaches. Section 3 explains the process of our new proposed multi-swarm knowledge and weighted velocity improvements. Finally, the experimental results and conclusion based on the new method are shown in Sects. 4 and 5 respectively.

2   Introduction to PSO Object Tracking Methods

Traditional particle swarm optimization [8] has been developed based of the research on natural behavior on birds flock and fish cluster in late 20th century. It produces flexible and robust simulation of natural creature's collective intelligence in solving real world problems. They consider problem solutions as particles within a defined domain; by randomly distributing the particles. The algorithm will calculate particles' location, direction and velocities; then particles will move to more proper place according to the combination of local best experience and global best experience iteratively. In each iteration, the following equation will be used to update the particle's movement pattern:

$$v^{i+1}_{[p][d]} = w \times v^i_{[p][d]} + C_1 R_1 \left(Pb^i_{[p][d]} - x^i_{[p][d]} \right) + C_2 R_2 \left(Gb^i_{[d]} - x^i_{[p][d]} \right) \tag{1}$$

$$x^{i+1}_{[p][d]} = x^i_{[p][d]} + v^{i+1}_{[p][d]} \tag{2}$$

In Eqs. (1) and (2), x represents the actual particle location and v represents the particle velocity; p and d stand for particle and dimension numbers while i represents iteration numbers. Pb and Gb are the particle's local best vector and global best vector locations. The value of $C1$ and $C2$ are acceleration constants for the swarm learning and normally pre-defined as $C1 = C2 = 2$. In addition, $R1$ and $R2$ are values randomly generated within $[0, 1]$. Equation (1) consists of three parts indicates the combination of particle velocity from last iteration and inertia weight value, cognition and social components of the entire population. After calculated the current velocity value, the current solution will update according to Eq. (2).

Although PSO has some leading advantages, it also encounters two major draw-backs. The first one is the particles "explosion" problem [6] which happens when some of the particles move across the border and spread away. Shi. Y and R. C. Eberhart introduced a dynamic inertia weight value w instead of the original static value of 1 to address this problem. This is shown in Eq. (3) where w_1 and w_2 are the possible largest weight value and lowest weight value. This value of w will change according to the number of iteration to ensure that particles can converge to solution in later iterations and they found that the most efficient setting for the value of w is between 0.4 and 0.9 [9].

$$w = (w_1 - w_2) \times \frac{i_{max} - i_{current}}{i_{max}} + w_2 \tag{3}$$

The other problems are unbalance of particle distribution diversity and conver-gence. The possible particle solution could bound to local best experience and cease to diffuse again. Researchers have proposed different improvements for PSO such as setting better parameter values [10], optimizing initial particle distribution, topology, optimizing parameters [6, 7], and combining with other algorithms.

3 Proposed Multi-swarm Particle Grid Optimization

In order to overcome the drawbacks of the traditional PSO for better object tracking, the new proposed Multi-Swarm Particle Grid Optimization (MSGO) can achieve more effective particle searching by configuring multi-swarm distribution [9, 10] into frame grids to increase the search domain and to perform better organized swarm system while maintaining population diversity. Besides, the proposed new weighted velocity value will be assigned to help particles to converge within each swarm.

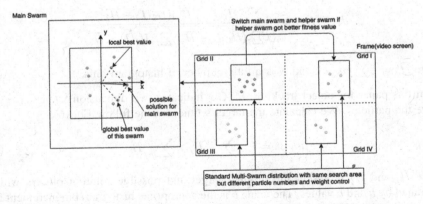

Fig. 1. Multi-swarm distribution and PSO particle movement within one swarm

The structure of possible multi-swarm grids distribution is shown in Fig. 1. When tracking starts, MSGO will equally divide the whole search screen into 2×2 or 3×3 grids, each grid will have its own swarm with different particles assigned to the population. The main swarm will locate at the grid where our selected or pre-defined target exists while other helper swarms will randomly distributed in other grids with similar size of search area. The population size differs from the main swarm and the helper swarms. The main swarm will have more particles to trace the target while helper swarms will have less size based on pre-defined parameters or will dynamically change size depending on the fitness values of each swarm.

For each frame, different swarms will calculate their own global solution and compare among each other. If the result shows that the best solution remains in current grid, the main swarm will continue tracking based on the best solution point from last frame. If the best solution appears in other grids, the main swarm will switch to target grid and replaces the helper swarm while a new helper swarm will be generated in the old grid.

To help particles converge to the true solution, our new MSGO assigns weight value for the velocity calculation. According to different particle numbers in each swarm, particles will move according to Eq. (4) instead of (1) so that the main swarm with more particles will have larger velocity to search for more possible solutions.

$$v_{[p][d]}^{i+1} = \frac{p}{\sum_{s}^{1} p} \times w \times v_{[p][d]}^{i} + C_1 R_1 \left(Pb_{[p][d]}^{i} - x_{[p][d]}^{i} \right) + C_2 R_2 \left(Gb_{[d]}^{i} - x_{[p][d]}^{i} \right) \qquad (4)$$

One important part for MSGO is to define the fitness value for tracking. In this paper, the calculation of likelihood after the comparison between original target and possible solution depends on the histogram values [2, 11]. There are three parameters include two array inputs $H1$, $H2$ and one method for different operation of comparison in color histogram definition to overcome complexity of the comparison processing [11]. The most important method is proven to be the Correlation [11]. The following correction process is used in MSGO:

$$d(H_1, H_2) = \frac{\sum_I (H_1(I) - \overline{H_1})(H_2(I) - \overline{H_2})}{\sqrt{\sum_I (H_1(I) - \overline{H_1}) \sum_I (H_2(I) - \overline{H_2})}} \qquad (5)$$

where $\overline{Hk} = \frac{1}{N} \sum_J Hk(J)$, and N is a total number of histogram bins.

In this paper for object tracking, MSGO indicates two dimension values (x, y) to locate the particles. We calculate the target's fitness value follow Eq. (6):

$$fitness = 1 - \sum_{b_1}^{h_1} H_1 \times \sum_{b_2}^{h_2} H_2 \qquad (6)$$

where H_1 and H_2 represent the original target and possible solution objects which calculated by h and b values. The value b is the appropriate histogram bin increment for each pixel and h is used to compute the sum of all bins and multiply each bin by the sum's inverse [11].

4 Experiment Results

In this paper, experiments have been designed to compare with traditional object tracking approach like Particle Filter [12] and other search algorithms like Genetic Algorithm and linearly decreasing PSO in real surveillance video and a self-made video with special movement patterns. The proposed MSGO has demonstrated excellent outcomes in complex environment and quick object retrieval.

All experiments were conducted using same system configuration as follows: Intel i7 CPU 2.5 GHz, 8 GM RAM, Nvidia Gefore GT 635 M and Windows 10 OS with Java and OpenCV library. All testing videos can be found in YouTube with following link: https://www.youtube.com/playlist?list=PLqTUSqrBpP-4f19AF1fJBVV-Qupr8RA7e

The testing videos' information are listed in Table 1:

Table 1. Experiment video Summary

	Name: complex blue ball motion tracking
	Challenge: fast moving speed; partially and fully cover by other objects; disappears for short period of time and re-appears again; other interference color and shape; motion blur.
	Frame: size: 1920*1080; No.: 562; Length: 22secs
	Name: surveillance video on elevator
	Challenge: glass and ground light reflection; target disappear inside elevator and re-appear again on second floor;
	Frame: size: 640*480; No.: 1679; Length: 33secs
Parameters: Particle Filter: 200 particles; Genetic Algorithm (GA): Standard configuration; inertia weight PSO: 30 particles, 20 iterations; MSGO: 2*2 Grids, main swarm has 15 particles, helper swarms have 5 particles each, iteration number = 20.	

4.1 Experiment 1: Surveillance Video in Building Hall

The first testing video was captured inside a building in front of four elevator doors. The target woman wearing pink coat and blue jeans tried to take bottom-left elevator lift from first level to second level and then walked towards to left exit in level two; she was covered by the white pillar and re-appeared again then walked through the second floor and leave the building on right side of the screen.

When she was waiting for the elevator in first part of the video, particle filter, inertia weight PSO and our MSGO approaches could always focus on the target because of linear trajectory while the genetic algorithm had already become unstable with some light reflection interference. Before the elevator door closed on frame 430, the wPSO lost track due to the light reflection on the elevator door while GA kept losing target.

After target disappeared inside the elevator, all methods lost target on frame 517. Particle filter and wPSO could not retrieve target at all due to their poor particle dispersion in small search area, while GA could only catch a few frames. However, MSGO could immediately restore tracking on frame 915 when target re-appeared in second floor and kept tracking until target disappeared again on frame 1175. Besides, MSGO could perform another quick retrieval on frame 1385 and maintain tracking until the last frame. MSGO had only lost a few frames due to the glass and light reflection.

4.2 Experiment 2: Complex Blue Ball Motion Tracking

This experiment conducted complex object movement patterns including motion blur caused by fast moving speed; interference deformation, and disappeared when target covered by other objects.

The particle filter, Genetic algorithm and inertia weight PSO performed badly with poor tracking outcomes. Particle filter and wPSO could not handle fast moving object and motion blur that wPSO lost track on frame 26 whereas GA had already lost target on frame 2. Furthermore, they could not retrieve target again in remaining part of video.

On the other hand, the MSGO could expand the search area by implementing multiple swarms in different grids, which led to much better tracking performance until frame 120 when target significantly changed shape and color and then quickly retrieved on frame 150 once the target dropped down; the same thing applied to frame 335 and 342, and frame 404 and 487 when a blue-color magazine covered and shaded the target, MSGO could retrieve target and kept tracking until last frame of video (Tables 2 and 3).

Table 2. Test result for video 1 - surveillance video on elevator

FrameNo.	Particle Filter	GA	wPSO	MSGO
430				
517				
915				
1175				
1385				

4.3 Screenshots for Experiment Results

Table 3. Test result for video 2 - complex blue ball motion tracking

FrameNo.	GA	wPSO	MSGO
2			
26			
120			
155			
335			
342			
404			
487			

5 Conclusion

Swarm intelligence algorithms produce excellent outcomes in image processing domain in recent years. The PSO algorithm demonstrates a faster and more accurate approach to solve both research and application problems. Many researchers continue to improve the process and try to obtain better performance for different applications. However, there is still no general method exists that can provide perfect reliable object tracking in wide variety of environments and adapt all kinds of changes.

In this paper, a novel multi-swarm particle grid optimization method has been proposed to overcome the drawback of small search region and particle convergence problem. This new approach performs better object tracking progress for non-linear target moving pattern and quick retrieval with color-based feature definition. It divides the entire frame into small grids and assigning weighted particles into main swarm and helper swarms. Thus it can allocate them into different grids efficiently.

When compare to traditional Particle Filter, Genetic Algorithm and inertia weight PSO, our proposed MSGO can fetch and retrieve desired target under complex environment and complicated motion pattern and produce stable tracking until last frame. In our future research, more efficient multiple swarm distribution and special particle update rules will be proposed, more testing will be conducted on different effects among swarms. Also reducing redundancy will be performed to increase accuracy and tracking speed.

References

1. Zhang, L., van der Maaten, L.: Structure preserving object tracking. In: Structure Preserving Object Tracking, pp. 1838–1845. IEEE (2013)
2. Culjak, I., Abram, D., Pribanic, T., Dzapo, H., Cifrek, M.: A brief introduction to OpenCV. In: A Brief Introduction to OpenCV, pp. 1725–1730. IEEE (2012)
3. Bradski, G., Kaehler, A.: Learning OpenCV: Computer Vision with the OpenCV Library. O'Reilly Media Inc, Sebastopol (2008)
4. Sha, F., et al.: A probability-dynamic particle swarm optimization for object tracking. In: 2015 International Joint Conference on Neural Networks (IJCNN), pp. 1–7. IEEE (2015)
5. Hsu, C., Dai, G.-T.: Multiple object tracking using particle swarm optimization. World Acad. Sci. Eng. Technol. **68**, 41–44 (2012)
6. Shi, Y., Eberhart, R.: A modified particle swarm optimizer. In: A Modified Particle Swarm Optimizer, pp. 69–73. IEEE (1998)
7. Clerc, M., Kennedy, J.: The particle swarm-explosion, stability, and convergence in a multidimensional complex space. IEEE Trans. Evol. Comput. **6**(1), 58–73 (2002)
8. Kennedy, J.: Particle swarm optimization. In: Encyclopedia of Machine Learning, pp. 760–766. Springer (2010)
9. Shi, Y., Eberhart, R.C.: Empirical study of particle swarm optimization. In: Empirical Study of Particle Swarm Optimization. IEEE (1999)
10. Langdon, W.B., Poli, R.: Evolving problems to learn about particle swarm and other optimisers. In: Evolving Problems to Learn About Particle Swarm and Other Optimisers, pp. 81–88. IEEE (2005)
11. OpenCV: OpenCV 2.4.9.0 documentation, opencv dev team, 21 April 2014. © Copyright 2011–2014
12. Wang, Q., Chen, F., Xu, W., Yang, M.-H.: An experimental comparison of online object-tracking algorithms. In: An Experimental Comparison of Online Objecttracking Algorithms (International Society for Optics and Photonics), pp. 81381A–81381A-81311 (2011)

Energy-Based Multi-plane Detection from 3D Point Clouds

Liang Wang[1,2]([✉]), Chao Shen[1], Fuqing Duan[3], and Ping Guo[4]

[1] College of Electronic Information and Control Engineering,
Beijing University of Technology, Beijing 100124, China
wangliang@bjut.edu.cn
[2] Engineering Research Center of Digital Community,
Ministry of Education, Beijing 100124, China
[3] College of Information Science and Technology,
Beijing Normal University, Beijing 100875, China
[4] School of Systems Science, Beijing Normal University, Beijing 100875, China

Abstract. Detecting multi-plane from 3D point clouds can provide concise and meaningful abstractions of 3D data and give users higher-level interaction possibilities. However, existing algorithms are deficient in accuracy and robustness, and highly dependent on thresholds. To overcome these deficiencies, a novel method is proposed, which detects multi-plane from 3D point clouds by labeling points instead of greedy searching planes. It first generates initial models. Second, it computes energy terms and constructs the energy function. Third, the point labeling problem is solved by minimizing the energy function. Then, it refines the labels and parameters of detected planes. This process is iterated until the energy does not decrease. Finally, multiple planes are detected. Experimental results validate the proposed method. It outperforms existing algorithms in accuracy and robustness. It also alleviates the high dependence on thresholds and the unknown number of planes in 3D point clouds.

1 Introduction

With the rapid development of emerging RGB-D sensors [1], 3D data of the natural world can be more flexibly captured. This provides powerful tools for analyzing and understanding natural scenes. However, in order to effectively make use of these 3D data, the raw 3D data should be enriched with abstractions and possibly semantic information. Detecting multi-plane from 3D point clouds can not only provide concise and meaningful abstractions of 3D data, but also give users higher-level interaction possibilities [2]. However, when performing multi-plane detection, the input 3D data are mixture of a huge number of points supporting an unspecified number of planes with outliers. Therefore, multi-plane detection from 3D point clouds is a challenging problem.

Several methods for multi-plane detection from 3D point clouds have been proposed, which can be roughly classified into three categories: the random sample consensus (RANSAC) paradigms [3], the Hough transform techniques [4] and

© Springer International Publishing AG 2016
A. Hirose et al. (Eds.): ICONIP 2016, Part II, LNCS 9948, pp. 715–722, 2016.
DOI: 10.1007/978-3-319-46672-9_80

the surface growing strategies [5]. The RANSAC [6] can robustly deal with out-liers, even if their percentage is more than 50 %. However, it usually can only detect one instance of one model. Moreover, it highly depends on thresholds of the distance from points to the model and manual adjustments. Nister [7] assumes that the number of candidates is fixed in advance to improve RANSAC. However, in practice, the number of planes to be detected is unknown and cannot be specified in advance. The Hough transform, even in the presence of high level noise and a high proportion of outliers, remains robust. However, high memory consumption is its major deficiency [4]. Algorithms of this type are only suitable for 2D image applications where the number of parameters typically is quite small. Surface growing for 3D point clouds is the extension of region growing [8] in 2D image space. It can detect multiple instances of multiple types of geomet-ric primitives. However, it lacks efficiency and robustness, and highly depends on manual selection of some parameters to avoid over-fitting and under-fitting.

In this paper, a novel method for detecting multi-plane from 3D point clouds is proposed. It first generates some initial models. Second, it computes the energy terms and constructs the energy function. Third, it solves the points labeling problem by minimizing the energy function. Then, it refines the labels and parameters of detected planes. This process is repeated until the energy does not decrease. Finally, the parameters and support sets of multiple planes are obtained. The main contributions are as follows. I. The multi-plane detection problem is solved by labeling points instead of traditional greedy search used by the Hough transform and RANSAC, which overcomes the greedy search's high dependence on distance thresholds. II. By introducing the labels and parameters refinement, the deficiency of the original energy-based multi-model fitting algo-rithm that the unknown number or a very large number of detected planes should be specified in advance, is overcome. III. The optimal results can be obtained by constructing and minimizing an energy function, and iterating the energy minimization. IV. The proposed algorithm has high accuracy and robustness.

The rest of this paper is organized as follows. Section 2 describes the proposed method in detail. Section 3 provides experimental evaluation on synthetic and real data. The last section concludes the paper.

2   The Proposed Method

2.1   Generate Initial Models

First, the proposed method uses random sampling of data points to generate an initial finite set of possible planes $\mathbf{M}_0$, which is borrowed from RANSAC [6,9]. A plane in a 3D point cloud can be described as

$$aX + bY + cZ + d = 0. \tag{1}$$

where (a, b, c) is the unit normal vector of the plane, and the distance from the original point of the coordinate system describing the 3D point cloud to the plane can be expressed as $|\frac{d}{\sqrt{a^2+b^2+c^2}}|$. Since (a, b, c) is a unit vector, the distance can also be expressed as $|d|$. To determine parameters (a, b, c, d), at least three different points on the plane should be known, i.e., the cardinality of the minimal support set of plane is three. So three sample points $S_k = \{p_{k1}, p_{k2}, p_{k3}\}$ are randomly selected in each sampling to compute parameters $M_k = (a_k, b_k, c_k, d_k)$ as follows. First, the unit directional vector of three lines formed by three sample points are computed by simply vector subtraction and normalization. Then, the unit normal vector of the plane determined by three sample points, (a_k, b_k, c_k), is computed with cross-product of any two of three computed unit directional vectors. Finally, three sample points are submitted into (1) to calculate the distance from the original point of the coordinate system to the plane, d_k. Therefore, a set of initial plane models $\mathbf{M}_0 = \{M_1, \ldots, M_K\}$ and the corresponding initial labels $\mathbf{L}_0 = \{L_1, \ldots, L_K\}$ are obtained.

The number of planes in 3D point cloud, K, is unknown in advance in practice. The initial value of K depends on the desired level of each plane model's confidence, number of data points and outliers, and cardinality of the minimal support set. It is very huge in theory [10]. Although experiments show that far fewer samples are needed than the theoretical value due to converging iterations, a large number has to be assigned [10]. In the proposed method, a small number, K, is specified in the initial step; then it is adaptively adjusted in the later optimization and refinement steps via labels merging and splitting. It finally converges to the real number of planes in the 3D point clouds.

2.2   Formulate Energy Terms

We extend the energy-based framework proposed by [10] to refine these labels and assign these labels to data points. Assume that p is a given point in data points set $\mathbf{P}$, i.e., $p \in \mathbf{P}$, and that L_p is a label assigned to point p. Then, the proposed method refines the labels and their spatial support (inliers) by minimizing the energy $E(\mathbf{L})$ of labeling $\mathbf{L} = \{L_p | p \in \mathbf{P}\}$

$$E(\mathbf{L}) = \sum_{p \in \mathbf{P}} D_p(L_p) + \sum_{(p,q) \in \mathcal{N}} V_{pq}(L_p, L_q) + f(\mathbf{L}). \qquad (2)$$

The first term $D_p(L_p)$ in (2) denotes the geometric error energy. For the plane detected from 3D point clouds, if a given point p is assigned a label L_p, the geometric error energy is measured by the distance from the point p to the plane $M(L_p)$ corresponding to the label L_p,

$$D_p(L_p) = \|p - M(L_p)\| = |\frac{ax_p + by_p + cz_p + d}{\sqrt{a^2 + b^2 + c^2}}| = |ax_p + by_p + cz_p + d|. \quad (3)$$

The shorter the distance $D_p(L_p)$ becomes, the less the penalty for assigning the point p with the label corresponding to the plane model $M(L_p)$ becomes. Otherwise, the point p can be taken as an outlier rather than an inlier of $M(L_p)$.

The second term $V_{pq}(L_p, L_q)$ in (2) denotes the smoothness energy, which describes the spatial smoothness constraints between neighboring points in 3D space. The neighbor relationship can be justified by Delaunay triangulation for all points $\mathbf{P}$, $\mathcal{N} = Delaunay(\mathbf{P})$. And every pair of vertices in each Delaunay triangle are neighboring points. The smoothness energy between neighbor points is described by the Potts model,

$$V_{pq}(L_p, L_q) = \lambda \cdot \omega_{pq} \cdot \delta(L_p \neq L_q). \tag{4}$$

where λ is the weight coefficient, whose value can be freely selected from 0.5 to 2.5 in practice. $\delta(\cdot)$ takes 1 if the specified condition holds, otherwise 0. ω_{pq} is the coefficient of discontinuity penalties for each pair of neighboring data points. In the case of multi-plane detection, the coefficient ω_{pq} is inversely proportional to the distance between points p and q, i.e., the closer two neighboring points without spatial smoothness become, the larger the penalty coefficient is:

$$\omega_{pq} = exp(-\frac{\|p - q\|^2}{\xi^2}), \tag{5}$$

where ξ is a constant coefficient, whose value usually varies from 3 to 5 for different point clouds in practice.

The third term $f(\mathbf{L})$ in (2) denotes the label energy, which is introduced to prevent too many labels due to over-fitting. It has the form

$$f(\mathbf{L}) = \beta \cdot |\ell_{\mathbf{L}}| + \tau, \tag{6}$$

where $\ell_{\mathbf{L}}$ is the set of distinct labels assigned to data points by labeling $\mathbf{L}$, $|\ell_{\mathbf{L}}|$ is the cardinality of $\ell_{\mathbf{L}}$ and equal to K. The initial value of K, i.e. $|\ell_{\mathbf{L}}|$, usually takes 5. Coefficient β sets the weight for the label energy. The method can obtain better performance when the value of β is freely chosen from 10 to 20. τ is introduced to take the label, outliers, into account, which has been ignored in the original energy-based algorithm [10]. It can help to adaptively determine the unknown number of planes in point cloud. Note that the real outlier rate is unknown in real applications. So τ is not a constant, but a variant during the energy optimization, which varies with the number of plane models and the desired level of each plane model's confidence. In practice, τ takes 10 times the actual outlier rate which is the percentage of the points that are not assigned any label in $\ell_{\mathbf{L}}$ relative to the whole 3D point cloud.

2.3   Optimize Labels

The energy function for multi-plane detection can be expressed as

$$E(\mathbf{L}) = \sum_{p \in \mathbf{P}} \|p - M(L_p)\| + \sum_{(p,q) \in \mathcal{N}} \lambda \cdot \omega_{pq} \cdot \delta(L_p \neq L_q) + \beta \cdot |\ell_{\mathbf{L}}| + \tau. \tag{7}$$

It has a form similar to the energy function of [11]. Thus, (7) can rapidly converge to the minimum, and the optimal planes $\mathbf{M} = \{M_1, \ldots, M_K\}$ and the corresponding labels $\mathbf{L} = \{L_1, \ldots, L_K\}$ are obtained.

2.4 Refine Labels and Parameters

For each detected plane, the standard deviation σ of the inliers is computed. Then, inliers are purified with the 3σ standard, whose deviations are larger than 3σ are classified as outliers. We also introduce the threshold on the minimal inlier number of each plane to prevent the case that few outliers occasionally form a plane, which generally takes 15. It can divide detected planes into two categories: planes with inliers more than the threshold and the others. We compare parameters of the latter with those of the former. If the difference between two groups of parameters is small, we merge the latter into the former. Otherwise, the latter plane is deleted, and its support points are classified as outliers. Then, the number of labels K, i.e. the number of plane models, which originally takes 5 in initial step, changes to $\hat{K}$. Thus, labels are refined by the label merging.

Then, parameters $\hat{M}$ of each detected plane are refined by solving

$$\hat{M} = arg\min_{M} \sum_{p \in \mathbf{P}(L)} \|p - M(L_p)\|, \tag{8}$$

with the least-squares method, where $\mathbf{P}(L) \subseteq \mathbf{P}$ denotes point sets with label L.

2.5 Iterate the Process

After refining labels and parameters, we have labels $\hat{\mathbf{L}} = \{\hat{L}_1, \ldots, \hat{L}_{\hat{K}}\}$, corresponding planes $\hat{\mathbf{M}} = \{\hat{M}_1, \ldots, \hat{M}_{\hat{K}}\}$, inliers and outliers. Then, we take current outliers as input to generate new initial models. The union of the current models $\hat{\mathbf{M}}$ and the new initial models is taken as initial values of energy function to perform energy optimization. By this operation, the labels splitting can be fulfilled. It can overcome the deficiency of the original energy-based algorithm that the number of initially generated models is too small to omit some planes due to the limited sample number. Repeat this process until the energy does not decrease. Finally, the parameters and support set of multiple planes are obtained.

3 Experiments

Many experiments are performed to validate the proposed method. Some of them are reported here to highlight the advantages of the proposed method.

3.1 Experiments with Synthetic Data

The synthetic scene shown in Fig. 1(a) contains three parallel planes shown in blue, green and red respectively, and another plane shown in magenta cutting across three parallel planes. Three parallel planes have the unit normal vector of $[\sqrt{3}/3, \sqrt{3}/3, \sqrt{3}/3]$ and the parameter d of $-100\sqrt{3}$, 0 and $100\sqrt{3}$ respectively. The plane in magenta has the unit normal vector of $[\sqrt{6}/6, \sqrt{6}/6, -\sqrt{6}/3]$ and d of $20\sqrt{6}$. Gaussian noise with a standard deviation $\sigma = 0.5$ is added to inliers of four planes. 400 outliers, shown in black, are uniformly distributed in the bounding box of 3D point cloud. The proposed method and RANSAC are performed. The key parameters of RANSAC are as follows: the distance threshold is 0.005 times the width of the bounding box, the angle threshold is $7°$, and the minimum number of inliers of a plane is 50, which are tuned via trial and error. The error of angle between the estimated unit normal vector and the ground truth, the error between estimated d and the ground truth, and estimated inliers are computed. Experimental results are shown in Fig. 1 and Table 1. The proposed method outperforms RANSAC in accuracy and robustness. It can deal well with overlapping between models (i.e., intersecting planes) due to exploiting the global and local constraints simultaneously, while RANSAC fails due to highly depending on threshold and need of manual adjustment.

(a) Original point cloud   (b) Result of proposed method   (c) Result of RANSAC

Fig. 1. Experiment with synthetic data.

Table 1. Configuration and result of experiment with synthetic data

Plane	Ground truth		Proposed method			RANSAC			
	Distance	Inlier	$error_{angle}$	$error_{distance}$	Inlier	$error_{angle}$	$error_{distance}$	Inlier
Blue	$-100\sqrt{3}$	100	0	-0.14	91	0	10.51	78
Green	0	300	0	0.05	281	2.99	-0.01	271
Red	$100\sqrt{3}$	200	0.27	0.51	197	0.70	62.20	170
Magenta	$20\sqrt{6}$	200	0	0.05	178	1.58	6.79	168

$error_{angle}$ is in degree

3.2   Experiments with Real 3D Point Clouds

For a real indoor scene as shown in Fig. 2(a), 3D point cloud is captured by a Microsoft Kinect. Multiple planes are in a more complicated configuration. The calibration pattern surface intersects the front face of the carton with a small dihedral angle, which makes distinguishing them challenging. The voxel grid-smoothing filter [12] is applied to deal with invalid data a is shown in Fig. 2(b), where the edge length of the voxel is 0.005 m. Then, the proposed method and RANSAC are performed. Results are shown in Fig. 2(c) and (d). We can see that the proposed method outperforms RANSAC. It succeeds in detecting and distinguishing two intersecting planes mentioned above, while the RANSAC fails.

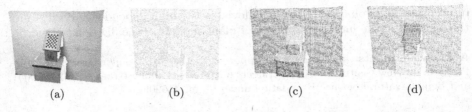

(a)   (b)   (c)   (d)

Fig. 2. Experiment with real data. (a) Original point cloud. (b) Smoothed point cloud. (c) Result of proposed method. (d) Result of RANSAC. In (c) and (d), outliers are in red, detected planes are in other colors (Color figure online)

(a)   (b)   (c)   (d)

Fig. 3. Experiment with more cluttered real data. (a) Original point cloud. (b) Smoothed point cloud. (c) Result of proposed method. (d) Result of RANSAC. In (c) and (d), outliers are in red, detected planes are in other colors (Color figure online)

Further, some points on the intersecting lines of intersecting planes are mistaken for outliers in RANSAC. It is because the RANSAC highly depends on distance thresholds. If the threshold value is small, intersecting planes with small dihedral angles can be distinguished, while many points on planes and intersecting lines are mistaken for outliers. Conversely, only few points are mistaken for outliers, while some intersecting planes are mistaken for one plane.

For a more cluttered scene [13,14] as shown in Fig. 3(a), multi-plane detection is also performed. First, the voxel filter [12] with edge length of 0.04 m is applied. Fig. 3(b) shows the result. Then, the proposed method and RANSAC are performed. As shown in Fig. 3(c) and (d), three key planes, the floor, vertical wall and table surface, are all detected by two algorithms. However, the false alarm rate of outliers for RANSAC is relatively greater. It is because RANSAC highly depends on thresholds. Although some false outliers can be removed by increasing the value of distance thresholds for RANSAC, many points of objects on the plane, for example points of the cup on the table, will be mistaken for points of the plane. Thus, the proposed method outperforms RANSAC.

4   Conclusions

A novel method for multi-plane detection from 3D point cloud is proposed, which formulates the multi-plane detection as an point labeling problem instead of the traditional greedy searching problem. Experiments with synthetic and real data validate the proposed method. It outperforms the existing robust algorithms in accuracy and robustness. It also alleviates the high dependence on distance thresholds and the unknown number of plane models in the 3D point cloud.

Acknowledgments. This work was supported by the NSFC under Grant 61572078, Program for New Century Excellent Talents in University under Grant NCET-13-0051, and Beijing Natural Science Foundation under Grant 4152027.

References

1. Zhang, Z.: Microsoft kinect sensor and its effect. IEEE Multimedia **19**(2), 4–10 (2012)
2. Gallup, D., Frahm, J., Pollefeys, M.: Piecewise planar and non-planar stereo for urban scene reconstruction. Proc. IEEE CVPR **II**, 803–806 (2010)
3. Liu, J., Wu, Z.: An adaptive approach for primitive shape extraction from point clouds. Optik **125**(9), 2000–2008 (2014)
4. Ogundana, O., Coggrave, C., Burguete, R., Huntley, J.: Automated detection of planes in 3-D point clouds using fast Hough transforms. Opt. Eng. **50**(5), 053609-1–053609-11 (2011)
5. Trevor, A., Gedikli, S., Rusu, R., Christensen, H.: Efficient organized point cloud segmentation with connected components. In: 3rd Workshop on Semantic Perception Mapping and Exploration, Karlsruhe, Germany (2013)
6. Fischler, M., Bolles, R.: Random sample consensus: a paradigm for model fitting with applications to image analysis and automated cartography. Commun. ACM **24**, 381–395 (1981)
7. Nister, D.: Preemptive RANSAC for live structure and motion estimation. Mach. Vis. Appl. **16**(5), 321–329 (2005)
8. Fan, M., Lee, T.: Variants of seeded region growing. IET Image Process **6**(9), 478–485 (2015)
9. Duan, F., Wang, L., Guo, P.: RANSAC based ellipse detection with application to catadioptric camera calibration. In: Wong, K.W., Mendis, B.S.U., Bouzerdoum, A. (eds.) ICONIP 2010, Part II. LNCS, vol. 6444, pp. 525–532. Springer, Heidelberg (2010)
10. Isack, H., Boykov, Y.: Energy-based geometric multi-model fitting. Int. J. Comput. Vis. **97**(2), 123–147 (2012)
11. Delong, A., Osokin, A., Isack, H., Boykov, Y.: Fast approximate energy minimization with label costs. Int. J. Comput. Vis. **96**(1), 1–27 (2012)
12. Rusu, R., Cousins, S.: 3D is here: point cloud library (PCL). In: Proceedings of IEEE ICRA, pp. 1–4 (2011)
13. Henry, P., Fox, D., Bhowmik, A., Mongia, R.: Patch volumes: segmentation-based consistent mapping with RGB-D cameras. In: Proceedings of IEEE 3DV, pp. 803–806 (2013)
14. Lai, K., Bo, L., Fox, D.: Unsupervised feature learning for 3D scene labeling. Proc. IEEE ICRA **II**, 803–806 (2014)

Bi-Lp-Norm Sparsity Pursuiting Regularization for Blind Motion Deblurring

Wanlin Gan, Yue Zhou$^{(\boxtimes)}$, and Liming He

Institute of Image Processing and Pattern Recognition,
Shanghai Jiao Tong University, Shanghai, China
{gwl008,zhouyue,heliming}@sjtu.edu.cn

Abstract. Blind motion deblurring from a single image is essentially an ill-posed problem that requires regularization to solve. In this paper, we introduce a new type of an efficient and fast method for the estimation of the motion blur-kernel, through a bi-lp-norm regularization applied on both the sharp image and the blur kernel in the MAP framework. Without requiring any prior information of the latent image and the blur kernel, our proposed approach is able to restore high-quality images from given blurred images. Moreover a fast numerical scheme is used for alternatingly caculating the sharp image and the blur-kernel, by combining the split Bregman method and look-up table trick. Experiments on both sythesized and real images revealed that our algorithm can compete with much more sophisticated state-of-the-art methods.

Keywords: Blind motion deblurring · Bi-Lp-Norm · Split Bregman method

1 Introduction

A motion blur is one of the common artifact that causes unpleasant blurred images with unavoidable information loss. During exposure, if the camera sensor or any object moves, a motion blurring will be generate.

Blind motion deblurring from a single image is essentially an ill-posed problem since there's more unknowns variables than observed data, leading to the result that a great change of the output might be influenced by a light undulation of the input variables. Blind motion deblurring has been extensively studied since the infusive work of Fergus *et al.* [1]. In a hypothesis, the input blurred image has a uniform blur kernel, which means that it's a spatial-invariant system, the degradation model is as follows

$$b = k * u + n, \tag{1}$$

where $*$ denotes the convolution operation, b is the observed blurry image, u is the latent image and k is called the point spread function (PSF) or the blur kernel, n is assumed to be random additive noise generally Gaussian noise. Blind

© Springer International Publishing AG 2016
A. Hirose et al. (Eds.): ICONIP 2016, Part II, LNCS 9948, pp. 723–730, 2016.
DOI: 10.1007/978-3-319-46672-9_81

motion deblurring is the problem of recovering u and k, only known the observation b. If k is known, the problem reduces to that of non-blind deconvolution [2,3].

Blind motion deblurring is ill-posed since none of the latent image u, the blur kernel h or the noise n are known. To solve the problem, prior assumptions on the structure of u and k must be explored. The heavy-tailed prior [3] on u is generally used, which is motivated from the observation that gradients of natural image is likely a hyper-Laplacian distribution [2]. But prior on the kernel k just received less attention, which is generally used the sparsity pursuiting form of the kernel for motion blurs, such as the l_1 norm $\|k\|_1$ [4,5], or the sparse coefficients under the curvelet transform [6].

Recent years, there are some top-performing methods using different forms of norm in image priors such as the $\frac{L_1}{L_2}$ normalized sparsity-based image prior in [5], the re-weighted L_2-norm-based image prior in [7], the recent approximate L_0-norm-based image prior in [8], the $L_{0.3}$-norm-based image prior in [9], the L_0-L_2-norm-based image prior in [10]. The L_1 regularizer is inefficient when the errors in data have heavy tail distribution [12]. Though there are some different forms of norm in image priors having good performance, is there any new regularization which is more sparse than the L_1 regularizer while it is still easier to be solved than the L_0 regularizer? In [11], it proposes L_p regularizer can achieve the requirement and L_p regularizer promotes sparsity more than L_1 regularizer which meets the conclusion [7] that spasity inducing regularizations are the key ingredient, irrespective of whether they provide good image gradient priors or not. Based on large number of experiments, the natural images have heavy tails [9] than the Laplacian (1-norm) and the Gaussian distribution (2-norm) shown in Fig. 1.

In this paper, we propose the method bi-L_p-norm regularization pursuiting the sparsity of the prior in both natural images and blur kernels according to what we find above. We use the split Bregman method to tackle this non-convex optimization problem. Since the ranges of the variables in the formulations are finitude, we use a look-up table trick to reduce the redundant calculating and accelerate the process. Experimental results on Levin *et al.*'s benchmark image dataset certify the effectiveness and robustness of our model and outperform state-of-the-art methods.

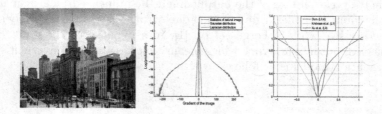

Fig. 1. Left: One sample of natural image. Center: a comparison between two different distribution and the statistics of the natural image. Right: plots of different penalty functions.

2   Model Using Bi-Lp-Norm Regularization

Most successful algorithms are classified either as Variational or Maximum A Posteriori (MAP). In practice the best MAP formulation techniques have proven as effective as variational methods and both the variational method and the MAP method lead to essentially the same framework [7]. We choose MAP framework for blind single-image deblurring. Those common MAP methods are formulated by the probabilisitc model. Based on the statement of the Baysian theorem, the process can be described as follows:

$$P(u,k|b) \propto P(b|u,k)P(u,k) = P(b|u,k)P(u)P(k), \qquad (2)$$

where $P(u,k|b)$ is the posterior probability, $P(u)$ and $P(k)$ are respective prior probability for the latent image and blur kernel, and $P(b|u,k)$ is the likelihood function. By taking negative logarithm on both sides, maximization of the posterior $P(u,k|b)$ equals to minimization of its negative logarithm $E(u,k|b) = -logP(u,k|b)$, i.e.,

$$E(u,k|b) = -logP(u,k|b) + const = \frac{\gamma}{2}||u*k-b||_2^2 + Q_0(u,k) + const, \qquad (3)$$

where the first term is called the error function or the data fidelity, γ is the weight of the fidelity term, the second term is the bi-L_p-norm regularization defined as follows,

$$Q_0(u,k) = \alpha_u Q_u(u) + \alpha_k Q_k(k) = \alpha_u||\nabla u||_{p_1} + \alpha_k||k||_{p_2}, 0 \le p_1, p_2 \le 1 \qquad (4)$$

where $Q_u(u) = -logP(u)$ and $Q_k(k) = -logP(k)$ can be regarded as regularizers that induce the optimization to the right and sparse solution and prevent a drift towards the trivial solution. The operation $||\bullet||_p$ denotes the L_p norm, α_u and α_k are the regularization weights. The red curve $L_{0.4}$ in Fig. 1 illustrates the form of L_p, and it is between the two function L_0 and L_1. Some other sparsity-pursuit functions which approach L_0 or L_1 used in deblurring [5,8] are also plotted in Fig. 1.

In [2], it demostrates the observation that gradients of natural images follow a hyper-Laplace distribution. In [7], it argues that enforcing sparsity of ∇u is a regularizer for k which is highly efficient, even when input images do not have sparse gradients. Therefore we use a sparsity pursuiting fomular of $Q_u(u)$ defined as

$$Q_u(u) = \Psi(\nabla_x u, \nabla_y u) = \sum_j (\sqrt{[\nabla_x u]_j^2 + [\nabla_y u]_j^2})^p, 0 \le p \le 1 \qquad (5)$$

where ∇_x and ∇_y are the partial derivative operators.

The estimation of k alone (ignore x) can recover the true kernel with an increasing accuracy for an increasing image size [13]. So the prior on k is just as important as the prior on u and motion blur kernel also pursues sparsity and refuses negative values. We concentrate on the case of spatially uniform blur of

Eq. (1), the prior of k leads to an improved estimation precision by sparsifying the blur-kernel. The fomular of $Q_k(k)$ and constraints on k are defined as

$$Q_k(k) = \sum_j \Phi(k_j)^p, 0 \leq p \leq 1 \tag{6}$$

$$\Phi(k_j) = \begin{cases} k_j & k_j \geq \varepsilon \\ 0 & k_j < \varepsilon \end{cases}, \quad (\varepsilon \text{ is a small positive threshold}) \tag{7}$$

$$\sum_j Q_k(k) = 1 \tag{8}$$

Thus the final loss function is as follows:

$$\min E(u, k) = \frac{\gamma}{2} ||u * k - b||_2^2 + Q_0(u, k) \tag{9}$$

3 Optimization

In order to solve the optimization problem Eq. (9), our practical implementation is alternating the estimations of the latent image u and the motion blur kernel k respectively while keeping the other one constant during one iteration, and the result estimated in the current iteration is taken as the input of the next iteration. The iteration not ends until it converges or reaches the setting maximum interation number. Therefore the problem is divided into two phases, and we explain the two phases process with respect to u and k separately. There are several schemes to solve the optimization problem, [14] refers to augmented Lagrangian method and Bregman iterative method, and both is equivalent when the constraints are linear. Here, we choose the split Bregman method (SBM) [15] and use the look-up table trick to solve the problem.

3.1 U Estimation

First we solve the latent image u while keeping the blur kernel k constant. The loss function of the optimization of the latent image u is expressed as follows,

$$\min_u \{ \frac{\gamma}{2} ||Ku - b||_2^2 + \alpha_u \Psi(\nabla_x u, \nabla_y u) \} \tag{10}$$

where K is the fixed convolution matrix corresponding to blur kernel k from the previous iteration. Then according to the method [15], to separate the minimization of data term and regularizer, the formulation of the problem can be transformed into the expression as follows,

$$\min_u \{ \frac{\gamma}{2} ||Ku-b||_2^2 + \alpha_u \Psi(d_x, d_y) + \frac{\alpha}{2} ||\nabla_x u - d_x - l_x||^2 + \frac{\alpha}{2} ||\nabla_y u - d_y - l_y||^2 \} \tag{11}$$

where introducing new variables $d_x = \nabla_x u$, $d_y = \nabla_y u$ to decouple two terms in Eq. (10), l_x and l_y are needing in Bregman formulation. According to the Bregman iteration, the problem in Eq. (10) can be divided into two sub-problems:

$$\min_u \{ \frac{\gamma}{2} ||Ku - b||_2^2 + \frac{\alpha}{2} ||\nabla_x u - d_x - l_x||^2 + \frac{\alpha}{2} ||\nabla_y u - d_y - l_y||^2 \} \tag{12}$$

$$\min_{d_x, d_y} \{ \alpha_u \Psi(d_x, d_y) + \frac{\alpha}{2} ||\nabla_x u - d_x - l_x||^2 + \frac{\alpha}{2} ||\nabla_y u - d_y - l_y||^2 \} \tag{13}$$

For the sub-problem Eq. (12), FFT (Fast Fourier Transformation) method is used to calculate u directly in Fourier domain. After differentiating Eq. (12) w.r.t u and setting the derivative to zero, we get the solution on u,

$$F(\mathbf{u}^{n+1}) = \frac{\overline{F(\mathbf{k}^n)} \circ F(\mathbf{b}) + \frac{\alpha}{\gamma}(\overline{F(\nabla_{\mathbf{x}})} \circ F(\mathbf{d_x} + \mathbf{l_x}) + \overline{F(\nabla_{\mathbf{y}})} \circ F(\mathbf{d_y} + \mathbf{l_y}))}{\overline{F(\mathbf{k}^n)} \circ F(\mathbf{k}^n) + \frac{\alpha}{\gamma}(\overline{F(\nabla_{\mathbf{x}})} \circ F(\nabla_{\mathbf{x}}) + \overline{F(\nabla_{\mathbf{y}})} \circ F(\nabla_{\mathbf{y}}))}, \quad (14)$$

$$u^{n+1} = F^{-1}(F(\mathbf{u}^{n+1})), \quad (15)$$

where $\circ$ denotes multiplication operator of the corresponding element-wise matrices, $F(\bullet)$ denotes the FFT operator of $\bullet$ and $\overline{F(\bullet)}$ denotes the complex conjugate of $\bullet$, and $F^{-1}(\bullet)$ denotes the IFFT (Inverse Fast Fourier Transform) of $\bullet$. Notice that $F(K^T)$, the FFT of the transposition of convolution matrix of k, equals to $\overline{F(k)}$, the complex conjugate of $F(K)$.

For the sub-problem Eq. (13) w.r.t l_x, l_y, we use a trick to solve it. Supposing j is the fixed pixel index, let $m = ([d_x]_j, [d_y]_j)$ and $w = (\nabla_x u_j - [l_x]_j, \nabla_y u_j - [l_y]_j)$, therefore the Eq. (13) is redescribed as follows,

$$\min_m \{\alpha_u ||m||^p + \frac{\alpha}{2}||m - w||^2\} \quad (0 \le q \le 1). \quad (16)$$

For the common use of $p = 1$, the solution of Eq. (16) is to use soft shrinkage operators, $m = softshrink(w, \frac{\alpha_u}{\alpha})$, where $softshrink(x, \alpha) = \frac{x}{|x|} * max(|x| - \alpha, 0)$. When $p = 0$, the solution of Eq. (16) is $m = hardshrink(w, \sqrt{\frac{2 * \alpha_u}{\alpha}})$, where $hardshrink(x, \alpha) = \frac{x}{|x|} * max(|x|, \alpha)$. When $0 < p < 1$, it has no closed form solution, but since p is set previously, it can be precalculated numerically to solve Eq. (16) in the form of lookup table (LUT). The update of $\{[d_x^{n+1}]_j, [d_y^{n+1}]_j\}$, l_x^{n+1}, l_y^{n+1} is as follows,

$$\{[d_x^{n+1}]_j, [d_y^{n+1}]_j\} = LUT_p([\nabla_x u^{n+1} + l_x^n]_j, [\nabla_y u^{n+1} + l_y^n]_j), \quad (17)$$

$$l_x^{n+1} = l_x^n + d_x^{n+1} - \nabla_x u^{n+1}, \quad (18)$$

$$l_y^{n+1} = l_y^n + d_y^{n+1} - \nabla_y u^{n+1}. \quad (19)$$

3.2 K Estimation

After calculating the latent image u_{n+1} at the n-th iteration, we calculate the blur kernel k in a similar way. According to Eqs. (6), (9), the loss function of the optimization of blur kernel k is expressed as follows,

$$\min_k \{\frac{\gamma}{2}||Uk - b||_2^2 + \alpha_k \Phi(k)\} \quad (20)$$

where U is the fixed convolution matrix constructed from the estimation of u_{n+1}. Then we make a substitution $d_k = k$ in $\Phi(k)$, and applying the split Bregman method in Eq. (20), the result is as follows,

$$\min_{k, l_k} \{\frac{\gamma}{2}||Uk - b||_2^2 + \alpha_k \Phi(d_k) + \frac{\beta}{2}||k - d_k - l_k||^2\}, \quad (21)$$

where l_k is the new variable in the quadratic penalty and also related to SBM method.

To do a similar operation in Sect. 3.1, the detailed expression are shown as follows,

$$F(\mathbf{k}^{n+1}) = \frac{\overline{F(\mathbf{u}^{n+1})} \circ F(\mathbf{b}) + \frac{\beta}{\gamma} F(\mathbf{d_k} + \mathbf{l_k})}{\overline{F(\mathbf{u}^{n+1})} \circ F(\mathbf{u}^{n+1}) + F(\frac{\beta}{\gamma})}, \tag{22}$$

$$k^{n+1} = F^{-1}(F(\mathbf{k}^{n+1})) \tag{23}$$

$$[d_k^{n+1}]_j = \mathrm{LUT}_p([k^{n+1} - l_k^n]_j), \tag{24}$$

$$l_k^{n+1} = l_k^n + d_k^{n+1} - k^{n+1}, \tag{25}$$

where after all the interation, we threshold small values(including-negative values) of the kernel K into zero, which can strengthen the robustness to noise.

3.3 Implementation

The two sub-parts, blur-kernel estimation and non-blind estimation is demonstrated in Algorithm 1.

Algorithm 1. Bi-L_p-norm deblurring

Require: : blurred image b
Ensure: : blur kernel k,deblurred image u
 1: initialize k^1 from the coarser-scale kernel estimate
 2: **for** level=1:max_level **do**
 3:     Initialization k and u in the current level
 4:     **for** m=1:max_iteration **do**
 5:         //U Estimation
 6:         **for** i=1 to max_i **do**
 7:             Update u by solving Eq. (14) (15)
 8:             Update $d_x, d_y; l_x, l_y$ by using Eq. (17)(18)(19)
 9:         **end for**
10:         //K Estimation
11:         **for** j=1:max_j **do**
12:             Update k by solving Eq. (22) (23)
13:             Update d_k, l_k by using Eq. (24)(25)
14:         **end for**
15:     **end for**
16: **end for**

---

4 Experimental Results

We test our algorithm on the benchmark dataset by Levin *et al.* [13].

The first experiment we introduce compares blind motion deblurring performance using the proposed bi-L_p-norm regularization (4). We calculate the

PSNR (peak signal to noise ratio) of the latent image and the deblurred one to make quality accessment of the deblurred image and, the higher value, the better performance. Figure 2 shows the result that we select p_1 ranging from 0 to 1 with a interval of 0.1 and a fixed $\cdot p_2 = 1$. In Fig. 2, it reveals that when $0 \leq p_1 \leq 0.5$, the $L_{0.4}$ regularizer pursuiting sparsity shows the best average performance, and when $0.5 < p_1 \leq 1$, the L_p regularizers perform a bit worse than $L_{0.4}$ regularizer. So we conclude that the $L_{0.4}$ regularizer can be taken as the representative of the $L_{p_1}(0 \leq p \leq 1)$ regularizers.

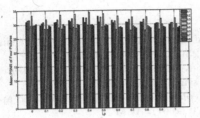 

Fig. 2. Left: Mean PSNR of four deblurring images blurred by eight kernels and Lp ranges from 0 to 1 with a interval of 0.1. Right: PSNR of deblurred images using three different methods in 32 test examples. Numbers on x-axis represent different images (u1 to u4) blurred by different kernels (kj)

Fig. 3. Every four figure, left to right: original; Xu's [8]; Krishnan's [5]; Ours.

In the other experiment, the proposed method is compared with two methods including Krishnan *et al.* [5] and Xu *et al.* [8] using the benchmark dataset by Levin *et al.* [13]. The final result is displayed in Fig. 2, where uikj means the blurred image is constructed from the i-th image blurring using the j-th kernel. The blue, green and red columns denote our, Krishnan [5]'s and Xu [8]'s method respectively. By comparing PSNR among these three methods, we can see that in most cases our method outperform than the other two methods. In Fig. 3, we can see that the deblurred images using our method are more flat and hold on more details and our estimated kernels are more sparse on the premise of high accuracy. All the three methods perform very well, but our method generates a bit more accurate results.

5   Conclusion

In this paper, we propose an effective model of bi-L_p-norm regularization pursuiting the sparsity of the prior in both natural image and blur kernels. As a conclude, we find the $L_{0.4}$ regularizer can be taken as the representative of the $L_{p_1}(0 \leq p \leq 1)$ regularizers which is more sparse than L_1 regularizers. Experiments shows that our proposed method has a competitive performance with the state-of-the-art method.

References

1. Fergus, R., Singh, B., Hertzmann, A., et al.: Removing camera shake from a single photograph. ACM Trans. Graph. (TOG) **25**(3), 787–794 (2006)
2. Krishnan, D., Fergus, R.: Fast image deconvolution using hyper-Laplacian priors. In: Advances in Neural Information Processing Systems (2009)
3. Levin, A., et al.: Image and depth from a conventional camera with a coded aperture. ACM Trans. Graph. (TOG) **26**(3), 70 (2007). ACM
4. Shan, Q., Jia, J., Agarwala, A.: High-quality motion deblurring from a single image. ACM Trans. Graph. (TOG) **27**(3) (2008). ACM
5. Krishnan, D., Tay, T., Fergus, R.: Blind deconvolution using a normalized sparsity measure. In: IEEE Conference on Computer Vision and Pattern Recognition (CVPR). IEEE (2011)
6. Cai, J.-F, et al.: Blind motion deblurring from a single image using sparse approximation. In: IEEE Conference on Computer Vision and Pattern Recognition, CVPR 2009. IEEE (2009)
7. Krishnan, D., Bruna, J., Fergus, R.: Blind deconvolution with non-local sparsity reweighting arXiv preprint arXiv:1311.4029 (2013)
8. Xu, L., Zheng, S., Jia, J.: Unnatural l0 sparse representation for natural image deblurring. In: Proceedings of the IEEE Conference on Computer Vision and Pattern Recognition (2013)
9. Kotera, J., Šroubek, F., Milanfar, P.: Blind deconvolution using alternating maximum a posteriori estimation with heavy-tailed priors. In: Wilson, R., Hancock, E., Bors, A., Smith, W. (eds.) CAIP 2013, Part II. LNCS, vol. 8048, pp. 59–66. Springer, Heidelberg (2013)
10. Shao, W.-Z., Li, H.-B., Elad, M.: Bi-l 0-l 2-norm regularization for blind motion deblurring. J. Vis. Commun. Image Represent. **33**, 42–59 (2015)
11. Xu, Z.B., et al.: L1/2 regularization. Sci. China Inf. Sci. **53**(6), 1159–1169 (2010)
12. Tibshirani, R.: Regression shrinkage, selection via the lasso. J. Royal Stat. Soc. Ser. B (Methodol.) 267–288 (1996)
13. Levin, A., et al.: Understanding and evaluating blind deconvolution algorithms. In: IEEE Conference on Computer Vision and Pattern Recognition, CVPR 2009. IEEE (2009)
14. Yin, W., et al.: Bregman iterative algorithms for L1-minimization with applications to compressed sensing. SIAM J. Imaging Sci. **1**(1), 143–168 (2008)
15. Goldstein, T., Osher, S.: The split Bregman method for L1-regularized problems. SIAM J. Imaging Sci. **2**(2), 323–343 (2009)

Author Index

Printed in the United States
By Bookmasters